Masahiko Utsuro and
Vladimir K. Ignatovich
Neutron Optics

Related Titles

Reimers, W., Pyzalla, A. R., Schreyer, A., Clemens, H. (Eds.)

Neutrons and Synchrotron Radiation in Engineering Materials Science

From Fundamentals to Material and Component Characterization

2008

ISBN 978-3-527-31533-8

Tsuji, K., Injuk, J., Van Grieken, R. (eds.)

X-Ray Spectrometry

Recent Technological Advances

2003

ISBN 978-0-471-48640-4

Masahiko Utsuro and Vladimir K. Ignatovich

Neutron Optics

WILEY-VCH

WILEY-VCH Verlag GmbH & Co. KGaA

The Authors

Masahiko Utsuro
Osaka University
Japan
masahiko.utsuro@rcnp.osaka-u.ac.jp

Vladimir K. Ignatovich
Joint Institute for Nuclear Research
Dubna
Russia
v.ignatovi@gmail.com

Library of Congress Card No.: applied for

British Library Cataloguing-in-Publication Data:
A catalogue record for this book is available from the British Library.

Bibliographic information published by the Deutsche Nationalbibliothek
The Deutsche Nationalbibliothek lists this publication in the Deutsche Nationalbibliografie; detailed bibliographic data are available on the Internet at http://dnb.d-nb.de.

Typesetting le-tex publishing services GmbH, Leipzig
Printing and Binding Strauss GmbH, Mörlenbach

Printed in the Federal Republic of Germany
Printed on acid-free paper

ISBN 978-3-527-40885-6

Contents

Preface *XV*

Textbooks *XIX*

1 Wave–Particle Duality of the Neutron *1*
1.1 Discovery of Matter Waves *1*
1.2 Proof of the Wave Nature of the Neutron *3*
1.2.1 Bragg Reflection *3*
1.2.2 Refractive Index and Total Reflection *5*
1.2.3 Fraunhofer Diffraction *8*
1.2.4 Fresnel Diffraction *11*
1.3 Coherence of Waves *13*
1.3.1 Coherence Lengths *13*
1.3.2 Pendellösung Interference *15*
1.4 Corpuscular Properties of the Neutron *16*
1.4.1 Time-of-Flight Analysis *16*
1.4.2 From a Particle to a Wave *19*
1.5 Magnetic Moment of the Neutron *22*
1.5.1 Spin Flip in Magnetic Resonance *22*
1.5.2 Adiabatic Spin Reversal in a Magnetic Gradient *25*

2 Reflection, Refraction, and Transmission of Unpolarized Neutrons *31*
2.1 Reflection and Transmission of Neutrons by a Single-Layer Mirror *31*
2.1.1 Measurement of Coherent Scattering Amplitude *31*
2.1.2 Neutron Guide Tube *33*
2.1.3 Total Reflection Potential and Surface Impurities *34*
2.1.4 Neutron Reflectometry *39*
2.2 Properties of Ultracold Neutrons *43*
2.2.1 Reflection Loss Probability *43*
2.2.2 Neutron Storage Bottle and Anomalous Loss Coefficient *45*
2.2.3 Limiting Velocity for Total Reflection and Neutron Absorption *48*
2.3 Reflection and Transmission of Neutrons by Multilayer Mirrors *51*
2.3.1 Developments of Multilayer Mirrors and Reflectivity Analyses *51*
2.3.2 High-Performance Supermirrors *55*

Neutron Optics. Masahiko Utsuro,Vladimir K. Ignatovich

ISBN: 978-3-527-40885-6

2.4 Reflection off and Transmission through Moving Matter *57*
2.4.1 Neutron Turbine *57*
2.4.2 Fizeau Experiment *62*
2.5 Neutron Resonator and Quasi-Bound Neutrons *64*
2.5.1 Neutron Reflection and Transmission Experiments with Macroscopic Resonators *64*
2.5.2 Nuclear Reaction Experiments of Neutrons *67*

3 Reflection, Refraction, and Larmor Precession of Polarized Neutrons *71*
3.1 Reflection from Magnetic Mirrors *71*
3.1.1 Magnetic Reflection and Neutron Polarization *71*
3.1.2 Polarized Neutron Reflectometry *74*
3.1.3 Magnetic Storage of Neutrons *77*
3.2 Reflection and Transmission of Neutrons by Magnetic Multilayers *81*
3.2.1 Reflectivity Analyses with the Transfer Matrix Method *81*
3.2.2 Analyses of Nonspecular Reflections *85*
3.3 Neutron Behavior in Magnetic Devices *87*
3.3.1 Magnetic Deceleration and Acceleration of Neutrons *88*
3.3.2 Magnetic Focusing *90*
3.3.3 Pseudomagnetic Prism *94*
3.4 Measurement of Larmor Precession *96*
3.4.1 4π Periodicity of Spinor *96*
3.4.2 Coherent Superposition of Spin States *99*
3.5 Neutron Spectroscopy with Larmor Precession *103*
3.5.1 Neutron Resonance Spin Echo Method *103*
3.5.2 Neutron Spin Maser *107*
3.6 Berry's Geometrical Phase *108*

4 Quantum Mechanics of Scalar Neutron Waves in One Dimension *113*
4.1 Some Descriptions of the Fundamental Theory *113*
4.2 Reflection and Refraction at a Potential Step *115*
4.2.1 Total Reflection *117*
4.2.2 Optical Potential and Ultracold Neutrons *117*
4.2.3 Digression on the Ultracold Neutron Anomaly *118*
4.2.4 Experience with an Experiment *121*
4.2.5 Surface Waves *126*
4.2.6 Goos–Hänchen Shift *127*
4.2.6.1 Total Reflection *128*
4.2.6.2 Nontotal Reflection *129*
4.3 Scattering by a Rectangular Potential Barrier *129*
4.3.1 The Other Method *130*
4.3.2 Asymptotical Behavior of Reflectivity from a Rectangular Barrier *131*
4.3.3 Wave Function inside the Barrier *132*
4.4 Neutron Scattering from Semitransparent Mirrors *132*
4.4.1 Bringing Two Mirrors Together *133*
4.4.2 Splitting of Rectangular Potentials *134*

4.4.3 A Film over a Substrate *134*
4.4.4 Experience with an Experiment *134*
4.4.5 Properties of the Reflection and Transmission Amplitudes *136*
4.4.5.1 Asymmetrical Potentials *136*
4.4.5.2 Relations between Phases of r and t *137*
4.5 Periodic Systems *138*
4.5.1 Asymmetrical Period *141*
4.5.2 Kronig–Penney Potential *141*
4.5.3 Supermirrors *144*
4.5.3.1 Proof of Eq. (4.102) *148*
4.5.3.2 Historical Remarks *150*
4.5.4 Experience with an Experiment *152*
4.6 Examples of Application of the Algebra *155*
4.6.1 Resonant Transmission *155*
4.6.1.1 A Resonance between Two Nonequal Barriers *156*
4.6.1.2 Breit–Wigner Formula *157*
4.6.1.3 Decay of Quasibound Systems *158*
4.6.1.4 Resonances at Total Reflection *159*
4.6.2 Channeling *159*
4.6.2.1 Wave Function inside a Channel *161*
4.6.2.2 Wave Function over the Nonilluminated Part of the Surface *163*
4.6.3 Bound States in Periodic Potentials *165*
4.6.3.1 Bound Levels in a Rectangular Potential Well [211, 212] *166*
4.6.3.2 Zones in Periodic Potentials *166*
4.7 Potentials of General Form *167*
4.7.1 Analytical Description of Reflection from a Combination of a Rectangular Barrier and Eckart Potentials *168*
4.7.1.1 Reflection from a Single Potential Eq. (4.180) *169*
4.7.1.2 Reflection from the Combination of Eq. (4.180) and a Potenital Step *171*
4.7.1.3 Reflection from the Full Potential Eq. (4.181) *174*
4.7.2 Continuous-Fractions Method *174*
4.7.3 Matrix Method for Scattering from an Arbitrary Potential *175*
4.7.4 Perturbation Theory *176*
4.7.5 Reflection from Mirrors in an External Field *177*
4.8 Some Experiments with Moving Mirrors *178*
4.8.1 Uniform Motion *179*
4.8.1.1 A Semiclassical Consideration *179*
4.8.1.2 Transformation of Reference Frames *180*
4.8.2 Accelerating Mirror *182*
4.8.2.1 Classical Approach *182*
4.8.2.2 Transformation to the Accelerating Reference Frame *184*
4.8.3 Vibrating Mirror *188*
4.8.4 Diffraction by a Moving Grating *189*
4.9 Conclusion *190*

5 One-Dimensional Scattering of Neutrons with Spin *191*
5.1 Reflection from a Magnetic Mirror *193*
5.1.1 Plane Wave in a Homogeneous Magnetic Field *193*
5.1.1.1 A Fundamental Question of Quantum Mechanics *194*
5.1.2 Solution of Eq. (5.9) *194*
5.1.2.1 Properties of the Pauli Matrices and Spinor States *195*
5.1.2.2 Matrix Elements of the Reflection Matrix *198*
5.1.3 Triple Splitting of the Reflected Beam *198*
5.1.3.1 A Fundamental Question *200*
5.1.3.2 Scalar Aharonov–Bohm Effect *201*
5.1.3.3 Comments on Dipole and Current Models of the Neutron Magnetic Moment *201*
5.2 Algebra of Magnetic Mirrors *201*
5.2.1 Magnetic Mirror of Finite Thickness *201*
5.2.2 System of Two Magnetic Mirrors *202*
5.2.3 Standing Waves *203*
5.2.4 Periodic System of Mirrors *203*
5.2.5 A Periodic Potential with Collinear Internal Fields *204*
5.2.6 Reflection from Helical Systems *205*
5.2.6.1 Reflection from a Semi-infinite Mirror *205*
5.2.6.2 Reflection from the Interface from within the Mirror *209*
5.2.6.3 Reflection from a Plate of Finite Thickness *210*
5.2.6.4 Violation of Fundamental Principles in Helical Systems *212*
5.3 Numerical Matrix Method for Multilayered Magnetic Systems *213*
5.4 Polarizers, Analyzers, and Spin Rotators *215*
5.4.1 Polarizers *215*
5.4.2 Analyzers *216*
5.4.3 Spin Rotators with Direct Current Fields *216*
5.4.3.1 Current Foil Method *216*
5.4.3.2 Zero Field Method *216*
5.4.3.3 Mezei Spin Rotator *217*
5.4.4 Resonant Spin Rotators *218*
5.4.4.1 The Rabi Formula *219*
5.4.4.2 Complete Conversion of an Unpolarized Beam into a Polarized One *220*
5.4.4.3 Linear Oscillating Field *220*
5.4.4.4 Broadband Adiabatic Spin Flippers *222*
5.5 The Berry Phase *224*
5.6 Inelastic Interaction of Neutrons with an RF Field *227*
5.6.1 Presumable Form of the Wave Function *227*
5.6.2 The Krüger Problem *229*
5.6.2.1 Solution to the Krüger Problem *230*
5.6.2.2 Angular Splitting of Reflected and Transmitted Neutrons *232*
5.6.2.3 Reflection and Transmission Matrices *233*
5.6.2.4 Approximate Expressions for the Transmission Matrix Eq. (5.169) *234*
5.7 Games with Polarized Neutrons *235*

5.7.1 Spin Wave and Intensity Modulated Wave after a $\pi/2$-Spin Flipper *235*
5.7.1.1 Classical Explanation of the Spin Wave. Wave–Particle Duality *236*
5.7.1.2 The First Miracle of St. Mark Street *238*
5.7.1.3 The Second Miracle of St. Mark Street *238*
5.7.2 π-Flipper *239*
5.7.2.1 The Ramsey Separated Fields Method *239*
5.7.2.2 Tuner *241*
5.7.2.3 Echo *241*
5.7.2.4 The Third Miracle of St. Mark Street *242*
5.7.2.5 Combination of Frequencies. The MIEZE Spectrometer *243*
5.7.2.6 The Fourth Miracle of the St. Mark Street *244*
5.7.2.7 Transformation of a Continuous Polarized Beam into a Pulsed One *244*
5.8 Neutron Holography *247*
5.8.1 General Principles of Holography *247*
5.8.1.1 A Distinguishing Feature of Our Approach *247*
5.8.1.2 Holographic Image *248*
5.8.1.3 Reproduction of the Scattered Field from the Hologram *249*
5.8.2 Spin Precession Wave *250*
5.8.3 Hologram of a Scattered SPW without a Reference Wave *252*
5.8.4 Magnetic Scattering of a SPW with Spin Flip *253*
5.8.5 Entangled States. A Fundamental Question of Quantum Mechanics *254*
5.9 Conclusion *255*

6 Optical Phenomena in Neutron Diffraction and Scattering *257*
6.1 Diffraction on Regular Lattices *257*
6.1.1 Borrmann Effect *257*
6.1.2 Darwin Table *260*
6.2 Inhomogeneity Scattering *265*
6.2.1 Reflection and Transmission Formulas for a Thick Sample *265*
6.2.2 Measurements of Inhomogeneity Scattering in the Long–Wavelength Region *266*
6.3 Small-Angle Scattering *268*
6.3.1 Analysis of the Guinier–Preston Zone *271*
6.3.2 Separation of Nuclear and Magnetic Structures *274*
6.3.3 Developments of Small Angle Scattering Spectrometers *275*
6.4 Diffraction in Time *277*
6.4.1 Quantum Shutter Analysis *277*
6.4.2 Fame-Overlapping Time-of-Flight Quantum-Chopper Spectrometer *280*
6.5 Neutron Holography Experiment *283*
6.5.1 Discovery of the Basic Idea of Holography *283*
6.5.2 Theoretical Principle of Wave Front Reconstruction *285*
6.5.3 Neutron Holography Experiment *287*

7 Dynamical Diffraction in Three-dimensional Periodic Media *291*
7.1 Diffraction on a Single Crystalline Plane *292*
7.1.1 Multiple Wave Scattering Theory *293*

7.1.2 Scattering on an Infinite Crystalline Plane *293*
7.1.3 Scattered Field of the Crystalline Plane *294*
7.1.4 An Idea for Calculation of S *296*
7.2 Diffraction from a Semi-infinite Single Crystal *299*
7.2.1 Properties of Dyadic Matrices *300*
7.2.2 Solution of Eq. (7.34) *301*
7.2.3 Bragg Diffraction *303*
7.2.3.1 Specular Bragg Diffraction *303*
7.2.3.2 Angular Width of the Specular Bragg Diffraction *305*
7.2.3.3 Nonspecular Bragg Diffraction *306*
7.3 Diffraction on a Crystal of Finite Thickness *307*
7.3.1 Eigenvectors and Eigenvalues of the Operator $\hat{X}$ *308*
7.3.2 All $e_n^2 = \exp(ik_{n\perp}a)$ Are Different *310*
7.3.3 Two Values of e_n Are Close to each other *310*
7.3.4 The Bloch Wave Vector at a Specular Bragg Diffraction *312*
7.3.5 Specular Bragg Diffraction *313*
7.3.6 Total Reflection of a Mosaic Crystal *314*
7.3.7 Extinction Coefficients *316*
7.3.7.1 Incoherent Reflectivity from N Equal Blocks *316*
7.3.8 The Laue Diffraction *318*
7.3.8.1 Simultaneous Fulfillment of Bragg and Laue Conditions *319*
7.3.8.2 Diffraction and Renormalization of the Scattering Amplitude *320*
7.4 The Standard Description of Diffraction in Single Crystals *321*
7.4.1 The Fermi Pseudopotential *321*
7.4.2 The Standard Dynamical Diffraction Theory *322*
7.4.2.1 Specular Bragg Diffraction *323*
7.4.2.2 Symmetrical Laue Diffraction *324*
7.5 Comparison of the Two Methods *325*
7.6 Kinematical Diffraction Theory and Practical Crystallography *326*
7.7 Unitarity and the Detailed Balance Principle *328*
7.7.1 Unitarity *329*
7.7.2 The Detailed Balance Principle *330*
7.8 The Optical Potential u_0 *332*
7.8.1 Kronig–Penney Corrections to the Optical Potential *333*
7.8.2 Diffraction Correction to the Optical Potential *334*
7.8.3 Digression on the Optical Potential *335*
7.8.3.1 Optics *335*
7.8.3.2 Neutrostriction in Neutron Stars *337*
7.8.4 The End of the Digression *338*
7.9 Conclusion *338*

8 Disordered Media. Incoherent and Small-Angle Scattering *339*
8.1 Wave Processes in Three-Dimensional Disordered Media *339*
8.1.1 Equations of MWS *339*
8.1.2 Solution for Coherent Values *341*

8.1.3 Renormalization of the Scattering Amplitude *343*
8.2 Scattering *344*
8.2.1 Propagation of Scattered Waves inside the Medium *346*
8.2.2 Scattering with the Modified Green Function *347*
8.2.2.1 Scattering on a Single Point Scatterer inside the Medium *347*
8.2.2.2 Scattering on Fluctuations of $\delta(\psi(\mathbf{r}_j)b_j)$ *348*
8.2.2.3 Yoneda Effect [368] *350*
8.2.2.4 Ultracold Neutrons *350*
8.2.2.5 Scattering on a Fixed Center outside the Medium *352*
8.2.2.6 Scattering because of Disorder *353*
8.2.3 Experience with an Experiment *355*
8.2.4 Small-Angle Scattering on a Single Inhomogeneity inside the Medium *356*
8.2.4.1 Asymptotic of the Structure Factor at Small κ *359*
8.2.4.2 Asymptotic of the Structure Factor at Large κ *359*
8.2.4.3 Porod Invariant for the Structure Factor *359*
8.2.5 Small-Angle Scattering on a Set of Inhomogeneities *360*
8.2.6 Scattering on Roughnesses at the Interface *362*
8.2.6.1 Common Approach to Roughness Scattering [372] *362*
8.2.6.2 Roughness of Interfaces in Multilayers *367*
8.3 Scattering on Magnetic Inhomogeneities *368*
8.3.1 The Spinor Green Function *369*
8.3.2 The Scattered Wave Function *371*
8.3.2.1 Scattering on a Helicoid in Matter *372*
8.3.2.2 General Calculation of Scattering Amplitude on an Inhomogeneity *373*
8.3.2.3 Calculations of Cross Sections *375*
8.4 Neutron Albedo *376*
8.4.1 Main Notions *376*
8.4.2 Albedo of an Infinitely Thick Wall *377*
8.4.3 Extinction of the Neutron Density in Matter *380*
8.4.4 Albedo of a Wall of Thickness L *380*
8.4.5 Comparison with Diffusion Formulas for the Albedo *381*
8.4.6 Albedo of Ultrathin Powdered Matter *382*
8.4.7 Angular Distribution of Reflected Neutrons for Isotropic Scattering *383*
8.4.8 Angular Distribution for Anisotropic Scattering *384*
8.5 Conclusion *387*

9 Neutron Interference and Phase Observation *389*
9.1 Neutron Interferometry with a Perfect Crystal Interferometer *390*
9.1.1 Triple Laue Diffraction Type Silicon Crystal Interferometer *390*
9.1.2 Experimental Verification of the Equivalence Principle on the Gravity Effect *392*
9.1.3 Coherence Length and Dispersive/Nondispersive Arrangements *396*
9.1.4 Stochastic and Deterministic Processes in the Absorbing Element *399*

9.1.5 Study of Quantum Entangled States with a Skew–Symmetrical Perfect Crystal Interferometer *402*
9.2 Experiments on Spin Behavior with a Perfect Crystal Interferometer *404*
9.2.1 4π Periodicity of Spin Precession in a Magnetic Field *404*
9.2.2 Complementary Principle in a Double-Resonance Experiment *405*
9.2.3 Experimental Test of the Einstein–Podolsky–Rosen Paradox *409*
9.3 Very Cold Neutron Interferometry with Grating Mirrors *413*
9.3.1 Development of a Grating Mirror Interferometer and Experiment on the Gravity Effect *413*
9.3.2 Experimental Verification of Aharonov–Bohm Effect *416*
9.4 Time-Dependent Interferometry *420*
9.5 Neutron Interferometry with Single-Layer and Multilayer Mirrors *422*
9.6 Spin Interferometry Experiments with Magnetic Multilayer Mirrors *426*
9.6.1 Development of a Phase-Spin-Echo Interferometer *426*
9.6.2 Observation of the Transverse Coherence Length *428*
9.6.3 Observation of Quasi-Bound States of the Neutron in a Magnetic Resonator *431*
9.6.4 Observation of Larmor Precession on Resonant Tunneling Neutrons *435*
9.6.5 Experimental Test of Wave Packet Theories with Very Cold Neutron Interferometry *437*

10 Contradictions of the Scattering Theory, and Quantum Mechanics *445*
10.1 How many Scattering Theories Do We Have? *445*
10.1.1 Theory of Spherical Waves *446*
10.1.2 The Standard Scattering Theory *447*
10.1.3 The Fundamental Scattering Theory *447*
10.2 Analysis of the Three Theories *447*
10.2.1 Scattering of Spherical Waves *447*
10.2.1.1 Contradiction 1 *449*
10.2.1.2 The Need for a Cross Section *449*
10.2.2 Cross Section in the Standard Theory *449*
10.2.2.1 Contradiction 2 *449*
10.2.3 Fundamental Scattering Theory [457, 458] *450*
10.2.3.1 Contradiction 3 *451*
10.3 General Definition of the Scattering Cross Section *452*
10.3.1 Phenomenological Definition of the Cross Section in the Self-Consistent Theory of Spherical Waves *452*
10.3.2 Relation between Cross Section and Probability *453*
10.3.2.1 A Consequence of the Introduction of the Parameter A *454*
10.3.2.2 Relation between the Cross Section and the Probability in the Fundamental Theory *454*
10.3.2.3 Scattering of Wave Packets *455*
10.3.2.4 Contradiction 4 *455*
10.4 Wave Packets *455*
10.4.1 Contradiction 5 *456*

10.4.1.1 The Singular de Broglie Wave Packet 456
10.4.2 Estimation of s 456
10.4.3 Contradiction 6 457
10.4.4 Resolution of Contradiction 6. Quantization of the Scattering Angle 457
10.4.4.1 The Value of the Angular Quantum and Resolution of Contradiction 6 459
10.4.5 Final Remarks 459
10.5 Details 459
10.5.1 Main Points of the Fundamental Scattering Theory According to [457, 458] 460
10.5.2 Elastic Scattering of a Wave Packet on a Fixed Center 462
10.5.3 Transition from Probabilities to Cross Sections 463
10.5.3.1 Transition to the Cross Section According to [457] 464
10.5.3.2 Transition to the Cross Section According to [458] 464
10.5.3.3 Analysis of Procedures in [457, 458] 464
10.6 The Hot Problems of Quantum Mechanics 465
10.6.1 The EPR Paradox, Its Logic, Error, and Consequences 465
10.6.1.1 Entangled States 466
10.6.1.2 Error of the EPR Paper and Definition of Physical Values 467
10.6.1.3 The Meaning of the Uncertainty Relations and of Dispersion 468
10.6.1.4 Locality, Nonlocality, Measurements, and Hidden Parameters 469
10.6.1.5 A Pair of Spin-1/2 Particles in an s-State 470
10.6.1.6 Nonlocality with Photon Pairs 472
10.6.1.7 Results of the First Experiment 473
10.6.1.8 The Experiments of Aspect *et al.* 474
10.6.2 Interpretation of Quantum Mechanics and Hidden Variables 476
10.6.2.1 Interference in Classical Physics 476
10.6.2.2 Nonlinear System of Classical Equations Instead of the Schrödinger Equation 477
10.6.3 Discussion of some Experiments with Wave Packets 478
10.7 Modern Art in Neutron Optics 479
10.8 Conclusion 482

11 Application of Neutron Optics to Material Science 485
11.1 Nuclear Scattering Density Profile Analyses of Layered Materials 485
11.2 Polarized Neutron Reflectometries 487
11.2.1 Penetration Depth in Superconductors 487
11.2.2 Reference Layer Methods for Multilayers 492
11.2.3 Nonspecular Reflection from Magnetic Multilayers 493
11.2.4 Magnetic Materials with Artificial Properties 494
11.3 Neutron Scattering Experiments in Various Wavelength Regions 498
11.3.1 Graphite Inhomogeneity Structure Observed with Very Cold Neutrons 498
11.3.2 Solid Water Inhomogeneity Structures Observed with Cold and Very Cold Neutrons 500

11.3.3 Mechanically Alloyed Graphite Structure Observed with Neutron Diffraction 504
11.4 Small Angle Scattering Experiments on Metallic Alloys 506
11.4.1 Observation of Spinodal Decomposition 506
11.4.2 Precipitation and Segregation in an Alloy 508
11.5 Defects in Material 509
11.5.1 Matter under Pressure. Stress and Strain 509
11.5.2 Texture of Matter 510
11.5.3 Small-Angle Neutron Scattering 513
11.5.4 Neutron Radiography 514
11.5.5 Phase-Contrast Neutron Radiography 517
11.5.5.1 The Theory of Phase Contrast 518
11.5.5.2 The Phase Contrast for a Plane Incident Wave 518
11.5.5.3 The Phase Contrast for a Point Monochromatic Source 520
11.5.5.4 Application of Phase-Contrast Neutron Radiography 521
11.6 An Example of Novel Applications of Spin Echo Spectrometry – Study on Dairy Products 522

12 Application of Neutron Optics in Biophysical/Biological Research 527
12.1 Contrast Variation 528
12.1.1 Isotope Substitution 529
12.1.2 Spin Contrast Variation 531
12.2 Scattering from a Dilute Ensemble of Monoparticles 533
12.2.1 Properties of $\gamma(\boldsymbol{r})$ 535
12.2.2 Properties of $I(q)$ 536
12.2.3 Some Special Forms of Particles 537
12.2.3.1 Spherical Particles: $\rho(\boldsymbol{r}) = \rho(r)$ 537
12.2.3.2 Spikes 538
12.2.3.3 Lamellar Particles 538
12.3 Multipole Expansion of the Scattering Density 539
12.3.1 Derivation of Eq. (12.54) 541
12.3.2 Isometric Particles 542
12.4 Small-Angle Scattering 542
12.4.1 Guinier Region. Scattering from a Dilute Monoparticle Solution 542
12.4.2 Small-Angle Scattering from a System of Two and more Different Particles 546
12.5 Reflectometry on Protein and Biomaterials 548
12.5.1 A New Tool to Study Protein Interactions 548
12.5.2 Membrane Investigations 549
12.6 Diffraction on Biological Macromolecules 551
12.7 Neutron Microscope for Biological Samples 553

Basic Formulas in Neutron Optics 557

References 559

Index 583

Preface

This book is one of the final products of very happy and close collaborations between us during the last 15 years.

A short message from Masahiko Utsuro:

In many previous textbooks, it has been written that the unique advantages of the neutron as a scientific probe are its electrical neutrality, its magnetic moment, and its mass, which is comparable to that of a hydrogen nucleus, and a further advantage is an average lifetime in the free state sufficiently long for physical experiments. However, I would like to mention accessibility to neutron studies. It is not an easy task for one in a small laboratory to prepare neutrons for use in physics experiments, in contrast to electrons and photons (X-rays and laser light). I think, however, this is not an unfortunate situation, but rather a happy one, because to start neutron studies we must go and work at some neutron laboratory, or must approach one of the neutron-user facilities that nowadays exist worldwide. Such efforts to gain access to neutrons lead us to join very nice neutron communities around such facilities, and together we can enjoy the scientific history of neutron research over about 50 years. Thus, my message to the readers:

> Let's speak of neutrons, then we will immediately become old friends.

I hope this book helps the readers to enjoy such good collaborations in neutron science.

I would like to express my deep gratitude to my research colleagues for their valuable collaborations with me, especially to Prof. Albert Steyerl for his kind collaboration on my experiments with very cold neutrons at the Munich reactor 30 years ago which formed the start of our fruitful communications thereafter, to Dr. Peter Geltenbort for his always well arranged and very effective directions for my experimental studies at the Grenoble reactor, to Dr. Vladimir Ignatovich, the coauthor of this book, for his stimulating discussions during my few visits to Dubna, and further to Dr. Mathias Hetzelt, who opened the door for me to the ultracold neutron experiment 30 years ago at the Grenoble reactor.

I would also like to express my sincere gratitude to the late Prof. Kazuhiko Inoue, my teacher at the start of my neutron studies at the Kyoto University Research Reactor Institute. Even now I recall his words on his thoughts about science: *True*

Neutron Optics. Masahiko Utsuro, Vladimir K. Ignatovich

ISBN: 978-3-527-40885-6

scientists were the scientists before the eighteenth century, who need not do science for the salary, nor for the professional position. He thereafter invented the excellent cold neutron spectrometer LAM for pulsed neutron sources at KENS.

Finally, I would like to thank my family for their continuous support for me during my scientific studies.

Now, I hand my pen to my coauthor, Vladimir Ignatovich

A message from Vladimir Ignatovich:

My chapters in this book summarize all my research in neutron optics. It is not a compilation from other textbooks. Some time ago, when I was studying the interaction of ultracold neutrons with the walls of a storage vessel, I was lucky to find an analytical method for the description of neutron scattering in layered media. I was excited to find a simple Fresnel-like formula for reflection of neutrons (and X-rays) from an *arbitrary* periodic potential. It seemed unbelievable, because periodic media after more than 100 years of study were so well understood! I thought that my approach would be accepted by the scientific community with enthusiasm. However, after more than 30 years, it still remains unnoticed.

Two years ago I was invited by Prof. V.L. Aksenov to give lectures on neutron optics to students of neutronography in the physics faculty of M.V. Lomonosov Moscow State University. It was a pleasure for me, because I like to talk. To be well prepared for every lecture, and to give the students an opportunity to read the lectures afterwards or in advance, I started this book. So, on the one hand it is a collection of lectures for students, and on the other hand it is a monograph, which, I hope, will be useful for neutron, optics, and solid-state physicists.

The title of the book may give the impression that it is devoted to a very narrow branch of neutron physics, but its scope is considerably broader. Here neutron optics is a tool which helps us explore all of quantum mechanics. It helps to reveal the beautiful features and weaknesses of quantum mechanics. It was a great pleasure for me not only to give the lectures, but also to write this book. I felt free to express my opinion and to justify it mathematically.

This book, to be brief, is about neutron reflection from and transmission through multilayered mirrors, diffraction in one-dimensional and three-dimensional periodic systems, elastic scattering in random media, interaction with stationary and nonstationary magnetic fields, and at the end it shows the inconsistency of the present-day standard scattering theory, which is the main mathematical tool used in the whole book. For me, as a coauthor, it is not surprising that science develops so fast, notwithstanding the weakness of some ideas; however, I believe that elucidation of weak points is very important for further progress, because it liberates our minds from dogmatic tenets of accepted ideology. After pointing out the contradictions of the present-day theory, I show a way to resolve them, and say why these contradictions do not cancel out the results presented earlier in the book.

I want to express my deep gratitude to V.L. Aksenov for his invitation, and to many colleagues at the Neutron Physics Laboratory of the Joint Institute for Nuclear Research, Dubna, for fruitful collaboration over so many years, especially to

A.V. Strelkov, Yu.N. Pokotilovskiy, and E.P. Shabalin. I would also like to mention several people who played a very important role in my life. They are as follows: my friend Mathias Hetzelt, the grandson of the famous Russian scientist S.P. Timoshenko; Igor Carron, who 20 years ago was a student at A&M Texas University and arranged wonderful trips for me to many scientific centers in the United States; Albert Steyerl, Steve Lamoreaux, Robert Golub, Peter Geltenbort, and Roland Gähler, with whom I had pleasure to collaborate; Prof. M. Utsuro, who invited me to Japan for a whole year (it was the greatest time of my life), and my colleagues and friends there. I am also grateful to my wife, who during our life together has helped me so much.

At the same time, I want to devote this book to the blessed memory of my father, who was shot dead in the sheer hell of the Stalin era, when I was only half a year old, and to my mother, who notwithstanding terrible difficulties, being branded a wife of the people's enemy, was able to raise me and give me an education. I also want to devote this book (who knows whether I will be able to write another one) to the dear memory of my teacher F.L. Shapiro. When he was alive, I felt like a court theoretician.

My fruitful collaboration with Japanese neutron scientists started at my invited stay for 1 year in the middle of the 1990s at the Kyoto University Research Reactor Institute in Kumatori, Osaka. This book is one of the outcomes of this collaboration on neutron optics. My coauthor (Masahiko Utsuro) prepared several chapters as reviews of the most important experimental works in the field of neutron optics, so my theoretical part is a good complement of use to experimenters. I hope that the result of our collaboration will be helpful not only to active participants in research in this field, but also to students who are choosing the direction of their scientific activity. I am also very grateful to the Ministry of Sport, Science, and Culture for giving me an opportunity to work for a whole year in the beautiful country of Japan.

> Genuine is relativity theory,
> Great is field theory and the physics of elementary particles,
> Stars capture imagination,
> But nothing in the world is as interesting as neutron optics.

Time changes everything, and the time given to me until the final submission of my last manuscript was used to improve the book even more. Some material which requires scrupulous calculations has been omitted. Some new material has been added, and some derivations have been made clearer. The part devoted to the fundamental questions of quantum mechanics has been enlarged.

In this edition, I would like to express my gratitude to George Soros, who helped Russian scientists and me too in the poor times of perestroika. I am also grateful to my son Filipp for his help in the preparation of this edition, and I would like to express my special thanks to Edward Kapuscik and the editorial board of the journal *Concepts of Physics. The Old and New* for publication of my articles, including discussions with referees. I think discussions with referees are an important part of scientific work, and they are worthy of publication.

During my scientific life, I have participated in many conferences and collaborated with scientists from many scientific centers and I am very grateful to them.

We are grateful to Elsevier, APS, IOP and Springer for permission to reproduce the figures from their journals.

As an additional remark for the reader, several fundamental textbooks on quantum mechanics, neutron physics, and neutron optics that are referred to throughout this book are listed at the front of the book.

November, 2009

Masahiko Utsuro (in Kumatori)
Vladimir Ignatovich (in Dubna)

Textbooks

A Pauli, W. (1980) *General Principles of Quantum Mechanics*, Springer-Verlag, Berlin, Heidelberg, New York (translated by P. Achuthan, K. Venkatesan).

B Landau, L.D., Lifshitz, E.M. (1977) *Quantun Mechanics (Nonrelativistic Theory)*, 3rd edn, Pergamon Press, Oxford.

C Schiff, L. (1968) *Quantum Mechanics*, 3rd edn, McGraw-Hill, New York.

D Byrne, J. (1994) *Neutron, Nuclei and Matter – An Eploration of the Physics of Slow Neutrons*, Inst. of Physics Pub., Bristol, Philadelphia.

E Gurevich, I.I, Tarasov, L.V. (1965) *Physics of Low Energy Neutrons* (eds R.I. Sharp, S. Chomet), Nauka, Moscow (translated by Scripts Technica Ltd., North-Holland Pub. Co., Amsterdam, 1968).

F Steyerl, A. (1977) *Neutron Physics*, in: Springer Tracts in Modern Physics, Vol. 80, Springer-Verlag, Berlin, Heiderberg, New York.

G Ignatovich, V.K. (1990) *The Physics of Ultracold Neutrons (UCN)*, Clarendon Press, Oxford.

H Golub, R., Richardson, D., Lamoreaux, S.K. (1991) *Ultra-Cold Neutrons*, Adam Hilger, Bristol, Philadelphia, New York.

I Sears, F.V. (1989) *Neutron Optics*, Oxford Univ. Press, New York, Oxford.

J Kawano, S., Kawai, T., Kawaguchi, A., Utsuro, M. (Eds) (1996) Proc. of Int. Symp. *Advance in Neutron Optics and Related Research Facitilies*, (Neutron Optics in Kumatori '96, NOK'96); J. Phys. Soc. Japan. Vol. 65, Suppl. A, Tokyo; Phys. Soc. Japan.

K Achiwa, N., Ebisawa, T., Kawai, T., Tasaki, S., Hino, M., Yamazaki, D. (2003) *Neutron Spin Optics* (eds N. Achiwa), Fukuoka, Kyushu Univ. Pub. Soc. (in Japanese).

Neutron Optics. Masahiko Utsuro, Vladimir K. Ignatovich

ISBN: 978-3-527-40885-6

1
Wave–Particle Duality of the Neutron

1.1
Discovery of Matter Waves

The fundamental fact that every particle with mass is at the same time a wave was discovered in 1925 by *de Broglie* [1]. He was honored with the award of the Nobel Prize for Physics in 1929 for this discovery. We begin our study of neutron optics with his proof of *the matter wave* following his Nobel lecture [2].

De Broglie was of the opinion that to solve the new serious question in physics that arose around 1900, the unification of matter and radiation was necessary, and more practically said that it should be possible to establish the equality of corpuscular motion and wave propagation. As the simplest case, he assumed a system consisting of a corpuscle at rest and completely free from all outside influence, and expressed the system as $\circ x_0 y_0 z_0$. In the sense of Einstein's relativity principle, we can consider this system being the "intrinsic" system of the corpuscle. Since the corpuscle is steady and at rest, the phase of the wave we are now considering must be the same at every point; that is, it can be expressed in the form $\sin[2\pi\nu_0(t_0-\tau_0)]$, where t_0 is the intrinsic time for the corpuscle and τ_0 is a constant.

As the next step, according to the principle of inertia, in every Galilean system we can make the corpuscle have linear motion and constant velocity. Let us consider such a Galilean system where the corpuscle has velocity $v = \beta c$. We will not lose generality by taking the x-axis as the direction of motion. According to the Lorentz transformation, the time t elapsing for the observer in this new system will be related to the intrinsic time t_0 defined above through the equation

$$t_0 = \frac{t - \frac{\beta x}{c}}{\sqrt{1-\beta^2}}\,, \tag{1.1}$$

and therefore the phase of the wave for the present observer will be given by

$$\sin\left[2\pi\frac{\nu_0}{\sqrt{1-\beta^2}}\left(t - \frac{\beta x}{c} - \tau_0\right)\right]. \tag{1.2}$$

Therefore, for the observer the wave will now have frequency

$$\nu = \frac{\nu_0}{\sqrt{1-\beta^2}}\,, \tag{1.3}$$

Neutron Optics. Masahiko Utsuro, Vladimir K. Ignatovich

ISBN: 978-3-527-40885-6

and will then propagate in the direction of the x-axis with *phase velocity*

$$V = \frac{c}{\beta} = \frac{c^2}{v} . \tag{1.4}$$

On the other hand, we can define *the group velocity* U for the wave as the velocity corresponding to the resultant amplitude from a group with very similar frequencies, and according to Rayleigh's definition for this velocity, $U = \partial\omega/\partial k$, where the wave number $k = 2\pi\nu/V$, it satisfies the equation

$$\frac{1}{U} = \frac{\partial\left(\frac{\nu}{V}\right)}{\partial\nu} = \frac{1}{v} . \tag{1.5}$$

In this way, we obtain the very important relation for the development of the present theory that the group velocity for the waves in the system $xyzt$ is equal to the velocity of the corpuscle.

To achieve our purpose to establish the equality of the corpuscle and the wave, we must combine the energy and the quantity of the motion. In the same way as in the previous Galilean transformation,

$$\text{Energy} = h \times \text{frequency} , \quad \text{or} \quad W = h\nu , \tag{1.6}$$

where h is Planck's constant. This relation reduces further according to the Einstein relation to its internal energy m_0c^2 in the intrinsic system as

$$h\nu_0 = m_0c^2 , \tag{1.7}$$

where m_0 is the rest mass. Since the quantity of movement, that is, the momentum, $\boldsymbol{p}$, has magnitude equal to $m_0v/\sqrt{1-\beta^2}$, then

$$p = |\boldsymbol{p}| = \frac{m_0 v}{\sqrt{1-\beta^2}} = \frac{Wv}{c^2} = \frac{h\nu}{V} = \frac{h}{\lambda} , \tag{1.8}$$

where the quantity λ is defined as the distance between two consecutive peaks of the wave (which corresponds to the phase velocity divided by the frequency), that is, *the wavelength*. In this way, we obtain the very important relation

$$\lambda = \frac{h}{p} . \tag{1.9}$$

This is *de Broglie's fundamental formula.*

De Broglie's matter wave was experimentally verified in the first place by Davisson and Thomson with *the discovery of the diffraction of electrons* [3, 4], and they were also awarded the Nobel Prize for Physics, in 1937. A few months after their experiment, Kikuchi reported the characteristic pattern of electron diffraction owing to the effects of thermal diffuse scattering from crystals (the so-called *Kikuchi pattern*) [5], and made an important contribution to the establishment of quantum mechanics developed by Heisenberg [6].

Pauli's textbook of quantum mechanics [A] starts with the following description: *The last decisive turning point of quantum theory came with de Broglie's hypothesis of matter waves, Heisenberg's discovery of matrix mechanics, and Schrödinger's wave equation, the last establishing the relationship between the first two sets of ideas.*[1)]

1.2 Proof of the Wave Nature of the Neutron

1.2.1 Bragg Reflection

Chadwick's discovery of the neutron in 1932 [7, 8] soon motivated the experimental verification of its *wave nature*. The approach is the same as that for the first proof on the matter wave of the electron, where *Bragg scattering* was observed [3, 4], but for the neutron, with a much larger mass and essentially no electric charge, it has significant advantages in view of the energy condition and of the electromagnetic effects in crystals. Actually, the wavelength calculated with Eq. (1.9) for neutrons with energy corresponding to room temperature (so-called *thermal neutrons*) is of the same order as the lattice spacing in most simple crystals, about 0.2 nm ($1\ \text{nm} = 10^{-9}\ \text{m} = 10\ \text{Å}$), and those neutrons that easily penetrate inside a crystal can be scattered directly by nuclei in the crystal, resulting in obvious *Bragg reflections*. As an example of such experiments, Mitchell and Powers [9] irradiated a single crystal of MgO with the neutron beam extracted from a paraffin moderator in which a Rn–Be neutron source was embedded, and they considered the contribution to the counting rate due to Bragg scattering as proof of the existence of a coherent component in nuclear scattering by the crystal nuclei.

However, the epoch-making event to initiate drastic developments of various kinds of neutron experiments was the realization of the first nuclear chain reaction in the reactor CP-1 conducted by Fermi in 1942. Furthermore, the first heavy water reactor, CP-3 in 1944, opened the door to precise neutron experiments by using a high-intensity neutron beam. Zinn [10] used the experimental devise shown in Figure 1.1a and reported for the first time a very clear distribution of Bragg-scattered neutrons, as shown in Figure 1.1b.

Thermal neutrons extracted from a reentrant hole in the graphite thermal column of the heavy water reactor are well collimated through an iron collimater in the reactor shielding and a couple of thick cadmium slits outside the shielding, and illuminate the single crystal on a rotating sample table. The neutrons scattered by the sample are registered by the BF_3 proportional counter at the end of the precisely rotating arm of a large mechanical device. The sample table and the counter arm are driven exactly with a 1 : 2 angular ratio. The results obtained on the sample of

1) Schrödinger followed de Broglie's idea of matter waves in setting up his equation. Later he proved the equivalence of his approach and that of Heisenberg.

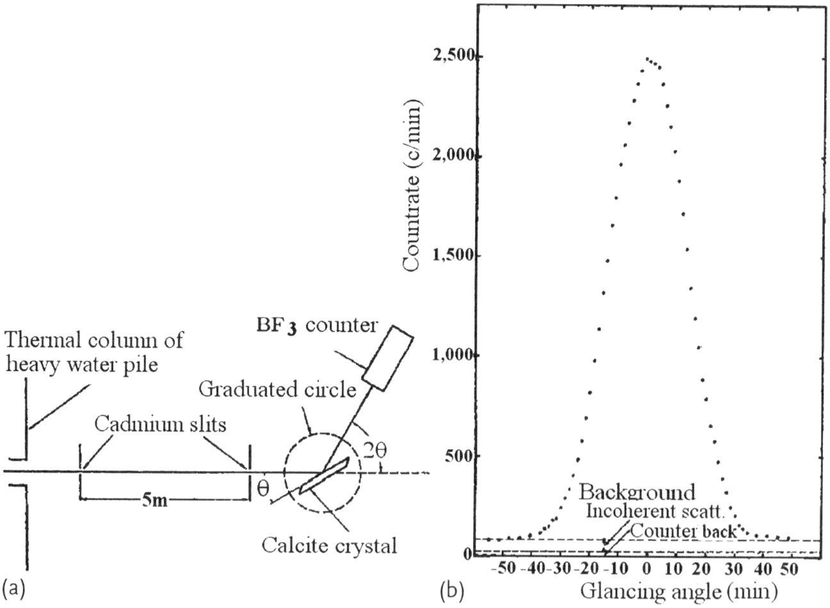

Figure 1.1 The first obvious measurement of Bragg-scattered neutrons. (a) Experimental setup; (b) a typical rocking curve of reflected neutrons measured on the (1,0,0) plane of a LiF crystal (Zinn [10]).

a LiF single crystal are given in Figure 1.1b. The distribution indicated is shown before the corrections for the detector efficiency and the energy resolution of the setup, and therefore a slight asymmetry is noticed. The apparatus was also used for measuring the energy distribution of incident thermal neutrons by replacing the sample with a larger single crystal of calcite, as shown in Figure 1.1a.

Furthermore, the whole apparatus was moved to a beam hole inserted directly in the reactor core, and was used for nuclear cross section measurements in a wider energy range, including *epithermal neutrons.*[2)]

Such neutron experiments, a typical setup for which is shown in Figure 1.1, proved that the diffraction experiment is possible using a device and setup quite similar to those for X-rays. However, in contrast to X-rays scattered mainly by atomic electrons, neutrons are mainly scattered or absorbed by the nucleus, and therefore quite different materials are utilized for radiation shielding for neutrons.

After these initial developments, neutron spectroscopy experiments made great progress, and nowadays various kinds of neutron spectrometers are used in experiments applied to a wide variety of samples, including crystalline solids, alloys,

2) Those neutrons with energy higher than the room temperature Maxwellian distribution but lower than the resonance region for nuclear reactions; that is, the energy region of about 0.1–1 eV.

complicated compounds, polymers, and biological material. Typical cases of these applications will be given in the last two chapters.

Among these spectroscopies, the method based on Bragg scattering with crystals called *crystal spectrometry* is one of the most popular and important approaches. The relation for the spectroscopic resolution can be derived from the well-known *Bragg law* on the neutron wavelength λ of Bragg scattering for lattice spacing d and neutron incident angle θ of the crystal:

$$n\lambda = 2d \sin\theta\,, \quad \text{or} \quad \tau = 2k \sin\theta\,, \tag{1.10}$$

where $\tau = 2\pi n/d$, $k = 2\pi/\lambda$, and n is an integer. Differentiation of both sides of the logarithm of Eq. (1.10) and the expression for the neutron energy E,

$$E = \frac{\hbar^2}{2m}k^2 = \frac{\hbar^2}{2m}\left(\frac{2\pi}{\lambda}\right)^2\,, \tag{1.11}$$

lead to

$$\Delta E/E \cong 2\Delta k/k \cong 2[(\Delta\tau/\tau)_{hkl} + \cot\theta\,\Delta\theta]\,, \tag{1.12}$$

where τ_{hkl} is the magnitude of the *reciprocal lattice vector* with *Miller indices* hkl for a three-dimensional crystal, and $\Delta\theta$ is the beam divergence.

This equation indicates that the highest energy resolution can be obtained by employing a Bragg angle of $\theta = \pi/2$. A novel technique based on the present principle was developed as *backscattering spectrometry* [11] for neutron scattering experiments with very fine energy resolution.

1.2.2
Refractive Index and Total Reflection

At the same time as these experimental developments, construction of an optical theory for neutrons was also carried out by taking into account the nuclear scattering, which has very different characteristics from scattering of X-rays. For the analysis of neutron optics, we must first describe the matter waves with the wavelength and the frequency given by de Broglie's fundamental formula (Eq. (1.9)) as a function of variables for practical situations. The wave function required to describe the matter wave and the related general fundamental equation in wave mechanics had already been by Schrödinger in 1926. Furthermore, in 1936 Fermi presented the simplest and most effective expression for the nuclear scattering potential to be inserted in the Schrödinger equation to obtain a solution [12, 13], the so-called *Fermi pseudopotential*, written in the form

$$V(\boldsymbol{r}) = 4\pi \sum_j b_j \delta(\boldsymbol{r} - \boldsymbol{r}_j)\,, \tag{1.13}$$

where b_j and $\boldsymbol{r}_j$ denote the scattering amplitude[3] and the position, respectively, of the jth nucleus.

3) In many textbooks it is called *the scattering length*, but here we denote it *the scattering amplitude*.

Starting from these fundamental arrangements, Foldy [14], Goldberger and Seits [15], and Lax [16] analyzed the interference phenomenon for neutrons under multiple scattering in media, then the preliminary theory of neutron optics was established. The details of the theory will not be discussed here since many standard textbooks on neutron scattering (e.g., [D]–[E]) have already been published, and here only one of the useful formulas will be given for the *index of neutron refraction n*, and derived by Goldberger and Seits [15] in the case of sufficiently weak absorption and without the effects of neutron spin:

$$1 - n^2 = \pm \frac{N(4\pi\sigma_s)^{\frac{1}{2}}}{k^2} \,. \tag{1.14}$$

In Eq. (1.14), N is the atomic density, σ_s the coherent scattering cross section, k the wave number of neutrons without the medium. To determine whether the right side should be positive or negative, information about the scattering nucleus, that is, the definition of the sign of the *coherent scattering amplitude (coherent scattering length)* becomes necessary. Foldy [14] carried out general analyses including randomly distributed scatterers, and Lax [16] discussed the effects of *incoherent scattering* and of possible anisotropy in the scattering amplitude.

According to Eq. (1.14), total reflection of neutrons will happen if we select the grazing angle θ between the incident neutrons and the surface of the medium with the positive sign on the right side such as to satisfy the condition

$$\sin^2\theta \le \sin^2\theta_c = 1 - n^2 \,. \tag{1.15}$$

In other words, whether total reflection from any material happens or does not happen for incident neutrons in a vacuum (or in atmospheric air) and, further, what the *critical angle for total reflection* θ_c is should give us information about the sign to be selected and the magnitude of the scattering amplitude $|b| = (\sigma_s/4\pi)^{1/2}$ on the right side of Eq. (1.14) for the element in the material.

Fermi and Marshall [17] modified the experimental setup shown in Figure 1.1 to the arrangement shown in Figure 1.2, where a monochromatic neutron beam with a wavelength of 0.1873 nm Bragg-reflected by the first crystal was incident on the front surface of a solid sample on the second turntable with a very small grazing angle.

The second turntable was provided with a detector arm, and the detector count rates were registered with precise alteration of the detector angle. Such a set of measurements was repeated for fine stepwise increments of the incidence angle to the sample, and clearly indicated the ending of total reflection with a sudden decrease of the reflected intensity. Thus, they obtained the values of the critical angle for total reflection on specimens of beryllium, graphite, iron, nickel, zinc, and others in the angular range of 7–12 minutes with the accuracy of about a tenth of a minute.

According to the present definition, the *refractive index formula* (1.14) and the *total reflection formula* (1.15) are reduced, respectively, to

$$n^2 = 1 - \frac{N b_{\mathrm{coh}} \lambda^2}{\pi} \tag{1.16}$$

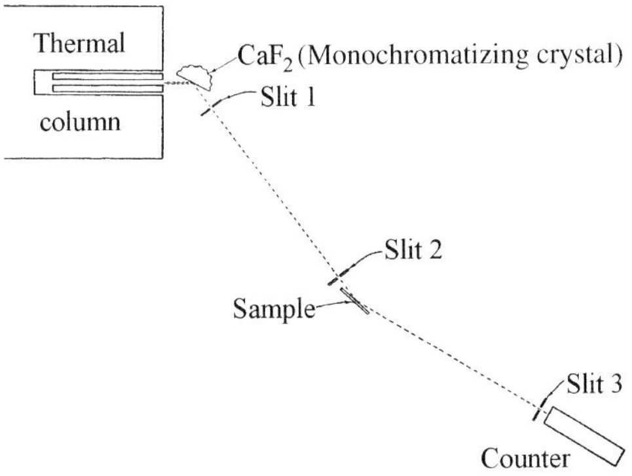

Figure 1.2 Arrangement of the total reflection experiment with a monochromatic neutron beam (Fermi and Marshall [17]).

and

$$\sin^2\theta \le \sin^2\theta_c = \frac{N b_{coh}\lambda^2}{\pi}\,. \tag{1.17}$$

Furthermore, the refractive index can be expressed in a more generalized form as

$$n^2 = 1 - \frac{U}{E}\,, \tag{1.18}$$

where U is the *optical potential* for the medium, given by

$$U = \frac{\hbar^2}{2m} 4\pi N b_{coh}\,. \tag{1.19}$$

Lax [16], who extended the approach of Foldy [14] to the case of an anisotropic scattering amplitude, showed that the coherent scattering amplitude appearing in the *total reflection formula* (1.17) corresponds to the amplitude for forward scattering, that is, $b_{coh}(\boldsymbol{a} \leftarrow \boldsymbol{a})$. Since this is a recoilless process, the value of the amplitude should not depend on the states of chemical binding. Nevertheless, it should take the value for the *bound atom scattering amplitude (scattering length)*, that is, the magnitude $b_{coh} = a_{coh}(A+1)/A$, where the *reduced mass factor* $(A+1)/A$ is a multiplication factor for an atom with mass number A bound to an infinite mass and a_{coh} is the amplitude of an isolated free atom.

Hughes, Burgy, and Ringo [18] proved the applicability of the total reflection formula (1.17) also for liquids; the details of their experiment are given in Section 2.1.

Furthermore, McReynolds [19] examined experimentally the applicability of the formula to samples in the gaseous state where atoms or molecules are distribute far apart from each other and are independently in free motion. The experiments were carried out at Oak Ridge and Brookhaven research reactors making use of the setup shown in Figure 1.3a. The observed total reflection intensity from the gas–liquid interface could be related to the criticality condition given by the difference

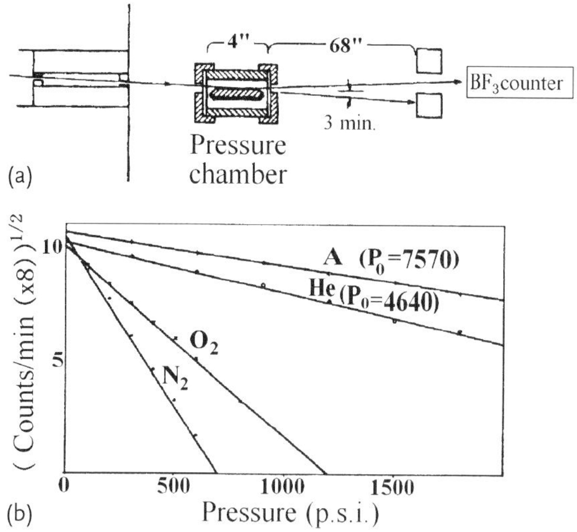

Figure 1.3 Neutron total reflection from a high-pressure gas and liquid interface: (a) experimental apparatus and (b) intensity of the neutron beam reflected from a surface of ethylene glycol at an angle of 3 minutes as a function of the surrounding gas pressure (McReynolds [19]).

between the right sides of Eq. (1.17) for the gas and the liquid, respectively. In the present experimental condition with the energy spectrum proportional to the neutron energy E, the square root of the reflected intensity I is expected to show a pressure dependence as

$$\left(\frac{I}{I_0}\right)^{\frac{1}{2}} = \frac{E_{c12}(P)}{E_{c1}} = \frac{N_1 b_{\text{coh}1} - N_2 b_{\text{coh}2}}{N_1 b_{\text{coh}1}} = 1 - \frac{P}{P_0}, \qquad (1.20)$$

where the subscripts 1 and 2 denote liquid and gaseous samples, respectively. Further, $E_{c12}(P)$ and E_{c1} represent the neutron energies satisfying the critical condition for the *total reflection from the gas–liquid interface* at sample pressures P and 0, respectively, whereas at pressure P_0 the refractive index for the gas becomes same as that for the liquid, and then the reflected intensity disappears. The experimental results shown in Figure 1.3b indicate the variation just as expected from Eq. (1.20).

1.2.3
Fraunhofer Diffraction

In addition to the diffraction experiments on crystal lattices and the total reflection experiments mentioned above, typical verification of the wave nature of neutrons is also possible with interference experiments using a *Fresnel biprism* or *Young's double slit* applied to neutrons. The former type of experiment was proposed and performed by Maier-Leibnitz and Springer at the FRM reactor, Technical University of Munich [20]. Their experimental results indicated an obvious interference pattern, which corresponded as a whole with their theoretical curves, but there remained locally some delicate disagreements which could be considered to indicate the difficulty of using a biprism interferometer for neutrons [21, 22].

The first trial of the slit interference experiments was carried out by Shull as a *single-slit interference experiment* [23]. In such a single-slit interference experiment, if the slit width is much smaller than the distance L between the beam source and the slit, and the distance L' between the slit and the observation point (i.e., the condition $a/2L + a/2L' \ll \lambda/a$ is satisfied for the neutron wavelength λ used in the experiment), then the interference pattern will reduce to the so-called *Fraunhofer diffraction* in which the curvature of the wave front can be neglected, and the intensity distribution $I(\theta)$ at the diffraction angle θ should be given by the equation $I(\theta) = I_0(\sin\beta/\beta)^2$, where $\beta = (\pi a/\lambda)\sin\theta$. Shull employed a neutron wavelength of 0.443 nm and a slit width of about 4–21 μm, and observed the apparent broadening of the diffraction peak, in good agreement with the calculated result of the Fraunhofer diffraction width. His result indicated that the wave front of neutrons entering the slit has coherency over a width of at least 20 μm in the direction transverse to the propagation direction of the neutron waves.

A more distinct slit interference for neutrons could be verified in the *double-slit interference experiment* corresponding to Young's optical experiment (1801). Zeilinger *et al.* [24] performed a precise double-slit interference experiment by making use of a beam of *very cold neutrons* extracted from the high-flux research reactor at the Institut Laue–Langevin, Grenoble, and with them being monochromatized to a wavelength of about 2 nm through a prism. The experimental result was compared with the numerical calculation simulating exactly the experimental procedure according to elementary wave mechanics. The width of slit S_1 in Figure 1.4a used in the experiment was carefully adjusted according to the neutron wavelength in the experiment, whereas the width of slit S_2 and that of the incident slit S_3 and the scanning slit S_4 were fixed at 100 and 20 μm, respectively. The object slit S_5 is a double slit consisting of two open channels with a width of about 22 μm each separated by a shielded part of boron wire with a width of about 104 μm as a neutron absorber, so that neutrons transmitted through slit S_5 are spatially *split* into two optical paths. Neutrons with a wavelength 1.845 ± 0.142 nm were used in the experiment.

It will be instructive for understanding the neutron optics in the present experiment to explain some details of their simulated calculation to obtain the double-slit interference. It starts from the *Huygens principle* (1690). Considering the vertical symmetry of the arrangement shown in Figure 1.4a, we can employ a two-dimensional structure in which the wave distribution at an arbitrary point in the object slit S_5 will be constructed by the interference with the phase distribution due to the products of the wave number and the optical path length for every point in the incident slit. On the other hand, the possible difference in the attenuation effect due to the optical path length difference can be neglected when considering the statistical accuracy of the experimental result because of the much smaller value for the slit width to the path length ratio. The wave propagation from the object slit to the detector slit can also be considered in a similar way. Therefore, the intensity

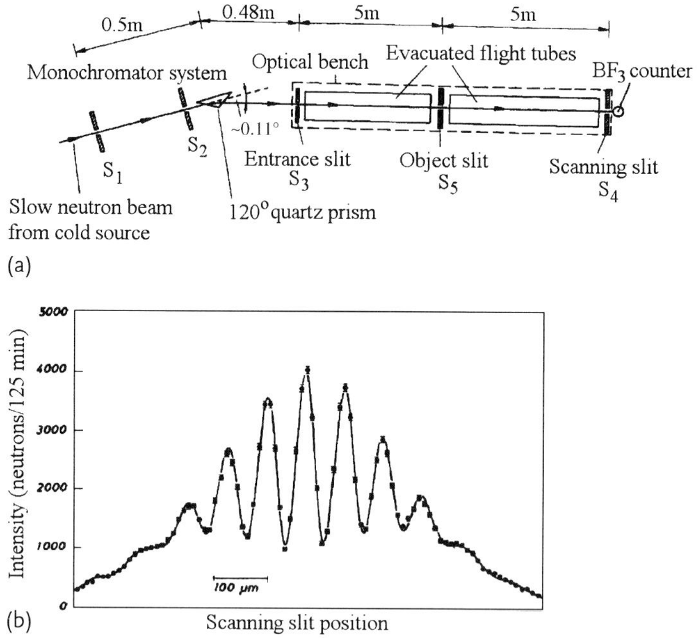

Figure 1.4 Double-slit neutron interference experiment. (a) Experimental arrangement (not to scale) and (b) experimental result compared with the theoretical calculation of the double-slit interference pattern (Zeilinger *et al.* [24]).

distribution I at an observation point P in the detector can be expressed by

$$I \propto \iiint |U(P)|^2 w(\lambda) w(\delta\vartheta) d\lambda d(\delta\vartheta) d S_4 \,, \tag{1.21}$$

where the amplitude $U(P)$ at point P is given by

$$U(P) \propto \iint f(\delta\vartheta) e^{ik(r+s)} d S_3 d S_5 \,, \tag{1.22}$$

k is the neutron wave number, r and s are the optical path lengths in the diffraction plane from the incident point to a point in the object slit and from there to the detection point, respectively, and $w(\lambda)$, $w(\delta\theta)$, and $f(\delta\vartheta)$ are the wavelength and angular distributions of incident neutrons, and the factor to take into consideration the relative phase at an incident point induced by the incident wave in the angle $\delta\vartheta$, respectively.

Equation (1.22) giving the amplitude $U(P)$ produced at point P by an incident plane wave is the integration as a coherent superposition, whereas Eq. (1.21), integrating over the incident wavelength and angular distributions, is simply the intensity integral as an incoherent superposition. The result of the present numerical calculation considering the slight asymmetry in two slit widths based on the actually observed result from optical microscopy shows very good agreement with the

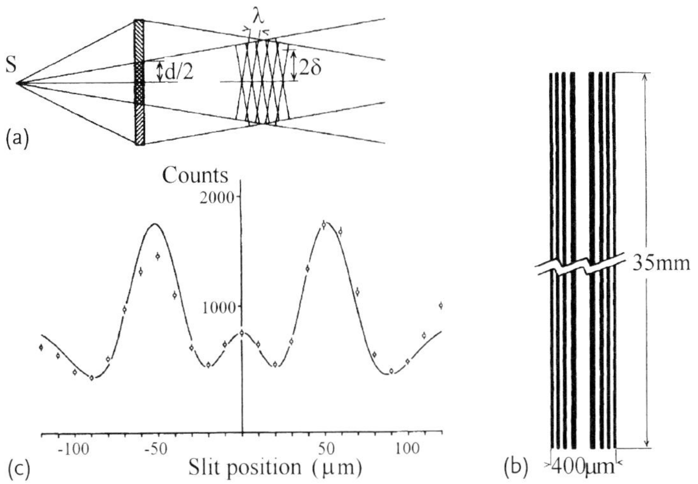

Figure 1.5 Split-lens interferometer configuration for neutrons: (a) principle of an overlapping split-zone plate working as a lens giving a constant fringe spacing; (b) split-lens geometrical pattern as a cylindrical zone plate preferable to a pinhole source for photomicrolithography; and (c) experimental interference pattern for a Cu zone plate fitted to the theoretical pattern for a zone thickness of 1.3 μm (Klein *et al.* [25]).

experimental result in Figure 1.4b, and the slit widths derived from the calculation gave values near those of the microscopy results. However, from the single-slit experiment carried out in a similar way with a slit width of about 95 μm, the slit width that gave the best fit was about 5% larger in comparison with the the microscopy result and the mechanical measurement with spacers, and this disagreement could not be explained well [24].

1.2.4 Fresnel Diffraction

All the slit interferences mentioned above are phenomena belonging the Fraunhofer diffraction; therefore, at the end of this section a few examples will be given for *Fresnel diffraction*, where the curvature of the wave front plays an important role in the interference. For the diffracting elements in such an experiment, a crystal restricts the neutron wavelength to below the Bragg cutoff, and a prism is also not preferable to focus on a short distance in the case of neutrons with a refractive index n very near unity ($|n-1| \sim 10^{-4}$ for a wavelength of 2 nm). Therefore, Klein *et al.* [25] employed a *Fresnel zone plate* and verified its applicability as an approach for the Fresnel diffraction of *cold neutrons* with a wavelength of 0.5–2 nm and *very cold neutrons* with a wavelength of about 2–30 nm.

In the *split-lens interferometer* configuration shown in Figure 1.5a, two kinds of zone plates, the central parts of where are overlapping, diffract the waves from the primary source S inward, and as the result is a fringe with a constant spacing in the region where the wave fronts overlap. They employed refractive zone plates to produce fringes with a *phase shift due to the refractive index* according to Eqs. (1.14)–(1.17), instead of ordinary absorption zone plates, for higher luminosity and applicability for neutrons with a refractive index very near unity. Further, they used the cylindrical zone plates illustrated in Figure 1.5b, which are preferable to a pinhole primary source as in the case of a narrow slit to be arranged in a neutron facility. They prepared such zone plates by means of photomicrolithographic techniques with UV light photography from the geometric patterns illustrated in Figure 1.5b, and then photoresists masked the electrolytic deposition. The required thickness $D(\lambda/2)$ for the wavelength λ to give a phase shift of 180°, that is, $\lambda/2$, can be estimated as

$$D\left(\frac{\lambda}{2}\right) = \frac{\lambda}{2(1-n)} \cong \frac{\pi}{N b_{\text{coh}} \lambda}, \tag{1.23}$$

from which they used $D(\lambda/2) = 2.4\ \mu\text{m}$ for copper at a wavelength of 2 nm. One of the experimental results of the *interference experiments with such split-zone plates* of electrodeposited copper carried out at the Grenoble high-flux reactor is shown in Figure 1.5c. The experimental fringe spacing δ agreed well with the theoretically estimated value of 51.8 μm from the equation

$$\delta = \frac{\lambda[f\rho - r(\rho - f)]}{\rho d}, \tag{1.24}$$

where the experimental wavelength $\lambda = 1.93 \pm 0.05$ nm, and f, ρ, and r are the focal length, the distances from the lens to the source, and the distance to the plane of detection, respectively, all being 5 m. Further detailed theoretical calculations indicated that the interference pattern corresponded to a copper thickness of 1.3 μm in the zone plates. This experiment verified the technical advantages of the interference experiments with split-zone plates, whereas the close spacing of the interfering beams would restrict the application of the method.

On the other hand, Steyerl *et al.* analyzed three-dimensional focusing with constructive interference of long-wavelength neutrons and designed an achromatic *concave Fresnel zone mirror* for ultracold neutrons [26]. The image formation experiment was carried out on such a circular zone ring mirror electrodeposited on an aspherical concave substrate with 60–80 nm *ultracold neutrons* at the inclined and highly curved guide facility PN5 of the Grenoble reactor. The experimental result agreed with the theoretical expectation at a magnification of 5 within the statistical error [27].

As mentioned above, the phenomenon of de Broglie matter waves of neutrons without consideration of the spin is considered to be precisely described by the law of *scalar optics*, where the amplitude and the phase of waves are exactly and uniquely decided by the position and the time.

However, a further advanced question that will arise quite naturally is if once split waves as considered in this section are superposed again at some position, how great a shift in the space or in the time is allowed to maintain the coherence of the waves. Such kinds of studies on the coherence of waves will be presented in the next section and in later chapters.

In addition to Bragg scattering and slit interferences or zone mirrors used in the historical experiments which verified the wave nature of neutrons, various types of neutron interferometers, such as Mach–Zehnder, Fabry–Pérot, and Jamin types, were also developed. These studies will also be introduced and discussed in Chapter 9. In Chapters 2 and 6, neutron optical experiments in various structures and systems will be described. As the neutron has spin, in some kinds of experiments the spin must be taken into consideration as a new variable, and such experiments relating to the neutron spin will be mainly discussed in Chapter 3 after a short preliminary study in the last section of this chapter.

1.3 Coherence of Waves

1.3.1 Coherence Lengths

In the theoretical calculation (Figure 1.4b) of Zeilinger *et al.* described in the previous section with the exact wave mechanical simulation of the double-slit interference [24], the integrals (1.21) and (1.22) were classified as the *coherent superposition*, Eq. (1.22), and the *incoherent intensity integral*, Eq. (1.21), over the wave components propagating downstream from each incident point. This classification was decided on depending on whether a certain correlation exists or does not exist between the phases of different partial waves, as the former belongs to the case without any accompanying random phase shift, whereas the latter belongs to the case with some accompanying random phase shifts. Since there is currently no coherent primary source for neutrons realized as there is laser light sources, superposition of waves for a couple of free neutrons should also belong to the latter case.

Since the monochromatization of incident neutrons in the experiment of Zeilinger *et al.* was performed with the combination of fine slits and a prism at a position sufficiently far from the primary source as shown in Figure 1.4a, the coherent superposition could be well approximated by the optical path integral, Eq. (1.22), of monochromatic plane *partial waves* owing to the prism having a different refraction angle for different wavelengths. In contrast, in the case of monochromatization with Bragg scattering, the correct relation between the wave components with different wavelengths and their coherence could not be represented so simply, but both of the quality of the monochromator crystal and the experimental setup must be taken into consideration.

For example, in the monochromatization setup for neutron scattering, a *mosaic crystal* which consists of a number of microcrystals with slightly different orientations is often used to obtain the optimum beam intensity by reflecting a rather wide wavelength range and over wide angular regions. *Pyrolytic graphite* is one such imperfect crystal that is nearly ideal for broad-angle monochromatization [28]. A recent study on pyrolytic graphite reported, in addition to such mosaicity, a distribution width in the lattice spacing, which induces additional broadening in the Bragg-scattered wavelength width [29]. When such a component is used in a neutron optics experiment such as interferometry, the wave phenomena cannot simply be represented by Eqs. (1.21) and (1.22) as the coherent superposition of monochromatic plane waves and the intensity integral over the wavelength and angular distributions, but the coherent superposition of partial waves over the reflected wavelength width, characteristic of the mosaic crystal, must be performed [30]. Further, the *two Gaussian distributions* of the incident neutrons observed in the interferometry experiment could be reduced to being caused by the two Gaussian distributions in the lattice spacing recently clarified for pyrolytic graphite [31].

As explained in the previous section on the double-slit experiment of Zeilinger *et al.* [24], the wave distributions were formed by the interference of partial waves with the phase difference due to the products of the wave number and optical path lengths from every incident point in the entrance slit. Then, it would be an interesting question to ask whether the coherence of waves holds or does not hold between the optical path lengths with a much larger difference, or in other words to ask how big is the path length difference or the time difference when the coherence finally disappears. Such a physical quantity expressing the maintenance of coherence in terms of spatial length is called the *coherence length*.

When we consider any experimental scheme to investigate the coherence length of a neutron, we must the first decide on whether on the incident neutrons for our device are sufficiently well approximated by monochromatic plane waves or whether they are have a wavelength distribution with a finite width. According to the principle of the Fourier integral, coherent superposition of plane waves over a finite wavelength width gives the resultant waves with a finite spatial broadening. A quite similar relation will also hold in time. Such waves localized in space or in time are called a *wave packet*. Therefore, the problem of how great is the coherence length is considered to be tightly related to the situation of how sharply the wavelength distribution is concentrated around the central value, or in other words how widely the wave packet is distributed in space and time.

However, a classical problem will arise that the propagation of waves composed of such a wavelength distribution, that is, the velocity distribution, with a finite width should be accompanied by an obvious dispersion of waves propagating to distant places. To get rid of such a dispersion, other concepts are possible in wave mechanics. One of the possible starting points is the thought that the wavelength distribution of a wave packet is the probabilistic concept of a neutron being sustained during propagation in space and time, and we observe the result of such probabilistic distributions. Experimental studies on such kinds of probabilistic concepts will be reported in the first section in Chapter 9. Another possible concept is a kind of singular wave packet inherently involving some physical structure persisting against the dispersion with the wavelength distribution. Such a singular wave packet will be studied in the last section in Chapter 9.

Therefore, here we do not go into detail on the problem of coherence and the concepts on the wave packet for a neutron, and we will study them in detail in Chapter 9.

1.3.2 Pendellösung Interference

Now, we introduce one of the early experiments estimating the *coherence length* of neutrons. Shull [32] carried out a very high precision experiment at Brookhaven National Laboratory with a neutron wavelength width of 0.0072 Å as shown in Figure 1.6a. This consisted of a thin silicon single crystal cut perpendicularly to the (111) reflection planes and fine entrance and exit slits with a width of 0.13 mm on each side of the crystal. The entrance side was illuminated with neutrons at incident angle θ very near the Bragg angle θ_B. This is the well-known setup of X-rays referred to as the *Pendellösung interference* experiment.

In this kind of *symmetry Laue diffraction* (refer to Section 6.1), the momentum component parallel to the reflecting lattice planes of incident neutrons is conserved during the propagation, whereas the perpendicular component experiences a number of Bragg reflections. As a result of the regular lattice effect and the potential effect of the crystal as a whole, with the refractive index given by Eqs. (1.14)–(1.17), and in the case of the incident angle θ with a very slight deviation from the Bragg angle θ_B, that is, at a very small deviation $\delta\theta = \theta - \theta_B$, the waves in the crystal become two split components with slightly different propagating velocities and are transported in two symmetrical directions with a small angle $\pm\epsilon$ to the Bragg angle. The transportation angle ϵ and the intensity I can be described by the equations ([32]; textbook [D] p. 431; textbook [I] Chapter 6, Section 6.3; [33] Chapter 6, Section 6.3; [34] pp. 201–202)

$$\gamma = \frac{\tan\epsilon}{\tan\theta_B} = \frac{\Delta\delta\theta/2d}{\sqrt{1+(\Delta\delta\theta/2d)^2}}, \tag{1.25}$$

$$I(\gamma) \cong (1-\gamma^2)^{-1/2}\sin^2\left[\frac{\pi}{4} + \frac{\pi t}{\Delta}(1-\gamma^2)^{1/2}\right], \tag{1.26}$$

where d is the lattice spacing in the crystal, t is the crystal thickness, $\Delta = \pi\cos\theta/NbF_{hkl}\lambda$, N is the number of unit cells per unit volume, b is the scattering amplitude of the nucleus, and F_{hkl} is the *crystal structure factor* per unit cell contributing the reflection with the Miller indices hkl.

In Eq. (1.25), we can see that at the exact Bragg incident angle, becoming $\pm\epsilon = 0$, the perpendicular components constitute standing waves, and therefore the waves are transported in parallel to the Bragg reflection planes. However, in all cases two split waves after being propagated through the crystal with slightly different velocities emerge from the exit surface of the crystal, thus a certain phase difference in proportion to the crystal thickness results between these two components. For a sufficiently narrow entrance slit, waves after the entrance slit extend over a wide angle, then the phenomena mentioned above spread widely inside the crystal and thus the Pendellösung interference due to the split-wave interference is observed. Figure 1.6b indicates such results of the Pendellösung interference for three different values of crystal thickness observed at the center of the exit surface of the crystal

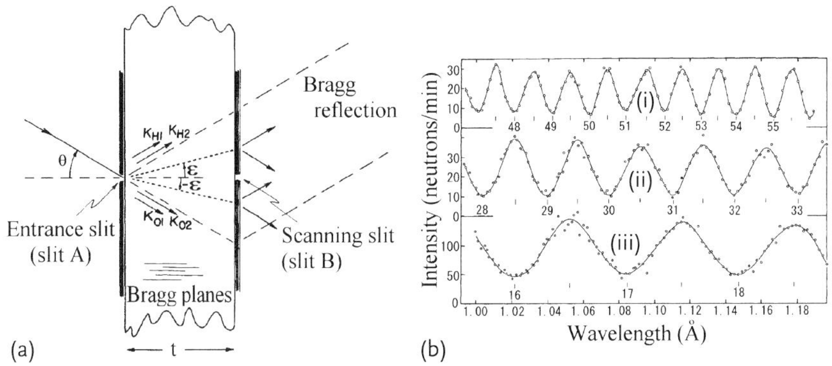

Figure 1.6 Pendellösung interference for neutrons. (a) The experimental setup; (b) fringe development at the center of the Bragg reflection as the wavelength increase for the crystal with the thickness; (i) 1.0000 cm, (ii) 0.5939 cm, (iii) 0.3315 cm (Shull [32]).

on minutely varying the incident neutron wavelength. From the order number of the *interference fringes* shown in Figure 1.6b, sufficiently high contrast is observed beyond 55th order. After the correction of the finite experimental resolution, the loss of contrast becomes of the order of 2% or even less, and thus the *coherence length* of neutrons in the present experiment was estimated to be larger than 2750λ, or 0.3 μm.

Further, interference variations induced by more effective and different kinds of phase shifts are observed for neutron waves by making use of various kinds of neutron interferometers, and the results will be described in Chapter 9.

1.4
Corpuscular Properties of the Neutron

1.4.1
Time-of-Flight Analysis

As early as a few years after the discovery of the neutron by Chadwick in 1932, it was experimentally assured that this neutral particle with mass much greater than that of the electron could be slowed down to the room temperature energy region by scattering by nuclei. For example, Dunning *et al.* [35] measured the time of flight of the neutrons emerging from the surface of a cylindrical paraffin block 16 cm in diameter and 22 cm long, in the center of which a 600 mCi Rn–Be neutron source was embedded. For the measurement they used the transmission through a couple of rotating absorber discs with slits. The two discs, each having a slit with an opening angle of 3.7° and provided with a similar fixed disc, were 54 cm apart and connected to each other with a lag angle of 3.5°. The device thus works as a *mechanical velocity selector* for neutrons in which neutrons with a velocity corresponding to the rotation speed are able to pass through the device. The count rates plotted against rotation speed agreed well with the Maxwellian distribution, having the maximum intensity at a neutron speed of about 2300 m/s.

The *flight velocity* of neutrons thus measured represents the particle motion of the neutrons, and at the same time it corresponds to the group velocity in the wave mechanics as mentioned in Section 1.1. The wavelength given by Eq. (1.9) for neutrons with energy corresponding to room temperature is on the order of interatomic distances, indicating possible applicability of these neutrons to spectroscopy experiments for material physics, but the neutron intensity as well as the energy resolution of the experimental arrangement of Dunning *et al.* were not sufficiently high for such applications. The precise time-of-flight spectroscopy for neutrons was carried out first by Alvarez, as shown in Figure 1.7a, by making use of a cyclotron to accelerate deuterons to be injected into a beryllium target and the neutrons pro-

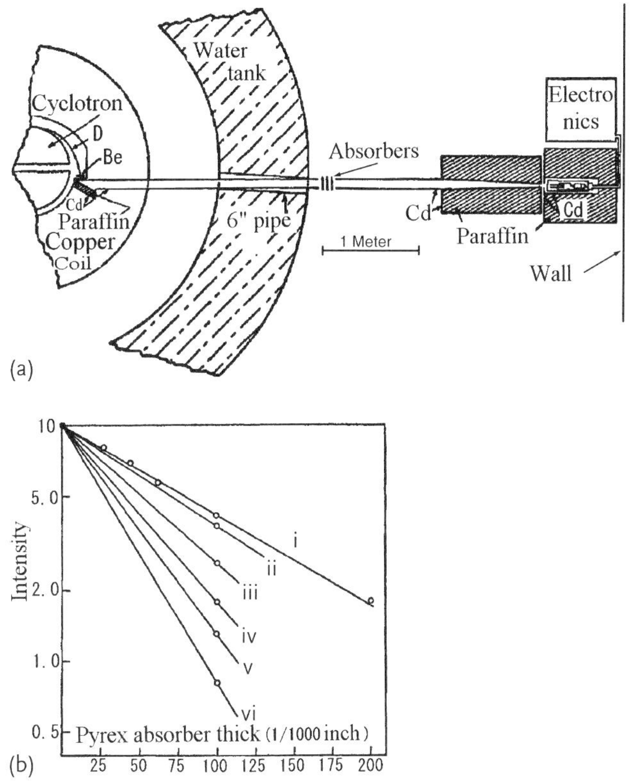

Figure 1.7 Time-of-flight spectroscopy experiment with time-modulated operation of a cyclotron; (a) Plan of cyclotron room; (b) Boron absorption variation with timing scheme, abscissa; absorver thickness, ordinate; transmitted neutron intensity (normalized), i: continuous measurement, ii–vi: delay time increased by about 1/240 s per step. (Alvarez [36]).

duced were moderated in a paraffin block. For the velocity analysis of moderated neutrons, he used the *electrical velocity selection* method with time-modulated operation for deuteron acceleration and synchronized control of the neutron measuring system [36].

There were several severe tasks and difficulties in the controlled operation of the cyclotron beyond 100 Hz to achieve thermal neutron spectroscopy, but finally the energy selection of neutrons around 300–10 K could be well achieved with the modulated operation at 120 Hz. From the measurements of the transmission ratio of such velocity-selected neutrons, the $1/v$ law, v being the neutron velocity, on the absorption cross section of boron was confirmed, as shown in Figure 1.7b.

The performance of the 1.5 m cyclotron used in the neutron experiment at Berkeley, the University of California, gave a deuteron energy of 8 MeV, a deuteron current of 50–60 μA in the steady-state operation, and 5–10 μA in the modulated operation at 60 Hz.

Later, in 1940, Baker and Bacher reported a *time-of-flight experiment* similar to the Alvarez one but with a much higher repetition frequency, that is, a period of 2500 μs, a neutron burst width of 50–100 μs, and a modulation time accuracy of 5 μs [37]. In the same year, Alvarez *et al.* carried out a magnetic resonance experiment for neutrons with the steady-state operation of the cyclotron; this will be described in the next section.

I would like to insert a short description of the activities in Japan during similar periods. Kikuchi *et al.* constructed a cyclotron with an accelerated deuteron energy of 4.2 MeV in the central Osaka campus opened in 1932, of Osaka University, and started nuclear physics studies with it. However, the outbreak of the Second World War unfortunately disturbed and finally interrupted the continuation of the research in the 1940s. Independently, at the Science and Chemistry Institute in eastern Japan, Nishina constructed a 1.5 m cyclotron in 1943 for nuclear physics studies. At Kyoto University, the construction of a cyclotron was started. All these experimental activities in nuclear and neutron physics in Japan were stopped at the end of the Second World War owing to the destructions of the facilities by the occupation forces. Thereafter, experimental neutron research in Japan recovered, accompanied by the development of research reactors and high-intensity accelerators after the 1960s.

Experimental research on neutrons after the 1950s progressed remarkably by making use of research reactors developed in the United States and in Europe. Further, after the 1970s, with the construction of high-flux reactors, various kinds of advanced neutron spectroscopies were proposed and practically applied to neutron optics and condensed matter studies. Furthermore, the successful developments of high-intensity accelerator sources in Europe and the United States promoted pulsed neutron experiments as a competitive approach to continuous beam experiments. Nowadays, the most intense pulsed neutron sources are constructed in the United States and in Japan and are starting their operations. Some of the wide variety of neutron optics and spectroscopy developments as well as their applications will be introduced in the following parts of this book.

1.4.2
From a Particle to a Wave

One of the very interesting investigations in Munich relating to the particle motion of neutrons reflected by a mirror will be described here. If the reader has some difficulty in understanding well the explanations, the information given in the next chapter will help gain a better understanding.

To understand deeply the wave–particle duality in the phenomenon of neutron reflection on a mirror, Felber *et al.* undertook a comparison between the results of the *classical mechanical analysis on the reflection of a particle beam* and of the *quantum mechanical analysis on the reflection of matter waves*, and studied the possible transition from the former to the latter under the developments of temporal and energy conditions in the reflection [38]. Practically, they considered the reflection of *very cold neutrons* with a wavelength of 2.4 nm on a surface vibrating at high frequency, and compared the theoretical analyses with the experimental results carried out at the Munich and at the Geesthacht research reactors. They succeeded in indicating the transition of the time-dependent reflection phenomenon from classical to quantum mechanical according to the developments of frequency and amplitude parameters with the surface vibration.

The vibrating mirror surface is represented by a *step potential* with height V_p under one-dimensional motion with amplitude a_p and angular velocity ω_p. The collimated neutrons with energy E_0 (the wave number k_0, velocity v_0) experiencing such a potential can be classified according to the values for three parameters $\alpha = 2k_0 a_p$, $\beta = V_p/E_0$, and $\gamma = a_p \omega_p / v_0$.

In the framework of the classical description, the neutron trajectory should be calculated iteratively by solving the equation of motion successively after every collision, but the neutron velocity change due to the collision necessary to obtain the neutron flux is given in an implicit form; therefore, the solution can only be found numerically.

In the framework of the quantum description, on the other hand, it is enough to introduce the step potential in the time-dependent Schrödinger equation, but our case is essentially different from the usual cases given in many textbooks, since our *potential is time-dependent.* Felber *et al.* therefore started with the expansion of the wave function into the *partial waves* with the number of transferred phonons n, and then investigated the behavior of the analytical solutions for the particle waves in the extreme cases for the parameters (small amplitude; $\alpha \ll 1$, or quasi-stationary; $\gamma \ll 1$) and further were able to calculate the approximate solutions for other general cases up to the order of the expansion $|n_{\max}| = N \cong 2\alpha$ to satisfy the required accuracy.

The results of these calculations are shown in Figures 1.8 and 1.9. From Figure 1.8a and b, we understand the quantum effects become significant for $\alpha < 10$, whereas for $\alpha \gg 10$ the results approach those of classical mechanics. Further, in Figure 1.8c, we recognize many black bars for the quantum reflections in the region where no reflections are expected in classical mechanics, that is, no black continuous spectrum is shown. Therefore, all of these reflections in $E' < 0.5$ in

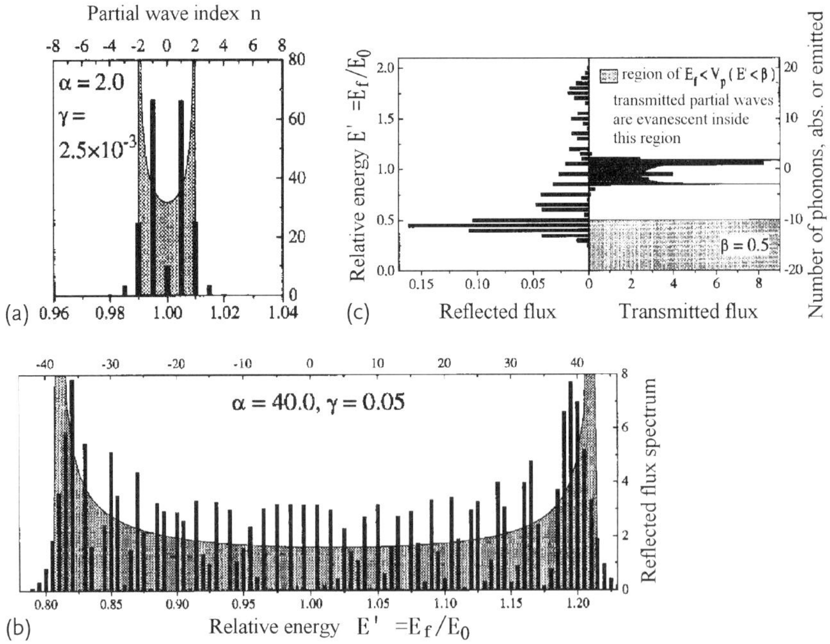

Figure 1.8 Comparison between the calculated results on a particle beam and on matter waves: (a) and (b) the case of the parameter value for $\beta = 10$, and the value for α varied (classical; gray patterns, quantum; black bars); (c) reflected and transmitted distributions for $\alpha = 15.0$, $\beta = 0.5$, $\gamma = 0.1875$ (classical; black pattern, actually no results for this parameter value, quantum; black bars, the gray region indicates the potential height) (Felber *et al.* [38]).

Figure 1.8c indicate quantum particle reflections. The comparison with their experiments are shown in Figure 1.9, where Figure 1.9a shows the experimental setup of the *reflection experiment with the vibrating mirror*.

Since the neutron velocity component parallel to the mirror surface $v_{0\parallel}$ is conserved, the situation can be reduced to a one-dimensional problem in which the previous parameter values are replaced by the corresponding relations for the perpendicular component $v_{r\perp}$. Thus, the energy and the velocity changes of neutrons reflected by the vibrating mirror described in Figure 1.8 are equal to the changes of the perpendicular components $E_{r\perp}$ and $v_{r\perp}$, respectively, and the resulting change of the reflection angle θ_r will be observed according to the relation

$$\tan\theta_r = \frac{v_{r\perp}}{v_{0\parallel}} = \tan\theta_0\sqrt{1 + \frac{\Delta E_\perp}{E_{0\perp}}}\,, \tag{1.27}$$

where $E_{0\perp} = E_0 \sin^2\theta_0$ and $\Delta E_\perp = E_{r\perp} - E_{0\perp}$. Figure 1.9b shows the experimental results for various values of the parameter α and compared with those of the matter wave calculations, where $1\,\text{neV} = 10^{-9}\,\text{eV}$. We can conclude that both results agree well by considering the effects of the finite resolution included in the experimental results

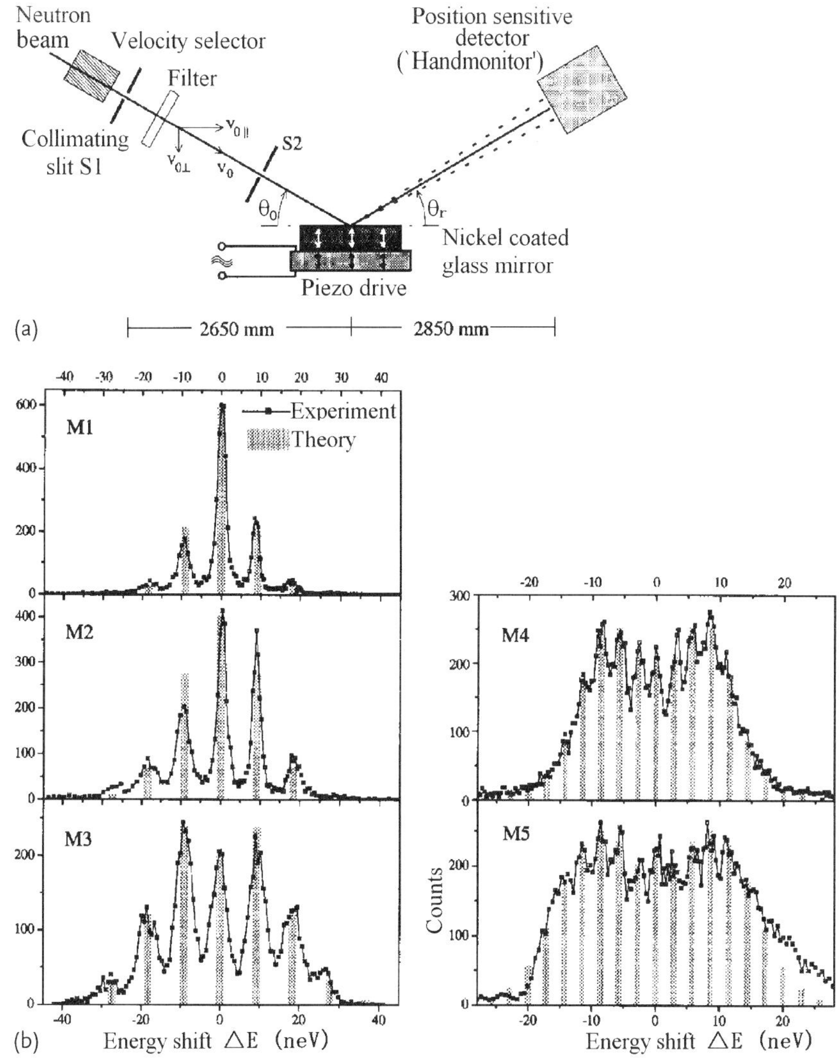

Figure 1.9 (a) Arrangement (incident wavelength $\lambda = 2.4$ nm, incident angle $\theta_0 = 2.00°$); (b) experimental results compared with calculations, Experimental results: Abscissa, energy transfer estimated from the change of reflection angle; Ordinate, reflected neutron counts; From M1 to M5, modulation index α increases, that is, M1–M3; $\alpha = 0.95$, 1.45, 1.98 (frequency $f = 2.2206$ MHz); M4, M5; $\alpha = 4.21$, 5.82 (frequency $f = 0.69295$ MHz), (Felber *et al.* [38]). (b) Reflection experiment of very cold neutrons on a vibrating mirror.

From the transition shown in Figure 1.8 from a few discrete spectra to the continuous spectrum for increasing the value for α, that is, in the transition from quantum mechanics to classical mechanics as shown in Figure 1.8, the analytical method appropriate for a given condition can be decided from the parameter values.

1.5 Magnetic Moment of the Neutron

The neutron has *spin* 1/2 and the *magnetic dipole moment* $\mu_n = -1.91\mu_{\mathrm{NB}}$, where μ_{NB} is the *nuclear magneton* $e\hbar/2Mc$, defined for the proton mass M. The half-integer neutron spin was predicted from the structure of a deuteron on the discovery of the neutron by Chadwick. Soon after, in 1937, it was confirmed, together with the spin dependence of the neutron–proton internucleon interaction, by the measurement and the analysis on the diffusion characteristics of thermal neutrons in liquid hydrogen samples with different ratios of two kinds of hydrogen molecules, orthohydrogen and parahydrogen [39, 40]. On the other hand, the magnitude of the neutron magnetic moment was determined from the measurement of neutron spin behavior in a magnetic field, which will be described next.

1.5.1 Spin Flip in Magnetic Resonance

A spin with magnetic moment μ precesses in a magnetic field with the *Larmor precession frequency* $\nu_L = 2\mu H/h$ around the direction of the magnetic field, the motion being called the *Larmor precession*. For a particle with a half-integer spin in a magnetic field, the theoretical formula for the spin reversal probability at time t in the situation of *nonadiabatic spin flip* was derived by Güttinger [41], who solved the Schrödinger equation for a spin-1/2 particle in a magnetic field rotating with frequency ν. His formula is given for the case where the rotational axis of the magnetic field makes an angle $\vartheta = \pi/2$ with the direction of the field. A more general formula was given by Rabi [42] for an extended condition of the *total magnetic field* H, to be decomposed into a *static magnetic field* H_0 and a *rotating magnetic field* with magnitude H_1 perpendicular to the direction of H_0. His formula, Eq. (1.28), applicable to an arbitrary value for the polar angle ϑ between the directions of H and H_0, is written as

$$P_{\left(\frac{1}{2},-\frac{1}{2}\right)} = \frac{\sin^2\vartheta}{1+q^2-2q\cos\vartheta}\sin^2\left[\pi\nu t\sqrt{1+q^2-2q\cos\vartheta}\right], \tag{1.28}$$

where $P_{\left(\frac{1}{2},-\frac{1}{2}\right)}$ denotes the probability that the spin state initially at $+1/2$, that is, polarized in the direction of the total field at time $t = 0$, becomes the $-1/2$ state at time t, and $q = \nu_L/\nu$. Putting $\vartheta = \pi/2$ in Eq. (1.28) gives the same result as with Güttinger's formula.

Since the terms as a function of q on the right side of Eq. (1.28) can be rewritten as $1+q^2-2q\cos\vartheta = (q-\cos\vartheta)^2+\sin^2\vartheta$, the *resonance condition for spin flip* in which $P_{\left(\frac{1}{2},-\frac{1}{2}\right)}$ takes the maximum value is given by $q_{\mathrm{res}} = \cos\vartheta$. Then, the spin flipping probability at that condition follows the time dependence

$$P_{\left(\frac{1}{2},-\frac{1}{2}\right)\mathrm{res}} = \sin^2(\pi\nu t\sin\vartheta)\,; \tag{1.29}$$

therefore, the probability will approach unity if we optimize the residence time of the particles inside the magnetic field. Furthermore, there being two kinds of rota-

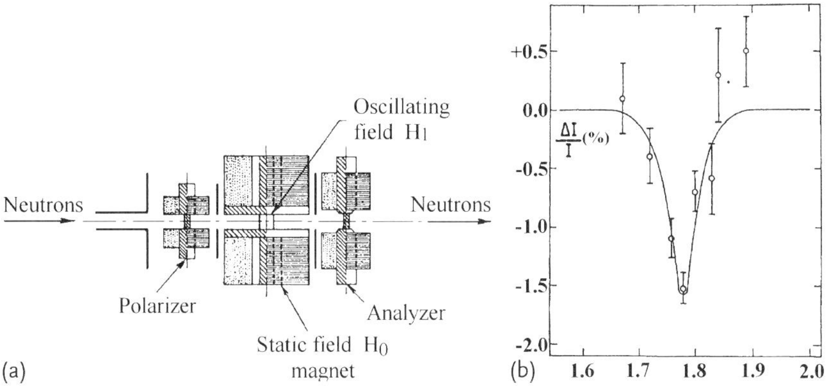

Figure 1.10 Magnetic resonance experiment for the neutron: (a) experimental apparatus; (b) count rate decrease observed at the resonance. The abscissa represents the magnetic field current in arbitrary units (Alvarez and Bloch [43]).

tion direction, that is, clockwise and counterclockwise, with the same magnitudes of frequencies ν_L and ν, the resonance condition in Eq. (1.28) requires not only the magnitude but also the sign of ν to match those of ν_L.

In other words, we can determine whether the magnetic dipole moment of the particle is positive or negative from the magnetic resonance experiment by making use of a *rotating magnetic field*, but this is not possible if we use the *reciprocating magnetic field*, which is the superposition of clockwise and counterclockwise rotating fields.

An experiment using this *nuclear magnetic resonance method* applied to neutrons was carried out by Alvarez and Bloch with the insertion of the resonance apparatus shown in Figure 1.10a in the cyclotron described earlier, and the *magnetic dipole moment of the neutron* was derived from their experiment [43]. In their experiment, the cyclotron was operated in a steady-state mode, which was advantageous from the viewpoint of the background effects in comparison with those in the modulated operation mode.

For the *polarization* of the incident neutrons and also for the *polarization analysis* of neutrons transmitted through the resonance apparatus, the *transmission method with magnetic scattering* through several centimeter thick iron plates magnetized by wires carrying electric currents was used. Further, they employed the method to find the resonance condition from the maximum of $P_{(\frac{1}{2},-\frac{1}{2})}$ by varying the static magnetic field H_0 under a constant frequency $\nu_n = \omega_n/2\pi$ of the oscillating magnetic field, assuming that the second term on the right side of Eq. (1.28) should be averaged over owing to the broad distribution of the residence time in the magnetic field of neutrons with various velocities. Furthermore, they used the resonance spin flip formula for the initial condition polarized in the direction of the static field, based on the small H_1/H_0 ratio, actually smaller than 2% for a static magnetic field H_0 of 600 G and an oscillating field strength H_1 of about 10 G.

The homogeneity of the magnetic field H_0 in the experiment of Alvarez and Bloch was 600 ± 1 G over the whole of the resonance region, and the voltage fluctuations of the magnet power supply were maintained below 0.1% by the control system compensating the effect due to the coil temperature variation. To determine precisely the number $\bar{\mu}_n$ which corresponds to the neutron magnetic moment μ_n expressed in the unit of *nuclear magneton*, $e\hbar/2Mc$ (M is the proton mass), they employed the method to compare the *resonance angular frequency* ω_n for the neutron in a *magnetic field* of strength H_n with the *resonance angular frequency* for the proton ω_p accelerated by the cyclotron, according to the equations

$$\omega_n = 2H_n\mu_n/\hbar = (2H_n\bar{\mu}_n/\hbar)(e\hbar/2Mc) = (eH_n/Mc)\bar{\mu}_n \,, \tag{1.30}$$

$$\omega_p = eH_p/Mc \,, \tag{1.31}$$

$$\bar{\mu}_n = (\omega_n/\omega_p)(H_p/H_n) \,. \tag{1.32}$$

One of their experimental results for the resonance curve is shown in Figure 1.10b for the magnetic field frequency $\nu_n = 1.843$ MHz.

As the final result of these measurements repeated many times, the *neutron magnetic dipole moment* in the unit of nuclear magneton was determined as

$$\bar{\mu}_n = -1.93_5 \pm 0.02 \,. \tag{1.33}$$

Although the sign of the magnetic moment for the neutron could not be determined in the experiment, where a reciprocating field was used, it had already been deduced to be negative from the *Stern–Gerlach magnetic deflection experiment* for protons and deuterons by Stern *et al.* [44, 45].

The main experimental efforts to determine precisely such a physical constant in these early neutron experiments by making use of accelerators were devoted to the stability and reproducibility of the measurements, especially with Alvarez *et al.* having had difficulty to achieve the required stable operation of the cyclotron (a fluctuation of the magnet source voltage below 0.1%). By the way, the experiment to determine directly the sign of the neutron magnetic moment by making use of two *mutually orthogonal magnetic fields* was later carried out by Rogers and Staub [46].

The phase difference between the incident and the exit neutron spins during the resonance will be accumulated over the period of Larmor precession and thus the accuracy of the phase determination will be improved in proportion to the number of precessions. *Ramsey's separated oscillating magnetic field method* [47] in which the oscillating field is split into two widely separated locations along the neutron flight path and a constant precession field is provided in the flight path between the separated fields for the continuation of the Larmor precession gave an epoch-making improvement in such resonance experiments for particles as mentioned above. Corngold *et al.* performed the first resonance experiment with this method applied to the neutron in the graphite reactor at Brookhaven National Laboratory [48, 49]. The results of this experiment are shown in Figure 1.11, and combined with the result of a similar measurement on a proton sample (H_2O) with the same

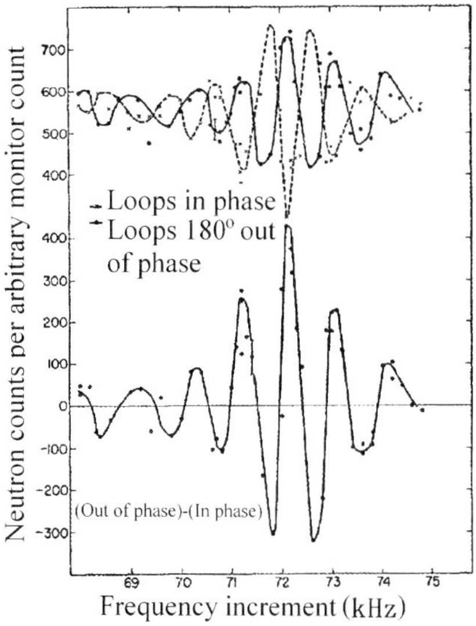

Figure 1.11 Neutron magnetic resonance experiment with separated oscillation fields method carried out by Corngold *et al.*: the upper two plots are the count rates as a function of the frequency (the resonance frequency being about 24.727 MHz) for the conditions in phase and 180° out of phase, respectively, and the lower one shows their difference (Corngold *et al.* [49]).

apparatus; this gave a much improved accuracy for the *neutron magnetic moment* as

$$\bar{\mu}_n = -1.913148 \pm 0.000066\,. \tag{1.34}$$

We add here one of the later results with further improved accuracy [50] also carried out with the separated field method:

$$\bar{\mu}_n = -1.91304308 \pm 0.00000058\,. \tag{1.35}$$

1.5.2
Adiabatic Spin Reversal in a Magnetic Gradient

The resonance phenomenon described above by making use of uniform fields has a very sharp resonance frequency, being very advantageous for precise measurements of a physical quantity of a particle. Therefore, besides the determination of magnetic moments, it is also applied to search for the *electric dipole moment* of the neutron as the most sensitive experimental method at present for verifying the possibility of the neutron having a finite electric dipole moment. However, from the viewpoint of efficient reversal of spins, Eq. (1.29) indicates the difficulty of attaining the resonance condition for neutrons with widely distributed velocities, spending different residence times in the precession field.

Another approach for spin reversal, named *adiabatic spin flip method with magnetic gradient*, is used for such cases to attain efficient performance for a wide velocity spectrum. This method of adiabatic spin reversal in a magnetic gradient consists of a static magnetic field with a gradient and an oscillating magnetic field in the middle of the static field, the working principle of which can be understood in the *rotating frame*.

Before considering the magnetic gradient method, it will be instructive to describe the experiment of Alvarez *et al.* in the *rotating frame* with the same frequency ν_n as the oscillating field around the axis in the direction of the static field. At the resonance condition where the frame rotation frequency is exactly the same as the Larmor precession frequency, the phenomenon of spin precession will disappear, that is, the effect of the static field H_n is canceled. On the other hand, the oscillating field seems to be at rest with a field strength of H_1. As a result of these transformations, the neutron spin will precess very slowly at the Larmor frequency in the field H_1 around the direction of the field H_1 now at rest in this rotating frame. The *spin flipping time*, defined as the time required for the complete spin reversal, that is, π spin turns, is just half of the Larmor precession period in this situation.

Now, we consider the magnetic field with a gradient as shown in Figure 1.12a, where the static field strength H_0 varies along the oscillating field region ($x = x_0 \sim x_1$) on the x-axis in the direction of the neutron flight. The equality $H_0(x) = H_n$ required in the previous case is satisfied only midway, $x = x_c$, in the path, and upstream the static field strength $H_0(x)$ becomes gradually larger than H_n, that is, $x_0 \le x < x_c$; $\Delta H(x) = H_0(x) - H_n \ge 0$, where $\Delta H(x_0) \gg H_1$. Downstream, however, H_0 gradually becomes smaller than H_n, that is, $x_c < x \le x_1$; $\Delta H(x) \le 0$, where $|\Delta H(x_1)| \gg H_1$. At the entrance ($x = x_p$) and the exit ($x = x_a$) positions of the device, there is a neutron polarizer and a polarization analyzer.

With a similar transformation to the rotating frame as previously considered, during the passage in the upstream region starting from the entrance to the mid-

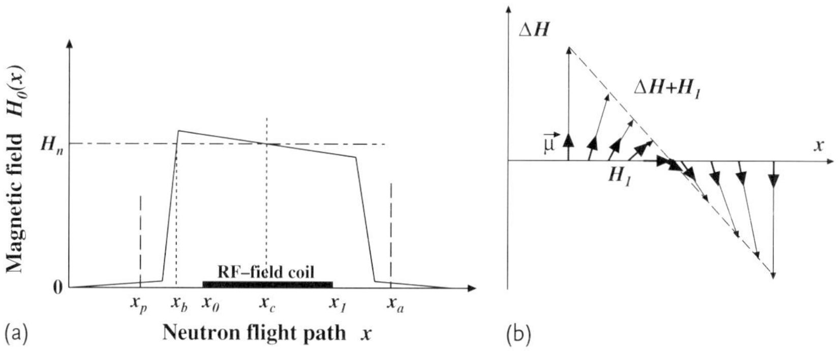

Figure 1.12 Description of the adiabatic spin flip method with a magnetic gradient: (a) setup and field distribution; (b) movement of the field direction followed by a spin vector adiabatically in a rotating frame synchronous with an oscillating magnetic field.

way position, where $\Delta H(x) > 0$, the neutron spin follows the motion of the magnetic field described as $\boldsymbol{H}_{\text{eff}}(x,t) = \Delta H(x)\boldsymbol{e}_z + H_1\boldsymbol{e}_{x'}$, where $\boldsymbol{e}_{x'}$ denotes the direction of H_1 at time t; thus, as shown in Figure 1.12b it results that the spin parallel (or the spin antiparallel) to H_0 at the entrance moves to the direction parallel (or antiparallel) to H_1 at the midway position, and further in the downstream region continues to move to antiparallel (or parallel) to H_0 as the sign of $\Delta H(x)$ reverses. In this way, regardless of the initial spin direction, the spin reversal is attained quite smoothly for a wide range of the neutron velocity. This is the principle of adiabatic spin flip with a magnetic gradient.

The velocity region of the neutrons for which the spin flipping mechanism mentioned above works well will be defined as follows.

First, the *adiabatic spin flip condition* requires the *adiabatic following after the magnetic field*, that is, the direction of the magnetic field in the device (now we consider it in the rotating frame), moves sufficiently slowly compared with the angular velocity of the Larmor precession in the magnetic field H_1. This can can be expressed as

$$\frac{v_{n\,\max}}{H_1}\left|\frac{dH(x)}{dx}\right| \ll \frac{2|\mu_n|H_1}{\hbar}\,,$$

$$\text{or}\quad v_{n\,\max}\left|\frac{dH(x)}{dx}\right| \ll \frac{2|\mu_n|H_1^2}{\hbar}\,,\qquad x_0 \le x \le x_1\,, \tag{1.36}$$

where μ_n is the magnetic dipole moment of a neutron, and $v_{n\,\max}$ denotes the maximum velocity for the neutrons performing the adiabatic spin flip.

Next, in addition to the midway position, there is another point, either at the entrance or at the exit (in the case in Figure 1.12a it is the entrance side), where the equality $H_0(x) = H_n$ holds at the position $x = x_b$. For successful working of the spin flip mechanism for the neutrons transmitted through the device, any additional spin reversal should not occur at this point of $x = x_b$, otherwise the spin would return back to the initial state. Since the oscillating field produced by the coil extending over the region $x = x_0 \sim x_1$ in the device should leak out somewhat to the point x_b, we have to make the field gradient around $x \cong x_b$ so steep that the turning speed of the local field direction becomes much faster than the Larmor precession of the spin at that location. This additional condition is written as

$$\frac{v_{n\,\min}}{H_1}\left|\frac{dH(x)}{dx}\right|_{x=x_b} \gg \frac{2|\mu_n|H_1}{\hbar}\,,$$

$$\text{or}\quad v_{n\,\min}\left|\frac{dH(x)}{dx}\right|_{x=x_b} \gg \frac{2|\mu_n|H_1^2}{\hbar}\,,\qquad x \cong x_b\,, \tag{1.37}$$

where $v_{n\,\min}$ denotes the minimum velocity for the neutron getting rid of such an additional spin reversal.

As a result of these requirements, the adiabatic spin flip will be attained successfully for neutrons in the rather wide velocity region defined by Eqs. (1.36) and (1.37).

The experimental study on the spin flipping characteristics with the gradient field adiabatic method was carried out by Egorov *et al.* [51], who measured the transmitted intensity and its static field dependence of very slow neutrons (*ultracold neutrons*) with a velocity below about 20 m/s through the spin flip device with a configuration similar to that shown schematically in Figure 1.12a. The ultracold neutrons were extracted from the curved guide tube at the VVR-M reactor of the St. Petersburg Nuclear Physics Institute. For the polarization and the polarization analysis, the transmission method of magnetized materials was employed (the setup was similar to that shown in Figure 1.10a), but in the present case for ultracold neutrons they used the *polarizer* and *analyzer* with a *magnetic reflection method* on a thin film, that is, a 1 μm thick evaporated iron layer, instead of the magnetic scattering method for thermal neutrons used by Alvarez. An oscillating field with a frequency 200 kHz, and a resonance field current of about 510 mA gave the resonant decrease of the neutron count rate down to about half with a resonance half width of about 3 mA [51].

Further, Ezhov *et al.* performed a comparison of the gradient field adiabatic method and Ramsey's resonance method in the case of bottled ultracold neutrons (refer to Section 2.2.2) experiencing Larmor precession in the bottle over a period of about 4.6 s on average. As mentioned already, the latter method is often used for the precise measurement of the resonance condition, such as for studying the existence of the electric dipole moment of the neutron. They carried out Monte Carlo calculations on the spin flipping performances in both methods mentioned above. The spin flip spectrometer had two separated oscillating fields at the entrance and the exit sides of the precession space in a homogeneous static field. Their comparisons of the calculated results for both methods are shown in Figure 1.13a, in which an about 1.8 times higher polarization ratio for the gradient field adiabatic

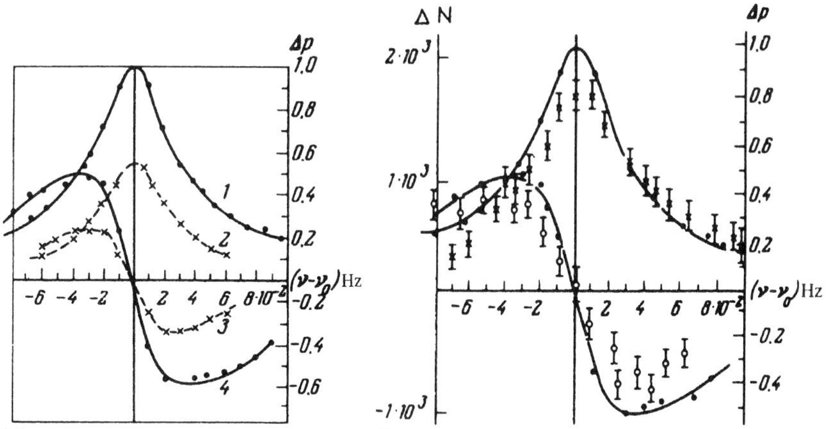

(a) (b)

Figure 1.13 Comparison between the spin flipping characteristics for ultracold neutrons with the gradient field method and with Ramsey's resonance method. (a) Results of numerical calculations; the solid curves represent the gradient field method, whereas the broken curves represent Ramsey's method. Curves 1 and 2 are for phase differences of 0° and 180° and curves 3 and 4 are those for phase differences of 90° and 270°. (b) Comparison with experimental results for the gradient field method (Ezhov *et al.* [52]).

method is indicated in comparison with Ramsey's resonance method. The feasibility of the correctness of the calculated values for the gradient field adiabatic method is also illustrated in Figure 1.13b as a comparison with the measured values, which show satisfactory agreement with the calculated values.

Herdin *et al.* also carried out the polarization experiment with the adiabatic spin flip method for ultracold neutrons with a velocity below about 8 m/s passing through a guide tube, in which a high spin-flipping efficiency of about 100% with a statistical error of 2% and a polarization efficiency of 95–98% for the neutron velocity region of 4.15–8.2 m/s were achieved [53].

The neutron spin precession in a magnetic field studied as described above can be applied for the development of neutron spectroscopies, and various kinds of high-resolution neutron spectrometers have been developed by utilizing these characteristics of Larmor precession. The first idea in this direction was suggested and developed by Mezei in 1972 [55], this new principle being named *neutron spin echo spectrometry* (see the schematic arrangement shown in Figure 1.14). It opened the door to the wide field of neutron spin spectroscopy. Generally speaking, neutron spectroscopy experiments are carried out to study the energy structures and their characteristics in a sample from the experimental results for the *transferred energy* as the difference between the incident and scattered neutron energies. Thus, for high-resolution investigations on the energy structures, similarly high-resolution analyses of the transferred energy are required.

The conventional spectroscopy method is based on wavelength or velocity spectrometry; time-of-flight spectrometry suffers from the serious intensity loss accompanying the severe selection of the incident neutrons needed to satisfy the requirement of such a high-energy resolution. In contrast, in the new principle of neutron spin echo spectroscopy, the information on the neutron energy does not require

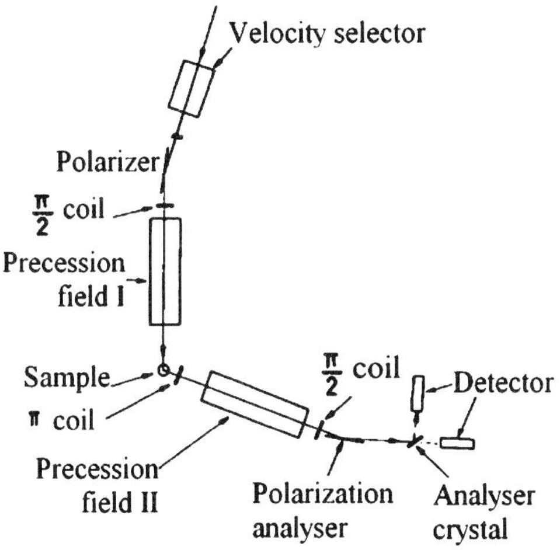

Figure 1.14 Schematic layout of neutron spin echo spectrometer. Typical lengths of the precession field regions are 2–4 m (Mezei [54]).

wavelength or velocity spectrometry, but is labeled on the neutrons as the number of Larmor precessions. A practical and basic configuration of the neutron spin echo spectrometer is shown in Figure 1.14. During their flights through the incident and scattered flight paths with, respectively, well-defined path lengths and well-defined magnetic field strengths, neutrons experience Larmor precessions; thus, after the flights the number of precessions is labeled on individual neutrons. It is quite possible to reverse the sign of the precession number, that is, label a negative number for the precession after the scattering by a sample, by reversing the direction of the magnetic field or by reversing the neutron spin. Then the resultant number of precessions observed at the end of the flight will be given by the difference between those of the incident and the scattered paths, $\Delta N \cong N\delta v/v$, where $\delta v = v' - v$ denotes the difference between the incident and the scattered neutron velocities. In this way, we need not select the incident neutron velocity within a very narrow width in proportion to the very high resolution of the velocity change δv, but we can obtain as high a velocity resolution δv as we wish by increasing the number of precessions N to satisfy the present relation under a given resolution ΔN in the experimental device. The present principle of echo spectrometry and the utilization of a much wider energy width than in conventional spectrometries are the important advantages of high-resolution spectroscopy with the neutron spin echo method.

We would like to add here a little advanced detail on another important characteristic of the neutron spin echo method. In conventional spectroscopy, the scattered intensity distribution $I(Q, \epsilon)$ is usually obtained for the transferred energy ϵ in inelastic and quasi-elastic neutron scattering experiments, whereas in the spin echo method the measured neutron intensity is given by the scattering function $S(Q, \epsilon)$ integrated over ϵ with the weight of the polarization analysis component $P_x = \cos(2\pi N\delta v/v)$ on the energy transfer structure in the sample. As supposed from the principle of the Fourier integral, the latter means that we can directly obtain data concerning the time-correlation function on the dynamical structure in the sample. Further details on the developments of spin echo spectrometers will be given in Chapter 3.

2
Reflection, Refraction, and Transmission of Unpolarized Neutrons

2.1
Reflection and Transmission of Neutrons by a Single-Layer Mirror

2.1.1
Measurement of Coherent Scattering Amplitude

The fundamental formulas for reflection, refraction, and transmission of neutrons are given by Eqs. (1.14)–(1.17) in Chapter 1, where a neutron beam with an angle incident to the surface of an infinite mirror with uniform thickness of homogeneous material was considered. It was also pointed out in Section 1.2 that these theoretical relations hold regardless of the physical states of the layer material, that is, even for liquid and gaseous states. Furthermore, the total reflection critical angle formula, Eq. (1.17), is a very simple expression as the product of the squared wavelength of neutrons, atomic density, and the coherent scattering amplitude. Since the present phenomenon of total reflection is recoilless forward scattering, essentially independent of the material temperature and the chemical bindings, it becomes one of the important methods to measure precisely the scattering amplitude of various materials for neutrons.

As one of the earliest experiments of total reflection for neutrons, Hughes, Burgy, and Ringo [18] carried out precise measurements under critical conditions for liquid-hydrocarbon mirrors at Argonne National Laboratory, and applied Eq. (1.17) to obtain the values for the amplitude of coherent scattering of neutrons by a proton, that is, a hydrogen nucleus, $b_{\mathrm{coh,H}}$, because the precise measurement of the amplitude of coherent scattering of slow neutrons by a proton was an important task at that time from the viewpoint of nuclear force theory. Other experimental approaches, such as neutron scattering on liquid hydrogen or hydrogen-containing crystals, were also performed, but their experimental error was greater than about 3% because of the statistical errors due to orthohydrogen in the sample in the former approach, or the correction errors for the diffuse scattering from thermal motions of atoms in the crystal. The present method of total reflection does not have these theoretical inaccuracies. Thus, Hughes *et al.* could obtain an accurate value for the coherent scattering amplitude for a proton with an error of less than 1%, from the total reflection experiment on trimethylbenzene, a liquid

Neutron Optics. Masahiko Utsuro, Vladimir K. Ignatovich

ISBN: 978-3-527-40885-6

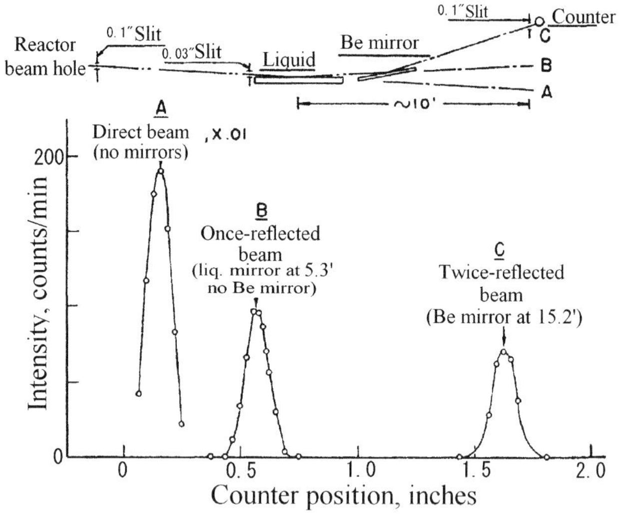

Figure 2.1 Experimental arrangement for neutron total reflection consisting of a liquid-sample mirror and a beryllium mirror for critical wavelength analysis (top) and the variation of the beam intensity with counter positions (bottom) (Hughes, Burgy, and Ringo [18]).

with a hydrogen to carbon ratio of 1.5 ($C_{12}H_{18}$), after subtracting the contribution from the carbon scattering amplitude $b_{\mathrm{coh,C}}$, which was known at that time with a precision of 0.5%. Their experimental accuracy for the proton scattering amplitude was effectively improved with the selection of the sample compound and they obtained a very small critical angle for total reflection as a result of balancing the positive scattering amplitude of carbon and the negative one of hydrogen.

A further consideration they employed to precisely measure the critical condition, Eq. (1.17), under the experimental situation of incident neutrons with a wavelength distribution is shown in the experimental arrangement in Figure 2.1. The angle of incidence with the liquid sample was held constant at 5.3 minutes, whereas the *critical wavelength for total reflection* was analyzed exactly by observing the critical angle for total reflection from a beryllium mirror after reflection by the sample by changing the detector position. The critical angle for total reflection of the beryllium mirror was determined in advance as a function of the wavelength with monochromatic neutrons. In this way, the precise value for the *coherent scattering amplitude for a proton* was derived as

$$b_{\mathrm{coh,H}} = -(3.75 \pm 0.03) \times 10^{-13}\,\mathrm{cm}\,, \tag{2.1}$$

where the error of about 0.9% mostly consists of statistical error.

2.1.2
Neutron Guide Tube

Applying the total reflection formula, Eq. (1.17), we can transport a neutron beam through a *neutron guide tube*, the inner wall of which is made of a material with a positive coherent scattering amplitude b_{coh}, and then the beam propagates through the inside hollow channel and experiences multiple total reflections. The principle and the design formula were proposed by Maier-Leibnitz [56, 57], and were further developed for practical uses. At present, various kinds of neutron guide tubes are installed in most neutron experiment facilities as indispensable devices for efficient and high-performance utilization of neutron beams. The main advantage of using a neutron guide tube is its capability to transport a neutron beam easily without significant intensity loss, that is, conserving, in principle, the phase space density of neutrons, to a place far distant from the neutron source where a wider area with lower background can be prepared for beam utilization. Furthermore, a *bent guide tube* instead of a *straight guide tube* transmits preferentially those neutrons with a wavelength longer than the *characteristic wavelength* given by the geometrical configuration of the guide tube, which reduces the components of shorter-wavelength neutrons and γ-rays unnecessary for neutron optics and slow neutron scattering experiments.

The characteristic wavelength λ^* for a bent guide tube satisfies, as shown in Figure 2.2a, the critical condition for the total reflection, Eq. (1.17), for the characteristic incident angle θ_c^* on the outer wall given by the characteristic flight path PQ tangent on the inner wall, as expressed by

$$\lambda^* = \sin\theta_c^* \sqrt{\frac{\pi}{N b_{\mathrm{coh}}}} \,. \tag{2.2}$$

For a circularly bent guide tube, as indicated in Figure 2.2a, the characteristic angle θ_c^* reduces to

$$\sin\theta_c^* = \sqrt{\frac{(\rho+d)^2-\rho^2}{\rho+d}} \cong \sqrt{\frac{2d}{\rho}}\,, \quad (\rho \gg d)\,, \tag{2.3}$$

where ρ and d are the radius of curvature and the channel width of the guide tube, respectively. After similar analyses on the critical conditions for all kinds of paths in the channel, the position-dependent transmission probability of the channel can be derived for every wavelength.

Since this optical performance of a neutron guide tube should become remarkable for neutrons with lower energy, or longer wavelength, the analytical results for a bent guide tube and the corresponding experimental results are compared in Figure 2.2b for very cold neutrons with long wavelength. The configuration of the guide tube used in the experiment is as follows: total length 1.8 m, channel width 4 cm channel height 5 cm, reflecting surface inside the guide tube of 3000-Å-thick nickel with *limiting velocity for total reflection* (refer to the next section) $v_l = 6.7\,\mathrm{m/s}$, evaporated on float glass substrates, radius of curvature of the guide tube of 9 m in a polygonal arrangement.

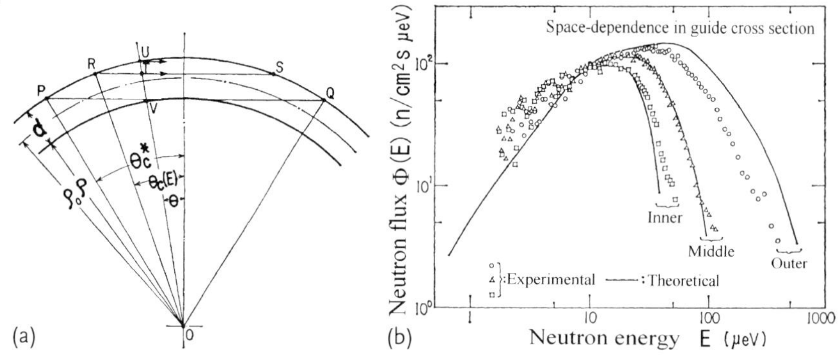

Figure 2.2 Characteristics of a bent guide tube: (a) geometrical analysis on the neutron flight paths and (b) experimental results obtained with the time-of-flight method for the position-dependent transmission characteristics in the cross section of a bent guide tube for pulsed very cold neutrons, compared with the numerical calculations (Utsuro *et al.* [58]).

The experimental results obtained with the time-of-flight method of pulsed very cold neutrons from a solid mesithylene cold source [58] at the linac facility at the Kyoto University Research Reactor Institute are consistent, as shown in Figure 2.2b, with the theoretical values from Eqs. (2.2) and (2.3) for a characteristic wavelength of about 56 Å, a *characteristic velocity* of about 71 m/s, and a *characteristic energy* of about 26 μeV for the guide tube configuration. This indicates that these characteristic parameters represent well the average and the central path performances of the neutron guide tube.

For the position dependence of the transmission probability, the following tendency can be pointed out. The inner portion among the various flight paths includes mainly the *zigzag* component, which experiences reflections off the inner and the outer walls alternatively. Thus, the incident angle on the outer walls becomes larger than the characteristic angle θ_c^*, and as a result becomes superior in the lower-energy component. In contrast, the outer portion includes mainly the *garland* component successively reflected only off the outer walls; thus, the incident angle is smaller than θ_c^*, and then significantly includes the component with energy higher than the characteristic energy of the guide tube.

Further applications of such a wavelength-selection performance and the beam-bending capability with the bent guide tube have been developed for novel devices such as a neutron bender [59, 60] and a neutron fiber or a neutron-focusing fiber consisting of a number of microguides [61, 62].

2.1.3
Total Reflection Potential and Surface Impurities

Now, we would like to go into some detail on the properties of the optical potential in the reflection of neutrons. In the fundamental analysis of neutron reflection, a single-layer mirror is most simply expressed by a single stepwise potential. Such a theoretical description of the optical potential is further extended to employ a more sophisticated form for the analyses of reflection and transmission in complicated situations we experience in practice.

One of the examples studied experimentally on such a sophisticated potential deformed from simple rectangular was reported by Schheckenhofer and Steyerl, who carried out reflection experiments of *ultracold neutrons* by making use of a very high resolution gravity diffractometer [63]. *Ultracold neutrons* were defined in Sections 1.2 and 1.5 as those with a velocity below the *limiting velocity for total reflection* v_l at the normal incidence to the mirror surface. In other words, their energy is less than the optical potential for total reflection off the mirror surface.

Schheckenhofer and Steyerl produced such low-energy ultracold neutrons at the Munich research reactor (FRM-I) with a special rotating device called a "*neutron turbine*" [64] to decelerate very cold neutrons with a velocity of about 50 m/s to ultracold neutrons with a velocity below about 10 m/s. Their neutron turbine consisted of a number of semicylindrical copper mirror blades rotating with a blade velocity of about 25 m/s, that is, about half the neutron velocity. During the passage through the blade with multiple reflections on the concave side, neutrons reverse their direction of relative motion and then the exit neutrons from the blade edge lose most of their momentum in the laboratory frame. Further details on the turbine structure and the operation characteristics will be given in Section 2.4. The ultracold neutrons produced in the turbine are guided to the *gravity diffractometer* shown in Figure 2.3a. After being collimated by the entrance slit, the neutrons experience gravitational fall and further those neutrons with a defined energy are selected by the reflection off a vertical nickel mirror. Then, they are reflected by a horizontal sample mirror and the reflected neutrons are counted after another vertical mirror and the exit slit.

The *gravity potential* for the neutron being about 10^{-7} eV/m induces a typical parabolic curve for the ultracold neutrons with a velocity of about 3 m/s, that is, with an energy of about 0.05 μeV, employed in the experiment. However, those neutrons with a velocity below about 3 m/s cannot pass through the aluminum wall (limiting velocity 3.2 m/s) of the neutron counter, and therefore they arrive at the counter after gravitational acceleration in the vertical fall. The measured count rates of the neutrons reflected from the sample shown in Figure 2.3b are plotted for the gravity fall height before the sample reflection, which selects the vertical component of the neutron velocity at the incidence at the sample.

The results indicate a typical reflectivity edge at a fall height of $h_{\text{cr}} = 93.6$ cm as expected for the low-alumina boron-free float glass, but the measured dependence of the count rate decease on the fall height $h > h_{\text{cr}}$ is obviously steeper than the theoretical curve of the reflectivity R for the *step potential* given by the equation[4)]

$$R = \left| \frac{k - k'}{k + k'} \right|^2 , \tag{2.4}$$

4) Refer to, e. g., textbook [B]; Chapter III, problem 1.

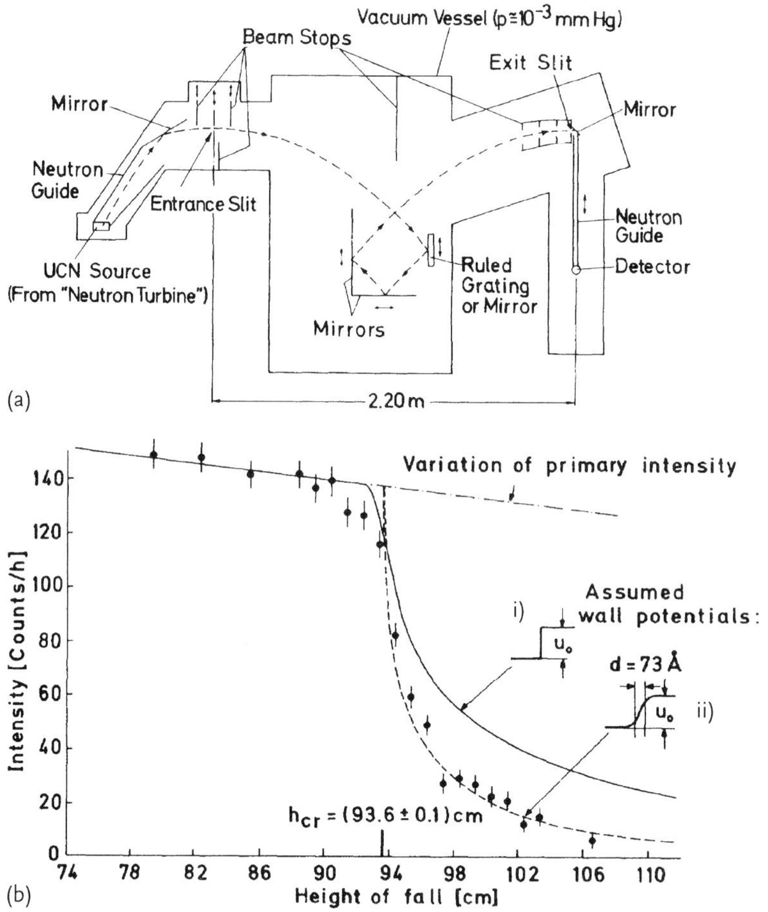

Figure 2.3 Mirror reflection experiment on ultracold neutrons: (a) arrangement of the gravity diffractometer and (b) measured distribution of the reflected intensity from a glass mirror (circles) compared with two kinds of theoretical curves, that is, (i) for the case of a step potential and (ii) for the case of a soft potential. UCN ultracold neutron. (Scheckenhofer and Steyerl [63]).

where k and k' are the wave vector components normal to the mirror surface of the neutrons in a vacuum and deep inside the mirror, respectively.

As a more sophisticated potential for the mirror surface, if we employ a *soft potential* with a *smoothed step* described by the equation

$$u(z) = \frac{u_0}{1 + \exp(z/d)}, \tag{2.5}$$

where z denotes the coordinate normal to the mirror surface and d is a parameter representing the thickness of the transient region in the potential, then the reflectivity R for a real value of u_0 reduces to

$$R = \left| \frac{\sinh[\pi d(k - k')]}{\sinh[\pi d(k + k')]} \right|^2 . \tag{2.6}$$

The calculated values for the present soft potential with a value of $d = 7.3 \pm 0.3$ nm plotted in Figure 2.3b agree well with the experimental values.

The existence of such a transient region in the reflection potential is considered to be explained by possible *hydrogenous impurity on the mirror surface.* Actually, the existence of a significant concentration of hydrogenous impurity on *clean* mirror surfaces was directly and commonly observed with two kinds of different experimental methods for precise element analyses as explained below [65, 66].

The first study was that by Lanford and Golub for the measurement of the *surface hydrogen concentration with resonant nuclear reaction* [65], the result of which is shown in Figure 2.4. The experiment was carried out by injecting a ^{15}N beam onto the surface of a sample mirror, and the γ-rays produced by the reaction $^{15}\text{N} + \text{H} \rightarrow {}^{12}\text{C} + {}^{4}\text{He} + 4.43$ MeV γ-ray were counted. Changing the energy of the incident beam enables one to determine the depth where the reacted hydrogen is located according to the relation for the thickness of the mirror necessary to moderate the incident beam down to just the resonance energy for the reaction. The

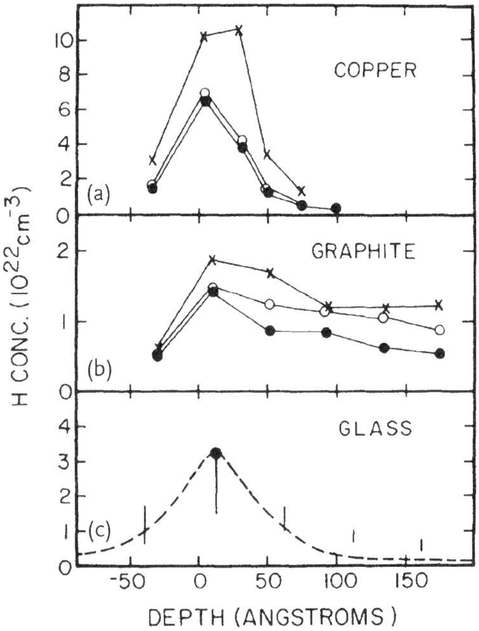

Figure 2.4 Distributions of surface hydrogen concentration measured on several mirror samples (mechanically polished copper (a), pyrolytic graphite baked at 400 °C and then exposed to air (b), and glass etched in HF (c)) (Lanford and Golub [65]).

thickness resolution in the present method varied with the mirror material within a range of about 3–7 nm (30–70 Å).

Among the materials studied as shown in Figure 2.4, the measurement on glass, an excellent insulator, was very difficult owing to the accumulation of electrical charge on the sample, thus inducing a change in the energy of the incident particle as it approached the surface. They tried to improve their experiment by lowering the beam intensity and repeating the measurements. The curve in Figure 2.4 for glass represents the minimum hydrogen content consistent with the data.

Anyway, all the results in Figure 2.4 obviously indicate a rather common peak concentration of about $(2-10) \times 10^{22}\ \mathrm{H/cm^3}$ with a depth width of about 5–10 nm, which means a significant amount of surface hydrogen; this needs to be considered when estimating the form of the neutron reflection potential.

Another measurement is that of the depth distribution profiles of the surface and interface hydrogen concentrations performed by Kawabata *et al.* with the method of *elastic recoil detection analysis* on neutron mirrors evaporated on substrates [66]. The experiments were carried out in a vacuum chamber by injecting an ^{40}Ar beam with an energy of 50 MeV onto the front surface of a sample mirror at an angle of 30°. The atoms that recoiled from the mirror with *Rutherford forward scattering* were registered with solid-state detectors.

The detectors were placed at angles of 37 ° and 47 ° to the incident direction, the former of which with a better depth resolution was used for the concentration profiling, whereas the latter was the monitor for the normalization. For the purpose of the present analysis for light-element profiles, an aluminum foil filter with a thickness of 10 μm was inserted in front of the detector.

An example of the measured energy spectrum of the recoiled atoms is shown in Figure 2.5a, and from the energy corresponding to the abscissa we can determine the energy loss for the transmission from the recoil position inside the mirror up to the mirror surface. The horizontal coordinates in the figure for each recoiled elements indicate the depth (Å) from the mirror surface reduced in such a way. The count rate shown on the ordinate can be converted for each element to the concentration using the formula for the Rutherford scattering cross section.

As clearly indicated in Figure 2.5a, a significant concentration of hydrogen is accumulated on the surface of the evaporated nickel and also in the Ni–Al interface in the nickel mirror processed in a high vacuum on an aluminum substrate with the surface polished by diamond machining. The sample was thermally insulated during the measurements, and the thermocouple on the backside surface indicated a temperature of about 100 °C. Thus, the sample could be considered to be rather under a vacuum baking state at a higher temperature.

Similar kinds of measurements were applied for various mirror samples as well as those after various trials to reduce the hydrogen concentration so as to expose the sample to a high vacuum or a deuterium gas atmosphere. The results are compared in Figure 2.5b, which reveals the existence of hydrogen in a concentration of $(2 \sim 5) \times 10^{15}\ \mathrm{H/cm^2}$ on the surface and in the interface, regardless of the various mirror materials and the surface processings. If we consider the results mentioned above to be the summation over the hydrogen depth width of 5–10 nm as shown

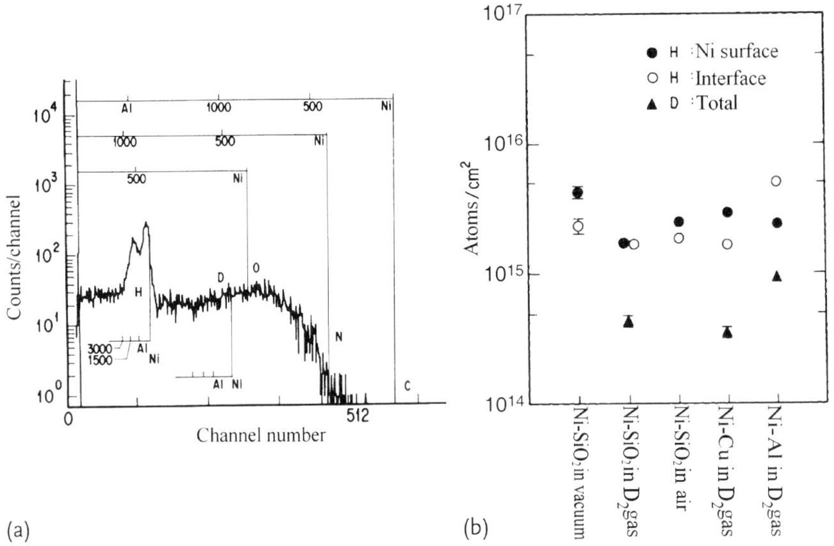

Figure 2.5 Hydrogen and deuterium concentration profiles: (a) spectrum measured by a solid-state detector with an aluminum filter of the recoiled atoms from 150-nm-thick nickel evaporated on an aluminum substrate and (b) comparison between the measured hydrogen/deuterium concentrations in various samples (Kawabata *et al.* [66]).

in Figure 2.4, then we can estimate the volume concentration of hydrogen to be higher than 5×10^{21} H/cm^3, which is sufficient to deform the optical potential for total reflection.

Thus, the two different experiments described above commonly reported a significant amount of hydrogen included in the important regions as a neutron mirror, and the coherent scattering amplitude of hydrogen being negative, the soft potential as a deformed step as shown in Figure 2.3b is considered to be a realistic description on the reflection potential for most kinds of neutron mirrors.

2.1.4
Neutron Reflectometry

As described in the present section, spectroscopy with neutron reflection is one of the most precise experimental methods; it is very sensitive and effective for analyzing the atomic and isotopic concentrations in the surface and interface regions. One of the advantages of the present approach compared with experiments with neutron inelastic scattering and small-angle scattering is the smaller angular dispersion factor, and it is thus suitable for discussing detailed structures based on experimental data with a higher statistical accuracy.

Furthermore, from the viewpoint of the theoretical analysis involving calculation of the neutron wave functions as the solution of the wave equation, the generally accepted assumption on the conservation of the wave vector component parallel to the interface allows one to employ the elementary solution in quantum mechanics for a one-dimensional potential (e. g., refer to textbook [B] Chapter III, problem 2, or textbook [C] Chapter 5, Section 17).

Practically, the *reflection coefficient* r and the *transmission coefficient* t for a single stepwise potential with thickness d are given by the next equations for a monochromatic incident wave with a unit amplitude:

$$r = \frac{(k^2 - k'^2)(1 - e^{2ik'd})}{(k + k')^2 - (k - k')^2 e^{2ik'd}} ,$$

$$t = \frac{4kk' e^{i(k'-k)d}}{(k + k')^2 - (k - k')^2 e^{2ik'd}} , \tag{2.7}$$

where k and k' denote the wave vector components normal to the interface for neutrons in a vacuum and in the potential, respectively.

Alternatively, Eq. (2.7) can be written as

$$r = \frac{-i(k^2 - k'^2)\sin(k'd)}{2kk'\cos(k'd) - i(k^2 + k'^2)\sin(k'd)} ,$$

$$t = \frac{2kk' e^{-ikd}}{2kk'\cos(k'd) - i(k^2 + k'^2)\sin(k'd)} . \tag{2.8}$$

The reflectivity for the layer thickness $d \to \infty$ in Eq. (2.7), that is, the reflectivity $R = |r|^2$ by putting the contribution from the backside surface, $e^{2ik'd}$ being zero, reduces to the previous Eq. (2.4). Further, this approach can easily be extended to the case of multilayers of the step potential as described in Section 2.3 Eq. (2.7). Owing to these advantages, *neutron reflectometry* has recently become widely applied to the study of various kinds of materials.

Surface and interface investigations with neutron *specular reflection* are now being extended to include various research fields, such as the working mechanism of surfactants at the air–solution interface, the structures of polymers and microphase-separated layers at the air–liquid interface, the surface chemistry on adsorption of a detergent layer at air–liquid and air–solid interfaces, and also studies on Langmuir–Blodgett films, solid films such as hard carbon and semiconductors, magnetic multilayers, ferromagnetic films, and so on [67].

The experimental accuracy at the present day of these specular reflectometries makes it quite standard to obtain the reflectivity plot over about 5 orders below the total reflection. The experimental procedures and their application guides are given in instructions (e. g., [69]) and reviews (e. g., [67]).

The example shown in Figure 2.6 is a comparison between two different measurements on the same sample [68], one of the measurements being carried out with a *horizontal type of reflectometer*, NG-7 (with a neutron wavelength of about 0.4 nm), at the National Institute of Standards and Technology (NIST) reactor, United States, and the other with a *vertical type of reflectometer*, C3-1-2 (with a neutron wavelength of 1.26 nm), at the JRR-3 reactor, Japan. The sample studied was a block

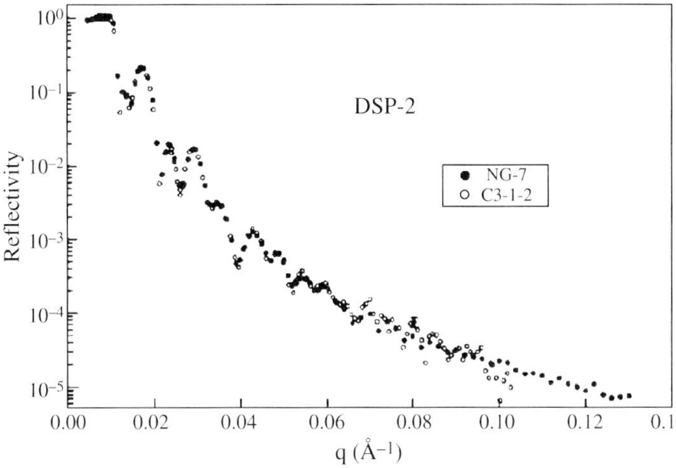

Figure 2.6 Measured cold neutron reflectivities compared from two different devices used for studying a thin film of triblock copolymer DSP (Ebisawa *et al.* [68]).

copolymer with a total thickness of about 250 nm, spin-cast from solution on a polished silicon substrate. The abscissa in the figure represents the vertical component q of the neutron wave vector[5)], and both results in Figure 2.6 agree well over the whole reflectivity region shown covering 5 orders of magnitude.

The analysis of these results from neutron reflectometry measurement gives the profile of the scattering density distribution in the sample. The reflectivity distributions in Figure 2.6 are different from the theoretical curve and the measured pattern for a single layer, and include many irregular structures. The data analysis of the reflectivity measurement for these kinds of multilayer samples will be discussed in Section 2.3.

Above we mainly discussed *specular reflection*, which is the reflection from a flat surface without any structures along the surface. In contrast, if the surface includes roughness and waviness, or very regular gratings or striped patterns, such a reflection component toward a shifted reflection angle from the specular angle or small-angle diffractions might be induced, and these components are denoted as *off-specular reflection*.

In specular reflections, the incident and the reflected angles are the same; therefore, the wave vector transfer occurs only for the normal component, and thus the description and the analyses of the phenomenon can be simplified. On the other hand, in the case of off-specular reflections, the description of the phenomenon requires at least two-dimensional variables to include also the event for the parallel component.

5) In specular reflection for neutron wavelength λ, wave number $k = 2\pi/\lambda$, and angle θ between the incident neutron and the mirror surface, the *wave vector transfer* is denoted as $Q = 2q = 4\pi \sin\theta/\lambda$. However, this notation of Q and q might be a little different in other references.

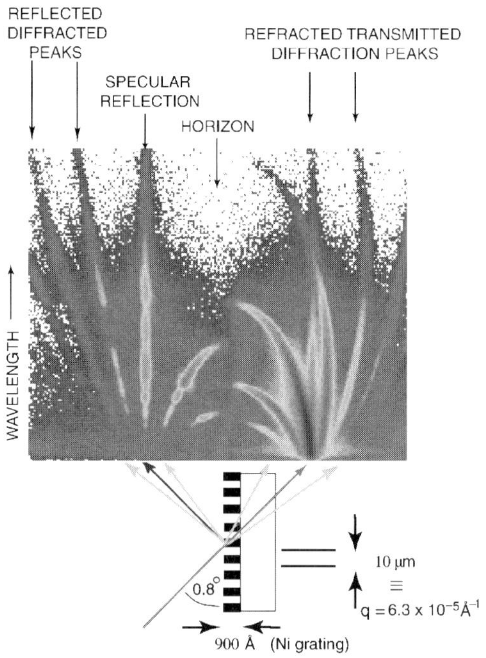

Figure 2.7 Time-of-flight data showing reflection, refraction, and off-specular diffraction from a nickel grating on a glass substrate (Cubitt [69]).

As an example of such cases, illustrated in Figure 2.7 are the results of neutron reflection and transmission for a nickel grating with several thousand lines with a spacing of 10 μm with a thickness of 90 nm and a width of 5 μm on a glass substrate, obtained from measurements [69] with the *chopper and time-of-flight reflectometer* D17 at the high-flux reactor at the Institut Laue-Langevin, Grenoble.

The abscissa in the reflection and transmission patterns corresponds to the pixels of the counter arranged normal to the sample surface, whereas the ordinate is the time of flight, that is, proportional to the neutron wavelength. The original picture shows the neutron intensity also with the color tones, which were, however, converted to a simpler colorless pattern for the present figure.

The figure indicates clearly, in addition to the specular components, with the straight vertical lines meaning the reflection angle is constant and the same as the incident angle independent of the neutron wavelength, several additional lines with the reflection angle dependent on the wavelength as the typical character of diffracted reflections, as well as similar additional lines with wavelength-dependent and wavelength-independent angles for refracted-transmission and diffracted-transmission components, respectively. In this way, the off-specular reflection and transmission method is expected to provide an abundance of information about the surface and interface of samples with complicated structures.

2.2 Properties of Ultracold Neutrons

2.2.1 Reflection Loss Probability

In Figures 2.3b and 2.6, the measured reflectivity in the total reflection region is usually considered to be 100%; however, in strict terms it is not exactly 100%, but is a little smaller than 100%. The expected magnitude of this reduction from 100%, that is, the theoretical value for the *reflection loss probability*, is given by the next equation for a step potential (refer to textbook [F], Section 5.1, or textbook [G], Appendix 1.1):

$$\mu = 1 - |\vec{r}_0|^2 = 2\frac{u_0''}{u_0'}\frac{k}{\sqrt{u_0' - k^2}} . \tag{2.9}$$

Since the above equation represents a one-dimensional system, the theoretical formula for a three-dimensional geometry applicable to usual ultracold neutron storage experiments with a neutron bottle should be written as

$$\mu = 2\frac{u_0''}{u_0'}\frac{k_\perp}{\sqrt{u_0' - k_\perp^2}} . \tag{2.10}$$

In Eqs. (2.9) and (2.10), k and $k_\perp$ denote the neutron wave number and its component normal to the mirror surface, respectively, and u_0' is the real part of the mirror step potential given by $u_0' = 4\pi N_0 b'$, whereas u_0'' is the imaginary part of the potential derived from the imaginary part b'' of the *complex scattering amplitude* $b = b' - ib''$.

Theoretical connections of b'' and u_0'' to the physical parameters of the mirror material are derived in Chapter 6 in textbook [G] and can be written as

$$b'' = \frac{k[\sigma_c(k) + \sigma_{ie}(k)]}{4\pi} , \tag{2.11}$$

$$\eta = \frac{u_0''}{u_0'} = \frac{b''}{b'} , \tag{2.12}$$

where η is the *reflection loss coefficient* and $\sigma_c(k)$ and $\sigma_{ie}(k)$ are the absorption and the inelastic scattering cross sections, respectively, of the mirror material for neutrons with wave number k.

For the incident angle θ of neutrons with the mirror surface normal, we can write $k_\perp = k\cos\theta$, and the angular average of Eq. (2.10) for an isotropic flux as assumed for neutrons in the storage experiments (neutron storage experiments will be described in the next subsection) reduces to (refer to textbook [G], Appendix 6.1, or [70])

$$\bar{\mu} = \frac{2\eta}{y^2}\left(\arcsin y - y\sqrt{1 - y^2}\right) , \quad y = v_0/v_l , \tag{2.13}$$

where v_0 is the neutron velocity and v_l the *limiting velocity for total reflection* of the mirror consisting the inside surface of the bottle.

Now, we can estimate the magnitude of the reflection loss probability for practical neutron mirrors according to Eq. (2.13). The *reflection loss coefficient* η can be rewritten as

$$\eta = \frac{b''}{b'} = \frac{k[\sigma_c(k) + \sigma_{ie}(k)]}{4\pi b_{\text{coh}}} = \frac{\sigma_c(k) + \sigma_{ie}(k)}{2\lambda(k) b_{\text{coh}}} = \frac{\sigma_{c0} + \sigma_{ie0}}{2\lambda_0 b_{\text{coh}}} , \qquad (2.14)$$

where b_{coh} is constant, and further the usual relations of the $1/v$ law for the absorption and the inelastic scattering cross sections of low-energy neutrons allow us to put $(\sigma_{c0} + \sigma_{ie0})/\lambda_0$ as a constant.

As an example of most idealized mirror conditions, we assume a beryllium mirror with a perfectly clean and flat surface, cooled down to a sufficiently low temperature, which leads to the theoretical values of $\sigma_{c0} + \sigma_{ie0} \cong \sigma_{c0} \cong 10$ mb for neutrons with wavelength $\lambda_0 = 0.18$ nm, and further $b_{\text{coh}} \cong 7.8$ fm.[6] Then, we obtain the expected value of $\eta \cong 7 \times 10^{-7}$. Similar theoretical estimations from Eq. (2.14) for other kinds of neutron mirrors also assumed with perfectly clean surface are $\eta \sim 10^{-4}$ for copper, nickel, and iron, and $\eta \sim 10^{-5}$ for boron-free glass, quartz glass, silicon, and aluminum, at room temperature.

On the other hand, we found in the previous section that the reflection potential for most neutron mirrors should be considered significantly deformed from the simple form of a step function, to a soft potential somewhat rounded in the region of the mirror surface and interfaces as shown in Figure 2.3b, because of the very common phenomenon of the adsorption and accumulation of *impurity hydrogens* on the surface and the interfaces.

The hydrogen concentration in this surface region, as indicated in Figures 2.4 and 2.5b, was about 5×10^{21} H/cm^3 with a depth width of about 5–10 nm, almost regardless of the materials and the surface treatments.

The penetration depth of the neutron wave function into the stepwise reflection potential under total reflection is theoretically estimated to be about 10 nm according to the solution of the Schrödinger equation. Experimental studies were also reported by Novopoltsev *et al.* [71] on the penetration depth of ultracold neutrons incident on a copper mirror with various thicknesses evaporated on a polished glass substrate. They measured the induced radioactivity of the copper layer by removing it from the substrate after irradiation with the ultracold neutron flux, and derived the dependence of the ultracold neutron absorption coefficient on the copper thickness, the result of which indicated the ultracold neutron penetration depth was consistent with the theoretical estimation, about 10 nm, as mentioned above.

On the basis of this information, here we want to try to calculate the expected reflection loss coefficient for a mirror surface which is assumed to include a hydrogen impurity with an atomic concentration of about 10% of the base material of the mirror. Since the physical and chemical forms of such an impurity are not

6) 1 fm (*fermi*) $= 1$ femtometer $= 10^{-15}$ m $= 10^{-13}$ cm.

known well, we also assume its state is rather near that of ordinary water. Then a very rough estimation for the constant magnitude $(\sigma_{c0} + \sigma_{ie0})/\lambda_0$ in the $1/v$ region leads to the increment of η_{surfH} in the reflection loss coefficient owing to the hydrogen impurity in the surface region: $\eta_{\mathrm{surfH}} \sim 10^{-4}$. The other part except η on the right side of Eq. (2.13) increases monotonously on varying y from 0 to 1 as $2(\arcsin y - y\sqrt{1-y^2})/y^2 = 0 \sim \pi$.

The experimental approach for the observation of the reflection loss coefficient and the practical results are described in the next subsection.

2.2.2
Neutron Storage Bottle and Anomalous Loss Coefficient

Because of very small magnitude expected for the reflection loss probability compared with unity, it is difficult to measure the magnitude in single reflection experiments such as those depicted Figures 2.3b and 2.6. The observation will become possible by precisely registering the decrease in the neutron number after thousands or tens of thousands of reflections have been experienced. *Ultracold neutron bottle experiments* make this possible.

In 1959, Zel'dovich proposed an idea to store for a long time such neutrons with very low energy to be totally reflected many times off the wall inside a vacuum vessel [72]. About 10 years after his proposal, Shapiro's group for the first time practically carried out the neutron storage experiment [73]. Since then, this new method for the ultracold neutron storage experiment has been applied for various kinds of physical experiments as a unique approach to hold neutrons in high-precision observations for a long period.

In such *ultracold neutron bottle experiments*, usually the temporal variation of the number of neutrons stored in the vacuum vessel experiencing total reflections is registered. As mentioned for Figure 2.3, *ultracold neutrons* are totally reflected even at normal incidence to the neutron mirror surface off the inside wall of the bottle, and then repeat free flights and total reflections until they are lost by any of the following causes: reflection loss, β decay to an electron, a proton, and an antineutrino, leakage from the vessel through an existing hole or gap, or any other possible loss mechanisms in the vacuum vessel.

As an example of the earliest neutron bottle experiments carried out at the Joint Institute for Nuclear Research, Dubna, and Kurchatov's Institute of Atomic Energy, the experimental results of Groshev *et al.* are shown in Figure 2.8 [74]. In Figure 2.8a, 1 and 2 indicate the entrance and exit shutters, respectively, for ultracold neutrons (the space between these two shutters thus works as the storage bottle) and 3 is the ultracold neutron detector. First, the entrance shutter is opened with the exit shutter closed. Then, the ultracold neutrons coming through an S-shaped guide tube installed in the research reactor fill the bottle. When the saturation density of neutrons is attained, the entrance shutter 1 is closed. After storage time t, the exit shutter is opened, and then the number of neutrons that flow out are registered. The results are plotted in Figure 2.8b, where experiment 1 was performed on a chemically polished copper bottle, giving the average lifetime T of the stored

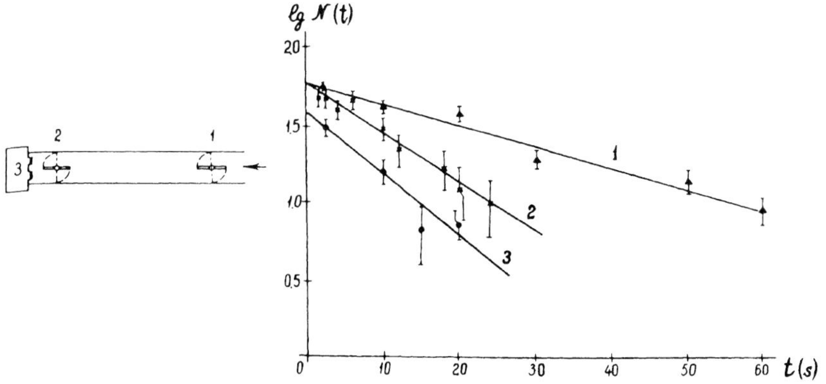

(a) (b)

Figure 2.8 Measurement of ultracold neutron storage lifetime: (a) storage bottle with an inner diameter of 14 cm and a length of 174 cm; (b) experimental results for the neutron number $N(t)$ after storage time t in various kinds of bottles; the ordinate shows the logarithm of $N(t)$ (Groshev *et al.* [74]).

neutrons of 33 s, and experiments 2 and 3 were for a copper foil bottle and a pyrographite bottle, both with unpolished surfaces, resulting in $T = 14\,\mathrm{s}$ and 11 s, respectively.

Applying the storage lifetime formula for T,

$$\frac{1}{T} = \frac{v}{\bar{\mu} d} + \frac{1}{\tau_n}, \tag{2.15}$$

to these experimental results, where v is the neutron velocity, d the bottle diameter, and τ_n the neutron decay lifetime, we can deduce the value for the average reflection loss probability per wall collision, $\bar{\mu}$, as $(0.7\text{–}2) \times 10^{-3}$. The present experimental values for $\bar{\mu}$ are 20–30 times larger than the theoretically expected values for these wall materials, and these experimental results are considered to indicate the existence of some unknown causes for ultracold neutron loss from the storage vessel.

As possible loss causes which are not considered in the above analysis, we can suppose reflection loss due to some defects on the inner surface and neutron leakage through some gaps in the shutter mechanism. Therefore, an experiment to diminish the possible effects of these loss causes was undertaken by Utsuro *et al.* [75] by preparing a larger-volume storage bottle of about 120 l which was equipped with tight contact type entrance and exit shutters with spring-pressing and the inner surface of the bottle was made of SUS-304 stainless steel treated with an *ultrafine electrochemical composite buffing polish process* [76].7) Ultracold neutrons for the storage lifetime experiments were fed to the bottle after gravitational deceleration through

7) The polished surface quality was an average roughness of about 50 Å and the peak-to-peak value about of 500 Å.

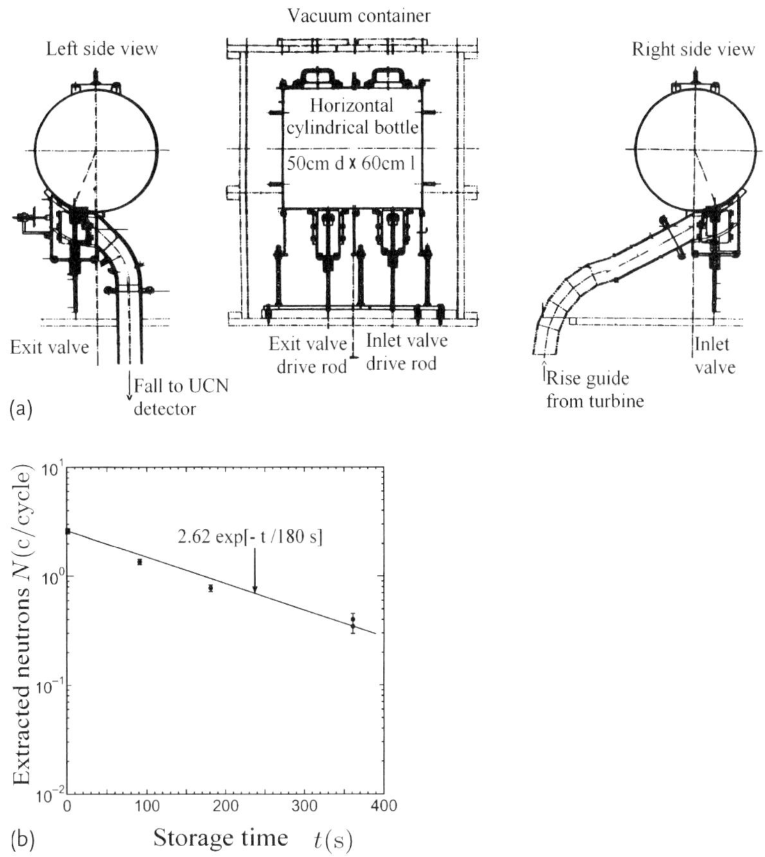

Figure 2.9 Ultracold neutron storage experiment with a large-volume ultrafine polished stainless steel bottle: (a) structure of the bottle; (b) experimental result for the storage lifetime (Utsuro and Okumura [75]).

a 1-m-rise guide tube connected to the exit port of the *supermirror turbine* [77] at the Kyoto University reactor described in Section 2.4.1. The experimental results are shown in Figure 2.9.

From the analysis of the measured storage lifetime of 180 s, that is, after subtracting the effect of the neutron decay according to Eq. (2.15), and by making use of the neutron mean free path $\bar{l} = 4V/S \cong 35$ cm, which replaces d in Eq. (2.15), (where V and S denote the storage volume and the inner surface area of the bottle, respectively), and the average velocity of the stored neutrons $\bar{v} = 5$ m/s estimated from the limiting velocity of total reflection $v_l = 5.4$ m/s for the present polished SUS-304 neutron mirror, the average reflection loss probability was obtained as $\bar{\mu} \cong 3.0 \times 10^{-4}$. The present value is near the expected value of 2×10^{-4} for the room temperature SUS bottle including the possible contribution of the surface hydrogen impurities.

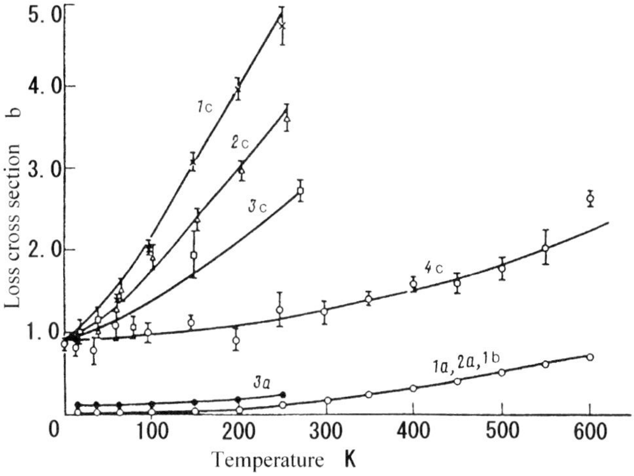

Figure 2.10 Neutron loss cross sections of beryllium after various purification treatments and at various temperatures, derived from an ultracold neutron storage experiment and from a very cold neutron transmission experiment, compared with the theoretical loss cross section ($\sigma_a + \sigma_{in}$) curve calculated in the Debye model. Transmission experiments: 1a, fused Be; 2a, quasi-single-crystal Be; 3a, pressed sintered Be. Storage experiments: 1c, deposited spherical trap, not degassed; 2c, deposited cylindrical trap, degassed at 250 °C; 3c, all-Be trap, degassed at 300 °C; 4c, deposited spherical trap, degassed at 350 °C and purification with He and D_2 gas. Theory: 1b. (Alfimenkov *et al.* [78]).

However, on the level of further lowered loss probability in bottled neutron experiments, significant disagreements have been reported between the experimental results and the theoretical expectation. As a typical example, the findings of the experimental study on the *beryllium bottle* by Alfimenkov *et al.* are shown in Figure 2.10 [78]. To elucidate the possible lower limit of reflection loss probability, they performed ultracold neutron storage and *very cold neutron transmission* experiments covering the wide temperature region from about 600 K down to 10 K on beryllium bottles and transmission samples after various purification treatments.

The value for the loss cross section extrapolated to absolute zero temperature from the results of the bottle experiment, that is, $\sigma_{ie}(k) \to 0$ in Eq. (2.12), is about 0.9 b larger than the results from the transmission experiment and the theoretical value based on the absorption cross section of beryllium; there are obvious disagreements. The reason for the present disagreements has not been sufficiently well clarified [79].

2.2.3
Limiting Velocity for Total Reflection and Neutron Absorption

Another disagreement between experimental results and theoretical expectation was observed by Kitagaki *et al.* [80, 81] in the development of a new kind of *solid-state ultracold neutron detector* at the Kyoto University Research Reactor Institute. Figure 2.11 shows the neutron velocity dependences of the count rate measured with the ultracold neutron beam at the high-flux reactor, Grenoble, on the ^{6}Li solid-state

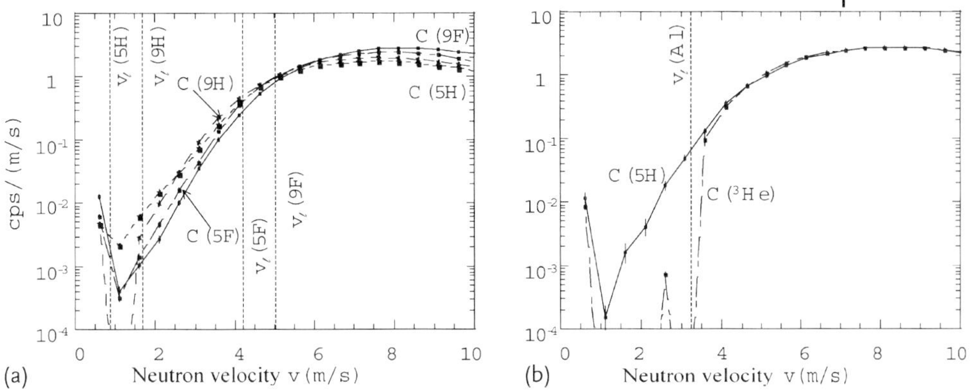

Figure 2.11 Velocity dependences of various kinds of ultracold neutron detectors: (a) ^{6}Li solid-state detectors with different limiting velocity for total reflection v_l and (b) a typical velocity dependence compared with that obtained with an aluminum window helium gas detector (Kitagaki *et al.* [81]).

detector [81] especially developed by Kitagaki with different values of the *limiting velocity for total reflection*, v_l. They compared the results with the similarly obtained results for a conventional aluminum window helium gas detector. The abscissa corresponds to the neutron flight velocity component perpendicular to the sensitive surface of the detectors, that is, $v_\perp$, which was derived from the flight time of the incident neutrons, and the ordinate is the registered count rate.

Since the isotope ^{6}Li has a large absorption cross section for neutrons ($\sigma_a \cong 945$ b at $v = 2{,}200$ m/s), the *contribution of the absorption to the limiting velocity* is taken into consideration according to the theory of Goldberger and Seits given in Chapter 1. According to their theory, the refraction phenomena are described by the *complex potential* related to the *complex scattering amplitude* for the isotope constituting the reflecting surface. As a result, b_{coh} on the right side of Eq. (1.16) should be replaced by the complex scattering amplitude b as

$$b_{\text{coh}} \to b = b' - i b'' , \tag{2.16}$$

$$b' = b_{\text{coh}} - \left(\frac{\sigma_a}{2\lambda}\right)^2 , \tag{2.17}$$

$$b'' = \frac{\sigma_a}{2\lambda} , \tag{2.18}$$

where the approximation $\sigma_c \cong \sigma_a$ (σ_a is the absorption cross section) was employed. Thus, we arrive at the formula for the reflection loss coefficient η, as previously described by Eq. (2.12).

The reliability of the present formula considering the *absorption* of neutrons was already assured by the consistency in cold neutron optical experiments on solid multilayers. An example is given in Section 9.6.4 for the reflection and transmission probabilities in experiments [445] with magnetic multilayer resonators including evaporated Gd with a giant absorption cross section.

We can estimate the contribution of the absorption cross section of ultracold neutrons $\sigma_a(v \sim 1\,\mathrm{m/s}) \cong 2 \times 10^6$ b for ^{6}Li in the Li-type solid-state detector according to Eqs. (2.16) and (2.17). The estimated contribution of the second term on the right side of Eq. (2.17) to the refractive index is smaller than 1%. Thus, the limiting velocity for total reflection of the solid-state detector is practically determined by the first term on the right side of Eq. (2.17), the derived values being shown by the vertical dotted lines in Figure 2.11a. In contrast to these apparently different values for the limiting velocity, the measured velocity dependences in Figure 2.11a indicate nearly the same and noticeably sensitive responses to ultracold neutrons even below the limiting velocity for total reflection.

Now, it is important for understanding these observations to point out the fact that the optical microscope inspection of the surface of the present solid-state detectors revealed very irregular surface structures with many cracks and waviness [80], that is, a not so simple flat surface as assumed in the theoretical formula given above. These irregularities should have significant effects on the velocity dependences reported in Figure 2.11a. For example, for a structure with size larger than the neutron wavelength, the perpendicular component $v_\perp$ of the incident neutron velocity varies locally from the supposed macroscopic value, whereas for the structure with the same size as the neutron wavelength, refractive and diffractive transmissions will be induced as already mentioned for Figure 2.7. These investigations and considerations will be important for clarifying various anomalous phenomena in ultracold neutron experiments, as mentioned in the previous section and in the last part of the present section.

Ultracold neutron optics in the case of a much larger absorption cross section will be more interesting. For the isotope with the largest absorption cross section, ^{157}Gd, the value of σ_a is about 250 times that of ^{6}Li, and for the complex scattering amplitude for such a nucleus in Eq. (2.16)–(2.18), it is actually possible that the imaginary component becomes dominant. There is a prediction [82] that the absorption cross section will saturate in such an extreme condition. Furthermore, concerning the neutron transmission ratio for a dilute solution of such an isotope, an additional term must be added to the usual formula for the transmission ratio represented by the average concentration, owing to the density fluctuations of the isotope with a *giant absorption cross section*. Actually, in the experimental result for the ultracold neutron transmission ratio for the dilute solution of ^{157}Gd and natural Gd, a deviation from the usual simple relation was observed in the velocity region below about 4 m/s [83].

By the way, we mentioned the reliability of Eqs. (2.16)–(2.18) for the reflection and transmission probabilities of the multilayer containing Gd, but we also notice a deviation from the simple theoretical curve at low grazing angles for the measured Larmor precession of the neutron spin. (This will be shown in Figure 9.27b.)

On the other hand, other kinds of various anomalous phenomena have also been reported in ultracold neutron experiments, such as abnormal transmission through thin films [84], minute energy changes of stored neutrons [85–87], and late neutrons emerging out of the bottle after a long discharge of about 100 s of stored neutrons on opening the exit shutter [88], in which the possibility of temporal adsorption of neutrons to the inner wall has been pointed out. A number of these singular phenomena introduced in this section are called *ultracold neutron anomalies*, and they can be considered to indicate that there are many unknown things in the interaction phenomena of ultracold neutrons on the surface of and the inside materials because of the extremely long wavelength, low velocity, and low energy of ultracold neutrons.

2.3 Reflection and Transmission of Neutrons by Multilayer Mirrors

2.3.1 Developments of Multilayer Mirrors and Reflectivity Analyses

When Schoenborn and his collaborators in the Biology Department and the Department of Applied Sciences at Brookhaven National Laboratory studied sciatic nerve fibers in D_2O, they noticed that a few repeats of relatively thick layers with alternative positive and negative scattering density should produce efficient monochromators, and suggested a new type of novel neutron monochromator [89]. Neutron scattering from such *multilayers* is not the same as simple Bragg scattering, but is given by a scattering factor proportional to the difference in the unit-volume scattering amplitudes of the two materials used. Therefore, in the case of bilayers with equal layer thickness, the even-order reflections will be absent, and if a sinusoidally varying scattering amplitude profile can be produced, higher odd orders will be absent.

To verify the applicability of the present method to produce *monochromatic neutrons with multilayers*, they prepared and tested Mn–Ge multilayer mirror monochromators deposited on glass substrates with a *vacuum evaporation method*. The scattering amplitudes of Mn and Ge are −3.6 and 8.2 fm, respectively, and the present combination is advantageous for producing a multilayer because of the low mutual diffusion coefficient. Ten Mn–Ge bilayers samples with a size of 1 cm × 4 cm were prepared and mounted at the position of the multilayer shown in Figure 2.12a in the neutron beam of the high flux beam reactor in their laboratory. The measured angular distributions of the reflected intensity are plotted in Figure 2.12b for fixed-wavelength monochromatic neutrons. For an angular region θ smaller than the critical angle for total reflection θ_c, the reflection should be total reflection, but actually the reflected intensity decreases for lower angles because the finite size of the sample could not cover the spreading beam. The letter M in the figure indicates the intensity level of the total reflection. In the angular region $\theta > \theta_c$ in Figure 2.12b, the reflection peaks corresponding to the multilayer period are clearly observed.

For the theoretical analysis of wave mechanics in these multilayers, the method of the one-dimensional potential approach described in Section 2.1 can be applied in principle by extending it to the increased number of the amplitude coefficients according to the increasing of the layer number.

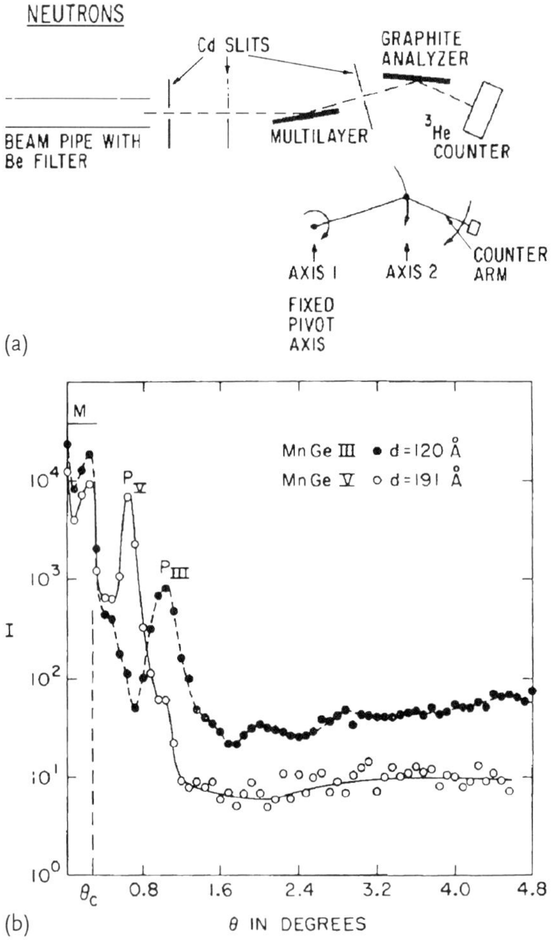

Figure 2.12 Neutron reflection with multilayers experiment at the high flux beam reactor: (a) experimental arrangement; (b) experimental results of the angular distribution of the reflected intensity for 4.2-Å fixed-wavelength neutrons (Schoenborn *et al.* [89]).

Practically, since Eq. (2.7) was obtained by solving the continuity conditions of the wave function and its derivative concerning the formula for the wave function $\Phi(x,z) = \exp(ik_x x)\phi(z)$ with the *reflection coefficient* r and the *transmission coefficient* t expressed as

$$\begin{aligned}\phi(z) &= e^{ikz} + re^{-ikz}, && z \le 0, \\ &= ae^{ik'z} + be^{-ik'z}, && 0 \le z \le d, \\ &= te^{ikz}, && d \le z, \end{aligned} \tag{2.19}$$

for multilayers the corresponding relation inside one of the layers can be written as

$$\phi(z) = a_n e^{ik_n z} + b_n e^{-ik_n z}, \quad z_{n-1} \le z \le z_n, \tag{2.20}$$

where $n = 1 \sim N$, $z_0 = 0$, $z_N = d$, and k_n denotes the component of the wave number vector perpendicular to the interface for the neutrons inside the region n, and so it is enough to solve these simultaneous equations of the layer number N.

After the initial study on the neutron multilayer mentioned above, a new field of the neutron multilayers, that is, artificially prepared material with periodic structures, has been developed, such as multilayer applications as various neutron optical devices and also neutron optics studies on various phenomena realized with multilayers. One of the greatest advantages of multilayers is the superior flexibility that allows us to design and control the separation of the layers according to our needs.

As a typical example to utilize this advantage, Mezei proposed a really new neutron optical element named a neutron supermirror, to be prepared by successively varying the separation of the layer according to the order of the layers; such a supermirror should provide a much wider wavelength region with a high neutron reflectivity compared with the usual crystalline devices and conventional simple multilayers. Further, applying the spin dependence of the neutron scattering amplitude for most isotopes, he proposed a *multilayer polarizer mirror* and a *supermirror polarizer*, and carried out numerical studies on their performance by extending diffraction theory [90, 91].

On the other hand, for theoretical studies on multilayer optics, Croce and Pardo developed the two lines to two arrays matrix method as a formalism to solve the wave functions in multilayers with the optical method [92]. According to their formalism, Eq. (2.20) can be expressed in the following matrix form after putting $A_n = a_n e^{ik_n z}$ and $B_n = b_n e^{-ik_n z}$, where the phase factors $e^{\pm ik_n z_n}$ at each interface $z = z_n$, $n = 1, 2, \ldots, N$, are contained in the respective coefficients:

$$\begin{pmatrix} A_n \\ B_n \end{pmatrix} = \frac{1}{2} \begin{pmatrix} \left[1 + \frac{k_{n+1}}{k_n}\right] e^{-i\varphi_{n+1}} & \left[1 - \frac{k_{n+1}}{k_n}\right] e^{i\varphi_{n+1}} \\ \left[1 - \frac{k_{n+1}}{k_n}\right] e^{-i\varphi_{n+1}} & \left[1 + \frac{k_{n+1}}{k_n}\right] e^{i\varphi_{n+1}} \end{pmatrix} \cdot \begin{pmatrix} A_{n+1} \\ B_{n+1} \end{pmatrix}, \tag{2.21}$$

where $\varphi_{n+1} = k_{n+1} d_{n+1}$. They also showed some calculated results of reflectivity for the multilayer example in which the absorption can be neglected.

Further, Yamada *et al.* [93] gave the formula for the j-layer derived from the two lines to two arrays matrix representation of Croce and Pardo's optical method, by eliminating the coefficients a_n and b_n in Eq. (2.20) and their derivatives at the interfaces, as follows:

$$\begin{pmatrix} \phi_{j+1}(0) \\ \phi'_{j+1}(0) \end{pmatrix} = \begin{pmatrix} \phi_j(\varphi_j) \\ \phi'_j(\varphi_n) \end{pmatrix} = \underline{\boldsymbol{M}}_j \begin{pmatrix} \phi_j(0) \\ \phi'_j(0) \end{pmatrix}, \tag{2.22}$$

where the matrix $\underline{\boldsymbol{M}}_j$ in the case that the absorption can be neglected, that is, as mentioned in the previous section, the scattering amplitudes are real, is given by the equation

$$\underline{M}_j = \begin{pmatrix} \cos n_j \varphi_j & n_j^{-1} \sin n_j \varphi_j \\ -n_j \sin n_j \varphi_j & \cos n_j \varphi_j \end{pmatrix}, \quad 4\pi N b_{j\,\text{coh}}/k^2 \leq 1 \tag{2.23}$$

$$= \begin{pmatrix} \cosh n'_j \varphi_j & n'^{-1}_j \sinh n'_j \varphi_j \\ n'_j \sinh n'_j \varphi_j & \cosh n'_j \varphi_j \end{pmatrix}, \quad 4\pi N b_{j\,\text{coh}}/k^2 > 1\,, \tag{2.24}$$

where $n_j = \sqrt{1 - 4\pi N b_{j\,\text{coh}}/k^2}$, $n'_j = i n_j = \sqrt{4\pi N b_{j\,\text{coh}}/k^2 - 1}$, and $\varphi_j = k d_j$.

In this way, in the case where the absorption is neglected, the solution for Eqs. (2.19) and (2.20) can be expressed as

$$\begin{pmatrix} t \\ i n_g t \end{pmatrix} = \underline{M}_T \begin{pmatrix} 1 + r \\ i n_0 (1 - r) \end{pmatrix}, \tag{2.25}$$

where $\underline{M}_T = \underline{M}_N \ldots \underline{M}_2 \underline{M}_1$, and n_0 and n_g are the refractive indices for the media before and after the multilayer, respectively. The formalism of Eqs. (2.22)–(2.25) is called the *transfer matrix method*, and the matrix $\underline{M}$ is the *transfer matrix*.

Putting $\underline{M}_T = \begin{pmatrix} A & B \\ C & D \end{pmatrix}$, we can solve Eq. (2.25) for r and t:

$$r = \frac{(n_0 n_g B + C) + i(-n_g A + n_0 D)}{(n_0 n_g B - C) + i(n_g A + n_0 D)}, \tag{2.26}$$

$$t = \frac{2 n_0}{(n_0 n_g B - C) + i(n_g A + n_0 D)}. \tag{2.27}$$

By making use of the present matrix formalism, Yamada *et al.* further carried out analyses and calculations on the reflectivity for various kinds of multilayers to be used for neutron monochromator mirrors, polarizer mirrors, and supermirrors [93]. The present transfer matrix method is a quite powerful approach to express and analyze very simply the reflection and transmission for very complicated optical systems consisting of a number of layers.

As the next step in multilayer developments, Ebisawa *et al.* prepared various Ni–Mn and Ni–Ti multilayers for monochromator mirrors and supermirrors with the vacuum evaporation method, and also measured the reflectivity at the Kyoto University reactor [94]. From the comparison of their results, they deduced the design formula and the optimum values for the parameters to obtain high reflectivity over a wide wavelength region of neutrons [93]. One of the experimental results for the reflectivity on the Ni–Ti supermirrors developed by them is shown in Figure 2.13. This example indicates an about 2.5 times wider region than for a nickel mirror in the *wave vector transfer*[8] for neutron reflection (refer also to Figure 2.15a).

8) For neutrons with wavelength λ, wave number $k = 2\pi/\lambda$, and reflected with angle θ between the mirror surface and the direction of reflection, the wave vector transfer is given by $Q = 4\pi \sin\theta/\lambda$.

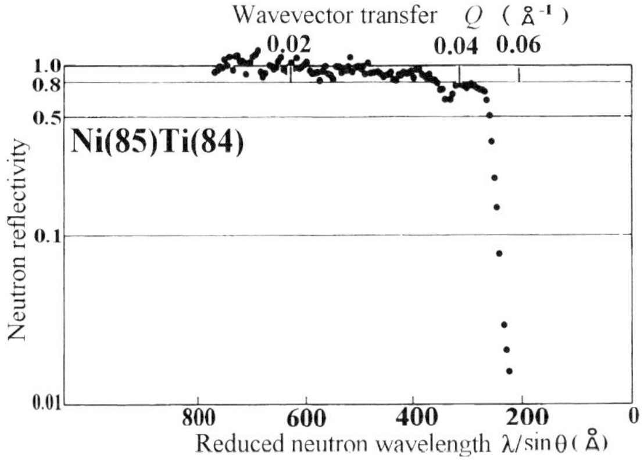

Figure 2.13 Typical example of the measured reflectivity of the supermirror (Ni 85 layers, Ti 84 layers) at the preliminary stage of the development at the Kyoto University reactor (Ebisawa *et al.* [94]).

On the basis of these theoretical analyses and practical developments of multilayers, the utilization of supermirrors for neutron guide tubes and other neutron optical devices was significantly promoted thereafter.

2.3.2
High-Performance Supermirrors

Active developments of high-performance multilayer neutron mirrors and supermirrors and also their applications are continuously promoted to this day. The method for preparing multilayers has been extended, in addition to vacuum evaporation, also to the *sputter method*, which is more appropriate than vacuum evaporation for preparing thin layers and controlling precisely the layer thickness. Saxena applied a high-frequency sputter method to realize *thinner layers* required for reflectivity improvement by eliminating the cause inducing layer imperfections and also for extension of the reflection region [95].

Recently, a very effective approach and practical success has been achieved by Soyama *et al.* [96, 97] and Hino *et al.* [98, 99] for reflectivity improvement and of the reflection region of multilayers (these developments are called *high-Q multilayer developments*).

With the increase of the layer number, and also the decrease of the minimum layer thickness required for the the reflection region, the effects of the interface roughness become fatal, and therefore an effective step to decrease the roughness is a crucial requirement. Soyama *et al.* employed an *ion-polishing* process during the sputtering process to improve the interface roughness, which inevitably increases in accordance with the progress of the ion-beam sputtering of Ni–Ti multilayers, and as a result they succeeded in drastically decreasing the resultant interface roughness in the completed multilayer.

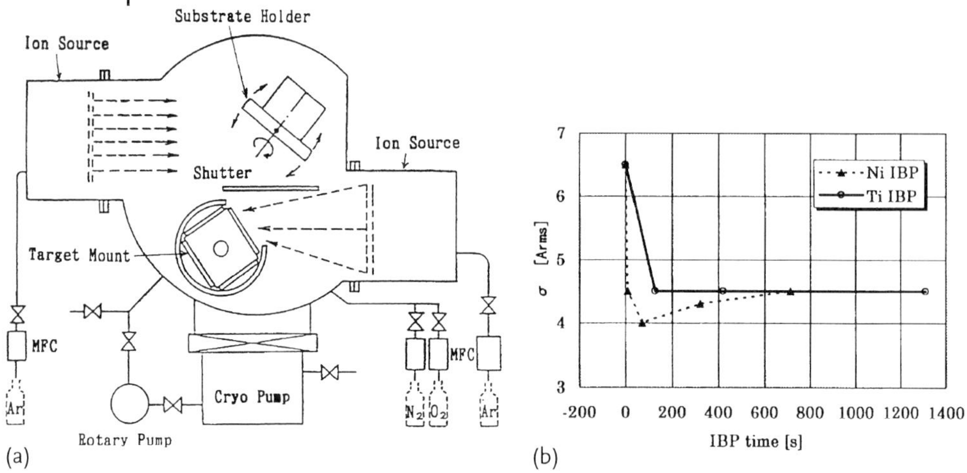

Figure 2.14 Improvements of multilayer interface roughness with the ion-polishing method: (a) the ion-beam sputtering apparatus ; (b) variation of interface roughness on Ni layers or Ti layers depending on ion-polishing time durations (ion energy 100 V, ion-injection angle 10°). IBP ion-beam polishing. (Soyama *et al.* [96]).

As shown in Figure 2.14a, the apparatus they used has two Ar^+ ion sources, one for the layer preparation and one for the ion-polishing. They looked for the optimum ion energy and polishing time duration which is most appropriate for obtaining the minimum interface roughness. Figure 2.14b indicates their solutions showing the resultant interface roughness for various polishing durations. The minimum interface roughness below 4 Å was obtained for an optimum polishing duration of about 69 s for Ni layers, whereas the interface roughness of about 4.5 Å for a duration longer than about 100 s was obtained for Ti layers. Both of these results are significantly improved with the application of ion-polishing.

The actual performance of neutron reflectivity was measured on the ion-polished multilayer by making use of a reflectometer with a wavelength of 4 Å and a resolution of about 5% at the JRR-3M reactor, Japan Atomic Energy Agency, and the result is shown in Figure 2.15 [97]. Figure 2.15a indicates an improved reflectivity of about 90% in the wave vector transfer of 0.03–0.06 $Å^{-1}$ for the ion-polished supermirror with a total layer number of 407, compared with about 80% [94] for the conventional supermirror without the ion-polishing. Further, Figure 2.15b illustrates the experimental result for an ion-polished monochromator mirror with an average layer spacing of 85 Å and a layer number of 130 [100], which can be well explained by the calculation for an interface roughness of 5.5 Å. We can see that the present reflectivity peak of monochromatization is much improved compared with the result shown in Figure 2.12b.

Further, very active development of high-performance supermirrors and various kinds of multilayer elements for neutron optical applications have been continued

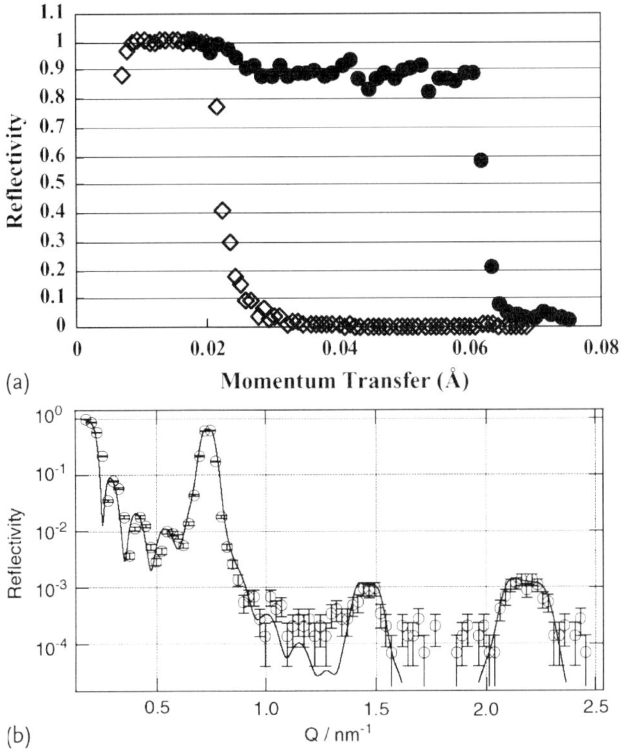

Figure 2.15 Neutron reflectivity performances of multilayers prepared with the ion-polishing method: (a) comparison between reflectivities of a Ni mirror and an ion-polished Ni–Ti supermirror with a 3 times wider reflection region (Soyama *et al.* [97]); (b) comparison between the measured reflectivity (○ mark) of an ion-polished Ni-Ti multilayer monochromator mirror and the calculated reflectivity (solid line) (Maruyama *et al.* [100]).

by Hino *et al.*, in which supermirrors with 5 times and 6 times wider extended reflection regions compared with a nickel mirror have been prepared [98, 99].

2.4 Reflection off and Transmission through Moving Matter

2.4.1 Neutron Turbine

As mentioned at the end of Section 1.4, in the case of the steady-state motion of an optical element, the reflection, diffraction, and transmission of neutrons can be analyzed classically. Suppose there is an optical element in translational motion with velocity v_T as shown in the lower part of Figure 2.16a, where, as illustrated in the upper part of Figure 2.16a, neutrons with velocity v_1 are incident with angle α to the direction of v_T. Then, the phenomena of neutron reflection and transmission

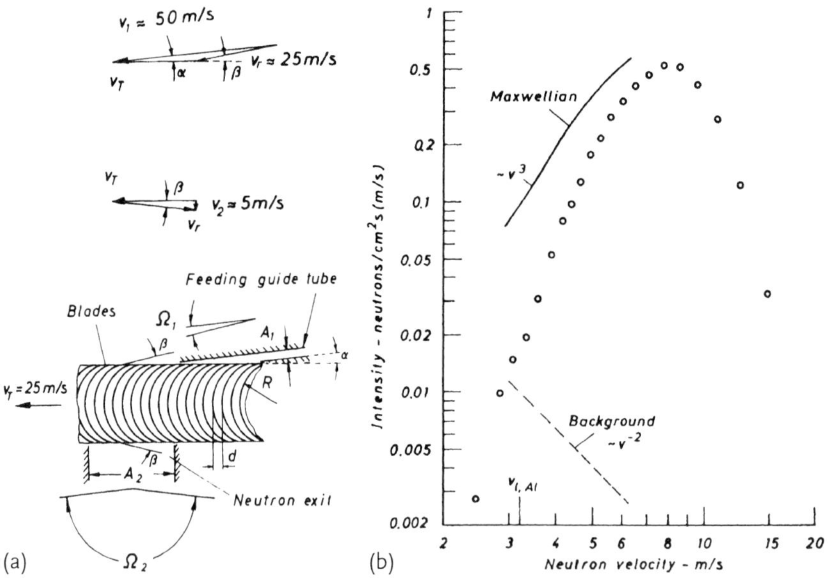

Figure 2.16 Neutron turbine: (a) working principle and (b) velocity spectrum of ultracold neutrons produced with Steyerl's turbine developed at the Technical University of Munich (Steyerl [64]).

can be analyzed in the translational frame of reference where neutrons with relative velocity v_r enter the element at rest with angle β. Further, the results of reflections in the optical element shown in the lower part Figure 2.16a can again be returned to the original laboratory frame as illustrated in the middle part Figure 2.16a. Our task in this subsection is to analyze the possible approach to arrive at the result we desire for a neutron beam or a group of neutrons by making use of the principle of the moving optical elements mentioned above.

However, it must be remarked that there exists an important fundamental law restricting the possible performance of such an optical element. The fundamental law is *Liouville's theorem*, which states that the phase space density of mutually noninteracting particles does not vary with the execution of any collective external forces. The *phase space density* ρ_{ps} is defined here as the differential number of particles per unit volume in the momentum space $\boldsymbol{k}$ per unit volume in the real space $\boldsymbol{r}$, that is, it is denoted as $\rho_{ps} = d^6 N/\hbar^3 d^3\boldsymbol{r} d^3\boldsymbol{k}$.

A further restrictive fact in practice is that the statement "does not vary" in the theorem applies to an ideal situation without any losses and imperfections in the execution of the collective external forces, and in the real conditions accompanying finite losses or imperfections it should be mentioned that one cannot reach the original density. Then, what is the meaning of executing such a collective external force as to convert the particle velocity illustrated above? The answer is that there exists really such a case where we can recover the lost density in the phase space by the conversion of the particle velocity from a high-density region to a region where the density is significantly low if we do not use the present principle.

As one of the actual cases, as mentioned in Section 2.2, ultracold neutrons have the attractive property of being reflected by neutron mirrors in a wide range of incident angles, but on the other hand their loss probability due to the absorption and inelastic scattering is very large compared with that of neutrons with higher energy. As the result, a significant density loss is accompanied in the course of the direct extraction of ultracold neutrons from a neutron source to an experimental apparatus. Therefore, it will be advantageous to transport cold or very cold neutrons to the neighborhood of the experimental apparatus and convert them there to ultracold neutrons by making use of a specially prepared device working on the principle shown in Figure 2.16a. The primary neutrons supplied to such a device will include high-density cold or very cold neutrons since they are much more easily extracted and transported, but the density of ultracold neutrons will be significantly diminished; thus, the recovery of the lost ultracold neutron density becomes an important task.

The optical element shown in the lower part of Figure 2.16a consists of a number of semicircular reflecting mirrors and rotates with a constant velocity in which the incident neutrons reverse the direction of the relative motion and experience multiple total reflections, as described in Figure 2.16a. The exiting neutrons are then decelerated in the laboratory frame to a velocity of about 1/10 of that of the incident very cold neutrons. Such a device with a special structure and performance as described above is called a *neutron turbine*. The neutron turbine designed and developed by Steyerl at the Technical University of Munich (called *Steyerl's turbine*) has nearly 700 semicircular blades of copper mirrors over the whole circumference of a big wheel, and rotates with a blade velocity of about 25 m/s. It then converts incident very cold neutrons with a velocity of about 50 m/s to ultracold neutrons with a velocity below about 10 m/s [64]. In Figure 2.16b the velocity spectrum measured with a time-of-flight method (circles) is compared with the expected spectrum (solid line) for the incident neutron intensity and an estimated turbine efficiency of 45%; the difference is attributed to the counting efficiency of the measuring system used for the ultracold neutrons.

To produce ultracold neutrons with such an axial flow turbine as described in Figure 2.16a, the turbine blades are rotated with a blade velocity of about half the incident neutron velocity. The velocity of the incident neutrons in Steyerl's turbine was designed to be about 50 m/s from the limiting velocity for total reflection of the copper mirror, $v_l = 5.6\,\text{m/s}$, the practical number of reflections in the blade (about 10), and the practical conditions for the construction (the width and number of the blades). In other words, if we can employ a turbine blade with a larger limiting velocity for total reflection, it is possible to decrease the required number of reflections in blades and also the number of blades under the same condition for the incident neutron velocity. Such an improvement will make the construction of the turbine easier.

A *supermirror neutron turbine* was developed by Utsuro *et al.* [77, 101, 102] at the Kyoto University reactor by making use of polygonal blades of Ni–Ti flat super-

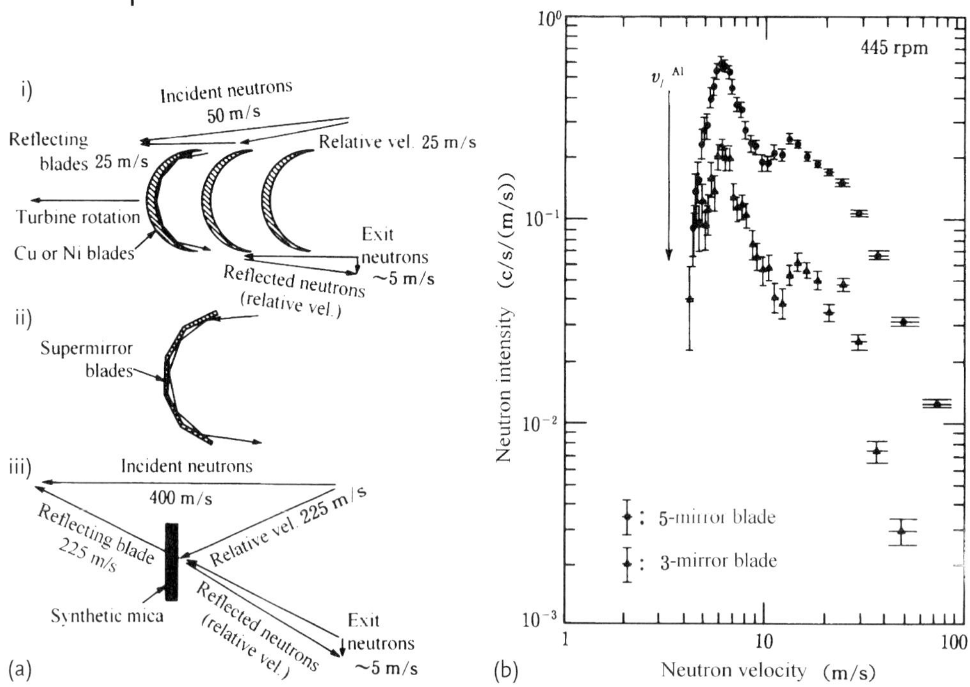

Figure 2.17 (a) Working principles of three kinds of neutron velocity conversion devices and (b) measured spectra of ultracold neutrons produced by the supermirror turbine at the Kyoto University reactor (comparison between the results from the three-mirror type and the five-mirror type of supermirror blades) (Utsuro *et al.* [77, 102]).

mirrors with a limiting velocity for the reflection of $v_l = 14.7$ m/s, with the blade shape shown in Figure 2.17a, configuration ii. The measured velocity spectra for the ultracold neutrons flowing out from the turbines with two different designs of baldes are given in Figure 2.17b, where the common decrease of the count rates below about 4 m/s is due to the drastic decrease of the counting efficiency of the ^{3}He counter used (as shown in Figure 2.11b) owing to the aluminum window reflection. The comparison of the two spectra in Figure 2.17b indicates that the turbine with the optimum design, that is, 96 blades of five flat supermirrors, gives about 3 times higher output intensity of ultracold neurtons compared with the turbine with 32 blades of three flat supermirrors. The structures of the spectra apparent in Figure 2.17b are caused by the reflected neutrons during passage through the polygonal blades.

The schematic arrangements of ultracold neutron facilities with these two kinds of turbines are illustrated in Figure 2.18a and b, respectively. The former facility [103] of Steyerl's turbine with nickel blades is installed on the top floor in the reactor room, and the incident very cold neutrons are supplied by a vertical-type nickel guide tube. It has five horizontal exit ports, four are for utilization of ultra-

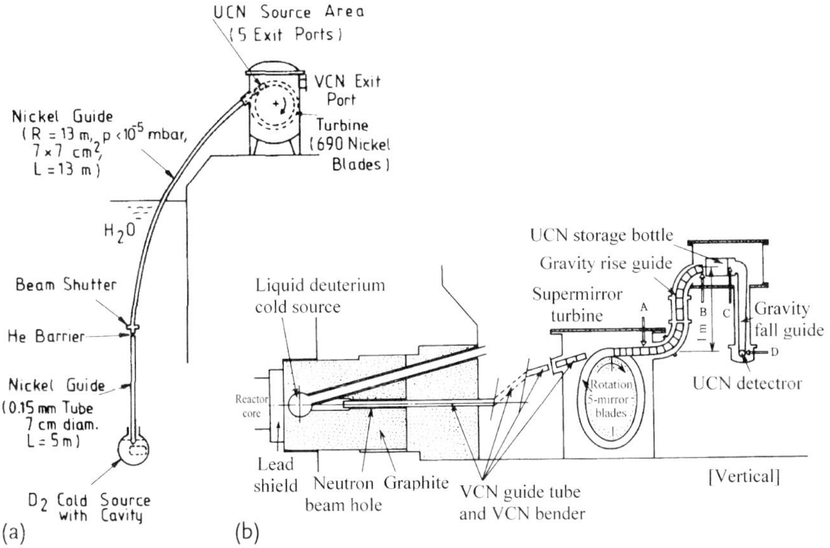

Figure 2.18 Two kinds of arrangements for the turbine type of ultracold neutron facilities: (a) facility at the Institut Laue-Langevin (Steyerl *et al.* [103]) and (b) facility at the Kyoto University Research Reactor Institute. VCN very cold neutrons (Utsuro *et al.* [77, 102]).

cold neutrons and one is for ultracold neutron test. The latter facility [77, 102] is installed on the experimental floor in the reactor room, and has a horizontal very cold neutron guide followed by an upward bender of nickel mirrors, and it has three horizontal exit ports for ultracold neutrons. Since the ultracold neutrons at the blade exit flow out with a wide divergent solid angle near 2π, the separation between the rotating blade edge and the fixed inlet port of the ultracold neutron guide must be minimized; in the above-mentioned facilities it is minimized to about 5 mm. Further, in the case of the facility shown in Figure 2.18b, there is a vertical rise guide for the *gravitational deceleration* of the output component with the maximum intensity in Figure 2.17b, that is, the velocity region of about 5.5–7 m/s, down to a velocity below about 5 m/s for ultracold neutron storage experiments, and a vertical fall guide from the experimental device to the detector, for *gravitational acceleration* up to a velocity beyond 3.2 m/s, the limiting velocity for total reflection of the aluminum detector window.

Both of these facilities utilize very cold neutrons from steady-state reactors; however, in recent years intense pulsed neutron sources have been constructed and utilized for various kinds of neutron experiments. To employ such a pulsed neutron beam for ultracold neutron production, the principle of multiple reflections for the velocity conversion as shown in Figure 2.17a, configurations i) and ii) is disadvantageous because of the dispersion of phase space density during the multiple reflections, and a different principle, a *Doppler shifter* as shown in Figure 2.17a, configuration iii), attaining the required velocity conversion by a single Bragg scattering, will become more appropriate. Dombech [104] and Brun *et al.* [105] developed and tested such a Doppler shifter with blades of synthetic mica (Thermica) by installing it at the pulsed source IPNS, at Argonne National Laboratory. Recently, a rather sophisticated concept of a *velocity-focusing type of supermirror Doppler shifter* was presented by Utsuro [106] in which the multilayer structure and the instantaneous velocity of the supermirror reflecting blade are controlled to synchronize the reflection condition with half the velocity of the incident neutrons depending on the arrival time from the pulsed source. A prototype machine constructed with a special mechanism for the blade motion was tested by Utsuro and Shima *et al.* [107] at the pulsed source KENS, the High Energy Physics Research Institute, Tsukuba.

2.4.2
Fizeau Experiment

In the previous subsection, we studied neutron optical elements in steady translational motion with a constant velocity or well approximated by such a motion. A typical contrary case of high-frequency reciprocal motion was also studied in Section 1.4. Therefore, in the present subsection, an another interesting case will be presented where the motion of an optical element induces different optical effects for different optical paths within the element. The *Fizeau experiment* in light optics is known as one of such experiments, and thus we consider the matter wave analogy to the Fizeau experiment in light optics.

As indicated by the formula for the refractive index, Eq. (1.16) given in Section 1.2,

$$n^2 = 1 - \frac{N b_{\text{coh}} \lambda^2}{\pi} \,,$$

and the definition of the refractive index, $n = k'/k$, where k and k' denote the neutron wave numbers in a vacuum and in a medium, respectively, the wave number in the medium differs from that in a vacuum owing to the effect of the *optical potential* of the medium.

Therefore, if we insert a neutron transmission rod between the object slit and the scanning slit in the double-slit experiment setup shown in Figure 1.4a, the phase of the transmitted neutrons changes by an amount proportional to the thickness. When the rod is stopped, the phase changes for the two optical paths for the double slit shown in Figure 2.19b are same and then the interference pattern will not change, whereas for the transmission through the rod in rotational motion, the interference pattern will change from that without the rod. A similar effect has already been shown in the Fizeau experiment in light optics. Klein *et al.* performed an analogous experiment by making use of the cold neutron beam at the high-flux reactor, Grenoble [108].

According to the theoretical analysis with the local wave number of neutrons in the transmission path through the moving element [108], the *phase change through*

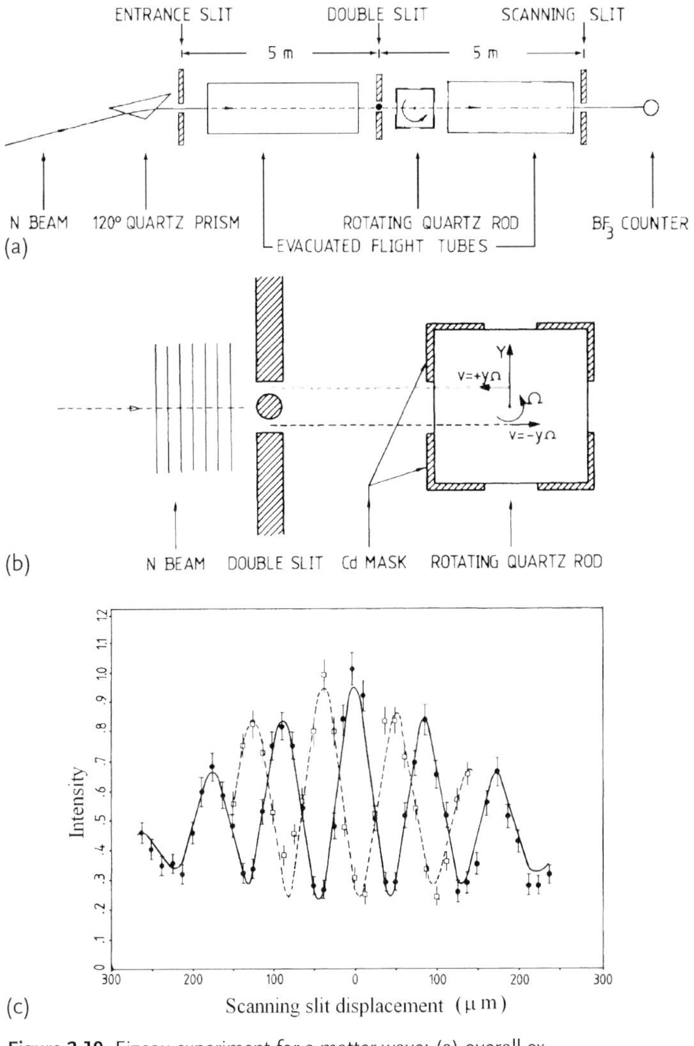

Figure 2.19 Fizeau experiment for a matter wave: (a) overall experimental layout (not drawn to scale); (b) a quartz rod rotating with angular velocity Ω and a double slit; (c) double-slit interference pattern, solid line $\Omega = 0\,\text{Hz}$, dotted line $\Omega = 100\,\text{Hz}$ (Klein *et al.* [108]).

the path in the rotating rod for thermal and cold neutrons is given by

$$\Delta\varphi = -k D \frac{U}{2E}\frac{v}{v_n}, \tag{2.28}$$

in which the inequality $U/E \ll 1$ holds for the neutron energy E, k and v_n are the wave number and the velocity of neutrons in a vacuum, respectively, and v is the local velocity of the rod material at a radius y from the rotation center, that is,

$\nu = \gamma \Omega$. The result plotted on the shift of the interference pattern as a function of the rotation speed agreed well with the theoretical straight line of Eq. (2.28).

As another example of related experiments, Hamilton *et al.* carried out a cold neutron diffraction experiment studying the effects of acoustic waves on a solid surface, and pointed out the possible difference from the description of light diffraction in a similar situation [109].

2.5
Neutron Resonator and Quasi-Bound Neutrons

2.5.1
Neutron Reflection and Transmission Experiments with Macroscopic Resonators

As described in the previous section, matter works as an optical potential for long-wavelength neutrons; therefore, as pointed out first by Kagan [110], it is possible to confine a neutron within a space between such kinds of positive potentials as given in textbooks on the state of a particle in a *potential well*.

The quantum states in such a potential well can be analyzed by solving the one-dimensional Schrödinger equation taking only the component k_z of the neutron wave number vector perpendicular to the potential into consideration. The solution for the quantum states is given, in the semiclassical Wentzel–Kramers–Brillouin approximation, by

$$k_{zn} \cong \frac{\pi}{d}\left(n + \frac{1}{2}\right) , \qquad (2.29)$$

where d is the potential width between the two potential hills and $k_{zn} = [2m(E_{zn} - U)]^{1/2}/\hbar$ by making use of the neutron energy E_{zn} corresponding to the perpendicular component (i. e., z-component) of the neutron velocity, where $U = 2\pi\hbar N b_{\text{coh}}/m$ is the optical potential of the material at the bottom of the well and N and b_{coh} are the atomic density and the coherent scattering amplitude of the material, respectively.

Steinhauser *et al.* carried out an experiment to demonstrate the existence of such kinds of quasi-bound states of neutrons [111] by utilizing the capability to vary the magnitude of k_z of neutrons incident on a sample with a potential well by changing the height of a sample put in place of the horizontal mirror in the central part of the *gravity diffractometer* shown in Figure 2.3a.

First, the reflected neutron intensity was measured on a sample put in place of the horizontal mirror in the diffractometer, the result of which is shown in Figure 2.20a. Next, the transmitted intensity was also measured in a slightly modified geometry by inserting the sample in the falling path between the horizontal mirror and the former vertical mirror, the result obtained also being shown in Figure 2.20b. The abscissa in this figure, that is, the fall height from the entrance slit

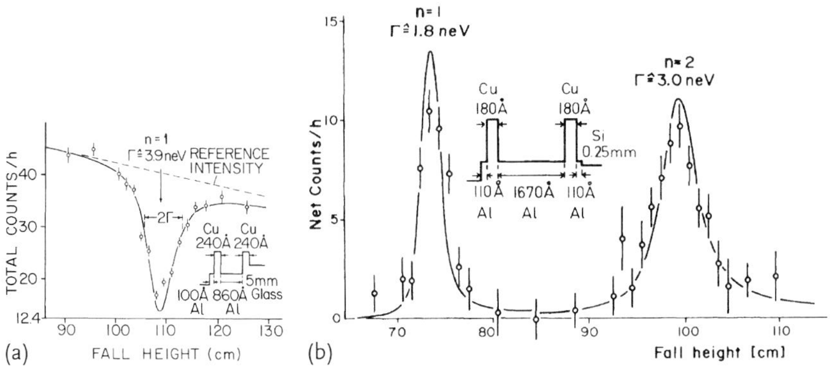

Figure 2.20 Experiments and analyses on the quasi-bound states of neutrons in a double-hump potential barrier studied with a gravity diffractometer: (a) reflected neutron intensity as a function of the the gravity fall height and (b) transmitted neutron intensity on a similar sample, where the solid curves correspond to the exact theoretical solutions (Steinhauser *et al.* [111]).

in Figure 2.3a, is related to the vertical wave number k_z in a vacuum as

$$k_z = \frac{m}{\hbar}(2gz)^{1/2}, \tag{2.30}$$

and the fall height resolution mainly defined by the slit widths was 3 cm (1 cm $\cong$ 1 neV).

The potential structures of these samples are of double-hump type as shown in the inserts, and the solid curves in the figure are the theoretical intensity distributions calculated from the exact solution for the Schrödinger equation applied to the potential and the experimental resolution. However, slight adjustments were necessary to match the intensity minimum of the calculated results with the experimental ones in Figure 2.20a, that is, the width of the potential well was adjusted to 881 ± 2 Å, and at the same time the width of the hill was adjusted to 225 ± 5 Å. Further, in the transmission experiments in Figure 2.20b where the sample was put at a position 16 cm higher than that of the horizontal mirror in Figure 2.3a, the calculation with the widths of the well and the hill, 1663 ± 3 and 190 ± 5 Å, respectively, agreed well with the experimental results. These typical widths of the resonance peaks are related to the lifetime of the *quasi-bound states of neutrons* in the well, which was derived as $\hbar/\Gamma = 2 \times 10^{-7}$ s from Figure 2.20a. The theoretical value for the resonance width in the exact solution, $2\Gamma = 5.1$ neV, is in reasonable agreement with the experimental result. However, in both cases, the measured heights of the resonance peaks are somewhat lower than the corresponding heights of the theoretical curves. Several reasons can be considered as the cause of these differences but none have yet been proven.

These *macroscopic resonators* for neutrons with double-hump or triple-hump potential structures are *Fabry–Pérot resonators* for neutrons. As pointed out in

Eq. (2.29), since only the perpendicular component k_z of the neutron wave number vector is effective in the experiments, it will be possible to use thermal or cold neutrons for similar experiments by arranging the grazing angle to be at a very small angle to the potential surface. Steyerl *et al.* [112] carried out such kinds of experiments at the cold neutron guide of the Kyoyo University reactor with the arrangement shown in Figure 2.21a, and compared the measured cold neutron reflectivities of double-hump and triple-hump resonators with the exact solutions for the one-dimensional Schrödinger equation.

The incident neutrons were collimated to a very small divergent angle of 0.2 mrad, the incident grazing angle was defined to be about 0.01 rad to the sample surface, and the incident neutrons were pulsed by a chopper with a wavelength resolution of $\Delta\lambda \cong 0.07$ Å in the wavelength region of $\lambda = 4$–8 Å. The time-of-flight measurements were performed on double-hump and triple-hump resonators, the results of which are given in Figure 2.21b in comparison with the theoretical calculations for the layer structures with an Al layer of 745 Å and a Cu layer of 181 Å

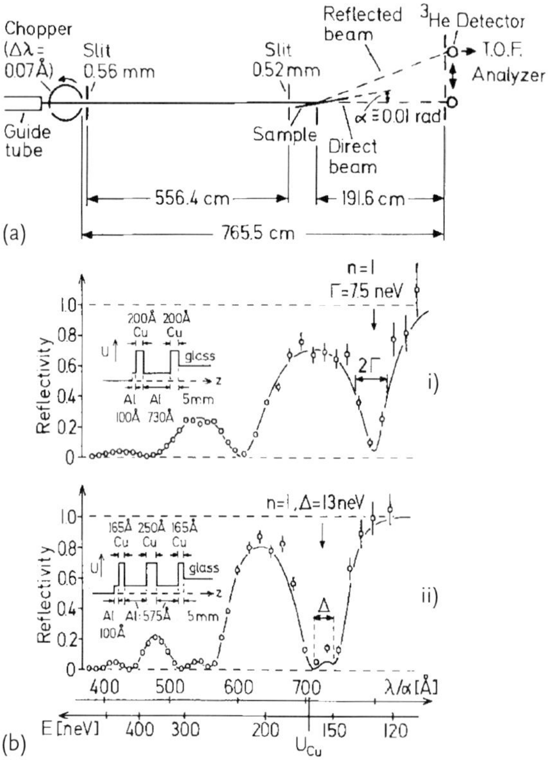

Figure 2.21 Cold neutron reflection experiments on macroscopic coupled resonators: (a) experimental arrangement of the time-of-flight reflectometry with the collimated and chopper-pulsed cold neutron beam at the Kyoto University research reactor and (b) neutron reflectivities of a) the double-hump and b) the triple-hump resonators with the potential structures shown as inserts, where the solid curves correspond to the exact theoretical solutions. (Steyerl *et al.* [112]).

(Figure 2.21b, plot i) and an Al layer of 609 Å and a central Cu layer of 257 Å with Cu layers on both sides of 144 Å (Figure 2.21b, plot ii).

Although the energy resolution of these measurements may be slightly worse ($\Delta E \cong 4\,\text{neV}$) than that of the gravity diffractometer previously described, the energy region to be covered is several times wider than that with the gravity diffractometer. As a result of the present advantage, it became possible to compare experiment with theory up to the excited quasi-bound state and for the continuous energy states in $E_z > U_{\text{Cu}}$, as illustrated in Figure 2.21b. The comparison between plots i) and ii) in Figure 2.21b indicates that the resonance state (the valley in the reflectivity) in the latter coupled twin wells is *split* into two levels, in contrast to the single level in the former single well. The energy splitting of the quasi-bound state for the $n = 1$ resonance was derived as $\Delta \cong 13\,\text{neV}$. In this way, these artificial macroscopic resonators for neutrons are used and applied for neutron optical studies for the observations of various quantum mechanical phenomena such as the quasi-bound states and the energy splitting.

2.5.2 Nuclear Reaction Experiments of Neutrons

As another approach for studying the neutron states in these resonators, it is also worth introducing such works observing the phenomena directly related to the spatial distribution of the wave functions of neutrons in the resonant state. Two kinds of measurements in such a category with the observation of the products from the *neutron reactions in a resonator* will be described below.

First, the experimental results obtained by Zhang *et al.* [113] are shown in Figure 2.22a,b, in which a thin layer of material inducing $(\boldsymbol{n}, \gamma)$ *reaction* was embedded in the gap layer of a resonator and the reaction γ-rays were measured. They also carried out theoretical calculations to compare the reaction rates and the reflectivities in the case with neutron absorption with the formalism of the *complex potential* explained in Section 2.2.

Practically, as explained concerning Eq. (2.16), making use of Eqs. (2.16)–(2.18) in place of b_{coh} in Eq. (1.19) gives us $U = U_{\text{Re}} - i\,U_{\text{Im}}$, where $U_{\text{Im}} = 2\pi\mu_a/\lambda$, and μ_a is the linear absorption coefficient. Applying to the wave equation with the complex potential our knowledge that the wave function in a multilayer can be expressed in the form of Eqs. (2.19) and (2.20), we arrive at the relation

$$d[\Phi^*(d\Phi/dx) - \Phi(d\Phi^*/dx)]/dx = -2i\,U_{Im}|\Phi|^2\,. \tag{2.31}$$

Spatial integration leads to the formula which combines the neutron reflectivity and the *neutron reaction rate* as

$$1 - |r|^2 = \frac{1}{k_x}\sum_i \int_{X_{ai}} |\Phi|^2 U^i_{\text{Im}}\,dx = \sum_i \tau_i\,, \tag{2.32}$$

where X_{ai} denotes the absorption region and τ_i is the absorption reaction rate of neutron-absorbent isotope i.

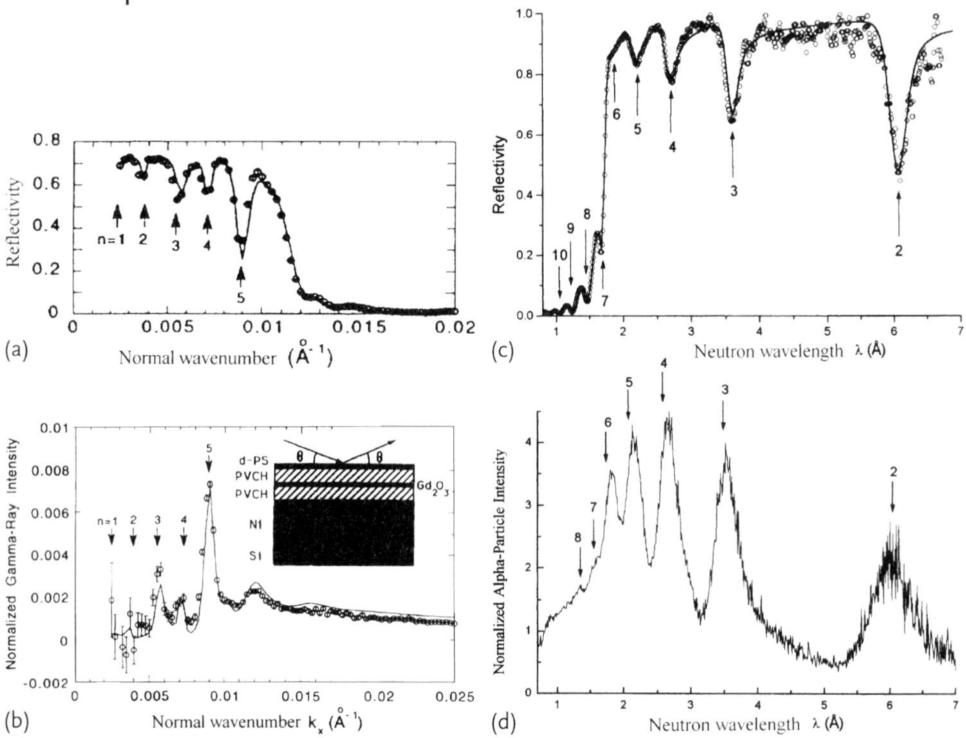

Figure 2.22 Experimental results compared with the theoretical calculations on the neutron reflections and reactions in multilayer resonators; the arrows indicate the positions of the resonance with the labeled modes: (a) neutron reflectivity and (b) (n, γ) reaction rate, where the insert shows the resonator structure and the solid curves are theoretical calculations (d-PS is predeuterated polystyrene, PVCH is polyvinylcyclohexane) (Zhang *et al.* [113]); (c) neutron reflectivity (the solid curve is the theoretical calculation) and (d) (n, α) reaction rate (Aksenov *et al.* [114]).

The resonator used is illustrated in the insert in Figure 2.22b. It was prepared by evaporating first about 0.6 μm thick nickel on a 5.1-cm-diameter polished silicon substrate, then spin-coating about 720 Å thick polyvinylcyclohexane, and next 50-Å-thick gadolinium was evaporated, and finally it was covered by an about 720 Å thick polyvinylcyclohexane film floating on water. Further, after it had been dried, it was covered by a predeuterated polystyrene film. The experiment was carried out at the NG7 reflectometer in the NIST reactor, and a 64-mm-diameter high-purity germanium crystal detector was used for the γ-ray measurements at a distance of 10 cm. The detection efficiency was known to be about 10% from the calibration.

The results of the theoretical calculations shown in Figure 2.22a,b were obtained by making use of the neutron wave function in the matrix formalism and the related equations given above. They agree well with the experimental results, and in the course of these calculations, consideration of the interface roughness was necessary. How the probability density of neutrons is amplified in the resonator in such resonance conditions is an important factor, and therefore the calculated results

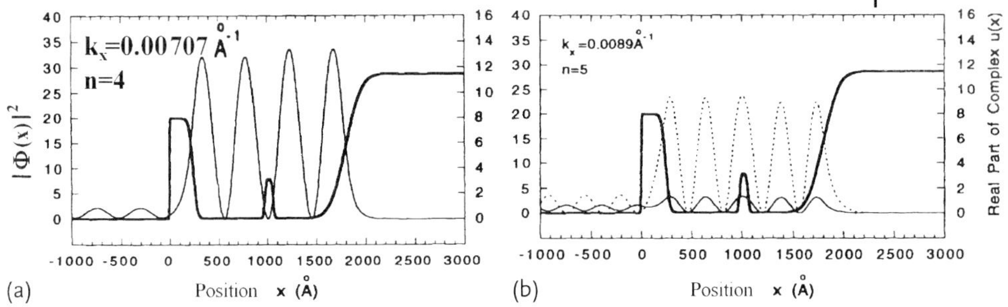

Figure 2.23 The calculated distributions of the real part of the one-dimensional complex potential $U(x)$ (thick solid curve) and the probability density of neutrons $|\Phi(x)|^2$ (thin solid curve). The dotted line is $|\Phi(x)|^2$ for the case where the imaginary part neglected. (a) The incident wave number corresponding to the $n = 4$ mode and (b) that corresponding to the $n = 5$ mode (Zhang *et al.* [113]).

are shown in Figure 2.23, which indicates the distribution of standing waves in a resonance condition compared with the real part of the potential $U_{\mathrm{Re}}(\times 10^{-5}\,\text{Å}^{-2})$. The remarkable amplification is recognized in the resonant state in mode $n = 4$ in Figure 2.23a, in contrast to the notable suppression owing to the effect of the absorption layer in the case in Figure 2.23b for mode $n = 5$.

Another kind of the neutron reaction experiment was performed by Aksenov *et al.* on $(\boldsymbol{n}, \alpha)$ *reaction*, where the multilayer resonator used has the layer structure ^{6}LiF(200 Å)/Ti(2000 Å)/Cu(100 Å)/glass substrate. First copper and titanium layers were prepared by vacuum sputtering on a substrate, then ^{6}LiF was spray-coated. The experiment was carried out at the polarized neutron spectrometer in the IBR-2 pulsed reactor, Joint Institute for Nuclear Research, Dubna. The α-particles and tritons from the reactions were detected by an ionization chamber containing the resonator in the ionizing gas with the copper layer as the cathode electrode [114]. The measured neutron reflectivity and α-ray intensity are shown in Figure 2.22c,d. The neutron reflectivity calculated with the matrix formalism is also shown in Figure 2.22c, where the effect of the possible interface roughness was taken into consideration to smooth the potential. In the present calculation, the amplification factor for the probability density for neutrons was about 35 for mode $n = 2$.

These two kinds of experiments and the corresponding calculations are consistently explained by the amplification of the resident density and the formation of standing waves of neutrons in the multilayer resonators.

At the end of this section, a very novel experiment will be briefly described. Suppose a one-dimensional potential well is arranged in the vertical direction, that is, instead of two flat reflecting mirrors consisting of a resonator, the upper one is replaced by the earth's gravity, as illustrated in Figure 2.24, which will also work as a macroscopic resonator for such incident neutrons as the vertical component of the energy E_{vert} is comparable to the energy difference of the gravity potential in the resonator. Therefore, quasi-bound neutrons, that is, quantum states in the

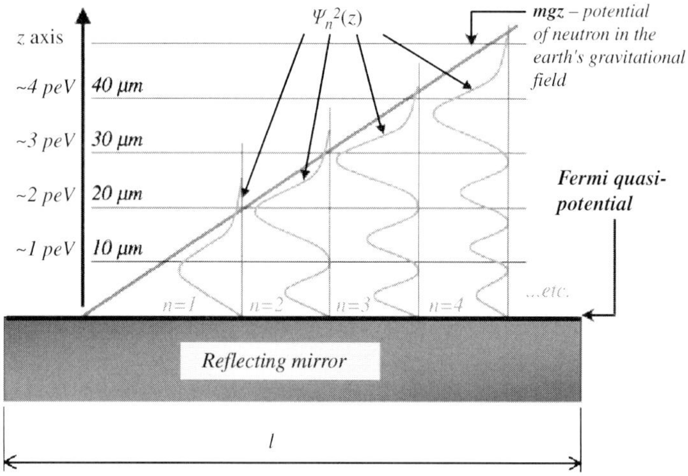

Figure 2.24 Quantum states of neutrons trapped between the potential of a horizontal flat reflecting mirror and that of the earth's gravity; in the ordinate, the z-axis is taken in the vertical direction, and the abscissa is the vertical component of the neutron energy, E_{vert}, that is, the energy corresponding to the vertical component of the neutron velocity. Then the gravity potential for the neutron is expressed with a straight line $E_{\text{vert}} = mgz$ as shown. In the energy region E_{vert} of picoelectronvolts (1 peV $=$ 10^{-15} eV), the solution $\Phi(z)$ for the one-dimensional Schrödinger equation is quantized as indicated for the states with quantum numbers $n = 1 \sim 4$ (Nesvizhevsky *et al.* [115]).

vertical component of the gravitational field, will be produced during the passage of neutrons through the resonator. Nesvizhevsky *et al.* carried out an ultracold neutron experiment to observe the characteristic feature of such quantum states in a gravitational field [115].

In their experiment performed at the high-flux reactor, Grenoble, incident neutrons with a velocity of about 5 m/s were supplied to the experimental apparatus consisting of a flat mirror with length of about 10 cm, and the intensity distribution of the transmitted neutrons was observed.

Although the experiment is very interesting in the sense of combining gravity and quantum mechanics, and also from the viewpoint of the extreme limit of the high-precision neutron experiment, we have no more space here to describe the details of the experiment as well as the result. Fortunately, some details were already reported in recent articles [116, 117], and so will not be described here.

3
Reflection, Refraction, and Larmor Precession of Polarized Neutrons

We already mentioned in Section 1.5 that a neutron has spin 1/2 and the magnitude of the magnetic dipole moment is also known precisely from neutron magnetic resonance experiments. Further, we also introduced the neutron spin precession in a magnetic field and the magnetic potential effects on neutron motion. In this chapter, we will study further the detailed behavior of neutrons as spin-1/2 particles, various types of optical experiments, and reflection, refraction, as well Larmor precession as phenomena which greatly concern neutron spin.

3.1
Reflection from Magnetic Mirrors

3.1.1
Magnetic Reflection and Neutron Polarization

In 1936, Bloch supposed that a neutron has a magnetic dipole moment with magnitude comparable to that of a proton but with the direction of the moment opposite to that of the angular momentum of the neutron. He formulated an expression for the magnetic scattering of neutrons as the dipole–dipole interaction and proposed an experiment to measure the polarization of neutrons scattered from or transmitted through a sample such as magnetized iron to directly determine the *magnetic dipole moment* of the neutron [118].

In the next year, Schwinger formulated another model on the magnetic interaction of neutrons as induced by the current density with the magnitude of the correct Dirac value to give the neutron magnetic dipole moment, and analyzed the magnetic scattering and polarization of neutrons [119].

In this situation, Halpern [120] pointed out that the experimental effect of the magnetic interaction should become maximum in an experimental arrangement where the magnetization of an iron plate sample is parallel to the projected direction of the neutron beam onto the sample surface. In addition he proposed that the total reflection experiment would be advantageous for elucidating the magnetic dipole moment of the neutron, because two critical angles would be observed for the reflection from the magnetized iron corresponding to two spin states of the

Neutron Optics. Masahiko Utsuro,Vladimir K. Ignatovich

ISBN: 978-3-527-40885-6

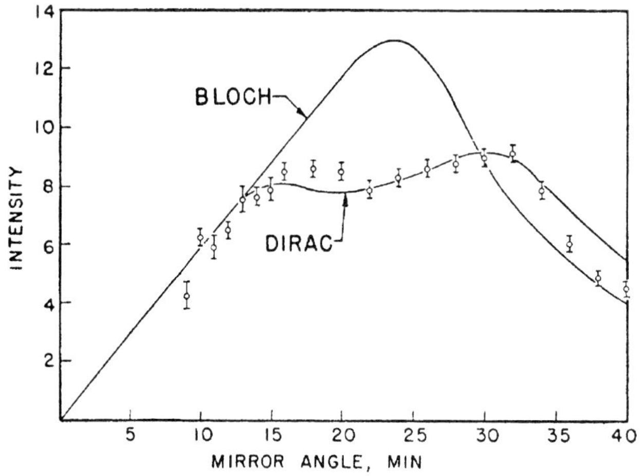

Figure 3.1 Intensity of cold neutrons reflected from a ferromagnetic mirror (Hughes and Burgy [121]).

neutron, above and below that of the unmagnetized one. Accordingly, Hughes and Burgy immediately carried out a magnetic reflection experiment of neutrons and obtained the result shown in Figure 3.1 [121].

Their experiment using a magnetized iron mirror was performed at the heavy water reactor, Brookhaven National Laboratory, with a cold neutron beam of wavelength longer than 4.4 Å extracted from the heavy water thermal column and transmitted through a beryllium oxide filter.

As mentioned in Chapter 1, the general expression for the refractive index of neutrons is given by

$$n^2 = 1 - \frac{U}{E}, \tag{3.1}$$

where U is the optical potential and E the neutron energy. Consideration of the magnetic interaction for neutrons leads to the relation $U = U_{\rm nuc} + U_{\rm mag}$, where $U_{\rm nuc}$ is given by Eq. (1.19).

Now, in the Bloch's *dipole model*, $U_{\rm mag}$ for neutrons with a magnetic dipole moment $\boldsymbol{\mu}$ becomes $U_{\rm mag} = -\boldsymbol{\mu} \boldsymbol{H}$, while Schwinger's *current model* leads to $U_{\rm mag} = -\boldsymbol{\mu} \boldsymbol{B} = -\boldsymbol{\mu}(\boldsymbol{H} + 4\pi \boldsymbol{M})$, where $\boldsymbol{M}$ is the magnetic dipole density of the medium.

The magnetic field $\boldsymbol{H}$ in the experimental arrangement proposed by Halpern varies continuously across the iron mirror surface, where the total reflection of neutrons occurs, and the magnetic interaction in the dipole model does not contribute significantly to the magnitude of the critical angle for total reflection. As a result, the dipole model should give a single value for the critical angle ($\theta_c = 23.8$ minutes for an iron mirror).

In contrast the magnetic induction $\boldsymbol{B}$ shows a discontinuous change with a significant magnitude across the mirror surface. Furthermore, its contribution differs in sign for parallel-spin and antiparallel-spin neutrons. As a result, in the current model there will be two significantly different values for the critical angle ($\theta_c = 14.2$ and 30.7 minutes for an iron mirror).

In this way, two kinds of the expected curve are drawn for the reflected intensity in Figure 3.1. It is obvious that the current model gives the correct prediction. In their experiment, by making use of a second iron mirror, they also ensured that the neutrons reflected from the first mirror were polarized. Thereafter, various applications of neutron magnetic reflection to studies on magnetic materials began as did theoretical developments of polarized neutron optics.

For polarized neutron experiments, firstly it is necessary to prepare a highly polarized neutron beam and an efficient and accurate polarization analyzer. In the magnetic resonance experiment of ultracold neutrons of Egorov *et al.* described in Section 1.5, the incident neutrons were polarized by transmission through a magnetized film as shown schematically in Figure 1.12, that is, in an arrangement similar to that shown in Figure 1.10a. After the resonance device, the polarization of exiting neutrons was analyzed again by transmission through the second magnetized film. The polarization efficiency of the *total reflection method with a magnetized iron thin film* for ultracold neutron beam was measured as $78 \pm 5\%$.

The polarizer was prepared by the evaporation of iron material with composition 20% ^{56}Fe, 80% ^{54}Fe, to a thickness of 1 μm on a copper substrate. Thereafter the substrate copper was removed by etching.

The polarization efficiency of the *magnetic scattering method* with an iron block employed by Alvarez *et al.* for thermal neutrons as shown in Figure 1.10a was far lower than the value mentioned above.

Concerning the polarization efficiency of the total reflection method being a little below 100%, Ignatovich [122] analyzed theoretically the reflection and refraction of neutrons in the situation where the magnetic inductions of two adjacent layers such as the polarizer film and its environment layer are not in the same direction. He explained the experimental results for the polarization efficiency and the polarization flipping rate of neutrons in such multilayers by the nonparallelism of the magnetic induction.

In the high-polarization experiment of Herdin *et al.* [53] briefly mentioned in Section 1.5, a single crystal of α-Fe with thickness 1500–3000 Å was employed. It was carefully grown epitaxially on a rock-salt substrate to improve the polarization efficiency by decreasing the magnetic inhomogeneity, thus preventing depolarization due to magnetic fluctuations, compared with the magnetized polycrystalline film used in the experiment of Egorov *et al.* They obtained a high polarization efficiency of 95–98% for neutrons in the velocity region of $4.15 < v < 8.2$ m/s, which corresponds to the region between the limiting velocities for total reflection v_l^- and v_l^+ of two spin states of neutrons. Such a polarizing film used in the ultracold neutron experiment can also be applied as a polarizer and analyzer for higher-velocity neutrons by locating the perpendicular component in the present velocity region.

3.1.2
Polarized Neutron Reflectometry

By making use of polarized neutrons prepared in this way it is possible to determine the depth profile of the magnetic property on a magnetic sample, similar to the case of unpolarized neutron reflectometry on nonmagnetic samples we studied in the previous chapter, from experiments measuring the polarized neutron reflectivity profiles. One of the simple examples of *polarized neutron reflectometry* applied to a magnetic single crystal by Felcher *et al.* [123] will be introduced in Figures 3.2 and 3.3.

The component of the magnetic induction perpendicular to the sample surface $\boldsymbol{B}_z$ continuously varies across the surface, and therefore we consider the magnetic interaction $U_{\text{mag}} = -\boldsymbol{\mu}\,\boldsymbol{B}$ in the situation of the reflectometry experiment shown in Figure 3.2a, in which the neutron polarization and the magnetic induction $\boldsymbol{B}$ both lie parallel to the sample surface. The component $\boldsymbol{B}_{\|}$ of the magnetic induction parallel to the neutron spin, to be determined by the direction of the external filed $\boldsymbol{H}$, induces the *magnetic potential* $U_{\text{mag}} = -\boldsymbol{\mu}(\boldsymbol{B}_{\|} - \boldsymbol{H})$, whereas the perpendicular component $\boldsymbol{B}_{\perp}$, if it exists, will contribute to possible local depolarization of the neutrons.

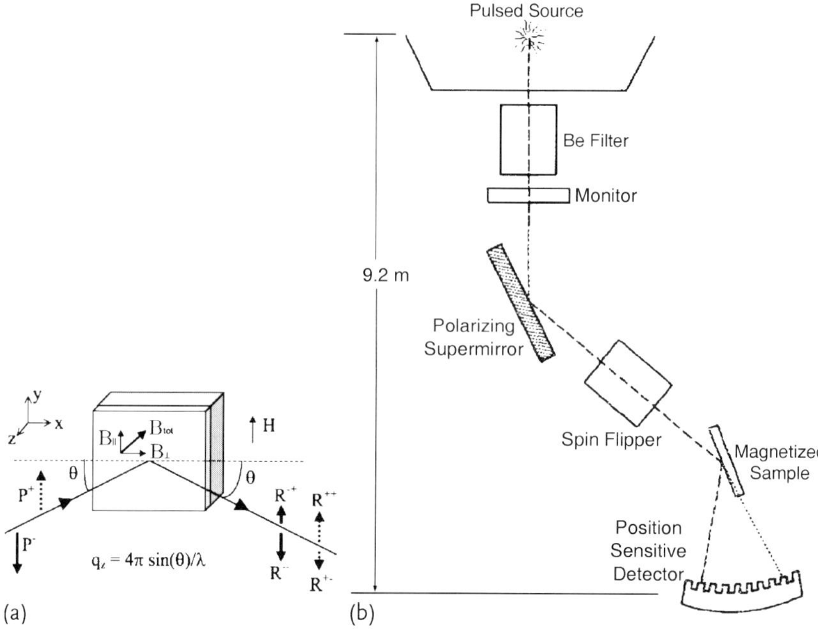

Figure 3.2 Experimental configuration for polarized neutron reflectometry. (a) Reflection situation of a polarized neutron beam; the incident neutrons are polarized in the direction of the external field **H**, where R is the reflected beam (Felcher *et al.* [124]). (b) Experimental arrangement for polarized neutron reflectometry (Felcher *et al.* [123]).

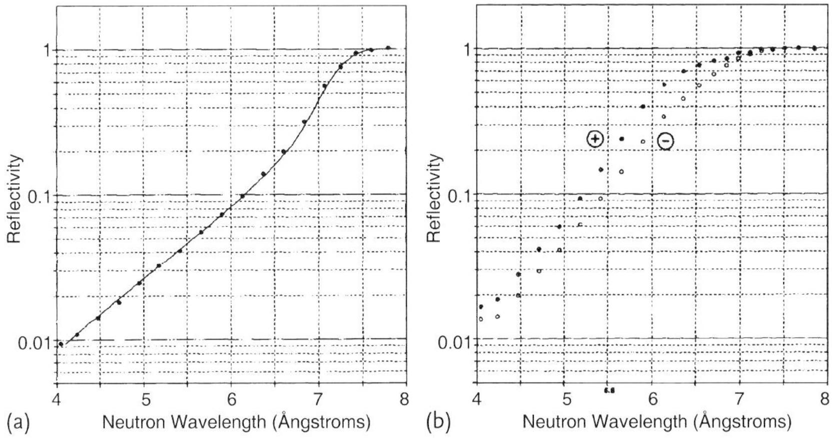

Figure 3.3 Neutron wavelength dependences of polarized neutron reflectivity: (a) measured reflectivity of silica glass with a high-quality polished surface compared with the theoretical curve for grazing angle $\theta_i = 0.435^\circ \pm 0.015^\circ$; (b) measured spin-dependent reflectivities for a nickel single crystal with the (110) face exposed at grazing angle $\theta_i \sim 0.7^\circ$, and with a magnetic field of 20 Oe applied in the $(1\bar{1}1)$ direction (Felcher *et al.* [123]).

The *Hamiltonian* $\mathcal{H}$ including the magnetic interaction for neutrons is expressed as $\mathcal{H} = -\hbar^2/(2m)\Delta + U(\boldsymbol{r}, t) - \underline{\boldsymbol{\mu}}\,\boldsymbol{B}(\boldsymbol{r}, t)$, and then the *magnetic potential* for neutrons is given by the term $-\underline{\boldsymbol{\mu}}\,\boldsymbol{B}(\boldsymbol{r}, t)$ in the Schrödinger equation:

$$i\frac{m}{\hbar}\frac{\partial}{\partial t}|\Phi(\boldsymbol{r}, t)\rangle = \left(-\frac{\Delta}{2} + \frac{m}{\hbar^2}U(\boldsymbol{r}, t) - \frac{m}{\hbar^2}\underline{\boldsymbol{\mu}}\,\boldsymbol{B}(\boldsymbol{r}, t)\right)|\Phi(\boldsymbol{r}, t)\rangle\,. \tag{3.2}$$

In relation to the magnetic interaction operator $-\underline{\boldsymbol{\mu}}\,\boldsymbol{B} = |\mu_n|\underline{\boldsymbol{\sigma}}\boldsymbol{B}$, the direction of the neutron magnetic moment can be written as $\boldsymbol{\mu} = 2\mu_n \boldsymbol{s} = -2|\mu_n|\boldsymbol{s}$, where, defining the coordinate system x', y', z' with z' in the direction of the quantization axis in $\boldsymbol{H}$ (i.e. $x' = x, y' = -z, z' = y$), $\boldsymbol{s} = (s_{x'}, s_{y'}, s_{z'})$ is the spin vector given by the expectation values for the respective spin components, $s_j = \langle\phi(t)|\hat{S}_j|\phi(t)\rangle$, $(j = x', y', z')$, and $\underline{\boldsymbol{\sigma}} = (\underline{\sigma_{x'}}, \underline{\sigma_{y'}}, \underline{\sigma_{z'}})$ is the *Pauli matrix* defined by

$$\underline{\sigma_{x'}} = \begin{pmatrix} 0 & 1 \\ 1 & 0 \end{pmatrix}, \quad \underline{\sigma_{y'}} = \begin{pmatrix} 0 & -i \\ i & 0 \end{pmatrix}, \quad \underline{\sigma_{z'}} = \begin{pmatrix} 1 & 0 \\ 0 & -1 \end{pmatrix}. \tag{3.3}$$

Furthermore, $U(\boldsymbol{r}, t)$ is the external potential and $\boldsymbol{B}(\boldsymbol{r}, t)$ is the magnetic induction, both being generally dependent on space and time.

Applying the property of the Pauli matrix to Eq. (3.2) leads to a couple of equations for two spinor components, $\phi_+(z), \phi_-(z)$, at depth z in a sample in a steady-state condition:

$$\begin{aligned} &\phi''_+ + \left[k_z^2 - 4\pi(N b_{\text{coh}} + c B_{\|})\right]\phi_+ - 4\pi c B_\perp \phi_- = 0\,, \\ &\phi''_- + \left[k_z^2 - 4\pi(N b_{\text{coh}} - c B_{\|})\right]\phi_- - 4\pi c B_\perp \phi_+ = 0\,, \end{aligned} \tag{3.4}$$

where $k_z = 2\pi \sin\theta_i/\lambda$ is the wave number vector component perpendicular to the sample surface, θ_i is the incident angle, and $c = 2\pi m\mu/h^2$.

From the solutions of Eq. (3.4), the perpendicular wave number components $k_\pm$ in the medium where $N b_{\text{coh}}$ and $\boldsymbol{B}$ are constant become

$$k_\pm = \sqrt{k_z^2 - 4\pi(N b_{\text{coh}} \pm c B_T)}\,, \tag{3.5}$$

where $B_T = \sqrt{B_\parallel^2 + B_\perp^2}$. This situation corresponds to the refractive indices for the medium given by the relation

$$n_\pm^2 = 1 - \frac{\lambda^2}{\pi}(N b_{\text{coh}} \pm c B_T)\,. \tag{3.6}$$

It must be remarked that, strictly speaking, if we consider a possible change of the magnetic potential on the sample surface in reflection and refraction, the term B_T in Eq. (3.6) should be replaced by $B_T - H$. However, in the ferromagnetic materials usually employed as a sample for magnetic reflection and refraction, the condition $B_T \gg H$ is well satisfied and then the correction mentioned above can be neglected. In contrast, in the case of superconductors with the property of perfect diamagnetism, the exact formula for the magnetic potential must be used as Eq. (11.1).

For the simplified case of $B_\perp = 0$ in Eq. (3.4), the coupling in the equation being resolved, the reflectivity R_{++} from the incident neutron spin state + to the reflected neutron spin state +, and that R_{--} from the incident neutron spin state − to the reflected neutron spin state −, reduce to the same form as the standard formula, Eq. (2.4), as

$$R_{\pm\pm} = \left| \frac{\sin\theta_i - \sqrt{\sin^2\theta_i - (\lambda^2/\pi)(N b_{\text{coh}} \pm c B_\parallel)}}{\sin\theta_i + \sqrt{\sin^2\theta_i - (\lambda^2/\pi)(N b_{\text{coh}} \pm c B_\parallel)}} \right|^2 . \tag{3.7}$$

Both reflectivities take the same value by scaling the wavelength with the relation

$$\lambda_+ \sqrt{N b_{\text{coh}} + c B_\parallel} = \lambda_- \sqrt{N b_{\text{coh}} - c B_\parallel}\,. \tag{3.8}$$

The polarized neutron reflectometry experiment was undertaken on the basis of such an expectation.

The experiment was carried out with the cold neutron beam from the solid methane moderator at the pulsed neutron source IPNS, Argonne National Laboratory. In the horizontal arrangement shown in Figure 3.2b, a guide field was applied over the whole neutron flight path after the polarizer to maintain the polarization.

The fundamental optical performance can be assured in Figure 3.3a, where the measured reflectivity of nonmagnetic silica glass with a high-quality polished surface is compared with the calculated curve of the standard formula, Eq. (2.4). Both results agree well, and the depth resolution $\Delta z = 1/2\Delta k_z$ in the present experimental condition was estimated to be about 40 Å. In the case of a sample with a lead layer of thickness 2 μm evaporated as usual on a silicon crystal, instead of the sample with a high-quality polished surface, the reflectivity decreased compared with the calculation using Eq. (2.4) owing to the effect of the surface roughness. The present result is consistent with the indication observed in Figure 2.3b.

Use of a nickel single crystal gave the experimental results for the polarized neutron reflectivities shown in Figure 3.3b. The sample magnetization was saturated in the $(1\bar{1}1)$ easy direction. During the measurements, the condition that magnetization direction is parallel to the neutron quantization axis, which is the premise for applying Eq. (3.7), otherwise the reflected beam would be depolarized, was routinely checked by the insertion of a polarization analysis leg. It can be easily checked that two reflectivity plots in Figure 3.3b for the spin states + and − satisfy well the relation for the wavelength scaling, Eq. (3.8). In this way, the present results from preliminary measurements of polarized neutron reflectometry on a sample of relatively weak magnetic material such as nickel gave an obvious difference between the reflectivities for the different neutron spin states, and indicated well the applicability of polarized neutron reflectometry.

As these experimental examples show, polarized neutron reflectometry was first applied to obtain magnetic depth profiling of magnetic materials and was then further developed for other fields in the middle of the 1980s. These developments of various applications in the early stage over about 10 years were reviewed by Felcher [125]. Not only was specular reflection with reflection angle $\theta_r = \theta_i$ applied to analyze the magnetic profiling in the direction of depth, but also experiments to measure and analyze nonspecular reflection with reflection angle $\theta_r \neq \theta_i$ were started to obtain possible information on the structure parallel to the surface. The application of specular reflectometry was also extended, in addition to single-layer samples, to sandwiched structures and multilayers. The fundamental principle of magnetic multilayer reflectometry will be described in the next section, and some examples of nonspecular reflectometry on a single-layer sample will be given in Chapter 11.

3.1.3
Magnetic Storage of Neutrons

As mentioned in Section 1.5, neutrons with spin parallel to the magnetic field are repelled by the field in a vacuum. Therefore, it is possible to use magnetic guides and magnetic storage bottles for polarized neutrons in a strong magnetic field. The idea of neutron magnetic storage was proposed by Vladimirskii in 1961 [126], and the first experimental realization of magnetic storage of ultracold neutrons was performed in late 1970 [127, 128].

As for the experimental principles for magnetic storage, two approaches are possible, that is, the *electromagnetic method* and the *permanent magnet method*. In the early experiments on magnetic storage, the former method was employed. In the electromagnetic apparatus, it is possible to attain a strong field over than 10 T by making use of a superconducting magnet, but it requires a relatively large scale experimental arrangement. In contrast, with permanent magnets, an apparatus with

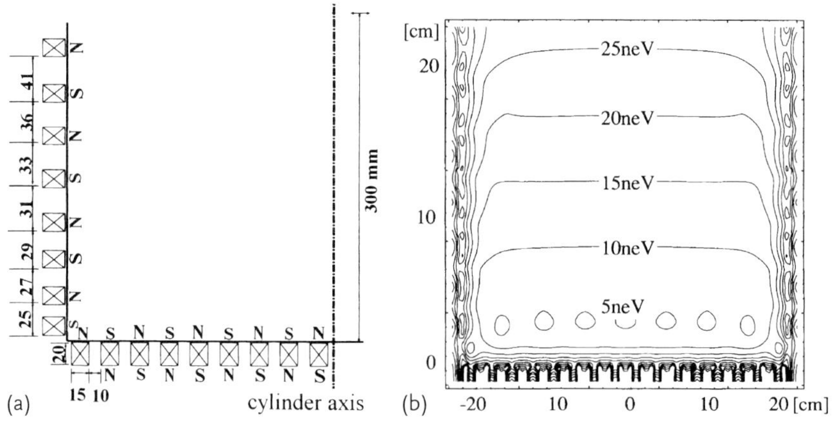

Figure 3.4 Magnetic reflection potential for ultracold neutrons produced by an array of rectangular rods of permanent magnet elements: (a) cross section of the magnetic element arrangement; (b) magnetic potential distribution for neutrons (1 neV = 10^{-9} eV) (Inoue *et al.* [129]).

a maximum magnetic field of about 1 T will become a practical concept by making use of recently developed high-performance magnetic materials. The magnetic design with an array of alternatively arranged poles in a fine pitch as shown in Figure 3.4a provides the large gradient of the magnetic field in the storage device which is an important requirement for the magnetic reflections of neutrons to be stored. In accordance with such a concept, a preliminary experiment on the magnetic storage of ultracold neutrons was carried out by Inoue *et al.*

As proposed by Vladimirskii, it is necessary for magnetic storage that the permanent magnet elements be arranged so as to realize a field distribution with strength increasing very steeply near the surface of the elements, whereas there should be almost no field away from the array of the elements. To meet this requirement, Inoue *et al.* arranged a number of fine elements of a Nd–Fe–B permanent magnet with dimensions of 15 mm × 20 mm, and 400 mm long, and with a surface field strength of about 0.4 T (Figure 3.4a), and thus realized a storage device for ultracold neutrons. The calculated magnetic field distribution in the device is shown in Figure 3.4b for the *magnetic potential* for neutrons [129].

As the preliminary experimental approach, the U-shaped magnetic chamber with the magnetic elements arranged as shown in Figure 3.4 was prepared and the experiments were performed with the ultracold neutrons from the supermirror turbine shown in Figure 2.17. The experimental results for the magnetic reflections of ultracold neutrons in the present device are plotted in Figure 3.5 [130]. Half of the unpolarized neutrons incident on the device guided by the turbine will be polarized with parallel spins and then repelled by the field, whereas the remaining half will be attracted by the magnetic field and absorbed by the structure materials such as the magnet elements and shielding materials. The chamber has a critical velocity of reflection for neutrons of about 2 m/s, and therefore it was positioned

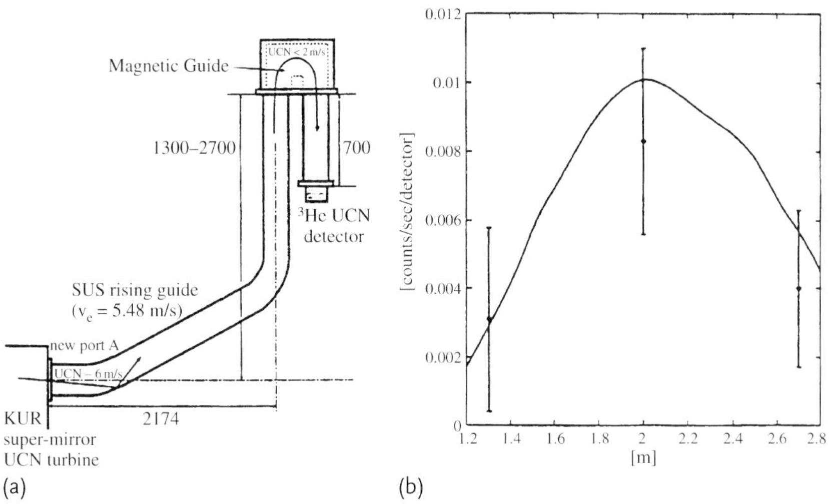

Figure 3.5 Ultracold neutron reflection experiment with a magnetic chamber consisting of permanent magnets: (a) experimental arrangement; (b) transmitted neutron count rates through the U-shaped magnetic chamber dependent on the gravity rise height. UCN ultracold neutrons; circles, experimental results; curve, calculated result (Inoue *et al.* [130]).

at a gravity rise height higher than the turbine exit port. This port provides the ultracold neutron spectrum shown in Figure 2.17b. The effect of the gravity rise was studied for three kinds of height, that is, 1.3, 2.0, and 2.7 m, in the present experiments. The count rates for the exiting neutrons were measured after the gravity fall from the chamber exit port as plotted in Figure 3.5b and were compared with the theoretical curve for the dependence on the gravity rise height calculated from the turbine spectrum shown in Figure 2.17b. The comparison indicates reasonable agreement in the rise height dependence of the neutron intensity, thus, this ultracold neutron storage chamber with a permanent magnet array has the magnetic reflection performance expected from the design.

Recently another development of the permanent magnet storage experiments was carried out by Ezhov *et al.*, in which the *permanent magnet* elements with a maximum field of 1 T are arranged in an array to form a vertical cylinder for the *long-time storage* of ultracold neutrons [131]. For the long-time storage, it is important to maintain the neutron spin parallel to the magnetic field during the period of a number of reflections in the storage device. As pointed out in Section 1.5, possible fluctuations such as a sudden change in the magnetic field direction faster than the Larmor precession frequency of neutrons induce depolarization of the neutrons and cause neutron leakage out of the storage space. The preliminary result of their experiments on ultracold neutron storage carried out at the high-flux reactor, Grenoble, is shown in Figure 3.6b, from which the storage lifetime for neutrons of 874.6 ± 1.6 s was deduced. This result indicates a small loss coefficient and

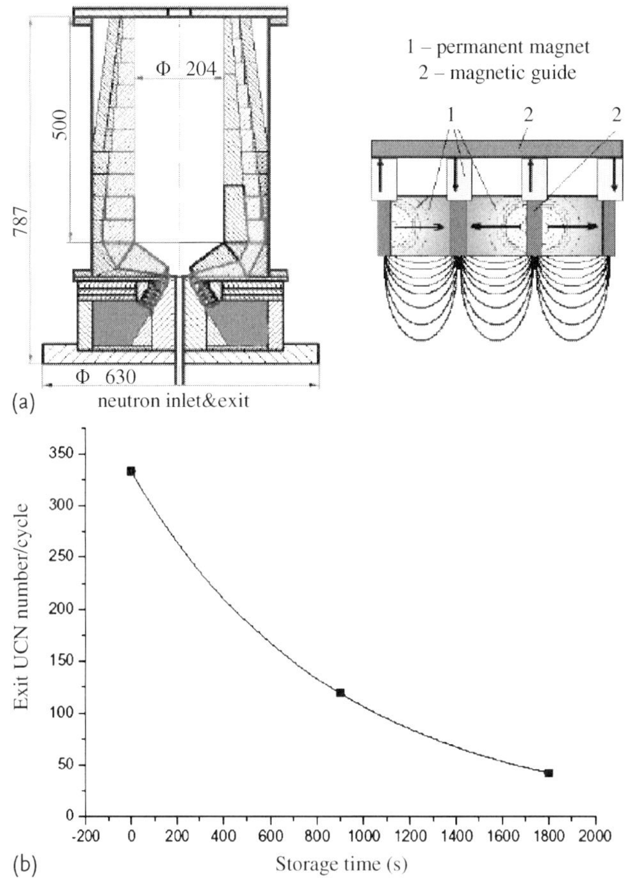

Figure 3.6 Ultracold neutron storage experiment with a storage chamber consisting of permanent magnets: (a) experimental arrangement (storage volume, inner diameter about 20 cm, height 50 cm) and the magnetic element configuration; (b) experimental dependence of exiting ultracold neutron counts on the storage time (Ezhov *et al.* [132]).

therefore their continued measurements will lead to valuable progress of neutron magnetic storage studies.

On the other hand, in the *electromagnetic storage* of neutrons, an experimental accuracy of about 10 s for the storage lifetime was reported [133, 134].

3.2
Reflection and Transmission of Neutrons by Magnetic Multilayers

3.2.1
Reflectivity Analyses with the Transfer Matrix Method

In Section 2.3 we studied the reflection and transmission of neutrons by nonmagnetic multilayers, and further various methods were introduced to derive the scattering amplitude density profile from experimental results of reflection and transmission by multilayer samples, in which it was predicted that polarized neutron reflectometry will become an effective approach for that purpose. In this section, such polarized neutron analyses on magnetic samples will be described.

As mentioned in Section 2.3, the extension of the analytical solution for reflection and transmission of unpolarized neutrons by a nonmagnetic single layer to nonmagnetic multilayers was the problem to solve simultaneous equations including an increasing number of amplitude coefficients according to the increase of the layer number, and it was rather a simple task to apply the analytical solution. However, in the case of polarized neutron analyses, the situation is completely different because of the property of the Pauli matrices included in the description of the magnetic potential, and this character appeared in Eq. (3.4). Practically, with regard to the two spinor components ψ_+ and ψ_-, the component of the magnetic induction $\boldsymbol{B}_\parallel$ parallel to the neutron spin induces the applied magnetic potentials $+U_{\text{mag}}$ and $-U_{\text{mag}}$, respectively, with opposite signs. Further, if the component perpendicular to the neutron spin, $\boldsymbol{B}_\perp$ exists, the transition between the spin components, that is, partial depolarization of neutrons, might be induced. For the analyses of such complicated situations in polarized neutron experiments, Blundell and Bland employed a *transfer matrix method* to obtain possible information on the magnetic profile on multilayers [135]. As mentioned in Section 2.3, the transfer matrix method was originally introduced by Croce and Pardo in the analyses of reflection and transmission of neutrons by multilayers [92].

In this matrix formalism, the wave functions in each region, that is, Eq. (2.20), are rewritten as follows by taking the y-axis as the direction of the layer thickness, in order to reserve the z-axis for the direction of magnetic induction lying on the layer surface,

$$\phi_n(y) = A_n e^{ik_n(y-y_n)} + B_n e^{-ik_n(y-y_n)}, \quad y_{n-1} \le y \le y_n, \tag{3.9}$$

and then $\phi_n(y)$ at the interface $y = y_n$ is expressed by a vector $\begin{pmatrix} A_n \\ B_n \end{pmatrix}$.

Defining $\phi_{n+1}(y)$, $y_n \le y \le y_{n+1}$ in a similar way, we can write the continuity conditions for these functions and their derivatives at $y = y_n$ as a whole as

$$\begin{pmatrix} 1 & 1 \\ k_n & -k_n \end{pmatrix} \begin{pmatrix} A_n \\ B_n \end{pmatrix} = \begin{pmatrix} e^{-ik_{n+1}d_{n+1}} & e^{ik_{n+1}d_{n+1}} \\ k_{n+1}e^{-ik_{n+1}d_{n+1}} & -k_{n+1}e^{ik_{n+1}d_{n+1}} \end{pmatrix} \begin{pmatrix} A_{n+1} \\ B_{n+1} \end{pmatrix}, \tag{3.10}$$

or

$$\underline{D}(k_n)\begin{pmatrix}A_n\\B_n\end{pmatrix}=\underline{D}(k_{n+1})\underline{P}(k_{n+1},d_{n+1})\begin{pmatrix}A_{n+1}\\B_{n+1}\end{pmatrix}, \tag{3.11}$$

where the *transmission matrix* $\underline{D}(k_n)$ and the *propagation matrix* $\underline{P}(k_n, d_n)$ are denoted as

$$\underline{D}(k_n)=\begin{pmatrix}1 & 1\\ k_n & -k_n\end{pmatrix}, \tag{3.12}$$

$$\underline{P}(k_n,d_n)=\begin{pmatrix}e^{-ik_nd_n} & 0\\ 0 & e^{ik_nd_n}\end{pmatrix}, \tag{3.13}$$

respectively.

As a result, the solutions of Eqs. (2.19) and (2.20) for the multilayer potential can be expressed with the *transfer matrix* $\underline{M}$ as follows:

$$\begin{pmatrix}1\\r\end{pmatrix}=\begin{pmatrix}M_{11} & M_{12}\\ M_{21} & M_{22}\end{pmatrix}\begin{pmatrix}t\\0\end{pmatrix}, \tag{3.14}$$

where r and t are the *reflection coefficient* and the *transmission coefficient*, respectively, and the transfer matrix $\underline{M}$ is given by

$$\underline{M}=\underline{D}^{-1}(k)\left[\prod_{n=1}^{N-1}[\underline{D}(k_n)\underline{P}(k_n,d_n)\underline{D}^{-1}(k_n)]\right]\underline{D}(k_N). \tag{3.15}$$

For the simplest case of a sufficiently thick substrate only, it becomes $N = 1$ and $\underline{M} = \underline{D}^{-1}(k)\underline{D}(k_1)$. Then the relation obtained, $r = M_{21}/M_{11} = (k - k_1)/(k + k_1)$, agrees with Eq. (2.4) as $R = |r|^2$. Further, the equation $t = 1/M_{11} = 2k/(k + k_1)$ agrees with the continuity condition $a = 1 + r$ for Eq. (2.19) by putting $d \to \infty$, that is, putting $b = 0$.

Next, we extend the matrix formalism to the case including a magnetic potential, that is, apply the magnetic potential in Eq. (3.2) to multilayers. For simplicity, we suppose the magnetic induction vector for each layer lies on the interface, that is, on the surface (x, z), and denote the vector for the $(n + 1)$th layer as rotated by an angle $\theta_{n,n+1}$ around y-axis relative to that for the nth layer. In this case, the rotation rule for a spin-1/2 system shown in Section 3.4, Eq. (3.48), and the formula for the Pauli matrices lead to the continuity condition, Eq. (3.11), being replaced by

$$\underline{D}\left(k_n^{\uparrow},k_n^{\downarrow}\right)\begin{pmatrix}\varphi_n^{\uparrow}\\ \varphi_n^{\downarrow}\end{pmatrix}=\underline{R}(\theta_{n,n+1})\underline{D}\left(k_{n+1}^{\uparrow},k_{n+1}^{\downarrow}\right)\underline{P}\left(k_{n+1}^{\uparrow},k_{n+1}^{\downarrow},d_{n+1}\right)\times\begin{pmatrix}\varphi_{n+1}^{\uparrow}\\ \varphi_{n+1}^{\downarrow}\end{pmatrix}, \tag{3.16}$$

where $\phi_n^s = \begin{pmatrix} A_n^s \\ B_n^s \end{pmatrix}$ is the solution for the spin $s = \uparrow, \downarrow$, and

$$\underline{\boldsymbol{R}}(\theta_{n,n+1}) = \begin{pmatrix} \cos(\theta_{n,n+1}/2)\underline{\boldsymbol{I}} & \sin(\theta_{n,n+1}/2)\underline{\boldsymbol{I}} \\ -\sin(\theta_{n,n+1}/2)\underline{\boldsymbol{I}} & \cos(\theta_{n,n+1}/2)\underline{\boldsymbol{I}} \end{pmatrix} \tag{3.17}$$

is the matrix for the rotation of the quantization axis with angle $\theta_{n,n+1}$ at the interface $n, n+1$, where $\underline{\boldsymbol{I}} = \begin{pmatrix} 1 & 0 \\ 0 & 1 \end{pmatrix}$.

Furthermore,

$$\underline{\boldsymbol{D}}(k_n^\uparrow, k_n^\downarrow) = \begin{pmatrix} \underline{\boldsymbol{D}}(k_n^\uparrow) & 0 \\ 0 & \underline{\boldsymbol{D}}(k_n^\downarrow) \end{pmatrix}, \tag{3.18}$$

$$\underline{\boldsymbol{P}}(k_n^\uparrow, k_n^\downarrow, d_n) = \begin{pmatrix} \underline{\boldsymbol{P}}(k_n^\uparrow, d_n) & 0 \\ 0 & \underline{\boldsymbol{P}}(k_n^\downarrow, d_n) \end{pmatrix} \tag{3.19}$$

are the 4×4 *transmission matrix* and *propagation matrix*, respectively, extended for the layers including the magnetic potential.

In this way, we obtain a 4×4 *transfer matrix*:

$$\begin{aligned} \underline{\boldsymbol{M}} &= \underline{\boldsymbol{D}}^{-1}(k, k)\underline{\boldsymbol{R}}(\theta_{1,2}) \\ &\times \left[\prod_{n=1}^{N-1} \left[\underline{\boldsymbol{D}}\left(k_{n+1}^\uparrow, k_{n+1}^\downarrow\right) \underline{\boldsymbol{P}}\left(k_{n+1}^\uparrow, k_{n+1}^\downarrow, d_n\right) \underline{\boldsymbol{D}}^{-1} \right.\right. \\ &\times \left.\left. \left(k_{n+1}^\uparrow, k_{n+1}^\downarrow\right) \underline{\boldsymbol{R}}(\theta_{n,n+1}) \right] \right] \underline{\boldsymbol{D}}(k_N, k_N). \end{aligned} \tag{3.20}$$

In the analysis of Blundell and Bland explained above, each magnetic induction vector is assumed for simplicity to lie on each interface, whereas Ignatovich and Radu gave the analysis in principle for arbitrary directions of the magnetic inductions [122, 136].

By making use of these analytical formulations, we can easily calculate reflection and transmission by multilayers with an arbitrary magnetic structure, and then we will be able to obtain the *reflection coefficient for a magnetic multilayer* and the *transmission coefficient for a magnetic multilayer*. However, the actual work involves the inverse process for practical requirements, that is, the reduction of the magnetic profile from experimental results on the reflection and transmission for an actual structure will not be so easy. Particularly, in such a case where the direction of the magnetic induction for each layer is different from layer to layer, many trials with various model calculations will be required to obtain a reliable magnetic profile.

As mentioned in Section 2.3, Majkrzak *et al.* developed the reference layer method with reflectivity measurements for the sample with known reference layers added in addition to the unknown layer to be studied, in order to determine the unique scattering amplitude density profile of the unknown layer. As an extension of the method, they proceeded to polarized neutron reflectometry by making use of magnetic reference layers. If the magnetic induction of the magnetic layers lies along a single direction, and incident neutrons have spin polarized parallel or antiparallel to the magnetic induction, then the off-diagonal elements in Eq. (3.17) become zero, and the coupling between the spinors is resolved. Then, the resultant relation reduces to the application of the 2×2 matrices, Eqs. (2.22)–(2.27). Therefore, the next equation can be derived for the case of negligible absorption and real scattering amplitudes [137]:

$$\begin{pmatrix} t \\ i n_b t \end{pmatrix} = \underline{\boldsymbol{M}} \begin{pmatrix} 1 + r \\ i n_f (1 - r) \end{pmatrix} , \tag{3.21}$$

where

$$\underline{\boldsymbol{M}} = \begin{pmatrix} A & B \\ C & D \end{pmatrix} = \prod_{j=N}^{1} \begin{pmatrix} \cos \theta_j & n_j^{-1} \sin \theta_j \\ -n_j \sin \theta_j & \cos \theta_j \end{pmatrix} . \tag{3.22}$$

The matrix elements $A \sim D$ are all functions of k, and $\theta_j = n_j d_j k$, where n_j for each layer is given by Eq. (3.6). Since it can be derived [138] that the determinants of each matrix on the right hand of Eq. (3.22) are always unity, that is, they are *unimodular matrices*, by making use of the relation $AD - BC = 1$ and of the assumption that the previous and the next layers of the set of multilayers are a vacuum, we obtain the reflection coefficient as

$$r = \frac{B + C + i(D - A)}{B - C + i(D + A)} = \frac{(B^2 + D^2) - (A^2 + C^2) - 2i(AB + CD)}{2 + (B^2 + D^2) + (A^2 + C^2)} . \tag{3.23}$$

Although the measured reflectivity corresponds to $|r|^2$, and from the above equations the relation

$$2 \left[\frac{1 + |r|^2}{1 - |r|^2} \right] = A^2 + B^2 + C^2 + D^2 \equiv \Sigma \tag{3.24}$$

holds, it seems by looking at these relations that we cannot determine $A \sim D$ from the measured reflectivities. However, employing the reference layer method, that is,

$$\begin{pmatrix} A & B \\ C & D \end{pmatrix} = \begin{pmatrix} a & b \\ c & d \end{pmatrix} \begin{pmatrix} w & x \\ y & z \end{pmatrix} , \tag{3.25}$$

where (a, b, c, d) refer to the reference layer, (w, x, y, z) to the unknown layer, then putting $\alpha = a^2 + c^2$, $\beta = b^2 + d^2$, $\gamma = ab + cd$, we obtain

$$\Sigma = (w^2 + x^2)\alpha + (y^2 + z^2)\beta + 2(wy + xz)\gamma . \tag{3.26}$$

Thus, if Σ_l, $l = 1$–3 can be derived from the experimental results for the conditions with three different known layers, and this yields three different relations corresponding to Eq. (3.26) for the cases $l = 1$–3. Then we can obtain α, β, and γ

as the common solutions for these three measurements, that is, we can determine the matrix elements $a \sim d$ in the unimodular matrices [137].

Further, from the unimodularity of these matrices, the restriction $\gamma^2 = \alpha\beta - 1$ can be easily derived, and finally only two measurements become sufficient to determine the matrix elements. These two kinds of measurements can be carried out by making use of neutrons polarized (+) and (−), respectively, without changing the sample. This method of reflectometry to obtain the *phase information on reflectivity* and to determine the scattering amplitude density profile of an unknown layer is named the *polarized neutron reference layer method* developed by Majkrzak *et al.*

An example of a practical application of the polarized neutron reference layer method will be introduced in Chapter 11.

3.2.2
Analyses of Nonspecular Reflections

Hitherto, studies on the specular reflections of polarized neutrons, that is, the experiments on and the analyses of the phenomena in the reflection angle the same as the incident one, have been described. Recently, further extended trials have been performed to obtain specific information from experiments on nonspecular reflections and also on spin-flipping reflections. As one of the analyses for such trials, the *supermatrix formalism* of Rühm *et al.* will be introduced below. Rühm *et al.* expected that neutron reflectometry will be further applied in the future to much more complex magnetic structures according to the progress of the experimental techniques because of its simple and powerful approach for magnetic structure studies. They proposed the *supermatrix formalism* as a generalized formalism with strictness and applicability directly to the layer structures with any kind of complexity, and demonstrated also an example of the practical analysis [139].

From the Schrödinger equation including the magnetic potential, Eq. (3.2), making use of the *Hamiltonian* for the mth layer, $\mathcal{H}_m = -\hbar^2/(2m_n)\Delta + U_m(z) - \underline{\boldsymbol{\mu}}\,\boldsymbol{B}_m(z)$, leads to the expression for multilayers as $\mathcal{H} = \Sigma\mathcal{H}_m$, where m_n is the neutron mass, and U_m, the nuclear potential, is given by Eq. (1.19). Equation (2.20) and its derivative, or Eqs. (2.22)–(2.25), or further Eq. (3.9) and its derivative, can be written by extending them to magnetic multilayers as

$$
\begin{aligned}
|\boldsymbol{\phi}(0)\rangle &= (1 + \underline{\boldsymbol{R}})|\boldsymbol{t}_0\rangle\,, \\
|\boldsymbol{\phi}'(0)\rangle &= ik_0(1 - \underline{\boldsymbol{R}})|\boldsymbol{t}_0\rangle\,, \\
|\boldsymbol{\phi}(z_n)\rangle &= \underline{\boldsymbol{T}}|\boldsymbol{t}_0\rangle\,, \\
|\boldsymbol{\phi}'(z_n)\rangle &= ik_s\underline{\boldsymbol{T}}|\boldsymbol{t}_0\rangle\,,
\end{aligned}
\tag{3.27}
$$

where the subscript s denotes the substrate.

These formulations were extended from Eq. (3.9) and its derivative to a system with the magnetization in an arbitrary direction, where $|\boldsymbol{\phi}(z)\rangle$ is a two-component eigenvector of the Hamiltonian $\mathcal{H}$ including the spin system, $|\boldsymbol{t}_0\rangle$ being the incident vector, and $\underline{\boldsymbol{R}}$ and $\underline{\boldsymbol{T}}$ are the 2×2 *reflectance matrix* and *transmittance matrix*, respectively.

Extending the formalism similar to Eqs. (3.14) and (3.15), or Eqs. (3.21) and (3.22), to the condition of the magnetic structure with an arbitrary magnetized direction leads to the expression with the *transfer supermatrix* $\mathcal{S}_m$ for each layer and the transfer supermatrix $\mathcal{S}_{\text{tot}}$ for the whole structure in place of these matrices as

$$\begin{pmatrix} |\varphi(z_m)\rangle \\ |\varphi'(z_m)\rangle \end{pmatrix} = \mathcal{S}_m \begin{pmatrix} |\varphi(z_{m-1})\rangle \\ |\varphi'(z_{m-1})\rangle \end{pmatrix}, \quad m = 1 \sim n , \tag{3.28}$$

$$\begin{pmatrix} |\varphi(z_n)\rangle \\ |\varphi'(z_n)\rangle \end{pmatrix} = \mathcal{S}_{\text{tot}} \begin{pmatrix} |\varphi(0)\rangle \\ |\varphi'(0)\rangle \end{pmatrix}, \tag{3.29}$$

where, from the condition of the continuity at the interface, similar to Eq. (3.22),

$$\mathcal{S}_m = \begin{pmatrix} \underline{S}_m^{11} & \underline{S}_m^{12} \\ \underline{S}_m^{21} & \underline{S}_m^{22} \end{pmatrix} = \begin{pmatrix} \cos \underline{k}_m d_m & \underline{k}_m^{-1} \sin \underline{k}_m d_m \\ -\underline{k}_m \sin \underline{k}_m d_m & \cos \underline{k}_m d_m \end{pmatrix}, \tag{3.30}$$

and $\underline{k}_m = \sqrt{k_0^2 - \underline{k}_{mc}^2}$, $\underline{k}_{mc}^2 = 2m_n/\hbar^2 \cdot \mathcal{H}_m$.

Substituting Eq. (3.27) into Eq. (3.29) and eliminating $\underline{T}$, we obtain

$$\underline{R} = \underline{\mathbf{\Delta}}^{-1} \cdot \{(\underline{S}_{\text{tot}}^{22} - i k_0^{-1} \underline{S}_{\text{tot}}^{21})\, k_0 - (\underline{S}_{\text{tot}}^{11} + i k_0 \underline{S}_{\text{tot}}^{12})\, k_s\} , \tag{3.31}$$

where $\underline{\mathbf{\Delta}} = (\underline{S}_{\text{tot}}^{22} + i k_0^{-1} \underline{S}_{\text{tot}}^{21})\, k_0 + (\underline{S}_{\text{tot}}^{11} - i k_0 \underline{S}_{\text{tot}}^{12})\, k_s$.

Now we use the decomposition formula (refer to textbook [B], Chapter 8, Section 58) which is useful for the analyses with the magnetization in an arbitrary direction:

$$\underline{R} = \frac{1}{2}(R_0 + \boldsymbol{R}\underline{\boldsymbol{\sigma}}) , \tag{3.32}$$

$$\underline{S}_m^{\mu\nu} = \frac{1}{2}\left\{\left[S_{m+}^{\mu\nu} + S_{m-}^{\mu\nu}\right] + (\boldsymbol{e}_{Bm}\underline{\boldsymbol{\sigma}})\left[S_{m+}^{\mu\nu} - S_{m-}^{\mu\nu}\right]\right\} , \tag{3.33}$$

where $R_0 = \text{Tr}\{\underline{R}\} = R_+ + R_-$, $\boldsymbol{R} = \text{Tr}\{\underline{R}\boldsymbol{\sigma}\} = (R_+ - R_-)\boldsymbol{b}_r$, $\boldsymbol{b}_r^2 = 1$, $\boldsymbol{e}_{Bm} = \boldsymbol{B}_m/|\boldsymbol{B}_m|$, and $\boldsymbol{B}_m$ is the magnetic induction in the mth layer, $S_{m\pm}^{\mu\nu}$ being the eigenvalue of the matrix $\underline{S}_m^{\mu\nu}$.

By making use of these relations, we finally obtain the next equation on the reflectivity $\mathcal{R}$ measured in experiments:

$$\begin{aligned}
\mathcal{R} &= \overline{|\langle \boldsymbol{r}|\underline{R}|\boldsymbol{t}_0\rangle|^2} = \text{Tr}\left\{\underline{\rho}\,\underline{R}\,\underline{\rho}_0\,\underline{R}^+\right\} \\
&= \frac{1}{8}\left\{|R_0|^2[1 + (\boldsymbol{P}_0\boldsymbol{P})] + |\boldsymbol{R}|^2[1 - (\boldsymbol{P}_0\boldsymbol{P})]\right\} \\
&\quad + \frac{1}{4}\,\text{Re}\left\{R_0^*\,\boldsymbol{R}(\boldsymbol{P}_0 + \boldsymbol{P}) + (\boldsymbol{R}^*\boldsymbol{P}_0)(\boldsymbol{R}\boldsymbol{P})\right\} \\
&\quad - \frac{1}{4}\,\text{Im}\left\{R_0^*\,\boldsymbol{R}(\boldsymbol{P}_0 \times \boldsymbol{P}) + \frac{1}{2}(\boldsymbol{R}^* \times \boldsymbol{R})(\boldsymbol{P}_0 - \boldsymbol{P})\right\} ,
\end{aligned} \tag{3.34}$$

where the density matrices $\underline{\rho}_0 = (1 + \boldsymbol{P}_0\underline{\boldsymbol{\sigma}})/2$, $\underline{\rho} = (1 + \boldsymbol{P}\underline{\boldsymbol{\sigma}})/2$, and further $\boldsymbol{P}_0 = P_0\boldsymbol{n}_0$, and $\boldsymbol{P} = P\boldsymbol{n}$ are the polarization efficiency of polarizer and the polarization analyzer, for the polarization directions $\boldsymbol{n}_0$ and $\boldsymbol{n}$ of incident and analyzed neutrons, respectively.

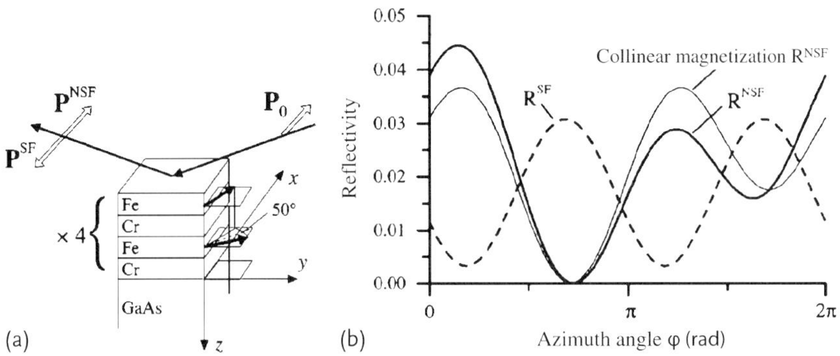

Figure 3.7 Analysis with the transfer supermatrix. (a) A multilayer with a magnetization profile; experimental setup for an FeCr multilayer to indicate the fundamental difference between collinear and noncollinear magnetic profiles, where $\mathbf{P}_0$, $\mathbf{P}^{SF}$, and $\mathbf{P}^{NSF}$ represent incident neutron polarization, spin-flip (SF) and non-spin-flip (NSF) polarization analyses of reflected neutrons, respectively. (b) Results of the analyses; variation of the dependence of the spin-flip and non-spin-flip reflectivity on the azimuthal rotation angle φ of the sample around the axis perpendicular to the sample surface, as calculated for the FeCr multilayer embracing an angle 50° in adjacent Fe layers, and for the experimental setup depicted in (a). The thin solid line represents the non-spin-flip calculation under the assumption of a collinear magnetization profile (Rühm *et al.* [139]).

An example of analyses with the present supermatrix formalism is shown in Figure 3.7. The sample studied consists of 16 layers in total, alternately of 53-Å-thick Fe layers and 17-Å-thick Cr layers, with the magnetizations in adjacent Fe layers embracing an angle of 50°.

The calculations were performed for an assumed measuring condition with the wave vector transfer $2k_\perp = 0.045\,\text{Å}^{-1}$. The effectiveness of the present method for analysis of the experimental data on a sample with *noncollinear magnetizations* is obviously indicted.

Practical polarized neutron experiments on a sample with a significant noncollinearity similar to that assumed in Section 3.2 will be introduced in Chapter 11.

3.3
Neutron Behavior in Magnetic Devices

In previous sections, we studied reflection and transmission of neutrons by magnetic mirrors and magnetic multilayers. Concerning the effects of the magnetic field, the fundamental explanations of the behavior of neutron spin in a magnetic field were also given in Section 1.5. Therefore, in this section, experimental studies on the particle motions of neutrons accompanied by spin motions in a magnetic field will be given. If we can control the particle motion of neutrons with a magnetic field, it will open an interesting area in physics and in applications in view of the fundamental difference from the utilization of neutron scattering.

3.3.1 Magnetic Deceleration and Acceleration of Neutrons

The Schrödinger equation including magnetic interaction of neutrons was given by Eq. (3.2). Therefore, incident neutrons with spin parallel (antiparallel) to the magnetic field in a vacuum and approaching from the left side as shown in Figure 1.12a experience a positive, that is, repulsive (negative, i. e., attractive) potential until the midpoint in the magnetic field, and thus the neutrons are decelerated (accelerated). If the neutron spin is reversed at the midpoint, the effect of the magnetic field is also reversed, and the neutrons are attracted (repulsed) during the passage through the later half of the magnetic field; thus, they decelerated (accelerated) again. The resultant effects of magnetic deceleration (acceleration) correspond to twice the magnitude of the magnetic potential for neutrons.

Such a *magnetic deceleration and acceleration* scheme for neutrons in a strong magnetic field was proposed and preliminarily tested by Utsuro [140] as an attractive approach available for a wide-velocity region by making use of the adiabatic spin reversal described in Section 1.5 with a high-frequency magnetic field simultaneously applied in the device.

The magnitude of the magnetic potential for neutrons in a magnetic field can be calculated from the magnetic dipole moment $\mu_n = -6.031 \times 10^{-12}$ eV/Oe. Thus, for a magnetic field of 1 T, corresponding to a neutron energy of about 0.06 μeV, the detection of the deceleration or acceleration effect will be possible only if incident neutrons with extremely low energy are used, that is, using *ultracold neutrons* with energy below about 1 μeV (velocity below about 10 m/s) or otherwise with the use of an experimental setup with an extremely high energy resolution to detect a very small fractional change below about 10^{-5} of the incident thermal neutron energy.

The former approach was employed by Utsuro in a preliminary experiment with the setup shown in Figure 3.8. The rotating supermirror blades of the turbine provide ultracold neutrons in the velocity region of about 4–30 m/s, as the velocity spectrum given in Figure 2.17b [77, 101]. In this magnetic acceleration experiment, the turbine was modified to remove the blades, except for two diagonal blade packets. It then produce two bursts of ultracold neutrons per revolution. The time-of-flight measurements of ultracold neutrons were performed through a short copper guide with an effective flight path length of about 28 cm and a cross section of $5 \times 2\,\text{cm}^2$. The static magnetic field with a strength of about 2.2 kG and the field gradient of about 53 G/cm with a couple of pole pieces of a permanent magnet and the high-frequency magnetic field of frequency 6.30 MHz and the coil current $I_{pp} = 15$ A satisfy the condition for the adiabatic spin reversal during the passage of ultracold neutrons through the guide.

In the case where the high-frequency field is not applied, all the neutrons in the velocity region beyond the critical velocity for total reflection of magnetized foils for parallel-spin neutrons (about 6 m/s), and half of the incident neutrons (antiparallel spin component) in the velocity region between about 3.5 m/s (critical velocity of the foil substrates) and about 6 m/s will pass through the guide. When the high-frequency field is applied, the former component will again be transmitted since

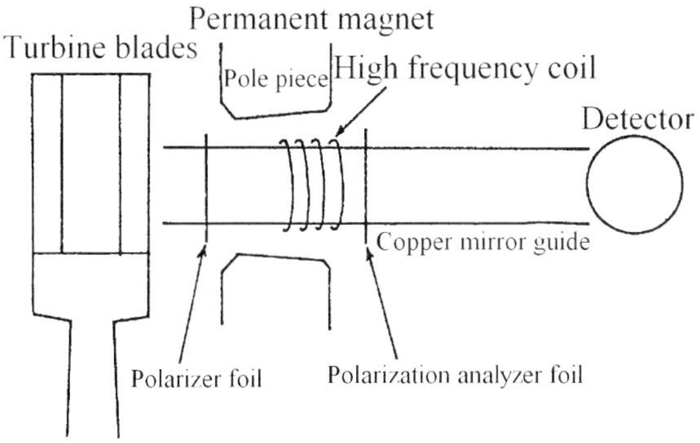

Figure 3.8 Configuration of a preliminary experiment on magnetic deceleration of neutrons (Utsuro [140]).

half of the neutrons (antiparallel spin component) are accelerated and the other half (parallel spin component) have their spin reversed to antiparallel, whereas the latter component will be accelerated with the spin reversed to parallel. A partial intensity decrease in the velocity region of about 3.5–6 m/s is thus induced owing to the reflection by the magnetized second foil. The experimental results [140] gave qualitatively reasonable agreement with these expectations but the statistical accuracy was not enough for further quantitative discussions.

Another experimental approach employing very high resolution spectroscopy on magnetic deceleration and acceleration effects on thermal neutrons was successfully carried out by Weinfurter *et al.* [141] at the Jülich Heavy Water Reactor. The experimental setup for the magnetic action on thermal neutrons is shown in Figure 3.9a. The working volume of the magnetic device is 5-cm diameter × 2.5-cm length, with a static magnetic field of 2.02 T and a gradient of 8 G/cm. The incident neutrons were monochromatized to a wavelength of $\lambda = 0.214$ nm by a copper mosaic crystal (refer to Section 1.3), and with regard to the magnetic device were very precisely analyzed with a *double-crystal spectrometer* of perfect silicon crystals (reflec-

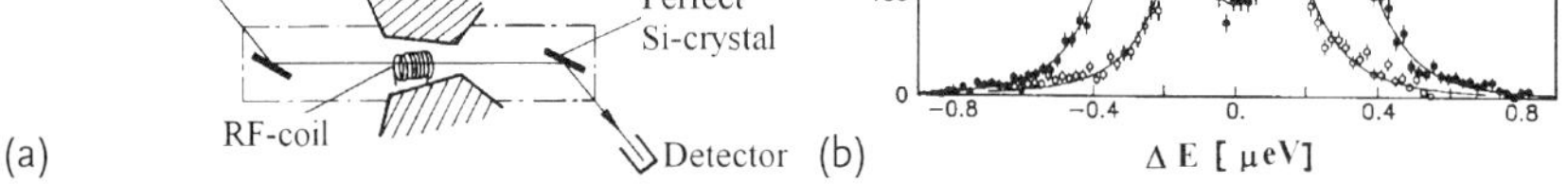

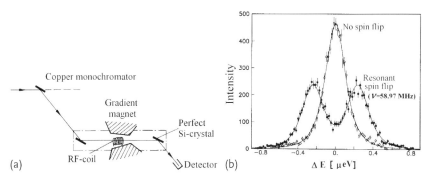

Figure 3.9 Experiment by Weinfurter *et al.* on the energy change of neutrons with adiabatic spin flip in a gradient magnetic field: (a) experimental setup; (b) comparison of energy change with and without spin flip (Weinfurter *et al.* [141]).

tion angle resolution $\sigma_{FWHM} = 6.9$ μrad). Thus, a very slight wavelength change ($\Delta\lambda/\lambda = 6.7 \times 10^{-6}$) could be evaluated from a slight change of the reflection angle of the second silicon crystal. The experimental results are given in Figure 3.9b, indicating the deceleration and acceleration effects with an energy change of about 0.240(5) μeV for parallel and antiparallel spin components, respectively, included in the unpolarized incident neutrons. The energy change measured agrees well with the theoretical expectation of 0.243 μeV. Thus, the phenomenon of neutron energy change induced by an adiabatic spin flip in a gradient magnetic field is well described by the theory, and can be applied to high-resolution spectroscopy and various kinds of energy control for neutrons.

3.3.2
Magnetic Focusing

In the famous *experiment of Stern and Gerlach* [44, 45], the flight path of particles with spin is split in the region with a magnetic field gradient. Application to neutrons was also carried out more than 50 years ago [142]. Such an approach to magnetic control of neutrons is not only very interesting in principle, but can also be expected to be applicable in various schemes very different from the applications of magnetic material mirrors and magnetic multilayers. In the present chapter, we studied the magnetic storage of ultracold neutrons in Section 3.1 and magnetic deceleration and acceleration in the previous subsection. Now, another attractive possibility is the magnetic bending and focusing of neutrons by making careful use of a magnetic potential without disturbing their polarized spin states. Such an experiment will be introduced in this subsection.

First, as mentioned in Section 1.5, a neutron spin in a magnetic field behaves with Larmor precession. The motion of a magnetic dipole in a magnetic field can be generally described by the Bloch equation [143], and the equation of motion for neutron spin was expressed as follows by Mezei [55], who opened the door for application of Larmor precession to neutron spectroscopy:

$$\frac{ds}{dt} = \frac{2|\bar{\mu}_n|\mu_N}{\hbar}[\boldsymbol{H} \times \mathrm{s}] , \tag{3.35}$$

where the spin vector $\mathbf{s} = (s_{x'}, s_{y'}, s_{z'})$, as explained in Section 3.1, represents the expectation value $s_j = \langle\phi(t)|\hat{S}_j|\phi(t)\rangle$ $(j = x', y', z')$ for each component of the neutron spin, and $\phi(t)$ is the wave function for a spin at time t, $\hat{S}_j$ being the operator with regard to the spin component. Further, $\bar{\mu}_n = \mu/\mu_N$ is the magnetic dipole moment of a neutron in units of the nuclear magneton μ_N, and the coefficient $\gamma = 2|\bar{\mu}_n|\mu_N/\hbar$ on the right side of Eq. (3.35) is the *gyromagnetic ratio*.

Therefore, if the change of direction of the magnetic field is sufficiently slow, the spin initially polarized in the direction of magnetic field follows well the magnetic field. Under such an adiabatic condition, Summhammer *et al.* analyzed the expected intensity gain of neutrons focused into a narrow-velocity region by making use of a magnetic potential traveling with the neutron flight and with controlled spatial distribution and temporal progress [144]. Shimizu *et al.* carried out analyses and experiments on the magnetic bending and focusing of neutrons in a static magnetic field with the spatial configuration satisfying the adiabatic condition [145]. Here the latter studies of Shimizu *et al.* will be explained below.

As the starting point, they expressed the motion of a neutron in a magnetic field by the equation [145]

$$\frac{d^2\boldsymbol{r}}{dt^2} = -\alpha\nabla(\boldsymbol{e}_s \cdot \boldsymbol{B}) , \tag{3.36}$$

where $\alpha = |\mu|/m = 5.77\,\mathrm{m^2\,s^{-2}\,T^{-1}}$, and $\boldsymbol{e}_s = \boldsymbol{s}/|\boldsymbol{s}|$ is the unit vector parallel to the neutron spin. Substituting the magnetic dipole moment of the neutron $\bar{\mu}_n = -1.913$ into Eq. (3.35), or the gyromagnetic ratio $\gamma = 1.833 \times 10^8\,\mathrm{rad\,s^{-1}\,T^{-1}}$ into the coefficient on the right-hand side, and further combining with Eq. (3.36), one can solve for both the motion of the neutron itself and the neutron spin.

The angular velocity of rotation of the magnetic field direction in the frame moving with the neutron is given by $\omega_B = |\partial \boldsymbol{e}_B/\partial q| \cdot \partial q/\partial t$, where q is the coordinate along the trajectory of the neutron and $\boldsymbol{e}_B = \boldsymbol{B}/|\boldsymbol{B}|$. If the magnetic field is so strong that the neutron spin *adiabatically follows the magnetic field*, then Eq. (3.36) is simplified to

$$\frac{d^2\boldsymbol{r}}{dt^2} = \mp\alpha\nabla|\boldsymbol{B}| , \tag{3.37}$$

where the upper and lower signs correspond to parallel spin and antiparallel spin, respectively, in the magnetic field.

As an example of a practical configuration, incident neutrons enter in the direction of the z-axis from an entrance pinhole in the arrangement shown in the upper part of Figure 3.10 into the *sextupole magnetic field* defined by the equation

$$\boldsymbol{B} = \frac{C}{2}\begin{pmatrix} y^2 - x^2 \\ 2xy \\ 0 \end{pmatrix} , \tag{3.38}$$

where C is a constant. Then, using the abbreviated notation as $dx/dt = \dot{x}$ and so on, Eq. (3.37) can be simplified as

$$\ddot{x} = \mp\omega^2 x , \quad \ddot{y} = \mp\omega^2 y , \quad \ddot{z} = 0 , \tag{3.39}$$

and the solution becomes

$$\begin{pmatrix} x & y \\ \xi & \eta \end{pmatrix} = \underline{\boldsymbol{M}} \begin{pmatrix} x_0 & y_0 \\ \xi_0 & \eta_0 \end{pmatrix} , \tag{3.40}$$

$$\underline{\boldsymbol{M}} = \begin{pmatrix} \cos\theta & \sin\theta \\ -\sin\theta & \cos\theta \end{pmatrix} , \quad \text{parallel spin} , \tag{3.41}$$

$$= \begin{pmatrix} \cosh\theta & \sinh\theta \\ \sinh\theta & \cosh\theta \end{pmatrix} , \quad \text{antiparallel spin} , \tag{3.42}$$

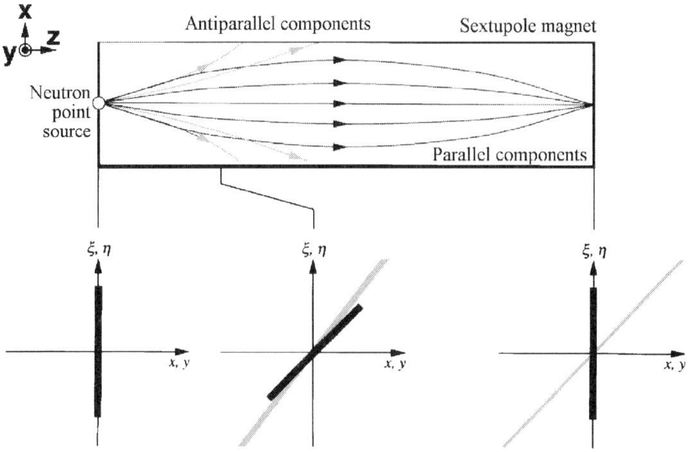

Figure 3.10 Principle of neutron beam focusing with a sextupole magnetic field. Upper part: trajectories of neutrons in the magnetic field emitted from a pinhole source at the center of the magnetic field entrance. Lower part: trajectories of neutrons described in the coordinate system of the $x\xi$-plane and the $y\eta$-plane at each section, where black and gray lines correspond to parallel and antiparallel components, respectively, to the local magnetic field. (Shimizu *et al.* [145]).

where $\omega^2 = C\alpha$, $\xi = \dot{x}/\omega$, $\eta = \dot{y}/\omega$, $\theta = \omega t$, and the subscript 0 denotes the quantity for $\theta = 0$.

As introduced in the previous section, the matrix $\underline{\boldsymbol{M}}$ for parallel-spin neutrons rotates the neutron beam by an angle $-\theta$ in the $x\xi$- and $y\eta$-planes. This situation is described in the lower part of Figure 3.10, and as illustrated in the figure, parallel-spin neutrons starting at the point source with $\theta = 0$ will be *focused* again on a point at the condition $\theta = \pi$ in the device. On the other hand, antiparallel-spin neutrons will be defocused and diverge in the direction of (1,1), whereas they will be focused in the direction of $(1, -1)$. However, according to *Liouville's theorem*, the phase space density of neutrons does not change owing to any operations of collective forces, and therefore the phase space area of the element in the $x\xi$- and $y\eta$-planes will not change. Even within the present restriction, the proposed scheme will give us a great gain in the neutron intensity compared with the usual divergent beam when we go far from the source point without such focusing devices if we can again focus the beam to a localized point at a position far from the primary source.

The experiment to test the focusing effect with a sextupole magnet was carried out by measuring the intensity of transmitted cold neutrons produced by a liquid-hydrogen moderator at the 46-MeV electron linac of Hokkaido University. The neutrons were introduced into the sextupole magnetic device with dimensions of 10-mm inner diameter, 48-mm outer square, and 2-m length consisting of a number of NEOMAX48 permanent magnet elements with a size of $5 \times 5 \times 50\,\text{mm}^3$. The magnetic field strength was measured as $C = (3.5 \pm 0.6) \times 10^4\,\text{T}\,\text{m}^{-2}$. The experimental results obtained with the setup of a 2-mm-diameter entrance hole

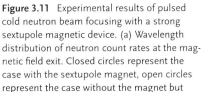

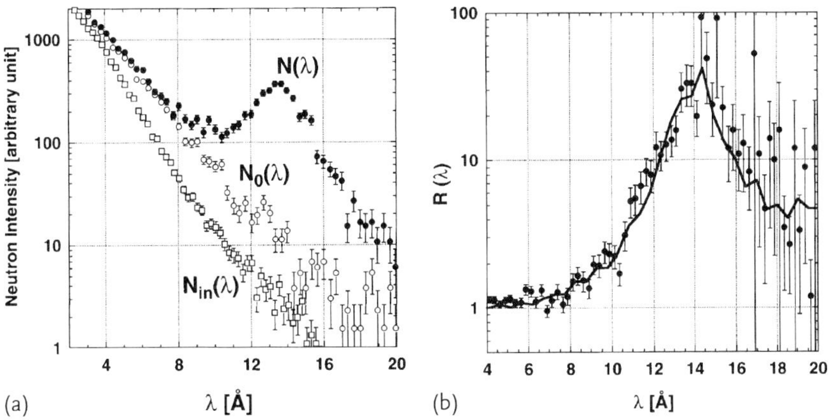

Figure 3.11 Experimental results of pulsed cold neutron beam focusing with a strong sextupole magnetic device. (a) Wavelength distribution of neutron count rates at the magnetic field exit. Closed circles represent the case with the sextupole magnet, open circles represent the case without the magnet but with the inner elements, and open squares represent the case without the inner elements. (b) Wavelength dependence of the magnetic gain factor denoted as $R(\lambda) = N(\lambda)/N_0(\lambda)$, compared with the result of the theoretical calculation (solid curve) (Shimizu *et al.* [145]).

about 1.7 m from the primary source and the detector at the position about 20 mm from the same 2-mm-diameter exit hole on the extension line of the center axis are plotted in Figure 3.11a as the wavelength dependence of the transmitted neutron intensity. The result indicates an obvious peak due to the magnetic gain at a wavelength of $\lambda \cong 14.4$ Å, and the gain factor shown in Figure 3.11b agrees well with the theoretical calculation over the whole wavelength region. The observed focusing performance for 14.4-Å neutrons with the 2-m-long device corresponds to the magnetic parameters $\omega = 432\,\mathrm{rad\,s^{-1}}$ and $C = 3.2 \times 10^4\,\mathrm{T\,m^{-2}}$, which are in reasonable agreement with the measured magnetic field strength. Further measurements on and analyses of the details of focusing characteristics and the degree of adiabaticity in neutron spin behaviors were also studied.

Further, these characteristics of magnetic focusing and control for neutrons can be generalized to other kinds of magnetic structures [146]. A generalized relation is derived as follows for the magnetic field configuration in a multipole magnet and the trajectory curvature for neutrons:

$$|\boldsymbol{B}| = B_n \left(\frac{\rho}{\rho_0}\right)^n , \tag{3.43}$$

$$\begin{cases} \frac{d^2 X}{d\theta^2} = \mp X(X^2 + Y^2)^{n/2-1} , \\ \frac{d^2 Y}{d\theta^2} = \mp Y(X^2 + Y^2)^{n/2-1} , \end{cases} \tag{3.44}$$

where $\theta = \omega t$, $\omega^2 = n B_n \mu / m\rho^2$, $X = x/\rho_0$, and $Y = y/\rho_0$, $n = 1, 2, 3, \ldots$, refer to quadrupole, sextupole, octupole, and so on, respectively, of the multipole magnetic field. The upper and lower signs correspond to parallel spin and antiparallel

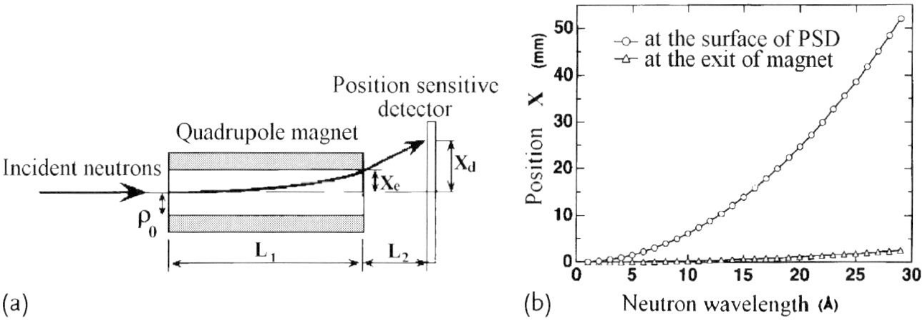

Figure 3.12 Neutron energy analysis with a quadrupole magnetic prism: (a) arrangement of experimental device, where $L_1 = 0.4$ m and $L_2 = 4$ m, and the magnet elements are essentially same as those in Figure 3.11; (b) measured wavelength dependence of neutron beam bending. PSD position-sensitive detector (Oku *et al.* [147]).

spin in the magnetic field. As an example, an experiment using a *magnetic prism* was performed by Oku *et al.* to bend a neutron beam to a position depending on the wavelength by making use of a *quadrupole magnetic field* [147]. The experimental setup and the result are shown in Figure 3.12. The effective separation of the neutron beam depending on the wavelength is obvious as expected from the relation given above. The energy resolution measured was about 0.5 μeV for a spatial resolution of 0.5 mm over the whole energy region of cold neutrons.

3.3.3
Pseudomagnetic Prism

In the ordinary optical prism, the incident beam is bent with an angle depending on the wavelength in transmission through the apex of a triangular rod, and can be used for analyzing the wavelength component in the beam. The magnetic prism for neutrons such as introduced in the previous Section provides spin- and wavelength-dependent separation of the optical path by making use of the effect of a magnetic field gradient or magnetic scattering. Now, spin-dependent separation of the optical path can be used for neutrons by making use of the effect of the spin-dependent nuclear scattering potential. Zimmer *et al.* carried out such an experiment to observe the *Stern–Gerlach effect without using a magnetic field gradient* [148] for neutrons passing through the apex of an angular rod with nuclear scattering in place of a magnetic field gradient.

Although the nuclear scattering potential U_m for a neutron was given by Eq. (1.19), this was a spin-independent formula. For the case including spin-dependent nuclear scattering, the formula becomes

$$U = \sum_{i=1}^{M} \frac{\hbar^2}{2m} 4\pi N_i b_0^i - \underline{\boldsymbol{\mu}} \cdot \left[- \sum_{i=1}^{M} \frac{\hbar^2}{2m\gamma} 4\pi N_i b_N^i I_i \underline{\boldsymbol{P}}_i \right] , \tag{3.45}$$

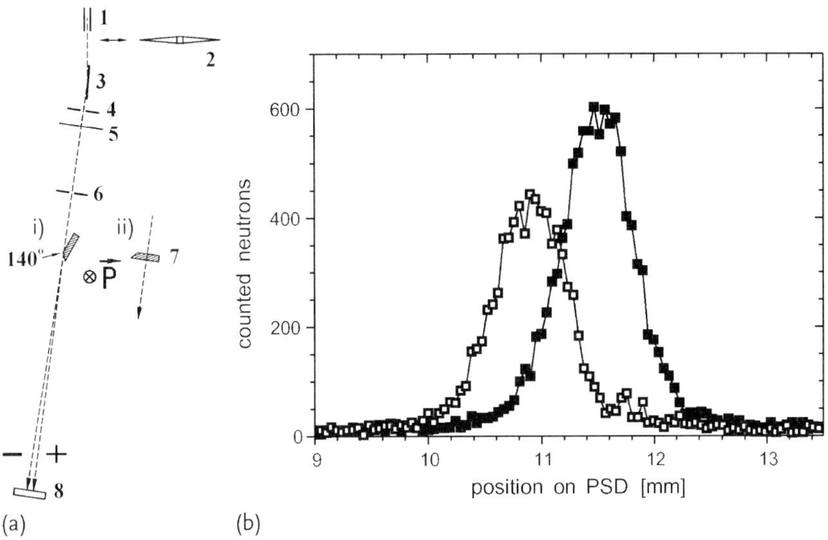

Figure 3.13 Stern–Gerlach effect observed with a pseudomagnetic prism. (a) Configuration of the experiment. 1 incident neutron beam, 2 chopper, 3 supermirror polarizer, 4 first slit, 5 spin flipper, 6 second slit, 7 nuclear polarized target plate, i) prism setting, ii) filter setting, 8 position-sensitive detector (PSD). (b) experimental results of spin-dependent splitting of a neutron beam due to a pseudo-magnetic field. Solid squares, parallel spin neutrons; open squares, antiparallel neutrons (Zimmer *et al.* [148]).

where the scattering amplitude operator extended to the spin-dependent expression $\hat{\underline{b}}_i = b_0^i + b_N^i \underline{s} \cdot \underline{I}_i$ was rewritten according to the magnetic potential formalism in the present chapter as composed of spin-independent term b_0^i and the term with the nuclear polarization $\underline{P}_i = \langle \underline{I}_i \rangle / I_i$ of a nuclear spin $\underline{I}_i$. Further, γ is the *gyromagnetic ratio* of the neutron.

Their experiment was carried out with the configuration shown in Figure 3.13a, where making use of a small refraction angle δ derived from the refractive index formula, Eq. (1.16), leads to the expression for the splitting angle, that is, the Stern–Gerlach effect, as

$$\Delta\delta = 2|n_+ - n_-| \tan\frac{\alpha}{2} = \frac{\lambda^2}{2\pi} N b_N |P| \tan\frac{\alpha}{2} , \tag{3.46}$$

where α is the apex angle of the prism and P is the nuclear polarization ratio. The spin-dependent scattering amplitude b_N is particularly large in the case of a proton, being $b_N = (5.8236 \pm 0.0014) \times 10^{-12}$ cm. Therefore, in their experiment a polarized hydrogenous compound was employed.

The prism used in the experiment was a mixture of 96 wt% propanediol and 4 wt% H_2O, doped with Cr(V) complexes, and frozen in the form of a plate with dimensions $18 \times 18 \times 3\,\text{mm}^3$. The latter component is needed to apply the technique of dynamical nuclear polarization. A special treatment was performed on the plate to obtain optically flat surfaces. The measurement was carried out with the thermal neutron beam F2 at the Technical University of Munich reactor. In the configuration shown Figure 3.13a, the incident beam is defined by the first slit, with width 200 μm, and the second slit 1.1 m away and with width 150 μm, and the detector was placed a further 3.59 m away. A supermirror polarizer selects polarized neutrons with a wavelength of 6 Å and resolution of $\Delta\lambda/\lambda \cong 30\%$ was used. The time-of-flight measurements indicated the resultant average wavelength and the wavelength width to be 5.80 ± 0.05 Å. From the experimental results on the Stern–Gerlach effect shown in Figure 3.13b and the distance between the prism and the detector of 287 cm, the split angle in the experiment is derived as $\Delta\delta = (2.063 \pm 0.035) \times 10^{-4}$ and the nuclear polarization ratio as $P = -35.9 \pm 1.0\%$. The negative sign in the nuclear polarization ratio comes from the polarized incident neutrons employed. Furthermore, the present results correspond to a *pseudomagnetic field* strength of about 1.5 T, and the difference between refractive indices is $|n_+ - n_-| = 3.8 \times 10^{-5}$.

3.4 Measurement of Larmor Precession

3.4.1 4π Periodicity of Spinor

One of the most important indications among the fundamental properties of *Larmor precession* is the result arising from *rotation by an angle of* 2π *radians* on a spin-1/2 system. In this subsection, one of the experiments in which the result was obtained by measurement will be described.

A theoretical introduction will be helpful before the description of the experiment. We define a vector angle $\boldsymbol{\alpha}$ to represent rotation by an angle $\alpha = |\alpha|$ in the right-handed rotation system around the axis directed toward the unit vector $\hat{\boldsymbol{\alpha}}$. The quantum mechanical *rotational operation* $R(\boldsymbol{\alpha})$ on an arbitrary system is given by

$$R(\boldsymbol{\alpha}) = \exp(-i\boldsymbol{\alpha} \cdot \boldsymbol{J}) , \tag{3.47}$$

where $\boldsymbol{J}$ is an angular momentum operator and determines the transfer property for rotation of the system [149]. In the case of a spin-1/2 system, according to the *Pauli matrices* $\underline{\boldsymbol{\sigma}} = (\underline{\sigma_x}, \underline{\sigma_y}, \underline{\sigma_z})$ given by Eq. (3.3), $R(\boldsymbol{\alpha})$ is a 2×2 matrix $\underline{r}(\boldsymbol{\alpha})$, and then it can be written as

$$\underline{r}(\boldsymbol{\alpha}) = \exp(-i\boldsymbol{\alpha} \cdot \underline{\boldsymbol{\sigma}}/2) . \tag{3.48}$$

Further, from the commutation relations on the Pauli matrices,

$$\begin{aligned} &\underline{\sigma_x \sigma_y} = i\underline{\sigma_z} , \quad \underline{\sigma_y \sigma_z} = i\underline{\sigma_x} , \quad \underline{\sigma_z \sigma_x} = i\underline{\sigma_y} , \quad \underline{\sigma^2_{x,y,z}} = 1 , \\ &\underline{\sigma_i \sigma_j} = -\underline{\sigma_j \sigma_i} , \end{aligned} \tag{3.49}$$

the next relation is derived for an arbitrary function $f(\boldsymbol{a}\underline{\boldsymbol{\sigma}})$ for an arbitrary vector $\boldsymbol{a}$:[9)]

$$f(\boldsymbol{a}\underline{\boldsymbol{\sigma}}) = \frac{1}{2}[f(a) + f(-a)] + \frac{\boldsymbol{a}\underline{\boldsymbol{\sigma}}}{2a}[f(a) - f(-a)] . \tag{3.50}$$

And as the typical case

$$\exp(i\boldsymbol{a}\underline{\boldsymbol{\sigma}}) = \underline{\mathbf{1}}\cos a + i\frac{(\boldsymbol{a}\underline{\boldsymbol{\sigma}})}{a}\sin a ; \tag{3.51}$$

therefore, we can write

$$\underline{r}(\boldsymbol{\alpha}) = \underline{\mathbf{1}}\cos\frac{\alpha}{2} - i\hat{\boldsymbol{\alpha}}\cdot\underline{\boldsymbol{\sigma}}\sin\frac{\alpha}{2} . \tag{3.52}$$

In this way, operating the rotation $\alpha = 2\pi$ on the wave function ϕ leads to the result

$$\phi(\boldsymbol{x}) \to \phi'(\boldsymbol{x}) = -\phi(\boldsymbol{x}) . \tag{3.53}$$

The possibility of experimental observation of the sign reversal with 2π rotation was proposed in 1967 [150, 151]. In 1976, Klein and Opat demonstrated it by measurement of *Fresnel diffraction* of a neutron beam [152, 153]. Since the observed quantities in many physical experiments are squares of the corresponding wave function, the present phase factor -1 could not be measured in such cases. However, if the wave function is divided into two partial waves and either one is rotated by an angle of 2π radians coherently to the other, thereafter both partial waves are superposed, and we will observe some noticeable difference. The experimental arrangement employed by Klein and Opat is shown in Figure 3.14a.

The experiment was performed with the high-flux reactor at the Institut Laue–Lanvegin to observe the Fresnel diffraction by magnetic domain boundaries in a ferromagnetic film. The high-flux beam of unpolarized monochromatic cold neutrons enters from the left side to the setup in the figure, where the lateral coherence of the beam was defined by passing it through a slit 5 μm wide and 2 mm high. The iron foil is a carefully oriented and aligned ferromagnetic crystal containing long and straight *Bloch walls*, the boundaries of which separate domains of opposite magnetization as illustrated in the figure. Therefore, in traversing the foil on either side of a domain boundary, the neutron spin precesses in opposite directions. The two parts of the wave function, as the trajectories drawn by thin lines in the figure, then acquire a relative phase shift, which leads to the appearance of a Fresnel diffraction pattern in the plane of observation. When the two parts traverse a foil of thickness d containing magnetic fields $+B$ and $-B$ on the left and right of the domain boundary, respectively, the relative angle α of precession of the neutron spin is given by

$$\alpha = \alpha_L - \alpha_R = 2\omega_L\frac{d}{v} = 2\mu B\frac{dm\lambda}{\pi\hbar^2} , \tag{3.54}$$

9) This fundamental relation can be easily derived, for example, separating the Tailor expansion of $f(\boldsymbol{a}\underline{\sigma})$ to even orders and odd orders, the former becomes $[f(a) + f(-a)]/2$, whereas the latter reduces to $[f(a) - f(-a)]/2 \times \boldsymbol{a}\underline{\sigma}/a$.

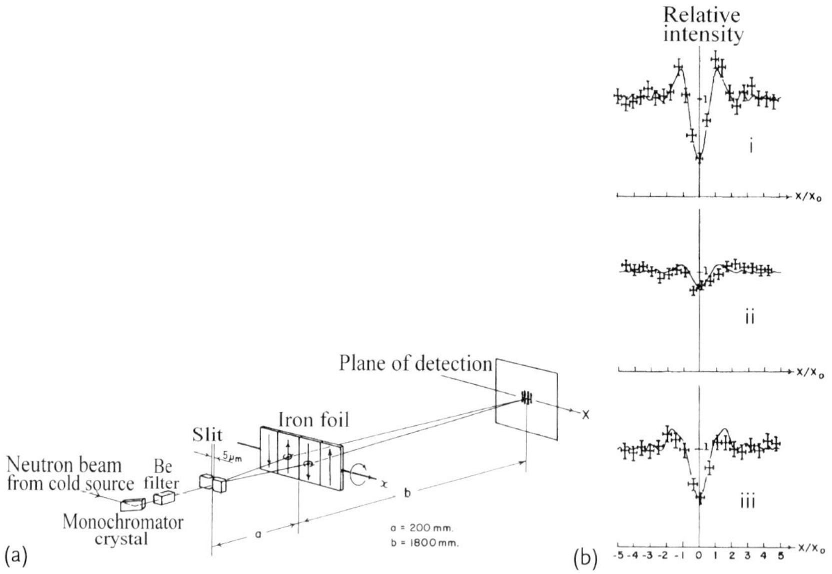

Figure 3.14 Neutron beam experiment indicating the result of rotation on a spin-1/2 system: (a) experimental arrangement; (b) Fresnel diffraction interference pattern for neutrons in the case of a relative precession angle for the spin: i $9.2 \times 2\pi$ rad, ii $10.2 \times 2\pi$ rad, iii $11.2 \times 2\pi$ rad (Klein and Opat [153]).

where ω_L, μ, m, v, and λ are the Larmor precession angular frequency, magnetic dipole moment, mass, velocity, and wavelength of the neutron, respectively. According to the theory, a destructive interference is expected when precession angle α in Eq. (3.54) is an odd multiple of 2π.

In the experiment, cold neutrons with wavelength $\lambda = 0.433$ nm and a wavelength resolution of $\Delta\lambda/\lambda = 2.3\%$ were used. The iron foil cut from a sheet of Fe-3% Si was 72 ± 1 μm thick and was mechanically and chemically polished. The domain boundaries were carefully aligned to be parallel with the slit to better than ± 2 mrad. The internal field B in the iron foil specimen was 1.98 T. A typical result of the measurements is shown in Figure 3.14b, part i, in comparison with the calculated result using standard Fresnel diffraction theory. Parts ii and iii in Figure 3.14b show similar results but for the foil tilted backward about the horizontal x-axis so as to present a greater effective thickness to the transmitted neutrons. From these results, the expected effect of destructive interference in the case of α being odd multiples of 2π was obviously recognized, and in the case of even multiples such a destructive interference tended to disappear.

3.4.2
Coherent Superposition of Spin States

Now, we would like to look at the experiment in which the principle of *superposition of spin states* for a fermion with spin 1/2 was studied. The fundamental interest in what would result from such an experiment is related to an experiment proposed by Wigner [154]. He presented a theoretical consideration on the observation in quantum mechanics concerning the neutron spin after the so-called *double Stern–Gerlach experiment,* that is, first making neutrons spin in the direction of x-axis, then splitting them into parallel and antiparallel states in the direction of the z-axis with a magnetic field gradient, and finally bringing them together again by a magnetic field produced by the current in a cable. The discussion is whether the final state is measured as a mixture of the two intermediate states, that is, parallel and antiparallel states in the z-direction, or as the recovery of the initial state polarized in the x-direction with no z-component. Summhammer *et al.* [155, 156] and Badurek *et al.* [155–157] carried out the experiment corresponding to that proposed by Wigner by making use of the experimental method of *interferometry with a perfect silicon single crystal.* Various kinds of physical experiments with neutron interferometry will be described in detail in Chapter 9, but the present experiment by Summhammer *et al.* and Badurek *et al.* will be appropriate to introduce here as the fundamental properties of Larmor precession.

In the experimental arrangement shown in Figure 3.15a, a three-wedge type of optical element prepared from a large silicon single crystal block was used as a triple-Laue interferometer. The three wedges were cut out exactly perpendicular to the lattice planes of the silicon crystal and with exactly equal thickness and equal separations. The incident neutrons were monochromatized with a wavelength of 1.835 Å and polarized by a magnetic prism. Proper angular setting of the interferometer crystal allowed both of the coherent partial waves Laue-reflected by the first wedge either in the spin-up or in the spin-down state with respect to the z-axis to be chosen to coincide with the direction of a magnetic guide field by a Helmholtz coil pair covering the whole setup. In the same way, the second wedge produces split four partial waves, and the third wedge brings two of them, labeled I and II, exactly together. Thus, each of the resultant two partial waves after the third wedge, that is, labeled as O-beam and H-beam in Figure 3.15a, is a superposition of two different intermediate states I and II. Rotating the phase shifter inserted between the first and second wedges makes it possible to adjust the optical path difference between states I and II, resulting a phase difference χ due to the refractive index given by Eq. (1.23).

Further, by actuating the spin flipper shown in the figure, the spin polarized in the direction of the z-axis rotates by 180° around the y-axis; thus, the state $|\mathrm{I}\rangle$ of the partial waves in path I becomes

$$|\mathrm{I}\rangle = \underline{r}(\pi, e_y)e^{i\chi}|\mathrm{II}\rangle = -i\underline{\sigma_y}e^{i\chi}|\uparrow_z\rangle = e^{i\chi}|\downarrow_z\rangle \,, \tag{3.55}$$

in contrast to that of path II, to be written as $|\mathrm{II}\rangle = |\uparrow_z\rangle$, where $\underline{r}$ is the rotation matrix defined by Eqs. (3.48)–(3.52) and $\underline{\sigma_y}$ is the Pauli matrix. Therefore, the O-

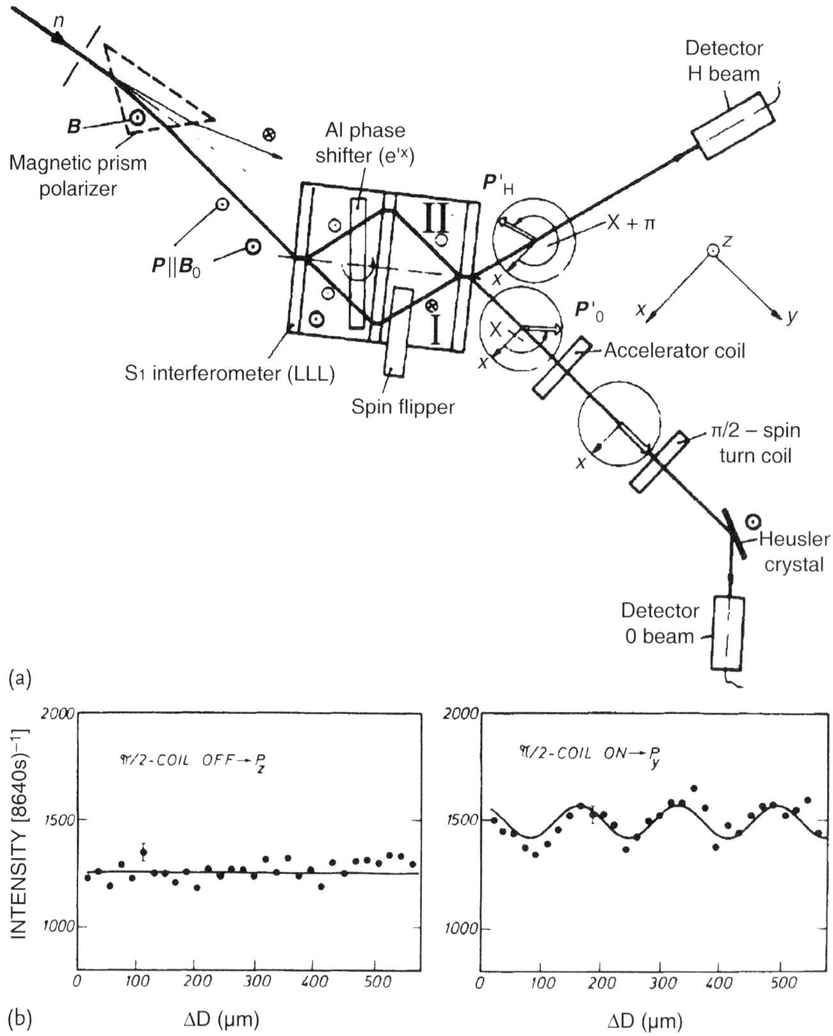

Figure 3.15 Experiment of coherent superposition of neutron spin states; (a) experimental arrangement; (b) experimental results of the forward beam (O-beam) intensity behind the analyzer versus the path difference between the two interfering partial waves I and II. For the oscillating interference pattern observed only in the case of the $\pi/2$-spin turn coil in action the y-component replaces the z-component. Solid lines are the results of least-squares fits of sinusoidal curves to the experimental values. (Summhammer *et al.* [155, 156]).

beam and H-beam arriving at the detector are represented as

$$
\begin{aligned}
|O\rangle &= \frac{1}{2}|\uparrow_z\rangle + \frac{1}{2}e^{i\chi}|\downarrow_z\rangle = \frac{e^{i\chi/2}}{\sqrt{2}}\left(\cos\frac{\chi}{2}\,|\uparrow_x\rangle - i\sin\frac{\chi}{2}|\downarrow_x\rangle\right) \\
&= 2^{-1/2}\left|\frac{\pi}{2},\chi\right\rangle,
\end{aligned}
\tag{3.56}
$$

$$|H\rangle = \frac{1}{2}|\uparrow_z\rangle + \frac{1}{2}e^{i\pi}e^{i\chi}|\downarrow_z\rangle = \frac{e^{i\chi/2}}{\sqrt{2}}\left(\cos\frac{\chi}{2}|\downarrow_x\rangle - i\sin\frac{\chi}{2}|\uparrow_x\rangle\right)$$
$$= 2^{-1/2}\left|\frac{\pi}{2}, \chi + \pi\right\rangle . \quad (3.57)$$

The present equations can be derived by making use of the expansion

$$|\uparrow_z\rangle = \frac{1}{\sqrt{2}}(|\uparrow_x\rangle + |\downarrow_x\rangle), \quad |\downarrow_z\rangle = \frac{1}{\sqrt{2}}(|\uparrow_x\rangle - |\downarrow_x\rangle), \quad (3.58)$$

on the eigenstates

$$|\uparrow_z\rangle = \begin{pmatrix}1\\0\end{pmatrix}, \quad |\downarrow_z\rangle = \begin{pmatrix}0\\1\end{pmatrix}, \quad (3.59)$$

of the Pauli matrix $\underline{\sigma_z}$ to the eigenstates $|\uparrow_x\rangle$ and $|\downarrow_x\rangle$ of $\underline{\sigma_x}$, where

$$\underline{\sigma_x}|\uparrow_x\rangle = |\uparrow_x\rangle = \begin{pmatrix}1\\1\end{pmatrix}, \quad \underline{\sigma_x}|\downarrow_x\rangle = -|\downarrow_x\rangle = -\begin{pmatrix}1\\-1\end{pmatrix}, \quad (3.60)$$

and considering the phase change by $\pi/2$ occurring for every reflection in the interferometer, with further representation of $|\theta,\varphi\rangle = \cos\frac{\theta}{2}|\uparrow_z\rangle + e^{i\varphi}\sin\frac{\theta}{2}|\downarrow_z\rangle$. Equations (3.56) and (3.57) indicate that the polarization of the beam emerging from the interferometer lies on the x–y plane, and rotates on the plane depending on the change of the effective thickness of the nonmagnetic phase shifter. In this way, the result obtained with the present experimental setup will give us an obvious demonstration of the properties of the coherent superposition of spin states.

The experiment was performed at the high-flux reactor in Grenoble, under special precautions to minimize possible temperature variation and a temperature gradient in the experimental apparatus, since only a slight deformation of the silicon crystal should induce a significant phase shift on the neutrons in the optical path. The polarization ratio in paths I and II was higher than 81 and −87%, respectively. The experimental results are plotted in Figure 3.15b, and agree well with the theoretical expectation of quantum mechanical superposition of spin states, given by Eq. (3.56), that is, a periodic pattern observed in the y-component depending on the phase difference caused by the nonmagnetic phase shifter, whereas there is no such pattern in the z-direction.

In the experiment of Summhammer *et al.* and Badurek *et al.* mentioned above, splitting of neutron waves was performed in the same spin state and thereafter the spin state was flipped for one of the waves. Another scheme for the spin state superposition experiment was employed by Ebisawa *et al.* They split neutron waves by the spin states [158], by making use of a *multilayer spin splitter* in place of the Laue-reflection splitter with a silicon single crystal. As shown in Figure 3.16a, the multilayer splitter was prepared by evaporating first a nonmagnetic reflection layer on a silicon substrate, next a gap layer of Ge with a designed thickness, and finally a reflection layer of magnetic material. Employment of the present multilayer spin splitter instead of the first and third wedges in the arrangement shown in

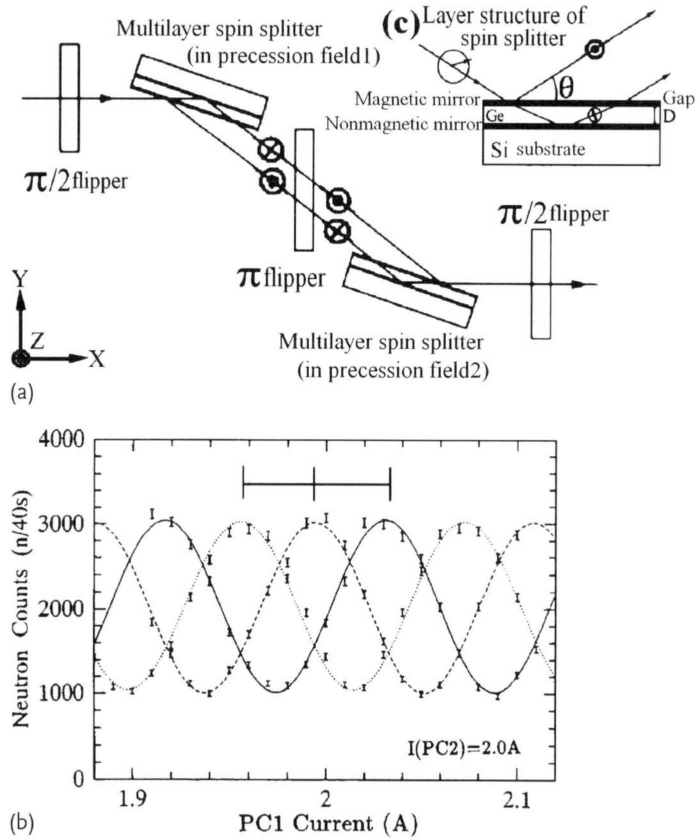

Figure 3.16 Experiment of coherent superposition of cold neutron spin states with multilayer spin splitters: (a) configuration of the spin splitter, π-flipper, and spin superposition; (b) count rates of neutrons from the polarization analyzer after spin superposition versus the current of precession field 1 (PC1), solid, broken, and dotted lines correspond to the least-squares sinusoidal fitting to the experimental data for the setting angle of the spin splitter successively changed by an increment of 0.03°, respectively, around about 1.76°; c) layer structure of multilayer spin splitters (Ebisawa *et al.* [158]).

Figure 3.15a provides spin splitting of incident neutrons as well as superposition of split partial waves, and insertion of the spin flipper into one of the split paths is not necessary. Therefore, by inserting the configuration shown in Figure 3.16a between the polarizer and the polarization analyzer in Figure 3.15a, one can vary the phase difference between the two partial waves by changing the *precession field* current or changing slightly the setting angle of the multilayer spin splitter. Such neutron spin interferometry experiments were developed by Ebisawa *et al.* at the cold neutron source facility of the Kyoto University reactor, and progressed further at the neutron guide hall of the JRR-3M reactor, Japan Atomic Energy Agency, Tokai. Some examples of the experimental results are shown in Figure 3.16b.

In the setup shown in Figure 3.16a, the neutrons with the spin rotated to lie on the x–y plane by the first $\pi/2$ flipper are decomposed into $+z$ and $-z$ components by the first spin splitter, and trace different optical paths, and next the polarizations are reversed by a π flipper working in a phase-echo scheme, and are then finally superposed by a second spin splitter by tracing the equivalent optical paths to the first splitter, to return again to the spin state lying on the x–y plane. Splitting to produce $\pm$ spin states in the region between the two spin splitters and superposition at the second splitter can be assured by the periodic pattern and the phase change of the pattern observed for slight angular shifts of the spin splitter, as shown in Figure 3.16b. Various kinds of physical studies making use of the present setup of the phase echo spin interferometer will be described in Section 9.6. A variety of neutron optical experiments based on Larmor precession will also be introduced in Chapter 9.

3.5
Neutron Spectroscopy with Larmor Precession

3.5.1
Neutron Resonance Spin Echo Method

In the present section, recent developments on the application of Larmor precession to neutron spectroscopies will be described. As already mentioned at the end of Chapter 1, the *neutron spin echo (NSE) method* [55] as a method of high-resolution spectrometry by analyzing a slight change in the Larmor precession angle was an epoch-making invention to attain very high energy resolution without undesirable sacrifice of the selected energy width of neutrons. In this subsection, one of the further recent developments, the *neutron resonance spin echo (NRSE) method*, will be introduced.

Gähler and Golub [159] proposed an advantageous spectrometric method for neutrons by making use of a completely different magnetic field structure as shown by solid lines in Figure 3.17, in contrast to that shown by broken lines used in the conventional spin echo method. In the case of the conventional NSE method, incident neutrons experience Larmor precession first in flight path l_1, then the neutron spin is reversed at the midpoint of the total path, and finally they attain Larmor precession during flight path l_2 in the opposite direction. Thus, the velocity change induced by the interaction with a sample put at the midpoint can be measured by detecting the change in the Larmor precession angle at the end of the flight path. In the proposed new NRSE method, in contrast, incident neutrons polarized parallel to the downward magnetic field with magnitude B_0 as shown in Figure 3.17 first pass through the incident spin rotator C_1. The magnetic field in spin rotator C_1 is directed perpendicular to $\boldsymbol{B}_0$ and rotates with the Larmor precession angular frequency ω_0 of neutrons in B_0. Further, the high-frequency coil current I_a of C_1 is adjusted so as to rotate the neutron spin by exactly $\pi/2$ in the frame rotating with angular frequency ω_0 according to Eq. (3.35); thus, the transmitted neutrons

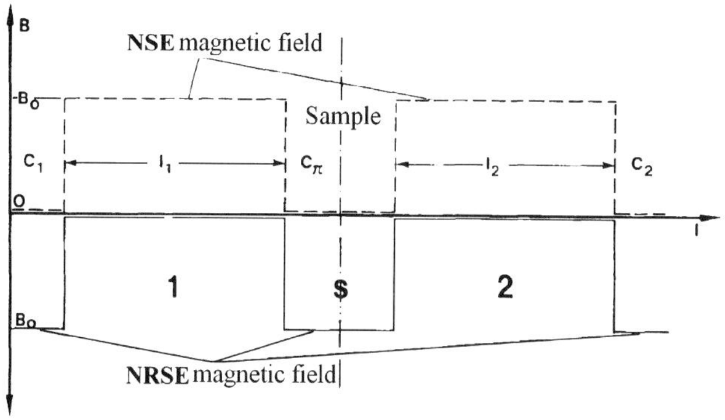

Figure 3.17 Comparison of working principles of the proposed neutron resonance spin echo (NRSE) method and the conventional neutron spin echo (NSE) method (Gähler *et al.* [159]).

are distributed in spin states $+1/2$ and $-1/2$ with equal probability, as indicated by Güttinger's formula, or the more general formula, Eq. (1.28). The high-frequency coil of the exit spin rotator C_2 works in the same way and the phase is also synchronized to that of C_1. The present operating condition corresponds to that of *Ramsey's separated oscillating field method* described in Section 1.5.

Further, in the present flight path between the separated oscillating magnetic fields, there are three magnetic fields, B_1, B_s, and B_2, all acting in the same direction as $\boldsymbol{B}_0$. In this setup, as the magnetic field in C_2 rotates with ω_0 during the flight period of neutrons in paths l_1, l_s, and l_2 which precess in fields B_1, B_s, and B_2, respectively, a phase difference will arise between the spin precession and the magnetic field rotation as

$$\begin{aligned}\Delta\varphi &= \omega_1 t_1 + \omega_s t_s + \omega_2 t_2 - \omega_0(t_1 + t_s + t_2)\\ &= (\omega_1 - \omega_0)t_1 + (\omega_s - \omega_0)t_s + (\omega_2 - \omega_0)t_2\\ &= \varphi_1 + \varphi_s + \varphi_2\,. \end{aligned} \tag{3.61}$$

Next, we consider the action of a π rotator C_π inserted just in front of the sample S at the midpoint of the total path. The rotator C_π in the NRSE method is different from that in the NSE method. C_π is a high-frequency magnetic field, the same as C_1 and C_2, and rotates with the phase $\pi/2$ delayed after C_1 and C_2. In this way it plays the same role as that in the conventional NSE method but now in the rotating frame; then the relation for the phase difference becomes

$$\Delta\varphi = -\varphi_1 + \varphi_s + \varphi_2\,, \tag{3.62}$$

and the modification of the conventional spin echo method to the resonance version is thus successfully achieved.

In the NRSE method, the operating conditions are settled as $\omega_s = \omega_0$ and at the same time $B_1 = B_2 = B_G$, where B_G is a weak guide field to prevent depolarization of neutrons during the flight. Then we can get rid of the effect of the flight path divergence, which was one of the noticeable problems deteriorating the energy resolution owing to the broadening of the precession angle distribution in the NSE method, and also of the effect of the scattering position distribution in the sample with a finite size. In this way, high-resolution spectroscopy for quasi-elastic and inelastic scattering can be performed [159, 160]. The resultant magnetic field configuration in the NRSE method is illustrated in Figure 3.17, which corresponds to that of the NSE method but as a whole lowered by B_0. In the case of an echo arrangement, $l_1 = l_2 = l$, the phase difference becomes

$$\Delta\varphi = \omega_0 l \left(\frac{1}{v_1} - \frac{1}{v_2}\right) \cong \frac{\omega_0 l}{2v} \cdot \frac{dE}{E} = \Omega \cdot \tau \,, \tag{3.63}$$

from which it arises that $\Omega = dE/\hbar = (E_2 - E_1)/\hbar, \tau = \hbar\omega_0 l/2vE$.

In the NSE method, serious effort must be devoted to attain high homogeneity of the precession field and also to restrict the divergence angle of the neutron flight path to realize highly homogeneous Larmor precession. In the NRSE method, the high accuracy of the high-frequency rotating field will play an important role.

Further details on the characteristics of and developments on devices used in the NRSE method will not be given here, but a typical example of the application of the NRSE method will be introduced here as a test experiment carried out by Klimko *et al.* on superfluid ^{4}He [161].

In superfluid ^{4}He, the dispersion relation of the excitation energy versus the momentum shows the well-known S-shaped curve, and the linear region near the origin represents phonon excitations with energy proportional to the momentum, whereas the elementary excitation around the bottom of the valley in the S-curve is called a *roton*. Figure 3.18 illustrates the line width Γ of the dispersion relation and the shift δ of the excitation energy for roton creation in superfluid ^{4}He measured by neutron inelastic scattering. The experiment was carried out with a new spectrometer, ZETA, which consists of a zero-field NRSE configuration installed in the spectrometer IN3 at the high-flux reactor in Grenoble. The temperature dependence of the line width reported previously indicated good agreement with the theory, whereas the previously reported temperature change of the excitation energy indicated disagreement with the theory, and further the present experiment gave another different result. Klimko *et al.* are also evaluating the advantage of the NRSE method in high-resolution measurements on a dispersive type of excitation beyond the roton minimum in superfluid ^{4}He or acoustic phonons in solids by making use of the configuration tilting the rotating field in the ZETA spectrometer with regard to the neutron beam.

The descriptions of the theory and experiments for the NRSE method given above belong the category of steady-state neutron utilizations, but in recent years pulsed neutron experiments have also become important. *Application of the NRSE method to pulsed neutrons* was performed by a group at the Kyoto University Research Reactor Institute [162]. Several things have to be considered to apply the NRSE method to pulsed neutrons. First, since the phase difference at the detector, τ, given by Eq. (3.63), is proportional to ω_0/v, control of ω_0, or the magnetic field B_0, in relation to the neutron flight time is necessary. Further, precautions for

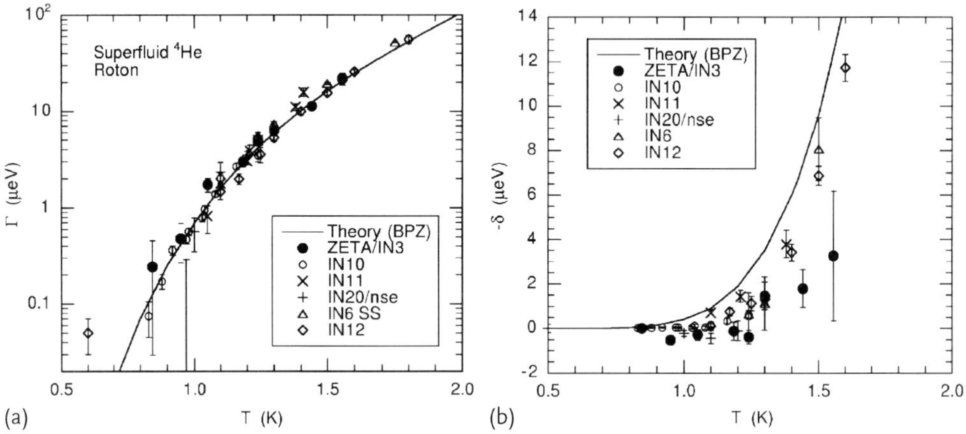

Figure 3.18 Temperature dependences of roton excitation in superfluid ^{4}He measured by the zero-field resonance spin echo spectrometer ZETA installed in IN3 in the high-flux reactor, Grenoble, compared with the results obtained with conventional spectrometers: (a) line width (half width and half maximum); (b) energy shift (Klimko *et al.* [161]).

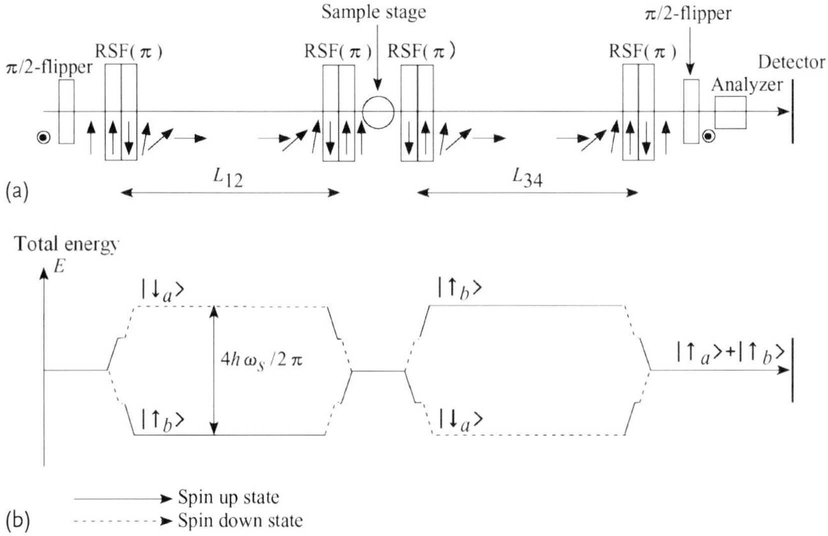

Figure 3.19 Principle of a resonance spin echo spectrometer proposed for installation in the intense pulsed neutron laboratory of the Japan Proton Accelerator Research Complex (J-PARC) facility: (a) system arrangement, where short arrows and double circles indicate the direction of magnetic fields; (b) development of the total energy for the two spin eigenstates (Maruyama *et al.* [163]).

the energy (velocity) change of neutrons owing to the effect of magnetic potential during their passage through the spin rotators C_1, C_π, and C_2 are also required. Furthermore, some auxiliary magnetic field is applied to compensate for the precession angle dispersion due to the broadening of the neutron flight path. The characteristic pair type of spin rotators and the auxiliary guide field illustrated by successively varying arrows in Figure 3.19a play such important roles in the pulsed neutron type of NRSE spectrometer under development by Ebisawa *et al.*, the working principle of which is described in Figure 3.19 [163, 164], and the pulsed type of NRSE spectrometer proposed for installation in the intense pulsed neutron laboratory at J-PARC, Tokai.

3.5.2
Neutron Spin Maser

As an another application of Larmor precession of neutrons, the concept of a *neutron spin maser* and its simulation experiment will be introduced below [165–167]. A maser is a kind of coherent electromagnetic amplifier which cooperatively produces coherent electromagnetic waves with the same phase and the same wavelength by irradiating an assembly of atoms or molecules excited to an unstable energy level with electromagnetic waves. A nuclear *spin maser* is an application of the present principle to a nuclear spin system in a magnetic field and holds the nuclear spins in the desired state for a long time. Conventional nuclear spin masers establish the maser condition, as shown in Figure 3.20a, by inducing an automatic oscillating current in a feedback coil and recovering the spin states by the effect of the rotating magnetic field if any transverse ($x-y$-) components arise in nuclear spins precessing in an assembly of atoms.

The direct application of the present principle of a nuclear spin maser to neutron spins is difficult because of the very low density of neutrons compared with the atomic density. Therefore, Yoshimi *et al.* proposed, as shown in Figure 3.20b, a new principle for a maser system by inserting a positive amplifying circuit between the detection of the spin precession and the induction of the rotating magnetic field, thus establishing the maser condition even in the case of very low spin density. Further, they carried out successfully a simulation experiment to operate

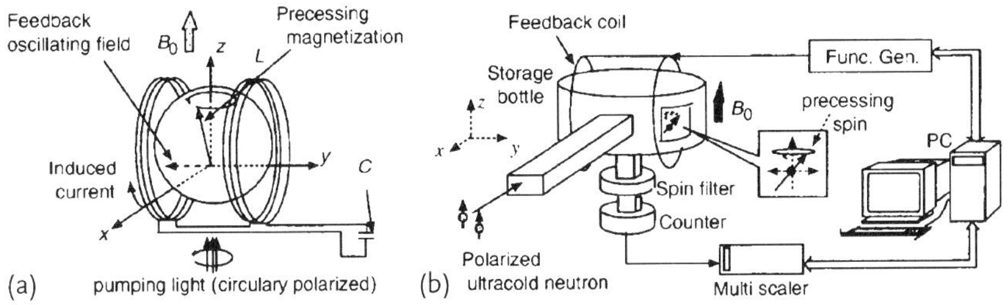

Figure 3.20 Operating principles of spin masers: (a) conventional spin maser; (b) neutron spin maser proposed to work by a new mechanism (Yoshimi *et al.* [166]).

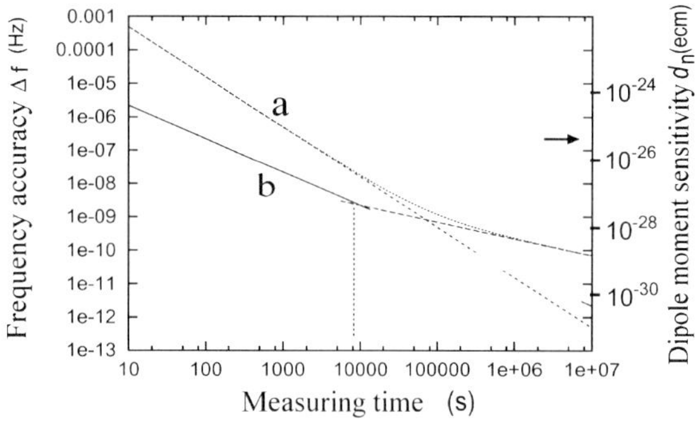

Figure 3.21 Expected experimental sensitivity of the neutron electric dipole moment with the new method of an ultracold neutron spin maser. The dotted line a is the expected experimental error due to statistical fluctuations and the solid line b is that due to resonance frequency fluctuations (Yoshimi *et al.* [167]).

nuclear maser oscillations and hold the condition for a long time by making use of polarized ^{129}Xe gas in place of low-density polarized neutron gas [165].

As already mentioned Section 1.5, continuation of neutron spin resonant precession for a long time for high-precision determination of the resonance frequency is one of the most important tasks in fundamental physics in relation to the search for the *electric dipole moment of the neutron*. Yoshimi *et al.* estimated the expected experimental precision for the determination of the electric dipole moment of a neutron by applying the spin maser method for ultracold neutrons with the system setup given in Figure 3.20b.

According to their estimation for a measuring time for about 30 days (2×10^6 s), the sensitivity of the experimental data at a level of 10^{-28} e for the neutron electric dipole moment will be attained as shown in Figure 3.21 [167] with the supposed experimental conditions of an ultracold neutron storage lifetime of 500 s and 10^4 ultracold neutrons being extracted per second to determine the precession conditions. The estimated sensitivity will surpass that of the conventional experiments by about 2 orders [166, 167].

3.6
Berry's Geometrical Phase

As an end to this chapter, we describe an interesting experiment on the observation of neutron spin behavior relating to an important viewpoint in quantum mechanics. In 1984, Berry [168] pointed out that when a quantum mechanical system at an eigenstate makes an excursion in the parameter space of the Hamiltonian $\mathcal{H}(\boldsymbol{R})$ for that system, such that the parameter $\boldsymbol{R}$ in the Hamiltonian is varied sufficiently slowly along a circuit C, the wave function of the state includes, in addition to the usual *dynamical phase* factor, a factor $\exp\{i\gamma(\mathrm{C})\}$ with a *geometrical phase* $\gamma(\mathrm{C})$, and also proposed an experiment to observe the effect. His recognition of the geometrical phase factor was received interest with corresponding concept of *gauge invariance*, which will be explained later in Section 9.3.

Following the outline of Berry's theory, we suppose that the state of the system $|\phi(t)\rangle$ satisfying the Schrödinger equation

$$\mathcal{H}(\boldsymbol{R})|\phi(t)\rangle = i\hbar|\dot{\phi}(t)\rangle , \tag{3.64}$$

is changed sufficiently slowly and stays in an eigenstate $|n(\boldsymbol{R})\rangle$, that is, is changed adiabatically satisfying

$$\mathcal{H}(\boldsymbol{R})|n(\boldsymbol{R})\rangle = E_n(\boldsymbol{R})|n(\boldsymbol{R})\rangle , \tag{3.65}$$

with energy $E_n(\boldsymbol{R})$ for the parameter $\boldsymbol{R} = \boldsymbol{R}(t)$ at the instant t. Then we arrive at the result that the state $|\phi\rangle$ can be written as

$$|\phi(t)\rangle = \exp\left\{-\frac{i}{\hbar}\int_0^t dt' E_n(\boldsymbol{R}(t'))\right\}\exp(i\gamma_n(t))|n(\boldsymbol{R}(t))\rangle , \tag{3.66}$$

that is, in addition to the *dynamical phase* factor of the first term on the right side, the second term, a factor including the phase $\gamma_n(t)$, appears. The important indication is that even though the parameter returns to the origin $\boldsymbol{R}(0)$ after time T has elapsed going around the circuit C, it results that $\gamma(T) \neq \gamma(0)$.

The expression for the function $\gamma_n(t)$ can be obtained by substituting Eq. (3.66) into Eq. (3.64). The expression becomes much simpler when the circuit C traces a loop close to some eigenstates $|m(\boldsymbol{R})\rangle$ in a degeneracy for the Hamiltonian $\mathcal{H}(\boldsymbol{R})$, that is, close to the region $\boldsymbol{R}^*$ where $E_m(\boldsymbol{R}^*) = E_n(\boldsymbol{R}^*)$. A typical example is a neutron precessing around a magnetic field $\boldsymbol{B}$, the two spin states of which are degenerate in the condition $\boldsymbol{B} = 0$. In this condition, Berry's phase is given by

$$\gamma_{\pm}(\mathrm{C}) = \mp\frac{1}{2}\Omega(\mathrm{C}) , \tag{3.67}$$

where $\Omega(\mathrm{C})$ is the solid angle for the circuit C extending around the origin in $\boldsymbol{B}$ space. In this way, Berry's phase is given by the phase of the geometrical circuit and therefore it is also called the *topological phase*. Further, it was revealed that the *Aharonov–Bohm effect* (to be introduced in Section 9.3) is a special case of the topological phase explained here.

The first experimental demonstration with neutrons on Berry's phase was performed by Bitter and Dubbers [169]. The key point of the experimental principle is given by the B_1 coil illustrated in Figure 3.22b, which, with the axial magnetic field coil not shown in the figure, induces a magnetic field $\boldsymbol{B}$ tracing the closed circuit C shown in Figure 3.22a, and at the same time changing adiabatically, during the flight of neutrons passing through the coil in the axial (z-) direction. The point $\boldsymbol{B} = 0$ in Figure 3.22a is the degenerate condition for the spin states mentioned above, and therefore Berry's phase defined by Eq. (3.67) will be realized in the region near such a degenerate point.

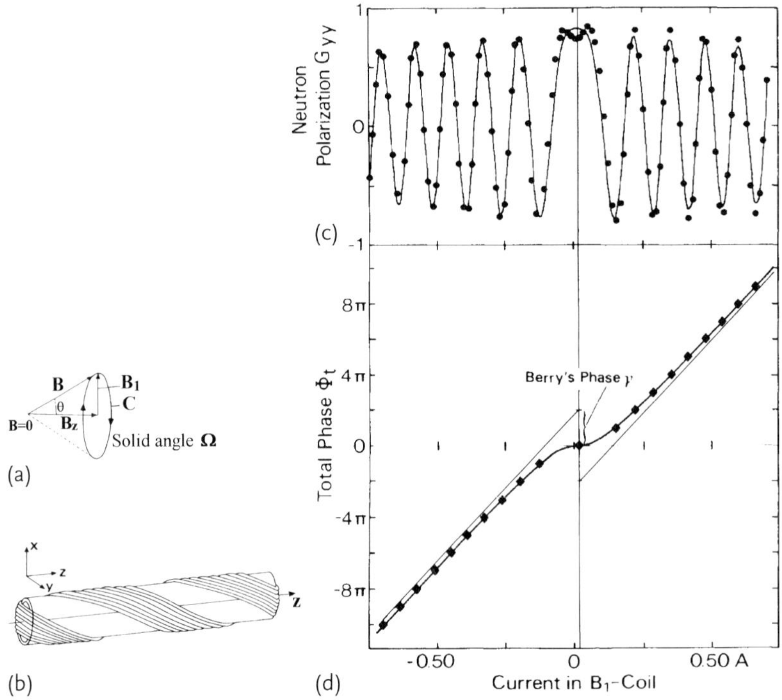

Figure 3.22 Neutron experiment on Berry's geometrical phase. (a) Adiabatic change of the magnetic field vector B tracing a closed circuit C. (b) A helical coil arrangement inducing the magnetic field B_1 in a right-handed system; neutrons are transmitted in the z-direction. (c) Spin rotation pattern for the transverse spin component $P_y(T) = G_{yy}(T)P_y(0)$ with regard to the helical magnetic field B_1 coil current (without Berry's phase the maxima of this curve should be equidistant), (d) observed and calculated phase shifts φ_t (Bitter and Dubbers [169]).

From the adiabaticity condition, the requirement $|\gamma| \ll |\varphi|$ must be satisfied by Berry's phase γ, where φ is the dynamical phase due to the total magnetic field including B_z. Thus, a high-precision measurement is necessary for the separation of γ. The experiment was carried out with a polarized beam (polarization ratio of about 97%) of cold neutrons with a velocity of about 500 m/s extracted through a guide tube at the high-flux reactor at the Institut Laue–Langevin, Grenoble. The incident neutrons enter in a direction perpendicular to the front end surface of the coil wound around an 8-cm diameter and 40-cm-long cylinder, then pass through the entrance magnetic field adiabatically. The magnetic field component B_1 rotates 2π for a coil length of 40 cm. The coil set was contained in a 30-cm diameter and 80-cm-long MuMetal cylinder.

The measured pattern of spin rotation with regard to the change of the B_1-coil current is plotted in Figure 3.22c, and the resultant total phase is plotted as shown in Figure 3.22d in comparison with the curves from the theoretical calculation, respectively. According to the theoretical derivation of Berry's phase mentioned above, by extrapolating to the adiabatic limit, Bitter and Dubbers obtained $\gamma = -\Omega = 2\pi$, in good agreement with the theoretical expectation.

The phenomenon of a topological phase, or an additional phase factor due to spin rotation in a helical magnetic field, is also observed in various experimental fields, such as nuclear magnetic resonance and polarized neutron scattering by helical magnetic materials. Further, measurement of Berry's phase in stored ultracold neutrons was also reported. In the research field of neutron interferometry, we will study in Chapter 9 several experiments on Berry's phase that are being carried out.

4
Quantum Mechanics of Scalar Neutron Waves in One Dimension

4.1
Some Descriptions of the Fundamental Theory

Optics deals with waves and their transformation by objects, that is, optics deals with scattering of waves. Likewise, neutron optics deals with scattering of neutron waves. However, such a definition is too broad, as it covers the whole field of neutronography. Here we limit ourselves to elastic reflection of neutrons from interfaces, and borrow from optics the notion of refractive index. We shall not consider devices such as lenses and microscopes; notwithstanding that these devices are familiar in the field of neutron optics. Though we do not consider them, neutron optics remains a very rich field of research. It is not a weak analog of the usual light optics, it has its own image, and it provides new results.

From the preceding chapters the reader already knows that the neutron is a wave, so we shall not dwell too long on this point. We only remind the reader that a free neutron is described by a plane wave

$$\psi_0(\boldsymbol{r}, \boldsymbol{k}, t) = \exp(i\boldsymbol{k}\boldsymbol{r} - i\omega t) , \tag{4.1}$$

containing wave vector $\boldsymbol{k}$, which is related to momentum $\boldsymbol{p}(\boldsymbol{k} = \boldsymbol{p}/\hbar)$, and frequency ω, which is related to neutron energy $E = p^2/2m(\omega = E/\hbar)$, where m is the neutron mass. The plane wave satisfies the free Schrödinger equation

$$i\hbar\frac{\partial}{\partial t}\psi = -\frac{\hbar^2}{2m}\Delta\psi , \tag{4.2}$$

and it is common to say that the plane wave Eq. (4.1) is the wave function of a free neutron with momentum p. However this is not so! A function is a wave function if it satisfies the Schrödinger equation and is normalizable. A plane wave is not normalizable; therefore, it must be replaced by a wave packet. However, at this point we enter a dense quantum mist. According to canonical quantum mechanics, a particle described by a wave packet cannot have momentum, because momentum is an eigenvalue of the momentum operator $\hat{p} = -i\hbar d/d\boldsymbol{r}$, and it exists only in states which are eigenfunctions of this operator, that is, in states described by plane waves, which are unacceptable for wave functions. So, at the very beginning we find

Neutron Optics. Masahiko Utsuro, Vladimir K. Ignatovich

ISBN: 978-3-527-40885-6

ourselves facing a terrible contradiction. We can forget about it without questioning the Holy Bible, or we should resolve it for to be honest with ourselves.

To resolve this contradiction we must reject the canonical definition of momentum. The full discussion of this point is postponed to Chapter 10. Here we only can say that the plane wave is a component of a wave packet, and, if we can predict what happens to every component, we can find what happens to their superposition. It is a great surprise to discover that the behavior of wave packets (this will be shown in Chapter 10) is very similar to that of plane waves. The difference is very small, but it is this difference which is the most important for development of the fundamental physics. We hope that some day physicists will realize this. Here and with some exceptions in all the other chapters we take a plane wave as a wave function, remembering that it is only one component of it.

Sometimes the wave packet is considered as a property of a neutron beam, and notions such as longitudinal and transverse coherence length and coherence time are introduced, as if the different neutrons were like different plane wave components of a single wave packet. This is not so, and it is misleading. We shall not use them. When needed, we shall use the natural characteristics of the neutron beam: energy and angular spread, which mean energy and angular resolution for a given experiment.

Equation (4.2) is valid only in empty space, and if some parts of the space are filled with matter the equation changes to

$$i\hbar\frac{\partial}{\partial t}\Psi(\boldsymbol{r},t)=\frac{\hbar^2}{2m}[-\Delta+u(\boldsymbol{r},t)]\Psi(\boldsymbol{r},t)\,, \tag{4.3}$$

where $\hbar^2 u/2m$ describes the interaction potential of the neutron with matter. The solution of this equation is not a plane wave and our task is to find it. In this book we shall consider mainly potentials which do not depend on time, so we can seek a stationary solution $\Psi(\boldsymbol{r},t)=\exp(-i\omega t)\Phi(\boldsymbol{r})$, and Eq. (4.3) is reduced to the form

$$[\Delta-u(\boldsymbol{r})+k^2]\Phi(\boldsymbol{r})=0\,, \tag{4.4}$$

where $k^2=2mE/\hbar^2=(p/\hbar)^2$. Moreover, in this chapter we shall consider only planar systems, which change only along one coordinate axis, say x, so the potential is $u(x)$. Therefore, the solution of Eq. (4.4) can be represented as $\Phi(\boldsymbol{r})=\exp(i\boldsymbol{k}_\parallel,\boldsymbol{r}_\parallel)\psi(x)$, where vectors $\boldsymbol{r}_\parallel$ are parallel to the (y,z) plane. Substitution into Eq. (4.4) gives

$$\left[d^2/dx^2-u(x)+k_x^2\right]\psi(x)=0\,, \tag{4.5}$$

where $k_x=\sqrt{k^2-k_\parallel^2}$ is the normal component of the wave vector $\boldsymbol{k}$. In the following we shall use this equation and omit the subscript x of k.

In this chapter we shall start by studying the reflection and refraction processes in semitransparent mirrors of finite thickness and multilayered systems by treating the neutron as a scalar particle. The one-dimensional systems are very convenient to study details of many different phenomena, such as resonances and Bragg diffraction. They are all in the realm of application of quantum mechanics, and we shall see how excellently it deals with them.

Now we Start the Main Task of this Chapter

We start by reviewing the simplest well-known problem in quantum mechanics – reflection and refraction at a rectangular potential step, and our main tool will be the Schrödinger equation:

$$\left[d^2/dx^2 - u(x) + k^2\right] \psi(x) = 0\,. \tag{4.6}$$

4.2 Reflection and Refraction at a Potential Step

Consider reflection from an infinitely thick wall filling the half-space $x > 0$ with an ideal flat interface at the point $x = 0$. The interaction between the wall and the neutron is a constant u_0 (what this constant is will be discussed later), so the whole potential $u(x)$ is represented by the function $\Theta(x > 0)u_0$, where $\Theta(x)$ is the step function equal to unity, when the inequality in its argument is satisfied, and to zero otherwise. So we have the potential step shown in Figure 4.1.

The solution of the Schrödinger equation with this potential step

$$\left[d^2/dx^2 - u_0\Theta(x > 0) + k^2\right] \psi(x) = 0\,, \tag{4.7}$$

is completely defined by an incident wave. For the incident wave e^{ikx} the solution is

$$\overrightarrow{\psi}(x) = \Theta(x \le 0)\left[e^{ikx} + \overrightarrow{r_0}\,e^{-ikx}\right] + \Theta(x > 0)\,\overrightarrow{t_0}\,e^{ik'x}\,, \tag{4.8}$$

with some yet unknown reflection, $\overrightarrow{r_0}$, and transmission (refraction), $\overrightarrow{t_0}$, amplitudes, where $\rightarrow$ above r and t means that the incident wave travels toward the interface from left to right. The refracted wave depends on k', which is the wave number inside the potential. Substitution of the refracted wave into Eq. (4.7) gives $-k'^2 - u_0 + k^2 = 0$, which defines $k' = \sqrt{k^2 - u_0}$.

If the incident wave originates inside the medium and propagates to the interface from right to left, then the solution is

$$\overleftarrow{\psi}(x) = \Theta(x \le 0)\,\overleftarrow{t_0}\,e^{-ikx} + \Theta(x > 0)\left[e^{-ik'x} + \overleftarrow{r_0}\,e^{ik'x}\right]. \tag{4.9}$$

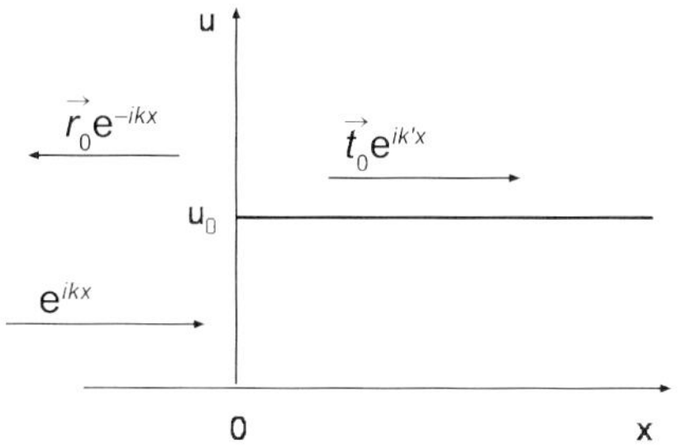

Figure 4.1 Reflection and refraction at a potential step of height u_0 located at $x = 0$.

To find the coefficients in Eq. (4.8), it is necessary to match the wave function at the interface, that is, at the point $x = 0$, to make the wave function and its first derivative continuous:

$$1 + \overrightarrow{r_0} = \overrightarrow{t_0}, \quad k[1 - \overrightarrow{r_0}] = k'\overrightarrow{t_0} \,. \tag{4.10}$$

The solutions of the above equations are

$$\overrightarrow{r_0} = \frac{k - k'}{k + k'}, \quad \overrightarrow{t_0} = 1 + \overrightarrow{r_0} = \frac{2k}{k + k'} \,. \tag{4.11}$$

Similarly, we can match the wave function (4.9); however, because of symmetry we can immediately obtain

$$\overleftarrow{r_0} = \frac{k' - k}{k' + k} = -\overrightarrow{r_0}\,, \quad \overleftarrow{t_0} = 1 - \overrightarrow{r_0} = \frac{2k'}{k' + k} \,. \tag{4.12}$$

In the above case, the space to the left of the step is a vacuum, that is, the potential to the left of the step is zero. However, we can consider a more general case, when the whole space is divided into two parts, with potential u_i to the left and u_f to the right. In such a case, if the incident wave travels from the left, that is, from u_i to u_f, the reflection and refraction amplitudes are

$$r_{if} = \frac{k_i - k_f}{k_i + k_f}, \quad t_{if} = 1 + r_{if} = \frac{2k_i}{k_i + k_f}, \tag{4.13}$$

where $k_{i,f} = \sqrt{k^2 - u_{i,f}}$.

For $k^2 \gg u_0$ we can approximate $k' = \sqrt{k^2 - u_0} \approx k - u_0/2k$ and the reflection amplitude $\overrightarrow{r_0}$ becomes approximately $u_0/4k^2$; therefore, at $k^2 \gg u_0$ the reflection coefficient $|\overrightarrow{r_0}|^2$, which is usually called reflectivity, decreases with increase of k as

$$|\overrightarrow{r_0}|^2 \approx \frac{1}{16}\frac{|u_0|^2}{k^4}. \tag{4.14}$$

Unitarity condition This is an appropriate place to mention unitarity. When considering the problem of reflection and refraction of plane waves at an interface, unitarity means that the total flux density of the reflected and refracted particles must be equal to the incident one:

$$k = k|\overrightarrow{r_0}|^2 + k'|\overrightarrow{t_0}|^2 \,. \tag{4.15}$$

This relation defines the refraction probability, $w = (k'/k)|\overrightarrow{t_0}|^2$, which can be written as

$$w = \frac{4kk'}{(k + k')^2} \,. \tag{4.16}$$

4.2.1
Total Reflection

Let us go back to the reflection amplitude $\overrightarrow{r_0}$ Eq. (4.11). Since $k' = \sqrt{k^2 - u_0}$, then for $k^2 < u_0$ the wave vector inside the medium, k', is imaginary, $k' = ik''$, where $k'' = \sqrt{u_0 - k^2}$, and the reflection amplitude becomes a complex number:

$$\overrightarrow{r_0} = \frac{k - ik''}{k + ik''} = \exp(-2i\gamma_0)\,, \quad \gamma_0 = \arctan\left(\frac{k''}{k}\right) = \arccos\left(\frac{k}{\sqrt{u_0}}\right). \tag{4.17}$$

This means that for $k^2 < u_0$ the reflection coefficient $|\overrightarrow{r_0}|^2$ is equal to 1, that is, this is the case of total reflection. The value $\sqrt{u_0}$ is called the critical wave number and it is denoted as k_c.

Recall that k in all expressions represents the perpendicular component, $k_\perp \equiv k_x$, of the total incident wave vector $\boldsymbol{k}$. The condition $k_\perp^2 < k_c^2$ means total reflection of the neutron from the wall. The stricter condition $|\boldsymbol{k}|^2 < k_c^2$ means that the neutron undergoes total reflection at any incident angle. Neutrons with energies $k^2 < k_c^2$ are called *ultracold neutrons*. They can be stored for quite a long time in closed vessels (textbooks [F,G,H]).

4.2.2
Optical Potential and Ultracold Neutrons

Above, we studied reflection and refraction for an abstract potential step. In this section we show how large this potential is in reality.

It will be deduced later, but now we simply declare, that the interaction of neutrons with materials is described by the optical potential

$$U = \frac{\hbar^2}{2m} 4\pi N_0 b_c\,, \tag{4.18}$$

where $\hbar = h/2\pi = 1.05 \cdot 10^{-34}$ J s is the reduced Planck constant, $m = 1.675 \cdot 10^{-24}$ g is the neutron mass, $N_0 \approx 10^{23}\,\text{cm}^{-3}$ is the number of atoms per unit volume of the medium, and b_c is the average (coherent) neutron scattering amplitude of a single atomic nucleus. Therefore, the reduced potential $u_0 = 2mU/\hbar^2$ used in our equations is equal to $4\pi N_0 b_c$.

The scattering amplitude b is different for different materials (textbook [E]), and it can be positive as well as negative, but it is on the order of $|b_c| \approx 10^{-12}$ cm. Substitution of numerical constants into Eq. (4.18) shows that the typical value of U is 10^{-7} eV. The energy of ultracold neutrons is $E = \hbar^2 k^2/2m \leq U \sim 10^{-7}$ eV, and their velocity is $v \sim 5$ m/s.

Historically, reflection from a potential step was described like in light optics with the help of a phenomenological constant, the index of refraction: $n \approx 1 - u_0^2/2k^2$. Fermi, who first observed specular and total reflection of neutrons from matter, used this index of refraction, which is valid only for $u_0 \ll k^2$ (see [170],

Chapter VIII), and he called total reflection total internal reflection, in accordance with light optics, whereas in reality the neutron reflection is an external one. He used an approximate expression, though at his time one could find in the literature a rigorous derivation of $n = \sqrt{1 - u_0/k^2}$ without the requirement for $u_0/2k^2$ to be small [15].

Fermi derived the refractive index with the help of the concept of a quasi-potential $u_1(\boldsymbol{r} - \boldsymbol{r}_i) = 4\pi b_c \delta(\boldsymbol{r} - \boldsymbol{r}_i)$, which he devised to describe the interaction of a neutron with a single nucleus at a point $\boldsymbol{r}_i$. This potential is called quasi because it does not describe the real neutron–nucleus interaction. However, in perturbation theory this potential gives the same scattering amplitude b as the real potential gives with the rigorous scattering theory.

With this quasi-potential it is easy to derive the optical potential u_0 by averaging the quasi-potential over many atoms:

$$u_0 = \sum_j u_1(\boldsymbol{r} - \boldsymbol{r}_j) = \int N_0 d^3r'\, u_1(\boldsymbol{r} - \boldsymbol{r}') = \int N_0 d^3r'\, 4\pi b_c \delta(\boldsymbol{r} - \boldsymbol{r}')$$
$$= 4\pi N_0 b_c\,. \tag{4.19}$$

This derivation, of course, is not a rigorous one. A rigorous derivation is given in Chapter 5.

It is interesting to note that the strong forces are short-range, that is, they act at distances on the order of 10 fm, and the interaction radius of the quasi-potential is even equal to zero. Nevertheless, the potential Eq. (4.19) turns out to be a long-range one and even macroscopic. This is clearly seen in everyday experiments with neutron reflectometers. Fermi clearly understood that the quasi-potential is not a real one. The real nuclear potential is attractive, that is, negative, and if we average it across the medium, we obtain a negative number, whereas Eq. (4.19) for most of the chemical elements is positive, that is, repulsive, because their scattering amplitude b_c is positive. The total reflection, and storage of neutrons in bottles, is possible because the potential Eq. (4.19) in many cases is repulsive.

Fermi, however, thought that the storage of neutrons is not possible, even though he and Zinn witnessed total reflection [171]. He, like Rutherford [172], thought that the neutron is a tight hydrogen atom, and he wrote [170] (p. 92): "There are no vessels capable of holding neutrons, because the neutrons diffuse in between the nuclei and freely permeate through matter".

Only in 1959 did Y.B. Zeldovich first point out that ultracold neutrons can be trapped in containers [72]. Such neutrons were first observed in 1968 [73], and the first storage experiment was performed in 1971 [74] under the supervision of F.L. Shapiro.

4.2.3
Digression on the Ultracold Neutron Anomaly

Since then, researchers have conducted numerous experiments with ultracold neutrons (see, for example, textbooks [F,G,H]), and they all reported one result, which has not been explained up to now. According to the measurements, conducted with very clean bottles where absorption by the nuclei and the probability of inelastic scattering are very low, the loss coefficient of ultracold neutrons in a single collision with the walls is almost two orders of magnitude higher than the theoretical estimate. This phenomenon is called the ultracold neutron anomaly.

Ultracold neutron losses (textbook [G]) can be described by the imaginary part u_0'' of the potential Eq. (4.19), caused by the imaginary part b_c'' of the scattering amplitude $b_c = b_c' - ib_c''$. For real potentials the refection coefficients of ultracold neutrons, $|\vec{r_0}|^2$, is unity. When the potential contains u_0'', the reflection coefficient is less than unity by the amount

$$\mu = 1 - |\vec{r_0}|^2 = 2\eta \frac{k_\perp}{\sqrt{u_0' - k_\perp^2}} \,, \tag{4.20}$$

where $\eta = u_0''/u_0' = b_c''/b_c'$, $u_0' = 4\pi N_0 b_c'$, and for the moment we restored the sign $\perp$ to denote the normal component of the neutron wave vector $\boldsymbol{k}$. The right side of Eq. (4.20) is the linear approximation of the left side in terms of u_0'', which is valid for $u_0'' \ll k^2 - u_0'$, that is, almost everywhere except the narrow range $k^2 - u_0' \approx u_0''$. The parameter η is called the loss coefficient. It is equal to $k\sigma_l/4\pi|b_c'|$, where $\sigma_l = \sigma_{ie} + \sigma_a$ is the loss cross section, comprising the inelastic, σ_{ie}, and absorption, σ_a, cross sections of neutrons for interactions with nuclei, and b_c' is the real part of the coherent scattering amplitude. σ_l is found from measurements of the transmission through the walls of neutrons with $k_\perp^2 > u'$. Since $\sigma_l \propto 1/k$, the coefficient η is a constant with a specific value for every substance. In Table 4.1 we give amplitude b_c' in femtometers, optical potentials $U'(1 \pm i\eta)$, where U' is in units of 10^{-7} eV, η is in dimensionless units of 10^{-5}, and for σ_l we used only σ_a, because usually $\sigma_{ie} \ll \sigma_a$ and it depends on temperature. For instance, the optical potential of Ti is $-0.51(1-0.0005i)\cdot 10^{-7}$ eV, because imaginary part is always negative.

The term ultracold neutron anomaly means that the experimentally found $\eta_{\rm exp}$ is considerably higher than theoretical one, $\eta_{\rm th} = k\sigma_l/4\pi b_c'$, with σ_l found from the transmission experiments. The difference can be of two orders of magnitude. For instance, in the case of Be [78], $\eta_{\rm th} = 3\cdot 10^{-7}$, whereas $\eta_{\rm exp} = 3\cdot 10^{-5}$ at low temperature (approximately 10 K).

At first sight it is very easy to explain the anomaly by contamination of surfaces with chemical elements of high absorption or inelastic cross sections, because in transmission experiments such contaminations are not seen. However, the required quantities of elements with large σ_a are too high compared with their abundances in the environment. In addition, elements with large σ_{ie} such as hydrogen, though very abundant, are improbable, because the inelastic cross section should have a steep temperature dependence, and should become ineffective at low temperatures, whereas the loss coefficient η does not show a steep temperature dependence. Nevertheless the role of hydrogen is discussed in the literature again and again.

The ultracold neutron anomaly was even considered to be a problem of quantum mechanics. In [173] it was suggested that the neutron can pass through the walls

Table 4.1 Values of b_c' in 10^{-13} cm and of the optical potential $U = U'(1 - i\eta \cdot 10^{-5})$ in 10^{-7} eV. The data were compiled with the help of Internet resources http://www.ncnr.nist.gov/resources/n-lengths/ and http://www.ktf-split.hr/periodni/en/. At the the latter Web page one can find densities of gases in solid or liquid states. The energy U' is calculated with a precision of two digits after the period, and η is calculated with a precision of two digits after period. It seems such precision is sufficient because there are always some additional factors, such as inelastic scattering and contamination, that increase η. With the help of U' it is easy to find the critical wave number $k_c = \sqrt{2mU'}/\hbar = 0.00695\sqrt{U'}$ in reciprocal angstroms if U' is in units of 10^{-7} eV. For instance, in the case of Ni we find $k_c = 0.00695\sqrt{2.45} = 1.088 \cdot 10^{-2}$ Å^{-1}. With the help of k_c we can find the reduced critical wavelength $\bar{\lambda}_c = \lambda_c/2\pi = 1/k_c$. For instance, in the case of Ni the reduced critical wavelength is approximately 92 Å.

N_Z	b_c'	U'	η	N_Z	b_c'	U'	η	N_Z	b_c'	U'	η
H_1	-3.74	-0.45	2.45	Ni_{28}	10.30	2.45	12.12	Ba_{56}	5.07	0.21	6.03
D_1	6.67	0.76	0.002	Cu_{29}	7.72	1.70	13.62	La_{57}	8.24	0.57	30.27
He_2	3.26	0.17	0.064	Zn_{30}	5.68	0.97	5.43	Ce_{58}	4.84	0.37	3.62
Li_3	-1.90	-0.22	10.3E2	Ga_{31}	7.29	0.97	10.49	Pr_{59}	4.58	0.35	69.82
Be_4	7.80	2.50	0.027	Ge_{32}	8.19	0.94	7.47	Nd_{60}	7.69	0.59	18.3E1
B_5	5.30	1.78	40.2E2	As_{33}	6.58	0.79	19.02	Pm_{61}	12.6	–	37.0E1
C_6	6.65	3.05	0.015	Se_{34}	7.97	0.76	40.82	Sm_{62}	0.80	0.06	32.9E5
N_7	9.36	1.07	5.64	Br_{35}	6.79	0.42	28.26	Eu_{63}	5.30	0.29	23.8E3
O_8	5.80	1.51	0.001	Kr_{36}	7.81	0.41	89.01	Gd_{64}	9.50	0.75	14.6E4
F_9	5.65	0.70	0.047	Rb_{37}	7.08	0.20	1.49	Tb_{65}	7.34	0.60	88.65
Ne_{10}	4.57	0.49	0.24	Sr_{38}	7.02	0.33	5.07	Dy_{66}	16.90	1.40	16.4E2
Na_{11}	3.63	0.24	4.06	Y_{39}	7.75	6.11	4.59	Ho_{67}	8.44	0.71	21.3E1
Mg_{12}	5.37	0.61	3.26	Zr_{40}	7.16	0.80	0.72	Er_{68}	7.79	0.66	56.8E1
Al_{13}	3.45	0.54	1.85	Nb_{41}	7.05	1.02	4.54	Tm_{69}	7.07	0.61	39.3E1
Si_{14}	4.15	0.55	1.14	Mo_{42}	6.72	1.12	10.27	Yb_{70}	12.40	0.78	78.04
P_{15}	5.13	0.47	0.92	Tc_{43}	6.80	1.25	81.78	Lu_{71}	7.21	0.64	28.5E1
S_{16}	2.85	0.29	5.17	Ru_{44}	7.02	1.35	10.14	Hf_{72}	7.70	0.91	37.5E1
Cl_{17}	9.58	0.82	97.23	Rh_{45}	5.90	1.11	68.2E1	Ta_{73}	6.91	1.00	82.90
Ar_{18}	1.91	0.12	9.83	Pd_{46}	5.91	1.05	32.46	W_{74}	4.77	0.78	10.7E1
K_{19}	3.67	0.13	15.91	Ag_{47}	5.92	0.90	29.73	Re_{75}	9.20	1.63	27.1E1
Ca_{20}	4.70	0.28	2.54	Cd_{48}	4.83	0.58	14.5E3	Os_{76}	10.70	1.99	41.58
Sc_{21}	12.10	1.26	63.19	In_{49}	4.07	0.41	13.3E2	Ir_{77}	10.60	1.95	11.2E2
Ti_{22}	-3.37	-0.50	50.25	Sn_{50}	6.23	0.60	2.80	Pt_{78}	9.60	1.65	29.83
V_{23}	-0.44	-0.08	31.9E1	Sb_{51}	5.57	0.48	24.51	Au_{79}	7.90	1.21	34.7E1
Cr_{24}	3.64	0.79	23.30	Te_{52}	5.68	0.44	23.01	Hg_{80}	12.60	1.34	82.2E1
Mn_{25}	-3.75	-0.80	98.62	I_{53}	5.28	0.32	32.42	Tl_{81}	8.78	0.80	10.86
Fe_{26}	9.45	2.09	7.53	Xe_{54}	4.92	0.21	13.54E1	Pb_{82}	9.40	0.81	0.51
Co_{27}	2.49	0.59	41.5E1	Cs_{55}	5.42	0.12	14.9E1	Bi_{83}	8.53	0.63	0.11

because its wave function is a wave packet, which does not spread and which is a neutron property. It was even estimated that to explain the ultracold neutron anomaly the width s of the packet in the momentum space should be

$$s \approx 4 \cdot 10^{-5} k\,, \tag{4.21}$$

where k is the average wave number of the packet. It was an exciting moment when soon after the publication of [173] transmission of ultracold neutrons through a thin Be foil was really found [84], but instead of the suggestion made in [173] a new theory with the notion of localization because of incoherent scattering was proposed [174]. However, very soon it was found [175, 176] that the transmitted neutrons are not really ultracold ones. Their energy E was in the range $U_c \leq E \leq 2U_c$, where U_c is the the optical potential of the foil. This means that the neutrons are heated a little bit. Such a small heating can be attributed [177] to interaction of ultracold neutrons with dust particles on the wall surfaces. It was hoped that the small heating would explain the anomaly, but the probability of small heating was found to be too low to do that, so to solve the problem of the anomaly the experimenters tested new materials. For instance, in [178] measurement of the loss coefficient in vessels with Be or Al walls covered by diamond-like carbon was reported. However, the loss coefficient was too high (approximately 10^{-4}) even at 70 K, and it was attributed to hydrogen and to holes in covering of diamond-like carbon.

Success was achieved with Fomblin-like hydrogen-free fluoropolymer, for which the loss coefficient $\eta_{\rm exp}$ at a temperature of approximately 110 K was reported to be $2 \cdot 10^{-6}$, which is higher than the theoretical value only by a factor of order 2. A review of recent research in this area can be found in [179].

4.2.4
Experience with an Experiment

Now we can compare with an experiment the theoretical predictions for the reflection coefficient

$$R(k) = |\overrightarrow{r_0}(k)|^2 \tag{4.22}$$

(it is usually called reflectivity) for $\overrightarrow{r_0}$ given by Eq. (4.11). Figure 4.2 shows the schematic of the experiment with a neutron reflectometer. If the mirror is a sufficiently thick plate (on the order of 1 cm) we can consider it as a semi-infinite one. There are three possible types of such an experiment:

Time-of-flight experiment In this type the detector and sample are fixed, the incident beam is well collimated, and its grazing angle θ is fixed but its spectrum is wide. The normal component of the wave vector of the incident neutrons $k_\perp = \sqrt{2mE/\hbar^2}\sin\theta$ varies with the energy E, which is determined by the time-of-flight method.

Experiment with monochromatic neutrons and a large stationary detector In this case variation of $k_\perp$ is achieved by rotation of the sample.

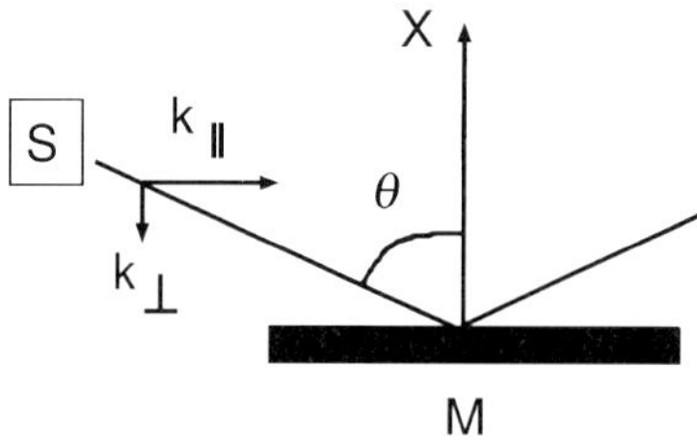

Figure 4.2 The reflectometry experiment. Neutrons emitted by a source S are reflected from a mirror M at an angle θ and are registered by a detector D.

$\theta - 2\theta$ experiment with monochromatic neutrons In this type of experiment the sample and detector rotate around the same axis, but the rotation angle of the detector is twice that of the sample, and the detector always registers the specularly reflected neutron beam.

Though the setup in Figure 4.2 looks simple, in reality the experiment is not so simple, because for thermal neutrons the grazing angle for total reflection is on the order of milliradians. At such an angle a mirror surface of 10-cm length looks like a filament of 0.1-mm thickness, and the incident neutron beam must have the same or a smaller width. We shall see what happens if the condition is not satisfied.

Figure 4.3 shows Japanese experimental data[10] for the reflectivity from a glass plate fitted by expression (4.22), where the fitting parameters are the real and imaginary parts of the optical potential $u' - iu''$ of the glass.

Before commenting on this figure, we want to warn the readers that our presentation of the results is different from the usual ones in several aspects. First, reflectivity is usually presented as its dependence on momentum transfer $\boldsymbol{q} = \boldsymbol{k}_0 - \boldsymbol{k}$, which for specular reflection is equal to $2k_\perp$. This seems strange, because the reflection amplitude depends on $k_\perp$. Second, the data are usually presented in dimensional units of reciprocal nanometers. We think it is much better to present the results in dimensionless units, $k = k_\perp / k_c$, where k_c is some characteristic wave number. Below we use the optical potential u_c of some selected material as a reference unit of energy, and determine k_c as the critical wave number corresponding to this reference energy. In Figure 4.3 the optical potential of Ni of 0.245 μeV is chosen for the unit of energy; therefore, the unit of the wave number is $k_c = 1/9.2\,\text{nm}^{-1}$. Third, sometimes when the fitted curve visually well reproduces the positions of the experimental points, researchers do not report the value of χ^2; we think that it is necessary to report not only χ^2, but the behavior of χ at all points. If we have fitted curve $f(x)$ and experimental data y_j at points k_j with statistical uncertainty

10) I am very grateful to Dr. Ryuji Maruyama from the Japan Proton Accelerator Research Complex (J-PARC) of the Japanese Atomic Energy Agency for sending me these data and for his explanations.

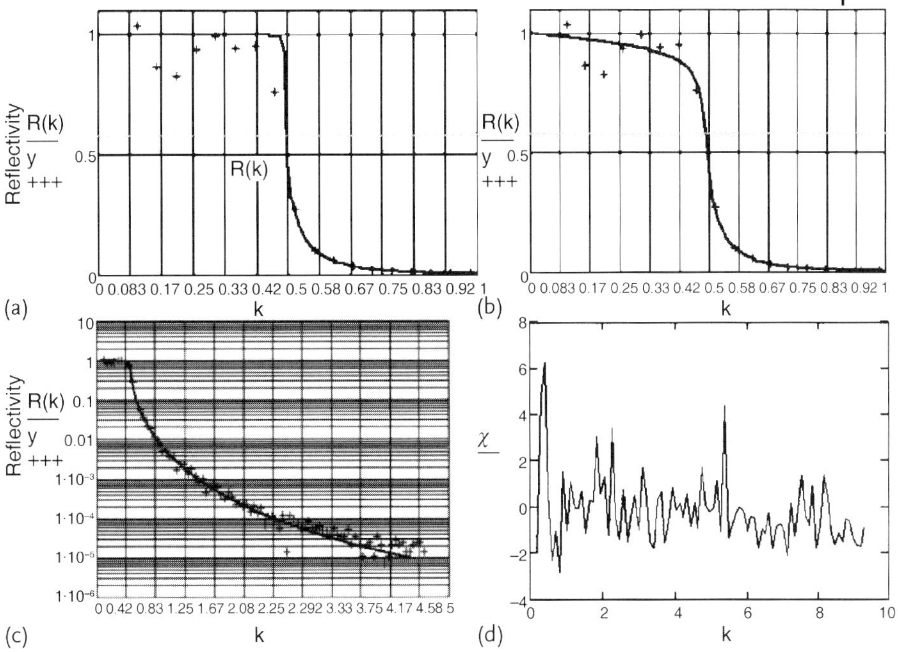

Figure 4.3 Fitting of the glass reflectivity data from Japan with $R(k)$ given by Eq. (4.22). The data were obtained with a $\theta - 2\theta$ reflectometer. The incident beam was wider than the surface area visible at small k. (a) One fitting parameter, u'; (b) two fitting parameters, u' and u''; (c) logarithmic scale of (b); (d) $\chi(k_j)$ distribution (Eq. (4.23)). The large variation seen at small k is explained by the properties of the setup.

σ_j, then the distribution

$$\chi(k_j) = \frac{f(k_j) - y_j}{\sigma_j} \tag{4.23}$$

gives information about the properties of the setup or about some missed physical phenomena. The single value

$$\chi^2 = \frac{1}{N - n - 1} \sum_{j=1}^{N} \chi(k_j)^2 \,, \tag{4.24}$$

where N is the number of experimental points and n is the number of parameters, is also informative, but without full $\chi(k_j)$ this information is very limited.

Figure 4.3a shows the result of fitting with a single parameter — the real part of the glass potential, when the imaginary part is fixed and equal to some more or less well estimated value 0.0002. Fitting gives $u' = 0.24$, in good agreement with the expected glass potential close to that of silicon, but $\chi^2 = 4.2$ in this case is large. Figure 4.3b shows the result of fitting with two parameters: the real and imaginary parts of the glass optical potential. The fitting gives the same real part $u' = 0.24$, but the imaginary part $u'' = 0.0094$ is too high compared with the expectation and makes the reflectivity curve smoother. However, $\chi^2 = 2.4$ is better than in Figure 4.3a. The high imaginary part can be attributed to scattering on inhomogeneities and surface roughness. In both plots we see some scattering of experimental points around the fitted curve. This scattering is due to the setup. The reflectivity was measured with a $\theta - 2\theta$ reflectometer. At small angles (small k up to approximately 0.08), part of the incident beam went over the mirror without touching the surface, and is registered by the the detector together with the reflected beam. In the interval $0.08 < k < 0.3$, the reflected beam separates from the direct beam, and this explains some dip in reflectivity near $k \approx 0.2$.

In Figure 4.3c the reflectivity is shown on a logarithmic scale. The fitting looks perfect, scattering at small k is not seen, and it can be overlooked, but this is not the case for the scattering of the points for curve $\chi(k)$ shown in Figure 4.3d. In this special case, when scattering at small k is well explained, we do not have a reason to seek an additional physical phenomenon, though a physical phenomenon could be masked by the properties of the setup.

Figure 4.4 shows the fitting of experimental data for glass reflectivity obtained in Hungary [188] with a reflectometer with a monochromatic beam, a rotating mirror,

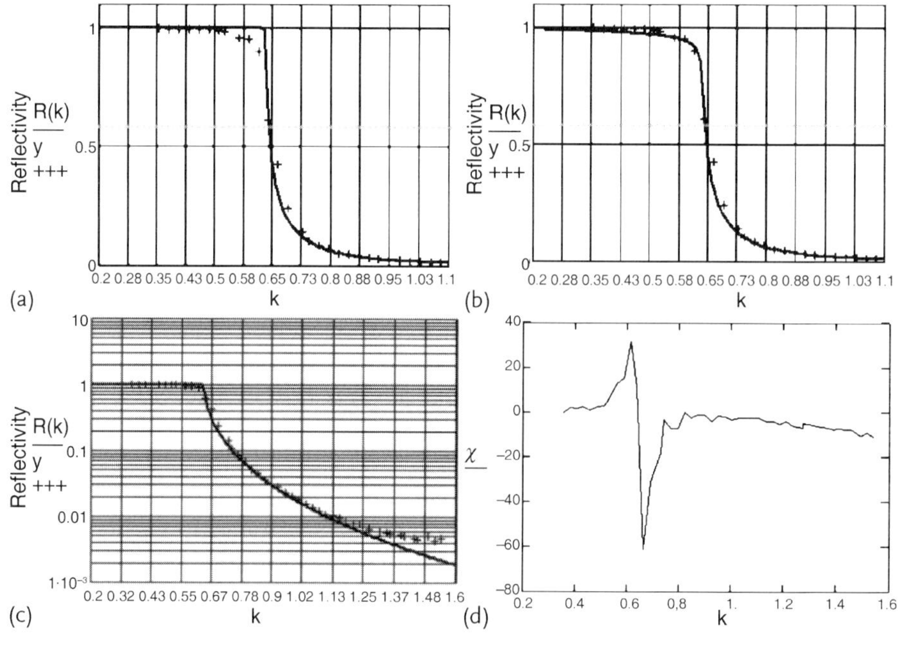

Figure 4.4 Fitting of the glass reflectivity data from Hungary [188] with $R(k)$ given by Eq. (4.22). The data were obtained with the reflectometer with a wide detector, which collects all the reflected neutrons at any angle θ of the sample. (a) One fitting parameter, u'; (b) two fitting parameters, u' and u''; (c) logarithmic scale of (a); (d) $\chi(k_j)$ distribution (Eq. (4.23)). The reason for the large variation seen at small k is not yet understood [188].

and a wide stationary detector, which collects all the reflected neutrons. Fitting with one parameter u' for $u'' = 0.0002$ with the same dimensional units as in the case of the Japanese data gives too high a potential $u' = 0.405$, which can be attributed to the special chemical formula of the glass used. On a linear scale the graph is shown in Figure 4.4a, and on logarithmic scale it is as shown in Figure 4.4c. We see a strong deviation of theoretical curve from the experimental points at large k. The parameter χ^2 for such a fitting is extremely high: $\chi^2 = 170$. The distribution $\chi(k)$ over all the experimental points is shown in Figure 4.4d. The result of fitting with two parameters u' and u'' on a linear scale is shown in Figure 4.4b. The parameter $\chi^2 = 132$ for such a fitting becomes a little bit less but still remains unacceptably high. Moreover, the fitted value of $u'' = -0.0046$ is completely unacceptable, because it is negative and therefore nonphysical. The experimental data can be fitted with more parameters. If we average the reflectivity over some interval 2δ, which determines the resolution of the system, and add background n_b, we obtain the formula

$$R(k) = \int_{k-\delta}^{k+\delta} |\vec{r_0}(p)|^2 \frac{dp}{2\delta} + n_b \,. \tag{4.25}$$

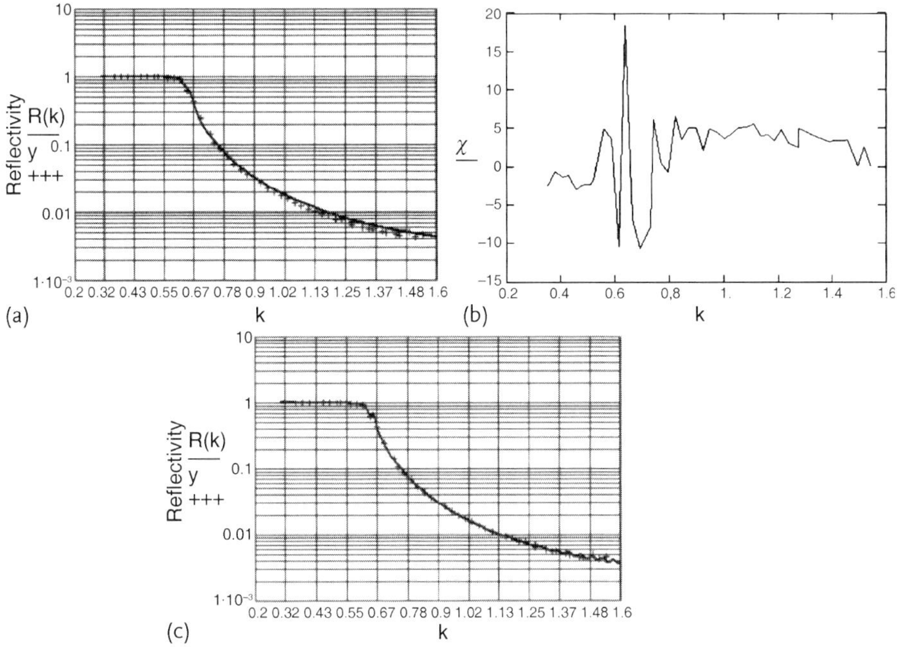

Figure 4.5 Fitting of the data from Hungary [188]: (a) with four parameters and $R(k)$ given by (4.25); (b) the distribution of $\chi(k)$ for such a fitting; (c) fitting with nine parameters, as shown in Eq. (4.26).

Fitting of this expression gives $u' = 0.408$ and $u'' = 0.003$, which is high for glass, but not negative, and $\delta = 4\%$ and $n_b = 0.0024$, which is quite reasonable. The result of fitting on a logarithmic scale is shown in Figure 4.5a. It looks quite reasonable; however, $\chi^2 = 29.1$ for this fitting is still too high. Distribution $\chi(k)$, shown in Figure 4.5b, demonstrates that there is still some peculiarity in the experimental data near the critical edge. After fitting with even more (nine) free parameters, one can achieve $\chi^2 = 4.5$ and even lower. The formula for the fitting and additional parameters is

$$R(k) = \int_{k-\delta}^{k+\delta} |\vec{r_0}(p)|^2 \left[a_0 + a_1|p - p'|^4 + a_2|p - p'|^6\right] \frac{dp}{2\delta} + n_b + a_3|k - k'|^4 + a_4|k - k'|^6 \,, \quad (4.26)$$

where $p' = \sqrt{p^2 - u}$. We do not list the fitted values for these parameters because their physical meaning is not yet clear. However, the result of the fitting, which is shown in Figure 4.5c, is interesting because it demonstrates that there is some unexpected detail near the critical edge. The problem is whether it is related to a physical phenomenon or whether it points out some instrumental peculiarity. Both cases are important for our understanding, so this question must be resolved.

4.2.5 Surface Waves

Surface waves propagate along the surface and decay exponentially in both directions: inside matter $x > 0$, and outside it. That is, the wave function is

$$\psi(x) = \Theta(x < 0)\exp(\kappa_1 x) + \Theta(x > 0)\exp(-\kappa_2 x) \,. \quad (4.27)$$

It is clear that such a function cannot be matched at the interface $x = 0$, because its derivative cannot be continuous. Therefore, surface states of particles do not exist. Nevertheless it was suggested [180] that there is some surfacelike solution of the Schrödinger equation for the case of a thin film with repulsive potential $u_0 > 0$ covering a substrate with an infinite potential. It is necessary to say that the solution found in [180] is not acceptable. Though it exists, the solution is nonphysical because it violates unitarity. According to it, particles are created in a vacuum [181].

In relation to this problem, the question arises of why surface waves exist in electrodynamics and in elasticity theory. The answer is simple. In quantum mechanics the wave equation is the main equation defined in all the space simultaneously. In electrodynamics the main equations are the Maxwell equations, which are used to derive wave equations. If the electromagnetic constants ε and μ are discontinuous, the wave equations in general can be derived separately in spaces with different ε and μ and the matching conditions are derived not from the wave equation but from the Maxwell equations. In particular, instead of continuity of the normal derivative of the wave function, one requires continuity of some field components.

Therefore, a discontinuous derivative of the wave function is not forbidden. The same is true for elastic waves [181].

4.2.6
Goos–Hänchen Shift

The Goos–Hänchen (G–H) shift was predicted for light by Newton. According to him, the light ray at total reflection dives in the reflecting medium and goes out of it at a point B (see Figure 4.6) shifted with respect to the point of entry A (see, for example, [182] and references therein). It was measured by Goos and Hänchen in 1947–1949 [183, 184]. This shift depends on the light wavelength and angle of incidence but its typical value was on the order of 1 μm. Since neutrons are also waves, they also should have a G–H shift.

It is impossible to define the entrance and exit points for plane waves. So we need to consider reflection of a wave packet.

Let the incident neutron be described by a wave packet

$$\psi_0(\boldsymbol{r}, t) = \int A(\boldsymbol{q}, \boldsymbol{k}_0) \exp(i\boldsymbol{q}\boldsymbol{r} - i\omega_q t) d^3q \,, \tag{4.28}$$

where $\boldsymbol{q} = (\boldsymbol{q}_{\|}, -q_z)$, $\boldsymbol{k}_0 = (\boldsymbol{k}_{0\|}, -k_{0z})$ is the average momentum of the incident particle, vectors with index $\|$ are parallel to the surface, and $\omega_q = q^2/2$ (we take units $\hbar = m = 1$). To be specific, we suppose that it is a Gaussian one with

$$A(\boldsymbol{q}, \boldsymbol{k}_0) = \frac{1}{(2\pi s^2)^{3/2}} \exp\left(-\frac{(\boldsymbol{q} - \boldsymbol{k}_0)^2}{2s^2}\right) , \tag{4.29}$$

where s is the width of the packet in the momentum space. We suppose the mirror to be the half-space $z < 0$ with optical potential u_0, and the neutron falls on it from the half-space $z > 0$.

The reflected wave function is

$$\psi_r(\boldsymbol{r}, t) = \int \vec{r_0}(q_z) A(\boldsymbol{q}, \boldsymbol{k}_0) \exp(i\boldsymbol{q}_r\boldsymbol{r} - i\omega_q t) d^3q \,, \tag{4.30}$$

where $\boldsymbol{q}_r = (\boldsymbol{q}_{\|}, q_z)$ and $\vec{r}(q_z)$ is given by Eq. (4.11).

We shall see that with the wave packet we can predict two effects: the G–H shift at total reflection and nonspecularity at nontotal reflection [185].

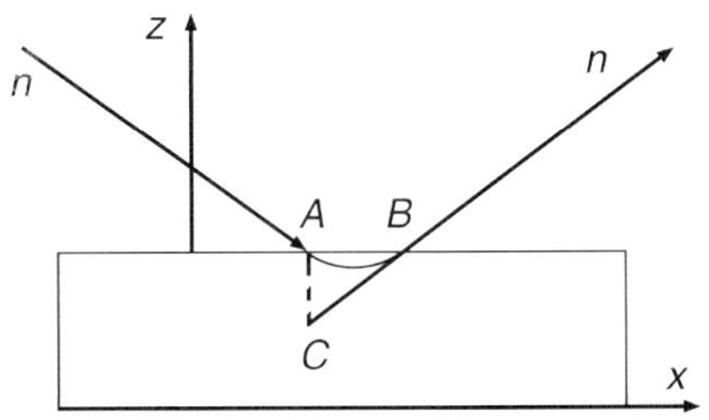

Figure 4.6 Explanation of the Goose–Hänchen shift.

4.2.6.1 **Total Reflection**

After substitution of Eq. (4.29) into Eq. (4.28) and integration over d^3q we obtain

$$\psi_0(\boldsymbol{r}, t) = \frac{1}{(1 + is^2t)^{3/2}} \exp\left(-\frac{s^2(\boldsymbol{r} - \boldsymbol{k}_0 t)^2}{2(1 + is^2t)} + i\boldsymbol{k}_0\boldsymbol{r} - i\omega_k t\right), \tag{4.31}$$

where s^2t characterizes spreading of the wave packet. We suppose that s is small and neglect spreading.

The cross section of the wave packet at the interface depends on time. We can choose the time origin $t = 0$, when the wave packet center crosses the surface. The footprint of the packet on the surface at that time is

$$\psi(\boldsymbol{r}_\|, z = 0, t = 0) = \exp\left(-\frac{s^2}{2}\boldsymbol{r}_\|^2 + i\boldsymbol{k}_\|\boldsymbol{r}_\|\right), \tag{4.32}$$

with its center at the point $\boldsymbol{r} = 0$. The center of the footprint of the reflected neutron is shifted along the surface by distance δ, and our task is to evaluate this distance.

According to Eq. (4.17), at total reflection we have

$$\vec{r_0}(q_z) = \exp(-i\gamma(q_z)), \tag{4.33}$$

where the phase $\gamma(q_z) = 2\arccos(q_z/\sqrt{u_0})$ can be expanded in a Taylor series near $q_z = k_{0z}$. In a linear approximation it is

$$\gamma(q_z) = \gamma(k_{0z}) - \zeta(q_z - k_{0z}), \tag{4.34}$$

where

$$\zeta = \frac{2}{k''_{0z}}, \quad k''_{0z} = \sqrt{u_0 - k_{0z}^2}. \tag{4.35}$$

After substitution of Eq. (4.34) into Eq. (4.30) and integration over d^3q we obtain

$$\psi_r(\boldsymbol{r}, t) \approx \exp\left(-\frac{s^2}{2}(\boldsymbol{r} + \boldsymbol{\zeta} - \boldsymbol{k}_{0r}t)^2 + i\boldsymbol{k}_{0,r}\boldsymbol{r} - i\omega_k t\right), \tag{4.36}$$

where $\boldsymbol{k}_{0r} = (\boldsymbol{k}_{0\|}, k_{0z})$ and $\boldsymbol{\zeta} = (0, 0, \zeta)$.

We see that at the entrance time $t = 0$ of the incident wave packet the center of the reflected wave packet is at the point $z = -\zeta$ under the surface. Therefore, at the exit time $t' = \zeta/k_{0z}$ its center is shifted with respect to $\boldsymbol{r} = 0$ by the amount

$$\delta = t'k_{0\|} = \frac{2}{k''_{0z}}\frac{k_{0\|}}{k_{0z}}. \tag{4.37}$$

To estimate the value of such a shift for thermal neutrons, we approximate $k_{0z}/k_{0\|} \approx 10^{-3}$, and $k''_{0z} \sim \sqrt{u}$, then $\delta \sim 10^6$ Å, or $\delta \sim 0.1$ mm.

We considered the simplest version of the neutron G–H shift from a passive mirror. Modification of this shift in the case of active media is discussed in [186].

4.2.6.2 Nontotal Reflection

In the case of nontotal reflection the reflection amplitude can be represented as

$$\vec{r_0}(q_z) = \exp(-\chi(q_z)) , \tag{4.38}$$

where

$$\chi(q_z) = -2\ln\left(\frac{q_z - \sqrt{q_z^2 - u_0}}{\sqrt{u_0}}\right) \tag{4.39}$$

can be expanded in a Taylor series near $q_z = k_{0z}$. In a linear approximation it is

$$\chi(q_z) = \chi(k_{0z}) + \varsigma(q_z - k_{0z}) , \tag{4.40}$$

where

$$\varsigma = \frac{2}{k'_{0z}}, \quad k'_{0z} = \sqrt{k_{0z}^2 - u_0} . \tag{4.41}$$

After substitution of Eq. (4.40) into Eq. (4.30) and integration over d^3q we obtain

$$\psi_r(\boldsymbol{r}, t) \approx \vec{r}(k_{0z})e^{\varsigma^2 s^2/2} \exp\left(-\frac{s^2}{2}(\boldsymbol{r} - \boldsymbol{k}'_r t)^2 + i\boldsymbol{k}'_r \boldsymbol{r} - i\omega_{k'} t\right) , \tag{4.42}$$

where $\boldsymbol{k}'_r = \boldsymbol{k}_{0r} - \boldsymbol{\kappa}$ and $\boldsymbol{\kappa} = (0, 0, s^2\varsigma)$. We see that the normal component of the wave vector $\boldsymbol{k}_0$ slightly decreased, because reflection of the waves with higher q_z is less than that of the waves with lower q_z.

The decrease of k_{0z} is equivalent to the decrease of grazing angle. If we denote $k_{0z} = k_0 \sin\theta$, then $\kappa \sim k_0 \Delta\theta$, that is,

$$\Delta\theta = 2\frac{s^2}{k_0 k'_{0z}} . \tag{4.43}$$

If the width s is given by Eq. (4.21), then for $k'_{0z} \sim s$ we get $\Delta\theta \sim 10^{-5}$ or on the order of an angular second.

4.3
Scattering by a Rectangular Potential Barrier

After the digression on the ultracold neutron anomaly and discussion of some specific properties of neutron reflection from an interface, let us return to one-dimensional scattering problems with a given optical potential u_0.

Consider reflection from and transmission through the potential barrier shown in Figure 4.7. Commonly, the reflection r and transmission t amplitudes are calculated by solving the Schrödinger equation with the potential $u_0\Theta(0 < x < d)$. The solution is given by

$$\psi(x) = \Theta(x < 0)\left[e^{ikx} + re^{-ikx}\right] + \Theta(0 < x < d)\left[Ae^{ik'x} + Be^{-ik'x}\right] + \Theta(x > d)te^{ik(x-d)} .$$

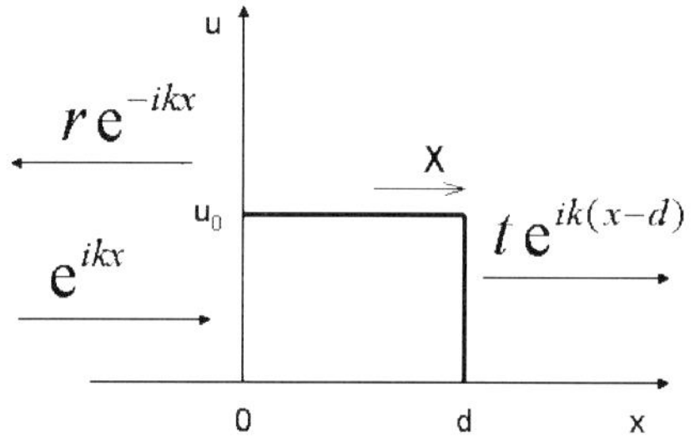

Figure 4.7 Rectangular potential barrier of height u_0 and width d.

(4.44)

Note that the wave function to the right of the barrier is written as $t\exp(ik(x-d))$, which is different from the familiar $t_{\text{usual}}\exp(ikx)$. Such a definition is more convenient, and it only redefines the phase of the transmission amplitude $t = \exp(ikd)t_{\text{usual}}$.

By matching the wave function Eq. (4.44) at $x = 0$ and $x = d$, we obtain a system of four equations for four unknown coefficients r, A, B, and t:

$$1 + r = A + B\,, \qquad k(1-r) = k'(A-B)\,,$$
$$Ae^{ik'd} + Be^{-ik'd} = t\,, \quad k'\left(Ae^{ik'd} - Be^{-ik'd}\right) = kt\,. \tag{4.45}$$

Solution of these equations gives the reflection and transmission amplitudes r and t. We now show how these amplitudes can be found without such a boring approach.

4.3.1
The Other Method

We know that the amplitude of the incident wave at the first interface is unity. If we could only know the amplitude of the wave incident inside the barrier (denote it by X) onto the interface at $x = d$, we would immediately obtain the transmission amplitude of the barrier

$$t = \overrightarrow{t_d}X\,. \tag{4.46}$$

and the reflection amplitude

$$r = \overrightarrow{r_0} + \overleftarrow{t_0}\exp(ik'd)\overrightarrow{r_d}X\,, \tag{4.47}$$

as a superposition of reflection from the first interface and the contribution of wave X, which after reflection from the second interface propagates toward the first one and passes through it into the vacuum. So we need to find X. But for it we can write a self-consistent equation:

$$X = \exp(ik'd)\overrightarrow{t_0} + \exp(ik'd)\overleftarrow{r_0}\exp(ik'd)\overrightarrow{r_d}X\,. \tag{4.48}$$

Here, the first term on the right describes the initial incident wave, which enters the barrier and travels to the right edge. The second term corresponds to the wave generated by X itself. The wave incident at the right edge first reflects from the edge with amplitude $\overrightarrow{r_d}$, then travels to the left edge, then reflects from it with amplitude $\overleftarrow{r_0}$, and travels back to the right edge. Both waves (or terms) on the right side of Eq. (4.48) combine to form the total amplitude of the wave incident at the right edge of the barrier. However, we already designated this amplitude by X, therefore the right side of Eq. (4.48) must be equal to X. The result is Eq. (4.48), which instantly leads to the solution

$$X = \frac{\overrightarrow{t_0} \exp(ik'd)}{1 - r_0^2 \exp(2ik'd)} , \tag{4.49}$$

where we denoted $\overrightarrow{r}(d) = \overleftarrow{r}(0) = -\overrightarrow{r_0} = -r_0$ according to Eqs. (4.11) and (4.12).

Substitution of X into Eqs. (4.46) and (4.47) gives

$$r = r_0 + \frac{\overrightarrow{t_0}\overleftarrow{t_0} \exp(2ik'd)(-r_0)}{1 - r_0^2 \exp(2ik'd)} = r_0 \frac{1 - \exp(2ik'd)}{1 - r_0^2 \exp(2ik'd)} ,$$

$$t = e^{ik'd} \frac{1 - r_0^2}{1 - r_0^2 \exp(2ik'd)} . \tag{4.50}$$

The same result was obtained by Godfrey [189], who used summation of the geometrical progression, which resulted from multiple reflections at two edges inside the barrier.

The expressions (4.50) reveal an interesting symmetry: amplitude r converts into t by permuting r_0 and $\exp(ik'd)$, and they show that at $u_0 = 0$ the transmission amplitude is $\exp(ikd)$.

4.3.2
Asymptotical Behavior of Reflectivity from a Rectangular Barrier

With Eq. (4.50) we can find the asymptotic of r for large k. According to Eq. (4.14) we see that r_0 at large k decreases; therefore, r_0^2 in the denominator of Eq. (4.50) can be neglected and we obtain

$$r(k \to \infty) = \frac{u_0}{4k^2}(1 - \exp(2ik'd)) , \quad |r(k \to \infty)|^2 = \frac{u_0^2 \sin^2(k'd)}{4k^4} . \tag{4.51}$$

We see that the reflectivity oscillates between zero and $u_0^2/4k^4$. This result is valid for ideal monochromatization. In the case of large d and some spread of k in the interval $\delta k > 2\pi/d$, we must average Eq. (4.51) over δ. As a result, the factor $\sin^2(k'd)$ is averaged to $1/2$, and we get for the reflectivity

$$|r(k \to \infty)|^2 = \frac{u_0^2}{8k^4} , \tag{4.52}$$

which is twice as high as the reflectivity Eq. (4.14) from the semi-infinite plate. This result is correct only when the imaginary part of the potential can be neglected. In the presence of the imaginary part, u'', which is always negative, the wave number k', in the case $\mathrm{Re}(k'^2) > 0$, acquires a positive imaginary part $k'' \approx u''/2\mathrm{Re}(k')$. Therefore, $\exp(ik'd)$ decays in proportion to $\exp(-k''d)$. For sufficiently large d we can neglect $\exp(2ik'd)$, and then the reflectivity of a layer becomes identical to Eq. (4.14). However, we must be cautious, because with increase of k' the imaginary part k'' decreases, therefore $\exp(-k''d) \to 1$, the reflectivity of the layer again becomes oscillating, and its averaged asymptotic again becomes equal to Eq. (4.52).

4.3.3
Wave Function inside the Barrier

Using X, the total wave function inside the barrier is immediately clear,

$$\psi(0 < x < d) = [\exp(ik'(x-d)) + \exp(-ik'(x-d))(-r_0)]X\,. \tag{4.53}$$

Indeed, the wave traveling to the right, $X\exp(ik'(x-d))$, must be equal to X next to the right barrier edge. The wave traveling to the left, $(-r_0)X\exp(-ik'(x-d))$, is equal to X reflected from the right edge. After substitution of Eq. (4.49) we get

$$\psi(0 < x < d) = \frac{\vec{t_0}\exp(ikd)}{1 - r_0^2\exp(2ikd)}\left\{\exp(ik'[x-d]) - r_0\exp(-ik'[x-d])\right\}\,. \tag{4.54}$$

4.4
Neutron Scattering from Semitransparent Mirrors

To describe reflection and transmission for arbitrary potentials, consider two semitransparent mirrors [193], separated by a distance l (Figure 4.8), Each mirror is described by some potential u_i $(i = 1, 2)$. The reflection and transmission amplitudes for each potential u_i are r_i and t_i. The task is to find the total amplitudes for reflection R_{12} and transmission T_{12} of the system of two mirrors. To solve this problem, we apply the same method as in the case of the rectangular barrier. Assume that a plane wave $\exp(ikx)$ is incident on the first mirror from the left. Its amplitude is 1 at the surface of the first mirror. The wave incident on the second mirror has amplitude X at its surface (Figure 4.8). The equation for X is similar to Eq. (4.48),

$$X = \exp(ikl)t_1 + \exp(ikl)r_1\exp(ikl)r_2X\,. \tag{4.55}$$

The solution of Eq. (4.55) is

$$X = \frac{t_1\exp(ikl)}{1 - r_1r_2\exp(2ikl)}\,. \tag{4.56}$$

The total transmission and reflection amplitudes can now be easily represented in terms of X:

$$R_{12} = r_1 + t_1\exp(ikl)r_2X\,, \quad T_{12} = t_2X\,. \tag{4.57}$$

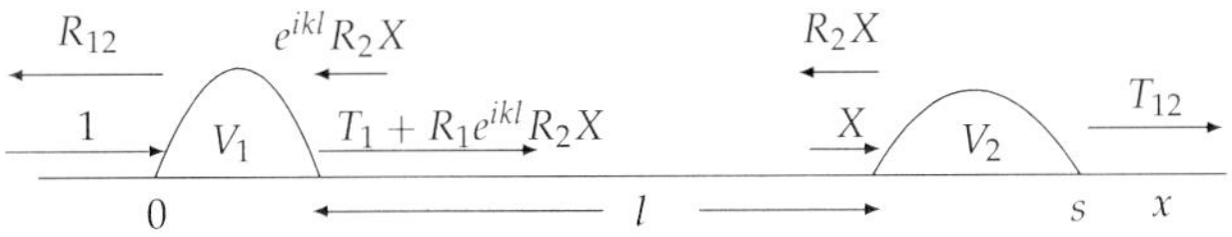

Figure 4.8 System of two semitransparent mirrors, represented by two potentials separated by a distance l.

Substitution of Eq. (4.56) gives

$$R_{12} = r_1 + t_1^2 \frac{\exp(2ikl) r_2}{1 - r_1 r_2 \exp(2ikl)}, \quad T_{12} = \frac{t_2 \exp(ikl) t_1}{1 - r_1 r_2 \exp(2ikl)}. \tag{4.58}$$

4.4.1 Bringing Two Mirrors Together

The distance l between two mirror is arbitrary and therefore, in particular, it can be set to zero. Equations (4.55) and (4.57) in this case are

$$X = t_1 + r_1 r_2 X, \quad R_{12} = r_1 + t_1 r_2 X, \quad T_{12} = t_2 X, \tag{4.59}$$

and amplitudes R and T become

$$R_{12} = r_1 + t_1^2 \frac{r_2}{1 - r_1 r_2}, \quad T_{12} = \frac{t_2 t_1}{1 - r_1 r_2}. \tag{4.60}$$

This result is very simple; however, it leads to an important discovery. We found that an arbitrary potential can be virtually split at any point into two parts, with the total reflection and transmission amplitudes expressed via the transmission and reflection amplitudes of the parts. However, we assumed that the mirrors and the respective subpotentials are symmetrical, which in general may not be true (Figure 4.9). In the case of asymmetrical potentials, the reflection and transmission amplitudes r_i and t_i depend on the direction of propagation. Therefore, Eq. (4.60) are to be rewritten as

$$\overrightarrow{R}_{12} = \overrightarrow{r_1} + \overleftarrow{t_1} \frac{\overrightarrow{r_2}}{1 - \overleftarrow{r_1}\,\overrightarrow{r_2}} \overrightarrow{t_1}, \quad \overleftarrow{R}_{21} = \overleftarrow{r_2} + \overrightarrow{t_2} \frac{\overleftarrow{r_1}}{1 - \overleftarrow{r_1}\,\overrightarrow{r_2}} \overleftarrow{t_2}, \tag{4.61}$$

and

$$\overrightarrow{T}_{12} = \frac{\overrightarrow{t_2}\,\overrightarrow{t_1}}{1 - \overleftarrow{r_1}\,\overrightarrow{r_2}}, \quad \overleftarrow{T}_{21} = \frac{\overleftarrow{t_1}\,\overleftarrow{t_2}}{1 - \overleftarrow{r_1}\,\overrightarrow{r_2}}, \tag{4.62}$$

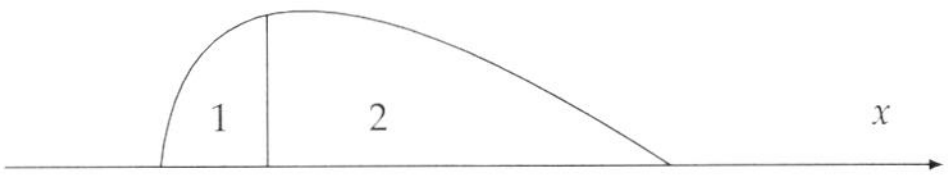

Figure 4.9 At any point every potential can be subdivided into two parts. Equations (4.61) and (4.62) describe the total reflection and transmission amplitudes, via the reflection and transmission amplitudes of the parts.

where the arrows indicate the directions of propagation for waves incident on the corresponding potential.

4.4.2 Splitting of Rectangular Potentials

It is a good exercise to verify the results of the previous section with a rectangular potential. Take a rectangular barrier of width d, and split it into two parts d_1, $d_2 = d - d_1$. Then,

$$r(d_1 + d_2) = \frac{r(d_1) - [r^2(d_1) - t^2(d_1)]r(d_2)}{1 - r(d_1)r(d_2)},$$

$$t(d_1 + d_2) = \frac{t(d_1)t(d_2)}{1 - r(d_1)r(d_2)}. \tag{4.63}$$

The validity of these relations can be directly checked by substitution of Eq. (4.50) into the last equations.

4.4.3 A Film over a Substrate

If we have a film with potential u_f and thickness l_f, evaporated over a thick (we can consider it as infinitely thick) substrate with potential u_s, we can find the reflectivity of the full system according to Eq. (4.60):

$$R_{fs}(k) = \left| r_f(k) + t_f^2(k)\frac{r_{0s}(k)}{1 - r_f(k)r_{0s}(k)} \right|^2, \tag{4.64}$$

where r_f and t_f are given by

$$r_f(k) = r_{0f}(k)\frac{1 - \exp(2ik'_f l_f)}{1 - r_{0f}^2(k)\exp(2ik'_f l_f)},$$

$$t_f(k) = e^{ik'_f l_f}\frac{1 - r_{0f}^2(k)}{1 - r_{0f}^2(k)\exp(2ik'_f l_f)}, \tag{4.65}$$

and r_{0f} and r_{0s} are given by

$$r_{0f}(k) = \frac{k - k_f}{k + k_f}, \quad r_{0s}(k) = \frac{k - k_s}{k + k_s}, \quad k_{f,s} = \sqrt{k^2 - u_{f,s}}. \tag{4.66}$$

4.4.4 Experience with an Experiment

Now we can check how well the theoretical reflectivity matches the experimental one. In Figure 4.10 we present the fitting of experimental data from Japan[11] ob-

11) A lot of thanks again to Dr Ryuji Maruyama from J-PARC of the Japanese Atomic Energy Agency for sending me these data.

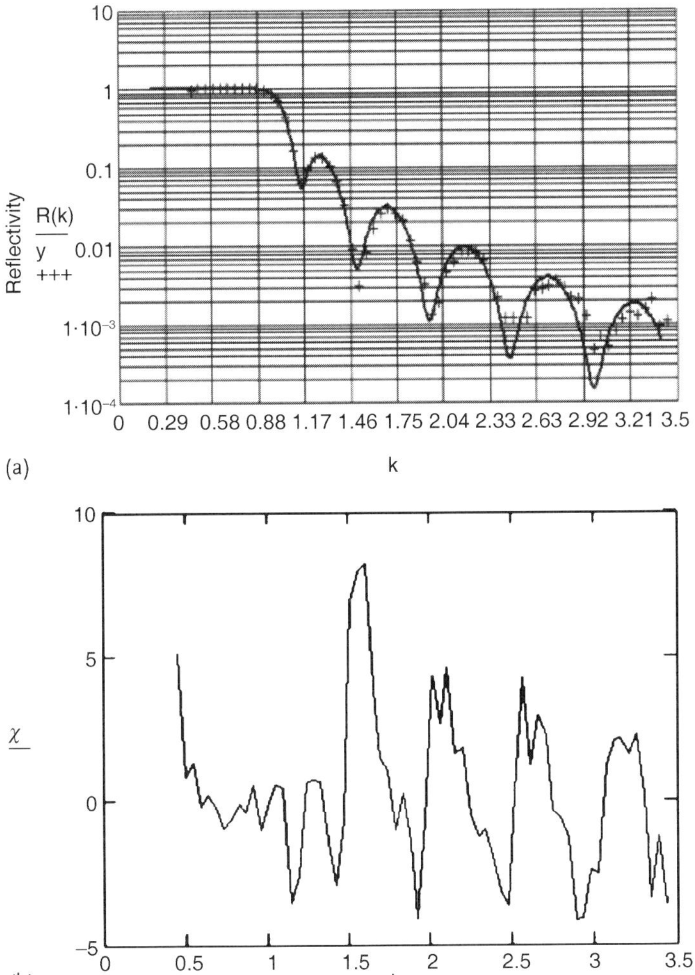

Figure 4.10 Reflectivity of a Ni film of thickness l over a Si substrate, fitted with the help of Eq. (4.67) and five fitting parameters. (a) Result of fitting on a logarithmic scale; (b) $\chi(k)$ distribution.

tained with a $\theta-2\theta$ reflectometer on the reflectivity from a Ni film of thickness 50 nm evaporated on a Si substrate. For the fitting the formula

$$R(k) = \int_{k-\delta}^{k+\delta} R_{fs}(k)\frac{dp}{2\delta}\,, \tag{4.67}$$

was used, where $R_{fs}(k)$ is given in Eq. (4.64). The fitting parameters were the real and imaginary parts of the Ni potential $u_{\mathrm{Ni}} = u'_{\mathrm{Ni}} - iu''_{\mathrm{Ni}}$, the thickness of the Ni film, l_f, the real part of the Si potential u'_{Si} (the imaginary part of the Si potential was fixed $u''_{\rm Si} = 2 \cdot 10^{-4}$), and the resolution δ. After fitting, we obtained $u_{\rm Ni} = 0.961 - 0.0022i$, $u'_{\rm Si} = 0.214$, $l = 50.35$ nm, and $\delta = 6\%$. All the values are quite realistic; however, $\chi^2 = 8.2$ for this fitting is too large. The distribution $\chi(k)$ is shown in Figure 4.10b. If we increase the number of fitting parameters, and include, for instance, the imaginary part of Si, $u''_{\rm Si}$, we shall decrease χ^2 to 5.6, but the resultant Si potential, $u_{\rm Si} = 0.6345{-}0.508i$, will be quite unrealistic.

4.4.5
Properties of the Reflection and Transmission Amplitudes

4.4.5.1 Asymmetrical Potentials

In general, the reflection and transmission amplitudes from the left and from the right of a potential can be different. To see what the difference can be we may consider some asymmetrical potential, which can be easily calculated. For instance, let us consider an asymmetrical potential constructed with two symmetrical ones, as shown in Figure 4.11.

Expressions for the total transmission and reflection amplitudes for such a potential are given in Eq. (4.60). We immediately see that $T_{12} = T_{21}$, and therefore for any arbitrary potential

$$\overleftarrow{t} = \overrightarrow{t} \,. \tag{4.68}$$

However, $R_{12} \neq R_{21}$, and therefore $\overrightarrow{r} \neq \overleftarrow{r}$, but for real potentials the difference is only in phase, that is,

$$\overleftarrow{r} = \exp(i\chi)\overrightarrow{r} \,, \tag{4.69}$$

where χ is a real number.

Indeed, the unitarity principle tells us that the number of reflected and transmitted particles is equal to the number of incident ones; therefore,

$$|\overleftarrow{r}|^2 + |\overleftarrow{t}|^2 = |\overrightarrow{r}|^2 + |\overrightarrow{t}|^2 = 1 \,. \tag{4.70}$$

Taking into account Eq. (4.68), we immediately obtain $|\overleftarrow{r}| = |\overrightarrow{r}|$, which means that there can only be a phase difference between the two amplitudes, as shown in Eq. (4.69). For a symmetrical potential the phase difference is zero, $\chi = 0$.

Note that Eq. (4.70) is only valid when there is no absorption inside a potential, that is, when u is real. If the potential has a nonzero imaginary part, then the phase χ has a nonzero imaginary part too. Therefore, the absolute values of the reflection amplitudes are not equal to each other. However, the transmission amplitudes from left and right are equal even for an absorbing potential.

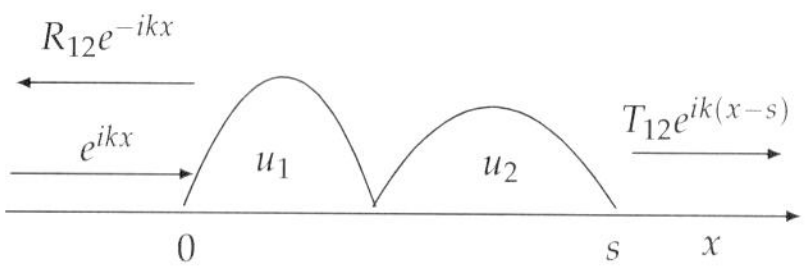

Figure 4.11 Using two symmetrical potentials to construct an asymmetrical potential.

4.4.5.2 Relations between Phases of *r* and *t*

Here we show that the difference between phases $\arg(r)$ and $\arg(t)$ is equal to $\pm\pi/2$ for real symmetrical potentials. To prove that, consider the gedankenexperiment, shown in Figure 4.12.

Assume that mirror M1 is described by an arbitrary, but symmetrical potential. The incident beam is split into two parts by mirror M1. The two beams then undergo total reflections on M2 and M3. The two beams then recombine on M4, which is identical to M1. Conservation of particle numbers requires that the total counts of the two detectors correspond to the number of incident neutrons. This means that

$$|rt|^2|e^{i\varphi} + e^{i\psi}|^2 + |r^2 e^{i\varphi} + t^2 e^{i\psi}|^2 = 1 \,. \tag{4.71}$$

If we write $r = |r|\exp(i\chi_r)$, $t = |t|\exp(i\chi_t)$ and assume for simplicity that $\varphi = \psi = 0$, then the last equation converts to

$$(|r|^2 + |t|^2)^2 + 2|rt|^2[1 + \cos(2\chi_r - 2\chi_t)] = 1 \,. \tag{4.72}$$

Without absorption, the first term on the left in (4.72) is equal to 1; thus, the rest of the expression is zero. Therefore,

$$2\chi_r - 2\chi_t = \pm\pi \,, \tag{4.73}$$

and $\chi_r = \chi_t \pm \pi/2$. End of proof. We leave it to the reader to check that the result is the same for arbitrary φ and ψ.

It is now clear that if $t = |t|e^{i\varphi_t}$, then $r = \pm i|r|e^{i\varphi_t}$. Therefore, $t \pm r$ are unit complex numbers $\exp(i\xi_\pm)$, where $\xi_\pm = \varphi_t \pm \arcsin(|r|)$ and $t^2 - r^2 = \exp(2i\varphi_t)$. According to this, the total reflection amplitude Eq. (4.60) of two symmetrical po-

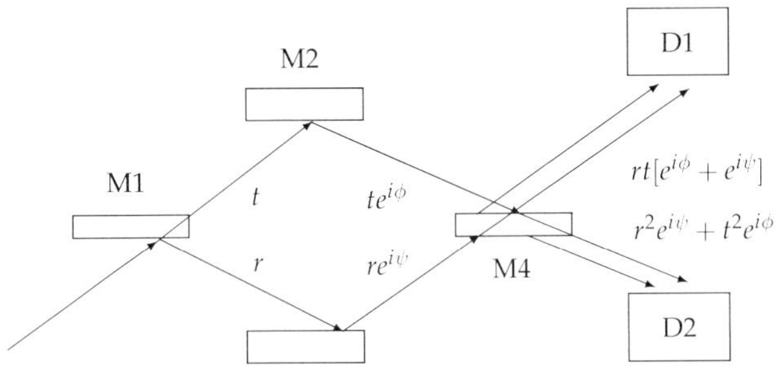

Figure 4.12 Gedankenexperiment aimed at finding the phase difference between *r* and *t*. Semitransparent mirrors M1 and M4 are identical. Mirrors M2 and M3 have 100% reflectivity, but different phases. Acquired phases φ and ψ correspond to two different propagation trajectories between mirrors M1 and M4. The total neutron counts from two detectors, D1 and D2, should be equal to the number of incident neutrons.

tentials can be written as

$$R_{12} = \frac{r_1 + (t_1^2 - r_1^2) r_2}{1 - r_1 r_2} = \frac{r_1 + \exp(2i\phi_{t1}) r_2}{1 - r_1 r_2} = \frac{r_1 - \exp(2i\phi_{r1}) r_2}{1 - r_1 r_2} , \quad (4.74)$$

where $\phi_{r,t1}$ are phases $\arg(r_1)$ and $\arg(t_1)$, respectively.

Thus, it is obvious that if one of the potentials has 100% reflectivity, that is, $r_j = \exp(i\chi_j)$ with $j = 1$ or 2, then the whole potential has 100% reflectivity too.

The above expression for the phase relations Eq. (4.73) can be easily generalized for an asymmetrical potential. If we write $\overrightarrow{r} = |\overrightarrow{r}| \exp(i\overrightarrow{\chi}_r)$ and $\overleftarrow{r} = |\overleftarrow{r}| \exp(i\overleftarrow{\chi}_r)$, then instead of Eq. (4.73) we get

$$\frac{1}{2}(\overrightarrow{\chi}_r + \overleftarrow{\chi}_r) = \chi_t \pm \frac{\pi}{2} . \quad (4.75)$$

4.5
Periodic Systems

> What novelty can we find in all that after excellent works by Parrait?
>
> A referee

The splitting method is very effective when applied to periodic systems. Imagine a periodic potential with N arbitrary periods. We want to find the reflection R_N and transmission T_N amplitudes of the total system. It turns out that the easiest way to tackle this problem is to find first the reflection amplitude R_∞ from a semi-infinite system $N \to \infty$.

To find R_N and T_N, we start [194–198] by examining a semi-infinite periodic potential, shown in Figure 4.13. The arrows indicate the incident and reflected waves, with the reflection amplitude denoted by R. The wave incident on the second period is denoted by X. Similarly to Eq. (4.59), we can immediately write a system of equations for X and R:

$$X = t + rRX, \quad R = r + tRX , \quad (4.76)$$

where r and t are the reflection and transmission amplitudes for a single isolated period (let it be symmetrical for simplicity), and we took into account that the semi-infinite potential does not change when its first period is removed. Note the symmetry: the second equation transforms into the first one after exchange of r and t.

To solve this system, determine X from the first equation and substitute it into the second one. As a result, we get

$$R = r + \frac{t^2 R}{1 - rR} . \quad (4.77)$$

This is a quadratic equation, which can be rewritten as $y^2 - 2py + 1 = 0$, where $y = R$ and $p = (r^2 + 1 - t^2)/2r$. The solution to such an equation is trivial,

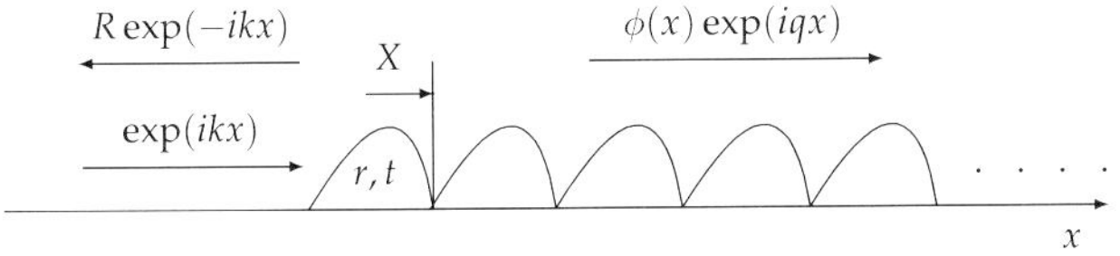

Figure 4.13 Semi-infinite periodic system.

$y = p - \sqrt{p^2 - 1}$, but it can also be represented as

$$y = \frac{\sqrt{p+1} - \sqrt{p-1}}{\sqrt{p+1} + \sqrt{p-1}}, \tag{4.78}$$

which immediately leads to [195–198]

$$R = \frac{\sqrt{(1+r)^2 - t^2} - \sqrt{(1-r)^2 - t^2}}{\sqrt{(1+r)^2 - t^2} + \sqrt{(1-r)^2 - t^2}}. \tag{4.79}$$

The last expression is very easy to remember and also easy to verify, as shown below.

1. Imagine that the potential is absent and all we have is free space. However, an empty space can also be thought of as a periodic system of empty periods. In this case a single period reflects nothing and therefore the whole system also reflects nothing, that is, $r = 0$ should lead to $R = 0$. It is easy to check using Eq. (4.79) that this is true.[12]
2. On the other hand, imagine that the height of the potential is very large or infinite and therefore the transmission t is zero. Then the total reflection R from the whole potential should be equal to the reflection from the first period r, since other periods effectively do not participate in the reflection processes. It is easy to see that this is also true.

Recall that the wave function inside a periodic potential (see Figure 4.13) can be represented by a product $\psi(x) = \exp(iqx)\phi(x)$ of a periodic function $\phi(x)$ and a phase factor $\exp(iqx)$ with the Bloch wave vector q. Therefore, if the wave incident on the first period is 1, the wave incident on the second period is $\exp(iqa)$, where a is the period thickness. But we denoted the latter wave by X, therefore $X = \exp(iqa)$!

It is easy to find X from the system (4.76). To do that, we need to find R from the second equation and substitute it into the first one. As a result, we again obtain a quadratic equation, this time for X. We can, however, immediately write down the solution using the r–t permutation symmetry of Eqs. (4.76). It follows from this symmetry that

$$X \equiv e^{iqa} = \frac{\sqrt{(1+t)^2 - r^2} - \sqrt{(1-t)^2 - r^2}}{\sqrt{(1+t)^2 - r^2} + \sqrt{(1-t)^2 - r^2}}. \tag{4.80}$$

12) Note that the above arguments are not true for $t^2 = 1$, because in this case we get an uncertain ratio 0/0 .

This formula, like (4.79), can also be easily checked.

1. Imagine again that the free space is a periodic system of empty periods. In this case $r = 0$ and $t = \exp(ika)$. By substituting these values into Eq. (4.80), we find that $\exp(iqa) = \exp(ika)$, that is, $q = k$, as expected.
2. On the other hand, imagine that the potential height of a single period is very large, and the transmission t is equal to zero. By substituting $t = 0$ into Eq. (4.80), we obtain $\exp(iqa) = 0$, that is, $\mathrm{Im}\,(q) = i\infty$, which means that nothing reaches the second period.

The formulas obtained can be modified in different ways. We consider here only one important modification valid for real potentials. Let us write $t = |t|\exp(i\phi_t)$ and $r = \pm i|r|\exp(i\phi_t)$. Then, $t^2 - r^2 = \exp(2i\phi_t)$. Therefore, $(1 \pm r)^2 - t^2 = 1 \pm 2r - \exp(2i\phi_t)$, and $(1 \pm t)^2 - r^2 = 1 \pm 2t + \exp(2i\phi_t)$. Substitution of these expressions into Eqs. (4.79) and (4.80) leads to

$$R = \frac{\sqrt{\sin\phi_t + |r|} - \sqrt{\sin\phi_t - |r|}}{\sqrt{\sin\phi_t + |r|} + \sqrt{\sin\phi_t - |r|}} ,$$

$$e^{iqa} = \mp\frac{\sqrt{\cos\phi_t + |t|} - \sqrt{\cos\phi_t - |t|}}{\sqrt{\cos\phi_t + |t|} + \sqrt{\cos\phi_t - |t|}} . \tag{4.81}$$

This representation can be very useful if amplitude t is calculated by perturbation theory. Then $|r| = \sqrt{1 - |t|^2}$, and formulas (4.81) take into account all the required properties of r and t automatically.

Amplitudes r and t depend on energy. At some energies for which $\sin^2\phi_t - |r|^2 < 0$, the amplitude R, according to Eq. (4.81), is a unit complex number, $|R| = 1$. At these energy values we have the total reflection from a semi-infinite periodic potential. This total reflection is called Bragg diffraction. It is important to note that at the points where $\sin^2\phi_t < |r|^2$ and $|R| = 1$, we have $\cos^2\phi_t = 1 - \sin^2\phi_t > |t|^2 = 1 - |r|^2$, and the absolute value of the Bloch phase factor is less than 1, that is, q is the complex number $q = \pi n + iq''$, with integer n and nonzero imaginary part iq''.

To find R_N and T_N we need only to take into account that the semi-infinite potential of a period a is also periodic with a period Na. So we can again consider the semi-infinite periodic potential, but instead of r and t we use R_N and T_N:

$$X^N = T_N + R_N R X^N , \quad R = R_N + T_N R X^N , \tag{4.82}$$

where instead of X we use $X^N = \exp(iqNa)$. The system (4.82) can be resolved for R_N and T_N. As the result we obtain

$$R_N = R\frac{1 - \exp(2iqNa)}{1 - R^2\exp(2iqNa)} , \quad T_N = \exp(iqNa)\frac{1 - R^2}{1 - R^2\exp(2iqNa)} . \tag{4.83}$$

It is easy to see that the above formulas have structure similar to the reflection and transmission amplitudes for a rectangular potential Eq. (4.50).

4.5.1
Asymmetrical Period

If the period is asymmetrical, the reflection amplitudes from the left and the right are different. We distinguish these reflections in the equations by using $\to$ and $\leftarrow$ above the letters. Equations (4.76) for a semi-infinite periodic structure then become

$$\overrightarrow{X} = t + \overleftarrow{r}\,\overrightarrow{R}\,\overrightarrow{X}\,, \quad \overrightarrow{R} = \overrightarrow{r} + t\,\overrightarrow{R}\,\overrightarrow{X}\,. \tag{4.84}$$

By solving the first equation for $\overrightarrow{X}$, and substituting the solution into the second equation, we can obtain the equation for $\overrightarrow{R}$, which can then be rewritten as

$$\overrightarrow{R}^2 - 2\frac{\overrightarrow{r}}{2r^2}(1 + r^2 - t^2)\overrightarrow{R} + \frac{\overrightarrow{r}^2}{r^2} = 0\,,$$

where $r = \sqrt{\overrightarrow{r}\,\overleftarrow{r}}$ is the symmetrized reflection amplitude. Denote $\overrightarrow{R} = (\overrightarrow{r}/r)R$, then the equation transforms to

$$R^2 - 2\frac{1}{2r}(1 + r^2 - t^2)R + 1 = 0\,,$$

and its solution coincides with Eq. (4.79). This means that the asymmetry of r propagates to R, that is,

$$\overleftarrow{R} = \frac{\overleftarrow{r}}{r}R\,, \quad \overrightarrow{R} = \frac{\overrightarrow{r}}{r}R\,, \quad R = \frac{\sqrt{(1+r)^2 - t^2} - \sqrt{(1-r)^2 - t^2}}{\sqrt{(1+r)^2 - t^2} + \sqrt{(1-r)^2 - t^2}}\,, \tag{4.85}$$

but

$$\overleftarrow{X} = \overrightarrow{X} = X = \frac{t}{1 - \overleftarrow{r}\,\overrightarrow{R}} = \frac{t}{1 - rR}\,,$$

that is, the Bloch phase factor is symmetrical.

Similarly, the asymmetry of $\overleftarrow{r}$ and $\overrightarrow{r}$ also propagates to $\overleftarrow{R_N}$ and $\overrightarrow{R_N}$, that is,

$$\overleftarrow{R_N} = \frac{\overleftarrow{r}}{r}R_N, \quad \overrightarrow{R_N} = \frac{\overrightarrow{r}}{r}R_N, \quad R_N = R\frac{1 - \exp(2iqNa)}{1 - R^2\exp(2iqNa)}\,. \tag{4.86}$$

Let us consider an example which shows the above equations at work.

4.5.2
Kronig–Penney Potential

Consider an ideal crystal consisting of an array of thin atomic planes, shown in Figure 4.14a.

The interaction with each plane is described by a δ-like potential $u(x) = 2p\delta(x)$, where p can be tied to the actual properties of the crystal, and the factor 2 is introduced for convenience. By averaging the potential of a single plane over a single

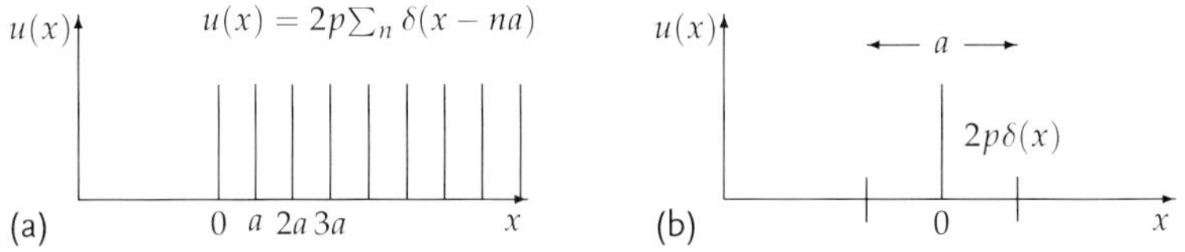

Figure 4.14 (a) A one-dimensional periodic potential, consisting of many planes parallel to the front interface. (b) A symmetrical potential representing a single period consisting of a single plane and two half-period wide gaps on each side of the plane.

period a, we obtain the average interaction potential, which should be equal to $u_0 = 4\pi N_0 b$:

$$\frac{1}{a}\int_0^a u(x)dx = \frac{1}{a}\int_0^a 2p\,\delta(x)dx = \frac{2p}{a} = u_0 = 4\pi N_0 b\,. \tag{4.87}$$

Therefore, $p = u_0 a/2 = 2\pi b N_s$, where $N_s = N_0 a$ is the number of atoms per unit surface area of a single atomic plane.

The scattering amplitude b is on the order of 10^{-12} cm, and a is about $3\cdot 10^{-8}$ cm. Therefore, according to Eq. (4.87), the parameter $p \approx 3\cdot 10^{-5}(2\pi/a)$ is almost 5 orders of magnitude less than the wave vector $k = 2\pi/\lambda$ of thermal neutrons with a wavelength of $\lambda = 1.8$ Å, which is important to know for future estimates.

The potential shown in Figure 4.14a is

$$u(x) = 2p\sum_n \delta(x - na)\,, \tag{4.88}$$

and we can identify a symmetrical single period shown in Figure 4.14b,

$$u_1(x) = 2p\,\Theta(-a/2 \le x \le a/2)\delta(x)\,. \tag{4.89}$$

It consists of an atomic plane and two empty spaces of width $a/2$ on both sides. To find the reflection r and transmission t amplitudes for such a period, we first need to find the reflection, r_1, and transmission, t_1, amplitudes for a single atomic plane. The Schrödinger equation in this case is

$$\left[\frac{d^2}{dx^2} + k^2 - 2p\,\delta(x)\right]\psi(x) = 0\,, \tag{4.90}$$

where the incident wave is $\exp(ikx)$ at $x < 0$. The solution to the above equation is

$$\psi(x) = \Theta(x < 0)[\exp(ikx) + r_1\exp(-ikx)] + \Theta(x > 0)t_1\exp(ikx)\,. \tag{4.91}$$

The unknown coefficients, r_1 and t_1, can be found by matching the wave function at $x = 0$. The wave function is continuous at that point, whereas the first derivative undergoes a step since the second derivative contains the δ-function. Therefore, the system of equations for r_1 and t_1 is

$$1 + r_1 = t_1\,, \quad ikt_1 - ik(1 - r_1) = 2p(1 + r_1)\,. \tag{4.92}$$

The solution to this system is

$$r_1 = \frac{-ip}{k+ip} = \mp i|r_1|e^{-i\gamma_1}\,, \quad |r_1| = \frac{|p|}{\sqrt{k^2+p^2}}\,,$$
$$t_1 = \frac{k}{k+ip} = |t_1|e^{-i\gamma_1}\,, \quad |t_1| = \frac{k}{\sqrt{k^2+p^2}}\,, \tag{4.93}$$

$$\gamma_1 = \arcsin\left(p/\sqrt{k^2+p^2}\right) = \pm\arcsin(|r_1|)\,, \tag{4.94}$$

where $\pm$ corresponds to the positive and negative values of p, respectively.

The amplitudes r and t are different from r_1 and t_1 only by the phase factor $\exp(ika)$, which describes propagation of the neutron in the two empty intervals $a/2$ before and after scattering on the δ-function potential. Thus,

$$r = \exp(ika)\frac{-ip}{k+ip} = \mp i|r_1|e^{i\phi}\,, \quad t = \exp(ika)\frac{k}{k+ip} = |t_1|e^{i\phi}\,,$$
$$\phi = ka - \gamma_1\,. \tag{4.95}$$

After substituting Eq. (4.95) into Eqs. (4.79) and (4.80) and performing some transformations, we obtain

$$R = \frac{\sqrt{k + p\tan(ka/2)} - \sqrt{k - p\cot(ka/2)}}{\sqrt{k + p\tan(ka/2)} + \sqrt{k - p\cot(ka/2)}}\,, \tag{4.96}$$

$$\exp(iqa) = \frac{\sqrt{p + k\cot(ka/2)} - \sqrt{p - k\tan(ka/2)}}{\sqrt{p + k\cot(ka/2)} + \sqrt{p - k\tan(ka/2)}}\,. \tag{4.97}$$

It is funny to see that the second formula can be also represented as

$$\exp(iqa) = \frac{\sqrt{1/k + 1/p\tan(ka/2)} - \sqrt{1/k - 1/p\cot(ka/2)}}{\sqrt{1/k + 1/p\tan(ka/2)} + \sqrt{1/k - 1/p\cot(ka/2)}}\,. \tag{4.98}$$

Equations (4.96) and (4.97) can be easily analyzed. For small $pa = u_0 a^2/2 \ll 1$ (this parameter is approximately equal to $2\pi b/a \approx 10^{-3}$ and therefore is always small) and small $ka/2 \ll 1$ we get

$$R = \frac{k - \sqrt{k^2 - u_0}}{k + \sqrt{k^2 - u_0}}\,, \quad e^{iqa} = \frac{1 + ia\sqrt{k^2 - u_0}/2}{1 - ia\sqrt{k^2 - u_0}/2} \approx e^{ia\sqrt{k^2 - u_0}}\,,$$
$$u_0 = 4\pi N_0 b\,, \tag{4.99}$$

which is in good agreement with the optical potential concept. In particular, from the second formula it follows that $q \approx \sqrt{k^2 - u_0}$. For $k^2 < u_0$ the Bloch wave number q becomes imaginary, and we obtain the total reflection.

According to Eq. (4.96) the total reflection takes place also at $k = k_{Bn} = \pi n/a$, where n is an integer and k_{Bn} is the Bragg wave number of nth order. The reflection

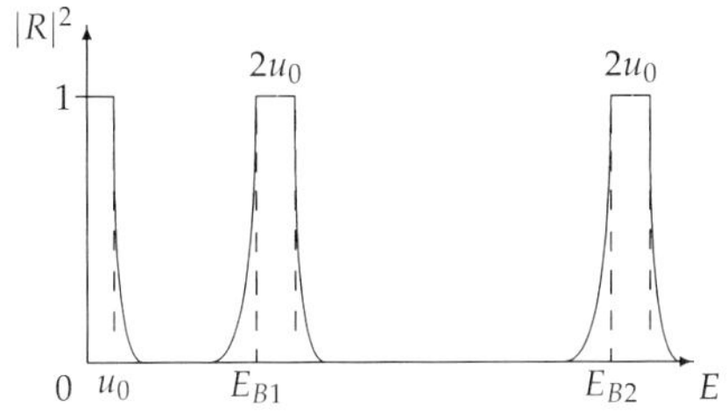

Figure 4.15 Dependence of the reflection coefficient on the energy of a neutron incident on the Kronig–Penney periodic structure. The curve shows that the the total Bragg reflections take place when $k_{Bn}^2 \le k^2 \le k_{Bn}^2 + 2u_0$.

coefficient $|R|^2$ is equal to 1 not just at one point, but within the whole interval. From Eq. (4.97) the nth order Bragg diffraction takes place within an interval $k_{Bn} \le k \le k_{Bn} + \delta k_{Bn}$. This interval is called the Darwin table. Figure 4.15 shows such Darwin tables for the dependence of the reflection on the neutron energy $E = k^2$. The reflection coefficient is equal to 1 when $k_{Bn}^2 < E < k_{Bn}^2 + 2u_0$ for $n \neq 0$, and in the interval $0 < k^2 < u_0$ for $n = 0$, where $u_0 = 4\pi N_0 b$ is the height of the potential barrier for the medium.

We considered the Kronig–Penney potential with a single atomic plane in a period, but we can also consider many atomic planes in the period and analyze when some of the Bragg peaks become forbidden [199, 200]. This means that at $k_{Bn} = \pi n$ the reflection amplitude is zero. It is a good exercise for the reader to consider two identical atomic planes in the period and to find the distance between them which makes the Bragg peak of the second order forbidden. After doing this exercise, the reader will find that the common assertion that the Bragg peak of the second order is forbidden when the distance d between two identical crystalline planes in a period a is equal to $a/4$ is not correct [199]. With such a period the Bragg peak of the second order is not forbidden, but is very narrow.

4.5.3 Supermirrors

In previous sections we learned about Bragg reflections in periodic potentials, where incident neutrons are totally reflected at energies located within a certain interval, called the Darwin table. The locations of the Darwin tables and their widths are defined by the properties of the potential. Therefore, it is possible to design a potential to get a mirror with the required properties [200, 201]. For instance, we can construct mirrors which reflect neutrons with energies higher than the optical potential u_0. Such mirrors are called supermirrors, and they are fabricated by depositing thin films with different optical potentials on a flat substrate.

The idea of a supermirror is demonstrated in Figure 4.16. We evaporate on a substrate a system of periodic chains with overlapping Bragg peaks. So the neutrons in some energy interval are Bragg-reflected by one of the chains. A period of

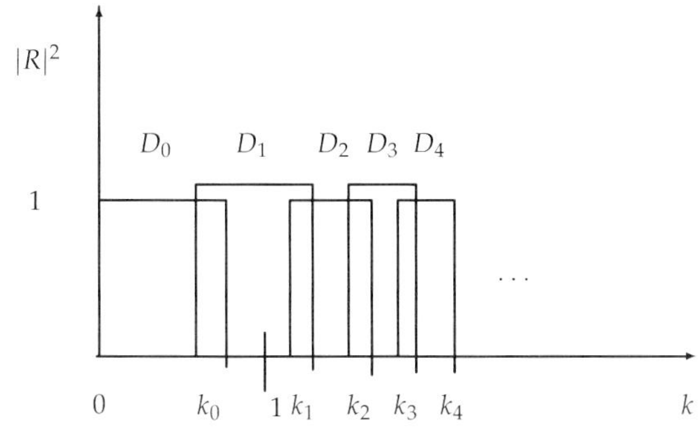

Figure 4.16 A system of periodic potentials with overlapping Bragg peaks. The system reflects neutrons with energies $k^2 > k_c^2 = u_b$, which are denoted here by unity. (I explain: by unity is denoted u_b)

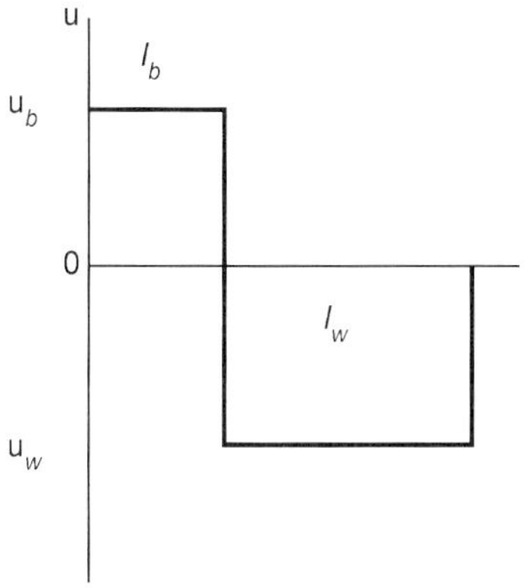

Figure 4.17 A single period of a multilayered system is fabricated using two different materials. The film of one material has a high optical potential u_b and a width d_b. The film of the other material has lower potential u_w and the width d_w.

a periodic chain is a bilayer composed of two layers with different potentials. The layer with the higher potential we call a barrier, and the one with the lower potential we call a well. Our goal is to find the positions k_i of the Bragg peak centers, the thicknesses of the barrier, $d_b(k_i)$, and the well, $d_w(k_i)$, layers in a single period (see Figure 4.17), and the optimal number, $N_i(k_B)$, of periods in the chains [201].

The algorithm is as follows. Suppose we want to have a mirror with critical wave number $k_m = 5k_c$, where k_c is the critical wave number of a single barrier. It is common to denote such a mirror by M5. Then we choose first $k_B = k_m$. The reflection amplitude of a periodic chain of N periods is

$$R_N(k) = R_\infty \frac{1 - \exp(2iqNa)}{1 - R_\infty^2 \exp(2iqNa)}, \tag{4.100}$$

where $\exp(iqa)$ is the Bloch phase factor and R_∞ is the reflection amplitude from the semi-infinite periodic potential. If we neglect imaginary parts of the potentials, then we can represent R_∞ as

$$R_\infty = \frac{\sqrt{\sin\psi + |r|} - \sqrt{\sin\psi - |r|}}{\sqrt{\sin\psi + |r|} + \sqrt{\sin\psi - |r|}}, \tag{4.101}$$

where ψ is the phase of the transmission amplitude of the single period (Figure 4.17) and $|r|$ is the modulus of the symmetrized reflection amplitude of the period. The total reflections take place at $\sin^2\psi < |r|^2$. The center of the Bragg peak of nth order is at $\psi = n\pi$, and its full width (width of the Darwin table) is equal to $2|r|$. Therefore, the widths d_b and d_w of the barrier and well layers are determined from two conditions: at the given position of the Bragg peak we must have $\psi = \pi$ and $|r|$ must be maximal. Solution of these two conditions (see the proof on page 148) gives

$$k_b d_b = k_w d_w = \frac{\pi}{2}, \qquad k_{b,w} = \sqrt{k_B^2 - u_{b,w}}. \tag{4.102}$$

When Eq. (4.102) is satisfied, we have

$$\begin{aligned} \overrightarrow{r} &= \frac{|r_b| - |r_w|}{1 - |r_b r_w|} = \frac{2r_{wb}}{1 + r_{wb}^2} = i\frac{k_w^2 - k_b^2}{k_w^2 + k_b^2} = \frac{u_b - u_w}{2k_B^2 - u_b - u_w}, \\ \overleftarrow{r} &= -\overrightarrow{r}, \end{aligned} \tag{4.103}$$

$$r = \sqrt{\overrightarrow{r}\overleftarrow{r}} = i\frac{2r_{wb}}{1 + r_{wb}^2}, \qquad r_{wb} = \frac{k_w - k_b}{k_w + k_b}, \tag{4.104}$$

and

$$R_\infty = i. \tag{4.105}$$

The Bloch phase factor at the middle of the Darwin table (when Eq. (4.102) is satisfied) becomes

$$\exp(iqa) = \frac{\sqrt{\cos\psi + |t|} - \sqrt{\cos\psi - |t|}}{\sqrt{\cos\psi + |t|} + \sqrt{\cos\psi - |t|}} = -\frac{\sqrt{1 + |t|} - \sqrt{1 - |t|}}{\sqrt{1 + |t|} + \sqrt{1 - |t|}}, \tag{4.106}$$

and

$$\begin{aligned} \exp(2iqa) &= \frac{1 - |r|}{1 + |r|} = \frac{(1 - |r_b|)(1 + |r_w|)}{(1 + |r_b|)(1 - |r_w|)} \\ &= \left(\frac{(1 - r_{0b})(1 + r_{0w})}{(1 + r_{0b})(1 - r_{0w})}\right)^2 = \frac{k_b^2}{k_w^2} < 1. \end{aligned} \tag{4.107}$$

Because of that

$$|R_N|^2 = \left| R_\infty \frac{1 - \exp(2iqNa)}{1 - R_\infty^2 \exp(2iqNa)} \right|^2 \approx 1 - 4\left(\frac{k_b^2}{k_w^2}\right)^N \equiv 1 - \zeta_N < 1. \tag{4.108}$$

The deviation from unity decreases with N, and the number N is defined from the acceptable deviation ζ:

$$\zeta_N = \zeta \to N = \frac{\ln(\zeta/4)}{\ln(k_b^2) - \ln(k_w^2)} . \tag{4.109}$$

To find the acceptable deviation we calculate R_∞ with the help of

$$R_\infty = \frac{\sqrt{(1+r)^2 - t^2} - \sqrt{(1-r)^2 - t^2}}{\sqrt{(1+r)^2 - t^2} + \sqrt{(1-r)^2 - t^2}} , \tag{4.110}$$

and find $|R_\infty(k_B)|^2$, which for complex potentials is less than unity: $|R_\infty(k_B)|^2 = 1 - \xi < 1$. It is evident that the optimal N is found from $\zeta = \xi$. Therefore,

$$N(k_B) = \frac{\ln(0.25[1 - |R_\infty(k_B)|^2])}{\ln(k_b^2) - \ln(k_w^2)} . \tag{4.111}$$

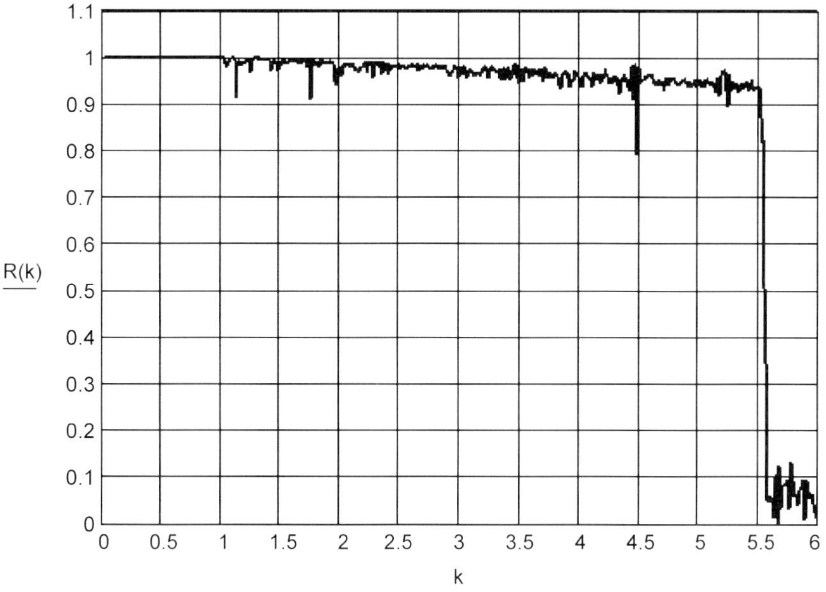

Figure 4.18 Supermirror M5.5 calculated with the algorithm explained in the text. The barrier has the potential $u_b = 1 - 0.0001i$ and the well has the potential $u_w = -0.5 - 0.0001i$. To improve the quality of the mirror the distance between the Bragg points was $2|r|$ in the interval $5.5 \geq k_B > 4.3$, $1.8|r|$ in the interval $4.3 \geq k_B > 3.8$, and $0.9|r|$ in the interval $3.8 \geq k_B > 1$. Moreover, the number of periods in the interval $1 < k_B < 2$ was twice that given by Eq. (4.112). One additional chain at $k_B = 1.05$ with six periods is also included to cure a defect in reflectivity. Atop of all the chains there is a single barrier of width 5. The total number of periodic chains is 48 and the total number of bilayers is 2323. For mirror M3.5 the total number of chains is 19 and the number of bilayers is 410. For mirror M3 the total number of chains is 14 and the number of bilayers is 256.

In practice, however, if $\xi < 0.01$, we should accept a deviation of order 1%. Therefore, practically

$$N(k_B) = \frac{\ln(0.25[1 - |R_\infty(k_B)|^2] + 0.01)}{\ln(k_b^2) - \ln(k_w^2)} . \tag{4.112}$$

Now, the algorithm is simple. We calculate the positions of all the Bragg points, the widths of the layers, and the number of periods for every periodic chain and after that recursively obtain the reflectivity from all the chains with some wide barrier atop of all them. The results of the calculations of mirror M5.5 with potentials $u_b = 1 - 0.0001i$ and $u_w = -0.5 - 0.0001i$ are shown in Figure 4.18. To get this result it was found empirically that it is better to modify the distance between the Bragg points and the number of periods. These modifications are pointed out in the figure caption.

4.5.3.1 **Proof of Eq. (4.102)**

The transmission of a period for real potentials is

$$t = \frac{t_b t_w}{1 - r_b r_w} = \exp(i\phi_b + i\phi_w)\frac{|t_b t_w|}{1 + |r_b r_w| \exp(i\phi_b + i\phi_w)} = |t| \exp(i\psi_t) , \tag{4.113}$$

where $\phi_i (i = b, w)$ are the transmission phases of separate barrier and well layers, respectively. From Eq. (4.113) it follows that

$$\psi_t = \phi_b + \phi_w - \arcsin\left(\frac{|r_b r_w| \sin(\phi_b + \phi_w)}{\sqrt{1 + |r_b r_w|^2 + 2|r_b r_w| \cos(\phi_b + \phi_w)}}\right) . \tag{4.114}$$

The condition $\psi_t = \pi$ is equivalent to

$$\sin(\phi_b + \phi_w) = -\frac{|r_b r_w| \sin(\phi_b + \phi_w)}{\sqrt{1 + |r_b r_w|^2 + 2|r_b r_w| \cos(\phi_b + \phi_w)}} , \tag{4.115}$$

and this can be satisfied only for $\sin(\phi_b + \phi_w) = 0$, or

$$\phi_b + \phi_w = \pi . \tag{4.116}$$

For separate layers we have

$$r_i = r_{0i} \frac{1 - \exp(2j k_i d_i)}{1 - r_{i0}^2 \exp(2j k_i d_i)} , \tag{4.117}$$

$$t_i = \exp(j k_i d_i) \frac{1 - r_{0i}^2}{1 - r_{i0}^2 \exp(2j k_i d_i)} , \tag{4.118}$$

where $i = b, w$, $k_i = \sqrt{k^2 - u_i}$, and

$$r_{0i} = \frac{k - k_i}{k + k_i} . \tag{4.119}$$

From Eq. (4.118) it follows that

$$|t_i| = \frac{1 - r_{0i}^2}{\sqrt{1 + r_{0i}^4 - 2r_{0i}^2 \cos(2k_i d_i)}} , \tag{4.120}$$

$$\phi_i = k_i d_i + \arcsin\left(\frac{r_{0i}^2 \sin(2k_i d_i)}{\sqrt{1 + r_{0i}^4 - 2r_{0i}^2 \cos(2k_i d_i)}}\right) , \tag{4.121}$$

and from Eq. (4.117) it follows that

$$r_i = -i \exp(j\phi_i)|r_i|, \quad |r_i| = \frac{2r_{0i} \sin(k_i d_i)}{\sqrt{1 + r_{0i}^4 - 2r_{0i}^2 \cos(2k_i d_i)}} . \tag{4.122}$$

The sine function is positive because we shall always have $k_i d_i < \pi$.

The reflection amplitudes of a single period are

$$\begin{aligned} \overrightarrow{r} &= r_b + \frac{t_b^2 r_w}{1 - r_b r_w} = \frac{r_b + \exp(2i\phi_b) r_w}{1 - r_b r_w} \\ &= -i \frac{|r_b| + \exp(i\phi_b + i\phi_w)|r_w|}{1 + |r_b r_w| \exp(i\phi_b + i\phi_w)} \exp(i\phi_b) , \end{aligned} \tag{4.123}$$

$$\begin{aligned} \overleftarrow{r} &= r_w + \frac{t_w^2 r_b}{1 - r_b r_w} = \frac{r_w + \exp(2i\phi_b) r_b}{1 - r_b r_w} \\ &= -i \frac{|r_w| + \exp(i\phi_b + i\phi_w)|r_b|}{1 + |r_b r_w| \exp(i\phi_b + i\phi_w)} \exp(i\phi_w) , \end{aligned} \tag{4.124}$$

so the symmetrized reflection amplitude of a single period is

$$r = \sqrt{\overrightarrow{r}\overleftarrow{r}} = -i \frac{\sqrt{|r_b|^2 + |r_w|^2 + 2|r_w||r_b| \cos(\phi_b + \phi_w)}}{1 + |r_b r_w| \exp(i\phi_b + i\phi_w)} \exp(i\phi_w + i\phi_b) . \tag{4.125}$$

Therefore,

$$|r| = \frac{\sqrt{|r_b|^2 + |r_w|^2 + 2\cos(\phi_b + \phi_w)|r_b||r_w|}}{\sqrt{1 + |r_b r_w|^2 + 2|r_b r_w| \cos(\phi_b + \phi_w)}} . \tag{4.126}$$

Now we must find the maximum of $|r|$ with the condition $\phi_b + \phi_w = \pi$. Therefore, we need to find the maximum of the combination $|r| - \lambda(\phi_b + \phi_w - \pi)$, where λ is the Lagrange multiplier. The condition for maximum gives three equations:

$$\frac{d}{dd_b}, \frac{d}{dd_w}, \frac{d}{d\lambda}[|r| - \lambda(\phi_b + \phi_w - \pi)] = 0 . \tag{4.127}$$

After exclusion of λ we obtain the equation

$$\frac{d}{dd_b}|r| \frac{d}{dd_w}\phi_w = \frac{d}{dd_b}\phi_b \frac{d}{dd_w}|r| . \tag{4.128}$$

The derivative of $|r|$ can be represented as

$$\frac{d}{dd_i}|r| = \frac{d}{dd_i}|r_i|\frac{d}{dr_i}|r| + \frac{d}{dd_i}\phi_i\frac{d}{d\phi_i}|r| \,. \qquad (4.129)$$

According to Eq. (4.126) the second term is proportional to $\sin(\phi_b + \phi_w)$ and is zero because of (4.116). Therefore, Eq. (4.128) is reduced to

$$\frac{d}{dd_b}|r_b|\frac{d}{d|r_b|}|r|\frac{d}{dd_w}\phi_w = \frac{d}{dd_b}\phi_b\frac{d}{dd_w}|r_w|\frac{d}{d|r_w|}|r| \,. \qquad (4.130)$$

Let us find first $d|r|/d|r_i|$ with account of (4.116) and (4.125):

$$\frac{d}{d|r_b|}|r| = \frac{d}{d|r_b|}\frac{|r_b| - |r_w|}{1 - |r_w||r_b|} = \frac{1 - |r_w|^2}{(1 - |r_w||r_b|)^2} \,. \qquad (4.131)$$

Now we calculate $d\phi_i/dd_i$ and $d|r_i|_i/dd_i$ using Eqs. (4.121) and (4.122):

$$\frac{d\phi_i}{dd_i} = k_i\frac{1 - r_{0i}^2}{1 + r_{0i}^4 - 2|r_{0i}^2|2\cos(2k_i d_i)} \,, \qquad (4.132)$$

$$\frac{d|r_i|}{dd_i} = k_i\frac{2r_{0i}\cos(k_i d_i)}{(1 + r_{0i}^4 - 2r_{0i}^2\cos(2k_i d_i))^{3/2}} \,. \qquad (4.133)$$

After substitution of Eqs. (4.131)–(4.133) into Eq. (4.130) we obtain

$$\frac{2r_{0b}\cos(k_b d_b)(1 - |r_w|^2)^2}{\sqrt{1 + r_{0b}^4 - 2r_{0b}^2\cos(2k_b d_b)}} = -\frac{2r_{0w}\cos(k_w d_w)(1 - |r_b|^2)^2}{\sqrt{1 + r_{0w}^4 - 2r_{0w}^2\cos(2k_w d_w)}} \,. \qquad (4.134)$$

This equation has the simplest solution, which is shown in Eq. (4.102). It also satisfies Eq. (4.116), which is seen after substitution of Eq. (4.121) in it.

4.5.3.2 Historical Remarks

As we know, the first article on supermirrors was published in 1967 by Turchin [202]. However, in this publication there was only an idea: if you prepare a periodic system of bilayers, you obtain Bragg reflection from it. If the thicknesses of the layers do vary, you can obtain almost total reflection in a wide interval of normal energies. As a candidate for the bilayer he considered two isotopes of Ni:^{62}Ni, with coherent scattering amplitude $b_c = -8.7$ fm, and ^{58}Ni, with $b_c = 14.4$ fm. He did not propose an algorithm to vary the thicknesses of the layers to get almost total reflection in a wide energy interval. The first prepared and experimentally measured periodic multilayer system was reported in 1974 [89]. In 1976, Mezei proposed the first algorithm for the preparation of a multilayer system with high reflectivity in a wide range of normal energies [90]. According to him the thickness of the jth layer in a multilayer system is proportional to $j^{-1/4}$.

In 1989, Hayter and Mook [204] proposed a different algorithm. They found a formula[13)] for the reflection amplitude from a semi-infinite periodic potential of

13) Equation (HM16) is incorrect because it gives $|R_\infty| > 1$ when $\cos^2\phi > r_{wb}$; however, it does not invalidate their results.

bilayers in our notation as (the numerations of equations with HM correspond to the numerations in [204])

$$R_\infty = e^{-i\phi} \frac{\sqrt{\cos\phi + r_{wb}} + \sqrt{\cos\phi - r_{wb}}}{\sqrt{\cos\phi + r_{wb}} - \sqrt{\cos\phi - r_{wb}}}, \tag{HM16}$$

where $\phi = k_b d_b \approx k_w d_w$, and

$$\exp(-q''a) = -\frac{k_b}{k_w}, \tag{HM18}$$

where $a = d_b + d_w$. If they were able to prepare a semi-infinite periodic system with the center of the Darwin table at some k_B, they would find the next Darwin point from the requirement $\Delta\phi \equiv d_b \Delta k_b \approx |r_{wb}|$. If they used N bilayers, then at the center of the Darwin table they would obtain reflectivity $1 - \exp(-2Nq''a)$. If a deviation from unity of order $\xi \sim 0.5\%$ is tolerable, they would need

$$N = \frac{\ln \xi}{2\ln(k_b/k_w)}. \tag{HM22}$$

However, instead of N bilayers they proposed taking only one. They supposed that a single bilayer also gives reflectivity of order unity but within a narrower Darwin table of the width $\tilde{r}_{wb}$, where

$$\tilde{r}_{wb} = \frac{1 - \tilde{k}_b/\tilde{k}_w}{1 + \tilde{k}_b/\tilde{k}_w}$$

with $\tilde{k}_b/\tilde{k}_w = (k_b/k_w)^{1/N}$. So they proposed preparing a nonperiodic system of bilayers, shifting the Bragg points of neighboring bilayers by an amount $\Delta k_b \sim \tilde{r}_{wb}$. Since at the center of the Darwin table $k_b d_b = \pi/2$, then $\Delta k_b d_b + k_b \Delta d_b = 0$, and $\Delta d_b = -|\tilde{r}_{wb}|/k_b$. This shift can be smaller than the interatomic distance,

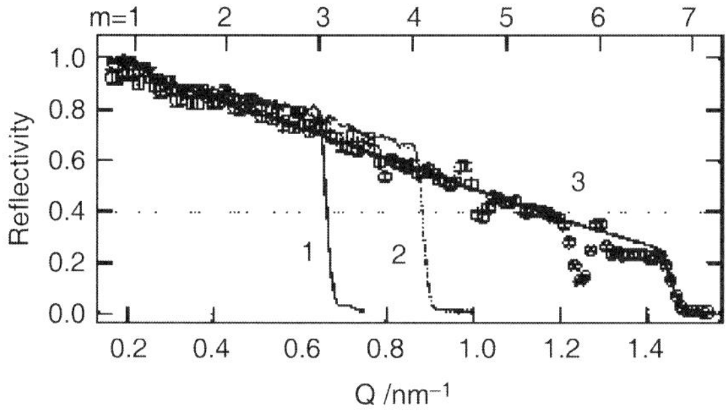

Figure 4.19 The measured neutron reflectivities of the supermirrors fabricated with the large-scale ion-beam sputtering instrument (IBS) for the spallation neutron source of J-PARC [205]. Reflectivities of mirror M3 with 403 layer(1), mirror M4 with 1201 layers (2), and mirror M6.7 with 4000 bilayers (3).

which leads to unavoidable roughness at interfaces, which deteriorates the quality of supermirrors.

Since then many supermirrors have been produced all around the world. Among the recent publications we cite [99], where a magnetic supermirror M5 preparation was reported with 1875 bilayers, and [100, 205], where production of mirror M6.7 with 4000 bilayers is reported. Experimentally measured reflectivities of such mirrors are shown in Figure 4.19.

Our approach here has an advantage because our shift of the Bragg peak positions is wider and we can make the thicknesses of the layers commensurate with interatomic distances. However, how practical this advantage is depends on the precision of technology.

4.5.4
Experience with an Experiment

To check how well the theoretical formulas work and how well the evaporation technique can be controlled, we consider experimental data on reflectivity from periodic chains of bilayers on a glass substrate. The samples were prepared at the Budapest Nuclear Center and they were measured with the KFKI reflectometer [188] with a wide position-sensitive stationary detector. Bilayers were composed from Ni (barrier $u_b = 1 - 0.00014i$) and Ti (well $u_w = -0.21 - 0.00012i$) of such thicknesses to get Bragg reflectivity at the point[14] $k = 2$. In accordance with Eq. (4.102), the thicknesses of the layers were chosen as $l_b = 0.91 \approx 84$ Å for Ni and $l_w = 0.76 \approx 70$ Å for Ti. In Figure 4.20 we present on a linear scale the results of fitting experimental data for two, four and eight periods with the formula

$$R(k) = \int_{k-\delta}^{k+\delta} |\overrightarrow{R}_{Ns}(p)|^2 \frac{dp}{2\delta} + n_b \,, \tag{4.135}$$

where

$$\overrightarrow{R}_{Ns}(p) = \overrightarrow{R_N}(p) + T_N^2(p) \frac{r_{s0}(p)}{1 - \overleftarrow{R_N}(p)\, r_{s0}(p)} \tag{4.136}$$

is the reflection amplitude from a periodic chain of N periods evaporated over a semi-infinite substrate, whose reflection amplitude is r_{s0}. The fitting parameters (in total seven) are real parts of all the potentials, the thicknesses of the Ni and Ti layers, the resolution δ, and the background n_b. The imaginary parts of the potentials were the same as the table ones. The results for the fitted parameters are presented in Table 4.2

14) We again use dimensionless units, where for the unit of energy and the wave number serve as the optical energy and the critical wave number of Ni: $u'_{\mathrm{Ni}} = 0.245$ µeV and $k_c = 1/92\,$Å^{-1}.

Table 4.2 Fitted values of real parts of potentials u' for Ni (b, barrier), Ti (w, well), and substrate (s), their thickness l, resolution δ, background n_b, and χ^2 for periodic chains with $N = 2, 4, 8$ bilayers. Imaginary parts u'' of the potentials were fixed. The results are given in dimensionless units. The unit of energy is equal to the real part of the Ni optical potential $u'_{\mathrm{Ni}} = 0.245$ µeV from Table 4.1, and the unit of length is the reduced critical wavelength $\lambda_c/2\pi = 92$ Å for Ni. The values in the fourth line were obtained for zero background, and the values in the fifth line were obtained for predetermined background $n_b = 0.003$. It is interesting to see that the fitting with a smaller number of parameters (six instead of seven) gave smaller χ^2.

N	u'_b	u'_w	u'_s	u''_b	u''_w	u''_s	l_b	l_w	δ	n_b	χ^2
2	0.964	−0.258	0.452	0.00014	0.00012	0.0001	1.121	0.59	0.036	0.003	25
4	0.934	−0.388	0.446	0.00014	0.00012	0.0001	1.182	0.525	0.033	0.003	114
8	0.993	−0.242	0.398	0.00014	0.00012	0.0001	1.061	0.649	0.035	0.0089	349
8	0.963	−0.421	0.415	0.00014	0.00012	0.0001	1.169	0.54	0.036	0	209
8	0.972	−0.349	0.408	0.00014	0.00012	0.0001	1.13	0.579	0.036	0.003	151

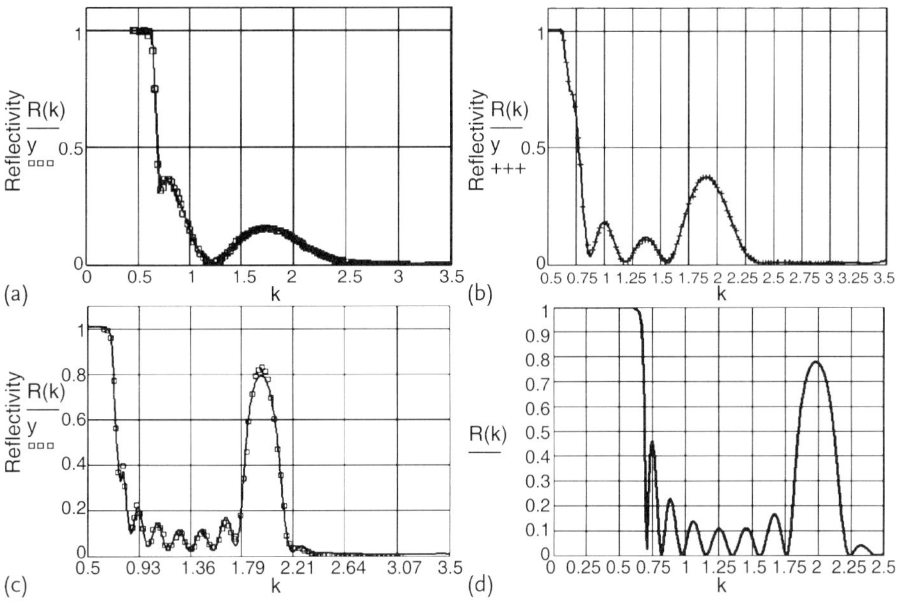

Figure 4.20 Fitting of the reflectivity data from periodic chains of bilayers evaporated on a thick float glass substrate. The fitting function is given by Eq. (4.135). The data were obtained in Hungary at a reflectometer with a wide stationary detector. The results are shown on a linear scale. The chains were (a) $(\mathrm{NiTi})_2$, (b) $(\mathrm{NiTi})_4$, and (c) $(\mathrm{NiTi})_8$. (d) Ideal theoretical reflectivity from $(\mathrm{Ni}(84\,\text{Å})\mathrm{Ti}(70\,\text{Å}))_8$ on glass with potential $u = 0.4 - i0.0001$.

In Figure 4.21 we represent on a logarithmic scale the results of fitting the chain of eight periods. The data shown on a logarithmic scale in Figure 4.21a correspond to the data shown on a linear scale in Figure 4.20c. We see that the background in the fitting is overestimated. The logarithmic scale fitting for zero background is

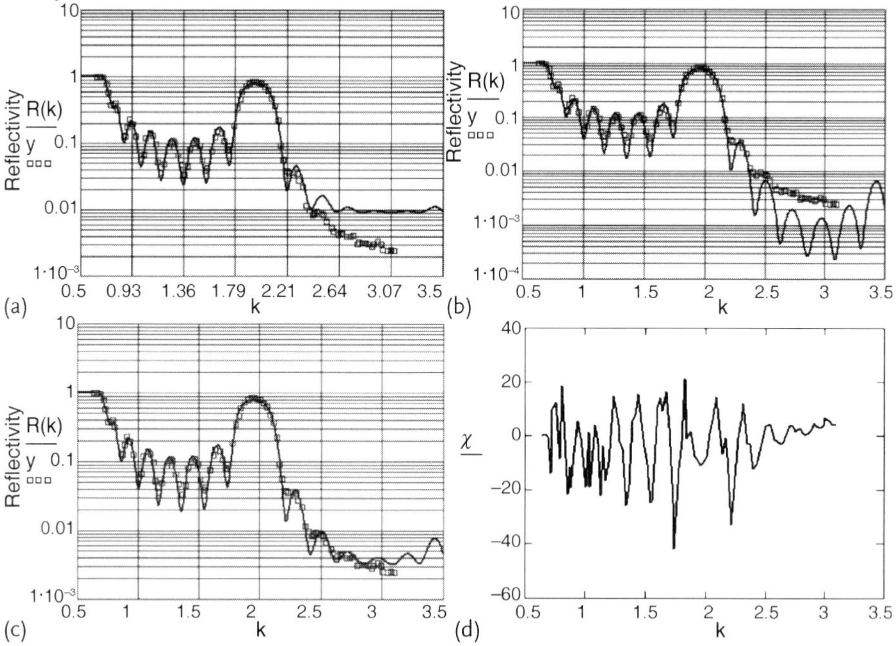

Figure 4.21 Logarithmic scale of different types of fitting of the reflectivity data from periodic chains of eight bilayers on a thick float glass substrate. (a) The fitting with seven parameters, corresponding to Figure 4.20c; (b) fitting with six parameters ($n_b = 0$); (c) fitting with six parameters ($n_b = 0.003$); (d) χ distribution for the case shown in (c).

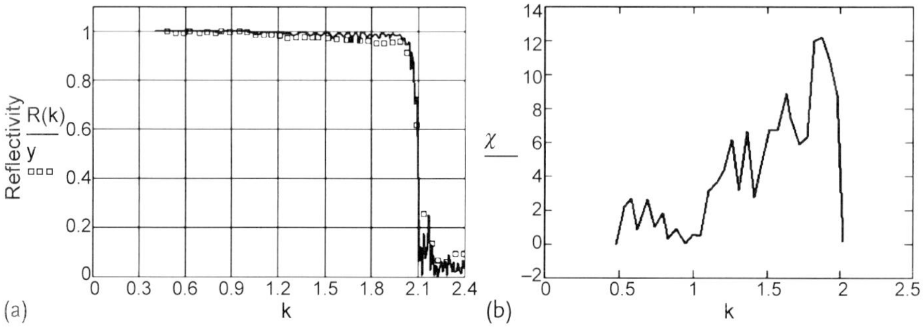

Figure 4.22 (a) Experimental data points and theoretical calculation of supermirror M2. (b) $\chi(k)$ distribution; it was obtained from direct comparison of experimental and theoretical data without fitting.

shown in Figure 4.21b. We see that the background is underestimated, and the better presentation of the experimental points is achieved when we set the background to 0.003, as shown in Figure 4.21c. Nevertheless the $\chi^2 = 151$ for this fit remains extremely high. The corresponding χ distribution is shown in Figure 4.21d.

Besides the samples described above, a supermirror M2 consisting of eight periodic chains and a total number of 59 bilayers was also prepared. The periods and the number of them were found according to the above prescription with some corrections like in the case of Figure 4.18. The result of measurements compared with calculations is shown in Figure 4.22a. We see good coincidence, however the χ distribution as shown in Figure 4.22b is wide, and $\chi^2 = 32$ is pretty high; nevertheless it is very acceptable because there was no fitting at all.

4.6 Examples of Application of the Algebra

We have already seen how fruitful the algebraic approach is. Now we shall apply it to some interesting multilayer systems.

4.6.1 Resonant Transmission

Let us look at the system with the optical potential shown in Figure 4.23.

It is a sandwich composed of two identical films of high potential u separated by a third film of lower potential. For simplicity we shall suppose that the intermediate film has zero potential.

In such a system a resonance can be seen, which means that at some energy of the incident neutrons the system of two barriers is transparent [206]. Indeed, the transmission amplitude of such a system is [207]

$$T = \exp(ikl)\frac{t^2}{1 - r^2\exp(2ikl)} = e^{2i\chi+ikl}\frac{|t|^2}{1 + |r|^2\exp(2i\chi + 2ikl)}, \quad (4.137)$$

where we denoted $t = |t|\exp(i\chi)$ and $r = \pm i|r|\exp(i\chi)$. The phase $\phi = 2\chi + 2kl$ depends on k, and at some k_n and sufficiently large l it can be equal to $\pi(2n + 1)$, where n is integer. At such values of the phase the absolute magnitude of the transmission amplitude becomes $|T| = |t|^2/(1 - |r|^2) = 1$, because $|r|^2 = 1 - |t|^2$. On the other side the reflection amplitude

$$R = r\frac{1 + (t^2 - r^2)\exp(2ikl)}{1 - r^2\exp(2ikl)} = r\frac{1 + \exp(2i\chi + 2ikl)}{1 + |r|^2\exp(2i\chi + 2ikl)} \quad (4.138)$$

at these k becomes zero.

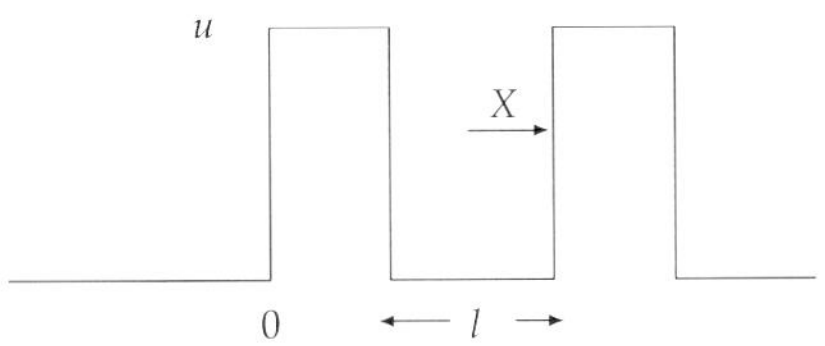

Figure 4.23 Resonant system of two barriers separated by a gap of width l.

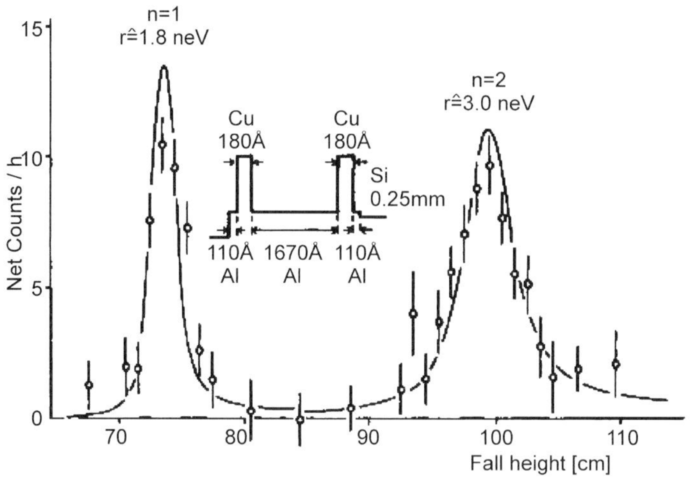

Figure 4.24 Experimentally measured transmission of ultracold neutrons though the multilayer resonant system shown in the inset (source [111]).

Figure 4.24 shows the result of an experiment [111] on transmission of ultracold neutrons through a resonant multilayer system. The two peaks correspond to two resonances. The resonant transmission is now used for monochromatization of neutrons.

The resonant transmission seems to be surprising because at these $k = k_n$ the transmission through a single barrier $|t|$ can be exponentially small. In effect, the transparency of two barriers is a result of accumulation. Because of the small transmission through both sides, the amplitude X of the wave incident from inside on the second barrier grows so much that even at small $|t|$ we have $|Xt| = 1$. This means that the amplitude of the wave function in-between the barriers is equal to $1/|t|$. Indeed, according to Eq. (4.56) we have

$$X = \frac{t\exp(ikl)}{1 - r^2\exp(2ikl)} = \exp(i\chi + ikl)\frac{|t|}{1 + |r|^2\exp(2i\chi + 2ikl)}, \quad (4.139)$$

and at $2\chi + 2kl = \pi(2n + 1)$ we obtain $|X| = 1/|t|$.

The growth of the wave function amplitude requires some time, like the increase at resonance of the amplitude of a mechanical pendulum. It would be very interesting to observe the transient process, that is, increase of the transmission probability with time or retardation of transmitted particles.

4.6.1.1 A Resonance between Two Nonequal Barriers

The resonant system can be transparent at resonance only if it is a symmetrical one, or if the transmissions through two barriers are equal. Let us consider the transmission for a resonant system with two different barriers separated by distance l.

The transmission amplitude T_{12} of the system is

$$T_{12} = \exp(ikl)\frac{t_2 t_1}{1 - r_1 r_2 \exp(2ikl)}; \tag{4.140}$$

therefore, at resonance (minimum of the denominator) we have

$$|T_{12}| = \frac{|t_2||t_1|}{1 - |r_1||r_2|}. \tag{4.141}$$

The resonance is high when $|r_1 r_2|$ is close to unity, or $|t_1|$ and $|t_2|$ are small. In that case $|r_i| = \sqrt{1 - t_i^2} \approx 1 - t_i^2/2$, and

$$|T_{12}| < \frac{2|t_2||t_1|}{|t_1|^2 + |t_2|^2} < 1. \tag{4.142}$$

This means that the transmission cannot be unity if $|t_1| \neq |t_2|$.

4.6.1.2 Breit–Wigner Formula

It is common in science to describe resonances by the Breit–Wigner formula:

$$F \propto \frac{\Gamma/2}{E - E_r + i\Gamma/2}, \tag{4.143}$$

where E_r is the resonance energy and Γ is its width. Let us show how this formula can be deduced in our case.

According to Eqs. (4.138) and (4.139), the resonance energy $E = E_n$, is determined by the equation $\phi(E) = 2\chi + 2kl = \pi(2n + 1)$, with integer n. At $E \approx E_n$ the phase $\phi(E)$ can be represented as

$$\phi(E) = \phi(E_n) + \phi'(E_n)(E - E_n) = \pi(2n + 1) + \phi'(E_n)(E - E_n), \tag{4.144}$$

where $\phi'(E) = d\phi(E)/dE$. If $|t|^2 \ll 1$, then $|r| \approx 1$, and the reflection amplitude near the resonance is

$$R = r\frac{-i\phi'(E_n)(E - E_n)}{1 - |r|^2 - i|r|^2\phi'(E_n)(E - E_n)} \approx \exp(i\chi)\frac{E - E_n}{E - E_n + i\Gamma_n/2}, \tag{4.145}$$

where

$$\Gamma_n = 2\frac{1 - |r|^2}{|r|^2\phi'(E_n)} \approx 2\frac{|t|^2}{\phi'(E_n)} \tag{4.146}$$

is the resonance width. To estimate it we suppose that $\phi'(E_n) \approx d(2k_n l)/dE_n = 2l/k_n$, where $k_n \approx n\pi/l$. Substitution into Eq. (4.146) gives $\Gamma_n \approx |t|^2 k_n^2/n\pi = 2|t(k_n)|^2 E_n/n\pi$. It is clear that $k_n \sim n$, and $|t(E_n)|$ increases with n; therefore, Γ_n also increases with n as is clearly seen in Figure 4.24.

At $E \approx E_n$, the transmission amplitude according to (4.137) is

$$T = e^{2i\chi}\frac{i\Gamma_n/2}{E - E_n + i\Gamma_n/2}. \tag{4.147}$$

We see that at $E = E_n$ it immediately follows that $R = 0$ and $|T|^2 = 1$.

4.6.1.3 Decay of Quasibound Systems

To every resonance in the potential shown in Figure 4.23 there corresponds a metastable quasibound state. The neutron in this state lives for a comparatively long time. Its lifetime between barriers can be calculated by stationary and nonstationary methods.

In the nonstationary approach one solves the Schrödinger equation

$$\left[i\frac{\partial}{\partial t}+\frac{\partial^2}{\partial x^2}-u(x)\right]\Psi(x,t)=0 \tag{4.148}$$

with initial value $\Psi(x,0)=A\Theta(x\in l)[\exp(ikx)+\exp(-ikx)]$, where A is a normalizing constant, and the Θ-function shows that the initial function is concentrated between two barriers. Solution of Eq. (4.148) is found with the help of the Green function, $G(x,t)$,

$$\Psi(x,t)=\int_0^l G(x-x',t)\Psi(x',0)dx' , \tag{4.149}$$

so we need to find a Green function which at $t\to 0$ becomes a δ-function: $\lim_{t\to 0} G(x-x',t)=\delta(x-x')$.

In the stationary approach we define the wave function in the form $\Psi(x,t)=\psi(x)\exp(-ik^2t)$, where $\psi(x)$ satisfies

$$[d^2/dx^2-u(x)+k^2]\psi(x)=0 , \tag{4.150}$$

and where k is a complex number $k=k'-ik''$ with a negative imaginary part. The negative sign is necessary to make the wave function stationary. Indeed, let us look at the wave between barriers going to the right. Suppose its amplitude there is unity. This wave is reflected from the right barrier with reflection amplitude $r=\pm i|r|\exp(i\chi)$ and propagates to the left. Near the left barrier it is equal to $r\exp(ikl)$. After reflection from the left barrier, the wave propagates to the right, and near the right barrier it becomes $r^2\exp(2ikl)$. But if the state is stationary, then the amplitude near the right barrier should be the same as at the very beginning, that is, unity. Therefore, we should have

$$|r|^2\exp(2i\chi+i\pi+2ikl)=1 , \tag{4.151}$$

which is a condition of the stationarity. But $|r|^2<1$ and therefore

$$\exp(2i\chi+i\pi+2ikl)=1/|r|^2>1 \tag{4.152}$$

and the phase $\phi(k)=2\chi+2kl$ have to be of the form $\pi(2n+1)-i\phi''$, that is, its real part should satisfy the resonance condition and the phase should also contain a negative imaginary part. It is possible to satisfy Eq. (4.152) when ϕ'' is

$$\phi''=-\ln|r|^2=-\ln(1-|t|^2)\approx|t|^2 . \tag{4.153}$$

The smaller is $|t|^2$, the better is the last approximation. The imaginary part $-i\phi''$ can appear only at complex $k = k_n - ik''$ or complex energy $E = E_n - iE''$. This means that the probability for a particle staying on the resonance level between two barriers is equal to $\int dx|\psi(x)|^2 \exp(-2E''t)$, that is, it decreases with time, and the lifetime on this level can be defined as $\tau = 1/2E''$.

To define E'' it is necessary to use expansion (4.144). Substitute $E = E_n - iE''$ in it. As a result, we get $\phi'' = E''\phi'(E_n)$. Substitution into Eq. (4.153) gives $E'' = |t|^2/\phi'(E_n) = \Gamma/2$, where Γ coincides with the width Eq. (4.146) of the resonance level, $\tau = 1/2E'' = 1/\Gamma$, which can be expected. We see that the theory of resonance decay is equivalent to the theory of α-decay.

4.6.1.4 **Resonances at Total Reflection**

Imagine the second barrier of the system in Figure 4.23 to be infinitely thick. Then our system becomes as shown in Figure 4.25. For it reflection at all energies below u will be total. Nevertheless, at some energies the system has resonances. The resonances correspond to maxima of the wave function amplitude between barriers:

$$X = \frac{t\exp(ikl)}{1 - rr_0\exp(2ikl)} = \frac{t\exp(ikl)}{1 - |r|\exp(i\chi_r + i\chi_0 + 2ikl)}, \tag{4.154}$$

where $r_0 = \exp(i\chi_0)$ is the reflection amplitude from the second potential step and $r = |r|\exp(i\chi_r)$ is the reflection amplitude from the left barrier. At resonance, where $\chi_r + \chi_0 + 2kl = 2\pi n$, we get

$$|X| = \frac{|t|}{1 - |r|} \approx \frac{2}{|t|}, \tag{4.155}$$

because at small t we have $|r| = \sqrt{1 - |t|^2} \approx 1 - |t|^2/2$.

The resonances can be seen in total reflections only if an absorber is placed between the barriers. The absorption probability is proportional to $|X|^2$, and has a maximum at resonances. As a result, reflection at resonances decreases [207].

4.6.2 **Channeling**

Let us consider the potential shown in Figure 4.26 It corresponds to a real system of films evaporated on a substrate, as shown in Figure 4.27. The upper film is partly

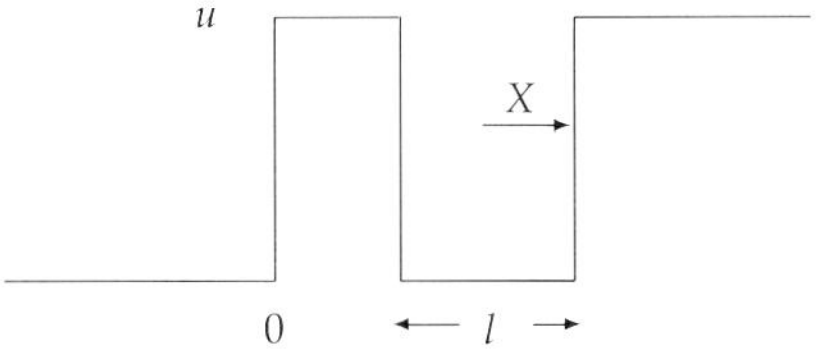

Figure 4.25 The resonant system with an infinitely thick right barrier.

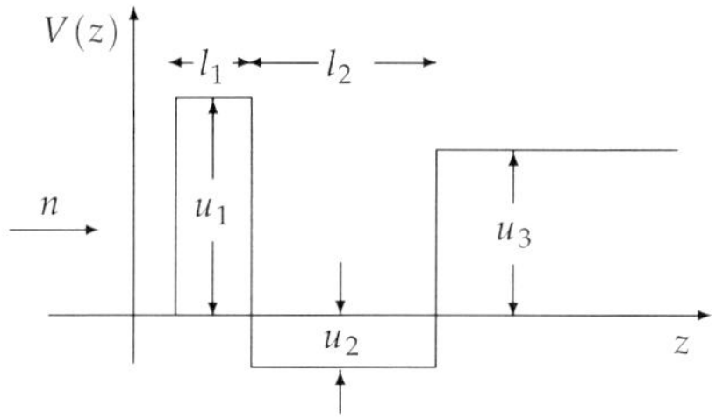

Figure 4.26 A resonant system for investigation of channeling.

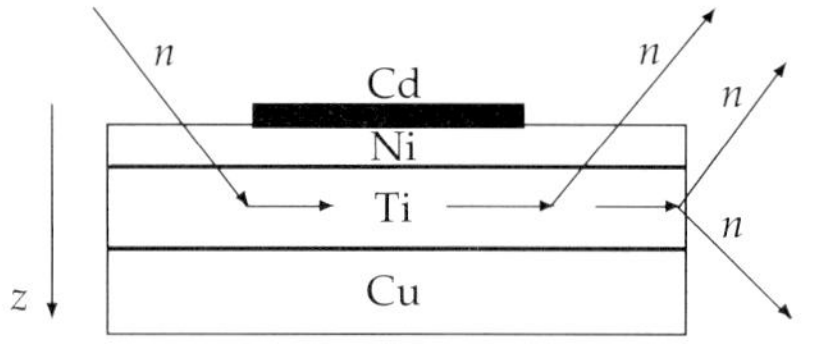

Figure 4.27 A real multilayered system corresponding to the potential in Figure 4.26.

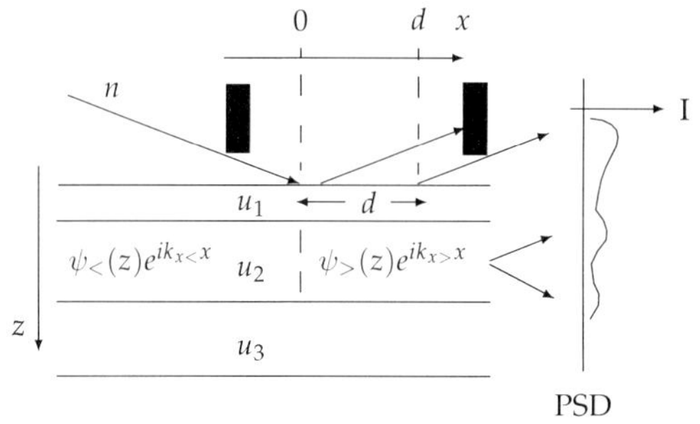

Figure 4.28 Simplified version of the channeling system. In the channel there is an interface between illuminated and nonilluminated parts, which is determined by the left absorber. The right absorber limits the area visible to the position-sensitive detector (PSD).

closed by Cd. We can imagine an experiment in which the part of the system to the left of Cd is illuminated by a neutron beam. The beam enters into resonance layer 2, propagates in this layer under Cd, and goes out on the other side of the Cd layer. Our task is to describe the neutron propagation under Cd and to estimate the intensity of the outgoing beam [208].

For simplicity we consider a slightly different system, as shown in Figure 4.28, where the Cd film makes only a shadow on the upper Cu film. In that case the potentials under the illuminated and nonilluminated parts of the upper film are identical. With Cd above Cu we should take into account also the Cd optical potential.

4.6.2.1 Wave Function inside a Channel

Let us denote x as the coordinate along the resonance layer with potential u_2 (call it a channel) and z the coordinate across it along the normal to interfaces. Consider the channel as consisting of two parts: one is under the illuminated part of the surface, $x < 0$, and the other is under the nonilluminated part of the surface, $x > 0$. The wave function in the channel is $\psi(x,z) = \psi_x(x)\psi_z(z)$. It satisfies the Schrödinger equation

$$\left[\partial^2/\partial x^2 + \partial^2/\partial z^2 - u(z) + 2(E_x + E_z)\right]\psi(x,z) = 0\,, \tag{4.156}$$

where $E = E_x + E_z$ is the total energy of the neutron, which is the same under the illuminated and nonilluminated parts of the surface. However, the splitting of this energy into two terms, E_x and E_z, can be different. This is because though the wave function in the whole channel is representable as the product

$$\psi(x,z) = \psi(z)\exp(ik_x x)\,, \tag{4.157}$$

under the part of the surface illuminated by the incident wave, $\exp(ik_x x + ik_z z)$, the component $k_{x<} \equiv k_x = \sqrt{2E_x}$ of the wave vector in Eq. (4.157) is the same as in the incident wave, but under the nonilluminated part of the surface $k_{x>}$ is different. Because of the outward flow of neutrons from the channel, the wave decays along the channel, and the component $k_{x>} = k_x + ik_x''$ should contain a positive imaginary part.

The z-component of the wave vector, $k_{2z} = \sqrt{2(E - E_x) + u_2}$, under the illuminated part of the surface is $\sqrt{k_z^2 + u_2}$, where k_z is the normal component of the wave vector of the incident wave, but under the nonilluminated part it contains a negative imaginary part $-ik_{2z}''$, which, because of energy conservation, $E = E_x + E_z = (k_x^2 + k_z^2)/2$, is naturally related to ik_x''.

Our task is to find the decay length along the channel $1/2k_x''$ and to predict the count rate of the detectors registering neutrons going out from the resonance channel on the right of the Cd shield. For that we have to find the wave function $\psi_>(z)$ under the nonilluminated part of the surface. This function is to be stationary there, that is, is to have the same form at all points under the nonilluminated part of the surface, and it must be continuously matched to $\psi_<(z)$ under the illuminated part:

$$\psi_<(z) = X(k_z)[\exp(-ik_{2z}(z + l_2)) + \exp(ik_{2z}(z + l_2))r_{23}(k_z)]\,, \tag{4.158}$$

where l_2 is the channel thickness (we consider that interface 23 is at the point $z = -l_2$) and $r_{23}(k_{2z})$ is the amplitude of the total reflection from the substrate with potential u_3 from inside the resonance channel. The factor function $X(k_z)$ is

$$X(k_z) = \frac{t_{02}\exp(ik_{2z}l_2)}{1 - r_{21}r_{23}\exp(2ik_{2z}l_2)}\,, \tag{4.159}$$

where t_{02} is the transmission amplitude from a vacuum into the channel through the first layer with potential u_1 and r_{21} is the reflection amplitude of layer u_1 from inside the channel layer. The denominator of the factor Eq. (4.159) determines the resonance behavior of the wave function amplitude in the channel. The absolute value of X is

$$|X| = \left|\frac{t_{02}\exp(ik_{2z}l_2)}{1-|r_{21}|\exp(i\chi_{23}+i\chi_{21}+2ik_{2z}l_2)}\right| = \frac{|t_{02}|}{1-|r_{21}|} \approx \frac{2|t_{02}|}{|t_{21}|^2} \approx \frac{2}{|t_{21}|}\,. \tag{4.160}$$

On the right of the point $x = 0$ the wave function $\psi_>(z)$ should be similar to $\psi_<(z)$, so we shall take it in the same form as Eq. (4.158). This way we achieve the continuity of the wave function in the channel at the point $x = 0$. Of course there is no total identity of the functions $\psi_<(z)$ and $\psi_>(z)$ at the point $x = 0$. To get perfect continuity of the function and its derivative at $x = 0$ we have to introduce reflected and scattered waves. We neglect them here.

The wave function $\psi_>(z)$, like Eq. (4.158), must be stationary along the channel. Ths is possible if a stationarity condition similar to Eq. (4.152),

$$r_{23}r_{21}\exp(2ik_{2z}l_2) = 1\,, \tag{4.161}$$

is satisfied. Now neglect absorption and assume that reflection from the potential u_3 is total. Then Eq. (4.161) is equivalent to

$$\exp(i\chi_{23}+i\chi_{21}+2ik_{2z}l_2) = \frac{1}{|r_{21}|} > 0\,. \tag{4.162}$$

It can be satisfied only if

$$\phi(E_{2z>}) = \chi_{23}+\chi_{21}+2k_{2z}l_2 = 2\pi n - i\phi''\,, \tag{4.163}$$

where n is an integer. We see that the phase has resonant real and negative imaginary parts. The last one can be represented as $-i\phi'' = (E_{2z>} - E_n)\phi'(E_n)$, where $\phi'(E) = d\phi(E)/dE$. It follows that $E_{2z>} = E_n - iE''$, that is, $k_{z>}$ has a negative imaginary part.

According to Eqs. (4.162) and (4.163), $\phi'' = -\ln|r_{21}| \approx |t_{21}|^2/2$; therefore,

$$E'' = \frac{|t_{21}|^2}{2\phi'(E_n)} \approx \frac{|t_{21}|^2 k_{2z}}{4l_2}\,, \tag{4.164}$$

where in the last term we approximated ϕ' by $(2k_{2z}l)' = 2l/k_{2z}$.

Now we can define the decay length $1/k_x''$ along the channel. Indeed, since $k_{x>} = \sqrt{2(E-E_{2z})} = \sqrt{k_x^2 + 2iE''}$, then for small E'' (it is natural to suppose it is small), we have

$$k_x'' = \frac{E''}{k_x} \approx \frac{|t_{21}|^2}{4l_2}\frac{k_{2z}}{k_x} = \frac{|t_{21}|^2}{4n\pi}\frac{k_{2z}^2}{k_x} \approx \frac{|t_{21}|^2}{4\pi}\frac{u}{k}\,. \tag{4.165}$$

For thermal neutrons the value of k'' can be estimated as $k'' \sim 10^2|t_{21}|^2$; therefore, for $|t_{21}| \sim 0.1$ we have $k'' \sim 1\ \mathrm{cm}^{-1}$. The losses because of escape through the barrier u_1 (escape losses) are much higher than the losses because of the absorption, σ_a, and inelastic, σ_{ie}, scattering cross sections, which can be estimated as $N_0(\sigma_a + \sigma_{ie}) < 0.1\,\text{cm}^{-1}$, where N_0 is the atomic density. However, if $|t_{21}| \ll 0.1$, the escape losses become negligible, and decay in the channel of length L is described by the usual exponent $\exp(-LN_0\sigma)$, where σ besides absorption and inelastic scattering includes also scattering from roughnesses at interfaces and absorption in layers 1 and 3, which affect the reflection amplitudes r_{23} and r_{21}.

4.6.2.2 Wave Function over the Nonilluminated Part of the Surface

For the known wave function inside the channel we can calculate the intensity of the neutrons going out through nonilluminated part of the surface on the right of the second absorber in Figure 4.28 and through the channel edge. We also can predict the distribution of the intensity on the position-sensitive detector shown on the right side in Figure 4.28.

The wave function over the nonilluminated interface can be calculated according to the Fresnel–Kirchhoff principle by an integral over the nonilluminated part of the surface ($\boldsymbol{r}' = (x' > L, y', z' = 0)$):

$$4\pi\psi(\boldsymbol{r}) = \int\limits_{x'>L} dx' \int\limits_{-\infty}^{\infty} dy' \left[\psi(\boldsymbol{r}')\frac{d}{dz'}G(\boldsymbol{r}-\boldsymbol{r}') - G(\boldsymbol{r}-\boldsymbol{r}')\frac{d}{dz'}\psi(\boldsymbol{r}')\right]_{z'=0}, \tag{4.166}$$

where the Green function $G(\boldsymbol{r} - \boldsymbol{r}')$ is $\exp(ik|\boldsymbol{r} - \boldsymbol{r}'|)/|\boldsymbol{r} - \boldsymbol{r}'|$. We take such a Green function because it transforms an incident plane wave into an identical plane wave.

Since the detector is usually very far from our system, we can use an approximation for G:

$$\frac{\exp(ik|\boldsymbol{r}-\boldsymbol{r}'|)}{|\boldsymbol{r}-\boldsymbol{r}'|} = \frac{\exp(ik|\boldsymbol{r}| - i\boldsymbol{p}\boldsymbol{r}' + iy'^2k/2r)}{|r|}, \tag{4.167}$$

where $\boldsymbol{p} = k\boldsymbol{r}/r$, and we have chosen the position of the detector to be at $\boldsymbol{r} = (x, 0, z)$.

Let us find $\psi(\boldsymbol{r}')$ and $d\psi(\boldsymbol{r}')/dz'$ of our resonant system at the upper surface $z' = 0$. The wave $\psi(\boldsymbol{r}')$ is generated by the wave $X(k_z)\exp(ik_{2z>}(z + l_2))r_{23}$ incident on the layer u_1 from inside the channel. Since the transmission through u_1 is determined by amplitude $t_{21}(k_{z>})$, we have

$$\psi(x', z' = 0) = t_{21}(k_{z>})X(k_z)\exp(ik_{2z>}l_2)r_{23}\exp(ik_{x>}x), \tag{4.168}$$

$$\frac{d}{dz'}\psi(x', z')|_{z'=0} = ik_{2z>}t_{21}(k_{z>})X(k_z)\exp(ik_{2z>}l_2)r_{23}\exp(ik_{x>}x), \tag{4.169}$$

and in all the amplitudes except X the wave vector k_z is to be replaced by $k_{z>} = k_z - ik_z''$.

Substitution of the boundary values into Eq. (4.166) with account of Eq. (4.167) after integration over $dx'dy'$ gives

$$\psi(\mathbf{r}) = t_{21}(k_{z>})\sqrt{\frac{2i\pi}{k}\frac{X(k_z)(p_z+k_{z>})}{p_x-k_x-ik_x''}}\exp(i(k_{x>}-p_x)L)\frac{\exp(ikr)}{4\pi\sqrt{r}}. \quad (4.170)$$

With this function we can find the flux to a detector of height Δz:

$$dw_1 = k\Delta z|\psi(\mathbf{r})|^2 = \frac{|t_{21}(k_{z>})X(k_z)|^2}{8\pi}\frac{(p_z+k_z)^2+k_x''^2}{(p_x-k_x)^2+k_x''^2}\frac{\Delta z}{L}. \quad (4.171)$$

After substitution of Eq. (4.160), $p_z = k\sin\theta$, $p_x = k\cos\theta$, $k_z = k\sin\theta_0$, $k_x = k\cos\theta_0$, and $\Delta z/r = d\theta$, and taking into account that $\theta \approx \theta_0 \ll 1$, we obtain

$$dw_1(\theta) = \frac{2}{\pi}\frac{d\theta\sin^2\theta_0\exp(-2Lk_x'')}{(\cos\theta-\cos\theta_0)^2+(k_x''/k)^2} = \frac{2}{\pi}\frac{d\theta\exp(-2Lk_x'')}{(\theta-\theta_0)^2+\Gamma^2}, \quad (4.172)$$

where $\Gamma = k_x''/k_z$. Integration over $d\theta$ gives

$$w_1 = 2\frac{k_z}{k_x''}\exp(-2Lk_x''). \quad (4.173)$$

For a given flux J_0 of the incident neutron beam, the count rate of the detector will be

$$J = J_0\frac{k_z}{k}|t_{02}|^2\frac{l_2}{D}w_1, \quad (4.174)$$

where D is the thickness of the beam, $k_z|t_{02}|^2/k$ is the probability of penetration of the incident beam into the resonance layer, and l_2/D is the fraction of the incident beam that is intercepted by the channel. For $k_z/k = 10^{-3}$, $l_2/D = 10^{-3}$, $t_{02} = 0.1$, $k_x'' = 1\,\text{cm}^{-1}$, and $k_z = 10^6\,\text{cm}^{-1}$ we obtain $J/J_0 = 0.02\exp(-2L)$, and for $L = 1\,\text{cm}$ we get $J/J_0 \approx 0.002$.

It is interesting to compare this result with experiments already performed [209, 210]. A direct comparison is impossible because the multilayer systems (see Figure 4.29) were different from that used here. Nevertheless, the numbers will not be very different. For instance, in the experiment reported in [209] the length L of the channel was not 1 cm, as here, but 2 cm. If we substitute this number into (4.173) we find $J/J_0 \approx 10^{-4}$, which is close to the result reported in [209]. In the experiment reported in [210], which was made with magnetic layers and polarized neutrons, the length L was 0.6 cm, but nevertheless the ratio J/J_0 was also on the level of 10^{-4}. It was attributed to roughnesses on interfaces. In the magnetic case, roughnesses not only decrease the reflection amplitudes r_{23} and r_{21}, but also depolarize neutrons. Therefore, the number 10^{-4} looks quite reasonable. It is important that in both experiments the angular distribution of outgoing neutrons was a Lorentzian just as is predicted here.

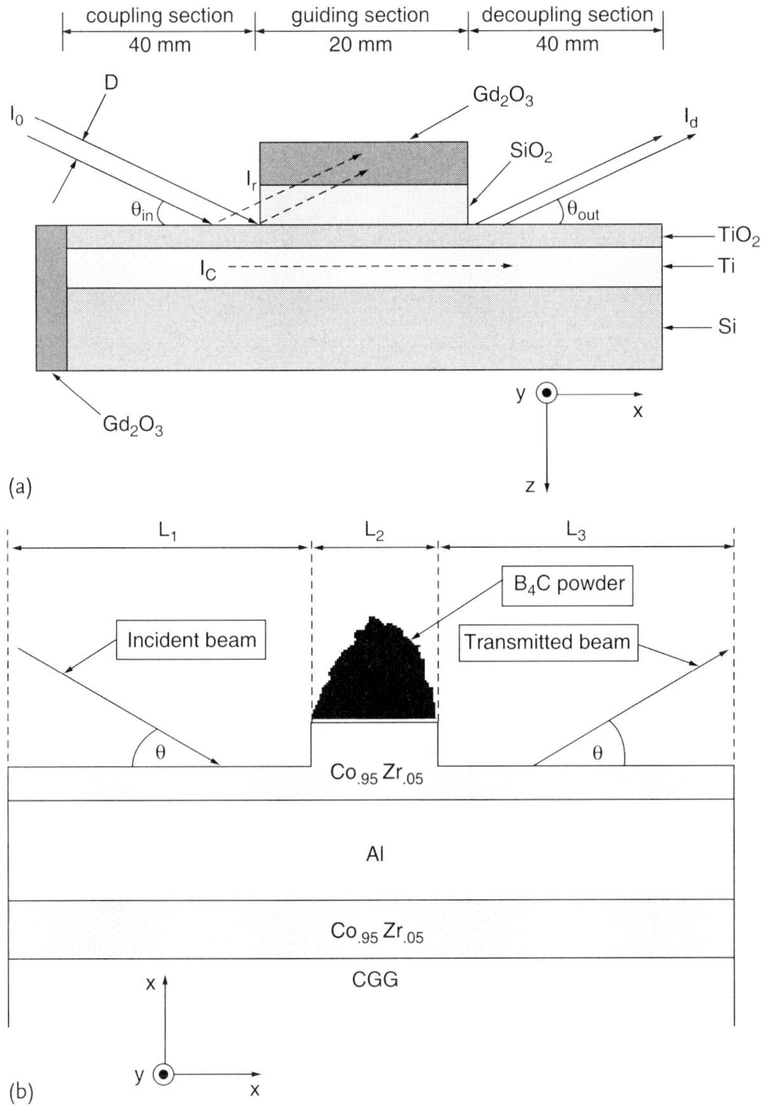

Figure 4.29 Multilayer systems for experiments on neutron channeling: nonmagnetic [209] (a) and magnetic [210] (b). In both experiments the transmitted intensity was 10^{-4} times that of the incident one, and the angular distribution of transmitted neutrons was described by a Lorenzian.

4.6.3
Bound States in Periodic Potentials

Now we want to show briefly how to apply the algebraic method to find bound states in rectangular and periodic potentials, and how a zone structure is created in periodic potentials.

4.6.3.1 Bound Levels in a Rectangular Potential Well [211, 212]

Let us find the bound levels in the potential well shown in Figure 4.30. To do that we shall take into account that the wave function of the bound level must be stationary. Therefore, we can use stationarity condition Eq. (4.151), which is now

$$r_0^2 \exp(2ik_w a) = 1\,, \tag{4.175}$$

where energy k_w^2 is counted from the bottom of the well and r_0 is the amplitude of the total reflection from the edges of the well:

$$r_0 = \frac{k_w - i\sqrt{u - k_w^2}}{k_w + i\sqrt{u - k_w^2}} = \exp(-2i\phi_0)\,, \quad \phi_0 = \arccos\left(\frac{k_w}{\sqrt{u}}\right)\,. \tag{4.176}$$

It follows from Eq. (4.175) with account of Eq. (4.176) that

$$k_w a - 2\phi_0 = \pi n\,, \tag{4.177}$$

where n is an integer. Equation (4.177) is equivalent to $\alpha z - n\pi = 2\arccos(z)$, where $z = k_w/\sqrt{u}$, and $\alpha = a\sqrt{u}$.

4.6.3.2 Zones in Periodic Potentials

For an infinite periodic potential with symmetrical periods we can split the potential at the beginning of any period into two semi-infinite periodic potentials. The stationarity condition Eq. (4.175) can be now represented in the form

$$R_\infty^2(k) = 1\,, \tag{4.178}$$

because the distance d between two potentials is now zero. The condition (4.178) is satisfied on two boundaries of the Darwin table, where $R_\infty(k) = \pm 1$. The stationarity condition is satisfied neither inside the Darwin table nor outside it. However, inside the Darwin table if the particle is put at some point inside the potential it cannot propagate and becomes localized, because its wave function decays exponentially in both directions. So the width of the Darwin table determines the forbidden zone. The permitted zone is the result of accumulation of bound states in a periodic potential with N periods when $N \to \infty$. Indeed, when we have a single potential well, a particle in the well can exist only on bound levels. When we have

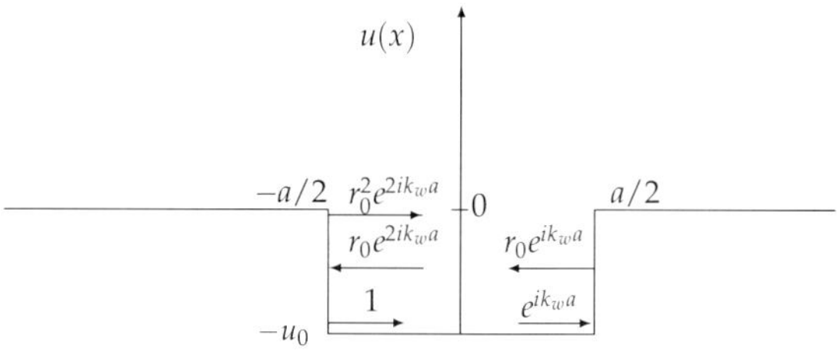

Figure 4.30 Bound states in the well are defined via the stationarity condition: $r_0^2 \exp(2ik_w a) = 1$.

two identical potential wells separated by a barrier, every bound level splits, and the particle on sublevels can propagate from one well to another one. When we have N identical potential wells, every bound level splits into N sublevels and a particle on every sublevel can propagate to any of these N wells. Every bound sublevel is a permitted zone of zero width. When $N \to \infty$, these zones of zero width accumulate to give a single zone of finite width. To see this process of splitting of energy levels and their accumulation into the permitted zone it is worth considering a simple model of a periodic potential with N periods surrounded on both sides by two potential barriers (Figure 4.31). Then we write the condition of stationarity in an infinitesimal gap between, say, the left potential step 1 with reflection amplitude $R_l = r_0$ and the right combination of the periodic potential and the other potential step 2. The reflection amplitude of this combination is $R_r = R_N + T_N^2 r_0/(1 - R_N r_0)$. The stationarity condition is now $R_l R_r = 1$, and it is reduced to $2r_0 R_N + (T_N^2 - R_N^2) r_0^2 = 1$. Substitution of Eq. (4.83) for R_N and T_N gives the equation

$$\exp(2iqNa) = \left(\frac{1 - r_0 R_\infty}{r_0 - R_\infty}\right)^2 . \tag{4.179}$$

For propagation we need the real wave number q. At these q the amplitude R_∞ is also real and the right-hand side of Eq. (4.179) is a unit complex number $\exp(2i\gamma)$ with real phase γ. If we define a bound state in a single period by the condition $q_1 a = \gamma + n\pi$, then for $N \neq 1$ we get

$$q_N a = \frac{\gamma}{N} + \pi l + \pi \frac{n}{N} .$$

So we see that every level is defined by the main quantum number l, which denotes the bound level of one period, and the secondary quantum number n, which runs from 0 to $N - 1$, and denotes the sublevels into which the main level is split. The distance between two sublevels is $\delta q \approx \pi/Na$, so the density of sublevels increases in proportion to N and in the limit $N \to \infty$ the system of zero-width sublevels accumulates into a finite range permitted zone.

4.7 Potentials of General Form

Up to now, when we considered concrete examples of some potential we used mainly rectangular barriers or potentials of the Kronig–Penney type. In real life we can meet more complex potentials. In particular, investigation of the experimental reflection curve from a Cu film over a glass substrate presented in Figure 4.32

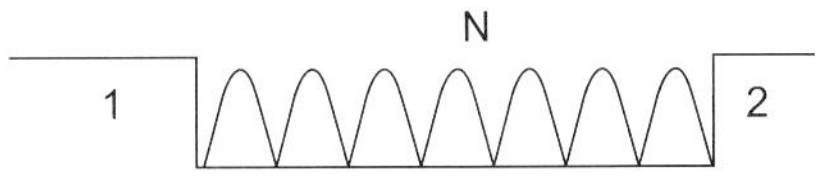

Figure 4.31 A periodic potential with N periods in a well. For derivation of the splitting of energy levels.

showed that description of films with a simple rectangular potentials is not correct. So it was supposed that the boundaries of the films are not sharp, but smoothed. The potential of a film on a substrate can be as shown In Figure 4.33. The edges of the film are smoothed on both sides and the smoothing is different.

4.7.1
Analytical Description of Reflection from a Combination of a Rectangular Barrier and Eckart Potentials

The smoothing can be described by many different functions; however, the most convenient are those for which it is possible to find an analytical solution of the Schrödinger equation. A good example is the Eckart potential (textbook [B]), which is represented by the function

$$u = \frac{u_0}{1 + \exp(-x/h)} . \tag{4.180}$$

It is a potential step of height u_0 with smooth edge of width h. The optical potential of a film evaporated over a substrate can be represented by the curve shown in Figure 4.33. It contains two smooth interfaces with different smoothing parameters. To find the reflection from such a potential one can split it into three parts [190]. The middle part is an ideal rectangular barrier of height u_0 and width b_2. The left

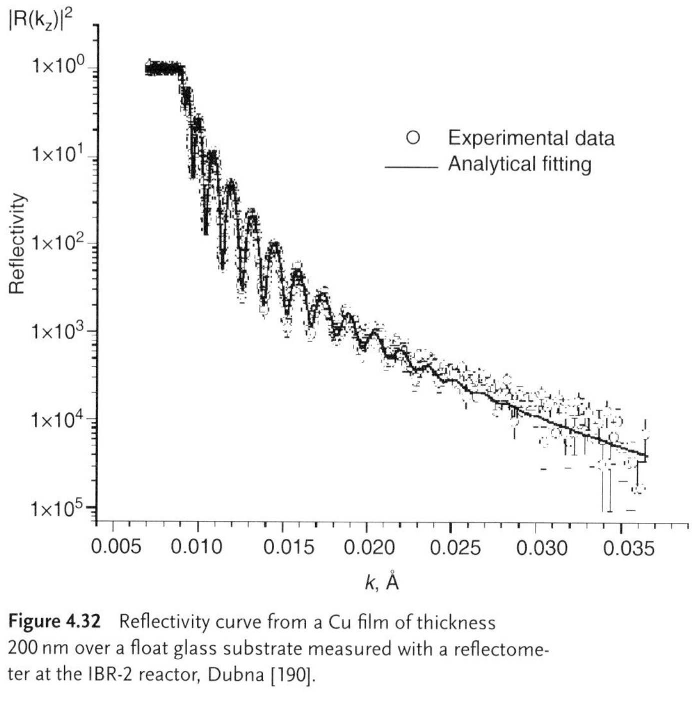

Figure 4.32 Reflectivity curve from a Cu film of thickness 200 nm over a float glass substrate measured with a reflectometer at the IBR-2 reactor, Dubna [190].

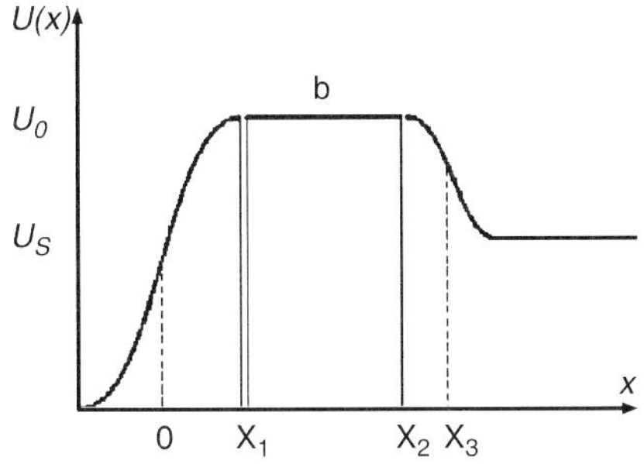

Figure 4.33 A film with smoothed boundaries on a substrate can be described by three potentials: a rectangular barrier for the film itself, and two Eckart potentials for smoothed boundaries.

and right parts are represented by the Eckart potentials, so the total potential is

$$u(x) = \theta(x < x_1) u_0 \frac{1 + e^{-x_1/h_1}}{1 + e^{-x/h_1}} + \theta(x_1 < x < x_2) u_0$$
$$+ \theta(x > x_2) \left(u_s + [u_0 - u_s] \frac{1 + e^{(x_2 - x_3)/h_2}}{1 + e^{(x - x_3)/h_2}} \right) . \quad (4.181)$$

The potential contains seven parameters: $u_{0,s}$, $h_{1,2}$, $b = x_2 - x_1$, x_1, and $x_3 - x_2$, where u_0 and u_s are the heights of the film and the substrate potential barriers, respectively, $h_{1,2}$ are smoothing parameters of the left and right edges of the film, b is the width of the film, and x_1 and $x_2 - x_3$ are parameters which determine how well the three pieces are matched at their interlocking. The larger these parameters are, the smaller the first derivatives of the function (4.181) at the points $x = x_1$ and $x = x_2$ are. Points $x = 0$ and $x = x_3$ correspond to the middle points of the function (4.180).

4.7.1.1 Reflection from a Single Potential Eq. (4.180)

To find the reflection of the whole potential we have to find the solution of the Schrödinger equation in three parts and match it at two points: $x = x_1$ and $x = x_2$. First we find the solution of the Schrödinger equation in the potential Eq. (4.180)

$$\left(\frac{d^2}{dx^2} + k_0^2 - \frac{u_0}{1 + e^{-x/h}} \right) \psi(x) = 0 . \quad (4.182)$$

After change of variables $z = x/h$, $k = k_0 h$, $u = u_0 h^2$,

$$y = -e^{-z}, \quad \psi(x) = y^{-\mu} w(\mu, y) , \quad (4.183)$$

where μ is an as-yet unknown parameter, we transform Eq. (4.182) into the equation for w

$$y^2(1-y)w'' + (1-2\mu)y(1-y)w' + [(\mu^2 + k^2)(1-y) - u]w(\mu, y) = 0. \quad (4.184)$$

Now we can define μ with the requirement $\mu^2 + k^2 - u = 0$. It follows that we can define $\mu = is = \pm ik'$, where $k' = \sqrt{k^2 - u}$.

$$y(1-y)w'' + [1 - 2is - (1-2is)y]w' + (s^2 - k^2)w(s,y) = 0\,. \tag{4.185}$$

This equation coincides with the one for hypergeometrical functions $F(\alpha, \beta, \gamma, y)$:

$$y(1-y)F'' + [\gamma - (1+\alpha+\beta)y]F' - \alpha\beta F = 0 \tag{4.186}$$

with $\alpha = i(k-s)$, $\beta = -i(k+s)$, and $\gamma = 1 - 2is$. Two linearly independent solutions of Eq. (4.184) correspond to the two opposite signs of $s = \pm k'$:

$$w_{1,2}(k,y) = F(\mp i[k'-k], \mp i[k'+k], 1 \mp 2ik', y)\,. \tag{4.187}$$

They both go to unity when $y \to 0$ or $x \to \infty$. Since $F(\alpha,\beta,\gamma,y) = F(\beta,\alpha,\gamma,y)$, then $w_{1,2}(k,y) = w_{1,2}(-k,y)$.

According to Eq. (4.183), both these functions correspond to linearly independent solutions of Eq. (4.182):

$$\psi_{1,2}(k,z) = e^{\pm ik'z} w_{1,2}(k,y)\,. \tag{4.188}$$

Since $F \to 1$, when $z \to +\infty$, that is, $y \to 0$, the wave functions $\psi_{1,2}$ at $z \to +\infty$ have asymptotics: $\psi_{1,2}(k,z) = e^{\pm ik'z}$. On the other hand, their asymptotics at $z \to -\infty$ are (textbook [B])

$$\psi_{1,2}(k,z) = C(k, \pm k')e^{ikz} + C(-k, \pm k')e^{-ikz}\,, \tag{4.189}$$

where

$$C(k,k') = \frac{\Gamma(-2ik)\Gamma(1-2ik')}{\Gamma[-i(k+k')]\Gamma[1-i(k+k')]}\,, \tag{4.190}$$

were $\Gamma(x)$ is the Euler Γ-function. From Eqs. (4.188) and (4.189) it follows that if we take the solution of Eq. (4.182) in the form

$$\vec{\psi}(k,z) = \frac{1}{C(k,k')}\psi_1(k,z)\,, \tag{4.191}$$

it will have asymptotics

$$\vec{\psi}(k,z) = \begin{cases} \frac{1}{C(k,k')}e^{ik'z}\,, & z \to +\infty\,, \\ e^{ikz} + \frac{C(-k,k')}{C(k,k')}e^{-ikz}\,, & z \to -\infty\,, \end{cases} \tag{4.192}$$

which gives immediately the reflection $\vec{r}(k)$ and refraction $\vec{t}(k)$ amplitudes for the plane wave incident onto the potential Eq. (4.180) from the left. They are

$$\begin{aligned} \vec{r}(k) &= \frac{C(-k,k')}{C(k,k')} = \frac{\Gamma(2ik)\Gamma[-i(k'+k)]\Gamma[1-i(k'+k)]}{\Gamma(-2ik)\Gamma[-i(k'-k)]\Gamma[1-i(k'-k)]} \\ &= \frac{k-k'}{k+k'}\,\frac{\Gamma(1+2ik)\Gamma^2[1-i(k'+k)]}{\Gamma(1-2ik)\Gamma^2[1-i(k'-k)]}\,, \end{aligned} \tag{4.193}$$

$$\overrightarrow{t}(k) = \frac{1}{C(k,k')} = \frac{\Gamma[-i(k'+k)]\Gamma[1-i(k'+k)]}{\Gamma(-2ik)\Gamma(1-2ik')}$$
$$= \frac{2k}{k+k'}\frac{\Gamma^2[1-i(k+k')]}{\Gamma(1-2ik)\Gamma(1-2ik')}, \tag{4.194}$$

where we used the relation $\Gamma(1+x) = x\Gamma(x)$. It is seen that these expressions can be represented as

$$\overrightarrow{r}(k) = \overrightarrow{r_0}(k_0)A_r(k_0,u_0,h)\,, \quad \overrightarrow{t}(k) = [1+\overrightarrow{r_0}(k_0)]A_t(k_0,u_0,h)\,,$$
$$\overrightarrow{r_0}(k_0) = \frac{k_0 - k_0'}{k_0 + k_0'}, \tag{4.195}$$

where $k_0' = \sqrt{k_0^2 - u_0}$.

We can also find the reflection and refraction amplitudes for a plain wave incident from inside the potential (4.180) and moving to the left. These expressions can be written down immediately by permutation of k and k' in (4.193) and (4.194), which follows from the symmetry of the space:

$$\overleftarrow{r}(k') = \frac{k'-k}{k'+k}\frac{\Gamma(1+2ik')\Gamma^2[1-i(k+k')]}{\Gamma(1-2ik')\Gamma^2[1-i(k-k')]}, \tag{4.196}$$

$$\overleftarrow{t}(k') = \frac{2k'}{k+k'}\frac{\Gamma^2[1-i(k+k')]}{\Gamma(1-2ik)\Gamma(1-2ik')}. \tag{4.197}$$

It is interesting to check the difference in reflectivity obtained with formula (4.193) from that of an ideal potential step. Since the arguments of all the Γ functions in Eq. (4.193) at small smoothness parameter h are very close to unity, and the deviation from unity is proportional to h, then at $h \to 0$ we asymptotically obtain $\overrightarrow{r}(k) \to \overrightarrow{r_0}(k)$ as the result for the ideal potential step. The difference at a finite quite high $h = 1$ is shown in Figure 4.34. The potential of the mirror was $u = 0.24 - 0.03i$. Its real part is close to Si, and the imaginary part was chosen to be high to show the difference clearly. The solid line gives the reflectivity of an ideal potential step. The dotted line shows the reflectivity with amplitude Eq. (4.193). The dashed line represents the reflectivity $|\overrightarrow{r_0}(k_0)A_{DW}(k_0,u_0,h)|^2$, where, according to [191], the smoothness of the interface is accounted for by the factor of Debye–Waller type $A_{DW} = \exp(-k_0k_0'h^2)$. On the linear scale of Figure 4.34a we see some difference between the three curves for the given $h = 1$. However, the dotted and dashed curves coincide, as is seen in Figure 4.34b, if h in A_{DW} is 2 times larger than in A_r of Eq. (4.195). However, the coincidence in this case is not complete, as is clearly seen on the logarithmic scale in Figure 4.34c.

4.7.1.2 **Reflection from the Combination of Eq. (4.180) and a Potenital Step**

Now we have to find the reflection from the potential

$$u(z) = \frac{u_0(1+e^{-x_1/h})}{1+e^{-x/h}}\Theta(x<x_1) + u_0\Theta(x>x_1)\,. \tag{4.198}$$

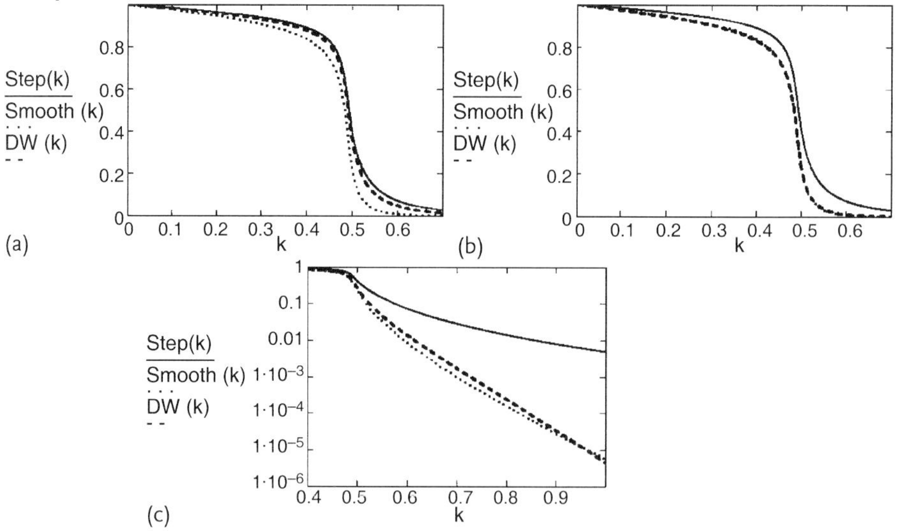

Figure 4.34 Comparison of reflectivity $|\overrightarrow{r_0}|^2$ of an ideal rectangular step (solid line) with reflectivity $|\overrightarrow{r}|^2$ (dotted line) and reflectivity $|\overrightarrow{r_0} A_{\mathrm{DW}}|^2$ proposed in [191] (dashed line), where $A_{\mathrm{DW}} = \exp(-kk') = \exp(-k_0 k_0' h^2)$. (a) The parameter h in A_{DW} is the same as in A_r of Eq. (4.195); (b) the parameter h in A_{DW} is 2 times higher than in A_r of (4.195); (c) The data in (b) on a logarithmic scale.

After change of variables $k = k_0 h$, $u_1 = u_0 h^2$, and $u = u_0 h^2(1 + \exp(-x_1/h))$, we obtain the Schrödinger equation in the form

$$\left(\frac{d^2}{dz^2} + k^2 - \frac{u}{1+e^{-z}}\Theta(z < z_1) - u_1\Theta(z > z_1)\right)\psi(z) = 0\,. \qquad (4.199)$$

To find the reflection amplitude from the left we match wave functions Eq. (4.188) on the left side of the point z_1 to plane waves on the right side of it.

First we consider the incident wave is going to the right at $z \to -\infty$. On the left side we take the wave function in the form $\psi_1(z) + a_1\psi_2(z)$, and on the right side we take it in the form $a_2\exp(-ik_1(z - z_1))$, where $k_1 = \sqrt{k^2 - u_1}$. The constants $a_{1,2}$ are obtained from matching equations

$$\psi_1(z_1) + a_1\psi_2(z_1) = a_2, \quad \psi_1'(z_1) + a_1\psi_2'(z_1) = ik_1 a_2\,, \qquad (4.200)$$

where $\psi' = d\psi/dz$. Their solution is

$$a_1 = \frac{\psi_1'(z_1) - ik_1\psi_1(z_1)}{ik_1\psi_2(z_1) - \psi_2'(z_1)}\,,$$

$$a_2 = \frac{\psi_2(z_1)\psi_1'(z_1) - \psi_1(z_1)\psi_2'(z_1)}{ik_1\psi_2(z_1) - \psi_2'(z_1)} = \frac{2ik'}{ik_1\psi_2(z_1) - \psi_2'(z_1)}\,. \qquad (4.201)$$

In the last equality we took into account that the Wronskian of two solutions (4.188) is equal to $2ik'$. With the help of asymptotical expressions (4.189), we can immediately find the reflection, $\overrightarrow{r_1}$, and refraction, $\overrightarrow{t_1}$, amplitudes. The wave function $z \to -\infty$ can be represented as

$$\psi(z \to -\infty) = \exp(ikz) + \frac{C(-k,k') + a_1 C(-k,-k')}{C(k,k') + a_1 C(k,-k')} \exp(-ikz)\,, \quad (4.202)$$

and at $z \to +\infty$ the wave function becomes

$$\psi(z \to \infty) = \frac{a_2}{C(k,k') + a_1 C(k,-k')} \exp(ik_1(z - z_1))\,. \quad (4.203)$$

Therefore

$$\overrightarrow{r_1} = \frac{C(-k,k') + a_1 C(-k,-k')}{C(k,k') + a_1 C(k,-k')}\,, \qquad \overrightarrow{t_1} = \frac{a_2}{C(k,k') + a_1 C(k,-k')}\,. \quad (4.204)$$

To find the reflection from the right we take the wave function in the form

$$\begin{aligned}\psi(z)) = {} & a_3 \Theta(z < z_1) \left[\frac{\psi_1(z)}{C(k,k')} - \frac{\psi_2(z)}{C(k,-k')} \right] \\ & + \Theta(z < z_1) \left[e^{-ik_1(z-z_1)} + \overleftarrow{r_1} e^{ik_1(z-z_1)} \right]\,, \end{aligned} \quad (4.205)$$

because the left combination at $z \to -\infty$ contains only outgoing wave $\propto \exp(-ikz)$. Matching conditions give

$$\begin{aligned} a_3 \left(\frac{\psi_1(z_1)}{C(k,k')} - \frac{\psi_2(z_1)}{C(k,-k')} \right) &= 1 + \overleftarrow{r_1}\,, \\ a_3 \left(\frac{\psi_1'(z_1)}{C(k,k')} - \frac{\psi_2'(z_1)}{C(k,-k')} \right) &= ik_1[-1 + \overleftarrow{r_1}]\,, \end{aligned} \quad (4.206)$$

from which we find

$$a_3 = \frac{2ik_1}{ik_1\left(\frac{\psi_1(z_1)}{C(k,k')} - \frac{\psi_2(z_1)}{C(k,-k')}\right) - \left(\frac{\psi_1'(z_1)}{C(k,k')} - \frac{\psi_2'(z_1)}{C(k,-k')}\right)}\,. \quad (4.207)$$

$$\overleftarrow{r_1} = \frac{ik_1\left(\frac{\psi_1(z_1)}{C(k,k')} - \frac{\psi_2(z_1)}{C(k,-k')}\right) + \left(\frac{\psi_1'(z_1)}{C(k,k')} - \frac{\psi_2'(z_1)}{C(k,-k')}\right)}{ik_1\left(\frac{\psi_1(z_1)}{C(k,k')} - \frac{\psi_2(z_1)}{C(k,-k')}\right) - \left(\frac{\psi_1'(z_1)}{C(k,k')} - \frac{\psi_2'(z_1)}{C(k,-k')}\right)}\,. \quad (4.208)$$

With a_3 we find the asymptotically outgoing wave $\overleftarrow{t_1}\exp(-ikz)$ with amplitude

$$\overleftarrow{t_1} = \frac{2ik_1\left(\frac{C(-k,k')}{C(k,k')} - \frac{C(-k,-k')}{C(k,-k')}\right)}{ik_1\left(\frac{\psi_1(z_1)}{C(k,k')} - \frac{\psi_2(z_1)}{C(k,-k')}\right) - \left(\frac{\psi_1'(z_1)}{C(k,k')} - \frac{\psi_2'(z_1)}{C(k,-k')}\right)}$$
$$= \frac{2ik'}{\psi_1'(z_1)C(k,-k') - \psi_2'(z_1)C(k,k') - ik_1(\psi_1'(z_1)C(k,-k') - \psi_2'(z_1)C(k,k'))}\,. \tag{4.209}$$

The derivatives over z are

$$\psi_{1,2}'(z) = e^{\pm ik'z}[\pm ik' w_{1,2}(y) - y\,dw_{1,2}'(y)/dy]\,, \tag{4.210}$$

and the derivative of the hypergeometrical function over y is defined via

$$\frac{d}{dy}F(\alpha,\beta,\gamma,y) = \frac{\alpha\beta}{\gamma}F(1+\alpha,1+\beta,1+\gamma,y)\,.$$

4.7.1.3 Reflection from the Full Potential Eq. (4.181)

Now we can construct the reflection and transmission of the whole potential Eq. (4.181). The formula is similar to that of the rectangular barrier:

$$\overrightarrow{R_t} = \overrightarrow{r_1} + \overleftarrow{t_1}\frac{\exp(2ik_0'b)\overrightarrow{r_2}}{1 - \overleftarrow{r_1}\,\overrightarrow{r_2}\exp(2ik_0'b)}\,, \tag{4.211}$$

where reflection amplitude $\overrightarrow{r_2}$ is easily expressed via $\overleftarrow{r_1}$. To do that we point out explicitly the dependence of $\overleftarrow{r_1}$ on all its parameters: $\overleftarrow{r_1}(k,k',k_1,u)$. Then

$$\overrightarrow{r_2} = \overleftarrow{r_1}(k_s, k_s', k_2, u_2)\,, \tag{4.212}$$

where $k_s = h_2\sqrt{k^2 - u_s}$, $u_2 = h_2^2[u_0 - u_s][1 + \exp(z_2 - z_3)]$, $z_{2,3} = x_{2,3}/h_2$, and $k_s' = \sqrt{k_s^2 - u_2}$, $k_2 = h_2 k_0'$. Reflectivity $|R_t|^2$ was used for fitting the experimental data shown in Figure 4.32, which correspond to reflection from a Cu film on a float glass substrate measured by time of flight at the IBR-2 reactor in Dubna [190].

4.7.2 Continuous-Fractions Method

Every smooth function can be represented as a histogram or a set of rectangular barriers [213]. Let us split the arbitrary smooth potential into n parts as shown in Figure 4.35. All the boundaries are indicated by numbers 1 to n from the left. Let us denote the reflection and transmission amplitudes of a single rectangle to the right of the kth boundary by r_k and t_k, respectively, and the reflection and transmission amplitudes of all the rectangles to the right of the kth boundary by R_k and T_k, respectively. Now we can write

$$R_1 = r_1 + \frac{t_1^2 R_2}{1 - r_1 R_2} = r_1 - \frac{t_1^2}{r_1 - 1/R_2}\,.$$

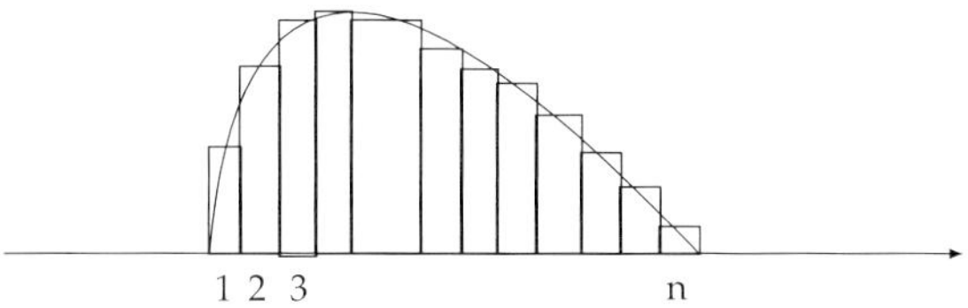

Figure 4.35 Approximation of an arbitrary potential by a set of rectangles.

Substitute here the similar expression for R_2, and so on, and as a result, we obtain the continuous fraction

$$R_1 = r_1 + \frac{t_1^2 R_2}{1 - R_2 r_1} = r_1 - \frac{t_1^2}{r_1 - \cfrac{1}{R_2}} = r_1 - \frac{t_1^2}{r_1 - \cfrac{1}{r_2 - \cfrac{t_2^2}{r_2 - \cfrac{1}{R_3}}}} = \cdots$$

4.7.3
Matrix Method for Scattering from an Arbitrary Potential

Your approach is good, but I know a better one.

A student

Analytical methods are very useful for analysis and fitting experimental data; however, sometimes they are very cumbersome, and then it is more preferable to use numerical methods. The most effective one is the numerical matrix method [214].

Let us again represent the potential by a set of rectangles, as shown in Figure 4.35, and count the jth rectangle to the right of the jth boundary. Inside the jth rectangle the wave function is

$$\psi_j(x) = A_j \exp(i k_j (x - x_j)) + B_j \exp(-i k_j (x - x_j)) , \tag{4.213}$$

where x_j is the coordinate of the jth boundary, $k_j = \sqrt{k_0^2 - u_j}$, and u_j is the potential of the jth rectangle. The wave functions behind the last $n + 1$ boundary and before the first x_1 boundary are of the same form as Eq. (4.213):

$$\psi_{n+1}(x) = A_{n+1} \exp(i k_{n+1} (x - x_{n+1})) + B_{n+1} \exp(-i k_{n+1} (x - x_{n+1})) , \tag{4.214}$$

$$\psi_0(x) = A_0 \exp(i k_0 x) + B_0 \exp(-i k_0 x) , \tag{4.215}$$

$k_{n+1} = k_0$, and we put $x_1 = 0$. Later we shall take into account that $A_0 = 1$, $B_0 = r$, $A_{n+1} = t$, and $B_{n+1} = 0$.

Now we have to match the wave functions at all the boundaries. Continuity of the function and its first derivative at the $(n+1)$th boundary gives the equations

$$\begin{aligned} A_n \exp(ik_n d_n) + B_n \exp(-ik_n d_n) &= A_{n+1} + B_{n+1}, \\ ik_n[A_n \exp(ik_n d_n) - B_n \exp(-ik_n d_n)] &= ik_{n+1}[A_{n+1} - B_{n+1}], \end{aligned} \tag{4.216}$$

where d_j is the width of the jth rectangle. The solution of the system can be represented in matrix form:

$$\begin{pmatrix} A_n \\ B_n \end{pmatrix} = M^{(n,n+1)} \begin{pmatrix} A_{n+1} \\ B_{n+1} \end{pmatrix}. \tag{4.217}$$

The matrix $M^{(n,m)}$ is a product of two matrices, $M^{(n,m)} = E^{(n)} K^{(n,m)}$, which are

$$\begin{aligned} E^{(n)} &= \begin{pmatrix} \exp(-ik_n d_n) & 0 \\ 0 & \exp(ik_n d_n) \end{pmatrix}, \\ K^{(n,m)} &= \frac{1}{2} \begin{pmatrix} 1 + k_m/k_n & 1 - k_m/k_n \\ 1 - k_m/k_n & 1 + k_m/k_n \end{pmatrix}. \end{aligned} \tag{4.218}$$

The matching of the wave function at all the other boundaries gives similar matrices; therefore,

$$\begin{pmatrix} A_0 \\ B_0 \end{pmatrix} = M^{(0,n+1)} \begin{pmatrix} A_{n+1} \\ B_{n+1} \end{pmatrix} = K^{(0,1)} \prod_{j=1}^{n-1} E^{(j)} K^{(j,j+1)} \begin{pmatrix} A_{n+1} \\ B_{n+1} \end{pmatrix}. \tag{4.219}$$

Substitute $A_0 = 1$, $B_0 = r$, $A_{n+1} = t$, and $B_{n+1} = 0$. As a result, we obtain

$$\begin{pmatrix} 1 \\ r \end{pmatrix} = M^{(0,n+1)} \begin{pmatrix} t \\ 0 \end{pmatrix} = \begin{pmatrix} M_{11}^{(0,n+1)} & M_{12}^{(0,n+1)} \\ M_{21}^{(0,n+1)} & M_{22}^{(0,n+1)} \end{pmatrix} \begin{pmatrix} t \\ 0 \end{pmatrix}, \tag{4.220}$$

which is equivalent to $1 = M_{11}^{(0,n+1)} t$, $r = M_{21}^{(0,n+1)} t$, or $t = 1/M_{11}^{(0,n+1)}$, and $r = M_{21}^{(0,n+1)}/M_{11}^{(0,n+1)}$. This numerical algorithm is very convenient.

4.7.4 Perturbation Theory

In many cases it is possible to use perturbation theory. The solution of the Schrödinger equation

$$\left(\frac{d^2}{dx^2} + k^2 - u(x) \right) \psi(x) = 0, \tag{4.221}$$

can be represented as

$$\psi(x) = \psi_0(x) + \int G(x - x') u(x') \psi(x') dx', \tag{4.222}$$

where $\psi_0(x)$ is a solution of the free equation, which is represented by a plane wave,

$$\psi_0(x) = \exp(ikx)\,, \tag{4.223}$$

and $G(x - x')$ is the Green function of the equation:

$$\left(\frac{d^2}{dx^2} + k^2\right) G(x) = \delta(x)\,. \tag{4.224}$$

The solution of it is

$$G(x) = -\frac{i}{2k} \exp(ik|x|)\,. \tag{4.225}$$

Substitution of Eq. (4.225) into Eq. (4.222) and replacement of $\psi(x')$ under the integral by $\psi_0(x)$ gives

$$r(k) = -\frac{i}{2k} \int_0^a u(x') \exp(2ikx') dx'\,,$$

$$t(k) = \exp(ika) \left(1 - \frac{i}{2k} \int_0^a u(x') dx'\right)\,, \tag{4.226}$$

where we assumed that perturbation is nonzero in the range $0 \le x \le a$.

Perturbation formulas do not satisfy unitarity. To improve them it is sufficient to choose only $r(k)$ in Eq. (4.226) and to put $t = \pm i(r/|r|)\sqrt{1 - |r|^2}$.

4.7.5
Reflection from Mirrors in an External Field

Above we everywhere supposed that outside the mirror the space is empty, and the wave function of the incident neutron is represented by a plane wave. Now we shall consider the general case of one-dimensional scattering in some external field, where outside the potential the wave function is not a plane wave [215]. For instance, we can consider vertical motion and reflection from mirrors in the presence of a gravity field, or we can consider radial motion in the presence of spherically symmetrical potentials (see also [216–218]). We shall present here only ideas.

The Schrödinger equation in presence of external field $F(z)$ is

$$\left(\frac{d^2}{dz^2} + k^2 - F(z) - v(z)\right) \Psi(z) = 0\,, \tag{4.227}$$

where $v(z)$ is the optical potential of a mirror.

Imagine that we have the uniform gravity field $F(z) = mgz$. In the absence of a mirror, the wave function (denote it as $f(z, k)$) describes motion, which is limited at $z > 0(z_m(k)$ is the maximal point, or turning point, of the vertical motion) and nonlimited at $z < 0$. The function $f(z, k)$ decays at $z \to \infty$. However, Eq. (4.227) also has another solution (denote it $g(z, k)$), which diverges at $z \to \infty$. At $z \to +\infty$ this solution is nonphysical; however, in the presence of mirrors we need it.

Let us consider a horizontal semi-infinite mirror with potential $u_0\Theta(z > 0)$. The solution of Eq. (4.227) can be represented as

$$\Psi(z) = \Theta(z < 0)\left(\frac{f(z,k)}{f(0,k)} + \vec{r}\,\frac{g(z,k)}{g(0,k)}\right) + \Theta(z > 0)\,\vec{t}\,\frac{f(z,k')}{f(0,k')}\,, \quad (4.228)$$

where $k' = \sqrt{k^2 - u_0}$. The requirement of continuity of the wave function and its derivative at the interface $z = 0$ gives the equations

$$1 + \vec{r} = \vec{t}\,, \qquad \frac{f'(z,k)}{f(0,k)} + \vec{r}\,\frac{g'(z,k)}{g(0,k)} = \vec{t}\,\frac{f'(z,k')}{f(0,k')}\,, \quad (4.229)$$

where the prime near the functions means their derivative over z. From these equations we can find the reflection r and refraction t amplitudes.

If we have two semitransparent mirrors, separated by distance l, and we know the amplitudes r_i, t_i $(i = 1, 2)$ for every one of them, then we can apply to them the same method as described on page 132. We can define the amplitude of the wave incident on the first mirror $f(z, k)/f(0, k)$, denote X the amplitude of the wave incident on the second mirror, and find the system of equations for X and the reflection r_{12} and transmission t_{12} amplitudes of both mirrors:

$$X = \vec{t}\,\frac{f(l+a,k)}{f(a,k)} + \frac{f(l+a,k)}{f(a,k)}\,\overleftarrow{r_1}\,\frac{g(a,k)}{g(a+l,k)}\,\overrightarrow{r_2}\,X\,,$$

$$\vec{r}_{12} = \overrightarrow{r_1} + \overleftarrow{t_1}\,\frac{g(a,k)}{g(a+l,k)}\,\overrightarrow{r_2}\,X\,, \quad \vec{t}_{12} = X\,\overrightarrow{t_2}\,, \quad (4.230)$$

where a is the end point of the potential $v_1(z)$. Solution of these equations gives all the required values. Note that transmissions in opposite directions are not equal now.

The scheme presented looks very reasonable and appropriate for any function $f(z, k)$; however, we must remember that reflection and transmission mean that we are dealing with currents or fluxes; therefore, the functions $f(x, k)$ and $g(x, k)$ cannot be arbitrary. For instance, they cannot be real, because real functions give zero current: $f^*\overrightarrow{\nabla} f - f^*\overleftarrow{\nabla} f = 0$. In that respect the Airy functions $Ai(x)$ and $Bi(x)$ are not appropriate for calculation of reflection and transmission in a gravity field. We shall discuss this again in the next section.

4.8 Some Experiments with Moving Mirrors

The considerations of one-dimensional systems will be not complete without discussion of moving mirrors. We shall consider here four types of motion: uniform, accelerating, vibrating, and diffraction of neutrons from moving gratings. For convenience we shall take unities $\hbar = m = 1$ so that the neutron speed can be denoted k, the speed of the mirror is denoted v, and difference in speed is $k - v$.

4.8.1 Uniform Motion

4.8.1.1 A Semiclassical Consideration

When a neutron enters a substance with optical potential $u > 0$, its speed k becomes slower $k' = \sqrt{k^2 - u}$; therefore, after transmission through a plate of thickness L its wave function acquires the phase factor $\exp(ik'L)$ instead of $\exp(ikL)$, which it would acquire if it were propagating through empty space of thickness L. The phase difference

$$\Delta\phi = (k - \sqrt{k^2 - u})L \approx uL/2k\,, \tag{4.231}$$

when $k^2 \gg u$, is well observed with the help of a neutron interferometer (Figure 4.36). If the mirror moves, say, along k with speed v, the neutron relative to it has speed $k - v$, and the phase (4.231) becomes

$$\Delta\phi = \left(k - v - \sqrt{(k-v)^2 - u}\right)L \approx \frac{uL}{2(k-v)} \approx \frac{uL}{2k} + \frac{uL}{2k}\frac{v}{k}\,, \tag{4.232}$$

where in the last equality we supposed that $v \ll k$. The last term in (4.232) is the result of the mirror motion.

Imagine that in the two paths of the interferometer (Figure 4.36) there are two plates of thickness L perpendicular to trajectories 1 and 2. One of them moves with speed v along k, and the other one moves against k. Then the phases of the wave functions on the two paths will change by the amount

$$\Delta\phi_{1,2} = \frac{uL}{2k} \pm \frac{uL}{2k}\frac{v}{k}\,, \tag{4.233}$$

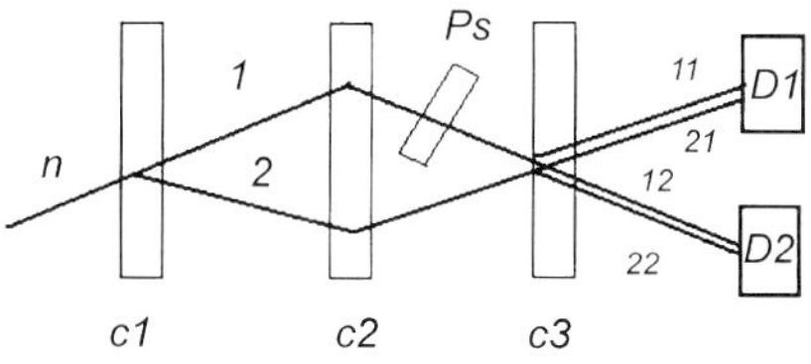

Figure 4.36 A neutron interferometer [219] consists of three crystals. The wave function of every neutron n in the incident beam splits in the first crystal $c1$ into two parts, which propagate along two paths 1 and 2. After diffraction in the second crystal $c2$, part 1 splits in the third crystal $c3$ into two subparts 11 and 12, and part 2 splits into two subparts 21 and 22. Subparts 11 and 21 recombine and go to detector $D1$, whereas the subparts 12 and 22 recombine and go to detector $D2$. The count rates of the two detectors depend on interference of the two subparts, which depends on the phase factors the wave function acquires on paths 1 and 2. The phase shifter Ps can change the phase in path 1. It is a rotatable plate of thickness l. When its angle changes, the effective thickness L of matter traversed by the neutron changes, and the phase along path 1 changes according to Eq. (4.231).

and the phase difference for them will be $\Delta\phi_1 - \Delta\phi_2 = (uL/k)(v/k)$. It is this phase difference which was measured in an experiment [220].

The above consideration is a semiclassical one. We want to consider the problem differently, purely quantum mechanically to see whether there will be a difference. In quantum mechanics the mirror of thickness L moving with speed v is described by the potential $u\Theta(vt < x < L + vt)$, which depends on time, and the Schrödinger equation becomes

$$\left(i\frac{\partial}{\partial t} + \frac{1}{2}\frac{\partial^2}{\partial x^2} - u\Theta(vt < x < L + vt)\right)\Psi(x,t) = 0\,. \tag{4.234}$$

With the time-dependent potential we can expect that if the incident wave is $\Psi(x,t) = \exp(ikx - ik^2t/2)$, then after scattering it will have a different energy and different wave number. In the case of uniform motion, everything is more or less evident. Reflected neutrons change energy and transmitted ones do not. This follows from semiclassical considerations; however, it is also interesting to see whether we can get this result rigorously.

4.8.1.2 **Transformation of Reference Frames**

In the case of uniform motion of the mirror we can make a transformation to a moving reference frame and find the transmission through the stationary mirror, but after such a transformation we get moving detectors. Therefore, after the interaction with the mirror we need to make a transformation back to the laboratory frame. The question is: what do "before" and "after" mean? Is there some special time interval? The uniform motion is a very good example to acquire experience which can be helpful in more complicated cases.

Transformation to the reference frame (coordinates x') moving with speed v is not a simple change of variables $x = x' + vt$. Change of variables transforms the neutron wave function $\Psi(x,t)$ to $\psi(x',t) = \Psi(x' + vt, t)$, and Eq. (4.234) to the distorted non-Schrödinger equation

$$\mathcal{L}\psi(x',t) \equiv \left(i\frac{\partial}{\partial t} - iv\frac{\partial}{\partial x'} + \frac{1}{2}\frac{\partial^2}{\partial x'^2} - u\Theta(0 < x' < L)\right)\psi(x',t) = 0\,. \tag{4.235}$$

If in the coordinates x' we want to have the canonical Schrödinger equation, we have to transform Eq. (4.235) with an operator $U(x',t)$:

$$U(x',t)\mathcal{L}U^{-1}(x',t)\,U(x',t)\psi(x',t) = \left(i\frac{\partial}{\partial t} + \frac{1}{2}\frac{\partial^2}{\partial x'^2} - u\Theta(0 < x' < L)\right)\phi(x',t) = 0\,, \tag{4.236}$$

where $\phi(x',t) = U\psi(x't)$. It is easy to check that the operator $U = \exp(-ivx' - iv^2t/2)$ does the job because

$$U(x',t)\left(i\frac{\partial}{\partial t} - iv\frac{\partial}{\partial x'} + \frac{1}{2}\frac{\partial^2}{\partial x'^2}\right)U^{-1}(x',t) = i\frac{\partial}{\partial t} + \frac{1}{2}\frac{\partial^2}{\partial x'^2}\,. \tag{4.237}$$

After such a transformation, the incident plane wave $\exp(ikx - ik^2t/2)$ is transformed to

$$\exp(-ivx' - iv^2t/2)\exp(ik(x' + vt) - ik^2t/2) = \exp\left(i(k-v)x' - i(k-v)^2t/2\right) . \tag{4.238}$$

With this incident wave we get a stationary solution of Eq. (4.236):

$$\phi(x', t) = \exp\left(-i(k-v)^2t/2\right)\left[\Theta(x' < 0)\left\{e^{i(k-v)x'} + R(k-v)e^{-i(k-v)x'}\right\} + \Theta(x' > L)T(k-v)e^{i(k-v)(x'-L)}\right] , \tag{4.239}$$

where $R(k-v)$ and $T(k-v)$ are the standard reflection and transmission amplitudes.

Now we want to make the back-transformation to the laboratory frame for both reflected and transmitted waves. To do that we first change variables $x' = x - vt$, obtain a distorted equation, and apply to it an operator $V(x, t)$ to restore the canonical form. In this case $V = \exp(ivx - iv^2t/2)$ does the job, and it transforms the wave function to

$$V(x, t)\phi(x - vt, t) = \Theta(x < vt)\left[e^{ikx - ik^2t/2} + R(k-v)e^{-i(k-2v)x - i(k-2v)^2t/2}\right] + \Theta(x > L + vt)T(k-v)e^{ivL}e^{ik(x-L) - ik^2t/2} . \tag{4.240}$$

We see that the energy of the transmitted wave does not change, but the transmission amplitude becomes

$$T(k-v)e^{ivL} = [1-\rho^2(k-v)]\frac{\exp\left(i\sqrt{(k-v)^2 - u}L + ivL\right)}{1 - \rho^2(k-v)\exp\left(2i\sqrt{(k-v)^2 - u}L\right)} , \tag{4.241}$$

where

$$\rho(p) = \frac{p - \sqrt{p^2 - u}}{p + \sqrt{p^2 - u}} . \tag{4.242}$$

From Eqs. (4.240) and (4.241) we can find all the changes associated with the transmitted wave. If we neglect ρ, we obtain only the change of phase, which for $(k-v)^2 \gg u_0$ completely coincides with Eq. (4.232).

It is interesting to note that the transformations to the moving reference frame and back are performed at the same time. They are made over incident and transmitted waves simultaneously because transmission through a mirror with multiple reflections at its edges takes no time.

4.8.2
Accelerating Mirror

Scattering on an accelerating mirror is more interesting. There we find that at transmission the neutron energy changes too [221, 222].

4.8.2.1 Classical Approach

In the classical approach, where nothing is left of quantum mechanics except the optical potential, we can follow the neutron history. Suppose that at the moment the neutron with speed k enters the mirror with optical potential u the mirror has speed v and acceleration a both along the neutron direction. Inside the mirror the neutron has speed $\sqrt{(k-v)^2-u}$ with respect to the entrance surface and continues to move with the same speed toward the exit surface. However, the exit surface moves away with acceleration a. Therefore, at the moment the neutron approaches the exit surface its speed with respect to the exit surface is $\sqrt{(k-v)^2-u-2aL}$, and after exiting into a vacuum this speed is $\sqrt{(k-v)^2-2aL}$. The travel time inside the mirror is $t_1 = \Delta v/a \equiv \left[\sqrt{(k-v)^2-u} - \sqrt{(k-v)^2-u-2aL}\right]/a$. During this time the mirror's speed becomes $v' = v + at_1 = v + \sqrt{(k-v)^2-u} - \sqrt{(k-v)^2-u-2aL}$. Therefore, in the laboratory frame the neutron's speed is

$$k_1 = \sqrt{(k-v)^2-2aL} + v + \sqrt{(k-v)^2-u} - \sqrt{(k-v)^2-u-2aL}. \quad (4.243)$$

It is easy to check that if $u = 0$, then $k_1 = k$, and if $u \ll (k-v)^2 - 2aL$, then

$$k_1 \approx k + \frac{u}{2}\left[\frac{1}{\sqrt{(k-v)^2-2aL}} - \frac{1}{k-v}\right]. \quad (4.244)$$

We see that the neutron speed is increased, that is, the mirror drags the neutron.

This effect was observed[222] at Institut Laue–Langevin, Grenoble, at the source of cold and ultracold neutrons. The scheme for the experiment is shown in Figure 4.37[15]. Ultracold neutrons were monochromatized at an energy of 107 neV by transmission through the resonant multilayer system described in Section 4.6.1. Below and near it was placed a Si plate of thickness 0.6 mm, which was harmonically vibrated by a driver with frequency 40 Hz. This vibration provided acceleration up to 75 m/s^2. Neutrons that passed a monochromator and accelerating mirror fell onto an analyzer multilayer system, which had a resonant transmission at an energy of 127 neV, and after the analyzer they were registered by a detector. Since in the free fall neutrons increase their energy by 1.026 neV every centimeter, the transmission through the analyzer varies with its height as shown in Figure 4.38a. The experimental points, shown in Figure 4.38b, correspond well to theoretical expectations.

15) I am grateful to A.I. Frank for permission to reproduce this and the next figures.

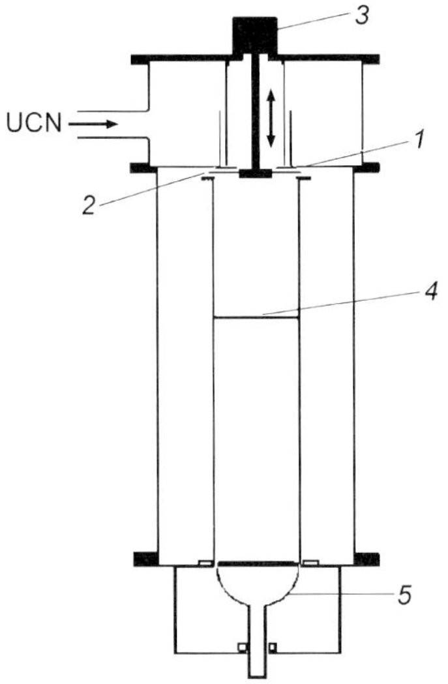

Figure 4.37 Experimental layout for measurement of the acceleration effect. UCN – ultracold neutrons; 1 – monochromator film; 2, Si sample of thickness 0.6 mm; 3 – vibrator; 4 – analyzer film; 5 – detector.

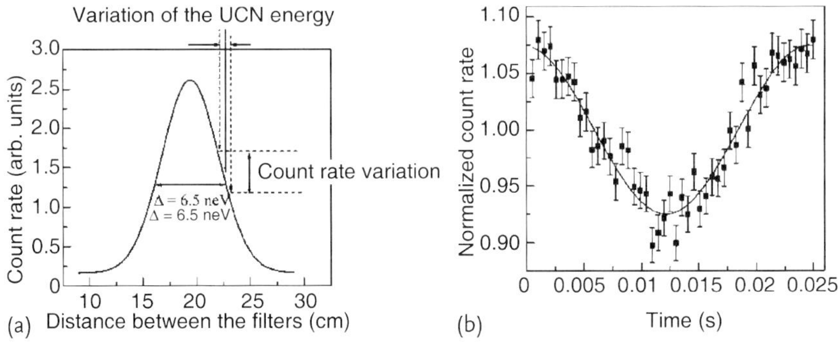

Figure 4.38 (a) Dependence of the transmission through the analyzer on its height. Its position was chosen to be at a height to give transmission at the steepest part of the side of the resonance peak. In such a position the variation of the transmission with a small variation of neutron energy is maximal. (b) The variation of the transmission with time, which corresponds to the variation of the neutron energy with the sample acceleration. The solid line is the theoretical curve fitted to experimental points.

However, this result is correct for a single passage of the neutron through the mirror. The probability of such a passage is

$$w_1 = \left|\tau(k-v)\tau'\left(\sqrt{(k-v)^2 - 2aL}\right)\right|^2,$$

where $\tau(p) = 2p/(p+p')$ and $\tau'(p) = 2p'/(p+p')$ are the refraction amplitudes at the edges, and $p' = \sqrt{p^2 - u}$. We can consider also n passages. The time for n

passages is $t_n = (2n-1)\Delta v/a$. Therefore, the neutron speed k_n after n passages is

$$k_n = \sqrt{(k-v)^2 - 2aL} + v + (2n-1) \times \left[\sqrt{(k-v)^2 - u} - \sqrt{(k-v)^2 - u - 2aL}\right], \tag{4.245}$$

and the probability of such a speed is

$$w_n = w_1 \left| r(k-v) r\left(\sqrt{(k-v)^2 - 2aL}\right)\right|^{2(n-1)},$$

where $r(p) = (p - p')/(p + p')$ is the reflection amplitude at the edges. We see that transmission through an accelerated mirror leads to a discrete spectrum of transmitted particles.

The above consideration is purely classical. To see how the effect can be described in quantum mechanics we must use the nonstationary Schrödinger equation with potential

$$u\Theta(\xi(t) < x < L + \xi(t)),$$

where $\xi(t) = vt + at^2/2$, to transform the incident wave to the accelerating reference frame, calculate the transmitted wave, and transform the transmitted wave back to the laboratory reference frame as in the case of the uniform motion.

4.8.2.2 Transformation to the Accelerating Reference Frame

To make the transformation to the accelerating reference frame we first make us of change of variables $x = x' + \xi(t)$. After that we obtain a distorted equation with the mirror at rest:

$$\mathcal{L}\psi(x',t) \equiv \left(i\frac{\partial}{\partial t} - i\dot{\xi}(t)\frac{\partial}{\partial x'} + \frac{1}{2}\frac{\partial^2}{\partial x'^2} - u\Theta(0 < x < L)\right)\psi(x',t) = 0, \tag{4.246}$$

where $\psi(x',t) = \Psi(x' + \xi(t), t)$, and $\dot{\xi}$ means the derivative over t. Now we need to find an operator $U(x',t)$ which transforms Eq. (4.246) to the canonical form of the Schrödinger equation in gravity potential ax':

$$U(x',t)\mathcal{L}U^{-1}(x',t)\,U(x',t)\psi(x',t) = \left(i\frac{\partial}{\partial t} + \frac{1}{2}\frac{\partial^2}{\partial x'^2} - \ddot{\xi}(t)x' - u\Theta(0 < x < L)\right)\phi(x',t) = 0, \tag{4.247}$$

where $\phi(x') = U\psi(x't)$. Let us check $U = \exp(if(t)x' + ig(t))$. With it we get

$$U(x',t)\mathcal{L}U^{-1}(x',t) = \left[i\frac{\partial}{\partial t} + \dot{f}(t)x' + \dot{g}(t) - i\dot{\xi}(t)\frac{\partial}{\partial x'} - \dot{\xi}(t)f(t) + \frac{\Delta'}{2} - if(t)\frac{\partial}{\partial x'} - \frac{f^2(t)}{2} - u\Theta(0 < x < L)\right], \tag{4.248}$$

so to have canonical form (4.247) we require $f(t) = -\dot{\xi}(t)$, and $\dot{g}(t) = -\dot{\xi}^2(t)/2$. Operator $U(x', t)$ transforms the incident plane wave $\exp(ikx - ik^2t/2)$ into the function

$$\phi_0(x', t) = \exp\left(i(k - v - at)x' - i\frac{(k-v)^2 t}{2} + i(k - v)\frac{at^2}{2} - i\frac{a^2t^3}{6}\right) . \tag{4.249}$$

In the following, for simplicity we shall put $v = 0$.

Now we need to find the transmission of such a wave through the mirror in gravity field ax'. However, we can find the transmission and the reflection only for a stationary wave function. Therefore, we must represent $\phi_0(x', t)$ in the form

$$\phi_0(x', t) = \int dE \exp(-i\omega t)\chi_0(\omega, x') , \tag{4.250}$$

find for every component $\chi_0(\omega, x')$ the transmission $T(E)$ through the mirror in the gravity field as shown on page 177, find transmitted wave $\int d\omega \exp(-i\omega t) T(\omega)$ $\chi_0(\omega, x')$, and then make the transfomation back to the laboratory frame by changing variables $x' = x - \xi(t)$ and applying an operator $V(x, t)$, which transforms the equation to its initial form.

From Eq. (4.249) with $v = 0$ and Eq. (4.250) it follows that

$$\chi_0(\omega, x') = \exp(ikx') \int \frac{dt}{2\pi} \times \exp\left(-i\left(ax' + \frac{k^2}{2} - \omega\right)t + ik\frac{at^2}{2} - i\frac{a^2t^3}{6}\right) . \tag{4.251}$$

After the change of variable $t \to t + k/a$, the cubic polynomial $\alpha t^3 - \beta t^2 + \gamma t$ in the exponent is transformed to the form $\alpha t^3 + \gamma' t + \delta$, and Eq. (4.251) becomes

$$\chi_0(\omega, x') = \exp\left(i\frac{k}{a}\omega - i\frac{k^3}{6a}\right) \int \frac{dt}{2\pi} \exp\left(-i(ax' - \omega)t - i\frac{a^2t^3}{6}\right) . \tag{4.252}$$

Change of the under-integral variable $t \to -t\sqrt[3]{2/a^2}$ gives

$$\begin{aligned}\chi_0(\omega, x') &= \exp\left(i\frac{k}{a}\omega - i\frac{k^3}{6a}\right) \sqrt[3]{\frac{2}{a^2}} \int \frac{dt}{2\pi} \exp\left(izt + i\frac{t^3}{3}\right) \\ &= \exp\left(i\frac{k}{a}\omega - i\frac{k^3}{6a}\right) \sqrt[3]{\frac{2}{a^2}} Ai(z) ,\end{aligned} \tag{4.253}$$

where

$$z = \sqrt[3]{\frac{2}{a^2}}(ax' - \omega) , \tag{4.254}$$

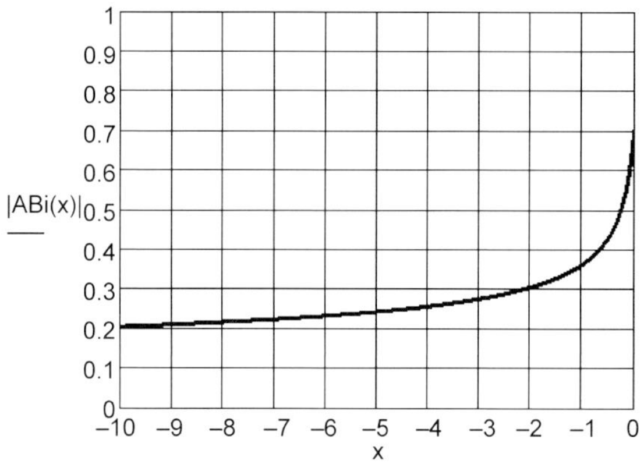

Figure 4.39 Dependence of $|Ai(x) + iBi(x)|$ on $x < 0$.

and

$$Ai(z) = 1/(2\pi) \int_{-\infty}^{\infty} e^{i(zt+t^3/3)} dt \tag{4.255}$$

is the Airy function. This Airy function is a real one. It satisfies the equation

$$\left(\frac{d^2}{dz^2} - z\right) Ai(z) = 0\,, \tag{4.256}$$

oscillates at $z < 0$, and decays exponentially at $z > 0$. The second solution of Eq. (4.256) is $Bi(x)$:

$$Bi(z) = \frac{1}{\pi} \int_0^{\infty} \exp\left(-\frac{t^3}{3} + zt\right) + \sin\left(\frac{t^3}{3} + zt\right) dt\,. \tag{4.257}$$

For the definition of the incident and reflected stationary waves, we cannot use real functions, so we try the combinations

$$ABi(z,\omega) = Ai(z,\omega) + iBi(z,\omega) \tag{4.258}$$

and

$$BAi(z,\omega) = Ai(z,\omega) - iBi(z,\omega)\,, \tag{4.259}$$

respectively, If these combinations are similar to $\exp(\pm ikz)$, then their modulus is monotonic for negative z. A direct calculation demonstrates (Figure 4.39) that this is indeed so.

Now the problem is: how do these combinations behave at $x > 0$? We suppose that it is not $Ai(x)$ but one of these combinations that decays exponentially at $x > 0$, but now we cannot prove it.

With these functions $ABi(z,\omega)$ and $BAi(z)$ we can find the reflection, $\overrightarrow{r_0}$, and transmission, $\overrightarrow{t_0} = 1 + \overrightarrow{r_0}$, amplitudes of the interface matching the wave function at the interface:

$$1 + \overrightarrow{r_0} = \overrightarrow{t_0}\,, \tag{4.260}$$

$$\frac{pABi(z_0,\omega)}{ABi(z_0,\omega)} + \overrightarrow{r_0}\frac{pBAi(z_0,\omega)}{BAi(z_0,\omega)} = \overrightarrow{t_0}\frac{pABi(z_0,\omega-u/2)}{ABi(z_0,\omega-u/2)}\,, \tag{4.261}$$

where $z_0 = -\sqrt[3]{2/a^2}\omega$ and letter p denotes a derivative with respect to z. From these two equations we find

$$\overrightarrow{r_0} = \frac{\dfrac{pABi(z_0,\omega)}{ABi(z_0,\omega)} - \dfrac{pABi(z_0,\omega-u/2)}{ABi(z_0,\omega-u/2)}}{\dfrac{pABi(z_0,\omega-u/2)}{ABi(z_0,\omega-u/2)} - \dfrac{pBAi(z_0,\omega)}{BAi(z_0,\omega)}}\,. \tag{4.262}$$

In a similar way we can find the reflection from inside the mirror $\overleftarrow{r_0}$,

$$1 + \overleftarrow{r_0} = \overleftarrow{t_0}\,, \tag{4.263}$$

$$\frac{pBAi(z_0,\omega-u/2)}{BAi(z_0,\omega-u/2)} + \overleftarrow{r_0}\frac{pABi(z_0,\omega-u/2)}{ABi(z_0,\omega-u/2)} = \overleftarrow{t_0}\frac{pBAi(z_0,\omega)}{BAi(z_0,\omega)}\,, \tag{4.264}$$

$$\overleftarrow{r_0} = \frac{\dfrac{pBAi(z_0,\omega-u/2)}{BAi(z_0,\omega-u/2)} - \dfrac{pBAi(z_0,\omega)}{BAi(z_0,\omega)}}{\dfrac{pBAi(z_0,\omega)}{BAi(z_0,\omega)} - \dfrac{pABi(z_0,\omega-u/2)}{ABi(z_0,\omega-u/2)}}\,, \tag{4.265}$$

and the reflection from inside at the interface at $x' = L$, $\overrightarrow{r_L}$,

$$1 + \overrightarrow{r_L} = \overrightarrow{t_L}\,, \tag{4.266}$$

$$\frac{pABi(z_L,\omega-u/2)}{ABi(z_L,\omega-u/2)} + \overrightarrow{r_L}\frac{pBAi(z_L,\omega-u/2)}{BAi(z_L,\omega-u/2)} = \overrightarrow{t_L}\frac{pABi(z_L,\omega)}{ABi(z_L,\omega)}\,, \tag{4.267}$$

$$\overrightarrow{r_L} = \frac{\dfrac{pABi(z_L,\omega-u/2)}{ABi(z_L,\omega-u/2)} - \dfrac{pABi(z_L,\omega)}{ABi(z_L,\omega)}}{\dfrac{pABi(z_L,\omega)}{ABi(z_L,\omega)} - \dfrac{pBAi(z_L,\omega-u/2)}{BAi(z_L,\omega-u/2)}}\,, \tag{4.268}$$

where $z_L = -\sqrt[3]{2/a^2}(\omega - aL)$. With these expressions we can write down the transmission through the plate:

$$\overrightarrow{T_L} = \frac{\dfrac{ABi(z_L,\omega-u/2)}{ABi(z_0,\omega-u/2)}[1+\overrightarrow{r_0}][1+\overrightarrow{r_L}]}{1-[1+\overleftarrow{r_0}][1+\overrightarrow{r_L}]\dfrac{ABi(z_L,\omega-u/2)}{ABi(z_0,\omega-u/2)}\dfrac{BAi(z_0,\omega-u/2)}{BAi(z_L,\omega-u/2)}}\,. \tag{4.269}$$

The dependences of $|rl0(\omega)| \equiv |\overrightarrow{r_0}|$ and $|Tl(\omega)| \equiv |\overrightarrow{T_L}|$ on ω are shown in Figures 4.40 and 4.41. With (4.269) we can construct the transmitted wave

$$\phi(x',t) = \int d\omega\,\overrightarrow{T_L}(\omega)\frac{ABi(z)}{ABi(z_L)}\exp(-i\omega t)\,, \tag{4.270}$$

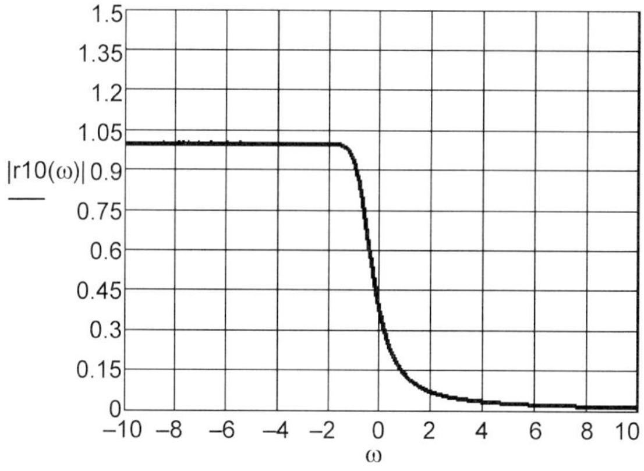

Figure 4.40 Dependence of $|\vec{r_0}|$ on ω.

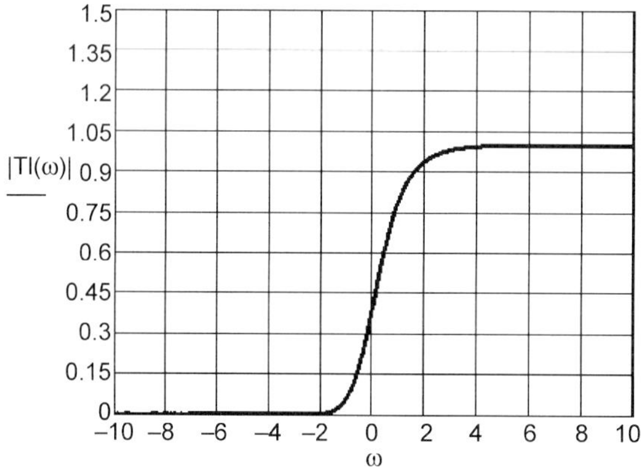

Figure 4.41 Dependence of $|\vec{T_L}|$ on ω.

make the back-transformation to laboratory reference frame, and find the function $\Psi(x, t)$ and its Fourier expansion

$$\Psi(x, t) = \int dE\,\Psi_F(E)\exp(ik_E x, -iEt)\,. \tag{4.271}$$

Then $|\Psi_F(E)|^2 k_E/k$ will give the spectrum of transmitted neutrons.

This program has not been realized yet. It represents a good start for research which is important both for mathematics and physics.

4.8.3
Vibrating Mirror

In practice, experiments on acceleration are performed with an oscillating mirror. If the oscillation period T is large compared with the flight time, t_f, of neutrons through the mirror, then for every neutron the mirror has some constant acceleration [223]. If not, we should use the full quantum theory for neutron scattering on an oscillating object [38].

In the experiment reported in [38] the total reflection of neutrons by a vibrating mirror was investigated. Let us consider a semi-infinite mirror of potential u vibrating with frequency ω. The position of the interface changes in time as $a(t) = a_0 \sin(\omega t)$, where a_0 is the oscillation amplitude. We cannot find a good analytical compact solution for such a problem, but it is clear what we can find. In the case $a_0 \ll \lambda_c \equiv 1/\sqrt{u}$, we can apply perturbation theory, which is linear in $a(t)$. This theory will give us three peaks in the spectrum of reflected neutrons. The middle one is the elastic peak with energy equal to the energy E_0 of the incident neutron, and the two side peaks (they are called Stokes and anti Stokes) have energies energies $E_0 \mp \hbar\omega$, respectively. When a_0 increases, we have to take into account the second-order perturbation, which is proportional to $a(t)^2$. This gives five peaks in the spectrum of reflected neutrons. The two new side peaks have energies $E_0 \mp 2\hbar\omega$. In the case of small oscillation frequencies $\omega \ll \omega_c = v_c/\lambda_c \approx 10^8\,\mathrm{s}^{-1}$, where v_c is the critical neutron speed and λ_c is the critical penetration depth, the spectrum of the reflected neutrons will be determined by the relative neutron speeds $v = v_0 - \dot{a}(t)$ with respect to the moving mirror, and we can calculate the reflection as if the mirror had uniform motion at time t. All that was calculated and experimentally checked in [38].

4.8.4
Diffraction by a Moving Grating

Some experiments were performed with moving gratings [224]. To calculate scattering from a moving grating we need only to transform the incident plane wave to the reference system moving with the grating. In this system scattering from the grating gives diffraction peaks[16]. This means that the component $k_\parallel$ of the neutron wave vector parallel to the grating plane will change after reflection to $k_\parallel + 2\pi n/d$, where n is an integer and d is the period of the grating. Since the diffraction on the grating is an elastic process, the normal component of the neutron wave vector after diffraction becomes $k_\perp = \sqrt{k^2 - (k_\parallel + 2\pi n/d)^2}$.

If the grating moves with speed v parallel to the grating plane, then the neutron wave vector component parallel to the grating plane relative to the grating is $k'_\parallel = k_\parallel - v$. After diffraction its wave components become $k'_{d\parallel} = k_\parallel - v + 2\pi n/d$ and

$$\begin{aligned} k'_{d\perp} &= \sqrt{k_\perp^2 + (k_\parallel - v)^2 - (k_\parallel - v + 2\pi n/d)^2} \\ &= \sqrt{k^2 - (k_\parallel + 2\pi n/d)^2 + 4\pi v n/d}\,. \end{aligned}$$

Transformation back to the laboratory reference frame gives $k_{d\parallel} = k_\parallel + 2\pi n/d$ as in diffraction from a nonmoving grating, but the normal component does not change after such transformation. Therefore, the total neutron energy after diffraction from the moving grating changes. The change is greater the smaller is the

16) Diffractions will be considered in Chapter 7.

grating period d. A grating with decreasing period can transform a continuous beam of neutrons into bunches, because it transfers greater energy to later neutrons than to earlier ones. The neutron density in the bunches at some moment of time can be considerably higher than in the previous continuous beam. This effect is called time focusing and it was demonstrated in [224].

4.9 Conclusion

The main topic in this chapter was the method, presented in Section 4.4, which was used for solution of one-dimensional problems. We illustrated the effectiveness of the method on several examples, but the field of application is much wider. We want to point out some unsolved problems. It is not known whether the multiple reflection used in this method is a real physical process. From the physical point of view, multiple reflections from two separated mirrors should proceed in time, though in stationary quantum mechanics they were considered as occurring with infinite speed or instantly. This seems unbelievable. Therefore, investigation of the transition processes in two-mirror system would be very important for our understanding of quantum mechanics.

We are dealing here with neutron optics, however the method developed here is applicable in many other areas. It is applicable to X-rays, to acoustics, elastic waves, and even corpuscular physics. Moreover, it is applicable not only to one-dimensional, but also to three-dimensional problems. This is what we shall talk about in the following pages.

5
One-Dimensional Scattering of Neutrons with Spin

In the previous chapter, the neutron was treated as a scalar particle, but it possesses spin and along with it a magnetic moment, because of which it can interact with a magnetic field. In this chapter we take this property into account.

A spinor particle is described by two functions, which can be arranged in a column,

$$|\psi(\mathbf{r})\rangle = \begin{pmatrix} \psi_u(\mathbf{r}) \\ \psi_d(\mathbf{r}) \end{pmatrix} = \psi_u(\mathbf{r})|\xi_u\rangle + \psi_d(\mathbf{r})|\xi_d\rangle\,, \tag{5.1}$$

where indices u and d point to the up and down components of the spinor, respectively, and we introduced two unit spinors,

$$|\xi_u\rangle = \begin{pmatrix} 1 \\ 0 \end{pmatrix}, \quad |\xi_d\rangle = \begin{pmatrix} 0 \\ 1 \end{pmatrix}, \tag{5.2}$$

which represent a basis in the spinor space.

With the introduction of the spinor we enter a dense mist of quantum ideology, which is marked by phrases and words such as wave-packet collapse, entangled states, and teleportation. We shall devote some pages to it in Chapter 10, but here we ignore it, and show how to work with a spinor, demonstrating that it is related to an absolutely classical object. In quantum physics it is a common to talk about observables represented by Hermitian operators, and we want to say that *no observable was ever observed!*

The two coefficients $C_{u,d}$ in the combination $|\psi\rangle = C_u|\xi_u\rangle + C_d|\xi_d\rangle$ determine a classical vector – the spin arrow, which is defined as

$$s = \langle\psi|\boldsymbol{\sigma}|\psi\rangle\,,$$

where $\boldsymbol{\sigma} = (\sigma_x, \sigma_y, \sigma_z)$ are Pauli matrices:

$$\sigma_x = \begin{pmatrix} 0 & 1 \\ 1 & 0 \end{pmatrix}, \quad \sigma_y = \begin{pmatrix} 0 & -i \\ i & 0 \end{pmatrix}, \quad \sigma_z = \begin{pmatrix} 1 & 0 \\ 0 & -1 \end{pmatrix}, \tag{5.3}$$

$\langle\psi(\mathbf{r})| = C_u^*\langle\xi_u| + C_d^*\langle\xi_d|$, the star means complex conjugate, and $\langle\xi_{u,d}|$ are transposed basic spinors: $\langle\xi_u| = (1,0)$, $\langle\xi_d| = (0,1)$, so $\langle\xi_{u,d}||\xi_{u,d}\rangle = 1$, $\langle\xi_{d,u}||\xi_{u,d}\rangle = 0$.

Neutron Optics. Masahiko Utsuro, Vladimir K. Ignatovich

ISBN: 978-3-527-40885-6

When we know the direction of the spin arrow, we also know the orientation of the magnetic moment $\boldsymbol{\mu} = \mu_n \boldsymbol{s} = -|\mu_n|\boldsymbol{s}$, where $\mu_n = -1.91\,\mu_{\rm NB}$, $\mu_{\rm NB} = e\hbar/2mc = 5\cdot 10^{-24}$ erg/G is the Bohr magneton. A neutron interacts with the magnetic field $\boldsymbol{B}$ via its magnetic moment. But this interaction is described with the operator $-\boldsymbol{\mu}\cdot\boldsymbol{B} = |\mu_n|\boldsymbol{\sigma}\cdot\boldsymbol{B}$, and not with the dot product $-\mu_n(\boldsymbol{s}\cdot\boldsymbol{B})$ [225]. This is an essential feature of quantum mechanics. The Schrödinger equation for the wave function $|\boldsymbol{\Psi}(\boldsymbol{r},t)\rangle$ of a neutron with spin is

$$i\frac{m}{\hbar}\frac{\partial}{\partial t}|\boldsymbol{\Psi}(\boldsymbol{r},t)\rangle = \left(-\frac{\Delta}{2} + \frac{m}{\hbar^2}U(\boldsymbol{r},t) - \frac{m}{\hbar^2}\boldsymbol{\mu}\cdot\boldsymbol{B}(\boldsymbol{r},t)\right)|\boldsymbol{\Psi}(\boldsymbol{r},t)\rangle\,, \tag{5.4}$$

where $U(\boldsymbol{r},t)$ is an external potential, which together with $\boldsymbol{B}$ can, in general, depend on time and coordinates. The equation can be simplified. For that we replace $t\hbar/m$ with new time variable t, $m|\mu_n|B/\hbar^2$ with new magnetic field B, and $2m\,U(\boldsymbol{r},t)/\hbar^2$ with $u(\boldsymbol{r},t)$[17]. As a result, we obtain

$$i\frac{\partial}{\partial t}|\boldsymbol{\Psi}(\boldsymbol{r},t)\rangle = \left(\frac{1}{2}[-\Delta + u(\boldsymbol{r},t)] + \boldsymbol{\sigma}\cdot\boldsymbol{B}(\boldsymbol{r},t)\right)|\boldsymbol{\Psi}(\boldsymbol{r},t)\rangle\,. \tag{5.5}$$

This chapter is devoted to one-dimensional problems: the potential u and the field $\boldsymbol{B}$ will depend only on a single coordinate x, and the magnetic field can also contain dependence on time. Therefore, the wave function $|\boldsymbol{\Psi}(\boldsymbol{r},t)\rangle$ is representable as

$$|\boldsymbol{\Psi}(\boldsymbol{r},t)\rangle = \exp(i\boldsymbol{k}_{\|}\boldsymbol{r}_{\|} - i|\boldsymbol{k}_{\|}|^2 t/2)|\boldsymbol{\Psi}(x,t)\rangle\,, \tag{5.6}$$

where $\boldsymbol{r}_{\|} = (0, y, z)$, and $\boldsymbol{k}_{\|}$ denotes components of the vector $\boldsymbol{k}$ parallel to the plane (y,z). Substitution of Eq. (5.6) into Eq. (5.5) leads to

$$i\frac{\partial}{\partial t}|\boldsymbol{\Psi}(x,t)\rangle = \left(-\frac{1}{2}\frac{\partial^2}{\partial x^2} + \frac{u(x)}{2} + \boldsymbol{\sigma}\cdot\boldsymbol{B}(x,t)\right)|\boldsymbol{\Psi}(x,t)\rangle\,. \tag{5.7}$$

If $\boldsymbol{B}$ does not depend on time, we can use a stationary solution, $|\boldsymbol{\Psi}(x,t)\rangle = \exp(-ik_x^2 t/2)|\boldsymbol{\Psi}(x)\rangle$. Substitution of this into Eq. (5.7) leads to

$$\left(\frac{\partial^2}{\partial x^2} + k^2 - u(x) - 2\boldsymbol{\sigma}\cdot\boldsymbol{B}(x)\right)|\boldsymbol{\Psi}(x)\rangle = 0\,, \tag{5.8}$$

where we removed index x in k_x^2 for simplicity.

From now on we will use Eqs. (5.8) and (5.7) most of the time. The stationary equation is used to find the reflection from and the transmission through magnetic mirrors and magnetic multilayered systems. The nonstationary equation is needed to study the interaction of neutrons with periodic fields, which are used to rotate the direction of the neutron spin arrow.

First, we will consider reflection from a semi-infinite magnetic mirror. This problem is similar to that of reflection of a scalar particle from a potential step; however,

17) It is useful to keep the factor 1/2, which always accompanies the kinetic energy.

now the reflection and transmission amplitudes are not numbers, r, t, but matrices, $\hat{\boldsymbol{r}}, \hat{\boldsymbol{t}}$. Because of that, the physical content of the process becomes richer: reflection and refraction can be accompanied by multiray beam splitting.

After that we will solve the problem of reflection from a magnetic mirror of a finite thickness (potential barrier analogy) and will find equations for a periodic multilayered system. Because of the matrix nature of the reflection and transmission amplitudes for one period, the equation for the full reflection amplitude $\hat{R}$ is more complex and in general cannot be solved. However, the solution can be found numerically with an algorithm similar to the matrix method described in previous chapter.

Then we will look at nonstationary processes, consider experimental methods of polarization, rotation and analysis of polarization of neutron beams, and investigate some interesting phenomena related to the dynamics of the neutron spin, such as particle–wave duality, spin echo, and neutron holography.

5.1 Reflection from a Magnetic Mirror

The reflection of a neutron in a magnetic field from a semi-infinite magnetized mirror is found similarly to the reflection of a scalar particle from a potential step. The stationary equation is

$$\left(-\frac{\partial^2}{\partial x^2} + u_0\Theta(x>0) + 2\boldsymbol{\sigma}\cdot\boldsymbol{B}_o\Theta(x<0) + 2\boldsymbol{\sigma}\cdot\boldsymbol{B}_i\Theta(x>0) - k^2\right) |\boldsymbol{\Psi}(x)\rangle = 0\,. \tag{5.9}$$

The interaction potential contains the optical potential of the mirror material, u_0, and the constant and homogeneous magnetic fields outside, $\boldsymbol{B}_o$ $(x<0)$, and inside, $\boldsymbol{B}_i$ $(x>0)$, the mirror. To solve Eq. (5.9) we need to define the incident wave.

5.1.1 Plane Wave in a Homogeneous Magnetic Field

The incident wave should satisfy the Schrödinger equation in the space outside the mirror, which is however not a free space as it contains the magnetic field $\boldsymbol{B}_o$,

$$\left(-\frac{\partial^2}{\partial x^2} + 2\boldsymbol{\sigma}\cdot\boldsymbol{B}_o - k^2\right) |\boldsymbol{\Psi}(x)\rangle = 0\,. \tag{5.10}$$

For arbitrary polarization $|\xi\rangle$

$$|\xi\rangle = \alpha_u|\xi_u\rangle + \alpha_d|\xi_d\rangle \tag{5.11}$$

where complex numbers $\alpha_{u,d}$ satisfy the normalization condition $|\alpha_u|^2 + |\alpha_d|^2 = 1$, and "up" and "down" are defined by the direction of the external field $\boldsymbol{B}_o$, the solution can be represented in the form

$$|\boldsymbol{\Psi}(x)\rangle = \alpha_u e^{ik_u x}|\xi_u\rangle + \alpha_d e^{ik_d x}|\xi_d\rangle = \begin{pmatrix} \alpha_u \exp(ik_u x) \\ \alpha_d \exp(ik_d x) \end{pmatrix} , \tag{5.12}$$

where $k_{u,d} = \sqrt{k^2 \mp 2B_o}$. If we take into account that any function $f(\sigma_z)$ has the same eigenvectors $|\xi_{u,d}\rangle$ as the matrix σ_z, and its eigenvalues are $f(\pm 1)$, then the above expression can be simplified to

$$|\boldsymbol{\Psi}(x)\rangle = \exp(i\hat{\boldsymbol{k}}_o x)\,(\alpha_u|\xi_u\rangle + \alpha_d|\xi_d\rangle) = \exp(i\hat{\boldsymbol{k}}_o x)|\xi\rangle , \tag{5.13}$$

where $\hat{\boldsymbol{k}}_o = \sqrt{k^2 - 2\boldsymbol{\sigma}\cdot\boldsymbol{B}_o}$. Since $|\xi\rangle$ is an arbitrary spinor, we can represent solution Eq. (5.13) in the form $|\boldsymbol{\Psi}(x)\rangle = \boldsymbol{\Psi}(x)|\xi\rangle$, where

$$\boldsymbol{\Psi}(x) = \exp(i\hat{\boldsymbol{k}}_o x) \tag{5.14}$$

is a plane wave matrix.

5.1.1.1 A Fundamental Question of Quantum Mechanics

Note that the incident wave has a well-defined energy $k^2/2$, and its two spin components propagate with different velocities. It is usually understood that the arbitrarily polarized neutron belongs to a coherent superposition of two states with different velocities. Because of the difference in velocities, the neutron polarization precesses in the external field. However, the following question arises: what happens to a neutron when the two components separate by a large distance? We know that the plane wave is only a single component of a wave packet, and the wave packet has a finite width, say, l which we can call "coherence length". So, what happens if the two components separate by a distance $> l$? Then n neutrons with their polarization given by Eq. (5.12) separate into $n|\alpha_u|^2$ neutrons polarized along the field and $n|\alpha_d|^2$ neutrons polarized against the field, at the distance $L = lk_u/(k_d - k_u)$ from the entry point. Moreover, the $n|\alpha_d|^2$ neutrons arrive at point L earlier than the $n|\alpha_u|^2$ neutrons, and the time difference is

$$\Delta t = l/k_d . \tag{5.15}$$

Such a possibility had never been checked. Instead, it is common to believe that the velocity of the neutrons with polarization Eq. (5.12) is given by the average velocity of the two components,

$$\overline{k} = |\alpha_u|^2 k_u + |\alpha_d|^2 k_d . \tag{5.16}$$

5.1.2 Solution of Eq. (5.9)

We are now ready to tackle the solution of Eq. (5.9) with the incident wave $\exp(i\hat{\boldsymbol{k}}x)$. We are looking for a solution in matrix form,

$$\boldsymbol{\Psi}(x) = \Theta(x<0)\left(\exp(i\hat{\boldsymbol{k}}_o x) + \exp(-i\hat{\boldsymbol{k}}_o x)\hat{\boldsymbol{r}}\right) + \Theta(x>0)\exp(i\hat{\boldsymbol{k}}_i x)\hat{\boldsymbol{t}} , \tag{5.17}$$

where $\hat{r}$ and $\hat{t}$ are the reflection and refraction amplitude matrices, and $\hat{k}_i = \sqrt{k^2 - u - 2\boldsymbol{\sigma} \cdot \boldsymbol{B}_i}$ is the wave vector operator inside the medium. To find $\hat{r}$ and $\hat{t}$, we need to match the wave function (5.17) and its first derivative at the interface. We then obtain

$$\hat{I} + \hat{r} = \hat{t}\,, \quad \hat{k}_o(\hat{I} - \hat{r}) = \hat{k}_i\hat{t}\,, \tag{5.18}$$

where $\hat{I}$ is the unit matrix. It follows that

$$\hat{r} = (\hat{k}_o + \hat{k}_i)^{-1}(\hat{k}_o - \hat{k}_i), \quad \hat{t} = \hat{I} + \hat{r} = (\hat{k}_o + \hat{k}_i)^{-1} 2\hat{k}_o\,. \tag{5.19}$$

These matrices can be used to obtain the probabilities of reflection with and without flip of the polarization. To find them let us first rewrite the reflection matrix (5.19) to show explicitly the dependence of the wave vectors $\hat{k}_{o,i}$ on the matrices $\boldsymbol{\sigma}$ and the fields $\boldsymbol{B}_{o,i}$:

$$\hat{r} = \left(\hat{k}_o(\boldsymbol{\sigma} \cdot \boldsymbol{B}_o) + \hat{k}_i(\boldsymbol{\sigma} \cdot \boldsymbol{B}_i)\right)^{-1} \left(\hat{k}_o(\boldsymbol{\sigma} \cdot \boldsymbol{B}_o) - \hat{k}_i(\boldsymbol{\sigma} \cdot \boldsymbol{B}_i)\right). \tag{5.20}$$

Consider reflection for neutrons polarized along or oppositely to the field $\boldsymbol{B}_o$, that is, if $|\xi_0\rangle = |\xi_u\rangle$ or $|\xi_d\rangle$, respectively. The diagonal elements of the reflection matrix $\langle\xi_{u,d}|\hat{r}|\xi_{u,d}\rangle$ define the probability amplitudes without spin flip, whereas the nondiagonal matrix elements define the probability amplitude of the reflection with spin flip. To find these elements, we need to define the properties of the Pauli matrices $\boldsymbol{\sigma}$ and the properties of spinor states.

5.1.2.1 Properties of the Pauli Matrices and Spinor States

The basic properties of the Pauli matrices are

$$\begin{aligned} &\sigma_x\sigma_y = i\sigma_z\,, \quad \sigma_y\sigma_z = i\sigma_x\,, \quad \sigma_z\sigma_x = i\sigma_y\,, \\ &\sigma^2_{x,y,z} = 1\,, \quad \sigma_i\sigma_j = -\sigma_j\sigma_i \quad \text{for} \quad i \neq j\,. \end{aligned} \tag{5.21}$$

These equations can be used to derive other relations, which include Pauli matrices:

1. For two arbitrary vectors $\boldsymbol{a}$ and $\boldsymbol{b}$,

   $$(\boldsymbol{\sigma} \cdot \boldsymbol{a})(\boldsymbol{\sigma} \cdot \boldsymbol{b}) = (\boldsymbol{a} \cdot \boldsymbol{b}) + i\boldsymbol{\sigma}(\boldsymbol{a} \times \boldsymbol{b})\,, \tag{5.22}$$

   where $(\boldsymbol{a} \times \boldsymbol{b})$ is the vector product of $\boldsymbol{a}$ and $\boldsymbol{b}$. This relation can be easily verified using Eq. (5.21).
2. Every spin state has a polarization along some unit vector $\boldsymbol{a}$, and we denote such a state as $|\boldsymbol{a}\rangle$.
3. Every state $|\boldsymbol{b}\rangle$ polarized along a unit vector $\boldsymbol{b}$ can be represented as

   $$|\boldsymbol{b}_a\rangle = \frac{\hat{I} + \boldsymbol{b} \cdot \boldsymbol{\sigma}}{\sqrt{2(1 + (\boldsymbol{b} \cdot \boldsymbol{a}))}}|\boldsymbol{a}\rangle \tag{5.23}$$

   for arbitrary $\boldsymbol{a}$. We will call $|\boldsymbol{a}\rangle$ a reference state. In cases where the nature of the reference state is not important, we will omit the index $\boldsymbol{a}$.

4. Every matrix $\boldsymbol{B}\cdot\boldsymbol{\sigma}$ with $\boldsymbol{B}/B=\boldsymbol{b}$ has eigenvalues $\pm B$ and eigenvectors $|\pm\boldsymbol{b}\rangle$:

$$\begin{aligned}\boldsymbol{B}\boldsymbol{\sigma}|\pm\boldsymbol{b}\rangle &= \boldsymbol{B}\boldsymbol{\sigma}\frac{\hat{I}\pm\boldsymbol{b}\boldsymbol{\sigma}}{\sqrt{2(1\pm(\boldsymbol{b}\cdot\boldsymbol{a}))}}|\boldsymbol{a}\rangle = \pm B\frac{\hat{I}\pm\boldsymbol{b}\boldsymbol{\sigma}}{\sqrt{2(1\pm(\boldsymbol{b}\cdot\boldsymbol{a}))}}|\boldsymbol{a}\rangle \\ &= \pm B|\pm\boldsymbol{b}\rangle\,. \end{aligned} \tag{5.24}$$

5. The eigenvectors of an arbitrary function $f(\boldsymbol{A}\cdot\boldsymbol{\sigma})$ with $\boldsymbol{A}/A=\boldsymbol{a}$ coincide with the eigenvectors of the matrix $\boldsymbol{a}\boldsymbol{\sigma}$, and the eigenvalues are equal to $f(\pm A)$:

$$f(\boldsymbol{A}\cdot\boldsymbol{\sigma})|\boldsymbol{a}\rangle = f(\pm A)|\boldsymbol{a}\rangle\,. \tag{5.25}$$

To prove this one can represent the function $f(\boldsymbol{A}\cdot\boldsymbol{\sigma})$ in terms of the Taylor series in powers of $(\boldsymbol{A}\cdot\boldsymbol{\sigma})$.

6. An arbitrary function $f(\boldsymbol{A}\cdot\boldsymbol{\sigma})$ can be represented as

$$\begin{aligned} &f(\boldsymbol{A}\cdot\boldsymbol{\sigma}) = f^{+}(A)+\frac{\boldsymbol{A}\cdot\boldsymbol{\sigma}}{A}f^{-}(A)\,, \\ &\text{where}\quad f^{\pm}(A)=\frac{1}{2}[f(A)\pm f(-A)]\,. \end{aligned} \tag{5.26}$$

For instance,

$$\exp(i\boldsymbol{A}\cdot\boldsymbol{\sigma}) = \cos A + i\frac{(\boldsymbol{A}\cdot\boldsymbol{\sigma})}{A}\sin A\,. \tag{5.27}$$

7. The diagonal matrix element of the matrix $\boldsymbol{A}\cdot\boldsymbol{\sigma}$ between states $|\boldsymbol{b}\rangle$ is

$$\langle\boldsymbol{b}|\boldsymbol{A}\cdot\boldsymbol{\sigma}|\boldsymbol{b}\rangle = (\boldsymbol{A}\cdot\boldsymbol{b})\,. \tag{5.28}$$

To prove it we substitute

$$|\boldsymbol{b}\rangle = \frac{1}{2}[\hat{I}+\boldsymbol{b}\cdot\boldsymbol{\sigma}]|\boldsymbol{b}\rangle \tag{5.29}$$

into (5.28) and use relations (5.22):

$$\langle\boldsymbol{b}|\boldsymbol{A}\cdot\boldsymbol{\sigma}|\boldsymbol{b}\rangle = \langle\boldsymbol{b}|\frac{1}{2}[\hat{I}+\boldsymbol{b}\cdot\boldsymbol{\sigma}]\boldsymbol{A}\cdot\boldsymbol{\sigma}\frac{1}{2}[\hat{I}+\boldsymbol{b}\cdot\boldsymbol{\sigma}]|\boldsymbol{b}\rangle = (\boldsymbol{A}\cdot\boldsymbol{b})\,. \tag{5.30}$$

8. The states $|\boldsymbol{b}_a\rangle$ and $|\boldsymbol{b}_{a'}\rangle$ for different reference states $|\boldsymbol{a}\rangle$ and $|\boldsymbol{a}'\rangle$ differ only by a numerical phase factor. For instance, suppose the state $|\boldsymbol{a}'\rangle$ is defined as $|\boldsymbol{a}'\rangle=\exp(i\alpha\boldsymbol{b}\cdot\boldsymbol{\sigma})|\boldsymbol{a}\rangle$. The phase difference of two states $|\boldsymbol{b}_a\rangle$ and $|\boldsymbol{b}_{a'}\rangle$ is

$$\begin{aligned}\langle\boldsymbol{b}_a||\boldsymbol{b}_{a'}\rangle &= \langle\boldsymbol{b}_a|\frac{\hat{I}+\boldsymbol{b}\cdot\boldsymbol{\sigma}}{\sqrt{2(1+(\boldsymbol{b}\cdot\boldsymbol{a}))}}\exp(i\alpha\boldsymbol{b}\cdot\boldsymbol{\sigma})|\boldsymbol{a}\rangle \\ &= \langle\boldsymbol{b}_a|\exp(i\alpha\boldsymbol{b}\cdot\boldsymbol{\sigma})|\boldsymbol{b}_a\rangle = e^{i\alpha}\,. \end{aligned} \tag{5.31}$$

9. The nondiagonal matrix element of the matrix $\boldsymbol{A}\cdot\boldsymbol{\sigma}$ between states $|\pm\boldsymbol{b}\rangle$ on one side is

$$\langle-\boldsymbol{b}|\boldsymbol{A}\cdot\boldsymbol{\sigma}|\boldsymbol{b}\rangle = \langle-\boldsymbol{b}|\left(\boldsymbol{A}-\boldsymbol{b}\,(\boldsymbol{b}\cdot\boldsymbol{a})\right)\cdot\boldsymbol{\sigma}|\boldsymbol{b}\rangle = \langle-\boldsymbol{b}|[\boldsymbol{b}\times[\boldsymbol{A}\times\boldsymbol{b}]]\cdot\boldsymbol{\sigma}|\boldsymbol{b}\rangle\,, \tag{5.32}$$

where we used the evident equality $\langle-\boldsymbol{b}|\boldsymbol{b}\cdot\boldsymbol{\sigma}|\boldsymbol{b}\rangle = 0$. On the other side

$$\langle-\boldsymbol{b}|\boldsymbol{A}\cdot\boldsymbol{\sigma}|\boldsymbol{b}\rangle = \langle-\boldsymbol{b}|(\boldsymbol{A}\cdot\boldsymbol{\sigma})\frac{\hat{I}+\boldsymbol{b}\cdot\boldsymbol{\sigma}}{2}|\boldsymbol{b}\rangle = \langle-\boldsymbol{b}|\frac{(\boldsymbol{A}\cdot\boldsymbol{\sigma})+i[\boldsymbol{A}\times\boldsymbol{b}]\cdot\boldsymbol{\sigma}}{2}|\boldsymbol{b}\rangle\,, \tag{5.33}$$

from which it follows that

$$\langle-\boldsymbol{b}|\boldsymbol{A}\cdot\boldsymbol{\sigma}|\boldsymbol{b}\rangle = \langle-\boldsymbol{b}|i[\boldsymbol{A}\times\boldsymbol{b}]\cdot\boldsymbol{\sigma}|\boldsymbol{b}\rangle\,. \tag{5.34}$$

Expressions (5.34) and (5.33) are different, though their absolute values are identical:

$$|\langle-\boldsymbol{b}|\boldsymbol{A}\cdot\boldsymbol{\sigma}|\boldsymbol{b}\rangle|^2 = |[\boldsymbol{A}\times\boldsymbol{b}]|^2 = |[\boldsymbol{b}\times[\boldsymbol{A}\times\boldsymbol{b}]]|^2\,. \tag{5.35}$$

10. For any function $f(\pm A)\neq 0$,

$$f^{-1}(\boldsymbol{A}\cdot\boldsymbol{\sigma}) = \frac{f(-\boldsymbol{A}\cdot\boldsymbol{\sigma})}{f(-\boldsymbol{A}\cdot\boldsymbol{\sigma})\,f(\boldsymbol{A}\cdot\boldsymbol{\sigma})} = \frac{f(-\boldsymbol{A}\cdot\boldsymbol{\sigma})}{f(A)\,f(-A)}\,, \tag{5.36}$$

because

$$f(\boldsymbol{A}\cdot\boldsymbol{\sigma})\,f(-\boldsymbol{A}\cdot\boldsymbol{\sigma}) = f(A)\,f(-A)\,. \tag{5.37}$$

As an example, let us consider a matrix $\hat{M} = 1/(1-\sigma_x-\sigma_y)$. It can be written as $1/(1-\boldsymbol{A}\cdot\boldsymbol{\sigma})$, where $\boldsymbol{A}=(1,1,0)$. Using (5.36), we obtain

$$\frac{1}{1-\sigma_x-\sigma_y} = \frac{1+\sigma_x+\sigma_y}{1-2} = -1-\sigma_x-\sigma_y\,.$$

11. For two arbitrary functions $f_1(A_1\boldsymbol{\sigma})$ and $f_2(A_2\boldsymbol{\sigma})$ we have

$$\frac{1}{I+f_1(A_1\cdot\boldsymbol{\sigma})\,f_2(A_2\cdot\boldsymbol{\sigma})} = \frac{I+f_2(-A_2\cdot\boldsymbol{\sigma})\,f_1(-A_1\cdot\boldsymbol{\sigma})}{1+f_1^{+}(A_1)f_2^{+}(A_2)+(\boldsymbol{a}_1\cdot\boldsymbol{a}_2)f_1^{-}(A_1)f_2^{-}(A_2)+f_1(A_1)f_1(-A_1)f_2(A_2)f_2(-A_2)}\,, \tag{5.38}$$

where $\boldsymbol{a}_i = \boldsymbol{A}_i/A_i$. It is easy to check the validity of this result.

12. For arbitrary function $f(x)$, the relation

$$\sigma_i\,f(A\sigma_j) = f(-A\sigma_j)\sigma_i \tag{5.39}$$

is valid for all $i = x, y, z \neq j$.

5.1.2.2 Matrix Elements of the Reflection Matrix

Using Eqs. (5.36) and (5.37), we can rewrite $\hat{\boldsymbol{r}}$ Eq. (5.20) as

$$\frac{1}{N}\Big(k_o(-B_o)k_o(B_o) - k_i(-B_i)k_i(B_i)\Big)$$
$$-\frac{1}{N}\Big(\hat{k}_o(-\sigma\cdot\boldsymbol{B}_o)\hat{k}_i(\sigma\cdot\boldsymbol{B}_i) - \hat{k}_i(-\sigma\cdot\boldsymbol{B}_i)\hat{k}_o(\sigma\cdot\boldsymbol{B}_o)\Big)$$
$$=\frac{1}{N}\Big(k_o(-B_o)k_o(B_o) - k_i(-B_i)k_i(B_i)\Big)$$
$$+\frac{2}{N}\Big((\sigma\cdot\boldsymbol{b}_o)k_o^-k_i^+ - (\sigma\cdot\boldsymbol{b}_i)k_o^+k_i^- - i(\sigma\cdot[\boldsymbol{b}_i\times\boldsymbol{b}_o])k_o^-k_i^-\Big)\,, \tag{5.40}$$

where $\boldsymbol{b}_{o,i} = \boldsymbol{B}_{o,i}/B_{o,i}$ are unit vectors, $k_{o,i}^{\pm} = [k_{o,i}(B_{o,i}) \pm k_{o,i}(-B_{o,i})]/2$,

$$N = \Big(\hat{\boldsymbol{k}}_o(\boldsymbol{\sigma}\cdot\boldsymbol{B}_o) + \hat{\boldsymbol{k}}_i(\boldsymbol{\sigma}\cdot\boldsymbol{B}_i)\Big)\Big(\hat{\boldsymbol{k}}_o(-\boldsymbol{\sigma}\cdot\boldsymbol{B}_o) + \hat{\boldsymbol{k}}_i(-\boldsymbol{\sigma}\cdot\boldsymbol{B}_i)\Big)$$
$$= (k_o(B_o) + k_i(B_i))\,(k_o(-B_o) + k_i(-B_i)) + 4k_o^-k_i^-\sin^2(\theta/2)\,, \tag{5.41}$$

and θ is the angle between $\boldsymbol{B}_o$ and $\boldsymbol{B}_i$.

With Eq. (5.40) we can immediately obtain all the matrix elements of the reflection matrix $\hat{\boldsymbol{r}}$. The diagonal ones are

$$\langle\pm\boldsymbol{b}_o|\hat{\boldsymbol{r}}|\pm\boldsymbol{b}_o\rangle = \frac{1}{N}(k_o(-B_o)k_o(B_o) - k_i(-B_i)k_i(B_i)$$
$$\pm 2[k_o^-k_i^+ - (\boldsymbol{b}_o\cdot\boldsymbol{b}_i)k_o^+k_i^-])\,, \tag{5.42}$$

and the nondiagonal matrix elements with account of Eqs. (5.33)—(5.35) are

$$\langle\mp\boldsymbol{b}_o|\hat{\boldsymbol{r}}|\pm\boldsymbol{b}_o\rangle = \mp 2i\langle\mp\boldsymbol{b}_o|[\boldsymbol{b}_i\times\boldsymbol{b}_o]\boldsymbol{\sigma}|\pm\boldsymbol{b}_o\rangle\frac{k_o^+k_i^- \pm k_o^-k_i^-}{N}$$
$$= \mp i\langle\mp\boldsymbol{b}_o|[\boldsymbol{b}_i\times\boldsymbol{b}_o]\boldsymbol{\sigma}|\pm\boldsymbol{b}_o\rangle\frac{k_o(\pm B_o)\,(k_i(B_i) - k_i(-B_i))}{N}\,. \tag{5.43}$$

It follows that the probability of reflection with spin flip is

$$w(\pm\to\mp) = \frac{k(\mp B_o)}{k(\pm B_o)}\,|\langle\mp\boldsymbol{b}_o|\hat{\boldsymbol{r}}|\pm\boldsymbol{b}_o\rangle|^2$$
$$= |[\boldsymbol{b}_i\times\boldsymbol{b}_o]|^2\frac{k_o(B_o)k_o(-B_o)\,(k_i(B_i) - k_i(-B_i))^2}{N^2}\,. \tag{5.44}$$

5.1.3 Triple Splitting of the Reflected Beam

The matrix elements (5.42) and (5.43) are linked to the phenomena of the triple reflection [226, 227] and quadruple refraction for neutrons incident on a magnetic mirror. Recall that the traditional optics deals maximum with double refraction.

Let us consider the process of reflection of unpolarized neutrons incident on a magnetic mirror which is inside an external magnetic field $\boldsymbol{B}$ (Figure 5.1).

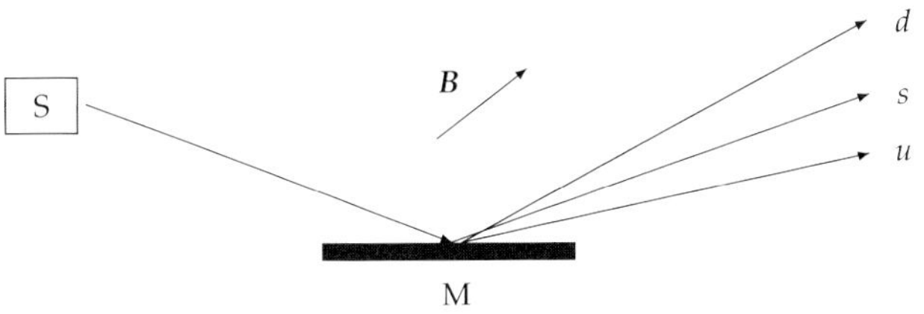

Figure 5.1 Experimental setup for measuring the triple splitting of the unpolarized beam of neutrons upon reflection from a magnetic mirror. Here, *S* is a neutron source, *M* is the magnetic mirror, ***B*** is the external field, *s* is the specularly reflected ray, and *u* and *d* are nonspecularly reflected rays, polarized along and oppositely to the external field, respectively. The external and the internal fields are noncollinear.

The component of the neutron velocity parallel to the mirror surface is conserved. If no spin flip occurs upon reflection, (the probability is defined by the square of the absolute values of the diagonal matrix elements (5.42)), then the magnitude of the normal component of the velocity is also conserved, and the reflection is specular.

The picture changes when the spin flip occurs. The total energy of the incident neutrons is equal to $E = k^2/2$. If the incident neutron is polarized along the magnetic field, then the reflected neutron is polarized oppositely to the field. The potential energy in this case is reduced by $2B$, whereas the total energy is unchanged since the reflection is elastic. Therefore, the kinetic energy increases up to $(k^2 + 4B)/2$, where B is the external magnetic field. In the other case, when the incident neutron is polarized oppositely to the magnetic field, and the reflected neutron is polarized along the field, the potential energy rises and the kinetic energy decreases down to $(k^2 - 4B)/2$.

Since the component of the velocity parallel to the interface is conserved, changes in the kinetic energy translate into changes in the perpendicular component of the velocity, $k_\perp \to \sqrt{k_\perp^2 \pm 4B}$. Therefore, the reflection angle is also changed, compared with the incidence angle. In general, $k_\perp^2 \gg 4B$, then the change in the normal component of the velocity is

$$\Delta k_\perp = \sqrt{k_\perp^2 \pm 4B} - k_\perp \approx \pm 2B/k_\perp ,$$

where the upper, +, sign corresponds to reflected neutrons polarized oppositely to the field, and the lower, −, sign corresponds to reflected neutrons polarized along the field. The change in the reflection angle can be found approximately as

$$\Delta\theta = \frac{\Delta k_\perp}{k} \approx \pm\frac{2B}{k^2}\frac{k}{k_\perp} \approx \pm\frac{B}{E\theta} ,$$

where θ is the grazing angle, which for thermal neutrons under total reflection is equal to a few milliradians. Thus, the change in the reflection angle due to a magnetic field of 1 kG is equal to few tenths of milliradians. At a distance of 1 m from the mirror, this angle translates into just a fraction of a millimeter deviation from the specular reflection.

In summary, the reflection of a nonpolarized neutron beam from a magnetic mirror, with its magnetization noncollinear to the external field, creates three reflected beams. Whereas the specularly reflected beam remains unpolarized, the adjacent beams are polarized. The beam which is located closer to the interface is polarized along the magnetic field, and the other beam is polarized against the magnetic field.

The intensity of the reflected beam polarized along the magnetic field, $I(\mu \| \boldsymbol{B})$, is less than the intensity, $I(\mu \| -\boldsymbol{B})$, of the other polarized beam. Indeed, for the same normal component $k_\perp$ of the incident neutrons for both polarizations, the flux $kw(+ \to -)$ from the mirror of beam d in Figure 5.1 according to Eq. (5.44) contains the factor $k_\perp^2 k_\perp(-2B)$, whereas that for beam u contains the factor $k_\perp^2 k_\perp(2B) < k_\perp^2 k_\perp(-2B)$. In the case of strong external fields and small grazing angles, the intensity of beam u can be suppressed. Normally, however, the field B is much less than the kinetic energy $k_\perp^2/2$, and both beams u and d are approximately equal in intensity:

$$I_{u,d} \approx I_0 \left([\boldsymbol{B}_i \times \boldsymbol{b}]/k_\perp^2\right)^2 ,$$

where I_0 is the incident beam intensity. Upon saturation, $B_{i\perp}$ is approximately 10 kG and the intensities of the polarized parts of the reflected triple can reach 10% of I_0.

Despite the small changes in the reflection angle, the reflected beam splitting has been observed experimentally [227–231]. These experiments confirmed the theoretical predictions [226] and created opportunities for applying the splitting effect to study magnetic materials [232].

5.1.3.1 A Fundamental Question

The process of reflection from a magnetic mirror is a coherent process, where the incident neutron state with a given velocity is transformed into a coherent superposition of spinor states with different velocities, different in values and in directions. Upon propagation, the reflected components can separate by a large distance (beam splitting is very visible in the experiment [227]), which leads to an important question: when does the coherent state decay? Experimentally, it is possible to estimate the coherence length of a neutron by placing some sort of analyzer, orientated perpendicular to the external field, in the reflected beam. The transmission through the analyzer will oscillate with the distance of the analyzer from the mirror owing to the spin precession. The amplitude of the oscillations will decrease with increasing separation of the reflected beam components. This rate of decrease of the oscillations depends on the difference in the angles and on the coherence length (or "size") of the neutron [192]. No experiment has yet been performed to see this effect.

5.1.3.2 Scalar Aharonov–Bohm Effect

When a polarized neutron with momentum k propagates along distance L in free space, its wave function acquires a phase factor $\exp(ikL)$ with phase kL. When this neutron enters magnetic field B parallel to its spin, its momentum k changes to $k' = \sqrt{k^2 - 2B}$, and propagation in this field along distance L will give the phase factor $\exp(ik'L)$ with phase $k'L$. The difference $\Delta\varphi = (k - k')L$ depends on k and B.

If the same neutron propagates in empty space, and the magnetic field is suddenly switched on for time T, then the neutron's speed does not change, but the neutron's energy E in the field changes to $E' = E + B$. The field changes the neutron's wave function from $\exp(ikx - iEt)$ to $\exp(ikx - i(E + B)t)$. Therefore, during time T of the field action the neutron acquires the additional phase factor $\exp(-iBT)$ with phase $-BT$, which does not depend on the neutron's speed. This phenomenon is called the scalar Aharonov–Bohm effect. The phase $-BT$ was measured, and its independence of the neutron's speed was demonstrated in [233].

5.1.3.3 Comments on Dipole and Current Models of the Neutron Magnetic Moment

In the discussion of the fundamental properties of the neutron, it is worth mentioning a question which was discussed some time ago [121]: whether the neutron magnetic moment is a dipole or a loop current. According to the experiment reported in [121] it should be a current, because reflection from magnetic mirrors shows that there are two limiting energies for two opposite spin polarizations, whereas for the dipole mode there should be a single limiting energy. However, this conclusion is wrong, because quantum mechanics predicts two limiting energies $u \pm B$ for both models, so for neutron optics the question is meaningless.

5.2 Algebra of Magnetic Mirrors

Let us generalize the scalar algebra developed in Chapter 4 to spinor particles with magnetic interactions.

5.2.1 Magnetic Mirror of Finite Thickness

Consider the problem with a rectangular potential barrier (Figure 4.7). This time, however, let us introduce a homogeneous magnetic field inside and outside the potential barrier u_0. The magnetic fields are different in both magnitude and orientation. Denote $\boldsymbol{B}_0$ as the external field, and $\boldsymbol{B}_i$ as the field inside the mirror (barrier) $0 \le x \le d$. The wave with spin state $|\xi_0\rangle$ is incident from the left. Denote $\hat{X}|\xi_0\rangle$ as the spinor amplitude of the wave incident on the right interface of the barrier. Analogously as we did for Eqs. (4.48) and (4.47), we can easily write equations for $\hat{X}$ and the reflection $\hat{R}$ and transmission $\hat{T}$ amplitudes,

$$\hat{X} = \exp(i\hat{k}_i d)\hat{t} + \exp(i\hat{k}_i d)\hat{r}' \exp(i\hat{k}_i d)\hat{r}' \hat{X} , \tag{5.45}$$

$$\hat{R} = \hat{r} + \hat{t}' \exp(i\hat{k}_i d)\hat{r}' \hat{X} , \quad \hat{T} = \hat{t}' \hat{X} , \tag{5.46}$$

where

$$\hat{r} = (\hat{k}_o + \hat{k}_i)^{-1}(\hat{k}_o - \hat{k}_i) = -\hat{r}' , \quad \hat{t} = I + \hat{r} , \quad \hat{t}' = I - \hat{r} . \tag{5.47}$$

The solution of Eq. (5.45) is

$$\hat{X} = \left(\hat{I} - \exp(i\hat{k}_i d)\hat{r} \exp(i\hat{k}_i d)\hat{r}\right)^{-1} \exp(i\hat{k}_i d)\hat{t} . \tag{5.48}$$

Substitution of Eqs. (5.48) into (5.46) gives $\hat{R}$ and $\hat{T}$,

$$\hat{R} = \hat{r} - \hat{t}' \exp(i\hat{k}_i d)\hat{r} \left(\hat{I} - \exp(i\hat{k}_i d)\hat{r} \exp(i\hat{k}_i d)\hat{r}\right)^{-1} \exp(i\hat{k}_i d)\hat{t} ,$$
$$\hat{T} = \hat{t}' \left(\hat{I} - \exp(i\hat{k}_i d)\hat{r} \exp(i\hat{k}_i d)\hat{r}\right)^{-1} \exp(i\hat{k}_i d)\hat{t} . \tag{5.49}$$

By using the properties of the Pauli matrices given on page 195, we can obtain all the matrix elements in analytical form, or we can obtain them numerically.

Using $\hat{X}$, we can also explicitly write the matrix wave function inside the mirror,

$$\Theta(0 \le x \le d)\boldsymbol{\Psi}(x) = \left[e^{i\hat{k}_i(x-d)} + e^{-i\hat{k}_i(x-d)}\hat{r}'\right]\hat{X}$$
$$= \left[e^{i\hat{k}_i(x-d)} - e^{-i\hat{k}_i(x-d)}\hat{r}\right]\left(\hat{I} - e^{i\hat{k}_i d}\hat{r} e^{i\hat{k}_i d}\hat{r}\right)^{-1} e^{i\hat{k}_i d}[I + \hat{r}] . \tag{5.50}$$

Multiplying it by an arbitrary spinor $|\xi\rangle$, we can represent it as a superposition $\alpha_u(x)|\xi_u\rangle + \alpha_d(x)|\xi_d\rangle$ of states $|\xi_{u,d}\rangle$ oriented along and against the internal magnetic field $\boldsymbol{B}_i$, and find their amplitudes.

5.2.2
System of Two Magnetic Mirrors

Consider a system of two potential barriers separated by distance l (see Figure 4.8), and add magnetic fields inside and outside the barriers. The external field is homogeneous, and the fields inside the mirrors can be arbitrary. Let $\exp(i\hat{k}x)|\xi_0\rangle$ be the arbitrary polarized wave incident on the first potential, and $\hat{X}|\xi_0\rangle$ is the wave incident on the second potential. The equation for $\hat{X}$ is then

$$\hat{X} = \exp(i\hat{k}l)\hat{t}_1 + \exp(i\hat{k}l)\hat{r}_1 \exp(i\hat{k}l)\hat{r}_2 \hat{X} . \tag{5.51}$$

The solution to this equation is

$$\hat{X} = \left(\hat{I} - \exp(i\hat{k}l)\hat{r}_1 \exp(i\hat{k}l)\hat{r}_2\right)^{-1} \exp(i\hat{k}l)\hat{t}_1 . \tag{5.52}$$

The reflection and transmission amplitudes for the full system are given by

$$\hat{R}_{12} = \hat{r}_1 + \hat{t}_1 e^{i\hat{k}l}\hat{r}_2\hat{X} = \hat{r}_1 + \hat{t}_1 e^{i\hat{k}l}\hat{r}_2\left(\hat{I} - e^{i\hat{k}l}\hat{r}_1 e^{i\hat{k}l}\hat{r}_2\right)^{-1} e^{i\hat{k}l}\hat{t}_1 , \quad (5.53)$$

$$\hat{T}_{12} = \hat{t}_2\hat{X} = \hat{t}_2\left(\hat{I} - e^{i\hat{k}l}\hat{r}_1 e^{i\hat{k}l}\hat{r}_2\right)^{-1} e^{i\hat{k}l}\hat{t}_1 . \quad (5.54)$$

When $l \to 0$, we obtain a more simplified version,

$$\hat{R}_{12} = \hat{r}_1 + \hat{t}_1\hat{r}_2\left(\hat{I} - \hat{r}_1\hat{r}_2\right)^{-1}\hat{t}_1 , \quad \hat{T}_{12} = \hat{t}_2\left(\hat{I} - \hat{r}_1\hat{r}_2\right)^{-1}\hat{t}_1 , \quad (5.55)$$

which can be obtained from equations

$$\hat{X} = \hat{t}_1 + \hat{r}_1\hat{r}_2\hat{X} , \quad \hat{R}_{12} = \hat{r}_1 + \hat{t}_1\hat{r}_2\hat{X} , \quad \hat{T}_{12} = \hat{t}_2\hat{X} . \quad (5.56)$$

5.2.3 Standing Waves

Methods used in the scalar case can be used to treat multiple magnetic mirror systems. We can then obtain the reflection amplitudes with and without the spin flip, the transmission amplitudes, and a full wave function. Normally, only the reflection and transmission coefficients are measured. However, it is also possible to measure the wave function [231]. In fact, we can easily do it in the case of standing waves [232, 247].

Standing waves appear because the incident and the reflected waves interfere with each other. They appear outside the mirror close to the reflecting surface, as well as inside a multilayered system. Standing waves are especially pronounced at resonances. Because of the interference, the wave function of a standing wave is a combination of fixed nodes and antinodes, parallel to the interface.

To measure such a wave function means measuring the location of the nodes and antinodes. For example, placing a thin slab of a material that captures neutrons in a node will decrease the neutron beam intensity and increase the products of the nuclear reaction, which accompanies neutron absorption. On the other hand, such a slab placed in an antinode is transparent for the beam and the nuclear reaction is suppressed. If we place a slab made out of polarization-sensitive material in the node, we will be able to observe the depolarization of the incident beam; and so on [207].

5.2.4 Periodic System of Mirrors

Just like in the scalar case, separate the first period from the rest of the system by an infinitesimal gap. The wave incident on the second period is then denoted by $\hat{X}$. The system of equations for $\hat{X}$ and for the reflection amplitude $\hat{R}$ for a semi-infinite potential is then

$$\hat{X} = \hat{t} + \hat{r}\hat{R}\hat{X} , \quad \hat{R} = \hat{r} + \hat{t}\hat{R}\hat{X} , \quad (5.57)$$

where $\hat{r}$ and $\hat{t}$ are the reflection and transmission amplitudes for a single period (for simplicity, consider symmetrical periods). Solving the first equation for $\hat{X}$ and substituting the solution into the second one, we get

$$\hat{R} = \hat{r} + \hat{t}\hat{R}(I - \hat{r}\hat{R})^{-1}\hat{t}\,. \tag{5.58}$$

This equation can be reduced to a quadratic matrix equation of type

$$ZAZ - ZB - CZ + D = 0\,, \tag{5.59}$$

where $Z = \hat{R}$, $B = \hat{t}^{-1}$, $A = \hat{t}^{-1}\hat{r}$, $C = \hat{r}\hat{t}^{-1}\hat{r} - \hat{t}$, and $D = \hat{r}\hat{t}^{-1}$. The last equation cannot be solved analytically in the general case of noncommuting matrices A, B, C, and D. Solution and analysis of such an equation is a challenge to mathematicians. It is worth noting that the solution of the matrix equation $\hat{X}^2 = \hat{I}$ in the case of 2x2 matrices is $\hat{X} = \boldsymbol{\sigma a}$, where $\boldsymbol{a}$ is an arbitrary 3-dimensional unit vector. So we see, that the number of solutions of such an equation is infinite. In more general case $\hat{X}$ can be represented as $\hat{X} = \hat{G}\hat{M}\hat{G}^{-1}$, where $\hat{M}$ is a diagonal matrix with ± 1 on the diagonal, and $\hat{G}$ is an arbitrary nonsingular matrix[18].

5.2.5
A Periodic Potential with Collinear Internal Fields

One particular case when an analytical solution can be obtained is the case where the magnetic field in the periodic system varies only in magnitude but not in direction, though its direction is not collinear to the external field. In that case it is possible to introduce an additional field $\boldsymbol{B}_0'$, collinear to the internal field and a virtual interface between fields $\boldsymbol{B}_0$ and $\boldsymbol{B}_0'$. Then the matrix reflection amplitude of the whole system, which includes the additional virtual interface, is given by

$$\hat{R} = \hat{r}_0 + (\hat{I} - \hat{r}_0)\hat{R}_\infty\left(\hat{I} + \hat{r}_0\hat{R}_\infty\right)^{-1}(\hat{I} + \hat{r}_0)\,, \tag{5.60}$$

where $\hat{R}_\infty$ is the matrix

$$\hat{R}_\infty = \frac{\sqrt{(\hat{I} + \hat{r})^2 - \hat{t}^2} - \sqrt{(\hat{I} - \hat{r})^2 - \hat{t}^2}}{\sqrt{(\hat{I} + \hat{r})^2 - \hat{t}^2} + \sqrt{(\hat{I} - \hat{r})^2 - \hat{t}^2}}\,, \tag{5.61}$$

which contains commuting diagonal amplitude matrices $\hat{r}$ and $\hat{t}$ of a single period,

$$\hat{r}_0 = \left(k(\boldsymbol{\sigma B}_0) + k(\boldsymbol{\sigma B}_0')\right)^{-1}\left(k(\boldsymbol{\sigma B}_0) - k(\boldsymbol{\sigma B}_0')\right)\,, \tag{5.62}$$

denotes the matrix amplitude of the reflection from the virtual interface between fields $\boldsymbol{B}_0$ and $\boldsymbol{B}_0'$, and $\boldsymbol{B}_0'$ is some field collinear to the internal field of the periodic potential. In particular, we can choose $\boldsymbol{B}_0'$ such that $|\boldsymbol{B}_0'| = |\boldsymbol{B}_0|$.

18) I am grateful to Rafael Sarkisyan for discussion of this problem.

In a similar way, introducing a virtual interface at the other side, we can calculate the reflection and transmission of a finite number N of periods. Since the potential with collinear internal fields can be treated as two separate potentials for scalar particles, we do not continue here and refer the reader to Chapter 4.

5.2.6
Reflection from Helical Systems

A helical periodic system is one where the analytical solution (5.59) exists. In a helical medium, the magnetization rotates clockwise or counterclockwise in space around the normal to the interface.

5.2.6.1 Reflection from a Semi-infinite Mirror

Consider first a semi-infinite mirror at $z > 0$, where the magnetization vector $\boldsymbol{b}$ rotates counterclockwise in the plane (x, y). The orientation of the external field $\boldsymbol{B}_0$ is arbitrary. The Schrödinger equation in the medium is

$$\left(\frac{d^2}{dz^2} - u_0 - 2b[\sigma_x \cos(2qz + 2\varphi) + \sigma_y \sin(2qz + 2\varphi)] + k^2\right)|\psi(z)\rangle = 0, \tag{5.63}$$

where u_0 is the optical potential of the medium, and the parameter π/q defines the rotational period of the field b. For generality we included the phase 2φ of the field $\boldsymbol{b}$ at the interface, and for convenience singled out the factor 2.

Using

$$\sigma_x \cos(2qz+2\varphi)+\sigma_y \sin(2qz+2\varphi) = \exp(-i\sigma_z(qz+\varphi))\sigma_x \exp(i\sigma_z(qz+\varphi)), \tag{5.64}$$

and substituting

$$|\psi(z)\rangle = \exp(-i\sigma_z(qz + \varphi))|\phi(z)\rangle \tag{5.65}$$

into Eq. (5.63), we get

$$\left(\frac{d^2}{dz^2} - 2iq\sigma_z \frac{d}{dz} - u_0 - 2b\sigma_x + k^2 - q^2\right)|\phi(z)\rangle = 0 . \tag{5.66}$$

The equation was solved first by Calvo [234] The solution was represented as

$$|\phi(z)\rangle = \exp(ipz)|\chi_p\rangle . \tag{5.67}$$

Substitution of Eq. (5.67) into (5.66) gives the equation

$$(-p^2 + 2qp\sigma_z - u_0 - 2b\sigma_x + k^2 - q^2)\,|\chi_p\rangle = 0 , \tag{5.68}$$

which is satisfied only if $|\chi_p\rangle$ is an eigenspinor of operator $2qp\sigma_z - 2b\sigma_x$ with eigenvalue $p^2 - k^2 + q^2 + u_0$.

> I am pretty convinced that all results beyond of those previously obtained by Calvo are incorrect and the paper should not be published.
>
> The referee of Physical Review B

To find the reflection even from a single interface at $z = 0$ with the help of eigenstates is a boring problem, because it requires solution of a system of four linear equations with four unknowns. Indeed, a polarized incident wave can be reflected with and without change of polarization. Therefore, reflection is characterized by two numbers. Inside the mirror the incident wave creates two eigenstates with amplitudes characterized by two other numbers. Matching of the wave function and its derivative for two outside spin states gives four equations. It is possible to solve them, but no one will do it without a computer.

Reflection from a mirror of finite thickness becomes an even more terrible problem, because it involves solution of eight linear equations. We do not like that, and below we show how to avoid this procedure and to find the solution in a compact analytical form.

The matrix solution When considering reflection from a single interface we can represent the neutron moving into matter away from the interface by the wave [235]

$$|\phi(z)\rangle = \exp(i[a + \vec{\boldsymbol{p}} \cdot \boldsymbol{\sigma}]z)|\chi\rangle \,, \tag{5.69}$$

with four unknown parameters a and $\vec{\boldsymbol{p}}$ and an arbitrary spinor state $|\chi\rangle$ at $z = 0$. Then, with account of Eq. (5.65), the total internal wave function becomes

$$|\psi(z)\rangle = \exp(-i\sigma_z(qz + \varphi))\exp(i[a + \vec{\boldsymbol{p}} \cdot \boldsymbol{\sigma}]z)|\chi\rangle \,. \tag{5.70}$$

At $z = 0$ we have $|\chi\rangle = \exp(i\sigma_z\varphi)|\psi(0)\rangle$, and therefore (5.70) is representable as

$$\begin{aligned}|\psi(z)\rangle &= \exp(-i\sigma_z(qz + \varphi))\exp(i[a + \vec{\boldsymbol{p}} \cdot \boldsymbol{\sigma}]z)\exp(i\sigma_z\varphi)|\psi(0)\rangle = \\ &\exp(-i\sigma_z qz)\exp(i[a + \vec{\boldsymbol{p}}_\varphi \boldsymbol{\sigma}]z)|\psi(0)\rangle \,,\end{aligned} \tag{5.71}$$

where $\vec{\boldsymbol{p}}_\varphi \cdot \boldsymbol{\sigma} = \exp(-i\sigma_z\varphi)\vec{\boldsymbol{p}} \cdot \boldsymbol{\sigma}\exp(i\sigma_z\varphi)$.

Substitution of Eq. (5.69) into Eq. (5.66) gives the equation

$$\begin{aligned}&\left[-a^2 - p^2 - 2a\vec{\boldsymbol{p}} \cdot \boldsymbol{\sigma} + 2q\sigma_z(a + \vec{\boldsymbol{p}} \cdot \boldsymbol{\sigma}) - u_0 - 2b\sigma_x + k^2 - q^2\right]|\phi(z)\rangle \\ &= 0 \,.\end{aligned} \tag{5.72}$$

> The authors suggest that the solution to the Schrödinger equation $M|\psi\rangle = 0$ is $M = 0$, assuming that here $|\psi\rangle$ is an arbitrary vector. All their following calculations are based on this assumption. This assumption, however, is principally wrong. Such a matrix

equation is nothing but a system of two equations, which enjoys nontrivial solutions provided that $\det(M) = 0$. It is principally wrong to try to force M to be zero.

A referee of Physical Review B

We find no reason to doubt our referees expertise or their ability to competently review your manuscript.

The editor of Physical Review B after our explanations.

An equation of the type $\hat{\boldsymbol{M}}|\phi(z)\rangle = 0$ with a constant matrix $\hat{\boldsymbol{M}}$ has three types of solutions. The first one is trivial: $|\phi(z)\rangle = 0$. We are not interested in it. The second type is obtained when $\det M = 0$. In that case $|\phi(z)\rangle$ is a constant spinor which does not depend on z. We are not interested in such a solution either. In fact, we should find the solution of the equation $\hat{\boldsymbol{M}}|\phi(z)\rangle = 0$ for an arbitrary z and an arbitrary state $|\phi(z)\rangle$. It is arbitrary because of the arbitrariness of $|\phi(0)\rangle$ and z.

The authors suggest that the solution to the Schrodinger equation $M|\phi\rangle = 0$ is $M = 0$, assuming that here $|\phi\rangle$ is an arbitrary vector. All their following calculations are based on this assumption. This assumption, however, is principally wrong. An equation $M|\psi\rangle = 0$, where $|\psi\rangle$ is not a wave function but an arbitrary vector, as the authors assume, is not the Schrodinger equation but something strange, having no relation to physics. Misinterpreting the Schrodinger equation makes all further authors' manipulations with formulas without physical meaning. The manuscript is basically wrong and has to be rejected.

Anonymous referee of Phys.Rev. B

In that case Eq. (5.72), which is of the form $\hat{\boldsymbol{M}}|\phi(z)\rangle = 0$ with a constant matrix $\hat{\boldsymbol{M}}$, can be satisfied only if $\hat{\boldsymbol{M}} = 0$, which gives

$$-[a^2 + p^2 + u_0 - k^2 + q^2]\hat{\boldsymbol{I}} - 2a\overrightarrow{\boldsymbol{p}}\cdot\boldsymbol{\sigma} + 2q\sigma_z(a + \overrightarrow{\boldsymbol{p}}\cdot\boldsymbol{\sigma}) - 2b\sigma_x = 0\,, \quad (5.73)$$

where $\hat{\boldsymbol{I}}$ is the unit matrix. This is equivalent to a system of four equations for parameters a and $\overrightarrow{p}$:

$$-a^2 - p^2 + 2q\overrightarrow{p_z} - u_0 + k^2 - q^2 = 0\,, \quad (5.74)$$

$$\begin{aligned} &-2a\overrightarrow{p_z} + 2qa = 0\,, \quad -2a\overrightarrow{p_x} - 2iq\overrightarrow{p_y} - 2b = 0\,, \\ &-2a\overrightarrow{p_y} + 2iq\overrightarrow{p_x} = 0\,. \end{aligned} \quad (5.75)$$

The three Eqs. (5.75) give

$$\overrightarrow{p_z} = q\,, \quad \overrightarrow{p_x} = \frac{ab}{q^2 - a^2}\,, \quad \overrightarrow{p_y} = i\frac{qb}{q^2 - a^2}\,. \quad (5.76)$$

Substitution of these expressions into Eq. (5.74) gives the equation for a:

$$-a^2 + q^2 - \frac{b^2}{a^2 - q^2} - u_0 + k^2 - q^2 = 0\,, \quad (5.77)$$

The solution to this equation is

$$a = \sqrt{q^2 + \left(\frac{\sqrt{K^2 + 2b} + \sqrt{K^2 - 2b}}{2}\right)^2}\,, \tag{5.78}$$

where $K^2 = k^2 - u_0 - q^2$. The sign inside the root is chosen such that the parameter a coincides with $k' = \sqrt{k^2 - u_0}$ when $b = 0$.

Let us now find the reflection and refraction amplitudes for the wave $\exp(i\hat{\boldsymbol{k}}_0 z)\,|\xi_0\rangle$ incident on the interface from a vacuum $z < 0$. The full wave function is

$$|\psi(z)\rangle = \Theta(z < 0)\left(\exp(i\hat{\boldsymbol{k}}_0 z) + \exp(-i\hat{\boldsymbol{k}}_0 z)\hat{\boldsymbol{r}}\right)|\xi_0\rangle$$
$$+\,\Theta(z > 0)\exp(-i\sigma_z q z)\exp(i[a + \overrightarrow{\boldsymbol{p}}_\varphi \cdot \boldsymbol{\sigma}]z)\hat{\boldsymbol{t}}|\xi_0\rangle\,, \tag{5.79}$$

where $\hat{\boldsymbol{r}}$ and $\hat{\boldsymbol{t}}$ are the matrix reflection and transmission amplitudes, $\hat{\boldsymbol{k}}_0 = \sqrt{k^2 - 2\boldsymbol{B}_0\boldsymbol{\sigma}}$, and $|\xi_0\rangle$ is an arbitrary spin state of the incident wave.

By matching the wave function (5.79) and its derivative at $z = 0$, we obtain

$$\hat{\boldsymbol{I}} + \hat{\boldsymbol{r}} = \hat{\boldsymbol{t}}\,, \quad \hat{\boldsymbol{k}}_0[\hat{\boldsymbol{I}} - \hat{\boldsymbol{r}}] = [a\hat{\boldsymbol{I}} - q\sigma_z + \overrightarrow{\boldsymbol{p}}_\varphi \cdot \boldsymbol{\sigma}]\hat{\boldsymbol{t}} \equiv (a\hat{\boldsymbol{I}} + \overrightarrow{\boldsymbol{p}}'_\varphi \cdot \boldsymbol{\sigma})\hat{\boldsymbol{t}}\,, \tag{5.80}$$

where

$$\overrightarrow{\boldsymbol{p}}'_\varphi \cdot \boldsymbol{\sigma} = \overrightarrow{\boldsymbol{p}}_\varphi \cdot \boldsymbol{\sigma} - q\sigma_z = \exp(-i\sigma_z\varphi)(\overrightarrow{p}_x\sigma_x + \overrightarrow{p}_y\sigma_y)\exp(i\sigma_z\varphi)\,. \tag{5.81}$$

The solution to Eq. (5.80) is

$$\hat{\boldsymbol{r}} = (\hat{\boldsymbol{k}}_0 + a\hat{\boldsymbol{I}} + \overrightarrow{\boldsymbol{p}}'_\varphi \cdot \boldsymbol{\sigma})^{-1}(\hat{\boldsymbol{k}}_0 - a\hat{\boldsymbol{I}} - \overrightarrow{\boldsymbol{p}}'_\varphi \cdot \boldsymbol{\sigma})\,, \quad \hat{\boldsymbol{t}} = \hat{\boldsymbol{I}} + \hat{\boldsymbol{r}}\,. \tag{5.82}$$

At $b = 0$ the amplitudes are reduced to

$$\hat{\boldsymbol{t}} = \frac{2\hat{\boldsymbol{k}}_0}{\hat{\boldsymbol{k}}_0 + k'}\,, \quad \hat{\boldsymbol{r}} = \frac{\hat{\boldsymbol{k}}_0 - k'}{\hat{\boldsymbol{k}}_0 + k'}\,, \quad k' = \sqrt{k^2 - u_0}\,, \tag{5.83}$$

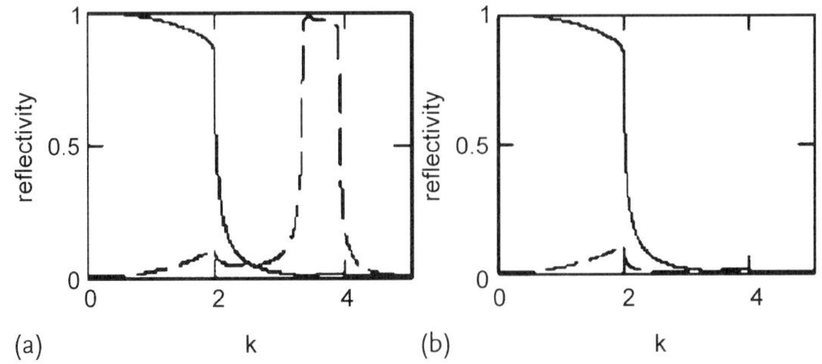

Figure 5.2 Wave-vector dependence of (a) the reflection coefficients without (solid line), $|r_{--}|^2$, and with (dashed line), $|r_{+-}|^2$, spin flip for the initial polarization (right subscript -) in the direction opposite the z-axis, which is parallel to the inward normal to the mirror, and (b) the respective reflection coefficients $|r_{++}|^2$ (solid line) and $|r_{-+}|^2$ (dashed line) for the initial polarization directed along the z-axis. The external field $\boldsymbol{B}_0$ is zero.

which is expected, since rotation does not matter when $b = 0$[19]. When $q = 0$, the above expressions become the reflection and transmission amplitudes for a mirror with constant internal field $\boldsymbol{b}$.

By means of Eq. (5.82), it is easy to obtain the reflection coefficients with and without spin flip. Figure 5.2 shows the calculation results for the simplest case $B_0 = 0$. The calculations were performed with a wave-vector length unit equal to $\sqrt{b}$, and with parameters $u_0 = 4 - 0.01i$, $q = 3$, and, correspondingly, $b = 1$. It is clearly seen that for one polarization there is a resonant reflection with spin reversal. It is a very important feature of helical systems.

> Two referees have attempted to explain to the authors what solving the Schrodinger equation entails. The authors refuse to accept these standard arguments. I cannot see any point in continuing the discussion.
> I do not recommend publication.
>
> Nonanonymous referee of the Phys.Rev. B editorial board

5.2.6.2 Reflection from the Interface from within the Mirror

Earlier we considered a wave which travels to the right and falls on the interface of the mirror from a vacuum. To find the reflection and transmission amplitudes for a mirror of finite thickness L, we also need to find the amplitudes for a wave incident on an interface from within the mirror. The wave traveling to the left inside the medium is

$$|\psi(z)\rangle = \exp(-i\sigma_z(qz+\varphi))\exp\left(-i[a+\overleftarrow{\boldsymbol{p}}\cdot\boldsymbol{\sigma}]z\right)\exp(i\sigma_z\varphi)|\psi(0)\rangle\,, \quad (5.84)$$

where $\overleftarrow{\boldsymbol{p}} \neq \overrightarrow{\boldsymbol{p}}$. By substituting $\exp(-i[a+\overleftarrow{\boldsymbol{p}}\cdot\boldsymbol{\sigma}]z)|\chi\rangle$ into Eq. (5.66), we obtain

$$-[a^2+p^2+u_0-k^2+q^2]\hat{\boldsymbol{I}}-2a\overleftarrow{\boldsymbol{p}}\cdot\boldsymbol{\sigma}-2q\sigma_z\left(a+\overleftarrow{\boldsymbol{p}}\cdot\boldsymbol{\sigma}\right)-2b\sigma_x = 0\,. \quad (5.85)$$

Notice that this equation differs from Eq. (5.72) only in the sign in front of q. Therefore, the solutions are

$$\overleftarrow{p_z} = -q = -\overrightarrow{p_z}\,,\quad \overleftarrow{p_x} = \frac{ab}{q^2-a^2} = \overrightarrow{p_x}\,,\quad \overleftarrow{p_y} = -i\frac{qb}{q^2-a^2} = -\overrightarrow{p_y}\,. \quad (5.86)$$

We can now write the total wave function when the incident wave originates from inside the mirror as

$$\begin{aligned}|\psi(z)\rangle = {}& \Theta(z<0)\hat{\boldsymbol{t}}'\exp(-i\hat{\boldsymbol{k}}_0z)|\xi_0\rangle \\ & + \Theta(z>0)\exp(-i\sigma_zqz)\left[\exp\left(-i[a+\overleftarrow{\boldsymbol{p}_\varphi}\cdot\boldsymbol{\sigma}]z\right)\right. \\ & \left.+\exp(i[a+\overrightarrow{\boldsymbol{p}_\varphi}\cdot\boldsymbol{\sigma}]z)\hat{\boldsymbol{r}}'\right]|\xi_0\rangle\,. \end{aligned} \quad (5.87)$$

19) In the following we shall frequently replace the unit matrix $\hat{\boldsymbol{I}}$ by 1

By matching this function and its derivative at the interface, we obtain

$$\hat{\boldsymbol{r}}' = \left[\hat{k}_0 + a + \overrightarrow{\boldsymbol{p}}'_{\varphi} \cdot \boldsymbol{\sigma}\right]^{-1} \left[a + \overleftarrow{\boldsymbol{p}}'_{\varphi} \cdot \boldsymbol{\sigma} - \hat{k}_0\right], \quad \hat{\boldsymbol{t}}' = I + \hat{\boldsymbol{r}}', \tag{5.88}$$

where

$$\overleftarrow{\boldsymbol{p}}'_{\varphi} \cdot \boldsymbol{\sigma} = \exp(-i\sigma_z\varphi)\left(\overleftarrow{p}_x\sigma_x + \overleftarrow{p}_y\sigma_y\right)\exp(i\sigma_z\varphi).$$

It is clear that if the internal magnetic field rotates clockwise instead of counterclockwise, the parameter q changes its sign, and $\overrightarrow{\boldsymbol{p}}$ and $\overleftarrow{\boldsymbol{p}}$ swap their places in the equations.

5.2.6.3 Reflection from a Plate of Finite Thickness

Let us find the reflection and transmission amplitudes for a plate of finite thickness L. First, we need to the find reflection and transmission amplitudes at the second interface. For convenience, let us place the origin at $z = L$. The wave function near it is

$$\begin{aligned} |\psi(z)\rangle = {} & \Theta(z < 0)e^{-iq\sigma_z z}\left[e^{i\left(a + \overrightarrow{\boldsymbol{p}_{\varphi}} \cdot \boldsymbol{\sigma}\right)z} + e^{-i(a + \overrightarrow{\boldsymbol{p}_{\varphi}} \cdot \boldsymbol{\sigma})z}\hat{\boldsymbol{r}}''\right]|\xi_0\rangle \\ & + \Theta(z > 0)e^{i\hat{k}_0 z}\hat{\boldsymbol{t}}''|\xi_0\rangle, \end{aligned} \tag{5.89}$$

where $\hat{\boldsymbol{t}}'' \neq \hat{\boldsymbol{t}}'$, $\hat{\boldsymbol{r}}'' \neq \hat{\boldsymbol{r}}'$, and φ is different from the angle at the first interface. By matching the function at the interface,

$$\hat{\boldsymbol{t}}'' = 1 + \hat{\boldsymbol{r}}'', \quad \hat{k}_0\hat{\boldsymbol{t}}'' = -\left[a + \overleftarrow{\boldsymbol{p}}'_{\varphi} \cdot \boldsymbol{\sigma}\right]\hat{\boldsymbol{r}}'' + a + \overrightarrow{\boldsymbol{p}}'_{\varphi} \cdot \boldsymbol{\sigma}, \tag{5.90}$$

we get

$$\hat{\boldsymbol{r}}'' = \left[\hat{k}_0 + a + \overleftarrow{\boldsymbol{p}}'_{\varphi} \cdot \boldsymbol{\sigma}\right]^{-1} \left[a + \overrightarrow{\boldsymbol{p}}'_{\varphi} \cdot \boldsymbol{\sigma} - \hat{k}_0\right]. \tag{5.91}$$

Note that at a distance $-L$ from the second interface the reflected wave is

$$\exp(iq\sigma_z L)\exp\left(i[a + \overleftarrow{\boldsymbol{p}}_{\varphi} \cdot \boldsymbol{\sigma}]L\right)\hat{\boldsymbol{r}}''. \tag{5.92}$$

We are now ready to consider reflection from a plate. Assume for simplicity that at the first interface $\varphi = 0$, then at the second interface $\varphi = qL$. Denote $\hat{X}$ the wave incident on the second interface. For $\hat{X}$ we can write down the equation

$$\hat{X} = e^{-iq\sigma_z L}e^{i(a + \overrightarrow{\boldsymbol{p}} \cdot \boldsymbol{\sigma})L}\hat{\boldsymbol{t}} + e^{-iq\sigma_z L}e^{i(a + \overrightarrow{\boldsymbol{p}} \cdot \boldsymbol{\sigma})L}\hat{\boldsymbol{r}}'e^{iq\sigma_z L}e^{i(a + \overrightarrow{\boldsymbol{p}_{qL}} \cdot \boldsymbol{\sigma})L}\hat{\boldsymbol{r}}''\hat{X}. \tag{5.93}$$

The solution to the above equation is

$$\begin{aligned} \hat{X} = {} & \left[1 - e^{-iq\sigma_z L}e^{i(a + \overrightarrow{\boldsymbol{p}} \cdot \boldsymbol{\sigma})L}\hat{\boldsymbol{r}}'e^{iq\sigma_z L}e^{i(a + \overleftarrow{\boldsymbol{p}_{qL}} \cdot \boldsymbol{\sigma})L}\hat{\boldsymbol{r}}''\right]^{-1} \\ & \times e^{-iq\sigma_z L}e^{i(a + \overrightarrow{\boldsymbol{p}} \cdot \boldsymbol{\sigma})L}\hat{\boldsymbol{t}}. \end{aligned} \tag{5.94}$$

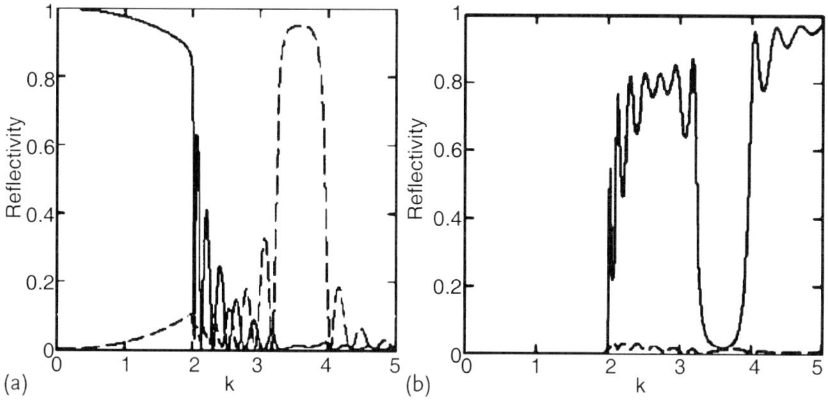

Figure 5.3 Wave-vector dependence of (a) the reflection coefficients without (solid line), $|R_{--}|^2$, and with (dashed line), $|R_{+-}|^2$, spin flip for the initial polarization (right subscript -) in the direction opposite the z-axis, which is parallel to the inward normal to the mirror, and (b) the respective transmission coefficients $|T_{--}|^2$ (solid line) and $|T_{+-}|^2$ (dashed line) for the initial polarization directed opposite the z-axis. The external field $\boldsymbol{B}_0$ is zero.

The reflection, $\hat{\boldsymbol{R}}$, and transmission, $\hat{\boldsymbol{T}}$, amplitudes are then

$$\hat{\boldsymbol{R}} = \hat{\boldsymbol{r}} + \hat{\boldsymbol{t}}' \exp(iq\sigma_z L) \exp\left(i[a + \overleftarrow{\boldsymbol{p}_{qL}} \cdot \boldsymbol{\sigma}]L\right) \hat{\boldsymbol{r}}'' \hat{\boldsymbol{X}}\,, \quad \hat{\boldsymbol{T}} = \hat{\boldsymbol{t}}'' \hat{\boldsymbol{X}}\,. \tag{5.95}$$

By means of Eq. (5.95), it is easy to obtain the reflection coefficients with and without spin flip. Figure 5.3 shows the results of the calculation for the simplest case $B_0 = 0$. The calculations were performed with a wave-vector length unit equal to $\sqrt{b}$ and with the parameters $u_0 = 4 - 0.01i$, $q = 3$, $L = 8$ and, correspondingly, $b = 1$. It is clearly seen that for one polarization there is a resonant reflection with spin reversal. The transmission coefficient for this polarization is strongly suppressed.

> This manuscript was handled appropriately and professionally. The editor clearly found competent referees with good publication and review histories. As is always the case with peer review, the editor made a well reasoned decision based on the material at hand. This manuscript did not meet the PRB criteria for publication and as such was rejected after a number of rounds of review. You were given opportunities to respond directly to the concerns of the referees and they remained unconvinced.
>
> Editorial Director of American Physical Society

Everything described in this paragraph is also valid in the case of the presence of a constant internal field $\boldsymbol{B}$ parallel to z. The main difference will be in the form of Eq. (5.77), which will become cubic with respect to a^2.

We discussed reflection from a helical system. The same approach can be applied to fanlike systems, where the internal field rotates not in the plane parallel to the interface, but in a perpendicular plane: $\boldsymbol{B}_{\text{fan}}(z) = (B_0, B_1\cos(2qz), B_1\sin(2qz))$.

The field $\boldsymbol{B}_{\text{fan}}(z)$ which models perpendicular exchange-spring multilayer [236, 237] and can be relevant to a pseudomagnetic field created by spin–spin neutron–nucleus interaction. The approach is identical to the above one, and the results are almost identical [238].

Magnets with a rotating field are called spring magnets. They can be prepared by evaporating a soft magnetic layer with small coercive force upon a hard magnetic layer with strong coercive force [241, 242]. After magnetization in a saturated magnetic field, the external induction is diminished to a small value and changes its direction. Then magnetization in the soft layer is on one side is tied to the magnetization of the hard layer, and near the vacuum it follows the external field. So it rotates along the depth coordinate. Spring magnets have been studied theoretically [243] and experimentally with neutron reflectometers [244].

5.2.6.4 **Violation of Fundamental Principles in Helical Systems**

It is important to notice that with helical systems we meet violation of time-reversal symmetry and violation of the detailed-balance principle [238]. Violation of the time parity is seen in the presence of correlation between neutron spin $\boldsymbol{s}$ and helicoidal vector $\boldsymbol{q}$:$\boldsymbol{sq}$ in transmission. On reversal of time, the vector $\boldsymbol{q}$ does not change notwithstanding that the field $\boldsymbol{b}$ changes its sign. This time parity violation is not a fundamental one, because we can imagine a change in the preparation of the mirror on time reversal.

The detailed balance is violated because with the helical mirror submersed in an isotropic unpolarized neutron gas there appears a cycle in phase space, which decreases the entropy. Indeed, imagine a vessel subdivided into two parts by a foil with helicoidal magnetization. The neutrons on the left side polarized parallel to the z-axis will go through the foil to the right side, and the neutrons with the opposite polarization will be reflected with reversal of their spin. After reflection from the vessel walls, the neutron with reversed spin return to the foil and this time pass through it to the right. It seems that all the neutrons from the left side become polarized along the z-axis and collect in the right side of the vessel. This, however, does not happen, because there is an opposite process from right to left. Nevertheless, an equilibrium is sustained by a cycle in the phase space. This cycle consists of reflection and transmission from the left, then reflection and transmission from the right. There is no detailed equality of the number of neutrons reflected from one and or the opposite direction. We can say that the presence of the helicoidal field with only one direction of rotation makes the space be in a nonequilibrium state.

Detailed balance is violated even in a system of two layers with noncollinear magnetization and three layers with noncomplanar magnetization. The last system also exhibits a property of one directional transmission for unpolarized neutron beam. This can be found in [239, 240]

5.3
Numerical Matrix Method for Multilayered Magnetic Systems

In addition to the analytical approach, it is important to know about numerical methods for finding the reflection and transmission effects in complex magnetic systems. The popular matrix method, applied for scalar particles in Section 4.7.3, can be generalized for spinorial particles [136].

Consider an arbitrary multilayered system (Figure 5.4) with n interfaces. Assume that the nuclear potentials and the magnetic fields are different in all layers. The wave function is

$$\psi_0 = \begin{pmatrix} \exp\left(i\hat{\boldsymbol{k}}_0^+ x\right) & 0 \\ 0 & \exp\left(-i\hat{\boldsymbol{k}}_0^+ x\right) \end{pmatrix} \begin{pmatrix} |\xi_0\rangle \\ \hat{\boldsymbol{\rho}}|\xi_0\rangle \end{pmatrix}, \tag{5.96}$$

to the left of interface 1 and

$$\psi_n = \begin{pmatrix} \exp\left(i\hat{\boldsymbol{k}}_n^+ (x - x_n)\right) & 0 \\ 0 & \exp\left(-i\hat{\boldsymbol{k}}_n^+ (x - x_n)\right) \end{pmatrix} \begin{pmatrix} \hat{\boldsymbol{\tau}}|\xi_0\rangle \\ 0 \end{pmatrix}, \tag{5.97}$$

to the right of interface n, where $|\xi_0\rangle$ is an arbitrary spinor and $\hat{\boldsymbol{\rho}}$ and $\hat{\boldsymbol{\tau}}$ are the matrices of the reflection and transmission amplitudes for the full system.

The wave functions (5.96) and (5.97) are related to each other through the block matrix $\hat{\boldsymbol{M}}$:

$$\Xi_0 = \hat{\boldsymbol{M}}\Xi_n \equiv \begin{pmatrix} |\xi_0\rangle \\ \hat{\boldsymbol{\rho}}|\xi_0\rangle \end{pmatrix} = \begin{pmatrix} \hat{\boldsymbol{M}}_{11} & \hat{\boldsymbol{M}}_{12} \\ \hat{\boldsymbol{M}}_{21} & \hat{\boldsymbol{M}}_{22} \end{pmatrix} \begin{pmatrix} \hat{\boldsymbol{\tau}}|\xi_0\rangle \\ 0 \end{pmatrix}, \tag{5.98}$$

where $\hat{\boldsymbol{M}}_{ij}$ are 2×2 matrices. From here,

$$\hat{\boldsymbol{\tau}} = \hat{\boldsymbol{M}}_{11}^{-1}, \quad \text{and} \quad \hat{\boldsymbol{\rho}} = \hat{\boldsymbol{M}}_{21}\hat{\boldsymbol{M}}_{11}^{-1}. \tag{5.99}$$

It is therefore necessary first to find the 4×4 matrix $\hat{\boldsymbol{M}}$, from which $\hat{\boldsymbol{\rho}}$ and $\hat{\boldsymbol{\tau}}$ are found elementarily. The matrix $\hat{\boldsymbol{M}}$ can be found by matching the wave function at all interfaces.

Rewrite the wave function inside the ith rectangle located to the right of the ith interface at the point a_i as

$$\left(\exp(i\hat{\boldsymbol{k}}_i(x - a_i))\hat{A}_i + \exp(-i\hat{\boldsymbol{k}}_i(x - a_i))\hat{B}_i\right). \tag{5.100}$$

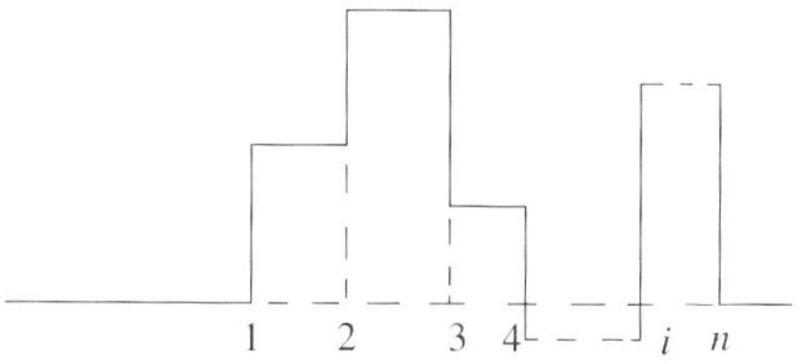

Figure 5.4 Multilayered system with a variety of different nuclear potentials and magnetic fields.

By matching the wave function and its first derivative at the ith interface, we obtain a system of two matrix equations:

$$\exp(i\hat{\boldsymbol{k}}_{i-1}d_{i-1})\hat{A}_{i-1} + \exp(-i\hat{\boldsymbol{k}}_{i-1}d_{i-1})\hat{B}_{i-1} = \hat{A}_i + \hat{B}_i\,,$$
$$\hat{\boldsymbol{k}}_{i-1}\left[\exp(i\hat{\boldsymbol{k}}_{i-1}d_{i-1})\hat{A}_{i-1} - \exp(-i\hat{\boldsymbol{k}}_{i-1}d_{i-1})\hat{B}_{i-1}\right] = \hat{\boldsymbol{k}}_i[\hat{A}_i - \hat{B}_i]\,,$$

where d_i is the width of the ith layer. From these two equations,

$$\hat{A}_{i-1} = \exp(-i\hat{\boldsymbol{k}}_{i-1}d_{i-1})\left(\frac{1}{2}\left[1 + \hat{\boldsymbol{k}}_{i-1}^{-1}\hat{\boldsymbol{k}}_i\right]\hat{A}_i + \frac{1}{2}\left[1 - \hat{\boldsymbol{k}}_{i-1}^{-1}\hat{\boldsymbol{k}}_i\right]\hat{B}_i\right)\,, \tag{5.101}$$

$$\hat{B}_{i-1} = \exp(i\hat{\boldsymbol{k}}_{i-1}d_{i-1})\left(\frac{1}{2}\left[1 - \hat{\boldsymbol{k}}_{i-1}^{-1}\hat{\boldsymbol{k}}_i\right]\hat{A}_i + \frac{1}{2}\left[1 + \hat{\boldsymbol{k}}_{i-1}^{-1}\hat{\boldsymbol{k}}_i\right]\hat{B}_i\right)\,. \tag{5.102}$$

Matrices $\hat{A}_i$ and $\hat{B}_i$ can be combined into a super spinor, or a four-dimensional vector,

$$\Xi_i = \begin{pmatrix}\hat{A}_i\\ \hat{B}_i\end{pmatrix}.$$

In this case, the two Eqs. (5.101) and (5.102) can be combined into one,

$$\Xi_{i-1} = \hat{Q}_i^{i-1}\Xi_i\,,$$

where

$$\hat{Q}_i^{i-1} = \hat{E}_{i-1}\hat{K}_i^{i-1}\,,\quad \hat{K}_i^{i-1} = \frac{1}{2}\begin{pmatrix}1 + \hat{\boldsymbol{k}}_{i-1}^{-1}\hat{\boldsymbol{k}}_i & 1 - \hat{\boldsymbol{k}}_{i-1}^{-1}\hat{\boldsymbol{k}}_i\\ 1 - \hat{\boldsymbol{k}}_{i-1}^{-1}\hat{\boldsymbol{k}}_i & 1 + \hat{\boldsymbol{k}}_{i-1}^{-1}\hat{\boldsymbol{k}}_i\end{pmatrix},$$
$$\hat{E}_i = \begin{pmatrix}e^{-i\hat{\boldsymbol{k}}_i d_i} & 0\\ 0 & e^{i\hat{\boldsymbol{k}}_i d_i}\end{pmatrix}.$$

By matching the wave function sequentially at all interfaces, the complete matrix is

$$\hat{\boldsymbol{M}} = K_1^0\prod_{i=2}^{n}\mathcal{N}\left[\hat{Q}_i^{i-1}\right] = K_1^0\prod_{i=2}^{n}\mathcal{N}\left[\hat{E}_{i-1}\hat{K}_i^{i-1}\right]\,, \tag{5.103}$$

where the symbol $\mathcal{N}$ next to the product sign means that the product elements should be arranged with increasing i from the left to the right. The approach described was used to prepare an original software program [246] for calculation of different magnetic configurations. I will stop the derivation here, since the reader has all the tools at his disposal to do this work on his own.

5.4
Polarizers, Analyzers, and Spin Rotators

This section is devoted to various methods of polarizing neutrons, analyzing them, and rotating their spin.

5.4.1
Polarizers

There are three ways to polarize thermal neutrons: via reflection from magnetized mirrors, via transmission through magnetized films, and via transmission through a polarized gas, ^{3}He. In the first two methods, the interaction between a neutron and a medium is described by an optical potential $u + 2\boldsymbol{\sigma} \cdot \boldsymbol{B}$, which contains the nuclear ($u = 4\pi N_0 b$) and the magnetic ($\boldsymbol{\sigma} \cdot \boldsymbol{B}$) parts. The potential is higher for neutrons polarized along the magnetic field ($u + 2B$) than for neutrons polarized against the magnetic field $u - 2B$. The medium reflects all the neutrons polarized along the field and some of the neutrons polarized against the field when the momentum $k_\perp$ of the motion perpendicular to the interface is within the interval $\sqrt{u - 2B} \leq k_\perp \leq \sqrt{u + 2B}$. The transmitted neutrons will be completely polarized against the field (external and internal fields are assumed to be collinear).

In many setups, neutrons are polarized inside a special neutron guide, where the walls totally reflect neutrons polarized along the field, and partially reflect neutrons with the opposite polarization. After multiple reflections, the number of oppositely polarized neutrons decays to zero.

The mirrors and films provide a high degree of polarization, but the output intensity in the case of thermal neutrons is low because of very small grazing angles. Alternatively, magnetic crystals work well for larger grazing angles, but only within a narrow spectral region and for well-collimated beams.

The spatially and spectrally wide beams can be obtained via transmission through a polarized gas, ^{3}He; however, a high degree of polarization is accompanied with strong losses of the transmitted beam. In this method, a polarized nucleus absorbs oppositely polarized neutrons and does not absorb the collinearly polarized ones. The polarization of the gas is achieved by the optical pumping of rubidium vapor. The polarization of the rubidium atoms is then transmitted to ^{3}He atoms upon collisions, or through the optical pumping of ^{3}He atoms when the collision excites helium atoms to a metastable state. The degree of polarization of the helium atoms can be as high as 30%. The degree of polarization of the transmitted neutrons depends on their energy – the lower is the energy, the higher is the degree of polarization. However, the higher is the degree of polarization, the lower is the intensity. The optimal polarization is about 80%. The polarized helium can maintain its polarization state for several days at room temperature.

5.4.2 Analyzers

The basic principle behind the analyzers is the same as for polarizers. An analyzer is a filter, which in case of films or mirrors transmits only neutrons polarized against the field, and reflects neutrons polarized along the field. In the case of helium, the analyzer transmits only neutrons polarized along the direction of polarization of the nuclei. Of course, the word "only" has a limited validity because analyzers and polarizers are not ideal filters.

5.4.3 Spin Rotators with Direct Current Fields

Spin rotators are the most important tool used in the majority of experiments involving polarized neutrons. Spin rotators are sometimes called spin flippers or simply flippers. The spin can be turned using a magnetic field constant in time but rapidly changing in space, a so-called direct current (DC) magnetic field (in fact, this is a field flipper not a spin flipper). It can also be turned with a combination of constant and oscillating magnetic fields, so-called radiofrequency (RF) magnetic fields [248]. The RF field not only changes the direction of the spin arrow, but also produces a lot of interesting phenomena, which are worth considering separately. Here we will consider first the DC flippers.

Upon neutron propagation in slowly changing space fields, the neutron spin arrow $\boldsymbol{s}$ follows the magnetic field vector, that is, the angle α between $\boldsymbol{B}$ and $\boldsymbol{s}$ is conserved. This conservation rule is stricter, the lower is the ratio between the angular velocity of the magnetic field rotation, $\Omega = (1/B)\boldsymbol{v} \cdot d\boldsymbol{B}/d\boldsymbol{r}$ ($\boldsymbol{v}$ is the neutron velocity) and the precession frequency $\boldsymbol{\mu} \boldsymbol{B}/\hbar$ of the spin around the field in the system coordinate tied to the moving neutron. The ratio Ω/ω is called the adiabatic parameter. The smaller is the parameter, the longer is the conservation of $\boldsymbol{s} \cdot \boldsymbol{B}$. If, in contrast, the angular velocity Ω is large, then the spin arrow $\boldsymbol{s}$ cannot keep up with $\boldsymbol{B}$, and nonadiabatically changes its direction with respect to the field.

Two methods are used to obtain the fast magnetic field rotation.

5.4.3.1 Current Foil Method

The spin rotator is a thin metal film with an electric current propagating parallel to its surface [249, 250]. The field on both sides of the foil has opposite direction. After the neutron has passed the film, its spin retains its direction in the flipped field.

5.4.3.2 Zero Field Method

The spin rotator is constructed using two coils with opposite currents. The configuration in Figure 5.5a is called a Drabkin flipper [249, 251]. The configuration in Figure 5.5b is a called Korneev flipper [252–256]. The letter A in figure denotes the space where the magnetic fields from the two coils cancel each other, and the total field is zero. Upon the neutron passing through this region of zero field, its spin is conserved, whereas the direction of the field changes [253].

These methods are effective in a broad spectrum of neutron energies, which is their main advantage.

5.4.3.3 Mezei Spin Rotator

There is also a third type of spin rotator, using time-constant fields, first proposed by Mezei [55, 249]. It can be used to rotate the spin by any amount around any axis. The Mezei spin flipper is shown in Figure 5.5c. The magnetic field is created by two oppositely oriented coils. The coils are positioned perpendicular to the neutron beam. The magnitude of the magnetic field inside the coils is equal to the external magnetic field, which is oriented perpendicular to both the beam axis and the fields inside the coils. The external field is marked by "+" in the figure. Suppose that the external field and the neutron polarization before the flipper are along the z-axis, and the fields of the coils are along the $-x$ and $+x$ directions, respectively. If in the first coil the neutron spin is rotated by an angle 2φ around the total field, then the spin state after the first coil is

$$|\xi_1\rangle = \exp(i[\sigma_z - \sigma_x]\varphi/\sqrt{2})|\xi_u\rangle\,, \tag{5.104}$$

and after the second coil it becomes

$$\begin{aligned} |\xi_2\rangle &= \exp(i[\sigma_z + \sigma_x]\varphi/\sqrt{2})|\xi_1\rangle \\ &= \exp(i[\sigma_z + \sigma_x]\varphi/\sqrt{2})\exp(i[\sigma_z - \sigma_x]\varphi/\sqrt{2})|\xi_u\rangle \\ &= \left(\cos^2\varphi + \sqrt{2}i\cos\varphi\sin\varphi\sigma_z - i\sigma_y\sin^2\varphi\right)|\xi_u\rangle\,. \end{aligned} \tag{5.105}$$

When $\varphi = \pi/2$, then after the flipper the spinor state is $|\xi_2\rangle = -|\xi_d\rangle$, which means that the spin is flipped up–down. The fast change in the field direction takes place in walls of the coils.

By choosing the size of the coils, their orientation, and the relative strength of the external and internal fields, one can rotate an initially arbitrarily oriented spin by an

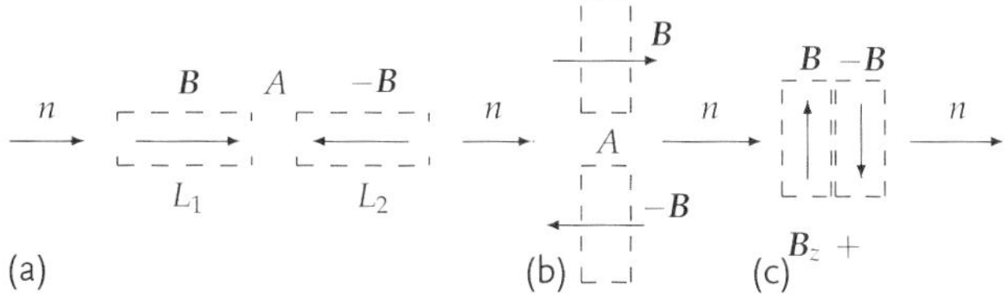

Figure 5.5 Three types of nonadiabatic direct current (DC) spin rotators. For simplicity, the rotators are placed in the same beam n. The Drabkin spin rotator (a) and the Korneev rotator (b) consist of two coils with opposite currents; thus, the field between the coils (in the space denoted by A) is zero. The Mezei spin rotator (c) consists of two coils oriented perpendicular to the beam. The fields B in the coils are equal in magnitude and perpendicular to the external field B_z (denoted by a cross). The neutron spin rotates a half circle around the sum of the internal and the external fields in each coil, $B_z + B$ and $B_z - B$, respectively.

arbitrary angle for neutrons with an arbitrary energy. However, since the amount of rotation in the coils depends on the time of flight of the neutron through the coils, and thus on the neutron speed, the Mezei spin rotator is effective only in a narrow spectral interval.

5.4.4
Resonant Spin Rotators

The review of the systems described above is rather brief. The resonant spin rotator will be considered in more detail. The distinctive feature of the RF flipper is the change of the neutron energy because of absorption or emission of RF-field quanta. In the presence of an RF field, the neutron wave function obeys the nonstationary Schrödinger equation,

$$i\frac{\partial}{\partial t}|\boldsymbol{\Psi}(t,x)\rangle = \left(-\frac{1}{2}\frac{\partial^2}{\partial x^2} + \frac{u(x)}{2} + \boldsymbol{\sigma}\cdot\boldsymbol{B}_0(x) + \boldsymbol{\sigma}\cdot\boldsymbol{B}_{\rm rf}(t,x)\right)|\boldsymbol{\Psi}(t,x)\rangle\,, \tag{5.106}$$

where $u(x)$ is the material potential. The field here is split into two parts. The first one is the time-constant DC field, B_0. The second part is the oscillating in time RF field, $B_{\rm rf}$.

Let us consider the ideal conditions, when the fields are confined to a finite space, $0 \le x \le D$, the so-called Krüger configuration (Figure 5.6). So,

$$\boldsymbol{B}_0(x) + \boldsymbol{B}_{\rm rf}(t,x) = \Theta(0 \le x \le D)[\boldsymbol{B}_0 + \boldsymbol{B}_{\rm rf}(t)]\,. \tag{5.107}$$

Furthermore, assume for simplicity that $u = 0$, and set the components of $\boldsymbol{B}_0$ and $\boldsymbol{B}_{\rm rf}$ as $(0, 0, B_0)$ and $B_{\rm rf}(\cos(2\omega t), \sin(2\omega t), 0)$, that is, the constant field is directed along z, and the RF field vector is rotating counterclockwise with angular

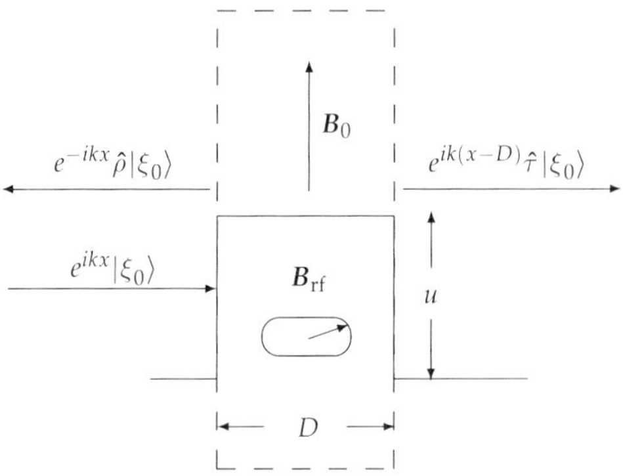

Figure 5.6 The reflection and transmission of the neutron through the magnetic field, localized in $0 \le x \le D$. The magnetic field contains a constant component, $\boldsymbol{B}_0 = (0, 0, B_0)$, and a radiofrequency component, $\boldsymbol{B}_{\rm rf} = B_{\rm rf}(\cos(2\omega t), \sin(2\omega t), 0)$. A medium characterized by an optical potential u may also be present in addition to the fields. The incident neutron spin (from the left) has an arbitrary orientation. The polarization of the neutron is described using a spinor, $|\xi_0\rangle$.

velocity 2ω (2 is chosen for convenience) in the (x, y) plane. The spin flip with respect to B_0 is accompanied by a change in the potential energy by $2B_0$ or a change in the total energy by the RF quantum 2ω.

We solve Eq. (5.106) for the incident wave

$$\Theta(x < 0)\exp(ik_0x - iE_0t)|\xi_0\rangle\,, \tag{5.108}$$

where $|\xi_0\rangle$ is an arbitrary spinor, and $E_0 = k_0^2/2$.

5.4.4.1 The Rabi Formula

Consider the most common case when $E_0 \gg B_0 \gg B_{\text{rf}}$, that is, there is almost no change in the neutron velocity when it enters B_0. Then we can represent the wave function as

$$|\boldsymbol{\Psi}(x,t)\rangle = \exp(ik_0x - iE_0t)|\boldsymbol{\Psi}(t)\rangle \tag{5.109}$$

and neglect the reflections at the interfaces. Substitution of Eq. (5.109) into Eq. (5.106) gives

$$id|\boldsymbol{\Psi}(t)\rangle/dt = [\boldsymbol{\sigma}\cdot\boldsymbol{B}_0 + \boldsymbol{\sigma}\cdot\boldsymbol{B}_{\text{rf}}(t)]|\boldsymbol{\Psi}(t)\rangle \tag{5.110}$$

with the initial condition $|\boldsymbol{\Psi}(0)\rangle = |\xi_0\rangle$, where $t = 0$ is the time when the neutron enters the area of the fields.

To solve Eq. (5.110), we use an easily verifiable equality (see Section 5.1.2.1)

$$\boldsymbol{\sigma}\cdot\boldsymbol{B}_{\text{rf}}(t) = B_{\text{rf}}[\sigma_x\cos(2\omega t) + \sigma_y\sin(2\omega t)] = B_{\text{rf}}e^{-i\omega\sigma_z t}\sigma_x e^{i\omega\sigma_z t}\,, \tag{5.111}$$

and write

$$|\boldsymbol{\Psi}(t)\rangle = e^{-i\omega\sigma_z t}|\boldsymbol{\phi}(t)\rangle\,. \tag{5.112}$$

Substitution of Eqs. (5.111), (5.112) into Eq. (5.110) gives

$$id|\boldsymbol{\phi}(t)\rangle/dt = (\boldsymbol{\sigma}\cdot\boldsymbol{\Omega})|\boldsymbol{\phi}(t)\rangle\,, \tag{5.113}$$

where $\boldsymbol{\Omega} = (B_{\text{rf}}, 0, B_0 - \omega)$.

The expression in the brackets on the right-hand side of Eq. (5.113) does not depend on time; thus, the solution $|\boldsymbol{\phi}(t)\rangle$ is obvious: $|\boldsymbol{\phi}(t)\rangle = \exp(-i\boldsymbol{\Omega}\cdot\boldsymbol{\sigma}t)|\boldsymbol{\phi}(0)\rangle$. The initial condition $|\boldsymbol{\Psi}(0)\rangle = |\xi_0\rangle$ is equivalent to $|\boldsymbol{\phi}(0)\rangle = |\xi_0\rangle$, and by substituting it into Eq. (5.112), we obtain

$$|\boldsymbol{\Psi}(t)\rangle = e^{-i\omega\sigma_z t}e^{-i\boldsymbol{\Omega}\cdot\boldsymbol{\sigma}t}|\xi(0)\rangle = e^{-i\omega\sigma_z t}\left[\cos\Omega t - i\frac{\boldsymbol{\Omega}\cdot\boldsymbol{\sigma}}{\Omega}\sin\Omega t\right]|\xi_0\rangle\,, \tag{5.114}$$

where $\Omega = |\boldsymbol{\Omega}| = \sqrt{B_{\text{rf}}^2 + (\delta\omega)^2}$, $\delta\omega = B_0 - \omega$.

Assume that the spin of the incident particle is parallel to the field $\boldsymbol{B}_0$, that is, $|\xi_0\rangle = |\xi_u\rangle$. The factor $\boldsymbol{\sigma} \cdot \boldsymbol{\Omega} = B_{\rm rf}\sigma_x + \delta\omega\sigma_z$ in Eq. (5.114) contains a nondiagonal matrix σ_x, which provides the spin flip. Therefore, the probability of the spin flip after flight time $t = t_1 = D/k_0$ through the fields is

$$w_{\rm fl} = \left|\frac{\Omega_x}{\Omega}\sin(\Omega t_1)\right|^2 = \frac{B_{\rm rf}^2}{B_{\rm rf}^2 + (\omega - B_0)^2}\sin^2\left(\sqrt{B_{\rm rf}^2 + (\omega - B_0)^2}\, t_1\right) . \tag{5.115}$$

This expression is the well-known Rabi formula. It shows that $w_{\rm fl} = 1$ when $\omega = B_0$ and $t_1 B_{\rm rf} = \pi n + \pi/2$, with integer n.

5.4.4.2 Complete Conversion of an Unpolarized Beam into a Polarized One

According to Eq. (5.115), the probability of the spin flip at $\omega = B_0$ is

$$w_{\rm fl} = \sin^2(B_{\rm rf} t_1) . \tag{5.116}$$

It depends on the time of flight through the coil, $t_1 = l/v_0$, where l is the coil thickness and v_0 is the neutron speed. It is natural to believe that an unpolarized beam will remain unpolarized after passing through the coil; however, this is true only if the velocities of the neutrons inside the field $\boldsymbol{B}_0$ do not depend on the polarizations of the neutrons. This assumption is valid when the kinetic energy $v_0^2/2$ is much higher than the magnetic potential energy B_0, and this is the assumption which we used to obtain Eq. (5.115). In the case of low kinetic energy neutrons, such as ultracold neutrons, the difference in velocities of differently polarized neutrons is significant. Neutrons polarized parallel to the field slow down (their velocity becomes $v_+ = \sqrt{v_0^2 - 2B_0} < v_0$). Neutrons polarized oppositely to the field $\boldsymbol{B}_0$ speed up ($v_- = \sqrt{v_0^2 + 2B_0} > v_0$). Therefore, the time of flight of the oppositely polarized neutrons is different, $t_+ = l/v_+$ and $t_- = l/v_-$, respectively.

One can exploit this time difference to completely polarize an unpolarized beam by passing it through a specially configured resonance RF spin flipper [257, 258]. If the coil thickness l is such that $B_{\rm rf}t_+ = \pi$ and $B_{\rm rf}t_- = \pi/2$, then the polarization of neutrons polarized along the field remains unchanged, whereas the polarization of the oppositely polarized neutrons is flipped. Therefore, all the neutrons become polarized along the field $\boldsymbol{B}_0$ after exiting the spin flipper.

In the case when the coil thickness is such that $B_{\rm rf}t_+ = 3\pi/2$ and $B_{\rm rf}t_- = \pi$, the polarization of neutrons polarized along the field is flipped, whereas it remain unchanged for the other neutrons. Therefore, all neutrons become polarized against the field $\boldsymbol{B}_0$.

5.4.4.3 Linear Oscillating Field

Above we considered a rotating RF field,

$$\boldsymbol{B}_{\rm rf} = B_{\rm rf}(\cos(2\omega t), \sin(2\omega t), 0) .$$

In practice, a simple linear oscillating field is used, $\boldsymbol{B}_{\text{rf}} = B_{\text{rf}}(\cos(2\omega t), 0, 0)$. Such an oscillating field can be split into a superposition of two fields, rotating in opposite directions,

$$\boldsymbol{B}_{\text{rf}} = B_{\text{rf}}(\cos(2\omega t), 0, 0) = \frac{1}{2} B_{\text{rf}}(\cos(2\omega t), \sin(2\omega t), 0) + \frac{1}{2} B_{\text{rf}}(\cos(2\omega t), -\sin(2\omega t), 0) . \quad (5.117)$$

Thus, in addition to the field, which rotates counterclockwise, that is, in the same direction as the precessing spin, there is another field rotating in the opposite direction.

The effect of the clockwise-rotating field on the spin can be estimated using Eq. (5.115) and $\omega = -B_0$. The probability of the spin flip in such a field is

$$w_{\text{fl}} = \left| \frac{\Omega_x}{\Omega} \sin(\Omega t_1) \right|^2 \approx \frac{B_{\text{rf}}^2}{16 B_0^2} \sin^2(2 B_0 t) . \quad (5.118)$$

It is easy to see that the probability of obtaining the opposite polarization is no larger than $B_{\text{rf}}^2/16 B_0^2 \ll 1$. Therefore, the clockwise-rotating field leads to a depolarization on the order of $\delta P = B_{\text{rf}}^2/16 B_0^2 \ll 1$. Such a depolarization is usually very small and can be neglected compared with the deviation of the polarization from the ideal one because of the nonmonochromaticity of the incident beam.

It is believed that the clockwise-rotating field also shifts the resonant frequency ω a little, which is known as Bloch–Siegert shift. This shift can be estimated as follows. We have seen that a counterclockwise field can give a probability of spin flip equal to unity. The clockwise field also gives a probability of spin flip Eq. (5.118). The total probability cannot be larger than unity; therefore, at $\omega = B_0$ the probability of spin flip is not unity and the factor before the sine function in Eq. (5.115) should be replaced by

$$\frac{B_{\text{rf}}^2/4}{B_{\text{rf}}^2/4 + (\omega - B_0)^2 + \delta^2} . \quad (5.119)$$

In this form it is not unity at resonance.[20] The value of δ is estimated from the requirement

$$\frac{B_{\text{rf}}^2/4}{B_{\text{rf}}^2/4 + (\omega - B_0)^2 + \delta^2} + \frac{B_{\text{rf}}^2/4}{B_{\text{rf}}^2/4 + (\omega + B_0)^2} = 1 , \quad (5.120)$$

which at resonance $\omega = B_0$ gives

$$\approx 1 - \frac{4\delta^2}{B_{\text{rf}}^2} + \frac{B_{\text{rf}}^2}{16 B_0^2} = 1 , \quad (5.121)$$

or $\delta / B_0 = B_{\text{rf}}^2/8 B_0^2$.

20) The factor 1/4 at B_{rf}^2 appears because of the factor 1/2 in Eq. (5.117)

21) Note that the sine factors are not important because by appropriate choice of t_1 and B_0 they can both be made equal to unity.

We see that the Bloch–Siegert shift is not a shift of the resonant frequency but an increase of the width of the resonance. If it were a shift $\omega - \Delta\omega$, then at $\omega = B_0 + \Delta\omega$ the counterclockwise-rotating component would give unit probability of spin flip and unitarity would be violated.[21)]

5.4.4.4 Broadband Adiabatic Spin Flippers

Resonant spin rotators are very sensitive to the spectrum of the incident neutrons. They are effective only for those neutrons which stay in the flipper (see Eq. (5.115)) during time $t = \pi/2B_{\rm rf}$. Now we consider a so-called adiabatic spin flipper, which is effective for the broad spectrum of the incident neutrons [259].

Suppose that the neutron propagates along the x-axis in the system of two fields similar to those in a resonant spin flipper. The initial neutron spin is parallel to the stationary field B_z directed along the z-axis. However, the field B_z now changes along the neutron path according to the law $B_z(x) = B_0 + b\cos(2px)$. The rotating RF field, $\boldsymbol{B}_{\rm rf}(t, x)$, is perpendicular to the z-axis and it also varies along the x-axis but according to a different law: $\boldsymbol{B}_{\rm rf} = -b(\cos(2\omega t), \sin(2\omega t), 0)\sin(2px)$. The Schrödinger equation (5.106) for the neutron wave function is now

$$i\frac{\partial}{\partial t}|\boldsymbol{\Psi}(t,x)\rangle = \left\{-\frac{1}{2}\frac{\partial^2}{\partial x^2} + \sigma_z[B_0 + b\cos(2px)] - b[\sigma_x\cos(2\omega t) + \sigma_y\sin(2\omega t)]\sin(2px)\right\}|\boldsymbol{\Psi}(t,x)\rangle\,. \tag{5.122}$$

To exclude the time dependence we use the relation (5.111), represent the wave function as

$$|\boldsymbol{\Psi}(t)\rangle = e^{-i\omega\sigma_z t - iEt}|\boldsymbol{\varphi}(x)\rangle\,, \tag{5.123}$$

where E is the total energy of the neutron, and suppose that $\omega = B_0$. Substitution of Eqs. (5.111) and (5.123) into Eq. (5.122) gives

$$\left(\frac{d^2}{dx^2} - 2\sigma_z b\cos(2px) + 2b\sigma_x\sin(2px) + k^2\right)|\boldsymbol{\varphi}(x)\rangle = 0\,, \tag{5.124}$$

where $k^2 = 2E$. To exclude the dependence on x in this equation we use the representations

$$\begin{aligned}\sigma_z\cos(2px) - \sigma_x\sin(2px) &= \exp(i\sigma_y px)\sigma_z\exp(-i\sigma_y px)\,,\\ |\boldsymbol{\varphi}(x)\rangle &= \exp(i\sigma_y px)|\xi(x)\rangle\,,\end{aligned} \tag{5.125}$$

and reduce Eq. (5.124) to the form

$$\left(\frac{d^2}{dx^2} + 2i\sigma_y p\frac{d}{dx} - 2b\sigma_z + k^2 - p^2\right)|\xi(x)\rangle = 0\,. \tag{5.126}$$

The solution of this equation is representable as

$$|\xi(x)\rangle = \exp(iqx + i\boldsymbol{\sigma}\cdot\boldsymbol{\rho}x)|\xi_0\rangle\,. \tag{5.127}$$

Substitution into Eq. (5.126) shows that the equation is satisfied for an arbitrary initial spin state, $|\xi_0\rangle$, if

$$-(q+\boldsymbol{\sigma}\cdot\boldsymbol{\rho})^2 - 2\sigma_y p(q+\boldsymbol{\sigma}\cdot\boldsymbol{\rho}) - 2b\sigma_z + k^2 - p^2 = 0\,. \tag{5.128}$$

This is equivalent to a system of four equations for four unknowns (the scalar q and the vector $\boldsymbol{\rho}$):

$$-2q\rho_x - 2ip\rho_z = 0\,,\quad -2q\rho_y - 2pq = 0\,,\quad -2q\rho_z + 2ip\rho_x - 2b = 0\,, \tag{5.129}$$

$$-q^2 - \rho^2 - 2p\rho_y + k^2 - p^2 = 0\,. \tag{5.130}$$

From Eq. (5.129) it follows

$$\rho_z = \frac{bq}{p^2-q^2}\,,\quad \rho_x = -i\frac{bp}{p^2-q^2}\,,\quad \rho_y = -p\,. \tag{5.131}$$

Substitution into Eq. (5.130) gives the equation $-q^2 - b^2/(q^2-p^2) + k^2 = 0$ for q, which has a solution

$$q^2 = \frac{1}{2}\left[k^2 + p^2 + \sqrt{(k^2-p^2)^2 - 4b^2}\right] \approx k^2 - \frac{b^2}{k^2-p^2}\,. \tag{5.132}$$

The approximation is valid when $k^2 \gg p^2$, $k^2 \gg 2b$, and $p^2 \gg b^2/k^2$, which is assumed.

If at the entrance point of the spin flipper the neutron spin state is $|\xi_u\rangle$, then after the path L in the flipper the state becomes

$$\exp(i\boldsymbol{\sigma}\cdot\boldsymbol{\rho}L)|\xi_u\rangle = \left(\cos\rho L + i\frac{\boldsymbol{\sigma}\cdot\boldsymbol{\rho}}{\rho}\sin\rho L\right)|\xi_u\rangle\,, \tag{5.133}$$

where

$$\rho = \sqrt{\frac{b^2}{q^2-p^2} + p^2} \approx p + \frac{b^2}{2p(k^2-p^2)}\,. \tag{5.134}$$

The probability amplitude f for spin flip is determined by the Pauli matrices $\sigma_{x,y}$, and from Eq. (5.133) we immediately find

$$f = i\frac{\rho_x + i\rho_y}{\rho}\sin(\rho L) = p\frac{q^2-p^2-b}{(q^2-p^2)\rho}\sin(\rho L) \approx \left(1 - \frac{b}{k^2-p^2}\right)\sin\left(pL - \frac{b^2 L}{2p(k^2-p^2)}\right)\,. \tag{5.135}$$

From this expression it follows that for a wide spectrum of neutrons, up to the lowest k_l, for which the inequality $p^2(k_l^2 - p^2) \gg b^2$ is violated, the probability of spin flip is

$$w = \sin^2(pL) = 1\,, \tag{5.136}$$

if $p = \pi/2L$. We see that this probability does not depend on the neutron speed; therefore, the flipper is really a broadband one.

5.5
The Berry Phase

A lot of papers in the literature have been devoted to the Berry phase, θ_B, which is defined as follows. Let us look at the neutron spin dynamics in the DC field B directed along the z-axis. It is determined by the Schrödinger equation:

$$i d|\boldsymbol{\Psi}(t)\rangle/dt = \sigma_z B|\boldsymbol{\Psi}(t)\rangle\,. \tag{5.137}$$

Its solution is

$$|\boldsymbol{\Psi}(t)\rangle = \exp(-i\sigma_z Bt)|\boldsymbol{\Psi}(0)\rangle\,. \tag{5.138}$$

If $|\boldsymbol{\Psi}(0)\rangle = |\xi_B\rangle$, that is, the neutron is polarized along B, then

$$|\boldsymbol{\Psi}(t)\rangle = \exp(-iBt)|\xi_B\rangle \equiv \exp(-i\phi_d(t))|\xi_B\rangle\,. \tag{5.139}$$

The value $\phi_d(t) = Bt$ is called the "dynamical phase".

If the field B itself rotates with small angular speed $2\omega \ll B$, as shown in Figure 5.7, then the spin state $|\boldsymbol{\Psi}(t)\rangle$ after one revolution, that is, after the period $T = 2\pi/2\omega$ becomes

$$|\boldsymbol{\Psi}(T)\rangle = \exp(-i\phi_d(T) - i\theta_B)|\xi_B\rangle\,, \tag{5.140}$$

that is, the phase factor acquires an additional term θ_B called the Berry phase. Berry proved a general theorem that this phase is equal to half of the solid angle subtended by the field $\boldsymbol{B}$ during its own motion. This solid angle, according to Figure 5.7, is equal to $\pi b^2/B^2$, where $b = \sqrt{B^2 - B_z^2}$, where B_z is the z-component of the field $\boldsymbol{B}$.

To understand how the Berry phase appears, it is sufficient to consider the dynamics of the neutron spin in the standard coil of an RF spin flipper [260], as shown on page 219. In this case we have a homogeneous DC field B_z directed along the z-axis and perpendicular to the RF field $\boldsymbol{B}_{\rm rf}(t)$ rotating with angular speed 2ω. This RF field plays the role of the component b of the field $\boldsymbol{B}$ in Figure 5.7. The exact solution shows that the Berry phase is not a fundamental or a "mystical" entity, but rather the result of a linear approximation in terms of ω of the total phase ϕ,

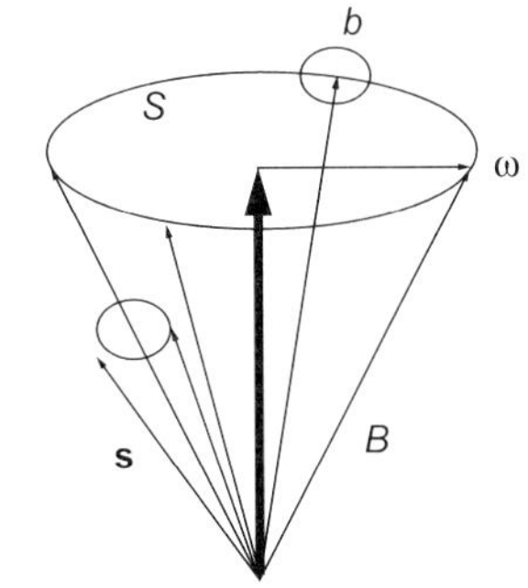

Figure 5.7 The definition of the Berry phase. The neutron spin arrow s precesses with the Larmor frequency $\omega_L = 2B$ around magnetic field $\boldsymbol{B}$, which slowly rotates around the z-axis with angular speed 2ω. After period $T = 2\pi/2\omega$ the magnetic field returns to its initial direction. The angle of the spin-arrow precession ϕ is $2(\varphi_d + \theta_B) = 2(BT + S/2B^2)$. The first term, BT, is the dynamical phase, which is related to the spin precession around standing field B. The second term, $\theta_B = S/2B^2$, is the Berry phase. The ratio S/B^2, where $S = \pi b^2$, where $b = \sqrt{B^2 - B_z^2}$, is the solid angle subtended by the field $\boldsymbol{B}$ during one revolution.

which can be found in this case absolutely rigorously. Below we shall find the precise phase ϕ, and show that its linear term in expansion in powers of ω gives the Berry phase.

The dynamics of the spin state for an arbitrary initial state $|\xi_0\rangle$ is given by Eq. (5.114)

$$|\xi(t)\rangle = e^{-i\omega\sigma_z t} e^{-i\boldsymbol{\Omega}_0\cdot\boldsymbol{\sigma} t}|\xi_0\rangle\,, \tag{5.141}$$

where $\boldsymbol{\Omega}_0 = (B_{\rm rf}, 0, B_z - \omega)$ is the vector of the effective field in the reference frame rotating with angular speed ω. The RF field in this reference frame is permanently directed along the$x$-axis, and the DC field B_z is partly compensated by the magnetic-like field ω.

For the derivation of the precise angle of rotation of the spin arrow in the rotating field, we suppose that initially the neutron is polarized along an arbitrary unit vector $\boldsymbol{a}$, that is, its state can be represented as

$$|\xi_0\rangle \equiv |\boldsymbol{a}(0)\rangle = \frac{I + \boldsymbol{\sigma}\cdot\boldsymbol{a}}{\sqrt{2(1+\boldsymbol{a}\cdot\boldsymbol{o})}}|\boldsymbol{o}\rangle\,, \tag{5.142}$$

where

$$|\boldsymbol{o}\rangle = \frac{I + \boldsymbol{\sigma}\cdot\boldsymbol{o}}{\sqrt{2(1+o_z)}}|\xi_u\rangle\,, \tag{5.143}$$

$\boldsymbol{o} = \boldsymbol{\Omega}/\Omega$, $|\xi_u\rangle$ is the eigenvector of the matrix σ_z: $\sigma_z|\xi_u\rangle = |\xi_u\rangle$, and $|\boldsymbol{o}\rangle$ is the eigenvector of the matrix $\boldsymbol{\sigma}\cdot\boldsymbol{\Omega}$ with eigen value Ω: $\boldsymbol{\sigma}\cdot\boldsymbol{\Omega}|\boldsymbol{o}\rangle = \Omega|\boldsymbol{o}\rangle$. After substitution of Eq. (5.142) into Eq. (5.141) we obtain

$$\begin{aligned}
|\xi(t)\rangle &= e^{-i\omega\sigma_z t} e^{-i\boldsymbol{\sigma}\cdot\boldsymbol{\Omega}(0)t}|\boldsymbol{a}(0)\rangle \\
&= e^{-i\omega\sigma_z t} e^{-i\boldsymbol{\sigma}\cdot\boldsymbol{\Omega}(0)t}\frac{I + \boldsymbol{\sigma}\cdot\boldsymbol{a}}{\sqrt{2(1+\boldsymbol{a}\cdot\boldsymbol{o})}} e^{i\boldsymbol{\sigma}\cdot\boldsymbol{\Omega}(0)t} e^{-i\boldsymbol{\sigma}\cdot\boldsymbol{\Omega}(0)t}|\boldsymbol{o}\rangle \\
&= e^{-i\Omega t} e^{-i\omega\sigma_z t}\frac{I + \boldsymbol{\sigma}\cdot\boldsymbol{a}_\sigma(t)}{\sqrt{2(1+\boldsymbol{a}\cdot\boldsymbol{o})}} e^{i\omega\sigma_z t} e^{-i\omega\sigma_z t} \\
&\quad \frac{I + \boldsymbol{\sigma}\cdot\boldsymbol{o}}{\sqrt{2(1+o_z)}} e^{i\omega\sigma_z t} e^{-i\omega\sigma_z t}|\xi_u\rangle \\
&= e^{-i\Omega t - \omega t}\frac{I + \boldsymbol{\sigma}\cdot\boldsymbol{a}_{\sigma(t)}(t)}{\sqrt{2(1+\boldsymbol{a}\cdot\boldsymbol{o})}}\frac{I + \boldsymbol{\sigma}\cdot\boldsymbol{o}(t)}{\sqrt{2(1+o_z)}}|\xi_u\rangle = e^{-i\phi(t)}|\boldsymbol{a}(t)\rangle\,,
\end{aligned} \tag{5.144}$$

where $\boldsymbol{o}(t) = \left(b\cos(2\omega t), b\sin(3\omega t), B_z - \omega\right)$ is vector of the effective field $\boldsymbol{\Omega}$ rotating around z-axis with angular speed 2ω. The state $|\boldsymbol{a}(t)\rangle$ corresponds to the spin arrow which rotates with angular speed 2Ω around vector $\boldsymbol{\Omega}$ and together with it around z-axis with the angular speed 2ω. During the time t the total rotation angle of $\boldsymbol{a}(t)$ around $\boldsymbol{\Omega}(t)$ and around z-axis is $2\phi(t)$, where

$$\phi(t) = (\omega + \Omega)t \tag{5.145}$$

is the total numerical phase of the spinor state (5.144) at time t.

The phase $\phi(t)$ can be expanded in powers of the small parameter ω, and in a linear approximation we obtain

$$\omega + \Omega \approx B + \frac{\omega}{B}(B - B_z) \approx B + \frac{B_{\rm rf}^2}{2B^2}\omega\,, \tag{5.146}$$

where $B = \sqrt{B_z^2 + B_{\rm rf}^2}$, and in the last term we assumed $B_{\rm rf}^2/B^2 \ll 1$, so $B_z = \sqrt{B^2 - B_{\rm rf}^2} \approx B - B_{\rm rf}^2/2B$. Therefore,

$$\varphi(t) \approx \left(B + \frac{B_{\rm rf}^2}{2B^2}\omega\right)t = \varphi_d(t) + \varphi_g(t)\,, \tag{5.147}$$

where the second term $\varphi_g(t) = B_{\rm rf}^2\omega t/2B^2$ is called the "geometrical phase" or the "topological phase". After the time of one cycle, $t = T = 2\pi/2\omega$, the geometrical phase becomes the Berry phase,

$$\theta_B = \varphi_g(T) = \frac{1}{2}\frac{\pi B_{\rm rf}^2}{B^2}\,. \tag{5.148}$$

It is equal to the solid angle subtended by the magnetic field during one cycle and multiplied by the spin quantum number 1/2.

If $\omega \gg B_z$ (we call it the "counteradiabatic" case), the phase $\varphi(t)$ can be expanded in powers of B_z. For $B_{\rm rf} \ll B_z$ in the linear approximation over $1/\omega$ we get

$$\varphi(t) \approx 2\omega t - B_z t + \frac{B_{\rm rf}^2}{2\omega}t\,. \tag{5.149}$$

Thus we see that in the above example we do not need such a notion as the Berry phase, because we can find the phase of the neutron wave function and the angle of rotation of its spin arrow absolutely precisely. However in the cases in which we cannot obtain a rigorous solution, we must use some approximation, and in such cases the notion of the Berry phase as a correction to the approximate dynamical phase and precession angle can be useful.

The Berry phase was found to be important, for instance, in experiments [261] with ultracold neutrons searching for the neutron electric dipole moment. It gives a false effect; therefore, it must be excluded. An exciting story about how it is excluded and how the rotation of Earth becomes involved in this exclusion can be found in [260].

5.6 Inelastic Interaction of Neutrons with an RF Field

Equation (5.115), though describing the spin flip, does not say anything about the changes in the energy and momentum of the neutron when it interacts with the field (which is expected, as we neglected the changes in the speed of the neutron when it enters and exits the fields). When we want think about what happens to the neutron energy upon the interaction with the fields, the following uncertainty arises. On one hand, we can assume that the neutron absorbs or emits an RF quantum, and the energy of the neutron changes by $\pm 2\omega$. On the other hand, we can assume that the kinetic energy remains unchanged upon the spin flip in the field B_0, and that the potential energy changes by $\pm 2B_0$; therefore, the total interaction energy changes by $\pm 2B_0$. In the resonant conditions, $\omega = B_0$, both assumptions lead to the same result. However, if $\omega \neq B_0$, the results are different. Below, I will show that the change in the total energy is defined entirely by the energy of the RF quantum, and not by B_0.

5.6.1 Presumable Form of the Wave Function

To prove the statement in the previous sentence, we need to study the complete dynamics of the interactions of neutron with fields, that is, we need to solve the Schrödinger equation (5.106) using the incident wave Eq. (5.108) with an arbitrary polarization $|\xi_0\rangle$. Before we start solving the complete equation, we can use some simple arguments to guess the final form of the wave function after the neutron exits the resonant spin flipper.

If initially the neutron with energy E_0 is polarized along the constant internal field, its incident wave function $\exp(ik_0x - iE_0t)|\xi_u\rangle$ can be written as

$$\exp(i\hat{\boldsymbol{k}}x - iE_0t - i\omega(\sigma_z - 1)t)|\xi_u\rangle\,, \tag{5.150}$$

where $\hat{\boldsymbol{k}} = \sqrt{k_0^2 + 2\omega(\sigma_z - 1)}$. The extra term $\propto \omega(\sigma_z - 1)$ does not matter, since $\sigma_z|\xi_u\rangle = 1$.

After the neutron has passed the rotator, the wave function becomes

$$|\Psi(x,t)\rangle = \exp\left(i\hat{\boldsymbol{k}}x - iE_0t - i\omega(\sigma_z - 1)t\right)\hat{\boldsymbol{\tau}}|\xi_u\rangle\,. \tag{5.151}$$

The transmission matrix may be nondiagonal; therefore, in Eq. (5.151) we can find the state $\tau_{du}|\xi_d\rangle$ with matrix element τ_{du}. For this part of the wave function $\sigma_z = -1$, and the extra term $\omega(\sigma_z - 1)$ becomes -2ω. Therefore, the energy of the state after the spin flip is $E_0 - 2\omega$, and its momentum is $k_d = \sqrt{k_0^2 - 4\omega}$, which means that after the spin flip the neutron becomes slower, that is, during the spin flip it emits an RF photon.

As follows from the Rabi formula, where $\boldsymbol{\Omega} = (B_{\rm rf}, 0, B_0 - \omega)$, and T is the time of flight through the rotator, in the case of resonance, the transmission amplitude matrix can be represented as

$$\hat{\boldsymbol{\tau}} = \exp(-i\sigma_x\phi) , \tag{5.152}$$

where $\phi = B_{\text{rf}}T$, and it converts $|\xi_u\rangle$ into a superposition of two states, $\hat{\boldsymbol{\tau}}|\xi_u\rangle = \cos\phi|\xi_u\rangle - i\sin\phi|\xi_d\rangle$. Substituting (5.152) into Eq. (5.151) shows that the wave function is transformed to

$$|\Psi(x,t)\rangle = \cos\phi\exp(ik_0x - iE_0t)|\xi_u\rangle - i\sin\phi\exp(ik'x - i(E_0-2\omega)t)|\xi_d\rangle , \tag{5.153}$$

where $k' = \sqrt{k_0^2 - 4\omega}$.

One can see that after the rotator the neutron with probability $\cos^2\phi$ retains its initial polarization and energy, and with probability $\sin^2\phi$, its spin is flipped and the energy is reduced to $E_0 - 2\omega$.

If initially the neutron is polarized against the field, its incident wave function is

$$\exp(ik_0x - iE_0t)|\xi_d\rangle \equiv \exp(i\hat{\boldsymbol{k}}x - iE_0t - i\omega(\sigma_z + 1)t)|\xi_d\rangle ,$$

where $\hat{\boldsymbol{k}} = \sqrt{k_0^2 + 2\omega(\sigma_z + 1)}$, and the extra term $\omega(\sigma_z+1)$ again does not matter, because in the $|\xi_d\rangle$ state $\sigma_z = -1$. After the neutron has passed the rotator, its wave function becomes

$$|\Psi(x,t)\rangle = \exp(i\hat{\boldsymbol{k}}x - iE_0t - i\omega(\sigma_z + 1)t)\hat{\boldsymbol{\tau}}|\xi_d\rangle , \tag{5.154}$$

and with the same transmission matrix, $\hat{\boldsymbol{\tau}}$ (5.152), this wave function is represented by the superposition of two states

$$|\Psi(x,t)\rangle = \cos\phi\exp(ik_0x - iE_0t)|\xi_d\rangle - i\sin\phi\exp(ik'x - i(E_0+2\omega)t)|\xi_u\rangle , \tag{5.155}$$

where in the up-state the extra term $\omega(\sigma_z + 1)$ was replaced by 2ω, which leads to an increase of the energy by 2ω and an increase of the momentum from k_0 to $k' = \sqrt{k_0^2 + 4\omega}$.

Thus, the neutron retains the same polarization and energy with probability $\cos^2\phi$, and its spin is flipped with a probability $\sin^2\phi$. In the flipped state the neutron energy and velocity increase, which means that during the spin flip the neutron absorbs an RF photon.

When the incident neutron is arbitrarily polarized,

$$|\xi_0\rangle = \alpha_u|\xi_u\rangle + \alpha_d|\xi_d\rangle$$

it is logical to assume that after the rotator the neutron becomes a superposition of three states. With probability $\cos^2\phi$ the neutron passes the flipper without a change of energy, with probability $|\alpha_u\sin\phi|^2$ its state becomes $|\xi_d\rangle$ with energy $E_0-2\omega$, and with probability $|\alpha_d\sin\phi|^2$ its state becomes $|\xi_u\rangle$ with energy $E_0+2\omega$ and speed $k_d' = \sqrt{k_0^2 - 4\omega}$.

5.6.2
The Krüger Problem

Above we guessed the wave function using the assumption that the RF field changes the neutron energy by $\pm 2\omega$ upon the spin flip in the rotator. Now we can verify that our guess was correct by solving the nonstationary Schrödinger equation (5.108) with $\boldsymbol{B}_0$ and $\boldsymbol{B}_{\mathrm{rf}}$ localized in the volume of the spin flipper.

Krüger [262] considered the problem with magnetic fields Eq. (5.107) to show that the RF field can be used to accelerate the neutron. The idea is the following. Imagine that the incident neutron is polarized oppositely to $\boldsymbol{B}_0$. Inside the fields, $0 \leq x \leq D$, the velocity of the neutron is increased, because the magnetic interaction for such a neutron is attractive. The RF field flips the spin, but does not affect the speed. Upon the spin flip, the interaction becomes repulsive. Thus, when the neutron exits the fields, its speed increases, that is, after the flipper the neutron is accelerated. Notice that according to this logic the energy of the neutron is increased by $2B_0$.

If the incident neutron is polarized along the constant field, then it is decelerated. Such deceleration was proposed 20 years before Krüger by Drabkin and Zjitnikov as a way to generate ultracold neutrons [263].

To find the probability of acceleration or deceleration for an arbitrary polarization of the incident neutron, we need to find the total wave function inside and outside the magnetic fields and to match it at the interfaces. For such an incident neutron, we obtain four reflected and four transmitted waves (total of eight unknown coefficients describing waves with different energies and polarizations). Two reflected and two transmitted waves have the same polarization as the incident wave, and the other two reflected and transmitted waves have their polarization flipped. One reflected and one transmitted waves correspond to the accelerated neutrons (they absorbed an RF quantum) and the other two correspond to the decelerated neutron (they emitted an RF quantum).

The wave function inside the volume $0 \leq x \leq D$ consists in total of eight waves propagating to the right or to the left. Upon matching this wave function at the interfaces, one obtains a system of 16 linear equations with 16 unknowns. There is no principal difficulty in solving such a system, but the analytical solution is quite cumbersome. Krüger himself found only an approximate solution, which helped him to prove that acceleration does really take place. Later, several authors tried to tackle this problem [264, 265], but they also limited themselves to some kind of an approximation.

Here we show how to find the rigorous analytical solution to the Krüger problem without cumbersome calculations. The solution will contain all the information about the transmitted and reflected waves, and we will have to use the Pauli matrix algebra (see Section 5.1.2.1) to find all the desired coefficients.

5.6.2.1 Solution to the Krüger Problem

Let us consider the Schrödinger equation again:

$$i\frac{\partial}{\partial t}|\boldsymbol{\Psi}(x,t)\rangle = \left\{-\frac{1}{2}\frac{\partial^2}{\partial x^2} + \left[\frac{u}{2} + \boldsymbol{\sigma}\cdot\boldsymbol{B}_0 + \boldsymbol{\sigma}\cdot\boldsymbol{B}_{\text{rf}}(t)\right]\Theta(0 \le x \le D)\right\} |\boldsymbol{\Psi}(x,t)\rangle\,, \tag{5.156}$$

where $\boldsymbol{B}_0 = (0, 0, B_0)$ and $\boldsymbol{B}_{\text{rf}}(t) = B_{\text{rf}}(\cos(2\omega t), \sin(2\omega t), 0)$ are localized inside $0 \le x \le D$, and u is the optical potential. Outside $x \in [0, D]$ there is an empty field-free space.

The solution to Eq. (5.156) is complicated by the time dependence of the field. This time dependence can be eliminated by the already-known transformation

$$\boldsymbol{\sigma}\cdot\boldsymbol{B}_{\text{rf}}(t) = B_{\text{rf}}[\sigma_x\cos(2\omega t) + \sigma_y\sin(2\omega t)] = B_{\text{rf}}e^{-i\omega\sigma_z t}\sigma_x e^{i\omega\sigma_z t}\,, \tag{5.157}$$

and substitution

$$|\boldsymbol{\Psi}(x,t)\rangle = e^{-i\omega\sigma_z t}|\boldsymbol{\phi}(x,t)\rangle\,. \tag{5.158}$$

As a result, Eq. (5.156) becomes

$$i\frac{\partial}{\partial t}|\boldsymbol{\phi}(x,t)\rangle = \left\{-\frac{1}{2}\frac{\partial^2}{\partial x^2} - \omega\sigma_z + \left[\frac{u}{2} + \sigma_z B_0 + \sigma_x B_{\text{rf}}\right]\Theta(0 \le x \le D)\right\} |\boldsymbol{\phi}(x,t)\rangle\,. \tag{5.159}$$

The above equation contains no time dependence; however, our manipulations introduced in the whole space an additional field $-\omega$, antiparallel to $\boldsymbol{B}_0$. The solution to Eq. (5.159) is equivalent to the solution to the problem with reflection and transmission by a magnetized mirror, with its magnetization noncollinear to the external field (Section 5.2.1).

The stationary solution to Eq. (5.159) for the incident plane wave with an arbitrary spinor $|\xi_0\rangle$ is

$$|\boldsymbol{\phi}(x,t,E)\rangle = e^{-iEt}\left\{\Theta(x<0)\left[e^{i\hat{k}x} + e^{-i\hat{k}x}\hat{\boldsymbol{\rho}}\right] + \Theta(x>D)e^{i\hat{k}(x-D)}\hat{\boldsymbol{\tau}}\right\} |\xi_0\rangle\,, \tag{5.160}$$

where E is the energy parameter (not the energy itself yet), $\hat{\boldsymbol{k}} = \sqrt{2(E+\sigma_z\omega)}$, and $\hat{\boldsymbol{\rho}}$, and $\hat{\boldsymbol{\tau}}$ are the matrix reflection and transmission amplitudes for a layer D containing the matter and the fields. We only included the solution outside the fields as we are not interested in the solution inside the fields for now.

Substitution of Eq. (5.160) into Eq. (5.158) gives

$$|\boldsymbol{\Psi}(x,t,E)\rangle = e^{-iEt-i\omega\sigma_z t}\Big\{\Theta(x<0)\left[e^{i\hat{\boldsymbol{k}}x}+e^{-i\hat{\boldsymbol{k}}x}\hat{\boldsymbol{\rho}}\right] + \Theta(x>D)e^{i\hat{\boldsymbol{k}}(x-D)}\hat{\boldsymbol{\tau}}\Big\}|\xi_0\rangle\,. \tag{5.161}$$

Let us analyze the solution obtained in the case of an arbitrary spinor:

$$|\xi_0\rangle = \alpha_u|\xi_u\rangle + \alpha_d|\xi_d\rangle\,. \tag{5.162}$$

By substituting Eq. (5.162) into Eq. (5.161), we see that the incident wave in Eq. (5.161),

$$|\boldsymbol{\Psi}_0(x,t,E)\rangle = e^{-iEt-i\omega\sigma_z t}\Theta(x<0)e^{i\hat{\boldsymbol{k}}x}|\xi_0\rangle\,, \tag{5.163}$$

is a superposition of two plane waves with opposite polarizations $|\xi_{u,d}\rangle$ and different energies,

$$|\boldsymbol{\Psi}_0(x,t,E)\rangle = \alpha_u e^{-i(E+\omega)t}e^{ik_+x}|\xi_u\rangle + \alpha_d e^{-i(E-\omega)t}e^{ik_-x}|\xi_d\rangle\,. \tag{5.164}$$

The energy and the momentum of the component $|\xi_u\rangle$ are $E+\omega$ and $k_+ = \sqrt{2(E+\omega)}$, respectively, and for $|\xi_d\rangle$ they are $E-\omega$ and $k_- = \sqrt{2(E-\omega)}$. To find a solution for a neutron with a fixed energy E_0 and an arbitrary polarization Eq. (5.162), we need to take a superposition of the two waves Eq. (5.160) with $E = E_0-\omega$ for $|\xi_u\rangle$ and with $E = E_0+\omega$ for $|\xi_d\rangle$. Let us rewrite the solution to Eq. (5.159) as

$$\begin{aligned}|\boldsymbol{\phi}(x,t)\rangle = {}& \alpha_u e^{-i(E_0-\omega)t}\Big\{\Theta(x<0)\left[e^{i\hat{\boldsymbol{k}}_u x}+e^{-i\hat{\boldsymbol{k}}_u x}\hat{\boldsymbol{\rho}}_u\right] \\ & + \Theta(x>D)e^{i\hat{\boldsymbol{k}}_u(x-D)}\hat{\boldsymbol{\tau}}_u\Big\}|\xi_u\rangle \\ & a_d e^{-i(E_0+\omega)t}\Big\{\Theta(x<0)\left[e^{i\hat{\boldsymbol{k}}_d x}+e^{-i\hat{\boldsymbol{k}}_d x}\hat{\boldsymbol{\rho}}_d\right] \\ & + \Theta(x>D)e^{i\hat{\boldsymbol{k}}_d(x-D)}\hat{\boldsymbol{\tau}}_d\Big\}|\xi_d\rangle\,,\end{aligned} \tag{5.165}$$

where $\hat{\boldsymbol{k}}_{u,d} = \sqrt{k_0^2+2\omega(\sigma_z\mp 1)}$, $k_0^2 = 2E_0$, and substitute it into Eq. (5.158). Then,

$$\begin{aligned}|\boldsymbol{\Psi}(x,t)\rangle = {}& \alpha_u e^{-iE_0t-i\omega(\sigma_z-1)t}\Big\{\Theta(x<0)\left[e^{i\hat{\boldsymbol{k}}_u x}+e^{-i\hat{\boldsymbol{k}}_u x}\hat{\boldsymbol{\rho}}_u\right] \\ & + \Theta(x>D)e^{i\hat{\boldsymbol{k}}_u(x-D)}\hat{\boldsymbol{\tau}}_u\Big\}|\xi_u\rangle \\ & + \alpha_d e^{-iE_0t-i\omega(\sigma_z+1)t}\Big\{\Theta(x<0)\left[e^{i\hat{\boldsymbol{k}}_d x}+e^{-i\hat{\boldsymbol{k}}_d x}\hat{\boldsymbol{\rho}}_d\right] \\ & + \Theta(x>D)e^{i\hat{\boldsymbol{k}}_d(x-D)}\hat{\boldsymbol{\tau}}_d\Big\}|\xi_d\rangle\,.\end{aligned} \tag{5.166}$$

This solution fixes the energy E_0 and the velocity k_0 for an arbitrary polarization $|\xi_0\rangle$ of the incident wave:

$$|\boldsymbol{\Psi}_0(x,t)\rangle = \alpha_u e^{-iE_0t - i\omega(\sigma_z-1)t} e^{i\hat{k}_u x}|\xi_u\rangle + \alpha_d e^{-iE_0t - i\omega(\sigma_z+1)t} e^{i\hat{k}_d x}\xi_d$$
$$= e^{ik_0x - iE_0t}|\xi_0\rangle\,. \tag{5.167}$$

Note that the change in the neutron energy upon the polarization flip is 2ω, not the expected $2B_0$. In addition, we guessed correctly the form of the outgoing wave function after the interaction with the spin rotator.

5.6.2.2 Angular Splitting of Reflected and Transmitted Neutrons

Here we show that if the space with fields is a plate, the reflection and transmission of an unpolarized neutron beam is accompanied by triple splitting [270] of the beam like in the case of reflection from a magnetic mirror with magnetization noncollinear to the external field. Indeed, the normal component, $k_\perp$, of the neutron wave vector after spin flip $d \to u$ increases from $k_{0\perp}$ to $\sqrt{k_{0\perp}^2 + 4\omega}$; therefore, the neutrons after such a spin flip will go at an angle closer to the normal to the interface, and they will be completely polarized along the internal field $\boldsymbol{B}_0$, as shown in Figure 5.8.

On the other side, after spin flip $u \to d$ the normal component of the neutron wave vector decreases to $\sqrt{k_{0\perp}^2 - 4\omega}$. Therefore, the neutrons after such a spin flip will go at an angle closer to the surface, and they will be completely polarized against the internal field $\boldsymbol{B}_0$.

Between these two beams there will be an unpolarized one consisting of the neutrons which passed the flipper without a spin flip.

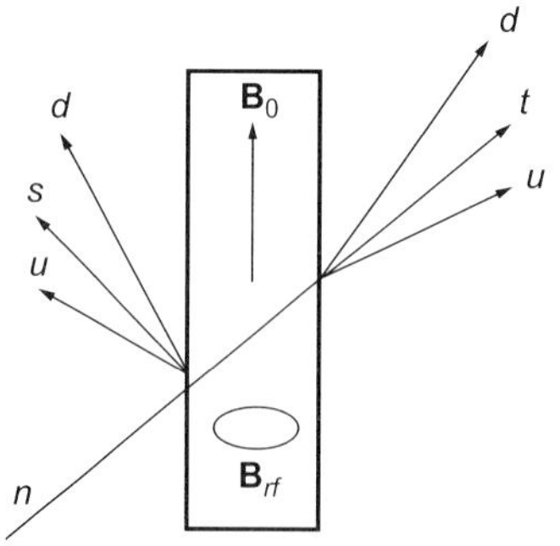

Figure 5.8 Triple splitting of the neutron beam n after interaction with a radiofrequency (RF) spin flipper. The normal component $k_\perp$ of the wave vector $\boldsymbol{k}$ of the incident beam will transform to $k_{u\perp} = \sqrt{k_\perp^2 + 4\omega}$ after spin flip $d \to u$, and will transform to $k_{d\perp} = \sqrt{k_\perp^2 - 4\omega}$ after spin flip $u \to d$. The beam components denoted by u, d are completely polarized along and opposite internal field $\boldsymbol{B}_0$ of the spin flipper. The specularly reflected, s, and directly transmitted, t, components contain those neutrons which did not change their polarization after interaction with RF field $\boldsymbol{B}_{\rm rf}$.

The spin flip is a coherent process; therefore, the neutron initially polarized along $\boldsymbol{B}_0$ after transmission will be in general in a superposition of states going along paths d and t in Figure 5.8. What the real distribution of neutrons after such a spin flipper will be is a fundamental question, which has not yet been tackled by an experiment.

5.6.2.3 Reflection and Transmission Matrices

The wave function (5.166) contains matrices with reflection, $\hat{\boldsymbol{\rho}}_{u,d}$, and transmission, $\hat{\boldsymbol{\tau}}_{u,d}$, amplitudes. We already derived them in Section 5.2.1 for an arbitrary polarization and fixed energy of the incident neutron. Currently, we need to derive these matrices separately for the two opposite polarizations of the incident neutron. We denote them $\hat{\boldsymbol{\rho}}_u$ and $\hat{\boldsymbol{\tau}}_u$ for initial polarization along $\boldsymbol{B}_0$, and $\hat{\boldsymbol{\rho}}_d$ and $\hat{\boldsymbol{\tau}}_d$ for initial polarization along $-\boldsymbol{B}_0$. Following the logic of Section 5.2.1, we obtain

$$\hat{\boldsymbol{\rho}}_{u,d} = \hat{\boldsymbol{\rho}}_{b0}^{u,d} + \hat{\boldsymbol{\tau}}_{0b}^{u,d} \frac{\exp\left(i\hat{\boldsymbol{k}}'_{u,d}D\right)\hat{\boldsymbol{\rho}}_{0b}^{u,d}}{1-\left[\exp\left(i\hat{\boldsymbol{k}}'_{u,d}D\right)\hat{\boldsymbol{\rho}}_{0b}^{u,d}\right]^2} \exp\left(i\hat{\boldsymbol{k}}'_{u,d}D\right)\hat{\boldsymbol{\tau}}_{b0}^{u,d}, \tag{5.168}$$

$$\hat{\boldsymbol{\tau}}_{u,d} = \hat{\boldsymbol{\tau}}_{0b}^{u,d} \frac{1}{1-\left[\exp\left(i\hat{\boldsymbol{k}}'_{u,d}D\right)\hat{\boldsymbol{\rho}}_{0b}^{u,d}\right]^2} \exp\left(i\hat{\boldsymbol{k}}'_{u,d}D\right)\hat{\boldsymbol{\tau}}_{b0}^{u,d}, \tag{5.169}$$

where

$$\hat{\boldsymbol{\rho}}_{b0}^{u,d} = \left(\hat{\boldsymbol{k}}_{u,d} + \hat{\boldsymbol{k}}'_{u,d}\right)^{-1}\left(\hat{\boldsymbol{k}}_{u,d} - \hat{\boldsymbol{k}}'_{u,d}\right) = -\hat{\boldsymbol{\rho}}_{0b}^{u,d}, \tag{5.170}$$

$$\hat{\boldsymbol{\tau}}_{b0}^{u,d} = I + \hat{\boldsymbol{\rho}}_{b0}^{u,d}, \quad \hat{\boldsymbol{\tau}}_{0b}^{u,d} = I + \hat{\boldsymbol{\rho}}_{0b}^{u,d}, \tag{5.171}$$

$$\hat{\boldsymbol{k}}_{u,d} = \sqrt{k_0^2 + 2\omega(\sigma_z \mp 1)}, \quad \hat{\boldsymbol{k}}'_{u,d} = \sqrt{k_0^2 \mp 2\omega - 2\boldsymbol{\sigma}\cdot\boldsymbol{\Omega} - u}, \tag{5.172}$$

and indices $b0, 0b$ correspond to the interface between the vacuum and the barrier from the vacuum and from the barrier (right index), respectively. Using the algebra of Pauli matrices, we can find all the matrix elements $\hat{\boldsymbol{\rho}}_{u,d}$ and $\hat{\boldsymbol{\tau}}_{u,d}$. The diagonal matrix elements define the probabilities of the reflection and transmissions without the spin flip, whereas the nondiagonal ones are for flipped spin states.

Any 2×2 matrix M can be written as $M = a\hat{\boldsymbol{I}} + \boldsymbol{b}\cdot\boldsymbol{\sigma}$, where $\hat{\boldsymbol{I}}$ is the unit matrix, and a and $\boldsymbol{b}$ are parameters. Transitions without the slip flip are described by $a\hat{\boldsymbol{I}} + b_z\sigma_z$, and the transitions with the spin flip are described by $b_x\sigma_x + b_y\sigma_y$.

Consider the structure of the reflected waves when the incident wave is Eq. (5.167) with polarization Eq. (5.162). The reflected wave, which corresponds to the elastic scattering (i. e., without the spin flip), contains the spinor

$$|\xi_r^{\mathrm{el}}\rangle = \alpha_u \rho_{uu}|\xi_u\rangle + \alpha_d \rho_{dd}|\xi_d\rangle, \tag{5.173}$$

where

$$\rho_{uu} = \langle\xi_u|\hat{\boldsymbol{\rho}}_u|\xi_u\rangle, \quad \rho_{dd} = \langle\xi_d|\hat{\boldsymbol{\rho}}_d|\xi_d\rangle.$$

The spinor Eq. (5.173) is different from the incident spinor $|\xi_0\rangle$, and thus the polarization of the reflected wave (without the spin flip) is different from the incident polarization, but the reflected wave propagates with the same energy and speed as the incident one:

$$\begin{aligned}|\boldsymbol{\Psi}_r^{\mathrm{el}}(x)\rangle &= \rho_{uu}\alpha_u e^{-iE_0t-i\omega(\sigma_z-1)t}e^{-i\hat{k}_u x}|\xi_u\rangle \\ &\quad + \rho_{dd}\alpha_d e^{-iE_0t-i\omega(\sigma_z+1)t}e^{-i\hat{k}_d x}\xi_d \\ &= e^{-ik_0x-iE_0t}|\xi_r^{\mathrm{el}}\rangle\,. \end{aligned} \tag{5.174}$$

The other part of the wave corresponds to inelastic scattering, that is, to scattering with spin flip,

$$\begin{aligned}|\boldsymbol{\Psi}_r^{in}(x)\rangle &= \rho_{du}\alpha_u e^{-iE_0t-i\omega(\sigma_z-1)t}e^{-i\hat{k}_u x}|\xi_d\rangle + \rho_{ud}\alpha_d e^{-iE_0t-i\omega(\sigma_z+1)t} \\ &\quad \times e^{-i\hat{k}_d x}|\xi_u\rangle \\ &= \rho_{du}\alpha_u e^{-i(E_0-2\omega)t-ik_-x}|\xi_d\rangle + \rho_{ud}\alpha_d e^{-i(E_0+2\omega)t-ik_+x}|\xi_u\rangle\,, \end{aligned} \tag{5.175}$$

where

$$\rho_{du} = \langle\xi_d|\hat{\boldsymbol{\rho}}_u|\xi_u\rangle\,,\qquad \rho_{ud} = \langle\xi_u|\hat{\boldsymbol{\rho}}_d|\xi_d\rangle\,,$$

$\boldsymbol{k}_\pm = \sqrt{2(E_0 \pm 2\omega)}$, $E_0 = k_0^2/2$. The first term in (5.175) corresponds to the spin flip $|\xi_u\rangle \to |\xi_d\rangle$, accompanied by emission of the RF quantum. Thus, the output neutron slows down. The second term corresponds to the spin flip $|\xi_d\rangle \to |\xi_u\rangle$ and to absorption of the RF quantum.

The wave function of the transmitted neutron has a similar structure. Note that the change in energy after emission or absorption of the RF quantum is equal to 2ω, that is, it does not depend on B_0. Only the probability of absorption and emission depends on B_0, but not the energy of the neutron spin components. This result resolves the problem which appears during the derivation of the Rabi formula (5.115).

5.6.2.4 Approximate Expressions for the Transmission Matrix Eq. (5.169)

We leave the explicit derivation of the matrix elements of Eqs. (5.168) and (5.169) to those who want to learn how to use the algebra of Pauli matrices and to those who need it for an experiment. We will limit ourselves to the simplest case, when the potential u is equal to zero, reflections can be neglected, and $B_{\mathrm{rf}} \ll k_0^2 - 2\omega$. In this case $\hat{\boldsymbol{\tau}}_{b0}^{u,d} = \hat{\boldsymbol{\tau}}_{0b}^{u,d} = \hat{\boldsymbol{I}}$ and expression (5.169) becomes

$$\hat{\boldsymbol{\tau}}_{u,d} \approx \exp\left(i\hat{\boldsymbol{k}}'_{u,d}(\boldsymbol{\sigma}\cdot\boldsymbol{\Omega})D\right)\,, \tag{5.176}$$

where $\hat{\boldsymbol{k}}'_{u,d}(\boldsymbol{\sigma}\cdot\boldsymbol{\Omega}) = \sqrt{k_0^2 \mp 2\omega - 2\boldsymbol{\Omega}\cdot\boldsymbol{\sigma}}$. Using Eq. (5.26),

$$\hat{\boldsymbol{k}}'_u(\boldsymbol{\sigma}\cdot\boldsymbol{\Omega}) = K_{u,d}^+ + (\boldsymbol{\Omega}\cdot\boldsymbol{\sigma}/\Omega)K_{u,d}^-\,,$$

where $\Omega = \sqrt{B_{\rm rf}^2 + (\omega - B_0)^2}$,

$$K_{u,d}^{\pm} = \frac{1}{2}\left[k'_{u,d}(\Omega) \pm k'_{u,d}(-\Omega)\right],$$

$$k'_u(\pm\Omega) = \sqrt{k_0^2 - 2\omega \mp 2\Omega}\,, \quad k'_d(\pm\Omega) = \sqrt{k_0^2 + 2\omega \mp 2\Omega}\,, \tag{5.177}$$

we represent the matrix $\hat{\boldsymbol{\tau}}_{u,d}$ as

$$\hat{\boldsymbol{\tau}}_{u,d} = \exp\left(i\hat{\boldsymbol{k}}'_{u,d}(\boldsymbol{\sigma}\cdot\boldsymbol{\Omega})D\right) = \exp\left(iK_{u,d}^{+}D\right)\exp\left(iK_{u,d}^{-}D\frac{\boldsymbol{\sigma}\cdot\boldsymbol{\Omega}}{\Omega}\right). \tag{5.178}$$

The phase factor $\exp(iK_{u,d}^{+}D)$ can be neglected because it does not affect probabilities, and at resonance $\omega = B_0$ the transmission matrix becomes

$$\hat{\boldsymbol{\tau}}_{u,d} = \exp(-i\varphi_{u,d}\sigma_x)\,, \tag{5.179}$$

where

$$\varphi_{u,d} = -K_{u,d}^{-}D, \qquad K_{u,d}^{-} \approx \frac{-B_{\rm rf}}{\sqrt{k_0^2 \mp 2\omega}}\,.$$

From Eq. (5.178) it is easy to derive the Rabi formula and to find that the flight time through the flipper is $T_{u,d} = D/\langle k_{u,d}\rangle$, where

$$\langle k_{u,d}\rangle = \frac{\sqrt{k_0^2 \mp 2\omega + 2B_{\rm rf}} + \sqrt{k_0^2 \mp 2\omega - 2B_{\rm rf}}}{2} \approx \sqrt{k_0^2 \mp 2\omega} \tag{5.180}$$

is the speed of each spin component inside the rotator.

5.7 Games with Polarized Neutrons

Now we can consider experiments with polarized neutrons, some of which are so nice that it is tempting to call them "games." They are useful not only for condensed matter research, but also as an illustration of some quantum mechanical phenomena. Moreover, they help to shed light on some quantum effects which are worth further research.

5.7.1 Spin Wave and Intensity Modulated Wave after a $\pi/2$-Spin Flipper

Let us look at the wave function after a $\pi/2$-flipper if before it the initial spinor state was $|\xi_u\rangle$. The $\pi/2$-flipper is one for which $\varphi_u = \pi/4$. According to Eq. (5.179), the transmitted wave function is

$$\begin{aligned}\psi_t(t,x) &= e^{-iE_0t - i\omega(\sigma_z - 1)t}e^{i\hat{k}_u(x-D)}\frac{|\xi_u\rangle - i|\xi_d\rangle}{\sqrt{2}} \\ &= \frac{1}{\sqrt{2}}\left(e^{ik_0(x-D)-iE_0t}|\xi_u\rangle - ie^{ik_-(x-D)-iE_-t}|\xi_d\rangle\right),\end{aligned} \tag{5.181}$$

where $k_- = \sqrt{k_0^2 - 4\omega}$ and $E_- = E_0 - 2\omega$. The direction of the spin arrow $\mathbf{s}(x,t)$ at any time t at an arbitrary point x is

$$\begin{aligned}\mathbf{s}(x,t) &= \langle\psi_t(x,t)|\boldsymbol{\sigma}|\psi_t(x,t)\rangle \\ &= -(\sin(k_s(x-D)-\omega_s t), \cos(k_s(x-D)-\omega_s t), 0)\,, \end{aligned} \tag{5.182}$$

where $k_s = k_0 - k_-$ and $\omega_s = 2\omega$. We see that at any point x the spin arrow rotates in the plane perpendicular to the z-axis with angular frequency ω_s, and at any moment t the direction of the spin arrow periodically changes in the x, y-plane with the distance along the x-axis. Note that there are no fields after the spin rotator; nevertheless we have a wavelike precession of the neutron spin arrow after it. This wave has all the attributes of a wave: frequency ω_s and wave vector k_s. The wave is a result of the superposition of two spin states $|\xi_u\rangle$, and $|\xi_d\rangle$, having two different energies.

The superposition can be used to prepare a neutron beam with periodically changing intensity. Indeed, if at some point x_1 on the beam path after the $\pi/2$-flipper we put an analyzer, which transmits only neutrons polarized along the y-axis, the wave function at the point x_1 after the analyzer will have polarization ξ_{+y} with amplitude

$$\psi_y(t,x_1) = \frac{1}{2}\left(e^{ik_0(x_1-D)-iE_0t} - e^{ik_-(x_1-D)-iE_-t}\right)\,. \tag{5.183}$$

To see this one needs only to expand spinors $|\xi_{u,d}\rangle$ over eigenspinors $|\xi_{\pm y}\rangle$ of the matrix $\sigma_y : \sigma_y|\xi_{\pm y}\rangle = \pm|\xi_{\pm y}\rangle$:

$$\begin{aligned}|\xi_{\pm y}\rangle &= \frac{1}{\sqrt{2}}\begin{pmatrix}1\\ \pm i\end{pmatrix}\,, \\ |\xi_u\rangle &= \frac{1}{\sqrt{2}}(|\xi_{+y}\rangle + |\xi_{-y}\rangle)\,, \quad |\xi_d\rangle = \frac{1}{i\sqrt{2}}(|\xi_{+y}\rangle - |\xi_{-y}\rangle)\,. \end{aligned} \tag{5.184}$$

According to Eq. (5.183) the intensity of the transmitted neutrons

$$\begin{aligned} I &= |\psi_y(t,x_1)|^2 = \frac{1}{2}[1-\cos(\phi_1 - \omega_s t)] = \sin^2(\phi_1/2 - \omega t)]\,, \\ \phi_1 &= k_s(x_1 - D)\,, \end{aligned} \tag{5.185}$$

oscillates in time, and we obtain the beam with modulated intensity, which can be used in all optical investigations, including holography.

5.7.1.1 Classical Explanation of the Spin Wave. Wave–Particle Duality

We found that the RF $\pi/2$-spin rotator puts the spin arrow into the plane perpendicular to the DC field $\boldsymbol{B}_0$ and creates a superposition of two states with different energies. As a result, a spin precession wave (SPW) is generated. In this wave, the phase speed

$$v_{\text{ph}} = \frac{\omega_s}{k_s} = \frac{2\omega}{k_0 - \sqrt{k_0^2 - 4\omega}} = \frac{k_0 + \sqrt{k_0^2 - 4\omega}}{2} = \langle v\rangle$$

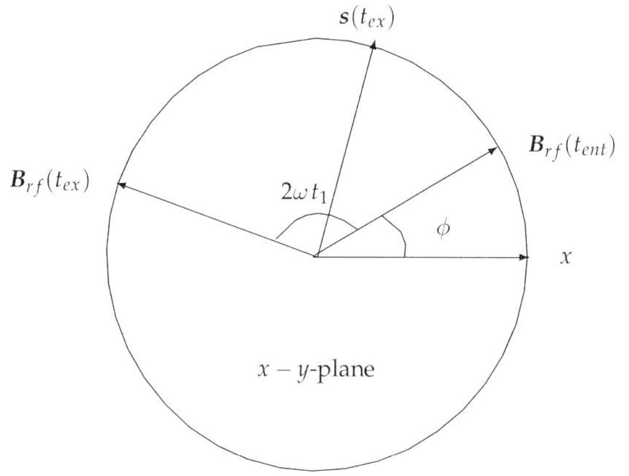

Figure 5.9 Direction of the neutron spin arrow $s(t_{ex})$ at the moment t_{ex} of its exit from the $\pi/2$-flipper. RF field B_{rf} rotates counterclockwise in the page plane, and at the moment t_{ent}, when the neutron enters the magnetic field area, the field B_{rf} has an angle ϕ with the x-axis. The permanent field B_0 and the spin arrow of the entering neutron are perpendicular to the page plane and directed toward the reader. During the flight time t_1 of the neutron through the flipper of thickness D the RF field B_{rf} turns to the position $B_{rf}(t_{ex})$, where $t_{ex} = t_{ent} + t_1$. At this moment the neutron spin arrow $s(t_{ex})$ is perpendicular to $B_{rf}(t_{ex})$. Therefore, the direction of the neutron spin arrow at the exit moment is determined by the direction of field B_{rf} at the neutron entrance moment t_{ent}.

is identical to the average speed of the two spin components, which at small $\omega \ll k_0^2$ is equal to the neutron velocity k_0. So, if we follow the single neutron moving with speed $\langle v \rangle$, we see that the spin arrow attached to it does not precess, because there is no magnetic field. But then, what does SPW mean?

The explanation is very simple, and to find it one needs to follow the neutron spin precession inside the flipper. When the neutron polarized along the DC field B_0 enters the flipper, its spin arrow starts to precess counterclockwise around RF field B_{rf}, looking down the arrow tip. After the spin arrow has deviated from the direction B_0, it starts to precess counterclockwise also around B_0. At resonance the angular precession speed around B_0 coincides with the angular speed of B_{rf} rotation. Therefore, the spin arrow remains perpendicular to B_{rf}, and when the spin falls into the plane perpendicular to B_0, in this plane it is also perpendicular to B_{rf}, as shown in Figure 5.9. Since the direction of the neutron spin arrow at the exit moment t_{ex} is determined by the position of $B_{rf}(t_{ex})$, it is determined also by the position of $B_{rf}(t_{ent})$ at the neutron entrance moment t_{ent} into the spin flipper. Different neutrons have different entrance moments t_{ent}; therefore, their spin arrows after exiting will also be different, as if the spin flipper were putting a mark on every neutron, which shows its time of its entrance into the flipper. The direction of the spin arrow of every neutron is fixed, but spin arrows of different neutrons are different. In fact, we do not have a wave, we have a periodic distribution of spin arrows of different neutrons. This can be illustrated in the following way [265].

5.7.1.2 The First Miracle of St. Mark Street

Let us imagine that at the beginning of St. Mark Street in a town N there is a gate. At the gate there is a clock with a single arm, which rotates with frequency ν. A miracle happens to every person entering the gate: a dim halo appears over his/her head with a bright spot on the halo's circumference at the place which is pointed out by the gate clock at the moment of entrance.

The people enter the street randomly, but go along it with the same speed v. If we look at the street from above, we can see the bright spots scattered along a sinusoid-like trajectory. We will call it the "bright-spot wave". The length of the wave is v/ν. Its amplitude is the radius of the halo. If we stay at some point on the street and glance at the halos of the passers-by, we will see that the bright spot (of different people) rotates with frequency ν equal to the clock's frequency, and this result does not depend on the people's speed (as long as the speed is the same for all the people going down the street). This example of the bright-spot wave illustrates the spin–wave property of the neutron beam after the $\pi/2$-flipper.

5.7.1.3 The Second Miracle of St. Mark Street

In the middle of St. Mark Street there is another gate with a clock, with single arm which points permanently at 12 o'clock. The gate opens only for those people whose bright spot in the halo is between 0 and 6 hours. The people with the bright spot between 6 and 12 hours are asked to go another way. If we look from above at the street, we will find that after the second gate the people would go in periodic bunches. These bunches are analogues of the intensity-modulated neutron beam wave after the analyzer. The superposition of two states here is the mutual action of the permanent direction of the clock hand at the middle gate and the periodic rotation of the clock hand at the first gate.

These examples illustrate not only SPW and intensity-modulated wave phenomena, but also the wave–particle duality in quantum mechanics. The main point is that there is no wave when you look at a single particle. To see a wave you need an experiment with many particles. A periodic process affects the neutron flux independently of how long an individual neutron is in the experimental device.

This illustration does not solve all the problems of quantum mechanics. It would be nice if the neutron speed were really equal to the average of the speeds of the two spin components. However, these components have different speeds and the question arises, what does it mean? Can the two components separate with time? If yes, how does it happen? What is the neutron with separated spin components? Moreover, we solved the Schrödinger equation for a single particle. How can we interpret the wave function of a single particle as an ensemble of many particles?

One more question makes all this system an enigmatic one. According to quantum mechanics, the neutron during the passage through the spin flipper can absorb or emit an RF photon. If the neutron wave function after the exit of the flipper is in a superposition of two states with different energies, what happened to the photon? Is it created in the RF coil, or not? We have some kind of entangled neutron–photon state. If we measure a neutron in a superposition, can we tell what happened to the photon? Or we should abandon the notion of the photon in such a case? We do not have answers to all these interesting questions.

5.7.2
π-Flipper

Let us go back to the transmission amplitude matrix (5.179) and suppose that $\varphi = \pi/2$. In this case the transmitted spin state is

$$\xi = \tau_u|\xi_u\rangle = -i|\xi_d\rangle\,, \tag{5.186}$$

which means a spin flip by 180°. It is necessary to note that the energy also changes. It becomes less by 2ω compared with previous one.

5.7.2.1 The Ramsey Separated Fields Method

Any π-flipper can be thought of as a combination of two consecutive $\pi/2$-flippers. Its transmission can be represented as[22)]

$$\hat{\tau}(\pi) = \hat{\tau}(\pi/2)\hat{\tau}(\pi/2)\,. \tag{5.187}$$

Let us separate two $\pi/2$-flippers by a distance L and find how the transmission will change. This separation is the main feature of the Ramsey separated fields (RSF) method. Before writing down the mathematics, let us think about what we can expect from such a separation. If no field is present between the two coils, then the spin of a neutron does not change its direction during the flight between the two flippers. The RF field inside the second coil, however, is rotating in phase with the first one with frequency 2ω during the neutron's flight. If during the time of flight T, the RF field turns by $2T\omega = 2\pi n$ degrees, where n is an integer, then the neutron entering the second $\pi/2$-flipper encounters the magnetic field $\boldsymbol{B}_{\mathrm{rf}}$ identical to the magnetic field in the first flipper at the time, t_{ex}, the neutron exited the first flipper. Thus, the second flipper works as if there is no separation, and both flippers act together as a single π-flipper.

If $2\omega T \neq 2\pi n$, the neutron spin arrow $\boldsymbol{s}(T + t_{\mathrm{ex}})$ (which is identical to $\boldsymbol{s}(t_{\mathrm{ex}})$) is not perpendicular to $\boldsymbol{B}_{\mathrm{rf}}(T + t_{\mathrm{ex}})$. If $2\omega T = 2\pi n - \alpha$, the neutron spinor is turned with respect to it by the operator $\exp(-i\sigma_z\alpha/2)$ and becomes

$$\xi = e^{-i\sigma_z\alpha/2}\frac{1}{\sqrt{2}}\left(|\xi_u\rangle - i|\xi_d\rangle\right) = e^{-i\alpha/2}\frac{1}{\sqrt{2}}\left(|\xi_u\rangle - ie^{i\alpha}|\xi_d\rangle\right)\,. \tag{5.188}$$

The action of the second $\pi/2$-flipper, which is represented by the operator $\exp(-i\sigma_x\pi/4)$, transforms Eq. (5.188) to

$$\exp(-i\sigma_x\pi/4)\xi = -i(\sin(\alpha/2)|\xi_u\rangle + \cos(\alpha/2)|\xi_d\rangle)\,. \tag{5.189}$$

22) In the following, for simplicity we omit index u when we consider only a single up-component of the incident wave.

This means that the second $\pi/2$-flipper flips the neutron spin only with probability $\cos^2(\alpha/2)$. If $\alpha = 0$, then both $\pi/2$-flippers act synchronously as a single π-flipper. If $\alpha = \pi$, then the second $\pi/2$-flipper completely cancels the effect of the first one.

The transmission operator of the entire system[23] is

$$\hat{\tau}(\pi/2, L, \pi/2) = \hat{\tau}(\pi/2)\exp(i\hat{k}L)\hat{\tau}(\pi/2)$$
$$\approx \exp(-i\sigma_x\omega_1 t_1/2)\exp[i(\sigma_z - 1)\omega T]\exp(-i\sigma_x\omega_1 t_1/2)\,, \tag{5.190}$$

where $T = L/k_0$. The common phase factor $\exp(ik_0[D + L])$ is excluded from this expression. We can exclude also the phase factor $\exp(-i\omega T)$. As a result, (5.190) is reduced to

$$\hat{\tau}_u(\pi/2, L, \pi/2) = \exp(ia\sigma_x)\exp(ib\sigma_z)\exp(ia\sigma_x)\,,$$

where $b = \omega T$. With the help of Eqs. (5.27) and (5.39) we get the state of the transmitted neutron:

$$\exp(-i\sigma_x\omega_1 t_1/2)\exp(i\sigma_z\omega T)\exp(-i\sigma_x\omega_1 t_1/2)|\xi_u\rangle$$
$$= [\exp(-i\sigma_x\omega_1 t_1)\cos(\omega T) + i\sigma_z\sin(\omega T)]|\xi_u\rangle$$
$$= [\cos(\omega_1 t_1)\cos(\omega T) + i\sin(\omega T)]|\xi_u\rangle - i\cos(\omega T)\sin(\omega_1 t_1)|\xi_d\rangle\,. \tag{5.191}$$

At $\omega_1 t_1 = \pi/2$ Eq. (5.191) is transformed to $\sin(\omega T)|\xi_u\rangle - \cos(\omega T)|\xi_d\rangle$, where we excluded the phase factor i.

We now see that if $2\omega T = 2\pi n$ or $\omega T = \pi n$ (with integer n), the initial $|\xi_u\rangle$ is flipped to $|\xi_d\rangle$. If $2\omega T = \pi(2n + 1)$, the action of the second $\pi/2$-flipper compensates the action of the first one, and the initial polarization $|\xi_u\rangle$ remains unchanged.

For arbitrary ωT, the intensity of an up-polarized neutron beam after two flippers is proportional to $\sin^2(\omega L/v)$, where T is replaced by the neutron time of flight, L/v, between the two separated coils. This means that the separated coils can be used for spectrometry. For large L/v, just a small variation of ω gives a large variation of the intensity, and thus L/v can be found very precisely. Because the value of L is usually well known and is large, the value of v can be found with high precision.

The RSF method is used in experiments to search for the neutron electric dipole moment (see, e. g., textbook [G], Chapter 8). In these experiments, a magnetic field $\boldsymbol{B}_0$ is added between the two coils, and thus the neutron's spin precesses along the flight path with frequency $2\omega_0$. To take this precession into account, we must replace $\exp(i\omega\sigma_z T)$ in Eq. (5.191) by $\exp(i(\omega-\omega_0)\sigma_z T)$. If $\omega = \omega_0$, then $\exp(i(\omega-$

23) Note there is no time, because the transmission and reflection amplitudes are defined for stationary states.

$\omega_0)\sigma_z T) = 1$, and both $\pi/2$-flippers work together as a π-flipper. The RSF method is a very sensitive method for searching for the electric dipole moment, because for a large T, we can see the effect even for a very small change in $\omega_0 - \omega$.

5.7.2.2 Tuner

We have seen that if $2\omega T = 2\pi n$, the two halves of the π-flipper act as if they were not separated. Suppose that $2\omega T$ is not exactly $2\pi n$. In this case, an additional coil of length l with a tunable magnetic field, B_t, parallel to the z-axis is placed between the two $\pi/2$-flippers. Such a coil is called a "tuner", because by variation of the field $\boldsymbol{B}_t$ it is possible to change the direction of the spin arrow in such a way that at the entrance of the second flipper it is perpendicular to RF field $\boldsymbol{B}_{\rm rf}$. Indeed, the magnetic field $\boldsymbol{B}_{rf}$ turns the spin by the angle $2\omega_t l/k_0$, where $\omega_t = B_t$.

With the tuner, the transmission amplitude $\tau(L)$ of the space L between the two $\pi/2$-flippers is $\exp(ik_o L)\exp(i\sigma_z[\omega T - \omega_t t_2])$, where $t_2 = l/k_o$. If $\omega T \neq 2\pi n$, ω_t can be tuned so that $\omega T - \omega_t t_2 = 2\pi n$. The tuner serves also as a detector of the change of $\omega - \omega_0$. Every change can be compensated by the tuner, and the required current is the measured signal.

5.7.2.3 Echo

If we split the distance L into two parts, L_1 and L_2, and place another π-flipper in-between, the transmission $\tau(L)$ of all the system becomes

$$\begin{aligned}\tau(L_1, \pi, L_2) = \tau(L_1)\tau(\pi)\tau(L_2) &= \exp(i\sigma_z\omega T_1)\exp(-i\sigma_x\pi/2)\exp(i\sigma_z\omega T_2)\\ &= -i\exp(i\sigma_z\omega[T_1 - T_2])\sigma_x\,. \end{aligned} \tag{5.192}$$

If $L_1 = L_2$, then the first exponential becomes equal to 1. This means that the phase accumulated along the first part of the path is canceled after the transmission through the second part; thus, the transmission through the full system of the flippers is $\tau(\pi/2)\tau(\pi)\tau(\pi/2) = -1$, as if all the flippers were absent. (A common phase factor of -1 in the wave function is not counted.) This phase cancellation is the main element in "echo" experiments.

Let us discuss why the two phases annihilate each other even if they are not equal to $2\pi n$, as in the RSF method. The role of the π-flipper here is crucial. Suppose that $L_1 = L_2 = L$. During the time T of flight through L_1, the RF field in all the coils turns by $\chi = 2\omega T$. If $\chi = 2\pi n - \alpha$, the angle between $\boldsymbol{s}$ and $\boldsymbol{B}_{\rm rf}$ is $\beta = \pi/2 - \alpha$ (see Figure 5.10). In the π-flipper the neutron spin is reflected with respect to $\boldsymbol{B}_\pi$ $(= \boldsymbol{B}_{\rm rf})$. The neutron proceeds to the second $\pi/2$-flipper with the same speed as after the first one. At the moment the neutron enters the second $\pi/2$-flipper, the RF field will turn again by $2\pi n - \alpha$. This time, however, the angle between the RF field and the neutron spin becomes $\beta + \alpha = \pi/2$, and the counterclockwise rotation around $\boldsymbol{B}_{\rm rf}$ in the second $\pi/2$-flipper brings the spin back to the initial state $|\xi_u\rangle$. Thus, the system of three flippers with arbitrary $L_1 = L_2$ is equivalent to free space.

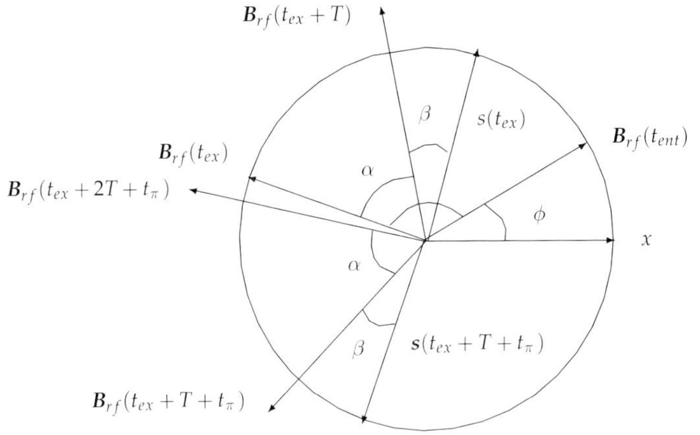

Figure 5.10 Phase compensation. $s(t_{\text{ex}})$ and $B_{\text{rf}}(t_{\text{ex}})$ are the directions of spin arrow s and RF field B_{rf} at the moment t_{ex} when the neutron leaves the first $\pi/2$-flipper (see Figure 5.9). $s(t_{\text{ex}})$ and $B_{\text{rf}}(t_{\text{ex}}+T)$ are the directions of the same vectors at the moment $t_{\text{ex}}+T$ when the neutron enters the π-flipper. The vector s does not change its direction, but B_{rf} turns by the angle $2\pi-\alpha$. In the π-flipper the spin arrow is reflected with respect to B_{rf} and turns with it by angle ωt_π, where t_π is the flight time through the π-flipper. After the flight time T to the second $\pi/2$-flipper, the field B_{rf} turns again by $2\pi-\alpha$ and points in direction $B_{\text{rf}}(t_{\text{ex}}+2T+t_\pi)$, which is now perpendicular to the arrow s. The spin arrow after counterclockwise rotation around B_{rf} in the third flipper by angle $\pi/2$ around B_{rf} returns to the initial position perpendicular to the figure plane and directed toward the reader.

5.7.2.4 The Third Miracle of St. Mark Street

The echo phenomenon can be illustrated in the following way. Let us imagine that there are three gates in the street: at the beginning, in the middle, and at the end. All these gates have identical clocks with a single arm in each clock rotating synchronously with the same speed ν. The effect of the first gate was described on page 238. When people enter the middle gate, the bright spot in their halos changes its position. The new position is obtained by reflecting the spot with respect to the direction of the clock arm at the entrance moment. At the end gate the halo disappears if the position of the spot coincides with the position of the clock arm. It is easy to see that all the halos disappear.

Suppose that the people go with such a speed that the positions of their halos coincide with the position of the clock arm when they arrive at the middle gate. This means that the clock arm turned by $2\pi n$ during the travel time T between the first two gates. In this case nothing happens with the bright spots at the middle gate. Because the people's speed does not change and the distances between the gates are equal, the people arrive at the end gate just when the third clock arm position coincides with the positions of their bright spots. So the halos will be removed.

If the speed of the people is slower, then at the moment they arrive at the middle gate the clock arm turns by $2\pi n+\alpha>2\pi n$. Thus, the people's bright spot will be behind the clock arm by the angle α. In that case the spot is reflected with respect to the clock arm and positioned before the clock arm by the same angle α. If the people's speed after the second gate is the same as before, they arrive at the end gate when the clock arm turns by the angle $2\pi n + \alpha$, and again it points just to the position of the spot, and the halo is extinguished. The same happens with people going faster.

The only way to save the halo is to change the people's speed after the middle gate. Many experiments with neutrons are aimed at measuring the change of the neutron's speed due to the interaction with a sample placed in the second half of the "echo" system.

Some details of spin echo experiments There are two types of echo methods. If the intervals $L/2$ are empty, the method is called neutron resonance spin echo [212]. If the intervals $L/2$ contain DC magnetic fields B_0 in opposite directions, the method is called neutron spin echo [55].

In the experiments using the effect of echo (see, e. g., [267]), two parts of the path L are first tuned with the tuner (in [267] it is called an accelerator), which is placed in one of the parts to achieve the phase cancellation. A sample is then placed in the second part. If there is a change in the neutron velocity $v \to v + \delta v$ (or in the length of the path δL), the time the neutron spends in the second part also changes:

$$\frac{L}{2v} \to \frac{L}{2v}\left[1 - \frac{\delta v}{v}\right] . \tag{5.193}$$

After this change the phases do not match any longer, and the tuner is used to cancel the phase difference again. The change of the current in the tuner characterizes the change in the neutron speed δv.

5.7.2.5 Combination of Frequencies. The MIEZE Spectrometer

Consider again an up-polarized neutron beam and the two $\pi/2$-spin flippers (it is unimportant whether they are close to each other or not). Now, however, assume that the DC fields $\boldsymbol{B}_0$ and therefore the resonant frequencies of the RF fields are different: ω in the first flipper, and $\omega' \neq \omega$ in the second one. A spectrometer with different flipper frequencies is called a MIEZE[24)] spectrometer [212].

The wave function of the neutron transmitted through the first flipper is

$$\psi_1 = \frac{1}{\sqrt{2}}[\exp(i k_0 x - i E_0 t)|\xi_u\rangle - i \exp(i k_- x - i E_- t)|\xi_d\rangle] , \tag{5.194}$$

where $E_- = E_0 - 2\omega$, $k_- = \sqrt{2E_-}$. The wave function of the same neutron after transmission through the second $\pi/2$-flipper (with different frequency ω') becomes

$$\psi_2 = \frac{1}{\sqrt{2}}\left[\psi_1 - \frac{i}{\sqrt{2}}\{\exp(i k'_- x - i E'_- t)|\xi_d\rangle - i \exp(i k'_0 x - i E'_0 t)|\xi_u\rangle\}\right] , \tag{5.195}$$

24) Note that "Mieze" is obtained by changing the position of letter i in the name Mezei, who was inventor of the spin echo spectrometer.

where $E_0' = E_0 - 2\omega + 2\omega'$, $E_-' = E_0 - 2\omega'$, and $k_{0,-}' = \sqrt{2E_{0,-}'}$. The first term, ψ_1, in Eq. (5.194) is related to elastic transmission through the second flipper. The other terms describe inelastic transmission. When the up-polarization analyzer is placed after the two spin flippers, the intensity of the transmitted neutrons is

$$I(x,t) = \left|\frac{1}{2}\left(e^{ik_0x - iE_0t} - e^{ik_0'x - iE_0't}\right)\right|^2 \approx \sin^2([\omega - \omega'][t - x/v]) \,. \quad (5.196)$$

So, at any point x, the intensity oscillates with frequency $\omega - \omega'$. We have again created a beam with modulated intensity, which can be considered as a coherent wave with wavelength $\lambda = 2\pi v/(\omega - \omega')$.

Such oscillations were first observed in [272]. The modulation frequency $\delta\omega = 2\times10^{-5}$ Hz was clearly seen. The observation of such a small frequency modulation seems remarkable because this frequency corresponds to a period of 12 h, while the frequencies ω and ω' of the flippers are on the order of 32 kHz, and the neutrons spend only a few milliseconds in the instrument. However, this observation was absolutely correct.

The oscillations characterize not the neutron properties, but properties of RF coils and the stability of their frequencies. The neutron was only a tool for checking the properties of the setup. However, the fact they are such good tool means that the experimenters understand their properties very well.

Considerably higher frequency modulations on the order of 200–400 kHz were reported in [273], and the observation was also in good agreement with the theoretical expectations and the equations above.

5.7.2.6 The Fourth Miracle of the St. Mark Street

The effect of low-frequency beating can also be illustrated with the help of the St. Mark Street gates. Let us imagine that the clock arm in the middle gate rotates with frequency $\nu' \neq \nu$. This gate opens only if the bright spot in the halo of the passers-by is inside angle α about the direction of the clock arm. If the frequency ν' is the same as ν, and νT is inside the interval $[2\pi n - \alpha, 2\pi n + \alpha]$, then all the people are permitted to pass the middle gate, and no beating is observed. If, however, νT is outside the interval, no one is permitted to go through the middle gate, and the intensity-modulated wave again cannot be observed. Only at $\nu' \neq \nu$ does the stream of people behind the middle gate consist of periodic bunches with frequency $|\nu - \nu'|$, which illustrates the intensity-modulated wave, as discussed above.

5.7.2.7 Transformation of a Continuous Polarized Beam into a Pulsed One

We saw above how to generate a SPW with an RF flipper and how to form an intensity-modulated wave with an analyzer. However, in this process because of elimination of one spin component, the intensity of the intensity-modulated wave is less than that of the incident one. Let us show how to transform a continuous polarized beam into a pulsed one without loss of intensity [274].

Let the continuous beam of neutrons with energy E_0 and polarization along the x-axis propagate along the y-axis. The wave function of such neutrons is

$$|\psi_0(\boldsymbol{r})\rangle = \exp(ik_0y - iE_0t)|\xi_{xu}\rangle \,, \quad (5.197)$$

where $k_0 = \sqrt{2E_0}$, and $|\xi_{xu}\rangle$ is the eigenvector of matrix $\sigma_x : \sigma_x|\xi_{xu}\rangle = |\xi_{xu}\rangle$.

Let us put on the beam path a π-spin flipper with internal DC field $\boldsymbol{B}_1$ parallel to the z-axis as shown in Figure 5.11a. Since

$$|\xi_{xu}\rangle = \frac{1}{\sqrt{2}}[|\xi_{zu}\rangle + |\xi_{zd}\rangle] ,$$

where $|\xi_{zu,d}\rangle$ are eigenspinors of the matrix σ_z:

$$\sigma_z|\xi_{zu,d}\rangle = \pm|\xi_{zu,d}\rangle ,$$

the neutrons after the RF flipper will become a superposition of the states $|\xi_{zu,d}\rangle$ with energies $E_0 \pm \omega_1$, respectively, as shown in Figure 5.11b, where $\omega_1 = 2B_1$. Their wave function is

$$\begin{aligned}|\psi_1(\boldsymbol{r})\rangle &= \exp\left(i\hat{k}_0 y - i(E_0 + \omega_1\sigma_z)t\right)\frac{-i}{\sqrt{2}}[|\xi_{zu}\rangle + |\xi_{zd}\rangle]\\ &= \frac{-i}{\sqrt{2}}\big(\exp(ik_{1+}y - i(E_0 + \omega_1)t)|\xi_{zu}\rangle\\ &\quad + \exp(ik_{1-}y - i(E_0 - \omega_1)t)|\xi_{zd}\rangle\big) ,\end{aligned} \tag{5.198}$$

where $\hat{k}_0 = \sqrt{k_0^2 + 2\omega_1\sigma_z}$, $k_{1\pm} = \sqrt{k_0^2 \pm 2\omega_1}$ and $\omega_1 = 2B_1$.

Immediately after the first π-flipper let us put the second one with DC magnetic field $\boldsymbol{B}_2$ parallel to the x-axis. Since $|\xi_{zu,d}\rangle$ are representable as a superposition of states $|\xi_{xu,d}\rangle$ with quantization axis along $\boldsymbol{B}_2$,

$$|\xi_{zu}\rangle = \frac{1}{\sqrt{2}}[|\xi_{xu}\rangle + |\xi_{xd}\rangle] , \quad |\xi_{zd}\rangle = \frac{1}{\sqrt{2}}[|\xi_{xu}\rangle - |\xi_{xd}\rangle] ,$$

then after transmission through the second π-flipper the neutron beam becomes a superposition of four states with different energies $E_0 \pm \omega_1 \pm \omega_2$, as shown in Figure 5.11b.

The wave function after the second flipper becomes

$$\begin{aligned}|\psi_2(\boldsymbol{r})\rangle &= -\frac{1}{2}\big(\exp(i\hat{k}_{1+}y - i(E_0 + \omega_1 + \omega_2\sigma_x)t)[|\xi_{xu}\rangle + |\xi_{xd}\rangle]\\ &\quad - \exp(i\hat{k}_{1-}y - i(E_0 - \omega_1 + \omega_2\sigma_x)t)[|\xi_{xu}\rangle - |\xi_{xd}\rangle]\big) ,\end{aligned} \tag{5.199}$$

where $\hat{k}_{1\pm} = \sqrt{k_0^2 \pm 2\omega_1 + 2\omega_2\sigma_x}$, or it can be represented as

$$\begin{aligned}|\psi_2(\boldsymbol{r})\rangle &= -\frac{1}{2}\big([\exp(ik_{++}y - i(E_0 + \omega_1 + \omega_2)t)\\ &\quad - \exp(ik_{-+}y - i(E_0 - \omega_1 + \omega_2)t)]|\xi_{xu}\rangle\\ &\quad + [\exp(ik_{+-}y - i(E_0 + \omega_1 - \omega_2)t)\\ &\quad + \exp(ik_{--}y - i(E_0 - \omega_1 - \omega_2)t)]|\xi_{xd}\rangle\big) ,\end{aligned} \tag{5.200}$$

where $k_{\pm,\pm} = \sqrt{k_0^2 \pm 2\omega_1 \pm 2\omega_2}$.

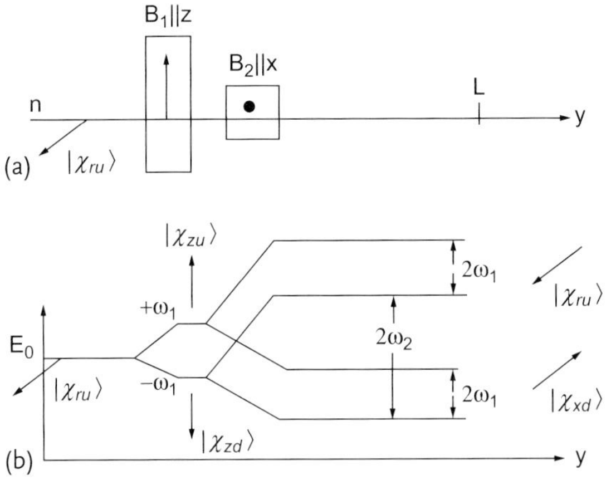

Figure 5.11 Principle of an experiment for total transformation of a continuous polarized beam into a nonpolarized pulsed one (a). After the first π-flipper with resonant frequency $\omega_1 = 2B_1$, where B_1 is a DC field parallel to the z-axis, the neutron becomes a superposition of two states with opposite polarizations on the z-axis and energies differing by $2\omega_1$ (b). After the second π-flipper with DC field $\boldsymbol{B}_2$ parallel to the x-axis, the neutron becomes a superposition of two spin states with opposite polarization along field $\boldsymbol{B}_2$, and every state is split in energy by $2\omega_2$. Because of this splitting, the intensity of every spin state oscillates in time, and the oscillation phase depends on the distance after the second flipper. At some distance L the oscillations of both spin components have identical phases. It is at this point that the previously continuous beam becomes completely transformed to a pulsed one.

The total intensity corresponding to this wave function at some distance L after the second flipper is equal to the sum of the intensities of the two spin components:

$$I(L) = \sin^2(K_+ L - \omega_1 t) + \cos^2(K_- L - \omega_1 t)$$
$$= 1 + \sin([K_+ + K_-]L - 2\omega_1 t)\sin([K_+ - K_-]L) , \qquad (5.201)$$

where

$$K_+ = \frac{1}{2}\left[\sqrt{k_0^2 + 2\omega_1 + 2\omega_2} - \sqrt{k_0^2 - 2\omega_1 + 2\omega_2}\right] \approx \frac{\omega_1}{\sqrt{k_0^2 + 2\omega_2}} , \qquad (5.202)$$

$$K_- = \frac{1}{2}\left[\sqrt{k_0^2 + 2\omega_1 - 2\omega_2} - \sqrt{k_0^2 - 2\omega_1 - 2\omega_2}\right] \approx \frac{\omega_1}{\sqrt{k_0^2 - 2\omega_2}} . \qquad (5.203)$$

Since $K_+ - K_- \approx -\omega_1\omega_2/k_0^3$, we have $\sin([K_+ - K_-]L) = -1$, when $L\omega_1\omega_2/k_0^3 = \pi/2$. At the point $L = \pi k_0^3/(2\omega_1\omega_2)$, the neutron intensity is

$$I(L) = 2\sin^2\left(\omega_1 t + \pi\left(k_0^2/2\omega_2 + 1/4\right)\right) .$$

Therefore, at the point L the previously continuous beam of polarized neutrons without any chopper is transformed into a pulsed beam of nonpolarized neutrons.

If $B_1 \sim 1$ kG, then $\omega_1 \approx 3$ MHz. The distance L, where transformation is achieved, is determined by

$$L = \lambda \frac{E_0^2}{4 B_1 B_2} .$$

If $B_2 = 2\,\text{kG}$ and $E_0 = 1\,\text{meV}$, then $L \approx 1\,\text{m}$.

On page 220, we discussed the method of transformation of a nonpolarized beam into a polarized one without loss of intensity. With this method it is possible to transform a nonpolarized continuous beam into a pulsed one without loss of intensity. However, that method is applicable only to very slow neutrons. At higher energies it is not feasible.

5.8 Neutron Holography

Here we shall discuss some ideas of the neutron holography [268], but first let us remind the main principles of the optical one and show some elements which distinguish our approach from the common one.

5.8.1 General Principles of Holography

If a plane wave $\exp(\boldsymbol{k}\boldsymbol{r} - i\omega t)$ illuminates an object, then the set of elastically scattered waves

$$\psi_s(\boldsymbol{r}, t) = \int d^3 r' N_0(\boldsymbol{r}') b(\boldsymbol{r}') \exp(\boldsymbol{k} \cdot \boldsymbol{r}') \frac{\exp(ik|\boldsymbol{r} - \boldsymbol{r}'| - i\omega t)}{|\boldsymbol{r} - \boldsymbol{r}'|} , \qquad (5.204)$$

focused on the retina gives a photo image of the object. Here $N_0(\boldsymbol{r}')$ and $b(\boldsymbol{r}')$ are the atomic density and the scattering amplitude of atoms at the point $\boldsymbol{r}'$ inside or on the surface of the object.

If the field ψ is focused onto a photo plate (focusing means concentration into a point of the field scattered by some point of the object), then the distribution of the intensities gives a photo image on the photo plate.

5.8.1.1 A Distinguishing Feature of Our Approach

Below we shall replace all spherical waves by a superposition of plane waves.

The spherical wave can be represented by a two-dimensional Fourier expansion

$$\frac{\exp(ikr)}{r} = \frac{i}{2\pi} \int \frac{d^2 p_{\|}}{\sqrt{k^2 - p_{\|}^2}} \exp\left(i\boldsymbol{p}_{\|} \cdot \boldsymbol{r}_{\|} + i\sqrt{k^2 - p_{\|}^2}|r_{\perp}|\right) ,$$

where $\boldsymbol{p}_{\|}$ are wave vector components parallel to some plane with coordinates $\boldsymbol{r}_{\|}$, and $r_{\perp}$ is the coordinate normal to this plane (Figure 5.12). The integration range

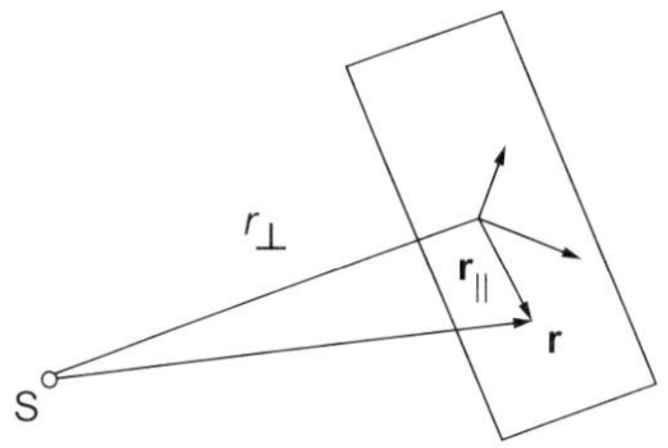

Figure 5.12 Geometry of the two-dimensional Fourier representation of a spherical wave: S is the scattering center radiating the spherical wave. At the observation point at a distance $r_\perp$ from S draw a plane perpendicular to $r_\perp$. Coordinates $\boldsymbol{r}_\parallel$ denote points in the plane in the vicinity of $r_\perp$.

of $d^2 p_\parallel$ includes all $\boldsymbol{p}_\parallel$ and in-between them are such $\boldsymbol{p}_\parallel$ that are larger than k; therefore, $k_\perp^2 = k^2 - p_\parallel^2 < 0$. These components belong to the waves which decay exponentially along $r_\perp$, and these waves can be neglected. As a result, the integral can be approximated by

$$\begin{aligned}\frac{\exp(ikr)}{r} &\approx \frac{i}{2\pi} \int\limits_{p_\parallel^2<k^2} \frac{d^2 p_\parallel}{\sqrt{k^2 - p_\parallel^2}} \exp\left(i\boldsymbol{p}_\parallel \cdot \boldsymbol{r}_\parallel + i\sqrt{k^2 - p_\parallel^2}|r_\perp|\right) \\ &= \frac{i}{\pi} \int d^3 p\, \delta(k^2 - p^2) \exp(i\boldsymbol{p} \cdot \boldsymbol{r}) \Theta(p_\perp r_\perp > 0) \\ &= \frac{ik}{2\pi} \int d\Omega \exp(i\boldsymbol{k}_\Omega \cdot \boldsymbol{r}) \Theta(\boldsymbol{k}_\Omega \cdot \boldsymbol{r} > 0)\,, \end{aligned} \tag{5.205}$$

where $|\boldsymbol{k}_\Omega| = k$, and we used the easily checked relation

$$\frac{d^2 p_\parallel}{2\sqrt{k^2 - p_\parallel^2}} \Theta(p_\parallel^2 < k^2) = d^3 p\, \delta(k^2 - p^2) \Theta(p_\perp > 0)\,.$$

Therefore, the scattered field is represented by

$$\psi_s(\boldsymbol{r}, t) = i \int d^3 r' N_0(\boldsymbol{r}') k b(\boldsymbol{r}') \exp(\boldsymbol{k}\cdot\boldsymbol{r}') \int \frac{d\Omega}{2\pi} \exp\left(i\boldsymbol{k}_\Omega \cdot (\boldsymbol{r} - \boldsymbol{r}') - i\omega t\right). \tag{5.206}$$

5.8.1.2 Holographic Image

The holographic image is obtained by registration of superposition of the scattered field $\psi_s(\boldsymbol{r}, t)$ with the field of a reference plane wave $\psi_0(\boldsymbol{r}, t) = \exp(i\boldsymbol{k}_0 \cdot \boldsymbol{r} - i\omega t)$ ($|\boldsymbol{k}_0| = k$) from the same coherent source, which also illuminates the object. The intensity $I(\boldsymbol{r})$ at a point $\boldsymbol{r}$ of the photo plate averaged over time,

$$I(\boldsymbol{r}) = \lim_{\tau\to\infty} \frac{1}{\tau} \int\limits_0^\tau |\psi_0(\boldsymbol{r}, t) + \psi_s(\boldsymbol{r}, t)|^2\, dt\,,$$

contains an interference term

$$I_i(\boldsymbol{r}) = \int d^3r'\, N_0(\boldsymbol{r}') k b(\boldsymbol{r}') \exp(\boldsymbol{k}\cdot\boldsymbol{r}') \times \int \frac{d\Omega}{\pi} \sin\left((\boldsymbol{k}-\boldsymbol{k}_\Omega)\cdot\boldsymbol{r}' + (\boldsymbol{k}_\Omega - \boldsymbol{k}_0)\cdot\boldsymbol{r}\right) . \tag{5.207}$$

This part of intensity creates the hologram.

5.8.1.3 Reproduction of the Scattered Field from the Hologram

To see the object we need to restore the scattered field (5.206) created by the object. For that the hologram, which is a transparency with transmission $T(\boldsymbol{r}_{\|}) \sim I_i(\boldsymbol{r})$ at points $\boldsymbol{r} = (\boldsymbol{r}_{\|}, r_{\perp} = 0)$, is illuminated by a reproducing wave $\exp(i\boldsymbol{q}\cdot\boldsymbol{r} - i\omega_q t)$, which is not necessary identical to the reference one. As a result, at the points $\boldsymbol{R}$ there appears the field transmitted through the transparency, which is calculated according to the Fresnel–Kirchhoff principle:

$$\psi_t(\boldsymbol{R}, t) = \frac{1}{4\pi} \int d^2 r_{\|} \frac{\exp(ik|\boldsymbol{R}-\boldsymbol{r}|)}{|\boldsymbol{R}-\boldsymbol{r}|} \left(\frac{\overleftarrow{d}}{dr_\perp} - \frac{\overrightarrow{d}}{dr_\perp} \right)_{r_\perp = 0} \times \exp(i\boldsymbol{q}\cdot\boldsymbol{r} - i\omega_q t) T(\boldsymbol{r}_{\|}) , \tag{5.208}$$

where the arrow over a derivative shows which factor has to be differentiated along the normal to the hologram plane. For a spherical wave, after substitution of its two-dimensional Fourier representation (5.205) we obtain

$$\psi_t(\boldsymbol{R}, t) = \frac{1}{4\pi^2} \int d^2 r_{\|} \int d^2 p_{\|} \frac{p_\perp + q_\perp}{2p_\perp} \exp(i\boldsymbol{p}\cdot(\boldsymbol{R}-\boldsymbol{r}_{\|}) + i\boldsymbol{q}\cdot\boldsymbol{r}_{\|} - i\omega_q t) T(\boldsymbol{r}_{\|}), \tag{5.209}$$

where $\boldsymbol{p} = (\boldsymbol{p}_{\|}, p_\perp)$ and $p_\perp = \sqrt{q^2 - p_{\|}^2}$. It is easy to check that at $T(\boldsymbol{r}_{\|}) = 1$ the field (5.209) completely reproduces the incident plane wave $\exp(i\boldsymbol{q}\cdot\boldsymbol{R} - i\omega_q t)$.

Assume that $T(\boldsymbol{r}_{\|}) = I_i(\boldsymbol{r}_{\|})$, substitute Eq. (5.207) into Eq. (5.209), and take into account that $\sin x = (1/2i)[\exp(ix) - \exp(-ix)]$. The first exponent after integration over the hologram plane (it is supposed to be infinite) gives

$$\psi_{t1}(\boldsymbol{R}, t) = \int d^3r'\, N_0(\boldsymbol{r}') k b(\boldsymbol{r}') \int d\Omega \frac{k_{\Omega\perp} + k_{0\perp}}{2k_{\Omega\perp}} \times \exp\left(i(\boldsymbol{k}-\boldsymbol{k}_\Omega)\cdot\boldsymbol{r}' + i(\boldsymbol{k}_\Omega - \boldsymbol{k}_0 + \boldsymbol{q})_{\|}\cdot\boldsymbol{R}_{\|} + ip_\perp R_\perp\right) , \tag{5.210}$$

where $p_\perp = \sqrt{q^2 - (\boldsymbol{k}_\Omega - \boldsymbol{k}_0 + \boldsymbol{q})_{\|}^2}$. (From now on we shall omit the temporal factor $\exp(-i\omega_q t)$.) At $\boldsymbol{q} = \boldsymbol{k}_0$, that is, when the reproducing wave is identical to the reference one, the field (5.210) is identical to the field scattered by the object and observed at the point $\boldsymbol{R}$. It is called a virtual image and is seen visually if the frequency ω_q of the field is in the visible range of electromagnetic radiation.

If $\boldsymbol{q} \neq \boldsymbol{k}_0$, the virtual image is shifted with respect to the object and becomes distorted, as if the object were in a medium with refractive index

$$n^2(\Omega) = \frac{k^2 - k_{\Omega\parallel}^2 + (\boldsymbol{k}_\Omega - \boldsymbol{k}_0 + \boldsymbol{q})_\parallel^2}{q^2} ,$$

and were illuminated by the wave with wave vector $\boldsymbol{k} - (\boldsymbol{k}_0 - \boldsymbol{q})_\parallel$.

The second exponent $\exp(-ix)$ of $\sin x$ after integration over the hologram plane gives the field which at some distance forms a real image. We shall not discuss it here.

5.8.2
Spin Precession Wave

Everything said above about electromagnetic waves is valid for neutron waves too. However, in neutron optics we do not have coherent sources of neutron radiation and the de Broglie wavelength is very small compared with the wavelength of visible light. Therefore, neutron holography seems to be impossible.

Nevertheless, the situation is not so hopeless, because with neutrons it is possible to generate waves with wavelengths considerably longer than the usual de Broglie wavelengths [275–277].

Let us consider the system of two π-spin flippers shown in Figure 5.13.

If the incident neutron is polarized along the x-axis, then its wave function is

$$|\psi_2(\boldsymbol{r}, t)\rangle = \exp(i\boldsymbol{k}\cdot\boldsymbol{r} - iEt)|\chi_{xu}\rangle , \tag{5.211}$$

where $E = k^2/2$ is the neutron kinetic energy and $|\chi_{xu}\rangle$ is its spin state with polarization along the x-axis. It is the eigenstate of the Pauli matrix σ_x: $\sigma_x|\chi_{xu}\rangle = |\chi_{xu}\rangle$. The state $|\chi_{xu}\rangle$ is a superposition of states with polarizations along and opposite

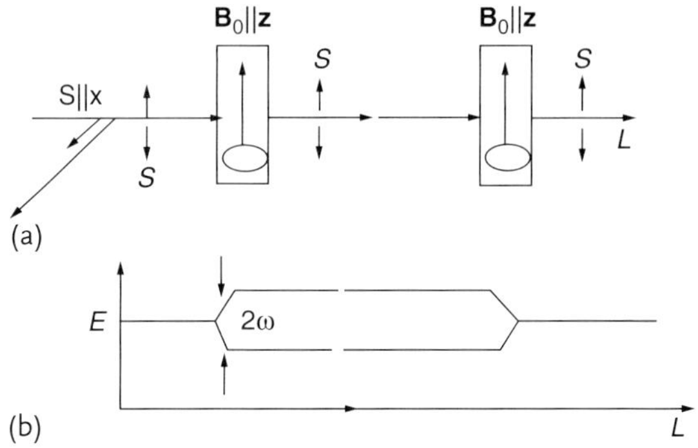

Figure 5.13 The generation of a spin precession wave (SPW) between two identical spin flippers with internal DC field $B_0 || z$-axis (a). The incident neutron is polarized along the x-axis. Outside the magnetic field is zero. After the first flipper every neutron is a superposition of two states with polarizations along and opposite the z-axis and with the two different energies shown in (b). After the second flipper the energies of the two components return back to the previous value E. A SPW exists between the two spin flippers.

the z-axis: $|\chi_{xu}\rangle = [|\chi_{zu}\rangle + |\chi_{zd}\rangle]/\sqrt{2}$, where $|\chi_{zu,d}\rangle$ are eigenstates of the Pauli matrix σ_z: $\sigma_z|\chi_{zu,d}\rangle = \pm|\chi_{zu,d}\rangle$.

In the incident wave the states $|\chi_{zu,d}\rangle$ have the same momentum k and energy E. After the first RF π-spin flipper the neutron wave function becomes

$$|\psi_1(\boldsymbol{r},t)\rangle = \exp\left(-i(E+\sigma_z\omega)t + i\hat{\boldsymbol{k}}\cdot\boldsymbol{r}\right)|\chi_{xu}\rangle$$
$$= \frac{-i}{\sqrt{2}}\exp\left(-i(E+\sigma_z\omega)t + i\hat{\boldsymbol{k}}\cdot\boldsymbol{r}\right)(|\chi_{zu}\rangle + |\chi_{zd}\rangle)\,, \qquad (5.212)$$

where $\hat{\boldsymbol{k}} = \sqrt{k^2 + 2\omega\sigma_z}\boldsymbol{k}/k$, and ω is the RF frequency of the flipper, which we take to be equal to $2B_0$. The two states $|\chi_{zu,d}\rangle$ are seen to now have different energies. Their difference is 2ω, as shown in Figure 5.13 (b). They also have different speeds. The difference leads to the wavelike variation of the neutron spin arrow in space and time, as discussed on page 235. Such a variation can be considered as a coherent SPW. Its coherence is provided by the RF field of the spin flipper.

After the second spin flipper the neutron state is described by the wave function

$$|\psi_2(\boldsymbol{r},t)\rangle = \frac{-1}{\sqrt{2}}\exp(-iEt + ik(y-L))$$
$$\times\left(\exp(ik_+L)|\chi_{zu}\rangle + \exp(ik_-L)|\chi_{zd}\rangle\right)\,, \qquad (5.213)$$

where $k_\pm = \sqrt{k^2 \pm 2\omega}$. It follows that after the second spin flipper the neutron has the same momentum and energy as before the first one, but the states $|\chi_{zu,d}\rangle$ have different phases. The phase difference can be observed in the experiment depicted in Figure 5.14. If we put an analyzer after the second spin flipper, which transmits only neutrons polarized along the x-axis, the intensity of the transmitted neutrons will be

$$I(L) = \frac{1}{4}|\exp(ik_+L) + \exp(ik_-L)|^2 = \cos^2(K_-L) = \frac{1}{2}[1+\cos(2\pi L/\Lambda)]\,, \qquad (5.214)$$

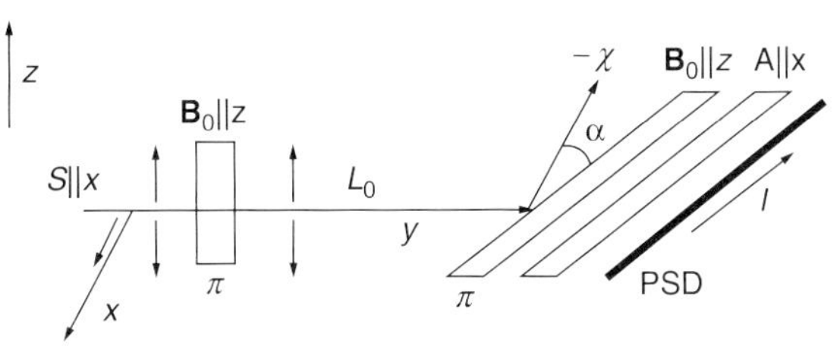

Figure 5.14 An experiment to observe the interference pattern produced by a SPW. The second spin flipper is turned in the plane (x, y) by the angle α with respect to the first one. After it there is an analyzer A, which transmits neutrons parallel to the x-axis, and a position-sensitive detector (PSD). The dependence of the count rate on the coordinate l along the PSD varies in a cosine-like manner as shown in Eq. (5.215).

where $K_- = (k_+ - k_-)/2 \approx \omega/k$, $\Lambda = \lambda E/\hbar\omega$, and $\lambda = 2\pi/k$ is the de Broglie wavelength. If the beam is wide enough along the x-axis, and the second spin flipper with magnetic field $\boldsymbol{B}_0 \| z$ is turned with respect to the second one by the angle α in the (x, y)-plane, then the dependence of the intensity on the coordinate l along the position-sensitive detector (PSD) (Figure 5.14) will vary as

$$I(l) = \frac{1}{2}\left[1+\cos(2\pi\{L_0+l\sin\alpha\}/\Lambda)\right] = \frac{1}{2}\left[1+\cos(\phi+2\pi l/\Lambda')\right], \quad (5.215)$$

where $\phi = 2\pi L_0/\Lambda$ and $\Lambda' = \Lambda/\sin\alpha$.

For neutrons of low energy (approximately 1 meV) with wavelength approximately 9 Å and magnetic field $B_0 \approx 2$ T, the wavelength is $\Lambda \approx 10\mu$.

5.8.3
Hologram of a Scattered SPW without a Reference Wave

Let us now investigate neutron scattering in a sample placed between two spin flippers as shown in Figure 5.15. The wave function of scattered neutrons before the second flipper is

$$|\psi_s(\boldsymbol{r},t)\rangle = \exp(-i(E+\omega\sigma_z)t) \times \int d^3r'\, N_0(\boldsymbol{r}')\frac{b(\boldsymbol{r}')}{|\boldsymbol{r}-\boldsymbol{r}'|}\exp(i\hat{k}|\boldsymbol{r}-\boldsymbol{r}'|)\exp(i\hat{\boldsymbol{k}}\boldsymbol{r}')|\chi_{xu}\rangle, \quad (5.216)$$

where $N_0(\boldsymbol{r}')$ and $b(\boldsymbol{r}')$ are the atomic density and the scattering amplitude at the point $\boldsymbol{r}'$ inside the sample, respectively, $\hat{k} = \sqrt{k^2+2\omega\sigma_z}$, and $\hat{\boldsymbol{k}} = \sqrt{k^2+2\omega\sigma_z}\boldsymbol{k}/k$. Substitute Eq. (5.205) for the spherical wave. Then the scattered field (5.216) transforms to

$$|\psi_s(\boldsymbol{r},t)\rangle = i\int d^3r'\, N_0(\boldsymbol{r}')\hat{k}b(\boldsymbol{r}') \times \int\frac{d\Omega}{2\pi}\exp\left(i\hat{\boldsymbol{k}}_\Omega\cdot(\boldsymbol{r}-\boldsymbol{r}')-i(E+\omega\sigma_z)t\right)\exp(\hat{\boldsymbol{k}}\cdot\boldsymbol{r}')|\chi_{xu}\rangle, \quad (5.217)$$

where $\hat{\boldsymbol{k}}_\Omega = \sqrt{k^2+2\omega\sigma_z}\boldsymbol{\Omega}$, $\boldsymbol{\Omega}$ is a unit vector in the direction Ω.

After the second flipper identical to the first one, and the analyzer, transmitting only neutrons polarized along the x-axis, the wave function (5.217) becomes

$$\langle\chi_{xu}||\psi_s(\boldsymbol{r},t)\rangle \approx \exp(-iE_0t) \times\int d^3r'\, N(\boldsymbol{r}')kb(\boldsymbol{r}')\exp\left(i\boldsymbol{k}\cdot\boldsymbol{r}'+\boldsymbol{k}_\Omega\cdot(\boldsymbol{r}-\boldsymbol{r}')\right)\cos\left(\boldsymbol{q}\cdot\boldsymbol{r}'+\boldsymbol{q}_\Omega\cdot(\boldsymbol{r}-\boldsymbol{r}')\right), \quad (5.218)$$

where $\boldsymbol{q} = (\omega/k^2)\boldsymbol{k}$, $\boldsymbol{q}_\Omega = (\omega/k)\boldsymbol{\Omega}$, and everywhere, where it is possible, we neglected ω with respect to k.

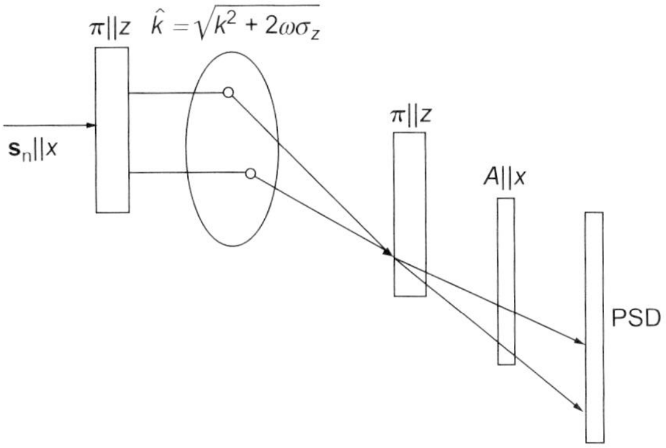

Figure 5.15 Experimental arrangement for recording of neutron hologram for incoherent scatterers. The coherent SPW created after the first π-spin flipper is scattered in a sample. After the second spin flipper and an analyzer, the hologram is registered by a PSD.

Assume that the scattering is incoherent. This means that according to wave function (5.218) the intensity of scattered neutrons at the PSD averaged over elements, isotopes, and spins is representable as

$$I(\boldsymbol{r}) = \langle\langle\, |\langle\chi_{xu}||\psi_s(\boldsymbol{r},t)\rangle|^2 \rangle\rangle = \int d^3r'\, N(\boldsymbol{r}')|kb(\boldsymbol{r}')|^2 \cos^2\left(\boldsymbol{q}\cdot\boldsymbol{r}' + \boldsymbol{q}_\Omega(\boldsymbol{r}-\boldsymbol{r}')\right)$$
$$= C + \frac{1}{2}\int d^3r'\, N(\boldsymbol{r}')|kb(\boldsymbol{r}')|^2 \cos\left(2\boldsymbol{q}\cdot\boldsymbol{r}' + 2\boldsymbol{q}_\Omega\cdot(\boldsymbol{r}-\boldsymbol{r}')\right) , \tag{5.219}$$

where $C = \int d^3r'\, N(\boldsymbol{r}')|kb(\boldsymbol{r}')|^2/2$. We see that the intensity depends on the coordinates $\boldsymbol{r}$ on the PSD, and it contains information about the distance between the point $\boldsymbol{r}$ on the PSD and the point $\boldsymbol{r}'$ of the scatterer.

If we put a photo plate on the PSD, we will get a hologram. After illumination of this hologram with a restoring light wave, we can get a visible image of the internal structure of the nontransparent object. However, for that it is necessary to overcome some difficulties related to the difference between Λ and the wavelengths of visible light. If the de Broglie neutron wavelength is too short, the SPW wavelength is too long.

5.8.4
Magnetic Scattering of a SPW with Spin Flip

Above we considered nonmagnetic scattering. In the case of magnetic samples, some scattering accompanies spin flip. In that case the state $|\chi_{zu}\rangle$ before scattering with higher energy $E+\omega$ transforms to the state $|\chi_{zd}\rangle$ with the same energy $E+\omega$ (we consider here only elastic scattering), and the state $|\chi_{zd}\rangle$ with lower energy $E-\omega$ transforms to the state $|\chi_{zu}\rangle$ with the same energy $E-\omega$. After the second spin flipper, identical to the first one, the spin components, which after the first flipper have energy difference 2ω, will interchange again, but their energies will not return to the previous E_0 but will separate apart again by 2ω. So, their energy difference will be 4ω, and this energy difference will make the intensity of the neutrons after the analyzer oscillate in time,

$$\begin{aligned} I(\boldsymbol{r},t) &= \int d^3r' N(\boldsymbol{r}')|kb(\boldsymbol{r}')|^2 \cos^2\left(2\boldsymbol{q}\cdot\boldsymbol{r}' + 2\boldsymbol{q}_\Omega\cdot(\boldsymbol{r}-\boldsymbol{r}') - 2\omega t\right) \\ &= C + \frac{1}{2}\int d^3r' N(\boldsymbol{r}')|kb(\boldsymbol{r}')|^2 \cos\left(4\boldsymbol{q}\cdot\boldsymbol{r}' + 4\boldsymbol{q}_\Omega\cdot(\boldsymbol{r}-\boldsymbol{r}') - 4\omega t\right), \end{aligned} \tag{5.220}$$

and without special time analysis this oscillation will produce a homogeneous background on a photo plate.

To avoid further splitting of energy, it is necessary to change the sign of the DC magnetic field in the second π-flipper. Then the intensity scattered without spin flip will oscillate and average to a homogeneous background, whereas the scattering with spin flip will create a steady-state interference pattern on the hologram. Illumination of this hologram with visible light will give a visual image of the magnetic structure of the sample, which gives scattering with spin flip. In this way we can separate scattering with and without spin flip.

5.8.5
Entangled States. A Fundamental Question of Quantum Mechanics

We found that after a spin flipper the neutron can occur as a coherent superposition of two states with different kinetic energies. However these states are related to the number of photons in the flipper. Suppose that initially the neutron is polarized perpendicular to the field $\boldsymbol{B}_0$ in the π-flipper, as shown in Figure 5.13, and the neutron enters the flipper at an angle, as shown in Figure 5.8. Then the transmitted neutron will be a superposition of two states. One of them corresponds to absorption of a photon, and the other one corresponds to emission of a photon. These two states constitute an entangled state with photons:

$$\begin{aligned} \Phi(\boldsymbol{k},n) = \frac{1}{\sqrt{2}}\big[&|n-1\rangle|\xi_u\rangle \exp(i\boldsymbol{k}_+\boldsymbol{r} - iE_+t) + \\ &+|n+1\rangle|\xi_d\rangle \exp(i\boldsymbol{k}_-\boldsymbol{r} - iE_-t)\big], \end{aligned} \tag{5.221}$$

where $|n \pm 1\rangle$ means the flipper state with one photon more or less and $E_\pm = E_0 \pm 2\omega$ and $k_\pm = \sqrt{2E_\pm}$ are the neutron energy and momentum after absorption or emission of a photon. The two components of the wave function propagate with speeds different in magnitude and direction. If we had a pulsed source and registered neutrons with a PSD, could we see two separate peaks coming with an appropriate time difference? Or would we see something different? In the first case we should conclude that after the flipper every neutron goes along a single path. Then what does coherent superposition mean? In the second case we should ask ourselves what process really takes place in the flipper: absorption or emission of a photon? We are inclined to expect the first version in a thought experiment, because it is similar to the Stern–Gerlach one. This means that every single neutron has a wave function equal to a single value of the two terms in the sum (5.221). Then, what meaning does superposition of the neutron states and their interference have? It is the most intriguing question of quantum mechanics.

5.9 Conclusion

Application of the theoretical methods presented here gives us an opportunity to investigate many phenomena observed in polarized neutron reflectometry, and not only phenomena related to condensed matter, but also ones related to the neutron itself. One of them is the problem of neutron speed, formulated on page 238: is the neutron speed really equal to the average speed of the two spin components, or can the spin components separate? If they separate, how does this happen? How does the coherence of two components then disappear? If these questions are be answered, our understanding of quantum mechanics will become one step better.

In this chapter we considered mainly uniform fields or fields with interfaces. We can also consider fields uniform along the interface, but changing along the normal to it. However, in all cases this external field is classical, that is, the particle does not change it, and we are not interested in the origin of this field. It is interesting when the field appears because of external electrical field $\boldsymbol{E}$. According to special relativity theory, in the neutron reference frame an external electrical field creates a magnetic one $\boldsymbol{B} = [\boldsymbol{v} \times \boldsymbol{E}]/c$, where $\boldsymbol{v}$ is the neutron speed and c is the speed of light. In reflectometry of thermal neutrons, relativistic effects are very small because of the smallness of the ratio v/c; however, in some special cases this effect can be the most important.

We only briefly discussed neutron holography with SPW. These waves have wavelengths much larger than light waves and can be used for imaging the macroscopic structure of matter. However, there is a holographic method where short wavelengths are used and atomic-size resolution is achieved. In this method a reference spherical wave $b_{\mathrm{Rf}} \exp(ikr)/r$ is generated by scattering on some reference atoms with scattering amplitude b_{Rf}. This reference wave is further scattered on surrounding atoms, and the interference of it with the scattered one

$$\left| \psi_0(\boldsymbol{k}_0, \boldsymbol{R}) b_{\mathrm{Rf}} \frac{\exp(ik|\boldsymbol{R}-\boldsymbol{r}|)}{|\boldsymbol{R}-\boldsymbol{r}|} \right.$$
$$\left. + \psi_0(\boldsymbol{k}_0, \boldsymbol{R}) b_{\mathrm{Rf}} \int \frac{\exp(ik|\boldsymbol{R}-\boldsymbol{r}'|)}{|\boldsymbol{R}-\boldsymbol{r}'|} d^3 r' b_{\mathrm{s}}(\boldsymbol{r}') \frac{\exp(ik|\boldsymbol{r}-\boldsymbol{r}'|)}{|\boldsymbol{r}-\boldsymbol{r}'|} \right|^2 \qquad (5.222)$$

is registered. Here $\boldsymbol{R}$ is the position of the reference atom, $\psi_0(\boldsymbol{k}_0, \boldsymbol{R})$ is the wave of the incident beam with wave vector $\boldsymbol{k}_0$, which illuminates the reference atom, and $b_{\mathrm{s}}(\boldsymbol{r}')$ is the scattering density of nearby atoms.

Usually experiments are performed on crystals, with one reference atom per elementary cell, and the interference pattern Eq. (5.222) is measured at different orientations of the crystal, which is equivalent to different orientations of the vector $\boldsymbol{k}_0$ of the incident beam with respect to it. The interference pattern is extracted from the full diffraction pattern of the crystal and numerical processing of the data obtained helps to extract the structure of the elementary cell. So this technique is similar to crystallography. We cannot dwell on this point, but for the interested reader we can recommend the papers [278, 279] and references therein.

The end of song is not the end of story,
The thought returns and wakes the brain to worry.

6
Optical Phenomena in Neutron Diffraction and Scattering

6.1
Diffraction on Regular Lattices

The foundation of the principles of the diffraction of waves by regular lattices was established in the 1910s through experiments on and analyses of X-ray diffraction. The framework can be classified into two categories as follows:

1. *kinematical theory*
2. *dynamical theory*

The fundamental difference between these two approaches lies in the treatment of multiple scattering, that is, in the former theory the effects of multiple scattering of waves are neglected, whereas in the latter they are taken into consideration. Therefore, the latter might become significantly complicated in a certain situation. The difference can also be expressed in the terminology usually used in scattering theory as the kinematical theory uses the Born approximation, whereas the dynamical theory does not use it.

The kinematical theory was established by Laue *et al.* [280], and their experiments on X-ray diffraction opened up the new research field to investigate material structures at the atomic level. On the other hand, the dynamical theory was initiated by Darwin [194, 281], and further refined by Ewald into a consistent diffraction theory [282, 283]. Some details of the differences between these two approaches will be introduced in the following subsections.

6.1.1
Borrmann Effect

As mentioned in Section 1.2, Bragg's law, Eq. (1.10), defines the angle between the directions of scattered waves and the incident waves in the phenomenon of coherent scattering of neutrons by a crystal. In such a phenomenon, two geometrical situations can be supposed relating to the emergence of diffracted and transmitted waves, as illustrated in Figure 6.1. One is the case shown in Figure 6.1a, where the diffracted waves emerge from the same side as the incident waves of the surfaces on a crystalline sample. The other case is shown in Figure 6.1b, where both

Neutron Optics. Masahiko Utsuro, Vladimir K. Ignatovich

ISBN: 978-3-527-40885-6

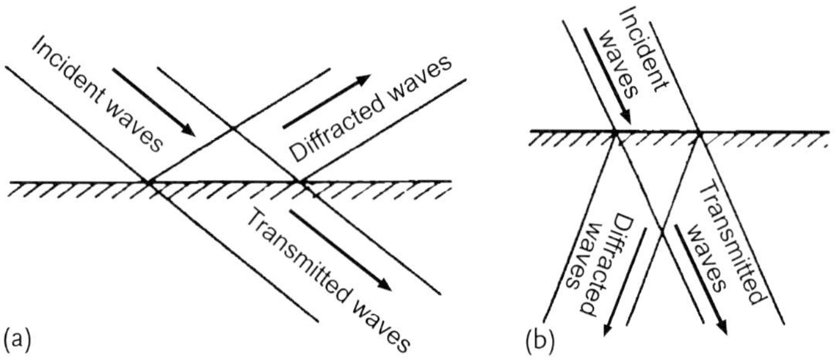

Figure 6.1 Two kinds of geometrical situations in Bragg scattering by a crystal: (a) Bragg case; (b) Laue case.

the diffracted and the transmitted waves exit from the opposite side of the sample against the incident surface. The former type of reflection geometry is called the *Bragg case* and the latter type of transmission geometry is called the *Laue case*.

As expected from the fundamental difference between the kinematical theory and the dynamical one mentioned above, both theories lead to essentially similar results in the case of a thin crystal, but for a thick crystal an important difference is possible between the results from the two theories. One of the most distinguished effects derived from the dynamical theory was shown for the first time in the 1940s by Borrmann [284, 285] as the anomalous behavior of X-ray diffraction for the Laue case, which is now called the *Borrmann effect*.

The effect is illustrated in Figure 6.2 according to the review by Batterman on the dynamical diffraction of X-rays [286]. In the case of Laue diffraction by a thin crystal as shown in Figure 6.2a, the angular scan of the intensity transmitted through the crystal indicates a decrease of the forward transmitted intensity around the Bragg angle as drawn by the upper curve in Figure 6.2c. In contrast, in the case of a thick crystal as shown in Figure 6.2b, an anomalous intensity increase should be observed in the direction of forward transmission as the lower curve in Figure 6.2c. If the details of these phenomena are analyzed for a X-ray-sensitive film inserted in the geometry as given in Figure 6.2d, an additional transmitted component should be recorded as indicating the energy flow due to multiple Bragg reflections different from the direct transmission of the incident beam. Such a flow of wave propagations due to multiple Bragg reflections forms a fan-shaped region of the diffracted beam which is called the *Borrmann fan*, and the intensity distribution in the region becomes the interference pattern named *Pendellösung* illustrated in Figure 1.6.

The first measurement with neutrons on the present anomalous transmission was performed by Knowles [287], and a little later a higher-resolution measurement was carried out by Sippel [288]. The experimental results obtained by Sippel *et al.* on the *rocking curve* as the angular distribution of neutron intensity are shown in Figure 6.3.

They used a double-crystal spectrometer with a neutron wavelength of 1.18 Å. For this wavelength, the InSb sample considered in Figure 6.3a has normal linear

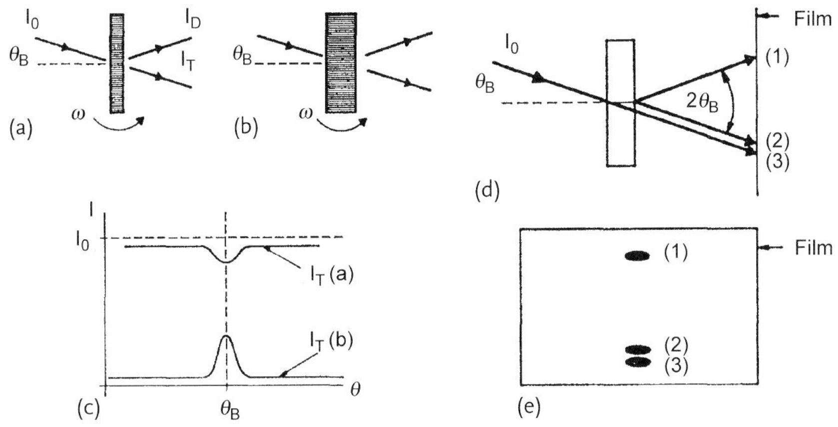

Figure 6.2 Anomalous transmission due to the Borrmann effect in X-ray diffraction by a crystal: (a) thin-crystal Laue diffraction; (b) thick-crystal Laue diffraction; (c) angular scans on the transmitted intensity, where the upper curve is the thin-crystal case and the lower curve is the thick-crystal case; (d) ray diagram explaining the spots on the film; (e) the diffracted beam (1) and the forward diffracted beam (2) on the film indicate the energy apparently flowed along the atomic planes in the crystal (Batterman [286]).

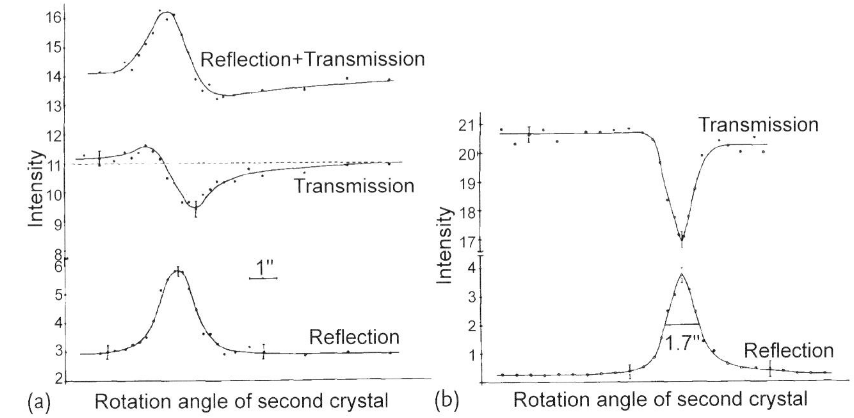

Figure 6.3 Rocking curves of Bragg-reflected and transmitted neutron beam intensities in a symmetrical Laue geometry from (a) InSb and (b) Ge crystal (Sippel *et al.* [288]).

absorption coefficient $\mu_{InSb} = 1.95\,cm^{-1}$ and thickness 7.2 mm, whereas the Ge sample considered in Figure 6.3b has $\mu_{Ge} = 0.08\,cm^{-1}$ and thickness 0.91 mm. As observed in the figure, the transmission curve of the InSb sample indicates obvious asymmetry, and the summation of reflected and transmitted intensities reveals an anomalous intensity increase, that is, the effect of an anomalous decrease of neutron absorption. In contrast, the transmission curve for the weakly absorbing Ge sample is as expected nearly symmetrical.

6.1.2
Darwin Table

As expected from the fundamental difference already mentioned between the kinematical and dynamical theories, the reflectivity R for a thin crystal (thickness $d \ll \Delta_h$) derived from the latter theory is similar to that derived from the former theory as

$$R(y, \alpha) = \left(\frac{\sin \alpha y}{y}\right)^2 , \tag{6.1}$$

where

$$\alpha = \pi d/\Delta_h \ll 1 ,$$
$$\Delta_h = \frac{\pi[\sin(\theta_B + \theta_S)\sin(\theta_B - \theta_S)]^{1/2}}{b_c e^{-W}|F_{hkl}|N\lambda} , \tag{6.2}$$

where θ_B is the Bragg angle and θ_S is the angle between the Bragg plane and the crystal surface. (Refer to, e. g., textbook [I], Eq. (6.2.21)).

In contrast, in a thick crystal ($d \gg \Delta_h$, i. e., $\alpha \gg 1$), the reflectivity from the dynamical theory, the so-called *Ewald formula*, leads to different equations for the Bragg case and the Laue case. In particular, for the Bragg case the formula of the *Darwin table* [194, 283] with saturated reflectivity as unity is derived as

$$R(y) = \begin{cases} 1 & y^2 < 1 \\ 1 - \sqrt{1 - \frac{1}{y^2}} & y^2 > 1 \end{cases} \tag{6.3}$$

(refer to, e. g., textbook [I], Eq. (6.2.23)), which illustrates an important difference from the kinematical theory.

The variable in these equations is the normalized deviation angle $y = (\theta - \theta_B)/\Delta\theta_B$, and $\Delta\theta_B$ is the *Darwin width* defined by

$$\Delta\theta_B = \frac{b_c e^{-W}|F_{hkl}|N\lambda^2}{2\pi\sqrt{f_{as}}\sin(2\theta_B)} , \tag{6.4}$$

where e^{-W} is the Debye–Waller factor, $|F_{hkl}|$ is the *crystal structure factor* for the indices h, k, l, N is the number of unit cells per unit volume in the crystal, and the asymmetry factor $f_{as} = \sin(\theta_B - \theta_S)/\sin(\theta_B + \theta_S)$.

Equation (6.3) gives a silk-hat type of reflectivity distribution, the plateau region, that is, the *Darwin region*, of which agrees qualitatively with the approximate result derived by Darwin [194]. This plateau shape of the reflectivity is called the *Darwin table*.

By making use of a specially arranged configuration of the double-crystal spectrometer shown in Figure 6.4a with the Laue case for the first crystal and the Bragg case for the second crystal, Kikuta *et al.* [289] obtained a very precisely parallel beam of thermal neutrons by the selection of the central portion as explained in

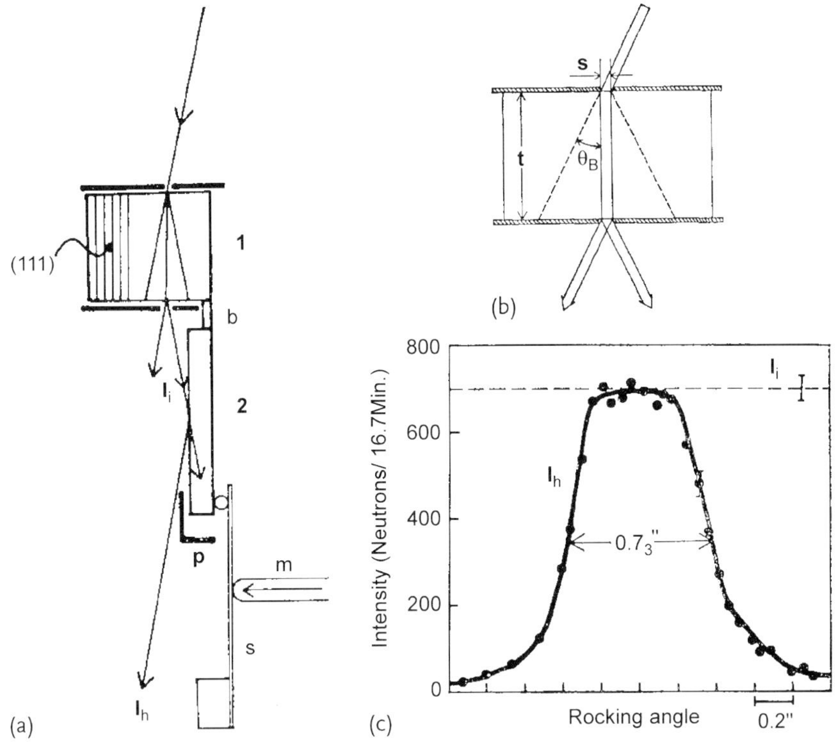

Figure 6.4 Intensity distribution studies on dyanamical diffraction of a precisely parallel thermal neutron beam: (a) double-crystal type of experimental arrangement with Laue-case (1) and Bragg-case (2) crystals, where crystal 2 is provided with a mechanism to rotate it through small angles, where b is a thin bending bridge, s is a plate spring, m is a micrometer, and p is a shielding plate to cut off the beam reflected from the backside of the second crystal; (b) detail explaining the method to obtain the precisely parallel beam by the selection of the central portion of the diffraction fan; (c) observed rocking curve of the intensity for Si(111) symmetrical Bragg-case diffraction (Kikuta *et al.* [289]).

Figure 6.4b of the *diffraction fan* through the first crystal. The parallel beam was further reflected by the second crystal in the setup of *symmetrical Bragg-case diffraction* (i. e., $f_{as} = 1$ in Eq. (6.4)), and thus they observed the silk-hat type of intensity distribution given in Figure 6.4c. The experiment was performed at the JRR-2 reactor at the Japan Atomic Energy Research Institute, Tokai, with a beam of thermal neutrons monochromatized at wavelength $\lambda = 0.86$ Å with a graphite crystal. In their experiment, therefore, (111) reflection of a silicon single crystal ($\theta_B = 7°51'$) was used with crystal thickness t of 13 mm and slit width s of 0.45 mm. In this situation, the angular divergence of the precisely parallel beam in the arrangement shown in Figure 6.4b, given by $\Delta\theta_s = \Delta\theta_B \Delta y_s$, where the selected angular width with the fine slit in the Laue-case diffraction is $\Delta y_s = s/(t \tan\theta_B)$, was estimated as $\Delta\theta_s = 0.065''$, this being sufficiently narrower than the broadening of the silk-hat curve. In this way, the observed intensity distribution illustrated in Figure 6.4c is considered to be very near to the intrinsic distribution in the Bragg-case diffraction.

Actually, the measured intensity distribution agreed well with the expected distribution from the incident intensity I_i indicated in the figure and the convolution of the angular distribution of the exploring beam and the silk-hat type intrinsic distribution in the Bragg-case diffraction, Eq. (6.3), except for a slight rounding on the Darwin table and a slight increase in the half width. This slight increase was considered to be due to the possible effect of geometrical instability in the experimental arrangement during the measuring time of 8 h. From these investigations, the experimental result of neutron diffraction from a Bragg-case crystal is considered to be well explained by the Darwin table formula derived from the dynamical diffraction theory.

After the clarification of the neutron reflectivity characteristics represented by the Darwin curve, neutron spectroscopy developed in two directions. One is the exploration of the reflected intensity increase by making use of a larger wavelength width or angular width with respect to the Darwin width defined by the crystalline parameters. The other is the approach to a higher resolution by realizing a narrower reflection width for neutrons. As the first category in these two directions, we introduce first, in Figure 6.5, the experiment by Klein *et al.* [290] on the neutron reflection in a wider wavelength region from crystalline lattices deformed locally in a thin vibrating crystal plate, and with subsequent local shift of the Bragg reflection wavelength.

A quartz bar with the dimensions indicated in Figure 6.5a vibrates at a frequency of 40 kHz with a peak strain amplitude close to 10^{-4} in the fundamental component of the modulation based on the known piezoelectric relationships. On account of the displacement-dependent strain gradient as shown in Figure 6.5b, more neutrons in a wider wavelength region $\Delta\lambda$ for an angular spread $\Delta\theta$ in the incident range can be accepted compared with the intrinsic Darwin width from an unperturbed crystal. The diffractive reflection is indicated by the ellipse for three slightly different lattice constants in Figure 6.5c.

The experiment was carried out with the thermal neutron beam monochromatized and collimated for a wavelength of 1.09 ± 0.01 Å and an angular divergence of $1.5°$ with a crystal with the dimensions shown in Figure 6.5a in the diffractometer at the high-flux reactor HIFAR, Australia. The experimental result obtained from the vibrating quartz crystal under these conditions gave a maximum diffracted beam intensity about 10 times higher than that from the nonvibrating crystal.

Further, Mikula *et al.* performed the observation and analyses of the intensity increase of a diffracted neutron beam according to the increase of the vibration amplitude of the crystal [291, 292]. Furthermore, as mentioned in Section 1.2, the *backscattering spectrometer* [11] with an especially high wavelength resolution was developed by making use of the combination of the complete backscattering condition with an extremely high wavelength resolution at the Bragg angle satisfying $\theta_B = \pi/2$ [293] and the shift of the Bragg wavelength with a vibrating crystal.

Another category of the studies relating to the Darwin region is the utilization of an assembly of crystals with slightly different Darwin regions and this results

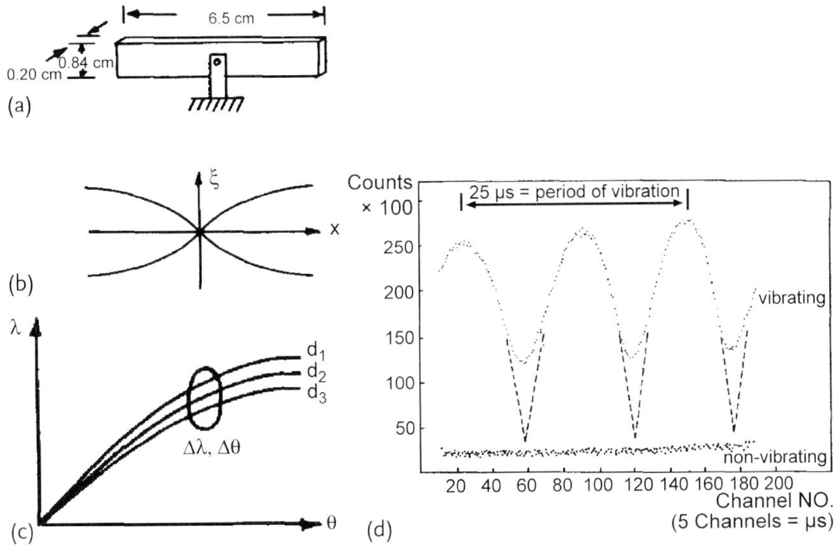

Figure 6.5 Neutron diffraction experiment on a vibrating quartz crystal: (a) dimensions and support; (b) distribution of the local strain gradient depending on the displacement; (c) the region satisfying the Bragg condition for three slightly different lattice constants in λ, θ space; (d) time variation of the neutron beam intensity diffracted by the (200) planes vibrating with a peak strain amplitude of 10^{-4} (Klein *et al.* [290]).

in a much narrower effective Darwin width to apply to ultra-small-angle neutron scattering and neutron optical experiments.

As an example, Treimer *et al.* [294] prepared two sets of sevenfold reflecting *channel cut crystals* (7CCC) cut out from a silicon single crystal, each set consisting of successive threefold and fourfold reflecting crystals, as shown in Figure 6.6a. They carried out a series of *double-crystal diffractometer* experiments by replacing the monochromator and the analyzer crystals in the diffractometer with the sets of 7CCC. Their arrangement of the double-crystal diffractometer corresponds essentially to that shown in Figure 1.2, but the sample in the figure is replaced by the analyzer crystal. Thus, in their configuration the first 7CCC were in the position of the monochromator and the second 7CCC were in the position of the sample working as the analyzer. (In the general configuration of a double-crystal diffractometer to study the structure, a sample with unknown structure is inserted between the monochromater and the analyzer.)

The reflectivity $R_E(y)$ in the case without the wedge in Figure 6.6a is given by $[R(y)]^7$ from Eq. (6.3), and then the Darwin width does not vary but the slope becomes much steeper. In contrast, in the case with the wedge inserted, the neutron waves are refracted by a small angle δ through the wedge, and then the reflectivity results as $R_E(y) = R^3_{\text{Shift}}(y)R^4(y)$, the Darwin width being reduced to $\theta_B - \delta$ [294, 295]. In this way, the rocking curve in the double-crystal diffractome-

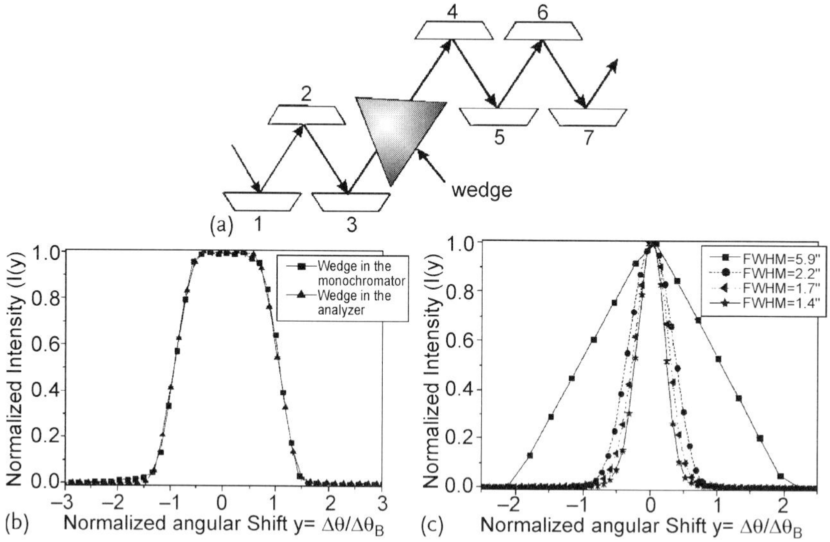

Figure 6.6 Experiment on the channel cut crystal for a tunable Darwin region: (a) sevenfold reflecting channel cut crystal (7CCC), to be used as a monochromator and as an analyzer in a double-crystal diffractometer; (b) measured rocking curves with a silicon wedge inserted in either of the 7CCC; (c) similar results but for the wedges inserted in both of the 7CCC with different Darwin shifts, where the full width at half maximum (FWHM) is indicated in seconds of arc (Treimer *et al.* [294]).

ter with same wedges inserted in both the monochromator and the analyzer is given by the next *convolution*:

$$I(y) = \int R_{E,\text{Monochromator}}(y') R_{E,\text{Analyzer}}(y - y') dy'$$
$$= \int_{-\infty}^{\infty} R^4(y') R^3_{\text{Shift}}(y') R^3_{\text{Shift}}(y - y') R^4(y - y') dy' , \tag{6.5}$$

where we must be careful that, in the case where R^4 is shifted with regard to R^3, R^4 and R^3 are exchanged in the equation, and a different result will be obtained.

They carried out the experiment with the double-crystal diffractometer V12b at the Hahn–Meitner Institute, Berlin, and obtained the results shown in Figure 6.6b for the case of the wedge inserted in either the monochromator or the analyzer. The corresponding Darwin width was sufficiently narrow, and the convolution resulted in essentially the Darwin curve for the other unit. For the wedges inserted in both units, the experimental result was a sharp triangle as shown in Figure 6.6c, and the full width at half maximum of the rocking curve could be varied by changing the refraction angle δ depending on the rotation angle of the wedge.

As an example of experiments in high-precision neutron optics applying the present tunable channel cut double-crystal diffractometer, Treimer *et al.* observed the interference pattern from a silicon phase grating with a periodicity of 16 or 32 μm inserted as the sample in the diffractometer, and compared the result with the theoretical calculations [296]. The comparison indicated the coherence length of a neutron in the experiment to be about 100 μm, assuming the theoretical model of a Gaussian distribution for the coherence of neutron waves with full width at half maximum Λ. The best correspondence with the experimental results was obtained for the case $\Lambda = 50$ μm. (Refer to the previous experimental result giving the neutron coherence length to be larger than 0.3 μm reported by Shull as described in Section 1.3, and also to the experimental result giving the neutron coherence length to be 100 μm or larger obtained by Scheckenhofer and Steyerl with the ultracold neutron gravity diffractometer shown in Figure 2.3a.)

6.2 Inhomogeneity Scattering

6.2.1 Reflection and Transmission Formulas for a Thick Sample

As a contrast to the previous section on the diffraction by regular lattices, neutron transmission and the scattering on a sample with an irregular spatial distribution of scattering nuclei will be introduced in this and the next sections, respectively.

We have already studied the formula for the neutron transmission coefficient on a single layer, Eq. (2.7). However, if we proceed to analyze a thick sample where the attenuation of the wave amplitude in transmission cannot be neglected but should be taken into consideration with the complex potential introduced in Section 2.2, then the wave number k' inside the sample must be replaced by a complex quantity $k'_r + ik''$. Then, we will arrive at the next equation including k'' by making use of Eqs. (1.11), (1.19), and (2.12)

$$k'_r + ik'' = \sqrt{k^2 - 4\pi N(b_{\text{coh}} - ib'')} \cong \sqrt{k^2 - 4\pi N b_{\text{coh}}} + i\frac{\Sigma_T}{2}\,, \tag{6.6}$$

and thus obtain the relations $k'_r = \sqrt{k^2 - 4\pi N b_{\text{coh}}}$, and $k'' = \Sigma_T/2$, where $\Sigma_T = \Sigma_a + \Sigma_{\text{ie}}$ with the approximation $\sigma_c \cong \sigma_a$ (σ_a is the absorption cross section).

The reflectivity R at the interface, that is, at the sample surface, is given by the formula for the reflectivity on a potential step, Eq. (2.4), and the transmission ratio T for the potential step similarly becomes $T = 1 - R = 4kk'_r/|k - k'|^2$. Such reflections and transmissions will take place in a sample with a finite thickness not only on the first arrival of the incident wave at the front surface, but also every time the transmitted waves arrive at the backside of the sample after passing through the sample, and also when the waves reflected by the backside arrive again at the front surface from inside, and so on to be continued repeatedly. Since the equations for R and T are the same for the exchange of k and k' with each other, and also for k and k'_r in the numerator of T, the values for R and T at the backside surface of the sample are the same as those at the front surface. In the case of a thick sample, the effects of these repeated phenomena of reflections and transmissions must be summed up incoherently. As the final result, the general formula for the total reflectivity ρ and the total transmission ratio τ can be obtained as follows:

$$\rho = R\frac{1+\alpha^2(T-R)}{1-\alpha^2R^2}, \qquad \tau = \frac{\alpha T^2}{1-\alpha^2R^2}, \tag{6.7}$$

where

$$\alpha = \exp(-\Sigma_T d) = \exp(-\sigma_T A_\nu w/A_w), \tag{6.8}$$

and d is the sample thickness, A_ν is the Avogadro number, A_w is the atomic weight, and $w = d\rho$ is the mass thickness for specific density ρ.[25]

In the present incoherent treatment, in addition to the attenuation due to the incoherent elastic scattering, in a thick sample with some inhomogeneity, coherent elastic scattering also induces attenuation of the wave being transmitted through the sample. Therefore, the formula for Σ_T includes the contributions from all these processes.

Solving Eq. (6.7) for $\alpha \geq 0$ leads to the result

$$\alpha = -\beta + \sqrt{\beta^2 + 1/R^2}, \tag{6.9}$$

where $\beta = T^2/(2\tau R^2)$. Further, in the case of the neutron energy E sufficiently larger than the optical potential U defined in Section 1.2, that is, in the case of $k \gg k_l = mv_l/\hbar$, with the limiting velocity of total reflection v_l, we can use the approximation $R \cong 0$, $T \cong 1$, then it reduces to $\alpha \cong \tau$. On the other hand, in the case of the velocity region in which the employment of the exact solution, Eq. (6.9), is required, the experimental result must be plotted in relation to the exact neutron velocity inside the sample, $v' = \sqrt{v^2 - v_l^2}$.

6.2.2
Measurements of Inhomogeneity Scattering in the Long-Wavelength Region

In the case of a sample including inhomogeneity on the order of several tens of angstroms, the effects of the inhomogeneity scattering will be observed most typically in an experiment using neutrons with wavelength of magnitude similar to the inhomogeneity size, that is, in very cold neutron scattering experiments. Steyerl *et al.* carried out transmission experiments for neutrons in the velocity region of about 5–500 m/s by preparing the experimental setup for very slow neutron spectrometry [297] shown in Figure 6.7a at the research reactor of the Technical University of Munich, and analyzed the results in view of the inhomogeneity scattering.

The very slow neutrons were extracted from the converter near the reactor core through a vertical guide tube, in which there was a chopper specially designed for very slow neutrons. The pulsed neutrons after the chopper were guided by gravitational deceleration up to the sample and detector assembly at the top of the guide

25) In the completely opposite case in which these multiple reflections and transmissions at the interfaces and also the transmissions through the sample occur completely coherently, that is, when we analyze by putting k' as a complex in Eq. (2.7), we arrive at the result $\tau = \alpha T^2/(1+\alpha^2R^2)$, where $\alpha = \exp(-2k''d) = \exp[-(\Sigma_a + \Sigma_{ie})]$.

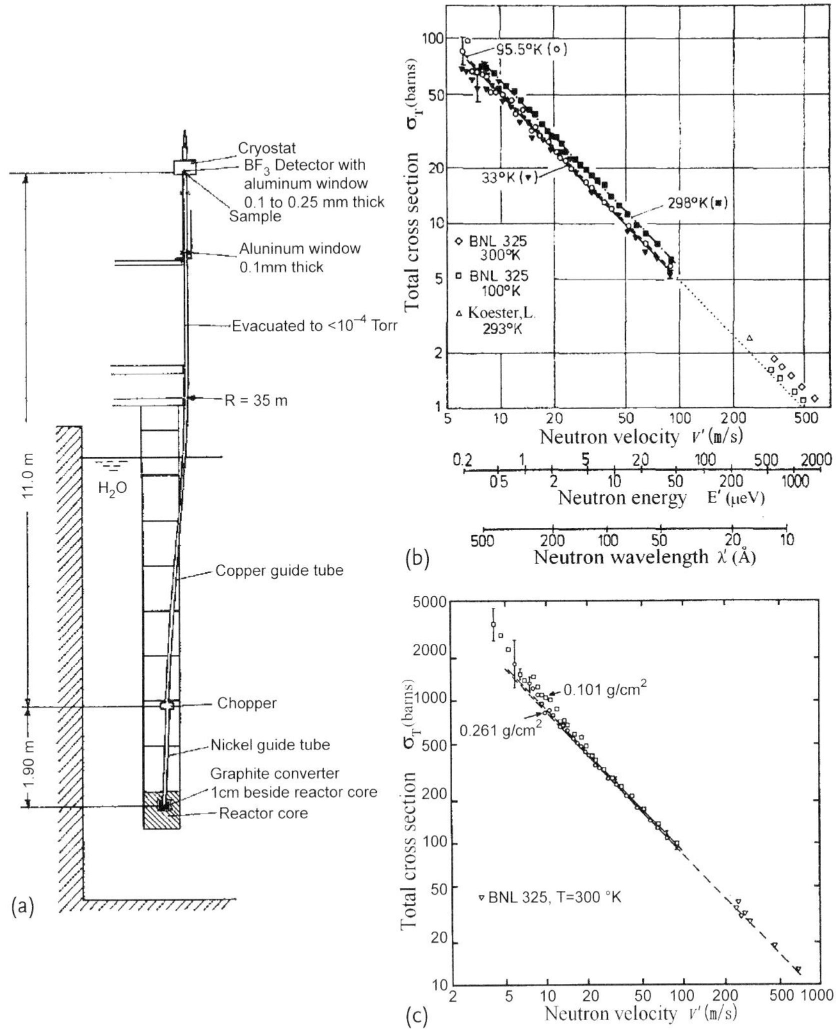

Figure 6.7 Time-of-flight spectroscopy experiment with very slow neutrons at the Technical University of Munich: (a) configuration of the vertical time-of-flight spectrometer (Steyerl [297]); (b) experimental results on the total cross section of a homogeneous aluminum sample; and (c) experimental results on the total cross section of two copper samples (Steyerl *et al.* [298]).

tube with a total length of about 13 m. The velocity resolution of neutrons at the sample position was about 10%, and the velocity-dependent transmission ratio was analyzed with the *time-of-flight method* taking the effect of the gravitational deceleration into consideration.

Prior to the investigation on the inhomogeneity scattering, an aluminum sample as a typical example without serious inhomogeneity was studied. The results are shown in Figure 6.7b [298] for sample temperatures of 33, 95.5, and 298 K. The sample size was $6 \times 6\,\text{cm}^2$, with mass thickness $w = 1.340\,\text{g/cm}^2 \pm 0.1\%$ and purity 99.999%. The detector looks at the sample with a solid angle of $\Omega = 0.53\,\text{sr}$ suspended for the detector window 2.5 cm from the sample, which reduces the additional correction due to the possible isotropic inelastic scattering smaller than 1% to obtain the exact magnitude for the total cross section σ_T. The experimental results on the transmission ratio τ were analyzed by making use of Eqs. (6.8) and (6.9), and the results derived for σ_T are plotted in Figure 6.7b with regard to the neutron velocity v', wavelength λ', and kinetic energy $E' = mv'^2/2$ in the sample with the limiting velocity for aluminum of $v_l = 3.21\,\text{m/s}$.

The velocity dependences shown in Figure 6.7b agree well for all of the sample temperatures studied with the respective straight lines proportional to $1/v'$. The possible contribution from inelastic scattering at a temperature of 33 K was theoretically estimated to be within the experimental errors, and therefore the extrapolation to the neutron velocity of 2200 m/s assuming $\sigma_T = \sigma_a$, where σ_a is the absorption cross section, gave the value $\sigma_{a,2200} = 227 \pm 6\,\text{mb}$, which is consistent with the reported value 230 ± 5 at BNL-325 [299]. In this way, it was confirmed that the velocity dependence of the total cross section for a homogeneous sample in the very slow neutron region can be well approximated by a straight line proportional to $1/v'$.

The next study was carried out on two kinds of thin copper plates of 99.99% purity which were prepared by reducing the thickness by cold-rolling from $1.8\,\text{g/cm}^2$ to $0.261\,\text{g/cm}^2 \pm 3\%$ and to $0.101\,\text{g/cm}^2 \pm 20\%$, respectively. The transmission measurements were performed at 80 and 300 K, and the experimental results were analyzed in the same way as in the previous study on aluminum but with the limiting velocity for copper of $v_l = 5.62\,\text{m/s}$. The results for σ_T at 80 K are plotted in Figure 6.7c for both samples for comparison. Although the velocity dependences seem to hold proportionality of σ_T to $1/v'$ except below $v' \cong 20\,\text{m/s}$, obvious deviations from the straight lines were observed below about 10 m/s for both samples. Furthermore, $1/v'$ fits in the velocity range $20 < v' < 90\,\text{m/s}$ yielded the values for the total cross section at $v = 10\,\text{m/s}$ as $\sigma_{T,10} = 831 \pm 30\,\text{b}$ for the thicker sample and $876 \pm 30\,\text{b}$ for the thinner sample.

The observed higher $\sigma_{T,10}$ for the more strongly deformed sample as well as the deviations from $1/v'$ lay outside the statistical error and were assumed to be caused by large inhomogeneities possibly due to surface impurities or large vacancies and dislocations produced by the rolling treatment.

Further studies on typical cases of inhomogeneity scattering with long-wavelength neutrons will be described in the next section and also in Section 11.3.

6.3
Small-Angle Scattering

Thermal and cold neutrons with a wavelength of 1–10 Å are scattered or diffracted by spatial structures much larger than a crystalline lattice cell or a simple molecule into small angles of a few degrees or below 1°. Such experimental investigations to observe the angular distribution of scattered neutrons in small angles named *small-angle neutron scattering* (SANS) have been used in recent years for structure studies on polymers and other macromolecules, metallic alloys, compounds, and other complex materials. Prior to the development of SANS, the experimental method of small-angle scattering was introduced and applied in the field of X-rays as *small-angle X-ray scattering* (SAXS). There are therefore many common theoretical approaches and subjects to be studied for these two fields of SANS and SAXS. For example, concerning the inhomogeneity studies mentioned in the previous section, various analytical approaches have been presented in addition to Rayleigh's formula for spherical zones [300], Eq. (6.10), described in the following, and most of the studies were started in the field of SAXS. as introduced in textbooks on X-ray scattering [300–304].

The earliest observation of small-angle scattering of neutrons was performed in neutron transmission through a nonmagnetized iron block. The effect observed was attributed to the small-angle scattering by *magnetic domain walls*, and was considered to be an undesirable problem complicating neutron polarization and experiments with polarized neutrons. In this viewpoint, Hughes *et al.* carried out an experiment to investigate in detail the effects occurring in neutron transmission through an iron block [305]. Their experimental arrangement is given in Figure 6.8a; a beam of thermal neutrons extracted from the thermal column of the heavy water reactor at Argonne National Laboratory was used. Prior to the sample measurements, the angular distribution of the direct beam without a sample was ensured to be proportional to the solid line shown in Figure 6.8b, and the distribution observed was well explained by the slit geometry in the experimental arrangement. Next, the measurement with a sample of a magnetized iron block with a thickness of 0.57 cm was carried out and the result is given by the open circles in Figure 6.8b. The distribution for the magnetized iron essentially agreed with that of the direct beam except for a constant attenuation factor of the transmission ratio due to the coherent scattering in the iron block. In contrast, in the case of a nonmagnetized iron block, the result observed as shown by solid circles in the figure indicated obvious dispersion of the angular distribution of the transmitted neutrons.

They analyzed the experimental results with the theoretical model for neutrons in a well-collimated beam experiencing broadening of the Gaussian angular distribution with the width increasing by σ_0 on average owing to the effect of small-angle refractive scattering for every transmission through the magnetic domain. The effects will be accumulated for the multiple transmissions through many domains contained in the nonmagnetized iron sample, leading to a final beam divergence $\sigma = \sigma_0(d/\delta)^{1/2}$ for the transmission of a sample with thickness d, where δ denotes the average size of the *magnetic domain*. Applying the present model to the distribution given by the solid circles in Figure 6.8b, and employing the known value for the domain size in nonmagnetized iron, $\delta = 3.4 \times 10^{-3}$ cm, leads to the estimation $\sigma_0 = 0.027'$. On the other hand, assuming a random distribution of the geometrical direction of the domain walls and also a randomly oriented magnetization direction in both sides of the walls, the theoretical value for σ_0 estimated from

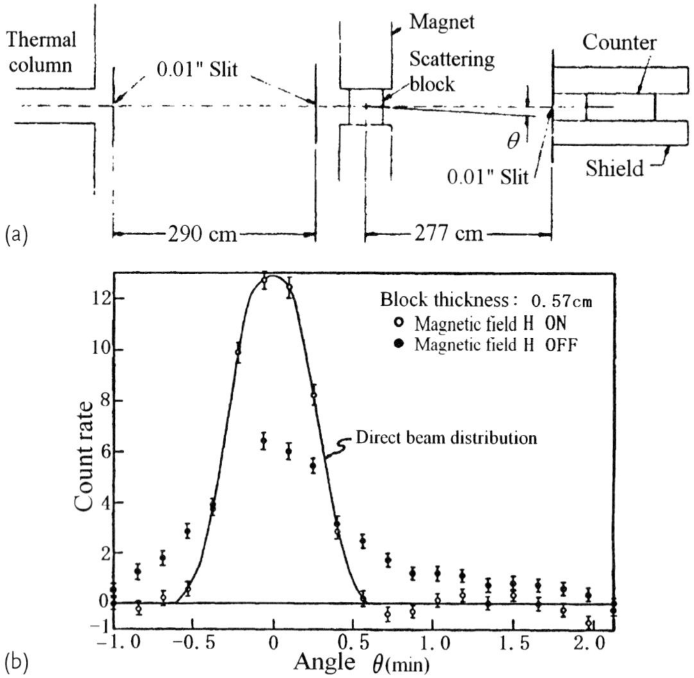

Figure 6.8 First experimental report on magnetic small-angle neutron scattering: (a) experimental setup with a combination of slits; (b) experimental results of transmitted neutrons compared for magnetized and nonmagnetized iron blocks, the fitted curve being proportional to the distribution of the direct beam (Hughes *et al.* [305]).

the refraction formula, Eq. (3.6), becomes $\sigma_0 = 0.029'$, in reasonable agreement with the result estimated from the observed effect of small-angle scattering in the nonmagnetized iron block.

Since the neutron scattering experiment is one of the most advantageous methods to study matter consisting of elements with similar atomic numbers or isotopes, but with very different neutron scattering amplitudes, or magnetic materials, the experimental methods of SANS and also of very cold neutron transmission (VCNT) introduced in the previous section are very powerful approaches to investigate macromolecular or inhomogeneity structures in solid matter, alloys, polymers, biological substances, and other complicated compounds. Actually, in recent years large-scale spectrometers dedicated uniquely to SANS studies have been developed and provided for user experiments in various research fields. In such SANS spectrometers, fundamentally similar configurations are employed as in Figure 6.8a for the first SANS experiments, but the incident neutrons are more precisely monochromatized to make high-resolution structure analyses possible. In addition, a one-dimensional or two-dimensional *position-sensitive detector* system is used for storing all the angular distribution of scattered neutrons at once, instead of the successive counting with slit scanning only in the direction in the arrangement shown in Figure 6.8a. Furthermore, it is worthwhile laboriously changing the distance between the sample and the position-sensitive detector defining the maximum range of the scattering angle θ, that is, the measuring range in the wave vector transfer for a given detector size, according to the requirements for the sample structures to be studied.

In the case of the VCNT method, on the other hand, the experimental apparatus is much simpler at the cost of only integrated information on the structure factor $S(Q)$ being obtained as indicated in Eq. (11.7).

In the following, some fundamental principles and examples of developments in SANS experiments will be described. Some typical results of SANS experiment applications will be given in Section 11.4.

6.3.1
Analysis of the Guinier–Preston Zone

First, one of the standard procedures of SANS experimental studies will be described in this subsection by introducing a typical case of material study on a binary alloy. Understanding the mixing procedure and the possible phase decomposition in binary alloys is a very important task for material science, and it also provides a typical example of the process for SANS experimental studies. Actually, in many cases of binary alloys, phenomena such as the creation of dense regions on one component, that is, precipitation, and the formation of highly concentrated zones on the component, the so-called *Guinier–Preston zone* (G–P zone), are observed [306]. As for the characterizing parameter on a G–P zone, the radius of gyration inertia, or simply the *gyration radius*, for the G–P zone, defined as the mean squared distance from the center of the scattering amplitude density distribution, can be derived from the distribution of the neutron scattering intensity, and is named the *Guinier radius*[26] [304].

The mechanical properties of alloys such as the strength and hardness vary depending on whether the material was *quenched* or gradually cooled down from a high temperature, and also on the aging time after the cooling. This suggests that the microscopic structure of materials such as the size and the constituent concentrations of the precipitated inhomogeneity regions should be different for different conditions during the treatments. Such a phenomenon depending on the time after cooling down is named *aging* [306], and investigations on the aging effects on the material properties and the related mechanism are important not only from the viewpoint of practical use of materials, but also from the viewpoint of scientific interest in nonlinear and nonequilibrium processes in a cooperative macrosystem being related.

26) The Guinier radius R_G can be obtained by applying the Guinier approximation $I(Q) \cong I(0)\exp(-R_G^2 Q^2/3)$ to the scattering intensity $I(Q)$ dependent on the wave vector transfer Q. Then, for spherical zones with radius R, $R_G^2 = 3R^2/5$, for ellipsoids of revolution with radii R, R, and νR, $R_G^2 = (2+\nu^2)R^2/3$, and for cylinders of revolution with diameter d and height L, $R_G^2 = d^2/8 + L^2/12$.

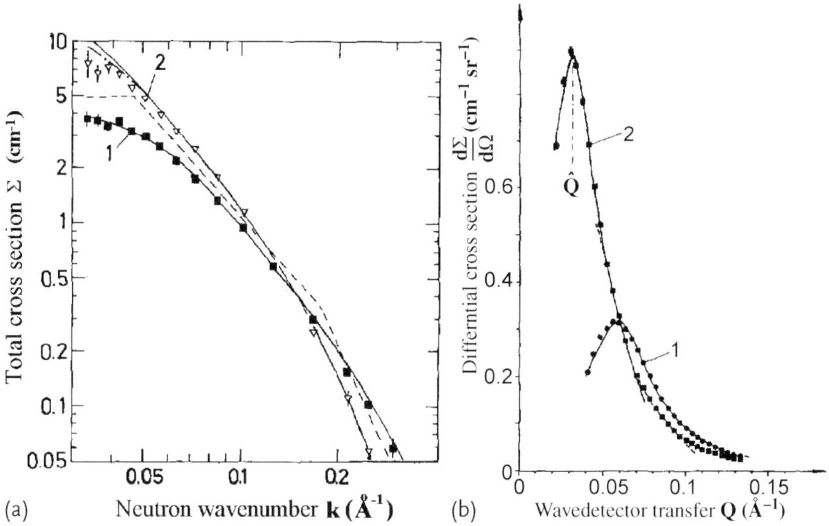

Figure 6.9 Comparison between the results of very cold neutron transmission (VCNT) and small angle neutron scattering (SANS) experiments on Al–5.5 at% Zn alloy; points 1 are the results after aging for several days at room temperature on a slowly quenched sample, and points 2 those after aging for 7 h on a rapidly quenched sample. (a) VCNT results (squares and triangles); the solid lines are from theoretical calculations on noncorrelated spherical precipitates, the dashed line indicates the asymptotic behavior of curve 1, whereas the interparticle interference was taken into consideration in the dash-dotted curve for points 2. (b) SANS results (symbols); the solid curves are for theoretical calculations considering interparticle interference effects in the hard-sphere model, where the wave vector transfer at the peak, $\hat{Q}$, gives an estimation of the interparticle interference parameter (Engelmann *et al.* [307]).

The comparison between the experimental methods of VCNT and SANS applied to same samples including G–P zones was performed by Engelmann *et al.*, the results of which are shown in Figure 6.9. The sample consisting of four electropolished discs of Al–5.5 at% Zn alloy was quenched from a homogenization temperature of 250–350°C, either rapidly in ice or slowly by N_2 vapor at 50°C/min. For such aging conditions, the phase separation in Al–Zn is known to be characterized by the growth of spherical G–P zones with a zinc concentration of 65–69% (depending only on aging temperature). The neutron experiments were carried out with the very cold neutron facility PN5-VCN and the small angle scattering spectrometer D11-SANS, at the Grenoble high-flux reactor.

The arrangement for the VCNT experiment consists of a very cold neutron guide, a chopper disc, and a 5.5-m-long time-of-flight path, with wavelength resolution $\Delta\lambda = 2.4\,\text{Å}$ for very cold neutrons of $10 < \lambda < 300\,\text{Å}$, and an incident neutron divergence angle of $2\theta_0 = 3.2°$, with detector aperture $\theta_1 = 15.2°$. The sample was maintained at liquid N_2 temperature for a measuring time of 0.5–1 day.

On the other hand, the SANS spectrometer consists of a helical slot velocity selector for cold neutrons with wavelength $\lambda = (6.6 \pm 0.6)\,\text{Å}$ and a two-dimensional position-sensitive detector with 64 × 64 elements of 1 cm^2 each. A measuring run had a duration of about 15 min.

For the analytical procedure for this kind of *diffractive scattering of waves in an inhomogeneous sample,* various models have been presented since the early years of the field of X-ray scattering. For the simplest case to be analyzed theoretically, *spherical particles* with radius r distributed in space sufficiently apart from each other, we obtain the angular distribution of the scattering per particle in the Born approximation given by Rayleigh's formula [300]:

$$\frac{d\sigma}{d\Omega} = |f(\theta)|^2 = V^2\rho_d^2 \frac{9\pi}{2}\left[\frac{J_{3/2}(\kappa r)}{(\kappa r)^{3/2}}\right]^2 = (4\pi r^3\rho_d)^2\left[\frac{j_1(\kappa r)}{\kappa r}\right]^2 , \qquad (6.10)$$

where V is the volume of the sphere, $\rho_d = \rho_z - \rho_m$ is the *coherent scattering power* of the spheres ($\rho_z = \langle N b_{\mathrm{coh}}\rangle_{\mathrm{zone}}$) immersed in the liquid-like homogeneous medium ($\rho_m = \langle N b_{\mathrm{coh}}\rangle_{\mathrm{matrix}}$), $\kappa = 2k \sin\theta/2$ is the wave vector transfer, $J_{n+\frac{1}{2}}(z)$ is the Bessel function of half-odd order, and $j_n(z)$ is the spherical Bessel function of the first kind.

The macroscopic cross section for a sample is given by

$$\frac{d\Sigma}{d\Omega} = \nu\frac{d\sigma}{d\Omega} , \qquad (6.11)$$

where ν is the number of zones (regions or particles) per unit volume in the sample and η (packing volume ratio) regarding the packing density of the inhomogeneity zones is defined as the fraction of the total volume occupied by the regions or particles and is given by $\eta = 4\pi r^3\nu/3$. In the present equations, the packing ratio was assumed to be so low that the zones are sufficiently far apart from each other. In contrast, in the case of a high packing ratio, interference effects between zones must be taken into consideration.

Applying the simplest model of spherical zones without interferences to the experimental results of VCNT in Figure 6.9a, one derives the zone radius $r = (22.7 \pm 1.0)$ Å, the packing ratio $\eta = (0.056 \pm 0.005)$, and the zone number per unit volume $\nu = (1.2 \pm 0.2) \times 10^{18}\ \mathrm{cm}^{-3}$ for sample 1. On the other hand, the analytical result for sample 2 was improved as shown by the dash-dotted line by considering the interzone interference effect with the value for the average hard-sphere radius R consistent with the SANS experimental results, although the numerical results converged in the analyses and gave somewhat different values according to the wavelength region applied.

With regard to the SANS results in Figure 6.9b, the next formula was applied as the product of the interzone interference term $F(\eta', QR)$ and the Rayleigh formula, Eq. (6.10), that is,

$$\frac{d\Sigma}{d\Omega} = 12\pi\eta\rho_d^2 r^3\left[\frac{j_1(Qr)}{Qr}\right]^2 F(\eta', QR) , \qquad (6.12)$$

and the *Percus–Yevick correlation function* for hard spheres [308, 309] was employed for the interzone interference structure factor $F(\eta', QR)$, where R is the sphere radius and the packing density $\eta' = (R/r)^3\eta$.

Such analyses of the SANS results for sample 1 gave the values as $r = (24.5 \pm 1.0)$ Å, $R = (44 \pm 4)$ Å, $\eta = (0.038 \pm 0.005)$, $\rho_d = (1.24 \pm 0.11) \times 10^{-6}$ Å^{-2}, and $\nu = (6.5 \pm 2.0) \times 10^{17}$ cm^{-3}. However, for sample 2, similar analyses did not give satisfactory agreement with the SANS experiment for a single set of the parameter values over the whole Q region with the present analytical model. As indicated in the figure, the values $r = 39$ Å and $R = 83$ Å were obtained in the low-Q region, whereas $r = 32$ Å and $R = 60$ Å were obtained in the intermediate-Q region. Such a situation was supposed to be caused by the zone size distribution or the zone shape deformed from spherical. An approximate measure for the average interzone distance could be derived from the wavenumber $\hat{Q}$ corresponding to the peak in Figure 6.9b.

As illustrated by the present comparison between the both experimental approaches and the analytical procedures, the very cold neutron transmission method and the cold neutron small angle scattering method give essentially same indications on the properties of G–P zones formed by the precipitation in binary alloys except the case in which the interzone interference effect plays an important role.

6.3.2
Separation of Nuclear and Magnetic Structures

A different situation regarding the precipitation of an inhomogeneous region will be possible in the case of alloys with a more complicated combination of elements. For example, in the case of ternary or quarternary alloys with a combination of magnetic and nonmagnetic elements, both the nuclear and the magnetic scattering processes occur, and thus there is such a possibility to derive two different inhomogeneity features from the experiments on a sample. The SANS experiment will become one of the most powerful approaches to study the precipitation phenomena and the analyses of the Guinier radius in such multielement alloys if we can separate the nuclear and magnetic scattering contributions in the analyses of SANS data.

Practically, as indicated in the textbook of Kunitomi *et al.* [310, 311], the possible approach to separate the nuclear and the magnetic components in the neutron scattering data is given by the next formula by making use of the very characteristic but simple angular dependence of the magnetic scattering amplitude, that is, its proportionality to $\sin\alpha$, where α is the angle between the wave vector transfer $\boldsymbol{Q}$ and the magnetization direction $\boldsymbol{H}$ in the sample:

$$\frac{d\Sigma}{d\Omega}(Q_x, Q_y) = \left(\frac{d\Sigma}{d\Omega}\right)_{\text{nuc}} + \left(\frac{d\Sigma}{d\Omega}\right)_{\text{mag}} \cdot \sin^2\alpha\,, \tag{6.13}$$

$$\left(\frac{d\Sigma}{d\Omega}\right)_{\text{nuc}} = \left(\frac{d\Sigma}{d\Omega}\right)_{H\|Q}, \tag{6.14}$$

$$\left(\frac{d\Sigma}{d\Omega}\right)_{\text{mag}} = \left(\frac{d\Sigma}{d\Omega}\right)_{H\perp Q} - \left(\frac{d\Sigma}{d\Omega}\right)_{H\|Q}. \tag{6.15}$$

A typical example of such a SANS experiment and the analyses performed by applying the present formula will be given in Section 11.4.

6.3.3
Developments of Small Angle Scattering Spectrometers

To end this section, a few examples of the development and experiments for advanced approaches in SANS will be briefly introduced. One is a new approach to develop *three-dimensional SANS* by Suzuki *et al.* [312], which is intended to obtain the anisotropic description on inhomogeneities surveyed from various directions in space, in contrast to the principal difficulty in deriving such a description with conventional SANS experiments already mentioned. The experimental principle is shown in Figure 6.10a. The SANS data are accumulated by successively changing the setting angle of a sample ω_s with regard to the two-dimensional detector. Although the amount of data to be stored will be huge, a three-dimensional description of inhomogeneity regions can be obtained. The detector at a higher angle is used to investigate the possible anisotropy on the inhomogeneous regions in the reciprocal lattice space in the case of a sample with regular lattices as a crystalline material. An experimental apparatus with the present configuration was installed in the SANS-J spectrometer at the JRR-3M reactor, Tokai.

The experimental result for a single crystal sample at room temperature of the superconductive compound $CeRu_2$ obtained by setting the $\langle 100 \rangle$-axis in the vertical direction is shown in Figure 6.10b for incident neutrons with wavelength $\lambda = 0.63$ nm, resolution $\Delta\lambda/\lambda = 0.14$, and a divergence angle of 0.17°. The two-dimensional detector at a position 10 m from the sample consists of 128×128 pixels in a diameter of 58 cm and accumulates the 360° data on a sample with an angular step of 3°. The corrected intensity distribution is displayed after the analyses by the data visualization software as the isosurface with equivalent intensity as illustrated in Figure 6.10b after being converted from the wave vector transfer $\boldsymbol{q}$ in the laboratory coordinates to the wave vector transfer $\boldsymbol{q}'$ in the single crystal lattice space. In

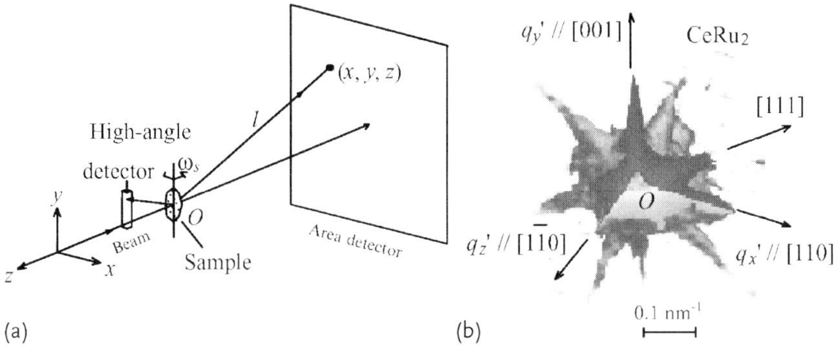

Figure 6.10 A three-dimensional SANS spectrometer and the experimental result. (a) Three-dimensional SANS configuration for a single crystal sample. (b) Profile measured with a three-dimensional SANS spectrometer on a superconductive single crystal of $CeRu_2$; intense diffuse streaks were obviously detected along symmetry directions of the crystalline lattices (Suzuki *et al.* [312]).

the figure, the region where $q_i > 0$, $i = 1$–3, is partly cut out to show the intensity distribution inside the isosurface.

The profile indicates the typical anisotropic distributions along the directions of $\langle 100\rangle$, $\langle 110\rangle$, and $\langle 111\rangle$ as well as the distinct diffuse features of the scattered intensity. Furthermore, the radial profile as the q' dependence averaged around the respective symmetry direction revealed the $(q')^{-4}$ dependence in the region somewhat distant from the peak, as already mentioned for the inhomogeneity formula in the present section. Therefore, the Guinier fitting was applied with an analytical model of the planar defects with the direction correlated to the crystalline structures. They arrived at the result [312] that the average thickness of the defects was 30 nm along the $\langle 111\rangle$ direction, this being about 1.5 times thicker than that along the $\langle 100\rangle$ and $\langle 110\rangle$ directions, 20 nm.

In this way, the experimental method of three-dimensional SANS and the analytical illustration become very powerful approaches to investigate such material with both inhomogeneity and anisotropy. A similar concept of three-dimensional SAXS was presented by Fratzl *et al.* for the study of anisotropic particles in a single crystal of an alloy [313].

Another important direction of SANS developments is the combination with the spin-echo method. Such a development of *spin-echo SANS* (SESANS) was carried out by Bouwman *et al.* [314]. The characteristic feature of SESANS spectroscopy is well understood from the analogy with the characteristic feature of neutron spin-echo (NSE) spectroscopy mentioned in Section 1.5, which directly gives the time correlation function on the dynamics in the sample studied [55]. The space correlation function $\tilde{S}(x)$ can be derived directly from SESANS as the next equation from the single scattered NSE profile on the wave vector transfer with a one-dimensional detector (x-direction) in a SANS configuration:

$$P_x/P_0 = 1 + \sigma d[\tilde{S}(x) - 1]\,,$$
$$\tilde{S}(x) = \frac{1}{\sigma k_0^2}\int_{\text{det}} dQ_x\, dQ_y\, S(Q_x, Q_y)\cos(Q_x x)\,, \qquad (6.16)$$

where $\sigma = \int_{\text{det}} dQ_x\, dQ_y\, S(Q_x, Q_y)/k_0^2$, d is the sample thickness, $x = c/k^2$, k is the neutron wave number, and c is a constant determined by the configuration and strength of the guide field.

Bouwman *et al.* carried out an evaluation experiment on the proposed SESANS spectroscopy by applying the method to a sample of colloidal silica particles under hard-sphere interactions [314].

As an end to this section, a few words are devoted to the development of a miniaturized SANS spectrometer (micro-focusing SANS) by Furusaka at Hokkaido University, in which there is scope to align a number of SANS spectrometers along a beamline exclusively oriented for time-consuming experimental studies.

6.4
Diffraction in Time

6.4.1
Quantum Shutter Analysis

The first situation which was analyzed as belonging to a diffraction in time was the shutter problem [315], that is, Moshinsky considered the geometry as shown in Figure 6.11, in which monochromatic neutrons propagate in the direction of the x-axis and the shutter at $x = 0$ interrupting the neutrons is opened at time $t = 0$. What will be observed by a detector at position x? This is the so-called the shutter problem.

For simplicity, the shutter is assumed to be a complete absorber of neutrons. Then, the wave function Φ for a neutron at time $t > 0$ is obtained by solving the time-dependent Schrödinger equation,

$$-i(\partial \Phi / \partial t) = (\hbar / 2m)(\partial^2 \Phi / \partial x^2) , \tag{6.17}$$

for the initial condition

$$\Phi(x, 0) = \begin{cases} \exp(ikx) & x < 0 , \\ 0 & x > 0 . \end{cases} \tag{6.18}$$

This kind of problem corresponds to the situation in which the stationary slit (or a stationary edge) in the simplest case of the Fresnel diffraction explained in Section 1.2 is replaced by a shutter according to the exchange of the space (z) for the time (t). Therefore, he wrote the solution for Eq. (6.17) as

$$\Phi(x, t) = \frac{1}{2} \chi(x, k, t) \tag{6.19}$$

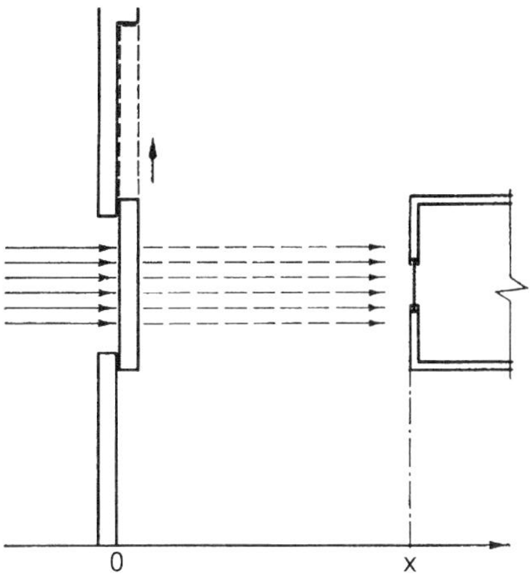

Figure 6.11 Geometry for a shutter problem (Moshinsky [315]).

by making use of

$$\chi(x, k, t) = \exp i\left(kx - \frac{\hbar}{2m}k^2 t\right)\mathrm{erfc}(y)\,, \tag{6.20}$$

where erfc(y) is defined by

$$\mathrm{erfc}(y) = \frac{2}{\sqrt{\pi}}\int_y^{\infty} \exp(-u^2)\,du\,, \tag{6.21}$$

with the complex variable

$$y = \frac{1-i}{2}\sqrt{\frac{m}{\hbar t}}\left(x - \frac{\hbar}{m}kt\right)\,, \tag{6.22}$$

and belongs to the confluent hypergeometrical functions [316–318] for the complex variable y by extending the Gauss error function erfc(x) for a real variable x [319–322]. In particular, the complex values for the integral erfc(y) can be obtained from the *Fresnel integral* [323–325]:

$$\begin{aligned} F(\eta) &= \sqrt{\frac{2}{\pi}}\int_0^{\eta} \exp(iw^2)\,dw \\ &= \int_0^{\tau} \exp\left(i\frac{\pi}{2}s^2\right) ds = C(\tau) + iS(\tau)\,. \end{aligned} \tag{6.23}$$

Since the asymptotic value of erfc(y) for $|y| \to \infty$ approaches 0 at $x > 0$, and to 2 at $x < 0$, the solution, Eq. (6.19), satisfies the initial condition, Eq. (6.18). Further, it can be easily proven by substitution that this solution also satisfies Eq. (6.17).

The exact numerical solutions for $C(\tau)$ and $S(\tau)$ obtained by numerical calculations were given by graphs [323] and tables [325]. The property of the present solution can also be observed to indicate an oscillatory behavior due to the diffraction effect in time, in contrast to the simple delayed stepwise response as the classical solution to the shutter problem, through the method of the *Cornu spiral* [326] by tracing the Fresnel integral on the two-dimensional $C(\tau)$-$S(\tau)$ plane.

However, the solution derived by Moshinsky is not sufficiently instructive for a comparison with neutron experiments on the shutter problem for the following reasons. First, the space was assumed to be one-dimensional, whereas for comparison with experiments the analysis must be extended to the geometry with a shutter with a finite width. Next, the shutter motion program should also be considered so as to make it possible to observe the diffraction effects in time for a practical condition of the neutron intensity. Gähler *et al.* performed these improvements and further carried out an optical experiment in the time domain.

As the starting point, Gähler and Golub [328] extended the space to two dimensions in (x, z). Further, in their analyses to obtain the solution $\Phi(L, d, v_0, t)$ at the observation point ($x = L, z = d$) for the neutron velocity v_0 (or the wavelength λ_0), by assuming the case of a shutter motion with a finite opening width

$z = (+a_0)-(-a_0)$, they solved the time-dependent Schrödinger equation with the Green function method. Furthermore, assuming the experimental conditions constitute the Fraunhofer limit, $a_0^2/2L \ll \lambda_0$; $d \ll L$, they used the *method of stationary phase*[27)] to approximate the integral in the solution by expansion of the phase factor with the time dependence (similar to γ in the previous Eq. (6.22)) around $t = t_\pm = (L^2 \pm 2a_0 d)^{1/2}/v_0 \cong L/v_0 = t_0$, and kept terms up to second order in $(t - t_0)^2$.

The solution obtained had the structure represented by the formula [328, 329]

$$\Phi(L, d, v_0, t) = -i\left(\frac{L}{\pi k_0}\right)^{1/2} \cdot \frac{\exp i(k_0 L - \omega_0 t)}{d} \cdot \phi_{xt}(L, d, v_0, t)\,, \tag{6.24}$$

where $\omega_0 = \hbar k_0^2/2m$ and ϕ_{xt} consists of the Fraunhofer diffraction in space and the Fresnel diffraction in time. In the case of square-pulse shutter openings (shutter opening and closing with infinite velocity), the diffractions in space and in time can be decomposed as

$$\begin{aligned} \phi_{xt}(L, d, v_0, t) &= \phi_x(L, d) \cdot \phi_t(L, t)\,, \\ \phi_x(L, d) &= \sin\left(\frac{k_0 a_0 d}{L}\right), \end{aligned} \tag{6.25}$$

where ϕ_t is given as follows according to shutter motion programs 1 and 2, whereas for case 3, with a triangular pulse opening to be programmed as $z = (+a(t))-(-a(t))$ at time t, the solution indicated a slightly complicated diffraction structure:

1. A single square pulse shutter opening;
   In this case, putting the time duration of opening as T,

$$\begin{aligned} a(t) &= a_0\,; \quad 0 < t < T \\ &= 0 \qquad \text{otherwise}\,, \end{aligned} \tag{6.26}$$

   then the solution becomes

$$\phi_t(L, t) = \{F[\gamma(t - t_0)] - F[\gamma(t - t_0 - T)]\}\,, \tag{6.27}$$

   where $\gamma = \hbar k_0/m\sqrt{k_0/\pi L}$, and $F[\gamma(t - t_i)]$ is the Fresnel integral.
2. Double square pulse shutter opening;
   Putting the shutter period T_2 as shown in Figure 6.12a,

$$\begin{aligned} a(t) &= a_0\,; \quad 0 < t < T \quad T_2 < t < T_2 + T \\ &= 0 \qquad \text{otherwise}\,, \end{aligned} \tag{6.28}$$

27) The method used for finding an approximate value of the integral on a rapidly oscillating function, due to Stokes and Kelvin, who argued that in a wave problem the contribution from the parts of the range of integration near a point of a stationary phase will be nearly in the same phase and add up, whereas those from other parts will interfere. This method is used especially in the theory of dispersion of waves and the theory of probability, including statistical mechanics. For mathematical details refer to [327].

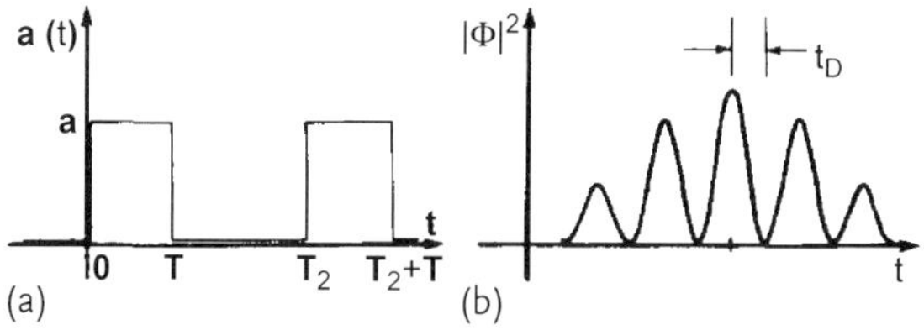

Figure 6.12 Double square pulse shutter opening; the opening function (a) and the calculated result for the neutron intensity (b) (Felber *et al.* [329]).

then the solution becomes

$$\phi_t(L,t) = \{F[\gamma(t-t_0)] - F[\gamma(t-t_0-T)] + F[\gamma(t-t_0-T_2)] - F[\gamma(t-t_0-T_2-T)]\} . \tag{6.29}$$

3. A triangular pulse shutter opening;
   In this case

$$\begin{aligned} a(t) &= \alpha t ; && 0 < t < T \\ &= \alpha(2T-t) ; && T < t < 2T \\ &= 0 && t < 0 \quad \text{or} \quad t > 2T , \end{aligned} \tag{6.30}$$

then the solution is written as

$$\phi_{xt}(L,d,v_0,t) = [\sin\omega_\alpha(t-t_0)\, W(t-t_0) - \sin\omega_\alpha(t-t_0-2T)\, W(t-t_0-T)] , \tag{6.31}$$

where $\omega_\alpha = k_0 \alpha d/L$ and $W(t-t_0) = F[\gamma(t-t_0)] - F[\gamma(t-t_0-T)]$.

The result calculated by Felber *et al.* according to Eq. (6.29) for the case 2 in the conditions of $\lambda = 30$ Å, $L = 1.5$ m, $T = 2 \times 10^{-8}$ s, and $T_2 \cong 6 \times 10^{-8}$ s is shown in Figure 6.12b.

The oscillatory behavior around the central peak of the intensity includes the effect of interference between two successive openings, or in other words *double-slit interference in time*, and the distance t_D in time from the central maximum to the first minimum can be estimated as $t_D = 1/(\gamma^2 T_2)$. They expect to realize $T_2 \cong 6 \times 10^{-8}$ s in their experimental apparatus, and then get $2t_D \cong 6$ µs, which was thought to be observable.

6.4.2
Fame-Overlapping Time-of-Flight Quantum-Chopper Spectrometer

On the basis of the analyses and calculations described in the previous subsection, Felber *et al.* prepared the *frame-overlapping time-of-flight* (FOTOF) quantum-chopper spectrometer as shown in Figure 6.13 for studying the time diffraction phenomena, and carried out the experiments at the GKSS research reactor, FRG-I, Geesthacht [330, 331].

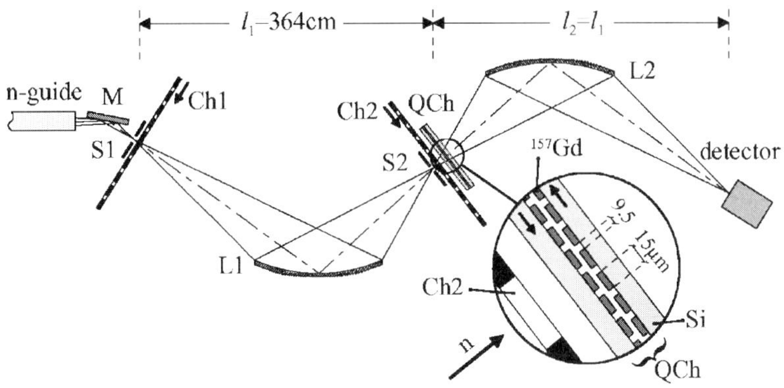

Figure 6.13 Arrangement on the experimental apparatus of the frame-overlapping time-of-flight (FOTOF) spectrometer; the symbols indicating the components are explained in the text (Hils *et al.* [330]).

In the experiment, very cold neutrons with a wavelength of about 34 Å extracted from an extension guide with a cross section of 30 × 3 mm^2, a radius of curvature of 5 m, and a characteristic wavelength 15 Å connected to the cold guide NL-1 were reflected by the monochromatizing mirror M with wavelength $\lambda_0 = 33.9$ Å and wavelength width $\Delta\lambda/\lambda = 7\%$. Then, the neutron beam was pulsed by the first chopper (Ch1) and the second chopper (Ch2) with a diameter of 270 mm rotating with a nominal speed of 347 Hz at the mutual distance l_1, in which Ch1 (Ch2) had effective slit widths of 0.6 mm (0.38 mm) and produced an effective pulse length of 2.0 μs (1.2 μs) full width at half maximum.

The low count rate of detected neutrons due to the severe selection of incident neutrons with long wavelength and in a very short time required special precautions to improve neutron efficiency. First, a neutron guide was used to enlarge the vertical divergent angle to about ±2° for neutrons perpendicular to the plane of

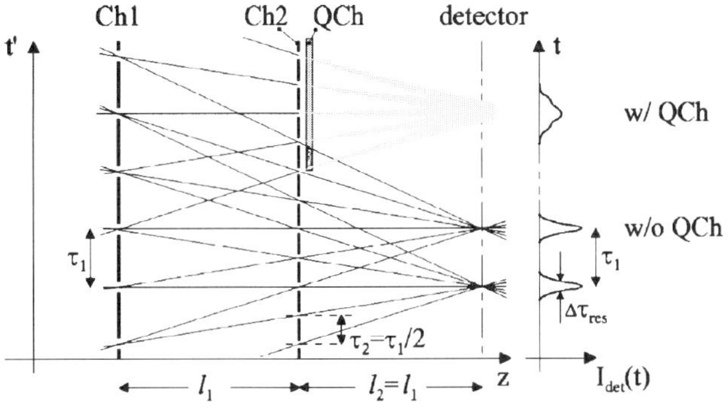

Figure 6.14 FOTOF operation principle: a wide-wavelength region indicated by a number of different gradients of neutron trajectories is focused on the detector, and the quantum chopper causes a characteristic defocusing (Hils *et al.* [330]).

Figure 6.13. Next, in the horizontal direction, two cylindrical mirror lenses L1 and L2 focus neutrons on chopper Ch2 and the detector, respectively, as shown in Figure 6.13. Furthermore, as illustrated in Figure 6.14 by the distance *vs.* time-of-flight diagram, the very precisely arranged FOTOF setup results in most of the transmitted neutrons being focused on the detector.

The *quantum chopper* QCh immediately after the second chopper corresponds to the diffraction sample in the usual diffraction experiments, and diffracts neutrons by the time lattices instead of the spatial lattices as in a crystal in the usual experiments. It consists of a couple of 150-mm-diameter, 3-mm-thick silicon crystal discs with a gap of about 1 mm, with 16 000 radial slits on a ^{157}Gd coating, with a mean distance of 24.9 μm with a width of 8.8 μm and a length of 20 mm. This produces an opening time of 33 ns full width at half maximum and a full period of $T = 186$ ns at the nominal rotation speed of 345 Hz. The operation of time diffraction by the quantum chopper is described by gray zones in Figure 6.14.

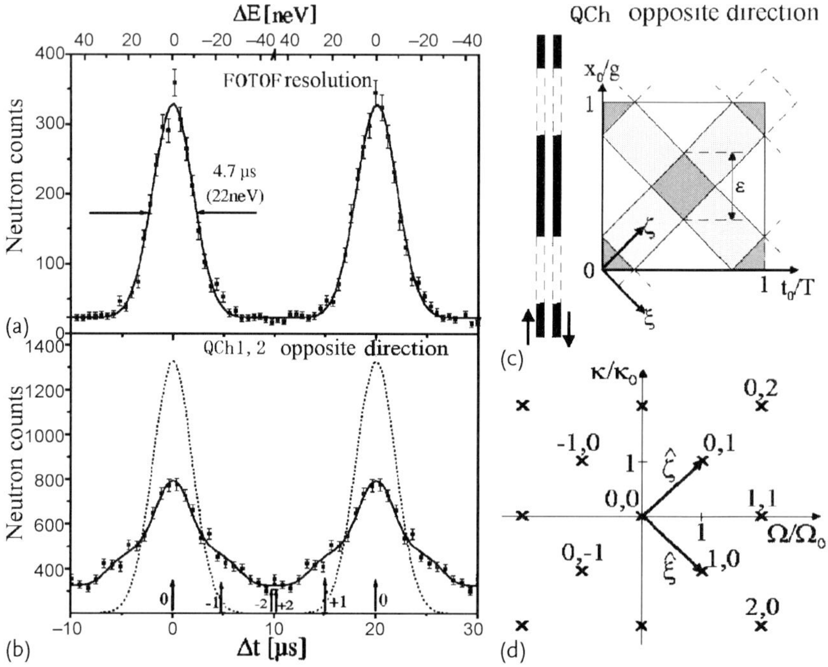

Figure 6.15 Examples of experimental results for time diffraction patterns: (a) FOTOF spectrometer resolution in the measured time (energy) spectrum; (b) neutron count distribution of the time diffraction pattern in quantum chopper QCh operated with both discs rotating in opposite directions, arrows indicate the expected position of sidebands; (c) time–space lattice structure of the transmission function in opposite direction mode, dark gray full transmission, light gray absorption by one grating, unshaded double absorption; (d) allocation of the respective Fourier components, the numbers denote the indices of reciprocal lattice points in the spatial wave number (momentum) κ and the time wave number (energy) Ω plane (Hils *et al.* [330]).

The experimental results and the analyses are shown in Figure 6.15. Figure 6.15, panel a indicates the resolution of the spectrometer in the measured time (energy) spectrum. For the operation of QCh, three kinds of rotation modes are possible, that is, only one disc rotating, both discs rotating in the same direction, and the discs rotating in the opposite directions. The result is given in Figure 6.15, panel b for the case of rotation in opposite directions. The effect of time diffraction convoluted with the spectrometer resolution can be recognized at the position of the expected sidebands ($\Delta E_{sb} = 22.2\,\text{neV}$, $t_{sb} = 4.82\,\mu\text{s}$). As help to understand the effect, Figure 6.15, panel c illustrates the *lattice structure in the time–space plane*, and Figure 6.15, panel d illustrates the *lattice points* and the indices of time diffraction *in the reciprocal space*, of the transmission function of QCh in the opposite direction operational mode. As shown in this figures, the time diffraction experiment with the FOTOF spectrometer gave results consistent with the theoretical expectation based on the time-dependent Schrödinger equation. In other words, an effect similar to the momentum transfer in the diffraction by regular spatial lattices was observed in the time domain. At the same time, the experiment achieved an energy resolution of 20 neV, a new limit for a time-of-flight experiment with slow neutrons.

6.5 Neutron Holography Experiment

6.5.1 Discovery of the Basic Idea of Holography

The basic idea of *holography* was discovered by Gabor[28] in 1948 as published in his articles on a new principle for the electron microscope by *wave-front reconstruction* [332–334]. Thereafter, the principle of holography was extensively developed and applied to such wide fields of image reproduction including not only electron and light optics, but also acoustics and gas dynamics, not only plane images, but also three-dimensional images, and not only a single focused image, but also successive multiple focused images on a single hologram. All these applications realized up to 1971 were reported in his Nobel lecture in 1971 [335].

His idea of holography originated from his consideration of the *resolution restriction of the electron microscope*. It was already known that the spherical aberration of electron lenses sets a limit on the resolving power of the electron microscope at about 5 Å. The prospects for the correction for the objectives are aggravated by the fact that the resolution limit is proportional to the fourth root of the spherical aberration. Thus, an improvement of the resolution by one decimal would require a correction of the objective to four decimals, a practically hopeless task.

28) Gabor called the interference pattern between the object beam and the coherent background (reference beam) produced by an electron microscope a "hologram", from the Greek word *holo* ("the whole"), because it contained the whole information.

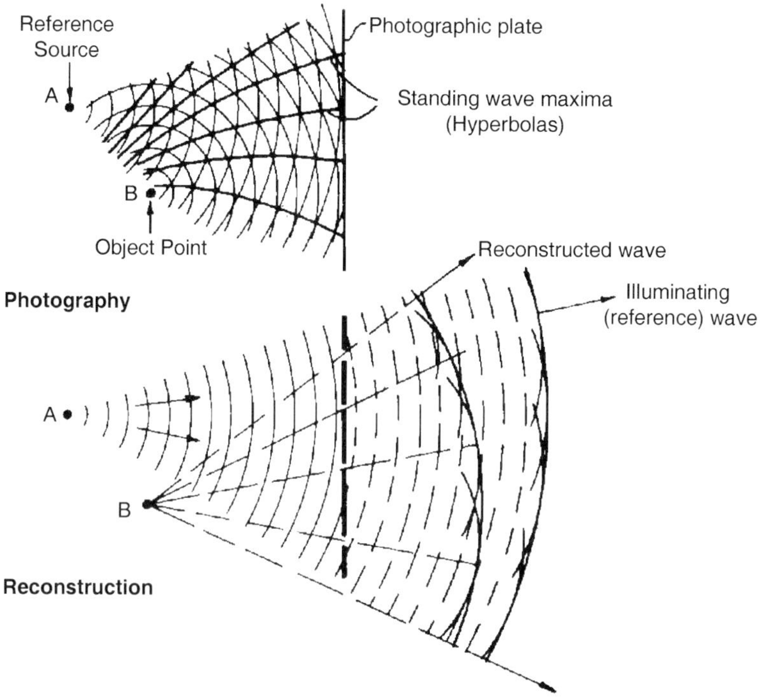

Figure 6.16 Basic idea of holography (Gabor [335]).

Gabor, however, was confident that such a diffraction diagram of an object, if it was taken with *coherent illumination* and then a *coherent background (coherent reference wave)* was added to the diffracted wave, would contain the full information on the modification which the illuminating wave had suffered in traversing the object, apart from an ambiguity of sign. To realize his confidence, he employed a *two-step process* to obtain a micrograph, by electronic analysis, followed by optical synthesis. His basic idea of holography is illustrated in Figure 6.16.

His idea was that, rather than take a bad electron picture, one would take a picture which contains the *whole* information, and correct it by optical means. It was clear to him that this could be done, if at all, only with coherent electron beams, with electron waves which have a definite phase. In contrast, an ordinary photograph loses the phase, since it records only the intensity. Of course, we lose the phase if there is nothing to compare it with. His argument is that, as illustrated in Figure 6.16 for the simplest case when there is only one object point, the interference of the object wave and of the coherent background or "reference wave" will produce interference fringes, with maxima whenever the phases of the two waves are identical. Now, let us make a hard positive record, so that it transmits only at the maxima, and illuminate it with the reference source alone. The phases are, of course, right for reference source A, but as at the slits the phases are identical, they must also be right for B; therefore, the wave of B must also appear *reconstructed*.

6.5.2
Theoretical Principle of Wave Front Reconstruction

Now, we deal with the wave-theoretical verification of the above-mentioned new method of *wave front reconstruction* [333, 334]. Consider a coherent monochromatic wave with a complex amplitude U striking a photographic plate. We write $U = A\exp^{i\varphi}$, where A and φ are real. U may be decomposed into a "background wave" $U_0 = A_0\exp^{i\varphi_0}$ and a remainder $U_1 = A_1\exp i\varphi_1$, which is due to the disturbance created by the object and may be called the secondary wave. Thus, the complex amplitude at the photographic plate is

$$U = U_0 + U_1 = A_0\exp^{i\varphi_0} + A_1\exp^{i\varphi_1} = \exp^{i\varphi_0}[A_0 + A_1\exp^{i(\varphi_1-\varphi_0)}], \quad (6.32)$$

and its absolute value is $A = [A_0^2 + A_1^2 + 2A_0A_1\cos(\varphi_1 - \varphi_0)]^{1/2}$.

The density of photographic plates plotted against the logarithm of exposure is an S-shaped curve, with an approximately straight branch between the two knees. In this region, the transmission of intensity is a power $-\Gamma$ of the exposure. The word "transmission" and the symbol t denote in this section the amplitude transmission, which is in general complex. For pure absorption, without a phase change, t is real number, the square root of the intensity transmission. Thus, we write for the negative process

$$t_n = (K_n A)^{-\Gamma_n}, \quad (6.33)$$

where K_n is proportional to the time of exposure. In the printing of the negative, the exposure is proportional to t_n; hence, the transmission of the positive print becomes

$$t_p = [K_p(K_n A)^{-\Gamma_n}]^{-\Gamma_p} = KA^{-\Gamma}, \quad (6.34)$$

where $\Gamma = \Gamma_n\Gamma_p$ is the "overall gamma" of the negative–positive processes. The same type of law applies if reversal development is used.

If now in the reconstruction process we illuminate the positive hologram with the background U_0 alone, a "substituted wave" U_s will be transmitted, which is given by, apart from a constant factor

$$U_s = U_0 t_p = A_0\exp^{i\varphi_0}[A_0^2 + A_1^2 + 2A_0A_1\cos(\varphi_1 - \varphi_0)]^{\Gamma/2}. \quad (6.35)$$

The simplest, and also the most advantageous choice, is $\Gamma = 2$, which gives

$$\begin{aligned} U_s = U_0A^2 &= A_0\exp^{i\varphi_0}\left[A_0^2 + A_1^2 + 2A_0A_1\cos(\varphi_1 - \varphi_0)\right] \\ &= A_0^2\exp^{i\varphi_0}\left[A_0 + \frac{A_1^2}{A_0^2} + A_1\exp^{i(\varphi_1-\varphi_0)} + A_1\exp^{-i(\varphi_1-\varphi_0)}\right]. \end{aligned} \quad (6.36)$$

Comparing Eq. (6.36) with Eq. (6.32), one sees that if $A_0 = $ const., that is, if the background is uniform, the substituted wave contains a component proportional to the original wave U (the first and third terms). This is not in itself a proof of the principle of reconstruction as any wave can be split into a given wave and a remainder. It remains to be shown that the remainder, that is, the spurious part of U_s, does not constitute a serious disturbance.

This remainder consists of two terms. One of these has the same phase as the background, with an amplitude $(A_1/A_0)^2$ times the amplitude of the background. This term can be made very small if the background is relatively strong, which does not mean that the contrast in the hologram must be poor. The contrast will fall below the observable limit of about 4% only for $(A_1/A_0)^2 \leq 0.0001$, that is, only if the flux scattered by the object into the area of the diagram is less than 0.01% of the illuminating flux.

The second term of the remainder has the same amplitude $A_1 A_0^2$ as the reconstruction of the original secondary wave, but it has a phase shift of opposite sign relative to the background. This wave produces a serious disturbance only in rather exceptional arrangements; in most cases the twin waves can be effectively separated. To make this plausible, one may think of Fresnel-zone plates. Zone plates act simultaneously as positive and negative lenses, producing two focal points, one at each side of the plate, at equal distance, which can be separately observed.

In Gabor's article [333], further possible approaches to eliminate, or at least effectively suppress, the spurious term and the distortion due to the remainder terms by a modification of the photographic process are described. It will be sufficient here to show an example of the practical success of the first holographic reconstruction he presented in his article [333], which is illustrated in Figure 6.17. In this figure, the original is a microphotograph of 1.5-mm diameter, and illuminated with radiation of $\lambda = 4358$ Å through a pinhole of 0.2-mm diameter, reduced by a microscope objective to 5-μm nominal diameter, at 50 mm from the object. The geometrical magnification is 12. The reconstruction was made with lens with an effective numerical aperture of 0.025, then the resolution limit was $0.6 \times 0.436/0.025 = 10$ μm, where 0.6 is the more accurate numerical factor. The noisy background in the reconstruction is chiefly due to imperfections of the illuminating objective.

The successful reconstruction with the holographic technique mentioned above and that became well known with the resolution in visible light region has been extended for images providing atomic resolution with electron waves and X-rays. There are two possible concepts for the generation of a reference wave and detection of interferences in such a holography approach with atomic resolution; that is, an "internal-source concept" [336] and an "internal-detector concept" [337].

In 1974 Bartell and Ritz produced atomic images of the electron cloud of rare-gas atoms by solving the requirement of coherent superposition of a strong reference wave and the weaker object wave by using strong scattering by atomic nuclei and weak scattering by individual atomic electrons in a gaseous sample for electron waves in holography [338]. Tonomura *et al.* developed a highly bright field emission electron beam realizing coherent illumination for imaging an atomic-resolution hologram in an electron microscope. In 1979 they presented a reconstructed image from the hologram of an evaporated gold particle generated with an external coherent source arrangement resulting in a magnification of 140 000 times by

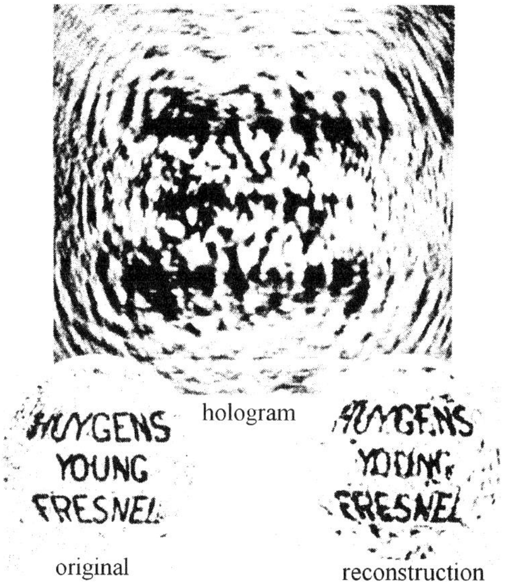

Figure 6.17 First holographic reconstruction in 1948 (Gabor [333]).

illumination with He–Ne laser light [339]. Harp *et al.* reported in 1990 the first successful experimental reconstruction of a three-dimensional crystal structure with atomic resolution from the *Kikuchi pattern* by electron holography with atomically localized incoherent electron sources [340]. X-ray holography with atomic resolution was demonstrated in 1996 by Tegze and Faigel on a single crystal platelet with hard X-ray photons based on the internal-source concept and two-dimensional holography measurement with a solid-state detector [341].

Recently, neutron holography was also demonstrated successfully on samples of an oxide mineral [342] and a single crystal [278, 343] by using thermal neutrons with a wavelength in the range of interatomic distance in solids. This will be described in the next subsection.

6.5.3
Neutron Holography Experiment

Neutrons have the capability to serve often as a unique probe among various scattering waves and particles in their interaction with materials because of the relative visibility of hydrogen and its isotopes, and this fact is a great advantage especially in the study of polymers and biological materials. The possibility of atomic-resolution neutron holography was indicated by Cser *et al.* in the middle of 2001 [344] as two different possible schemes. In the first approach, a *pointlike source of neutron waves* is produced inside the crystal investigated owing to the extremely large value of the incoherent-scattering cross section of the proton and thus hydrogen atoms embedded in a single crystal may serve as pointlike sources when the sample is irradiated by a monochromatic beam of low energy neutrons. In the second approach, the

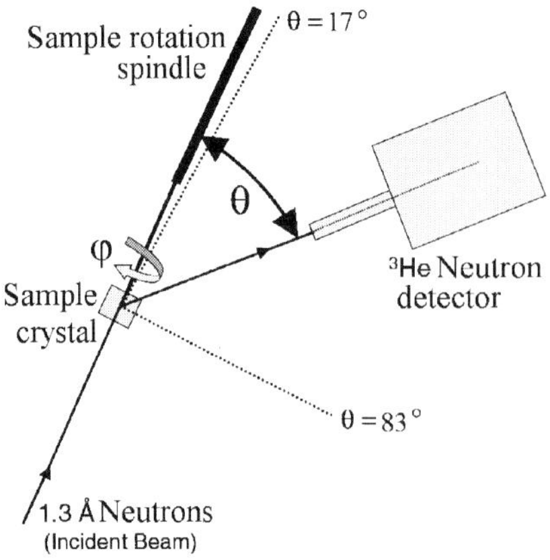

Figure 6.18 Plan view of the experimental setup for thermal neutron holography by Sur *et al.* The incident neutron beam is parallel to the sample rotation axis (φ). The hologram data were obtained by rotating the sample through 2π radians in optimal steps for the given detector coverage ($2^\circ \times 2^\circ$), at a given detector angle θ. This was repeated in 2° steps in θ for $17^\circ \leq \theta \leq 83^\circ$ (Sur *et al.* [342]).

interference between the incident and scattered waves may be registered by means of a *pointlike detector* inserted in the lattice of the crystal under investigation.

Immediately after Cser *et al.* had indicated the possibility of atomic-resolution neutron holography, practical results of atomic-structure holography on the oxide mineral simpsonite with a size of about 6.5 mm long and 3.9 mm thick using thermal neutrons were reported by Sur *et al.* [342] by adopting the *internal-source concept*. The experiment was performed at the N5 instrument located at the National Research Universal Reactor, Chalk River Laboratories, with the experimental setup illustrated in Figure 6.18 using neutrons with wavelength $\lambda = 1.3$ Å and estimated wavelength resolution $\Delta\lambda/\lambda \cong 1.5\%$ from a germanium monochromator.

The raw intensity data covering the sample orientation (φ) and detector angle (θ), after excluding evident intense Bragg peaks from the data analysis on a statistical basis, contains the hologram data plus a θ-dependent background, the latter of which could be subtracted by using the measured data without the sample. By employing a *hologram modulation function* for a scattering potential consisting of a strong incoherent scatterer at the origin surrounded by a coherent scattering amplitude distribution of a pure s-wave weak scatterer, and further after the process to eliminate the conjugate image, they could obtain the results of reconstructions from single-wavelength data as illustrated in Figure 6.19.

In Figure 6.19a, the central spot is the oxygen atom located directly above the origin, and the triangles are drawn to indicate the two triplets of oxygen atoms located slightly below this central oxygen atom. The distortion of the triangles is attributed to limitations in the quality of the present demonstration data. Similarly, in Figure 6.19b, the center of the triangle of the nearest-neighbor oxygen atoms is located directly below the origin in simpsonite.

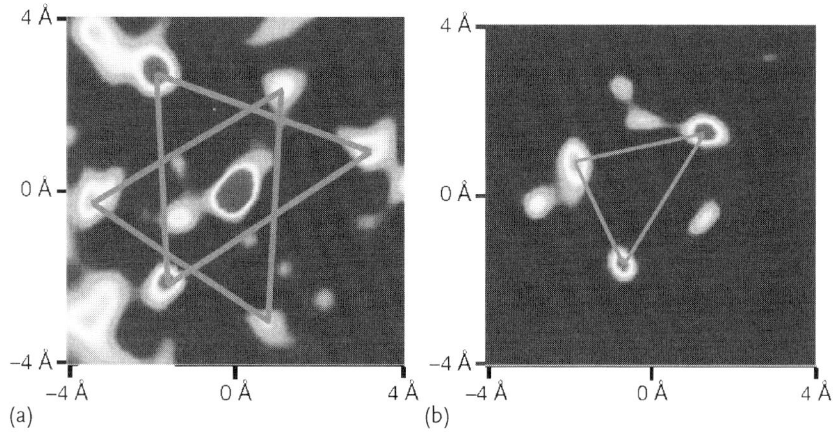

Figure 6.19 Reconstructed planes from the hologram data on a single crystal of natural simpsonite, a rare oxide mineral, $Al_4Ta_3O_{13}(OH)$ (originals are in color). (a) The plane located approximately +0.9 Å from the hydrogen atom (the origin); (b) the same result for the plane located approximately −1.4 Å from the hydrogen atom (the origin) (Sur *et al.* [342]).

Another of the earliest reports on atomic-resolution neutron holography was given by Cser *et al.* [278]. They used *internal-detector concept*, which requires strongly neutron absorbing isotopes acting as pointlike detectors in a sample. The feasibility experiment for this second concept was performed at the D9 four-circle diffractometer installed at the high-flux reactor, Grenoble, with neutrons of wavelength $\lambda = 0.8397$ Å from a copper monochromator. By recording a holographic image of lead nuclei in a Pb(Cd) spherical single crystal with a diameter of about 7 mm, they obtained the results illustrated in Figure 6.20.

Figure 6.20a shows the hologram data after normalization with the monitor counter and corrected for the slowly varying background level. For reconstruction of the object, computer-generated spherical waves were applied to the recorded hologram, as the convolution of converging *s* waves with the data matrix of the scanned (χ, φ) surface. The reconstructed hologram is shown in Figure 6.20b, where the intensity variation of the spots is due to the limited accuracy of the background approximation procedure. The positions of all first-neighbor lead nuclei are clearly visible. Since in a face centered cubic lattice the first 12 neighbors are equidistant from any lattice cite, the 12 lead atoms surrounding a given cadmium atom can be put on a surface of a sphere. This representation is given in Figure 6.20c. The value for the radius of this sphere, 4.93 Å, is in very good agreement with the values determined in the usual X-ray and neutron diffraction measurements.

In this way, both concepts of the internal-source arrangement and the internal-detector arrangement for atomic-resolution neutron holography were successfully realized in the very early 2000s. Holography experiments provide much structural information [344]. The crystal lattice can be restored from the hologram without

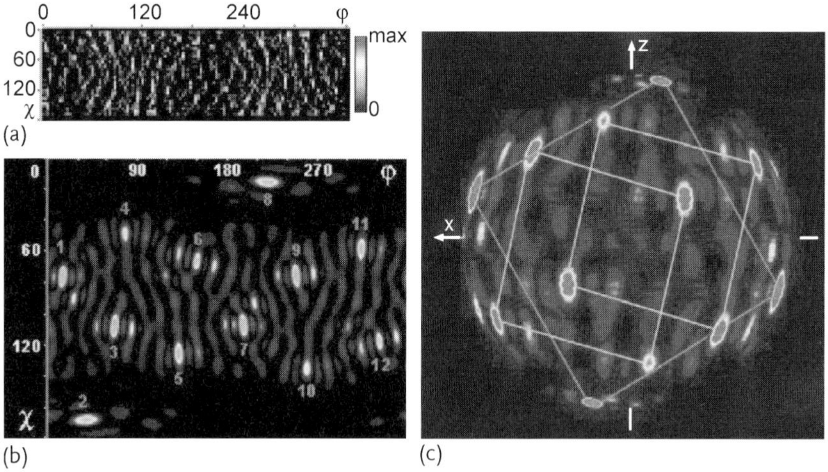

Figure 6.20 Results of the holography experiment on a spherically shaped single crystal of $Pb_{0.9974}Cd_{0.0026}$ (originals are in color): (a) raw hologram data plotted as a pixel map, each pixel covering $3° \times 3°$ in the angular coordinates χ and φ, respectively; (b) reconstructed hologram of 12 Pb neighbors of a Cd nucleus in χ and φ coordinates; (c) the spots representing the positions of the 12 Pb atoms forming the first neighbors of the Cd nucleus displayed on the surface of a sphere of radius $3.49\,\text{Å} = a/\sqrt{2}$, a being the lattice parameter, where the x-axis is the incident beam direction (Cser *et al.* [278]).

making use of any a priori knowledge about the orientation of the sample. Further, interatomic distances can be measured with extremely high accuracy, which is presently limited by the collimation and the wavelength spread of the neutron beam used. The above-mentioned pioneering experimental results can be considered to indicate a certain promise of the neutron holography technique for wide applications in the future.

7
Dynamical Diffraction in Three-dimensional Periodic Media

In this chapter we consider diffraction on an ideal monatomic single crystal, then recall standard dynamical diffraction theory, and finally consider fundamental problems of unitarity and detailed balance, after which we show how to derive the optical potential.

An ideal monatomic crystal can be represented by a system of equidistant atomic planes parallel to the interface and perpendicular to the z-axis. Atoms on atomic planes have discrete periodic positions. A neutron plane wave, $\exp(i\boldsymbol{k}\boldsymbol{r})$, with some wave vector $\boldsymbol{k}$ after scattering on such a crystalline plane situated at $z = 0$ splits into many reflected, $\sum_n \alpha_n \exp(i\boldsymbol{k}_{nr}\boldsymbol{r})$, and many transmitted, $\sum_n \beta_n \exp(i\boldsymbol{k}_{nt}\boldsymbol{r})$, plane waves. This splitting is called diffraction. The difference between wave vectors $\boldsymbol{k}_{nr}$ and $\boldsymbol{k}_{nt}$ is in the sign of their z-components. Wave vectors $\boldsymbol{k}_{nr}$ have negative z-components because they describe propagation to the left of the crystalline plane, and $\boldsymbol{k}_{nt}$ have positive z-components because they describe propagation to the right of the crystalline plane. The set of these wave vectors is infinite but countable, and this set is invariant with respect to scattering on other crystalline planes. The scattering of a plane wave $\exp(i\boldsymbol{k}_n\boldsymbol{r})$ on the crystalline plane can be described by reflection, $\hat{\boldsymbol{\rho}}$, and transmission, $\hat{\boldsymbol{\tau}}$, matrices. With matrices $\hat{\boldsymbol{\rho}}$ and $\hat{\boldsymbol{\tau}}$ we can describe diffraction [345] on a set of crystalline planes like scattering of a spinor particle on a one-dimensional magnetic system. In this chapter we do not take spin into account, treating the neutron as a scalar particle, but make use of matrix technique elaborated for the spinor particle. It is not difficult to include the spin property of the neutrons, but we shall not do it because here we are interested mainly in the diffraction problems.

To see what the wave vectors $\boldsymbol{k}_{nr}$ and $\boldsymbol{k}_{nt}$ of reflected and transmitted diffracted waves are, we describe the positions of atoms on a crystalline plane at $z = 0$ by vectors $\boldsymbol{r} = a\boldsymbol{n}$, where $\boldsymbol{n} = (n_x, n_y)$ is a two-dimensional vector with integer coordinates $n_{x,y}$ and a is the lattice parameter, which for simplicity we take to be the same as the distance between neighboring planes.

Diffraction of a plane wave $\exp(i\boldsymbol{k}_0\boldsymbol{r})$ incident on a crystalline plane means that after scattering the component $\boldsymbol{k}_{0\|}$ parallel to the plane becomes $\boldsymbol{k}_{n\|} = \boldsymbol{k}_{0\|} + \boldsymbol{\kappa}_n$, where $\boldsymbol{\kappa}_n = (2\pi/a)\boldsymbol{n}$ is a vector of the reciprocal lattice of the plane. The $k_{n\perp}$ component, because of the energy conservation (conservation of k^2), becomes $k_{n\perp} =$

Neutron Optics. Masahiko Utsuro, Vladimir K. Ignatovich

ISBN: 978-3-527-40885-6

$\pm\sqrt{k^2 - \boldsymbol{k}_{n\|}^2}$. The plus sign is related to transmitted diffracted waves, and sign minus is related to reflected ones.

It is important to note that the infinite countable discrete set of waves with wave vectors $\boldsymbol{k}_n = (\boldsymbol{k}_{n\|}, \pm k_{n\perp})$ created by diffraction on a single crystalline plane remains invariant after diffraction on other crystalline planes, and because of that we can consider diffraction like a one-dimensional scattering in a magnetic field of a particle with a high even infinite spin.

The states of a particle can be represented by a multidimensional (in fact, infinitely dimensional) vector:

$$|\psi_r\rangle = \begin{pmatrix} \vdots \\ a_{-2} \\ a_{-1} \\ a_0 \\ a_1 \\ \vdots \end{pmatrix} = \hat{\rho}|\psi_0\rangle\,, \quad |\psi_t\rangle = \begin{pmatrix} \vdots \\ b_{-2} \\ b_{-1} \\ b_0 \\ b_1 \\ \vdots \end{pmatrix} = \hat{\tau}|\psi_0\rangle\,, \quad |\psi_0\rangle = \begin{pmatrix} \vdots \\ 0 \\ 1 \\ 0 \\ 0 \\ \vdots \end{pmatrix}\,, \tag{7.1}$$

where the line with number zero is defined, and an element on the nth line shows the amplitude of the wave with wave vector $(\boldsymbol{k}_{n\|}, \pm k_{n\perp})$. With such notation, the incident wave can have arbitrary numbers on all the lines, but in the example shown in Eq. (7.1) it has only one unity on the line of number $n = -1$.

The matrices $\hat{\boldsymbol{\rho}}$ and $\hat{\boldsymbol{\tau}}$ are multidimensional matrices with matrix elements which relate components of the vector $|\psi_0\rangle$ of the incident wave to the components a_j and b_j of reflected and transmitted waves.

To calculate the diffraction we need to find matrices $\hat{\boldsymbol{\rho}}_s$ and $\hat{\boldsymbol{\tau}}_s$ for a single crystalline plane, and matrices $\hat{\boldsymbol{\rho}}$ and $\hat{\boldsymbol{\tau}}$ for a single period, containing the crystalline plane and the two empty spaces of width $a/2$ on both sides of it, as we did in the case of the Kronig–Penney potential (see Figure 4.14). With these matrices we can find the reflection matrix $\hat{\boldsymbol{R}}$ of a semi-infinite ideal crystal and matrix $\exp(i\hat{\boldsymbol{q}}a)$ of the Bloch phase factors for shifting along the z-axis between two adjacent crystalline planes. With $\hat{\boldsymbol{R}}$ and $\exp(i\hat{\boldsymbol{q}}a)$ we can find the reflection $\hat{\boldsymbol{R}}_N$ and transmission $\hat{\boldsymbol{T}}_n$ matrices for a set of N crystalline planes, discuss all the effects related to Bragg and Laue diffraction on single crystals, and compare our results with those obtained with the standard dynamical diffraction theory.

Thus, first, we need to describe diffraction of a plane wave on a single crystalline atomic plane.

7.1
Diffraction on a Single Crystalline Plane

We consider all atoms as fixed-in-space point scatterers with scattering amplitude b. Scattering on a set of scatterers is found with the help of the multiple wave scattering theory [14, 16].

7.1.1 Multiple Wave Scattering Theory

According to the standard scattering theory (its foundation will be considered in Chapter 10), isotropic scattering on a center fixed at a point $\boldsymbol{r}_1$ is described by the wave function

$$\psi(\boldsymbol{r}) = \psi_0(k, \boldsymbol{r}) - \psi_0(k, \boldsymbol{r}_1)\frac{b}{|\boldsymbol{r} - \boldsymbol{r}_1|} \exp(ik|\boldsymbol{r} - \boldsymbol{r}_1|) , \tag{7.2}$$

where $\psi_0(\boldsymbol{r})$ corresponds to the incident neutron with energy $k^2/2$ and the second term, which contains spherical wave $\exp(ikr)/r$, corresponds to the scattered part of the wave function. It contains factor b with dimension of length, called the scattering amplitude, and the factor $\psi_0(k, \boldsymbol{r}_1)$ corresponding to the field illuminating the scatterer. In the following we shall take $\psi_0(k, \boldsymbol{r}) = \exp(i\boldsymbol{k}\boldsymbol{r})$.

If we have N scattering centers, the total wave function is

$$\psi(\boldsymbol{r}) = \exp(i\boldsymbol{k}\boldsymbol{r}) - \sum_{j=1}^{N} \psi_j \frac{b_n}{|\boldsymbol{r} - \boldsymbol{r}_n|} \exp(ik|\boldsymbol{r} - \boldsymbol{r}_n|) , \tag{7.3}$$

where b_n is the scattering amplitude of the center at the point $\boldsymbol{r}_n$ and ψ_n is its illuminating field. For ψ_n we can write down the equations

$$\psi_n = \exp(i\boldsymbol{k}\boldsymbol{r}_n) - \sum_{n' \neq n} \psi_{n'} b_{n'} \eta_{n'n} , \tag{7.4}$$

where $\eta_{n'n} = \exp(ik|\boldsymbol{r}_{n'} - \boldsymbol{r}_n|)/|\boldsymbol{r}_{n'} - \boldsymbol{r}_n|$. In the case of an ideal crystalline plane it is possible to find an analytical solution of system (7.4).

7.1.2 Scattering on an Infinite Crystalline Plane

We assume that the elementary cell of the plane lattice is square with side a (Figure 7.1). The scattering amplitudes of all the atoms are the same: $b_n = b$. The incident plane wave illuminates all the atoms equally. Therefore, the illuminating fields ψ_n of different atoms differ only in the phase $\exp(i\boldsymbol{k}_\parallel \boldsymbol{r}_n)$ of the incident wave, where $\boldsymbol{k}_\parallel$ are the components of the vector $\boldsymbol{k}$ along the crystalline plane. Thus, from symmetry considerations it follows that the solution of system (7.4) can be represented as

$$\psi_n = C \exp(i\boldsymbol{k}_\parallel \boldsymbol{r}_n) , \tag{7.5}$$

with some constant C. Note that from now on we shall enumerate atomic positions with a couple of integers, represented by vector $\boldsymbol{n}$. Substitution of Eq. (7.5) into Eq. (7.4) gives the equation for C: $C = 1 - bSC$, where S is the sum

$$S = \sum_{n \neq 0} \exp(i\boldsymbol{k}_\parallel \boldsymbol{r}_n) \frac{\exp(ikr_n)}{r_n} . \tag{7.6}$$

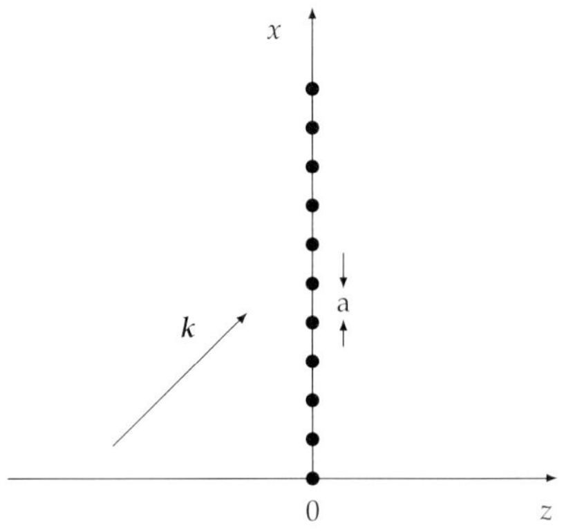

Figure 7.1 An infinite atomic plane and the incident plane wave with a wave vector $\boldsymbol{k}$. Atoms are in the plane x, y at $z = 0$. The elementary cell of the atomic lattice is a square with side a. The sides of the square are directed along the x- and y-axes. The y-axis is perpendicular to the figure plane and directed toward the reader.

Therefore,

$$C = \frac{1}{1 + bS} . \tag{7.7}$$

It is interesting that for some k, $\boldsymbol{k}_{\|}$ the sum S diverges; therefore, $C = 0$ and $b_c = bC = 0$. The neutron with such a wave vector does not notice the atomic plane and propagates like in a vacuum. This happens when $\boldsymbol{k}_{\|} = 2\pi \boldsymbol{m}/a$, where $\boldsymbol{m} = (m_x, m_y)$ is a vector with integer components $m_{x,y}$, and $k = 2\pi m_k/a$, where m_k is some other integer, which is larger than $|\boldsymbol{m}| = \sqrt{m_x^2 + m_y^2}$. For such a wave vector there are infinitely many points $\boldsymbol{r}_n$, where

$$\exp(i\boldsymbol{k}_{\|}\boldsymbol{r}_n)\exp(ikr_n) = 1 . \tag{7.8}$$

Therefore, the sum over such points is infinite. The equality (7.8) takes place at all points $a\boldsymbol{n}$ with such integer components $\boldsymbol{n} = (n_x, n_y)$, which are of the type $n_{x,y} = \nu p_{x,y}$, where ν is an arbitrary integer and $p_{x,y}$ together with $p = \sqrt{p_x^2 + p_y^2}$ constitute a Pythagorean triple.

7.1.3 Scattered Field of the Crystalline Plane

It is not necessary to calculate S for now. We can accept C as a parameter which renormalizes (redefines) the scattering amplitude $b \to b_c = bC$. So we start the calculation of the scattered field itself.

Substitution of Eq. (7.5) into Eq. (7.3) with $b_n = b$ gives

$$\psi(\boldsymbol{r}) = \exp(i\boldsymbol{k}\boldsymbol{r}) - b_c \sum_n \exp(i\boldsymbol{k}_{\|}\boldsymbol{r}_n)\frac{1}{|\boldsymbol{r} - \boldsymbol{r}_n|}\exp(ik|\boldsymbol{r} - \boldsymbol{r}_n|) . \tag{7.9}$$

The spherical function $\exp(ikr)/r$ has the two-dimensional Fourier representation

$$\frac{\exp(ikr)}{r} = \frac{i}{2\pi}\int \frac{d^2 q_\|}{q_\perp} \exp(i\boldsymbol{q}_\| \boldsymbol{r} + iq_\perp |z|)\,, \tag{7.10}$$

where $q_\perp = \sqrt{k^2 - q_\|^2}$. Substitution of it into Eq. (7.9) gives

$$\psi(\boldsymbol{r}) = \exp(i\boldsymbol{k}\boldsymbol{r}) - b_c \frac{i}{2\pi}\int \frac{d^2 q_\|}{q_\perp} \exp(i\boldsymbol{q}_\| \boldsymbol{r} + iq_\perp |z|) \sum_{\boldsymbol{n}} \exp(i[\boldsymbol{k}_\| - \boldsymbol{q}_\|]\boldsymbol{r}_{\boldsymbol{n}})\,. \tag{7.11}$$

For calculation of the sum on the right-hand side there is a useful rule valid for an arbitrary sum:

$$\sum_{n=N_1}^{N_2} f(n) = \sum_{m=-\infty}^{\infty} \int_{N_1}^{N_2} f(n) \exp(2\pi i m n)\, dn\,. \tag{7.12}$$

This rule is easily checked because twice applied it gives the original sum. Indeed, after the first application we obtain the new sum of new functions $F(m) = \int_{N_1}^{N_2} f(n) \exp(2\pi i m n) dn$. If we apply it again in the same way, we obtain

$$\begin{aligned}
\sum_{m=-\infty}^{\infty} F(m) &= \sum_{M=-\infty}^{\infty} \int_{-\infty}^{\infty} F(m) \exp(2\pi i M m)\, dm \\
&= \sum_{M=-\infty}^{\infty} 2\pi \int_{N_1}^{N_2} f(n) \delta(2\pi M + 2\pi n)\, dn \\
&= \sum_{M=-\infty}^{\infty} \int_{N_1}^{N_2} \delta(M - n) f(n)\, dn \\
&= \sum_{M=-\infty}^{\infty} f(M) \Theta(N_1 \le M \le N_2) \\
&\equiv \sum_{M=N_1}^{N_2} f(M)\,.
\end{aligned} \tag{7.13}$$

The rule (7.12) can be generalized to many dimensions. To apply it to the sum on the right-hand side of Eq. (7.11) we put $\boldsymbol{r}_{\boldsymbol{n}} = a\boldsymbol{n}$, where $\boldsymbol{n}$ is a two-dimensional vector with integer components $n_{x,y}$. Then we get

$$\begin{aligned}
\sum_{\boldsymbol{n}} \exp(i[\boldsymbol{k}_\| - \boldsymbol{q}_\|]\boldsymbol{r}_{\boldsymbol{n}}) &= \sum_{n_{x,y}=-\infty}^{\infty} \exp(i[\boldsymbol{k}_\| - \boldsymbol{q}_\|]\boldsymbol{n} a) \\
&= \sum_{m_{x,y}=-\infty}^{\infty} \int d^2 n \exp(i[\boldsymbol{k}_\| - \boldsymbol{q}_\|]\boldsymbol{n} a + 2\pi i \boldsymbol{m}\boldsymbol{n})\,,
\end{aligned} \tag{7.14}$$

where $\boldsymbol{m}$ is a two-dimensional vector with integer components $m_{x,y}$, and integration goes over all the infinite continuous plane of vectors $\boldsymbol{n}$. After change of variables $\boldsymbol{r}' = a\boldsymbol{n}$, and $2\pi\boldsymbol{m}/a = \boldsymbol{\kappa}_m$, where $\boldsymbol{\kappa}_m$ is a vector of the reciprocal lattice of the crystalline plane, we obtain

$$\sum_n \exp(i[\boldsymbol{k}_\| - \boldsymbol{q}_\|]\boldsymbol{r}_n) = \sum_m \frac{(2\pi)^2}{a^2}\delta(\boldsymbol{k}_\| - \boldsymbol{q}_\| + \boldsymbol{\kappa}_m) . \tag{7.15}$$

Substitution into Eq. (7.11) gives

$$\psi(\boldsymbol{r}) = \exp(i\boldsymbol{k}\boldsymbol{r}) - 2\pi i N_2 b_c \sum_\kappa \frac{1}{k_\perp(\boldsymbol{\kappa})} \exp(i\boldsymbol{k}_{\kappa\|}\boldsymbol{r}_\| + ik_\perp(\boldsymbol{\kappa})|z|) , \tag{7.16}$$

where $N_2 = 1/a^2$ is the two-dimensional density of atoms on the crystalline plane, $k_\perp(\boldsymbol{\kappa}) = \sqrt{k^2 - \boldsymbol{k}_{\kappa\|}^2}$, $\boldsymbol{k}_{\kappa\|} = \boldsymbol{k}_\| + \boldsymbol{\kappa}$, and summation over $\boldsymbol{m}$ is replaced by summation over all possible vectors $\boldsymbol{\kappa}$.

At negative z, the term $k_\perp(\boldsymbol{\kappa})|z|$ is equal to $-k_\perp(\boldsymbol{\kappa})z$, which corresponds to waves going to the left from the atomic plane. We call these waves reflected ones. At $z > 0$ we have $k_\perp(\boldsymbol{\kappa})|z| = k_\perp(\boldsymbol{\kappa})z$, which corresponds to waves going to the right. We call them transmitted waves.

Vector $\boldsymbol{\kappa} = 0$ of the reciprocal lattice corresponds to specularly reflected and to forwardly transmitted waves. Vectors $\boldsymbol{\kappa} \neq 0$ correspond to diffracted waves. The number of vectors $\boldsymbol{\kappa}$ is infinite; therefore, the number of diffracted waves is also infinite. However, if $|\boldsymbol{k}_{\kappa\|}| > k$, then the normal component $k_\perp(\boldsymbol{\kappa})$ becomes imaginary and diffracted waves with such a component decay away exponentially from the scattering plane. At sufficient a distance such waves can be neglected and we arrive at a finite number of the diffracted waves with $k_\perp^2(\boldsymbol{\kappa}) > 0$.

In the following we shall represent Eq. (7.16) in the form

$$\psi(\boldsymbol{r}) = \exp(i\boldsymbol{k}\boldsymbol{r}) - i\sum_n \beta_n \exp(i\boldsymbol{k}_{n\|}\boldsymbol{r}_\| + ik_{n\perp}|z|) , \tag{7.17}$$

where

$$\beta_n = p/k_{n\perp}, \quad p = 2\pi N_2 b_c , \quad \boldsymbol{k}_{n\|} = \boldsymbol{k}_\| + \boldsymbol{\kappa}_n , \quad k_{n\perp} = \sqrt{k^2 - \boldsymbol{k}_{n\|}^2} , \tag{7.18}$$

and $\boldsymbol{n}$ numerates vectors $\boldsymbol{\kappa}$ of the reciprocal plane lattice.

7.1.4
An Idea for Calculation of S

The sum S is a complex number $S = S' + iS''$, where

$$S' = \sum_{n\neq 0} \frac{\cos(ka|\boldsymbol{n}| + a\boldsymbol{k}\boldsymbol{n})}{a|\boldsymbol{n}|} , \quad S'' = \sum_{n\neq 0} \frac{\sin(ka|\boldsymbol{n}| + a\boldsymbol{k}\boldsymbol{n})}{a|\boldsymbol{n}|} . \tag{7.19}$$

The imaginary part can be calculated analytically because it can be extended over all points $\boldsymbol{n}$:

$$S'' = \operatorname{Im}\left(\sum_{n} \frac{\exp(ika|\boldsymbol{n}| + iak\boldsymbol{n})}{a|\boldsymbol{n}|} - ik\right)$$
$$= \operatorname{Im}\left(\sum_{m} \int d^2 n e^{2\pi i m n} \int \frac{i d^2 p_{\parallel} \exp(ia\boldsymbol{p}\boldsymbol{n})}{2\pi\sqrt{k^2 - p_{\parallel}^2}} - ik\right), \qquad (7.20)$$

where we used the two-dimensional Fourier representation Eq. (7.10) for the spherical function and the rule (7.12) for summation over the lattice points. Integration on the right-hand side is easily performed and we obtain

$$S'' = -k + \sum_{k_{\perp m}^2 > 0} \frac{2\pi N_2}{k_{\perp m}}, \quad k_{\perp m} = \sqrt{k^2 - (\boldsymbol{k}_{\parallel} + \boldsymbol{\kappa}_m)^2}, \quad \boldsymbol{\kappa}_m = \frac{2\pi}{a}\boldsymbol{m}. \qquad (7.21)$$

The summation goes over a finite number of reciprocal lattice points; therefore, S'' has a finite value for all $\boldsymbol{k}$ except the points where $k_{\perp m} = 0$. At these points $S'' = \infty$ and the renormalization constant $C = 1/(1 + bS)$ is zero. This means that the amplitudes of all the diffracted waves in Eq. (7.16) become zero except for the single wave with amplitude $\propto 1/k_{\perp m}$. At the points $k_{\perp m} = 0$ only the single diffracted wave with amplitude $2\pi N_2 b/k_{\perp m}$ survives, and its amplitude becomes equal to unity. This is a prediction for an experiment with single crystals, and this prediction has not yet been verified.

The real part S' of S is equal to

$$S' = \frac{1}{a} f(\boldsymbol{p}), \quad f(\boldsymbol{p}) = \sum_{n \neq 0} \frac{\cos(p|\boldsymbol{n}| + \boldsymbol{p}\boldsymbol{n})}{|\boldsymbol{n}|}, \quad \boldsymbol{p} = p\boldsymbol{k}/k, \quad p = ak. \qquad (7.22)$$

In textbook [G] it was shown that $f(\boldsymbol{p})$ can be represented as

$$f(\boldsymbol{p}) = \sum_{m=m'}^{\infty} \frac{2\pi}{p''_{\perp m}}, \quad p''_{\perp m} = \sqrt{(\boldsymbol{p}_{\parallel} + \boldsymbol{\kappa}_m)^2 - p^2}, \quad \boldsymbol{\kappa}_m = 2\pi\boldsymbol{m}, \qquad (7.23)$$

where $\boldsymbol{m}'$ is found from the minimum of $k''^2_{\perp m'} > 0$. This sum does not converge, but in textbook [G] it was suggested that Eq. (7.23) is a formal representation of a good function:

$$f(\mu, \boldsymbol{p}) = \sum_{m=m'}^{\infty} \frac{2\pi}{[(\boldsymbol{p}_{\parallel} + 2\pi\boldsymbol{m})^2 - p^2]^{\mu}}, \quad \boldsymbol{p} = \boldsymbol{k}a, \qquad (7.24)$$

which can be calculated for any $\mu > 1$ and then analytically continued to $\mu = 1/2$ like the series $\sum x^n$ representing the function $1/(1 - x)$ can be analytically continued from $x < 1$ to $x > 1$. With this suggestion, $S'(1/2, 0) = 1/a$ was obtained (textbook [G]).

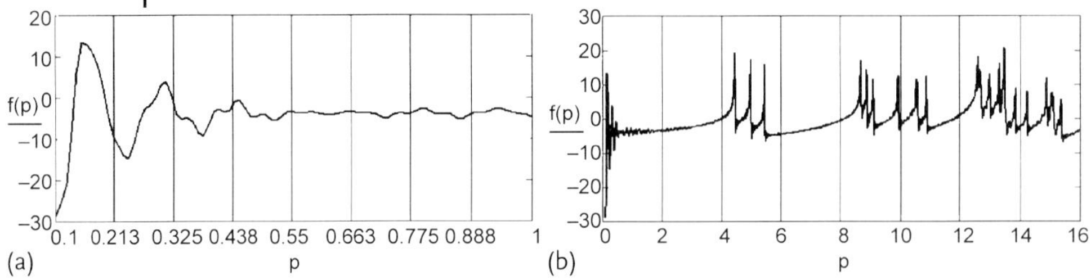

Figure 7.2 Dependence of the function $f(\boldsymbol{p})$ Eq. (7.22) on p for $\boldsymbol{p}$ shown in Eq. (7.25). The function $f(\boldsymbol{p})$ is calculated on a finite lattice with number of scattering points $2N + 1 \times 2N + 1$ and $N = 100$. (a) The function $f(\boldsymbol{p})$ in the range $0 < p < 1$. At small p and $N \to \infty$ it is proportional to $1/p$. (b) The function $f(\boldsymbol{p})$ in the range $0 < p < 16$. At $p > 2$ it has many narrow maxima. Their height is proportional to N. Therefore, at $N \to \infty$ the sum $f(\boldsymbol{p})$ at these points diverges and b_c becomes zero.

However, such a suggestion is not well justified. The dependence of the function $f(\boldsymbol{p})$ on $p = ak$ for a dimensionless wave vector

$$\boldsymbol{p} = p(0.5, 0.4, \sqrt{0.59}) \tag{7.25}$$

is shown in Figure 7.2. The summation in $f(\boldsymbol{p})$ was limited to $-N \le n, m \le N$, where $N = 100$. At small p the function $f(\boldsymbol{p}) \propto 1/p$ when $N \to \infty$. At some other p the function $f(\boldsymbol{p})$ has sharp maxima, which are proportional to N and therefore diverge at $N \to \infty$. At these points the renormalization constant C is zero and the incident wave does not notice the lattice of the crystalline plane. Their positions can be found from number-theoretical considerations and can be checked in an experiment.

To avoid the divergences we must limit the summation over $\boldsymbol{n}$ in Eq. (7.6) by introducing some losses along the path $a\boldsymbol{n}$. For instance, we can represent Eq. (7.6) in the form

$$S = \sum_{n \neq 0} \exp(i\boldsymbol{k}_{\parallel}\boldsymbol{r}_n)\frac{\exp(ikr_n - \mu r_n)}{r_n}, \tag{7.26}$$

where $\mu = \nu\sigma_l(k)$, where $\sigma_l(k)$ is a loss cross section at the neutron momentum k and ν is the atomic density. The loss cross section is the sum of two parts: $\sigma_l(k) = \sigma_0 + k_T\sigma(k_T)/k$. The first one appears because of some incoherent scattering, and the second one is due to inelastic scattering and absorption. This part is expressed via cross section $\sigma(k_T)$ measured at thermal point k_T. The atomic density ν for waves propagating along a crystalline plane should be equal to $1/a^2 r_N$, where r_N is the nuclear radius. With the loss factor, the function $f(p)$ instead of Eq. (7.22) becomes

$$f(\boldsymbol{p}) = \sum_{n \neq 0} \frac{\cos(p|\boldsymbol{n}| + \boldsymbol{p}\boldsymbol{n})}{|\boldsymbol{n}|}\exp(-\mu a|\boldsymbol{n}|). \tag{7.27}$$

With such an expression we obtain $f(p) = 0$ when $p \to 0$, and the maximal value of $f(p)$ is determined by range N of summation, which is not larger than

$N \sim 1/a\mu = ar_N/\sigma_l(k) \approx 100$, and because of that the value of $f(p)$, as seen in Figure 7.2 is not larger than 100; therefore, $bS' = f(p)b/a \le 0.01$, which means that $(b_c - b)/b \approx bS'$ is of the order of 1%. Though this correction is small it is not well determined and varies with energy, therefore it needs further investigation.

7.2 Diffraction from a Semi-infinite Single Crystal

The wave function (7.17) describing scattering on a crystalline plane is a discrete superposition of plane waves:

$$\psi(\boldsymbol{r}) = e^{i\boldsymbol{k}\boldsymbol{r}} - i\sum_n \beta_n \exp(i\boldsymbol{k}_{n\|}\boldsymbol{r}_\| + ik_{n\perp}|z|)$$
$$\equiv \Theta(z<0)\left[\psi_0(\boldsymbol{r}) + \psi_r(\boldsymbol{r})\right] + \Theta(z>0)\psi_t(\boldsymbol{r})\,, \tag{7.28}$$

which consists of two parts. On the left side $(z < 0)$ it contains the incident, ψ_0, and reflected, ψ_r, waves, and on the right it contains transmitted, ψ_t, waves.

Combination of the diffracted waves can be represented by a column vector $|\boldsymbol{\Psi}\rangle$ (7.1). The states $|\boldsymbol{\Psi}\rangle$ create a linear vector space. In this linear space we can choose basis vectors $|n\rangle$, every one of which contain unity on the line n and zeros at all other places. The state $|\boldsymbol{\Psi}\rangle$ represented in such a basis is $|\boldsymbol{\Psi}\rangle = \sum_n \alpha_n|n\rangle$. Complex numbers α_n corresponding to amplitudes of the plane waves with wave vectors $\boldsymbol{k}_n$ are the coordinates of the state vector $|\boldsymbol{\Psi}\rangle$ in the basis $|n\rangle$.

With the above notation, the wave function (7.28) is representable as

$$|\psi(\boldsymbol{r})\rangle = \Theta(z<0)\left[\exp(i\hat{\boldsymbol{k}}\boldsymbol{r}) + \exp\left(i\hat{\boldsymbol{k}}_r\boldsymbol{r}\right)\hat{\boldsymbol{\rho}}_s\right]|\boldsymbol{\Psi}_0\rangle$$
$$+ \Theta(z>0)\exp(i\hat{\boldsymbol{k}}\boldsymbol{r})\hat{\boldsymbol{\tau}}_s|\boldsymbol{\Psi}_0\rangle\,, \tag{7.29}$$

where the reflection, $\hat{\boldsymbol{\rho}}_s$, and transmission, $\hat{\boldsymbol{\tau}}_s$, matrix amplitudes of a single crystalline plane are introduced. We also introduced here the operators $\hat{\boldsymbol{k}}$ and $\hat{\boldsymbol{k}}_r$, which have eigenvectors $|n\rangle$ and eigennumbers $(\boldsymbol{k}_{n\|}, k_{n\perp})$ and $(\boldsymbol{k}_{n\|}, -k_{n\perp})$, respectively. The state vector $|\boldsymbol{\Psi}_0\rangle = |\boldsymbol{\kappa} = 0\rangle \equiv |0\rangle$ corresponds to the incident wave $\exp(i\hat{\boldsymbol{k}}\boldsymbol{r})|\boldsymbol{\Psi}_0\rangle = \exp(i\boldsymbol{k}\boldsymbol{r})|0\rangle$.

The basis $|n\rangle$ can be used for representation of matrices. A matrix $\hat{M}$ with a matrix element M_{ln}, which transforms the state vector $|n\rangle$ into $|l\rangle$, can be represented as $\hat{M} = \sum_{l,n}|l\rangle M_{ln}\langle n|$. Therefore, the matrix elements of the matrix $\hat{M}$ are $\langle l|\hat{M}|n\rangle = M_{ln}$.

The matrix elements of matrices $\hat{\boldsymbol{\rho}}_s$ and $\hat{\boldsymbol{\tau}}_s$ are, respectively,

$$\langle l|\hat{\boldsymbol{\rho}}_s|n\rangle = -i\beta_l\,, \quad \langle l|\hat{\boldsymbol{\tau}}_s|n\rangle = \delta_{ln} - i\beta_l\,, \tag{7.30}$$

where $\beta_l = 2\pi N_2 b_c/k_{l\perp}$ and δ_{ln} is the Kronecker symbol. Note that the matrix elements $\langle l|\hat{\boldsymbol{\rho}}_s|n\rangle$ of $\hat{\boldsymbol{\rho}}_s$ do not contain a dependence on the initial index n, and the matrix $\hat{\boldsymbol{\tau}}_s$ is equal to $\hat{\boldsymbol{I}} + \hat{\boldsymbol{\rho}}_s$, where $\hat{\boldsymbol{I}}$ is the unit matrix.

A period of the monatomic crystal with a cubic elementary cell and crystalline planes parallel to the interface can be represented by an atomic plane and two empty spaces of width $a/2$ on both sides of it like in the Kronig–Penney potential (Figure 4.14b).

The reflection, $\hat{\boldsymbol{\rho}}$, and transmission, $\hat{\boldsymbol{\tau}}$, matrices of a single period can be represented as $\hat{\boldsymbol{\rho}} = \hat{\boldsymbol{E}}\hat{\boldsymbol{\rho}}_s\hat{\boldsymbol{E}}$, $\hat{\boldsymbol{\tau}} = \hat{\boldsymbol{E}}\hat{\boldsymbol{\tau}}_s\hat{\boldsymbol{E}} = \hat{\boldsymbol{E}}^2 + \hat{\boldsymbol{\rho}}$, where $\hat{\boldsymbol{E}}$ is the diagonal matrix with matrix elements $\hat{\boldsymbol{E}}_{ln} = \langle l|\hat{\boldsymbol{E}}|n\rangle = \delta_{ln}e_l$, with $e_l = \exp(ik_{l\perp}a/2)$ describing propagation of waves along the empty space of width $a/2$.

It is worth introducing a diagonal matrix $\hat{\boldsymbol{\beta}}$ with elements $\hat{\boldsymbol{\beta}}_{ln} = \delta_{ln}\beta_n$, and a vector $|\boldsymbol{n}\rangle = \sum_n |n\rangle$. With them the matrix $\hat{\boldsymbol{\rho}}$ becomes

$$\hat{\boldsymbol{\rho}} = -i\hat{\boldsymbol{\beta}}\hat{\boldsymbol{E}}|\boldsymbol{n}\rangle\langle\boldsymbol{n}|\hat{\boldsymbol{E}}\,. \tag{7.31}$$

To find the reflection matrix $\hat{\boldsymbol{R}}$ from a semi-infinite crystal for the incident wave at the first period $|\boldsymbol{\Psi}_0\rangle$ we need to find the wave $\hat{\boldsymbol{X}}|\boldsymbol{\Psi}_0\rangle$ incident on the second period, where $\hat{\boldsymbol{X}} = \exp(i\hat{\boldsymbol{q}}a)$ is a matrix of the Bloch phase factors. Matrices $\hat{\boldsymbol{R}}$ and $\hat{\boldsymbol{X}}$ like in the one-dimensional case are related by the equations

$$\hat{\boldsymbol{X}} = \hat{\boldsymbol{\tau}} + \hat{\boldsymbol{\rho}}\hat{\boldsymbol{R}}\hat{\boldsymbol{X}}\,, \quad \hat{\boldsymbol{R}} = \hat{\boldsymbol{\rho}} + \hat{\boldsymbol{\tau}}\hat{\boldsymbol{R}}\hat{\boldsymbol{X}}\,. \tag{7.32}$$

The first equation can be resolved with respect to $\hat{\boldsymbol{X}}$:

$$\hat{\boldsymbol{X}} = (\hat{\boldsymbol{I}} - \hat{\boldsymbol{\rho}}\hat{\boldsymbol{R}})^{-1}\hat{\boldsymbol{\tau}}\,. \tag{7.33}$$

Substitution of Eq. (7.33) into Eq. (7.32) gives the equation for $\hat{\boldsymbol{R}}$:

$$\hat{\boldsymbol{R}} = \hat{\boldsymbol{\rho}} + \hat{\boldsymbol{\tau}}\hat{\boldsymbol{R}}(\hat{\boldsymbol{I}} - \hat{\boldsymbol{\rho}}\hat{\boldsymbol{R}})^{-1}\hat{\boldsymbol{\tau}}\,. \tag{7.34}$$

It is impossible to solve such an equation analytically. However, we can find an approximate solution and a peculiarity of the matrix $\hat{\boldsymbol{\rho}}$ helps for this. The peculiarity of the matrix $\hat{\boldsymbol{\rho}}$ is that its matrix elements are representable in the form of a product $\rho_{ln} = \langle l|\hat{\boldsymbol{\rho}}|n\rangle = a_l b_n$ of components $a_l = -ie_l\beta_l$ and $b_n = e_n$ of two state vectors $|\boldsymbol{a}\rangle$ and $|\boldsymbol{b}\rangle$. Matrices $\hat{\boldsymbol{M}}$ with matrix elements $M_{ln} = p_l q_n$ equal to the product of coordinates p_l, q_n of two vectors $|\boldsymbol{p}\rangle$ and $|\boldsymbol{q}\rangle$ are called dyadic ones. The dyadic matrices have the following properties.

7.2.1
Properties of Dyadic Matrices

If $\hat{\boldsymbol{A}}$ is a dyadic matrix, that is, a matrix representable as $\hat{\boldsymbol{A}} = |\boldsymbol{K}\rangle\langle\boldsymbol{N}|$ with the help of some vectors $|\boldsymbol{K}\rangle$ and $|\boldsymbol{N}\rangle$ with coordinates K_n, N_n, then:

1. For an arbitrary matrix $\hat{\boldsymbol{B}}$, the product $\hat{\boldsymbol{A}}\hat{\boldsymbol{B}}$ is also the dyadic matrix

$$\hat{\boldsymbol{A}}\hat{\boldsymbol{B}} = |\boldsymbol{K}\rangle\langle\boldsymbol{\mathcal{B}}^r|\,, \quad \hat{\boldsymbol{B}}\hat{\boldsymbol{A}} = |\boldsymbol{\mathcal{B}}^l\rangle\langle\boldsymbol{N}|\,,$$
$$\mathcal{B}_n^r = \sum_m N_m B_{mn}\,, \quad \mathcal{B}_n^l = \sum_m B_{nm}K_m\,; \tag{7.35}$$

2. For an arbitrary $\hat{B}$ we have

$$\hat{A}\hat{B}\hat{A} = c_{AB}\hat{A}\,, \quad c_{AB} = \mathrm{Sp}\hat{A}\hat{B} = \langle \boldsymbol{N}|\hat{B}|\boldsymbol{K}\rangle = \sum_{m,n} N_m B_{mn} K_n\,; \tag{7.36}$$

3. The product of the matrix A and an arbitrary vector $|\boldsymbol{q}\rangle = \sum_n Q_n|n\rangle$ gives the vector

$$\hat{A}|\boldsymbol{q}\rangle = c_{Nq}|\boldsymbol{K}\rangle = c_{Nq}\sum_n K_n|n\rangle\,, \quad c_{Nq} = \langle \boldsymbol{N}||\boldsymbol{q}\rangle = \sum_{n=-\infty}^{\infty} N_n Q_n\,. \tag{7.37}$$

   Therefore, the dyadic matrix $\hat{A}$ is proportional to a projection operator because it projects any vector onto the vector $|\boldsymbol{K}\rangle$.
4. The square of $\hat{A}$ is

$$\hat{A}^2 = |\boldsymbol{K}\rangle\langle \boldsymbol{N}||\boldsymbol{K}\rangle\langle \boldsymbol{N}| = c_A\hat{A}\,,$$
$$c_A = \mathrm{Sp}\hat{A} = \langle \boldsymbol{N}||\boldsymbol{K}\rangle = \sum_{m=-\infty}^{\infty} K_m N_m\,. \tag{7.38}$$

5. For unit matrix $\hat{I}$,

$$(\hat{I} - \hat{A})^{-1} = \hat{I} + C_A\hat{A} \tag{7.39}$$

   is valid, where $C_A = 1/(1 - c_A)$, $c_A = \langle \boldsymbol{N}||\boldsymbol{K}\rangle$. Indeed

$$(\hat{I} - \hat{A})^{-1}(\hat{I} - \hat{A}) = (\hat{I} + C_A\hat{A})(\hat{I} - \hat{A}) = \hat{I} - (1 - C_A + c_A C_A)\hat{A} = \hat{I}$$
$$\to C_A(1 - c_A) = 1\,.$$

7.2.2
Solution of Eq. (7.34)

The matrix $\hat{R}$ defines waves reflected to the vacuum. To find $\hat{R}$ we substitute $\hat{\boldsymbol{\tau}} = \hat{E}^2 + \hat{\rho}$ into Eq. (7.34) and use relation (7.39) for transformation of $(\hat{I} - \hat{\rho}\hat{R})^{-1}$. As a result, we obtain

$$\begin{aligned}\hat{R} &= \hat{\rho} + (\hat{E}^2 + \hat{\rho})\hat{R}(\hat{I} + C_R\hat{\rho}\hat{R})(\hat{E}^2 + \hat{\rho})\\ &= \hat{\rho} + (\hat{E}^2\hat{R} + C_R\hat{\rho}\hat{R} + C_R\hat{E}^2\hat{R}\hat{\rho}\hat{R})(\hat{E}^2 + \hat{\rho})\,,\end{aligned} \tag{7.40}$$

where in the second equality we multiplied first three factors and used the relations $C_R = 1/(1 - c_R)$ with

$$c_R = \mathrm{Sp}\hat{\rho}\hat{R} = -i\mathrm{Sp}(\hat{\boldsymbol{\beta}}\hat{E}|\boldsymbol{n}\rangle\langle\boldsymbol{n}|\hat{E}\hat{R}) = -i\langle\boldsymbol{n}|\hat{E}\hat{R}\hat{\boldsymbol{\beta}}\hat{E}|\boldsymbol{n}\rangle\,.$$

After multiplication of two factors on the right-hand side with account of rules (7.35) and (7.36), we reduce Eq. (7.40) to

$$\hat{R} - \hat{E}^2\hat{R}\hat{E}^2 = \frac{(\hat{I} + \hat{E}^2\hat{R})\hat{\rho}(\hat{I} + \hat{R}\hat{E}^2)}{1 + i\langle n|\hat{E}\hat{R}\hat{\beta}\hat{E}|n\rangle}$$
$$\equiv -i\frac{(\hat{E} + \hat{E}^2\hat{R}\hat{E})\hat{\beta}|n\rangle\langle n|(\hat{E} + \hat{E}\hat{R}\hat{E}^2)}{1 + i\langle n|\hat{E}\hat{R}\hat{\beta}\hat{E}|n\rangle} . \tag{7.41}$$

Let us represent $\hat{R}$ as $\hat{R} = -i\hat{E}\hat{\beta}\hat{R}^s\hat{E}$. Substitute it into Eq. (7.41) and divide both sides of the equation by diagonal matrices $-i\hat{E}\hat{\beta}$ on the left and by $\hat{E}$ on the right. As a result, (7.41) reduces to

$$\hat{R}^s - \hat{E}^2\hat{R}^s\hat{E}^2 = \frac{(\hat{I} - i\hat{E}^2\hat{R}^s\hat{\beta}\hat{E}^2)|n\rangle\langle n|(\hat{I} - i\hat{E}^2\hat{\beta}\hat{R}^s\hat{E}^2)}{1 + \langle n|\hat{E}^2\hat{\beta}\hat{R}^s\hat{\beta}\hat{E}^2|n\rangle} , \tag{7.42}$$

where $|n\rangle = \sum|n\rangle$. It follows that the matrix $\hat{R}^s$ is a symmetrical one. If we multiply this equation by $\langle l|$ on the left and by $|l'\rangle$ on the right, we find an equation for matrix elements $R^s_{ll'}$:

$$R^s_{ll'}\left(1 - e_l^2 e_{l'}^2\right) = x_l x_{l'} , \tag{7.43}$$

from which it follows that

$$R^s_{ll'} = \frac{x_l x_l'}{\left(1 - e_l^2 e_{l'}^2\right)} , \tag{7.44}$$

where

$$x_l = \frac{1 - ie_l^2\langle l|\hat{R}^s\hat{\beta}\hat{E}^2|n\rangle}{\sqrt{1 + \langle n|\hat{E}^2\hat{\beta}\hat{R}^s\hat{\beta}\hat{E}^2|n\rangle}} = \frac{1 - ie_l^2 x_l\sum_m x_m\beta_m e_m^2/\left(1 - e_l^2 e_m^2\right)}{\sqrt{1 + \sum_{i,j} x_i\beta_i e_i^2 x_j\beta_j e_j^2/\left(1 - e_i^2 e_j^2\right)}} . \tag{7.45}$$

The factor $\beta_l = 2\pi b_c N_2/k_{l\perp} \approx b_c\lambda/a^2$, where λ is the neutron wavelength, is usually small, and it can be neglected everywhere except in terms with small denominators $1 - e_l^2 e_m^2$.

If no denominator is small, we can put all the $x_l = 1$, then the matrix elements of the reflection matrix $\hat{R} = -i\hat{E}\hat{\beta}\hat{R}^s\hat{E}$ are

$$R_{ll'} = -i\frac{\beta_l e_l e_{l'}}{1 - e_l^2 e_{l'}^2} , \tag{7.46}$$

and probability of reflection with transition from state $|l\rangle$ to state $|l'\rangle$ is

$$w_{l'l} = \frac{k_{l'\perp}}{k_{l\perp}}|R_{l'l}|^2 = \frac{k_{l'\perp}}{k_{l\perp}}\left|\frac{2\pi N_2 b_c}{2k_{l'\perp}\sin[(k_{l\perp} + k_{l'\perp})a/2]}\right|^2 . \tag{7.47}$$

For specular reflection $l = l'$ we get

$$w_{ll} = \frac{u_c^2 a^2}{16 k_{l\perp}^2 \sin^2(k_{l\perp} a)} , \tag{7.48}$$

where $u_c = 4\pi N_0 b_c$ is the optical potential and $N_0 = N_2/a$ is the three-dimensional atomic density. At small $k_{l\perp} a$ we get the standard expression $u_c^2/16k_{l\perp}^4$ for specular reflection far from the total reflection range.

At $l' \neq l$ we get from Eq. (7.47) the expression

$$w_{l'l} = \frac{1}{16 k_{l\perp} k_{l'\perp}} \frac{u_c^2 a^2}{\sin^2[(k_{l\perp} + k_{l'\perp})a/2]} , \tag{7.49}$$

which is symmetrical with respect to interchange of $k_{l\perp}$ and $k_{l'\perp}$, and it shows that the reflectivity in a direction with $k_{l'} < k_l$ is higher than the specular one. This prediction can be checked by an experiment.

At $k_{l\perp} a = \pi n$ the reflectivity Eq. (7.48) diverges, which means that close to these points the total reflection takes place. At these $k_{l\perp} a$ the approximation $x_l = 1$ is not valid because some denominators in Eq. (7.45) are small.

7.2.3 Bragg Diffraction

Let us multiply the two equations of the system (7.45) with numbers l and l'. As a result, we obtain the new system

$$x_l x_{l'} \left(1 + \sum_{m,m'} \frac{x_m e_m^2 \beta_m x_{m'} e_{m'}^2 \beta_{m'}}{1 - e_m^2 e_{m'}^2} \right)$$
$$= \left(1 - i e_l^2 x_l \sum_m \frac{x_m e_m^2 \beta_m}{1 - e_l^2 e_m^2} \right) \left(1 - i e_{l'}^2 x_{l'} \sum_m \frac{x_m e_m^2 \beta_m}{1 - e_{l'}^2 e_m^2} \right) . \tag{7.50}$$

In the case when one of the products $e_l^2 e_m^2$ is close to unity, the most essential terms in the brackets are those with $x_l x_m$. In the simplest case of specular reflection we have $l = m = 0$ and $e_0^4 \approx 1$.

7.2.3.1 Specular Bragg Diffraction

In this case the most important terms are those proportional to x_0^2. Let us denote x_0^2 by z. For z we get the equation

$$z \left(1 - \frac{z r^2}{1 - e_0^4} \right) = \left(1 + \frac{e_0^2 z r}{1 - e_0^4} \right)^2 , \tag{7.51}$$

where we introduced the notation $r = -i e_0^2 \beta_0$. Solution of this quadratic equation gives z, from which we obtain the amplitude R of the specular reflection:

$$R = \frac{z r}{1 - e_0^4} = \frac{\sqrt{(1+r)^2 - t^2} - \sqrt{(1-r)^2 - t^2}}{\sqrt{(1+r)^2 - t^2} + \sqrt{(1-r)^2 - t^2}} , \tag{7.52}$$

where $t = e_0^2 + r$. The result completely coincides with Eq. (4.85), because $r = -i(p/k_\perp)e_0^2$ and t are the reflection and transmission amplitudes of the single period of the Kronig–Penney potential, as if calculated within the perturbation theory. We did not use the perturbation theory, and the result becomes perturbational only if we replace $b_c = Cb$ in p Eq. (7.18) by b, that is, put $C = 1$. If we had calculated the sum (7.6), then in our case we would get $C = 1/(1 + ip/k_\perp)$, and then r and t are precisely the amplitudes of the Kronig–Penney potential without perturbation theory.

Let us note that $(1 \pm r)^2 - t^2 = 1 - e_0^4 + 2i\beta_0(e_0^4 \mp e_0^2)$. The condition $e_0^4 \approx 1$ means that $2k_\perp a \approx 2\pi n$, where n is an integer. At $n = 0$ we can approximate $1 - e_0^4 \approx -2ik_\perp a$. As a result, the expression (7.52) becomes

$$R \equiv R_{B0} \approx \frac{k_\perp - \sqrt{k_\perp^2 - u_c}}{k_\perp + \sqrt{k_\perp^2 - u_c}}\,, \tag{7.53}$$

where the optical potential $u_c = 2p/a = 4\pi N_0 b_c$ appears again.

In the case $n = 1$, we have $2k_\perp a \approx 2\pi$. This means that $k_\perp \approx \pi/a$ and $e_0^2 \approx -1$. Let us introduce the Bragg wave number $k_{B1} = \pi/a$. Then $1 - e_0^4 \approx -2i(k_\perp - k_{B1})a$, and expression (7.52) becomes

$$R \equiv R_{B1} = \frac{\sqrt{k_\perp^2 - k_{B1}^2 - 2u_c} - \sqrt{k_\perp^2 - k_{B1}^2}}{\sqrt{k_\perp^2 - k_{B1}^2 - 2u_c} + \sqrt{k_\perp^2 - k_{B1}^2}}\,, \tag{7.54}$$

where we used the approximate equality $2k_\perp(k_\perp - k_{B1}) \approx k_\perp^2 - k_{B1}^2$. It is seen that reflection becomes total in the interval $k_{B1}^2 < k_\perp^2 < k_{B1}^2 + 2u_c$. The interval $2u_c$ is called the Darwin table because it was Darwin who was the first to find that the reflection is total not only at the Bragg wave number, which follows from the perturbation theory, but also in some finite energy interval.

At $k_\perp^2 < k_{B1}^2$ the reflection amplitude is a positive real number:

$$R_{B1} = \frac{\sqrt{k_{B1}^2 + 2u_c - k_\perp^2} - \sqrt{k_{B1}^2 - k_\perp^2}}{\sqrt{k_{B1}^2 + 2u_c - k_\perp^2} + \sqrt{k_{B1}^2 - k_\perp^2}}\,. \tag{7.55}$$

At $k_\perp^2 > k_{B1}^2 + 2u_c$ it, according to Eq. (7.54), is a negative real number. In the range of total reflection it is representable as

$$R_{B1} = -\frac{\sqrt{k_\perp^2 - k_{B1}^2} - i\sqrt{k_{B1}^2 + 2u_c - k_\perp^2}}{\sqrt{k_\perp^2 - k_{B1}^2} + i\sqrt{k_{B1}^2 + 2u_c - k_\perp^2}}\,. \tag{7.56}$$

The amplitude R becomes a unit complex number which with increase of $k_\perp$ runs counterclockwise along the upper semicircle.

The case $n = 1$ can be generalized to an arbitrary integer n. The Bragg reflection is total in all the intervals $k_{Bn}^2 < k_\perp^2 < k_{Bn}^2 + 2u_0$, where $k_{Bn} = n\pi/a$, where the reflection amplitude is a unit complex number

$$R_{Bn} = (-1)^n \frac{\sqrt{k_\perp^2 - k_{Bn}^2} - i\sqrt{k_{Bn}^2 + 2u_c - k_\perp^2}}{\sqrt{k_\perp^2 - k_{Bn}^2} + i\sqrt{k_{Bn}^2 + 2u_c - k_\perp^2}} . \tag{7.57}$$

Between neighboring Bragg wave numbers $k_{Bn}^2 + 2u_c < k_\perp^2 < k_{Bn+1}^2$ the amplitude R is real and positive for even n, and it is real and negative for odd n. In the range of total reflection it is a unit complex number which with increase of $k_\perp$ runs counterclockwise along a semicircle in the upper-half complex plane for even n and in the lower part for odd n.

However, we should recall that in the calculation of R we chose a symmetric period with a crystalline plane and two empty spaces of width $a/2$; therefore, our crystal contains the empty space of width $a/2$ before it. If we delete this space and look at the crystal starting with the atomic plane, then we get reflection matrix $\hat{\boldsymbol{R}}' = \hat{\boldsymbol{E}}^{-2}\hat{\boldsymbol{R}}$, and our formula (7.57) is replaced by

$$\begin{aligned} R'_{Bn} &= (-1)^n \exp(-ik_\perp a) \frac{\sqrt{k_\perp^2 - k_{Bn}^2} - i\sqrt{k_{Bn}^2 + 2u_c - k_\perp^2}}{\sqrt{k_\perp^2 - k_{Bn}^2} + i\sqrt{k_{Bn}^2 + 2u_c - k_\perp^2}} \\ &= \frac{\sqrt{k_\perp^2 - k_{Bn}^2} - i\sqrt{k_{Bn}^2 + 2u_c - k_\perp^2}}{\sqrt{k_\perp^2 - k_{Bn}^2} + i\sqrt{k_{Bn}^2 + 2u_c - k_\perp^2}} . \end{aligned} \tag{7.58}$$

The factor $\exp(-ik_\perp a)$ in the interval $k_{Bn}^2 + 2u_c < k_\perp^2 < k_{Bn+1}^2$ turns the phase of the amplitude R'_B clockwise by π, and because of that the amplitude R'_B at the start of the Bragg reflection always becomes -1. On the Darwin table it runs counterclockwise along the lower semicircle and at the end of the Darwin table almost reaches the value +1. After that it turns back and always begins with -1 at the start of the next Bragg reflection. In the following we shall use the last formula and omit the prime and index n.

7.2.3.2 Angular Width of the Specular Bragg Diffraction

According to Eq. (7.58) the reflection amplitude can be rewritten in the form

$$R_B = \frac{\sqrt{1-y} - i\sqrt{1+y}}{\sqrt{1-y} + i\sqrt{1+y}} . \tag{7.59}$$

where a new variable $y = [k_B^2 + u_c - k_\perp^2]/u_c$ was introduced. The center of the Bragg peak is at $y = 0$, and the Darwin table is in the range $-1 \le y \le 1$. The angle of incidence ϑ is defined via the relation $k_\perp = k\cos\vartheta$. The Bragg angle ϑ_B, corresponding to $y = 0$, is defined as $k^2\cos^2\vartheta_B = k_B^2 + u_c$. When $\vartheta = \vartheta_B + \varepsilon$, we have

$$y = \frac{k_B^2 + u_c - k^2\cos^2(\vartheta_B + \varepsilon)}{u_c} \approx \frac{k^2}{u_c}\sin(2\vartheta_B)\varepsilon . \tag{7.60}$$

The total reflection takes place in the range $-1 \leq y \leq 1$; therefore, the angular width of the Bragg reflection at given energy k^2 of the incident neutrons is

$$\Delta\varepsilon = 2u_c/k^2 \sin(2\vartheta_B)\,. \tag{7.61}$$

For $\hbar^2 u_c/2m = 10^{-7}$ eV and $\hbar^2 k^2/2m = 2.5 \cdot 10^{-2}$ eV, the full angular width of the Bragg peak is on the order of $\sim 10^{-5}$ rad, or approximately 2 angular seconds.

7.2.3.3 **Nonspecular Bragg Diffraction**

Let us consider now the case when $e_l^2 e_0^2$ is close to unity for $l \neq 0$. Then denoting in Eq. (7.50) $x_l x_0 = z$ and $-i\beta_l e_l e_n = r_{ln}$, we instead of Eq. (7.51) obtain the equation

$$z\left(1 - 2z r_{0l} r_{l0}/\left(1 - e_0^2 e_l^2\right)\right) = \left(1 + e_0^2 z r_{ll}/\left(1 - e_0^2 e_l^2\right)\right) \times \left(1 + e_l^2 z r_{00}/\left(1 - e_0^2 e_l^2\right)\right)\,. \tag{7.62}$$

It can be represented in the form

$$z^2 \frac{r_{0l} r_{l0}}{\left(1 - e_0^2 e_l^2\right)^2}\left[1 + \left(1 - e_0^2 e_l^2\right)\right] - z\left(1 - \frac{e_0^2 r_{ll} + e_l^2 r_{00}}{1 - e_0^2 e_l^2}\right) + 1 = 0\,. \tag{7.63}$$

The value $1 - e_0^2 e_l^2$ in brackets in the numerator of the first term can be neglected, and the equation is simplified to

$$z^2 \frac{r_{0l} r_{l0}}{\left(1 - e_0^2 e_l^2\right)^2} - z\left(1 - \frac{e_0^2 r_{ll} + e_l^2 r_{00}}{1 - e_0^2 e_l^2}\right) + 1 = 0\,. \tag{7.64}$$

Its solution is

$$z = \frac{1 - e_0^2 e_l^2}{\sqrt{r_{0l} r_{l0}}} \frac{\sqrt{(1 + \sqrt{r_{00} r_{ll}})^2 - t_{00} t_{ll}} - \sqrt{(1 - \sqrt{r_{00} r_{ll}})^2 - t_{00} t_{ll}}}{\sqrt{(1 + \sqrt{r_{00} r_{ll}})^2 - t_{00} t_{ll}} + \sqrt{(1 - \sqrt{r_{00} r_{ll}})^2 - t_{00} t_{ll}}}\,, \tag{7.65}$$

where $t_{jj} = e_j^2 + r_{jj}$. It follows that the amplitude of the Bragg reflection of the wave with $\boldsymbol{\kappa} = 0$ to the wave with $\boldsymbol{\kappa} = \boldsymbol{\kappa}_l$ is

$$R_{l0} = \sqrt{\frac{k_{0\perp}}{k_{l\perp}}} \frac{\sqrt{(1 + \sqrt{r_{00} r_{ll}})^2 - t_{00} t_{ll}} - \sqrt{(1 - \sqrt{r_{00} r_{ll}})^2 - t_{00} t_{ll}}}{\sqrt{(1 + \sqrt{r_{00} r_{ll}})^2 - t_{00} t_{ll}} + \sqrt{(1 - \sqrt{r_{00} r_{ll}})^2 - t_{00} t_{ll}}}\,, \tag{7.66}$$

and the reflection coefficient is

$$W_{l0} = \frac{k_{l\perp}}{k_{0\perp}} |R_{l0}|^2 = \left|\frac{\sqrt{(1 + \sqrt{r_{00} r_{ll}})^2 - t_{00} t_{ll}} - \sqrt{(1 - \sqrt{r_{00} r_{ll}})^2 - t_{00} t_{ll}}}{\sqrt{(1 + \sqrt{r_{00} r_{ll}})^2 - t_{00} t_{ll}} + \sqrt{(1 - \sqrt{r_{00} r_{ll}})^2 - t_{00} t_{ll}}}\right|^2\,, \tag{7.67}$$

where the factor $k_{l\perp}/k_{0\perp}$ characterizes the ratio of fluxes.

Since $(1 \pm \sqrt{r_{00}r_{ll}})^2 - t_{00}t_{ll} = 1 - e_0^2e_l^2 \mp 2ie_0e_l\sqrt{\beta_0\beta_l} + ie_0^2e_l^2(\beta_0 + \beta_l)$, then expression (7.66) near $e_0^2e_l^2 \approx 1$ is representable in the form

$$R_{l0} \approx \sqrt{\frac{k_{0\perp}}{k_{l\perp}}}$$
$$\frac{\sqrt{k_{0\perp}+k_{l\perp}-2k_B - \frac{u_c}{2}\left(\frac{1}{\sqrt{k_{0\perp}}} - \frac{1}{\sqrt{k_{l\perp}}}\right)^2} - i\sqrt{2k_B - k_{0\perp} - k_{l\perp} + \frac{u_c}{2}\left(\frac{1}{\sqrt{k_{0\perp}}} + \frac{1}{\sqrt{k_{l\perp}}}\right)^2}}{\sqrt{k_{0\perp}+k_{l\perp}-2k_B - \frac{u_c}{2}\left(\frac{1}{\sqrt{k_{0\perp}}} - \frac{1}{\sqrt{k_{l\perp}}}\right)^2} + i\sqrt{2k_B - k_{0\perp} - k_{l\perp} + \frac{u_c}{2}\left(\frac{1}{\sqrt{k_{0\perp}}} + \frac{1}{\sqrt{k_{l\perp}}}\right)^2}} . \tag{7.68}$$

Ifwe approximate

$$2\sqrt{k_{0\perp}k_{0l\perp}} \approx 2k_B \approx k_B + \frac{k_{0\perp} + k_{0l\perp}}{2} \equiv k_B + \overline{k}_{0l\perp} ,$$

then (7.68) transforms to

$$R_{l0} \approx \sqrt{\frac{k_{0\perp}}{k_{l\perp}}} \frac{\sqrt{\overline{k}_{0l\perp}^2 - k_B^2 - u_c(k_{0\perp} - k_{l\perp})^2/2k_B^2} - i\sqrt{k_B^2 - \overline{k}_{0l\perp}^2 + 2u_c}}{\sqrt{\overline{k}_{0l\perp}^2 - k_B^2 - u_c(k_{0\perp} - k_{l\perp})^2/2k_B^2} + i\sqrt{k_B^2 - \overline{k}_{0l\perp}^2 + 2u_c}} . \tag{7.69}$$

At $l = 0$ it coincides with Eq. (7.58).

7.3 Diffraction on a Crystal of Finite Thickness

To find the diffraction on a finite crystal we need besides $\hat{\boldsymbol{R}}$ also to know Bloch wave vectors, which are found from eigenstates of the operator $\hat{\boldsymbol{X}}$. If the crystal contains N crystalline planes, we take the semi-infinite crystal and separate from it the first N periods. Their reflection, $\hat{\boldsymbol{R}}_N$, and transmission, $\hat{\boldsymbol{T}}_N$, can be considered as amplitudes of a single period containing N planes. If the state vector of the incident wave on the first period is $|\boldsymbol{\Psi}_0\rangle$, then the state vector of the wave incident on the $N+$ first period is $\hat{\boldsymbol{X}}_N|\boldsymbol{\Psi}_0\rangle$, where $\hat{\boldsymbol{X}}_N = \hat{\boldsymbol{X}}^N = \exp(i\hat{q}Na)$.

The system of Eqs. (7.32) is replaced by

$$\hat{\boldsymbol{X}}_N = \hat{\boldsymbol{T}}_N + \hat{\boldsymbol{R}}_N\hat{\boldsymbol{R}}\hat{\boldsymbol{X}}_N, \quad \hat{\boldsymbol{R}} = \hat{\boldsymbol{R}}_N + \hat{\boldsymbol{T}}_N\hat{\boldsymbol{R}}\hat{\boldsymbol{X}}_N . \tag{7.70}$$

For known matrices $\hat{\boldsymbol{R}}$ and $\hat{\boldsymbol{X}}$ the system is resolved with respect to $\hat{\boldsymbol{R}}_N$ and $\hat{\boldsymbol{T}}_N$. Substitution of $\hat{\boldsymbol{R}}_N = \hat{\boldsymbol{R}} - \hat{\boldsymbol{T}}_N\hat{\boldsymbol{R}}\hat{\boldsymbol{X}}_N$ into the first equation gives

$$\hat{\boldsymbol{T}}_N = (\hat{\boldsymbol{I}} - \hat{\boldsymbol{R}}^2)\hat{\boldsymbol{X}}_N[\hat{\boldsymbol{I}} - \hat{\boldsymbol{R}}\hat{\boldsymbol{X}}_N\hat{\boldsymbol{R}}\hat{\boldsymbol{X}}_N]^{-1} , \tag{7.71}$$

and substitution of $\hat{\boldsymbol{T}}_N = \hat{\boldsymbol{X}}_N - \hat{\boldsymbol{R}}_N\hat{\boldsymbol{R}}\hat{\boldsymbol{X}}_N$ into the second equation gives

$$\hat{\boldsymbol{R}}_N = [\hat{\boldsymbol{R}} - \hat{\boldsymbol{X}}_N\hat{\boldsymbol{R}}\hat{\boldsymbol{X}}_N][\hat{\boldsymbol{I}} - \hat{\boldsymbol{R}}\hat{\boldsymbol{X}}_N\hat{\boldsymbol{R}}\hat{\boldsymbol{X}}_N]^{-1} . \tag{7.72}$$

To proceed we need to find $\hat{\boldsymbol{X}}$. First we suppose that the reflection is small. In this case diffraction is related to transmission through the crystal and it is called Laue diffraction. After that we consider the case of strong reflection, which is related to Bragg diffraction.

7.3.1 Eigenvectors and Eigenvalues of the Operator $\hat{X}$

To understand what waves propagate inside the crystal we need to find the eigenvalues and eigenvectors of the matrix $\hat{\boldsymbol{X}}$. The eigenvalues of the operator $\hat{\boldsymbol{X}}$ are complex numbers $\exp(iqa)$ with different q. An eigenvector corresponding to some q will be denoted $|\boldsymbol{q}\rangle$. It can be normalized and represented in the basis $|n\rangle$:

$$|\boldsymbol{q}\rangle = \sum_j Q_n(q)|n\rangle\,, \tag{7.73}$$

where $Q_n(q)$ are its coordinates in the basis $|n\rangle$.

Normalized vectors $|\boldsymbol{q}\rangle$ themselves also constitute a basis; therefore, the incident wave $|\psi_{\text{ins}}(\boldsymbol{r})\rangle = \exp(i\hat{\boldsymbol{k}}\boldsymbol{r})|\boldsymbol{\Psi}_0\rangle$ can be represented also in the basis $|\boldsymbol{q}\rangle$:

$$|\psi_{\text{ins}}(\boldsymbol{r})\rangle = \exp(i\hat{\boldsymbol{k}}\boldsymbol{r})|\boldsymbol{\Psi}_0\rangle = \exp(i\hat{\boldsymbol{k}}\boldsymbol{r})\sum_q \psi_q|\boldsymbol{q}\rangle\,, \tag{7.74}$$

where ψ_q are coordinates of the state vector $|\boldsymbol{\Psi}_0\rangle$ in the basis $|\boldsymbol{q}\rangle$. In this basis the wave field $|\psi_{\text{ins}}(\boldsymbol{r})\rangle$ inside the crystal becomes

$$\begin{aligned}|\psi_{\text{ins}}(\boldsymbol{r})\rangle &= \exp\left(i\hat{\boldsymbol{k}}_{\|}\boldsymbol{r}_{\|}\right)\sum_q \psi_q \exp(iqz)|\boldsymbol{q}\rangle \\ &= \exp\left(i\hat{\boldsymbol{k}}_{\|}\boldsymbol{r}_{\|}\right)\sum_q \psi_q \exp(iqz)\sum_n Q_n(q)|n\rangle \\ &= \sum_n \exp(i\boldsymbol{k}_{n\|}\boldsymbol{r}_{\|})\left(\sum_q Q_n(q)\psi_q\exp(iqz)\right)|n\rangle\,. \end{aligned} \tag{7.75}$$

If the crystal has thickness D, and reflection $\hat{\boldsymbol{R}}$ can be neglected, then the field of neutrons transmitted through the crystal is

$$\begin{aligned}\exp(i\hat{\boldsymbol{k}}\boldsymbol{r})|\boldsymbol{\Psi}_t\rangle &= \sum_n \exp(i\boldsymbol{k}_{n\|}\boldsymbol{r}_{\|} + ik_{n\perp}(z-D))\left(\sum_q Q_n(q)\psi_q\exp(iqD)\right) \\ &= \sum_n \psi_{\text{tn}}\exp(i\boldsymbol{k}_{n\|}\boldsymbol{r}_{\|} + ik_{zn}(z-D))\,, \end{aligned} \tag{7.76}$$

where coordinates ψ_{tn} determine the intensity of the transmitted neutrons $k_{n\perp}|\psi_{\text{tn}}|^2$ with wave vector $\boldsymbol{k}_n = (\boldsymbol{k}_{\|} + \boldsymbol{\kappa}_{n\|}, k_{n\perp})$. According to Eq. (7.76) the amplitudes $\psi_{\rm tn}$ are

$$\psi_{\rm tn} = \sum_q Q_n(q)\psi_q \exp(iqD) \,. \tag{7.77}$$

To calculated them we need the eigenvalues q and eigenvectors of $\hat{\boldsymbol{X}}$. It is important to notice that the amplitude $\psi_{\rm tn}$ shows interference of many Bloch eigenstates at the exit from the crystal. This interference depends on the crystal thickness D and leads to many interesting phenomena that can be observed with single crystals.

It follows from Eq. (7.33) that when $\hat{\boldsymbol{R}}$ can be neglected, then $\hat{\boldsymbol{X}} \approx \hat{\boldsymbol{\tau}} = \hat{\boldsymbol{E}}^2 + \hat{\boldsymbol{\rho}}$, where

$$\hat{\boldsymbol{\rho}} = |\boldsymbol{a}\rangle\langle\boldsymbol{b}| \equiv -i\hat{\boldsymbol{E}}\hat{\boldsymbol{\beta}}|\boldsymbol{n}\rangle\langle\boldsymbol{n}|\hat{\boldsymbol{E}} \,, \tag{7.78}$$

$|\boldsymbol{n}\rangle = \sum_n |n\rangle$ is a vector with all the coordinates equal to unity, and $\hat{\boldsymbol{E}}$ and $\hat{\boldsymbol{\beta}}$ are diagonal matrices

$$\hat{\boldsymbol{E}} = \sum_n |n\rangle e_n\langle n| \,, \quad \hat{\boldsymbol{\beta}} = \sum_n |n\rangle\beta_n\langle n| \,, \tag{7.79}$$

with matrix elements

$$e_n = \exp(ik_{n\perp}a/2) \,, \quad \beta_n = p/k_{n\perp} \,, \quad \text{and} \quad k_{n\perp} = \sqrt{k^2 - (\boldsymbol{k}_\parallel + \boldsymbol{\kappa}_n)^2} \,.$$

The equation for the eigenvectors $|\boldsymbol{q}\rangle = \sum_n Q_n(q)|n\rangle$ and eigenvalues λ of the matrix $\hat{\boldsymbol{X}}$ is

$$\hat{\boldsymbol{X}}|\boldsymbol{q}\rangle \equiv \hat{\boldsymbol{E}}^2|\boldsymbol{q}\rangle + \hat{\boldsymbol{\rho}}|\boldsymbol{q}\rangle = \lambda|\boldsymbol{q}\rangle \,. \tag{7.80}$$

Since $\hat{\boldsymbol{\rho}}$ is the dyadic matrix (7.78), then, according to Eq. (7.37), it projects any state (see Eq. (7.78) vector $|\zeta\rangle$ onto $|\boldsymbol{a}\rangle = \sum_n a_n(q)|n\rangle$, so Eq. (7.80) becomes

$$\hat{\boldsymbol{E}}^2|\boldsymbol{q}\rangle + C_q|\boldsymbol{a}\rangle = \lambda|\boldsymbol{q}\rangle \,, \tag{7.81}$$

where

$$C_q = \langle\boldsymbol{b}||\boldsymbol{q}\rangle \,. \tag{7.82}$$

It follows from Eq. (7.81) that

$$|\boldsymbol{q}\rangle = C_q(\lambda\hat{\boldsymbol{I}} - \hat{\boldsymbol{E}}^2)^{-1}|\boldsymbol{a}\rangle \equiv -iC_q(\lambda\hat{\boldsymbol{I}} - \hat{\boldsymbol{E}}^2)^{-1}\hat{\boldsymbol{E}}\hat{\boldsymbol{\beta}}|\boldsymbol{n}\rangle \,. \tag{7.83}$$

Substitution of Eq. (7.83) into Eq. (7.82) gives the equation for λ:

$$\langle\boldsymbol{b}|(\lambda\hat{\boldsymbol{I}} - \hat{\boldsymbol{E}}^2)^{-1}|\boldsymbol{a}\rangle \equiv -i\langle\boldsymbol{n}|\hat{\boldsymbol{E}}^2(\lambda\hat{\boldsymbol{I}} - \hat{\boldsymbol{E}}^2)^{-1}\hat{\boldsymbol{\beta}}|\boldsymbol{n}\rangle = 1 \,. \tag{7.84}$$

In coordinates of state vectors $|\boldsymbol{a}\rangle$ and $|\boldsymbol{b}\rangle$ or $|\boldsymbol{n}\rangle$ Eq. (7.84) takes the form

$$\sum_n \frac{a_n b_n}{\lambda - e_n^2} \equiv \sum_n \frac{-i\beta_n e_n^2}{\lambda - e_n^2} \equiv -i\sum_n \frac{p}{k_{n\perp}}\frac{e_n^2}{\lambda - e_n^2}$$
$$\equiv -2\pi i N_2 b_c \sum_n \frac{e_n^2}{k_{n\perp}(\lambda - e_n^2)} = 1 \,. \tag{7.85}$$

Now we need to find all the solutions for λ. With their help we find all the eigenvectors $|\boldsymbol{q}\rangle$ from (7.83), where C_q can be accepted as normalization constants.

7.3.2
All $e_n^2 = \exp(ik_{n\perp}a)$ Are Different

If all $e_n^2 = \exp(ik_{n\perp}a)$ are different, then, since all the values $\beta_n \equiv 2\pi N_2 b_c/k_{n\perp}$ are small (for thermal neutrons they are in general on the order $b_c/a \approx 10^{-4}$), the left side of Eq. (7.85) can be equal to unity only at such λ which are close to one of the exponent: $\lambda = \lambda_n \approx e_n^2$. In general, when all the e_n^2 are different, every λ_n is close only to one of e_n^2. Then Eq. (7.85) is reduced to

$$-i\beta_n \frac{e_n^2}{\lambda_n - e_n^2} \approx 1\,, \tag{7.86}$$

and it follows that

$$\begin{aligned}\lambda_n &= e_n^2(1 - i\beta_n) = \exp(ik_{n\perp}a)(1 - iu_c a/2k_{n\perp}) \\ &\approx \exp(ia(k_{n\perp} - u_c/2k_{n\perp}))\,,\end{aligned} \tag{7.87}$$

or, if we rewrite Eq. (7.87) as $\lambda_{n1} = \exp(iq_{n1}a)$ (the additional index 1 means that only one exponent is essential), then $q_{n1} \approx \sqrt{k_{n\perp}^2 - u_c}$, which is equivalent to a wave vector inside a continuous medium with the optical potential $u_c = 4\pi N_0 b_c$. The eigenvector for eigenvalue λ_{n1}, according to Eq. (7.83), has the single component $|\boldsymbol{q}_n\rangle = |n\rangle$. Therefore, the incident wave in the state $|\boldsymbol{\Psi}_0\rangle = |0\rangle$ has the same coordinate in basis $|q\rangle$ as in basis $|n\rangle$, $\psi_n = \delta_{n,0}$, and inside the substance there propagates a single wave with wave vector $q_0 = \sqrt{k_{0\perp}^2 - u_c}$.

It is also possible to find corrections to the wave number q if instead of using Eq. (7.86) we approximate Eq. (7.85) by the equation

$$-i\beta_n \frac{e_n^2}{\lambda - e_n^2} = 1 + i\sum_{m\neq n} \frac{\beta_m e_m^2}{\lambda_{n1}^2 - e_m^2}\,. \tag{7.88}$$

7.3.3
Two Values of e_n Are Close to each other

The case when two or more values of e_n are close to each other is more interesting. Suppose that $k_{n\perp} \approx k_{m\perp}$. Then for one of the λ, which is close to e_n^2, Eq. (7.85) is reduced to

$$-i\beta_n \frac{e_n^2}{\lambda - e_n^2} - i\beta_m \frac{e_m^2}{\lambda - e_m^2} \approx 1 \tag{7.89}$$

because all other terms can be neglected. The solutions, $\lambda_{n2}^{\pm}$, of Eq. (7.89) (the additional index 2 means that two exponents are essential) are

$$\lambda_{n2}^{\pm} = \frac{1}{2}\left(\lambda_{n1} + \lambda_{m1} \pm \sqrt{(\lambda_{n1} - \lambda_{m1})^2 - 4e_m^2 e_n^2 \beta_m \beta_n}\right)\,. \tag{7.90}$$

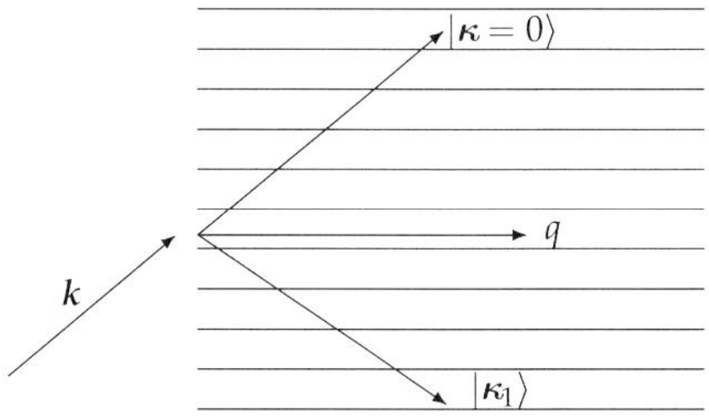

Figure 7.3 In the case when two waves have approximately equal wave vectors, $|\boldsymbol{k}_{\|}| \approx |\boldsymbol{k}_{\|} + \kappa|$, a superposition of two waves appears in the crystal, which propagates along crystalline planes with a single Bloch wave number q.

At $k_{m\perp} = k_{n\perp}$ we get $\lambda_{n2}^{+} = e_n^2$ and $\lambda_{n2}^{-} = e_n^2(1 - 2i\beta_n)$. These eigenvalues correspond to the normalized eigenvectors (7.83)

$$|\boldsymbol{q}_{n2}^{+}\rangle = \frac{1}{\sqrt{2}}(|n\rangle - |m\rangle)\,, \quad |\boldsymbol{q}_{n2}^{-}\rangle = \frac{1}{\sqrt{2}}(|n\rangle + |m\rangle)\,, \tag{7.91}$$

which are shown schematically in Figure 7.3.

The first eigenvalue $q_{n2}^{+} = k_{n\perp}$ means that the wave inside the crystal does not "see" nuclei and propagates like in empty space. Its eigenvector $|\boldsymbol{q}_{n2}^{+}\rangle$ corresponds to the wave function

$$\psi_1(\boldsymbol{r}) = \frac{1}{\sqrt{2}}\left(e^{i(\boldsymbol{k}_{\|}+\boldsymbol{\kappa}_n)\boldsymbol{r}_{\|}} - e^{i(\boldsymbol{k}_{\|}+\boldsymbol{\kappa}_m)\boldsymbol{r}_{\|}}\right)\exp\left(iq_{n2}^{+}r_{\perp}\right)\,. \tag{7.92}$$

It is zero at the locations of nuclei because $\boldsymbol{\kappa}_n\boldsymbol{r}_j = 2\pi N$, where N is an integer.

The second solution, $q_{n2}^{-} \approx \sqrt{k_{n\perp}^2 - 2u_c}$, corresponds to a continuous medium with a doubled optical potential. Its eigenvector $|\boldsymbol{q}_{n2}^{-}\rangle$ corresponds to the wave function

$$\psi_2(\boldsymbol{r}) = \frac{1}{\sqrt{2}}\left(e^{i(\boldsymbol{k}_{\|}+\boldsymbol{\kappa}_n)\boldsymbol{r}_{\|}} + e^{i(\boldsymbol{k}_{\|}+\boldsymbol{\kappa}_m)\boldsymbol{r}_{\|}}\right)\exp\left(iq_{n2}^{-}r_{\perp}\right)\,. \tag{7.93}$$

At nuclei the two exponents add, that is, $\psi_2(\boldsymbol{r}) = 2\exp(i\boldsymbol{k}_{\|}\boldsymbol{r}_{\|})$, which results in doubling of the potential. If the wave incident upon the crystal is described by the state $|\boldsymbol{\Psi}_0\rangle = |n\rangle$, then in the basis $|\boldsymbol{q}^{\pm}\rangle$ Eq. (7.91) it is represented by Eq. (7.74) with coordinates $\psi^{\pm} = 1/\sqrt{2}$. We see that the incident wave creates two pairs of plane waves inside the crystal. One of the pairs interferes negatively on nuclei and propagates through the crystal like in free space. The other one interferes positively on nuclei and propagates more slowly in the crystal. Interference of the two pairs on their exiting the crystal gives the phenomenon (see page 309), called Pendellösung and will be discussed later.

We considered the ideally symmetrical case $k_{m\perp} = k_{n\perp}$. To describe the diffraction lines in detail it is necessary to solve equations when there is some difference $\Delta k = k_{m\perp} - k_{n\perp}$ of two wave vectors, and to take into account the imaginary part of the scattering amplitude b_c. It requires some elementary algebra and we leave it to a researcher who will need it for processing his experimental data.

7.3.4 The Bloch Wave Vector at a Specular Bragg Diffraction

Near the Bragg reflections the matrix $\hat{\boldsymbol{R}}$ in $\hat{\boldsymbol{X}}$ Eq. (7.33) cannot be neglected. In that case with the rules Eq. (7.35)–(7.39) for the reciprocal matrix $(\hat{\boldsymbol{I}} - \hat{\boldsymbol{\rho}}\hat{\boldsymbol{R}})^{-1}$ we can transform $\hat{\boldsymbol{X}}$ to the form

$$\hat{\boldsymbol{X}} = (\hat{\boldsymbol{I}} + C_R\hat{\boldsymbol{\rho}}\hat{\boldsymbol{R}})(\hat{\boldsymbol{E}}^2 + \hat{\boldsymbol{\rho}})\,, \tag{7.94}$$

where

$$C_R = 1/(1 - c_R)\,, \quad c_R = \mathrm{Sp}\hat{\boldsymbol{\rho}}\hat{\boldsymbol{R}} = -i\langle \boldsymbol{n}|\hat{\boldsymbol{E}}\hat{\boldsymbol{R}}\hat{\boldsymbol{E}}\hat{\boldsymbol{\beta}}|\boldsymbol{n}\rangle = -\langle \boldsymbol{n}|\hat{\boldsymbol{E}}^2\hat{\boldsymbol{\beta}}\hat{\boldsymbol{R}}^s\hat{\boldsymbol{\beta}}\hat{\boldsymbol{E}}^2|\boldsymbol{n}\rangle\,. \tag{7.95}$$

Opening the brackets in Eq. (7.94) with account of Eqs. (7.35) and (7.36) gives

$$\hat{\boldsymbol{X}} = \hat{\boldsymbol{E}}^2 + \hat{\boldsymbol{\rho}} + C_R c_R\hat{\boldsymbol{\rho}} + C_R\hat{\boldsymbol{\rho}}\hat{\boldsymbol{R}}\hat{\boldsymbol{E}}^2 = \hat{\boldsymbol{E}}^2 + C_R\hat{\boldsymbol{\rho}}(\hat{\boldsymbol{I}} + \hat{\boldsymbol{R}}\hat{\boldsymbol{E}}^2)\,. \tag{7.96}$$

The equations for eigenvectors $|\boldsymbol{q}\rangle = \sum_n Q_n(q)|n\rangle$ and eigenvalues λ of the matrix $\hat{\boldsymbol{X}}$ are now

$$\hat{\boldsymbol{X}}|\boldsymbol{q}\rangle \equiv \hat{\boldsymbol{E}}^2|\boldsymbol{q}\rangle + C_R\hat{\boldsymbol{\rho}}(\hat{\boldsymbol{I}} + \hat{\boldsymbol{R}}\hat{\boldsymbol{E}}^2)|\boldsymbol{q}\rangle = \lambda|\boldsymbol{q}\rangle\,. \tag{7.97}$$

Since $\hat{\boldsymbol{\rho}}$ is dyadic matrix (7.78), then, according to Eq. (7.37), it projects any vector state $|\zeta\rangle$ onto $|\boldsymbol{a}\rangle = -i\hat{\boldsymbol{E}}\hat{\boldsymbol{\beta}}|\boldsymbol{n}\rangle = -i\sum_n e_n\beta_n|n\rangle$, and Eq. (7.97) becomes identical to Eq. (7.80)

$$\hat{\boldsymbol{E}}^2|\boldsymbol{q}\rangle + C_q|\boldsymbol{a}\rangle = \lambda|\boldsymbol{q}\rangle\,, \tag{7.98}$$

where

$$C_q = C_R\langle \boldsymbol{b}|(\hat{\boldsymbol{I}} + \hat{\boldsymbol{R}}\hat{\boldsymbol{E}}^2)|\boldsymbol{q}\rangle = C_R\langle \boldsymbol{n}|\hat{\boldsymbol{E}}(\hat{\boldsymbol{I}} + \hat{\boldsymbol{R}}\hat{\boldsymbol{E}}^2)|\boldsymbol{q}\rangle\,. \tag{7.99}$$

From Eq. (7.98) it follows that

$$|\boldsymbol{q}\rangle = C_q(\lambda\hat{\boldsymbol{I}} - \hat{\boldsymbol{E}}^2)^{-1}|\boldsymbol{a}\rangle = -iC_q(\lambda\hat{\boldsymbol{I}} - \hat{\boldsymbol{E}}^2)^{-1}\hat{\boldsymbol{E}}\hat{\boldsymbol{\beta}}|\boldsymbol{n}\rangle\,. \tag{7.100}$$

Substitution of Eq. (7.100) into Eq. (7.99) gives the equation for λ,

$$C_R\langle \boldsymbol{b}|(\hat{\boldsymbol{I}} + \hat{\boldsymbol{R}}\hat{\boldsymbol{E}}^2)(\lambda\hat{\boldsymbol{I}} - \hat{\boldsymbol{E}}^2)^{-1}|\boldsymbol{a}\rangle \equiv -i\frac{\langle \boldsymbol{n}|\hat{\boldsymbol{E}}(\hat{\boldsymbol{I}} + \hat{\boldsymbol{R}}\hat{\boldsymbol{E}}^2)(\lambda\hat{\boldsymbol{I}} - \hat{\boldsymbol{E}}^2)^{-1}\hat{\boldsymbol{E}}\hat{\boldsymbol{\beta}}|\boldsymbol{n}\rangle}{1 + i\langle \boldsymbol{n}|\hat{\boldsymbol{E}}\hat{\boldsymbol{R}}\hat{\boldsymbol{E}}\hat{\boldsymbol{\beta}}|\boldsymbol{n}\rangle} = 1\,, \tag{7.101}$$

which is a little bit different from Eq. (7.84). In the case of a specular Bragg reflection it is reduced to

$$\frac{r\left(1 + Re_0^2\right)}{(1 - rR)\left(\lambda - e_0^2\right)} = 1\,, \tag{7.102}$$

where $r = -ie_0^2\beta_0$. It follows that

$$\lambda \equiv \exp(iqa) = \frac{e_0^2 + r}{1 - rR} \approx e_0^2\left(1 - i\beta_0\left(1 + Re_0^2\right)\right) . \tag{7.103}$$

In the range of the total reflection near $k_\perp \approx k_{B1}$ the factor e_0^2 near R in the brackets can be replaced by -1. Substitution of R from Eq. (7.56) gives

$$\exp(iqa) \approx e_0^2\left(1 - i\frac{k_\perp^2 - k_{B1}^2}{2k_\perp}a - \frac{\sqrt{\left(k_\perp^2 - k_{B1}^2\right)\left(k_{B1}^2 + 2u_c - k_\perp^2\right)}}{2k_\perp}a\right) . \tag{7.104}$$

Since $k_\perp^2 - k_{B1}^2 \approx (k_\perp - k_{B1})2k_\perp$, we obtain

$$q = k_{B1} + i\frac{\sqrt{\left(k_\perp^2 - k_{B1}^2\right)\left(k_{B1}^2 + 2u_c - k_\perp^2\right)}}{2k_\perp} , \tag{7.105}$$

and we see that the imaginary part of the Bloch wave number is positive and the wave function inside the crystal decays exponentially at the total reflection.

Now we can make a generalization to an arbitrary Bragg wave number k_B. In the range of the total reflection near any Bragg number $k_B = \pi n/a$ with integer n the Bloch wave number is

$$\begin{aligned} q &\approx k_B + i\frac{\sqrt{\left(k_\perp^2 - k_B^2\right)\left(k_B^2 + 2u_c - k_\perp^2\right)}}{2k_B} \\ &\approx \sqrt{k_B^2 + \sqrt{\left(k_\perp^2 - k_B^2\right)\left(k_\perp^2 - k_B^2 - 2u_c\right)}} \end{aligned} \tag{7.106}$$

when $k_\perp^2 > k_B^2$, and it is

$$q \approx \sqrt{k_B^2 - \sqrt{\left(k_B^2 - k_\perp^2\right)\left(k_B^2 + 2u_c - k_\perp^2\right)}} \tag{7.107}$$

when $k_\perp^2 < k_B^2$. Far from the Bragg reflections, when $|k_\perp^2 - k_B^2| \gg u_c$, we get $q = \sqrt{k_\perp^2 - u_c}$, which is in accord with the case considered on page 310.

7.3.5
Specular Bragg Diffraction

At a specular Bragg diffraction we have in fact a one-dimensional result because the most essential term is the single matrix element $(R_N)_{00}$. Substitution of Eq. (7.57) and (7.106) into (7.72) in the range of the total reflection gives

$$\begin{aligned} (R_N)_{00} &= R_{00}\frac{1 - \exp(2iqD)}{1 - R_{00}^2\exp(2iqD)} \\ &= \frac{u_c\sinh(q''D)}{\left(k_\perp^2 - k_B^2 - u_c\right)\sinh(q''D) + i\sqrt{\left(k_\perp^2 - k_B^2\right)\left(k_B^2 - k_\perp^2 + 2u_c\right)}\cosh(q''D)} , \end{aligned} \tag{7.108}$$

where $q'' = \sqrt{(k_\perp^2 - k_B^2)(k_B^2 - k_\perp^2 + 2u_c)}/2k_\perp$, and $D = Na$. If we denote $y = (k_B^2 - k_\perp^2 + u_c)/u_c$ and $A = u_c D/2k_\perp$, then the probability of reflection $W = |(R_N)_{00}|^2$ in the range of the total reflection $y^2 < 1$ is representable in the form

$$W(y) = \frac{\sinh^2(A\sqrt{1-y^2})}{1 - y^2 + \sinh^2(A\sqrt{1-y^2})} . \tag{7.109}$$

Outside the total reflection range ($y^2 > 1$), the reflection probability is

$$W(y) = \left| R_{00} \frac{1 - \exp(2iqD)}{1 - R_{00}^2 \exp(2iqD)} \right|^2 = \frac{\sin^2(A\sqrt{y^2-1})}{y^2 - 1 + \sin^2(A\sqrt{y^2-1})} . \tag{7.110}$$

Both formulas can be represented for all y by the unique one

$$W(y) = \frac{|\sin^2(A\sqrt{y^2-1})|}{|y^2-1| + |\sin^2(A\sqrt{y^2-1})|} . \tag{7.111}$$

We shall use this expression when comparing the results with those of the standard dynamical diffraction theory.

7.3.6
Total Reflection of a Mosaic Crystal

The intensity of the Bragg-reflected neutrons is obtained by integration of Eq. (7.111) over all angles ε. To do this we approximate y according to Eq. (7.60) by

$$y = \frac{k_B^2 + u_c - k^2\cos^2(\vartheta_B + \varepsilon)}{u_c} \approx \frac{k^2}{u_c}\sin(2\vartheta_B)\varepsilon .$$

Then the total reflectivity for any D or any $A = u_c D/2k_B$ is

$$R(A) = \int W(y)d\varepsilon = \int_{-\infty}^{\infty} W(y)\frac{d\varepsilon}{dy}dy = F(A)\frac{u_c}{k^2\sin(2\vartheta_B)} , \tag{7.112}$$

where

$$F(A) = \int_{-\infty}^{\infty} W(y)dy . \tag{7.113}$$

Integration over y was extended to infinity because in fact only a narrow range of $y \approx 1$ gives the main contribution to it. The function $F(A)$ is shown in Figure 7.4. It reaches saturation $F(A) \approx \pi$ at $A = 3$. Therefore, for $D > D_c = 6k_B/u_c \approx 0.1$ mm the crystal can be considered to be infinitely thick, and its reflectivity is

$$R(A) = \frac{\pi u_c}{k^2\sin(2\vartheta_B)} \approx \pi \cdot 10^{-5} , \tag{7.114}$$

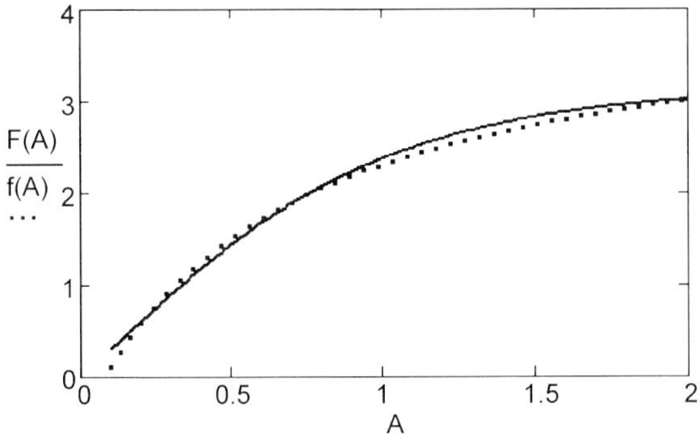

Figure 7.4 Dependence of the function $F(A)$ in Eq. (7.112) on the value of A (solid line), and its approximation by Eq. (7.115) (dotted line).

which means that the full angular width of the total reflection is $\Delta\varepsilon \approx 6.5''$ angular seconds.

For $A < 2$ the function $F(A)$ can be approximated by

$$f(A) = \frac{2\pi A - 0.5}{1 + 1.5A} \,. \tag{7.115}$$

It is not a fit but an approximation by eye. Usually for small $A < 1$ it is accepted that $f(A) \approx \pi A$, then

$$\begin{aligned} R(A) &\approx \pi A \frac{u_c}{k^2 \sin(2\vartheta_B)} \approx \frac{\pi u_c^2 D}{2k^3 \cos(\vartheta_B)\sin(2\vartheta_B)} = \frac{|N_0 b_c|^2 \lambda^3 D}{\cos(\vartheta_B)\sin(2\vartheta_B)} \\ &= QD\mathcal{L} \,, \end{aligned} \tag{7.116}$$

where in the last equality two parameters are introduced: (1) $Q = |N_0 b_c|^2 \lambda^3$ has dimension of reciprocal centimeters and is called the coefficient of the secondary extinction and (2) $\mathcal{L} = 1/\cos\theta_B \sin(2\theta_B)$, usually called the Lorentz factor.

Let us consider a mosaic crystal of thickness $L \gg D_c$ consisting of crystalline blocks with the orientations of their normals inside some solid angle ξ of the order of $1'$ angular minute. The reflectivity of a single block is given by Eq. (7.114), which can be estimated as an angular range ε, where the reflection is total. If the angular width of the incident neutrons after collimation is $\chi \gg \Delta\varepsilon$, then the single block reflects only a small fraction $\Delta\varepsilon/\chi$ of the incident intensity. In fact, however, the reflectivity of a real mosaic crystal is of the order

$$R_m = \xi/\chi \gg \Delta\varepsilon/\chi \tag{7.117}$$

because every neutron inside the angular range ξ finds on its path a sufficiently thick crystalline block, which will reflect it totally.

A mosaic crystal consisting of blocks with thickness $D \ll D_c$ can also reflect many more neutrons than an ideal single crystal. The reflectivity of a single block according to Eq. (7.111) is small and proportional to A^2; however, the angular width in terms of Δy is proportional to $1/A$ and is large. Therefore, the product of the two factors gives the result shown in Eq. (7.116). This total reflectivity is nevertheless smaller ($A \ll 1$) than Eq. (7.115). However, reflection of many small blocks can give the total reflectivity of the mosaic crystal of the same order as Eq. (7.117). To show it let us discuss two widely used notions: primary and secondary extinctions.

7.3.7 Extinction Coefficients

Extinction coefficients define how fast the intensity decays along the depth of the crystal. The primary extinction coefficient at the Bragg reflection is $2q''$ – double the imaginary part of the Bloch wave number. It defines the depth $l_0 = 1/2q''$ of the decay of the squared modulus of the wave function away from the interface. According to Eq. (7.105) this depth can be estimated as

$$l_0 \approx \frac{k}{u_c} = \frac{\lambda}{2\pi}\frac{k^2}{u_c}\,.$$

In the case of thermal neutrons, $l_0 \approx 7\,\mu$.

Let us note that the stronger is this extinction, the higher is the reflectivity of the crystal. For the majority of crystallographers this sentence looks contradictory. Usually diffraction of small crystals is calculated by perturbation theory in the first Born approximation. In this approximation all parts of the crystal are illuminated by the same amplitude of the incident wave. Extinction decreases the amplitude in the depth of the crystal; therefore, the total reflectivity decreases and this decrease is accounted for by a factor which is less than unity.

In dynamical diffraction theory, the reflectivity at the Bragg points is equal to unity, and it is possible only because the wave function decays exponentially in the crystal away from the entrance point. The faster it decays, the thinner can be the crystal whose reflectivity remains close to unity. So the primary extinction is nothing but a positive and very important factor for the diffraction.

The secondary extinction plays a similar role. It determines the decay of the beam inside a real crystal because of reflection by separate blocks, and it increases the reflectivity. The stronger is the secondary extinction, the higher is the reflectivity by separate blocks. Only if one deals with perturbation theory does the secondary extinction decrease the reflectivity and give rise to a factor (like primary extinction) in the reflected intensity which is less than unity.

7.3.7.1 Incoherent Reflectivity from *N* Equal Blocks

If we have *N* identical blocks with similar parameter $A < 1$ and with angular orientation within the angular range $\xi \sim 1/A$, but randomly distributed along the depth of the crystal, then the intensity reflected by every next block will be lower than reflected by the previous one because the number of neutrons which can reach it is smaller. Moreover, the intensity reflected by the next block is partly screened by the previous ones.

To estimate the reflectivity of a large mosaic crystal let us find the reflectivity of a group of N identical blocks of thickness D randomly distributed over the thickness L of the crystal, so we have no interference between waves reflected by different blocks.

To calculate the incoherent reflectivity of such a system we apply the same method as used for the calculation of the reflection amplitude in the case of one-dimensional potentials. First we shall find the reflectivity R of an infinite set of such blocks, when the reflectivity and the transmissivity of a single block are ρ and $\tau = 1 - \rho$. The intensity incident on the first block is supposed to be unity, and the intensity incident on the second block will be denoted by X. For R and X we have the system of equations

$$R = \rho + \tau R X\,, \qquad X = \tau + \rho R X\,. \tag{7.118}$$

However, if we solve this system, we shall find the trivial result $R = X = 1$, which is physically understandable because the infinite system of blocks reflects everything. We need to find the reflection of a finite number of N blocks. It is also possible to do this with the help of our method: $R_N = R(1 - X^{2N})/(1 - R^2 X^{2N})$. But now the result becomes indefinite, because with $X = R = 1$ we obtain the indefinite ratio $0/0$.

To resolve this ambiguity we can do the following. Let us introduce a small absorption by a single block. The absorption always exists, but here it is an artificial trick. So we suppose that $\tau = 1 - \rho - \alpha$, where α is an infinitesimal parameter. Now R and X are not unities:

$$R \approx \frac{\sqrt{4\rho} - \sqrt{2(1-\rho)\alpha}}{\sqrt{4\rho} + \sqrt{2(1-\rho)\alpha}}\,, \qquad X \approx \frac{\sqrt{4(1-\rho)} - \sqrt{2\rho\alpha}}{\sqrt{4(1-\rho)} + \sqrt{2\rho\alpha}}\,. \tag{7.119}$$

To resolve the ambiguity $0/0$ in R_N it is sufficient to differentiate the numerator $1 - X^{2N}$ and the denominator $1 - R^2 X^{2N}$ over $\sqrt{\alpha}$ and put $\alpha = 0$. As a result, we obtain

$$R_N = \frac{N\rho}{1 + N\rho}\,. \tag{7.120}$$

Now we can substitute for the reflectivity of a single block its value $Q D \mathcal{L}$. The total number of blocks in a crystal of total thickness L is $N_L = L/D$. If some of these blocks $N(\vartheta) = N_L W(\vartheta)$ have orientation distributed in the angular range ξ, and the function $W(\vartheta)$ is the probability density of finding blocks with a given average orientation, then the reflectivity of this system of blocks is [357]

$$R_N = \int\limits_{\chi} d\vartheta \frac{Q L \mathcal{L} W(\vartheta)}{1 + Q L \mathcal{L} W(\vartheta)}\,. \tag{7.121}$$

If $Q L \mathcal{L} \gg 1$, the reflectivity R_N is close to unity in the large fraction of the range ξ. Therefore, the reflectivity is again close to ξ/χ.

However, everything here depends on chance. The distribution $W(\vartheta)$ and the mosaicity (angular width of $W(\vartheta)$) are phenomenological values. For their description one uses some model and finds parameters of the model by fitting model calculations to experimental data.

7.3.8 The Laue Diffraction

At Laue diffraction the neutrons pass through the crystal. This means that reflection can be neglected, that is, one can put $\hat{\boldsymbol{R}} = \hat{\boldsymbol{R}}_N = 0$. According to Eq. (7.71) the transmission in this case is completely determined by $\hat{\boldsymbol{X}}_N$:

$$\hat{\boldsymbol{T}}_N = \hat{\boldsymbol{X}}_N \equiv \hat{\boldsymbol{X}}^N . \tag{7.122}$$

The wave function of the transmitted neutrons is

$$|\psi_t(\boldsymbol{r})\rangle = \Theta(z > D) \exp\left(i\hat{\boldsymbol{k}}_{\|}\boldsymbol{r}_{\|} + i\hat{\boldsymbol{k}}_{\perp}(z - D)\right)\hat{\boldsymbol{X}}^N |\boldsymbol{\Psi}_0\rangle . \tag{7.123}$$

If we know the eigenvectors $|\boldsymbol{q}_n\rangle$ of the matrix $\hat{\boldsymbol{X}}$, we can find an analytical expression for the wave function of the transmitted neutron

$$|\psi_t(\boldsymbol{r})\rangle = \exp\left(i\hat{\boldsymbol{k}}_{\|}\boldsymbol{r}_{\|} + i\hat{\boldsymbol{k}}_{\perp}(z - D)\right) \sum_{n=-\infty}^{\infty} e^{iq_n D}|\boldsymbol{q}_n\rangle\langle\boldsymbol{q}_n||\boldsymbol{\Psi}_0\rangle . \tag{7.124}$$

With the help of representation Eq. (7.83) for $|\boldsymbol{q}_n\rangle$, we obtain

$$\psi_t(\boldsymbol{r}) = \sum_{m=-\infty}^{\infty} \exp(i\boldsymbol{k}_{m\|}\boldsymbol{r}_{\|} + ik_{m\perp}(z - D)) \sum_{n=-\infty}^{\infty} e^{iq_n D}\frac{-i\beta_m}{\lambda_n - e_m^2}\langle\boldsymbol{q}_n||\boldsymbol{\Psi}_0\rangle . \tag{7.125}$$

This expression shows that the main contribution to the wave function gives those terms in the sum over m which have a small denominator, that is, $\lambda_n \approx e_m^2$.

In the case when all the e_m^2 are different, we have $\langle\boldsymbol{q}_n||\boldsymbol{\Psi}_0\rangle = \langle\boldsymbol{q}_n||0\rangle \equiv \delta_{n0}$, and the main contribution to the wave function gives the single term with $n = 0$ in the double sum (7.125):

$$\psi_t(\boldsymbol{r}) = \exp(i\boldsymbol{k}_{\|}\boldsymbol{r}_{\|} + ik_{0\perp}(z - D))\exp(iq_0 D) , \tag{7.126}$$

where $q_0 = \sqrt{k_{0\perp}^2 - u_c}$.

At Laue diffraction several $k_{n\perp}$ and e_n have similar values. The most common case is when two values, say, $k_{0\perp}$ and $k_{1\perp}$ are similar. Let $k_{1\perp} = k_{0\perp} + \delta k$, then $e_1^2 = e_0^2(1 + i\epsilon)$, where $\epsilon = \delta k a$. To find eigenvalues λ, we have to solve the system of Eqs. (7.89). Their solution is given in Eq. (7.90). Let us consider an ideal case: $\boldsymbol{k}_{1\|} = \boldsymbol{k}_{0\|} + \boldsymbol{\tau}_1 = -\boldsymbol{k}_{0\|}$. Then $k_{1\perp} = \boldsymbol{k}_{0\perp}$, $\delta k = 0$, and we obtain $q_1 =$

$\sqrt{k_{0\perp}^2 - 2u_c}$, and $q_2 = k_{0\perp}$. The eigenvectors are given in Eq. (7.91), and the total wave function of the transmitted neutrons consists of two waves:

$$\psi(\boldsymbol{r}) = \frac{1}{2} e^{i\boldsymbol{k}_{0\|}\boldsymbol{r}_{\|} + ik_{0\perp}(z-D)} \left(e^{iq_1 D} + e^{iq_2 D}\right) + \frac{1}{2} e^{-i\boldsymbol{k}_{0\|}\boldsymbol{r}_{\|} + ik_{0\perp}(z-D)} \left(e^{iq_1 D} - e^{iq_2 D}\right) . \tag{7.127}$$

One of them, with $\boldsymbol{k}_{0\|}$, propagates in the forward direction (we call it the direct wave), and the second one, with wave vector $-\boldsymbol{k}_{0\|}$, propagates in the diffraction direction (we call it the diffracted wave). The intensity of the direct wave is $I_1 = \cos^2(\delta q D/2)$, and that of the diffracted one is $I_2 = \sin^2(\delta q D/2)$, where $\delta q = q_1 - q_2$. When the thickness D or wave number k_0 of the incident wave changes, the intensity of the transmitted neutrons oscillates between two directions. This phenomenon is called Pendellösung. It is natural that the diffracted wave behaves oppositely to the direct one. Because of that the total intensity of the transmitted particles is equal to that of the incident ones.

Inside the crystal the wave function matched to Eq. (7.127) is

$$\psi(\boldsymbol{r}) = \frac{1}{2}\left(e^{i\boldsymbol{k}_{0\|}\boldsymbol{r}_{\|}} + e^{-i\boldsymbol{k}_{0\|}\boldsymbol{r}_{\|}}\right) e^{iq_1 z} + \frac{1}{2}\left(e^{i\boldsymbol{k}_{0\|}\boldsymbol{r}_{\|}} - e^{-i\boldsymbol{k}_{0\|}\boldsymbol{r}_{\|}}\right) e^{iq_2 z} . \tag{7.128}$$

Since at the lattice points

$$\exp(-i\boldsymbol{k}_{0\|}\boldsymbol{r}_j) = \exp(i(\boldsymbol{k}_{0\|} + \boldsymbol{\tau}_1)\boldsymbol{r}_j) = \exp(i\boldsymbol{k}_{0\|}\boldsymbol{r}_j) ,$$

we see that the first pair of waves in Eq. (7.128) enhance each other at the lattice points and the second pair of waves annihilate each other at the lattice points. Therefore, the second pair propagate inside the crystal as if in free space with $q = k_{0\perp}$, while the first pair see the crystal as a medium with doubled potential u_0.

In the presence of losses the potential u_0 contains an imaginary part; therefore q_1 also contains an imaginary part and $\exp(iq_1 D)$ decays exponentially with thickness D. In the case of sufficiently thick crystals the part of transmitted wave function (7.127) proportional to $\exp(iq_1 D)$ can be neglected and oscillation of the intensity with change of D or k_0 ceases.

Since q_2 does not contain the potential, the part of the wave function proportional to $\exp(iq_2 D)$ does not decrease in the crystal. Therefore half of the intensity passes through the thick crystal even if the absorption is pretty high. This anomalous transmission of waves through an absorbing crystal at Laue diffraction is known as the Borrmann effect in X-ray optics and the Kagan–Afanasiev effect in neutron optics.

7.3.8.1 Simultaneous Fulfillment of Bragg and Laue Conditions

The Bragg diffraction is the total reflection, which in the specular case takes place when condition $k_{n\perp} = k_B = \pi n/a$ or $e_n^4 = 1$ is satisfied. The Laue diffraction takes place when for some vectors of the reciprocal lattice, say, $\boldsymbol{\kappa}_m$ and $\boldsymbol{\kappa}_n$, the condition $k_{m\perp} \approx k_{n\perp}$ or $e_m^2 \approx e_n^2$ is satisfied. However, we can imagine the case when

$e_m^2 = e_n^2$ and simultaneously the condition $e_n^4 = 1$ is also satisfied. In such a case we should observe the Bragg diffraction, that is, total reflection from the crystal, and the Laue diffraction, that is, splitting of the beam into two parts and transmission through the crystal. It looks paradoxical, and the question arises: what will be seen in such a case in reality? Will the neutrons be reflected or transmitted?

The answer to this question happens to be pretty simple. We know that at Laue diffraction the neutron wave function inside the crystal contains two pairs of waves. One of them interferes in such a way that it becomes zero at lattice points. The other one doubles at lattice points. The first pair does not notice the crystal and passes through it freely. The second one sees the crystal and it is this pair which is totally reflected by the crystal. In this pair the direct wave will be reflected in the specular direction and the diffracted wave will be reflected in the nonspecular direction. It can be shown mathematically [352], but we leave it as an exercise for the reader.

It is interesting also to consider nonspecular Bragg reflection, when there are two states m and n for which $e_m^2 e_n^2 = 1$ is satisfied, and simultaneously there is the state l for which $e_l^2 = e_n^2$ is satisfied. What shall we observe in such a case? This problem is a good starting point for a reader who wants to conduct his own research.

7.3.8.2 **Diffraction and Renormalization of the Scattering Amplitude**

In Section 7.1.4 we showed that the the normalization constant C of the scattering amplitude b depends on the neutron wave vector $\boldsymbol{k}$ and for some of them, which we denote $\boldsymbol{k}_{j\infty}$, the constant C becomes zero. This conclusion is correct for an infinitely wide crystal, in which the crystalline planes extend to infinity in all directions. For a finite crystalline plane the constant $C(\boldsymbol{k})$ does not become zero; however, at these points $C(\boldsymbol{k}_{j\infty})$ becomes small, and it is very interesting to notice the decrease of the renormalized scattering amplitude b_c. It can be observed in Bragg diffraction from a single crystal. If one changes the energy of the incident neutrons and every time adjusts the crystal to observe a Bragg reflection, then at some energy the reflection will be strongly suppressed, though all the conditions for the Bragg reflection are satisfied and the structure factor for $b_c \neq 0$ is not equal to zero, that is, the reflection is not forbidden. If we were able to find the points $\boldsymbol{k}_{j\infty}$ and to see how much b_c at these points do decrease, we could understand how many lattice points do contribute to the sum (7.22), and therefore have a hint at what the neutron wave function is.

Besides the points $\boldsymbol{k}_{j\infty}$, the constant C becomes zero because of the imaginary part S'' Eq. (7.21) of the sum S. It becomes zero at such $\boldsymbol{k}$ for which there appears diffraction with $k_{n\perp} = 0$. For these $\boldsymbol{k}$ one can observe a strong diffraction in the direction along the interface and suppression of all the other scatterings. This effect has not been observed yet because no one has looked for it.

7.4 The Standard Description of Diffraction in Single Crystals

I hate diffraction... always had, always will

Anonymous. (From a lecture by R. Golub)

In the standard approach the neutron–crystal interaction is described by a periodic potential, which is represented by the Fourier series

$$u(\boldsymbol{r}) = \sum_{\kappa} v(\boldsymbol{\kappa}) \exp(i\boldsymbol{\kappa r}) , \tag{7.129}$$

where $\boldsymbol{\kappa}$ is a vector of a three-dimensional reciprocal lattice. In the case of a real potential the coefficients $v(\boldsymbol{\kappa})$ and $v(-\boldsymbol{\kappa})$ are complex-conjugated: $v(-\boldsymbol{\kappa}) = v^*(\boldsymbol{\kappa})$. The component $v(0)$ is an average potential which is identical to the optical one u_0.

To describe diffraction it is necessary to solve the Schrödinger equation with potential (7.129)

$$\left(\Delta + k^2 - \sum_{\kappa} v(\boldsymbol{\kappa}) \exp(i\boldsymbol{\kappa r})\right) \psi(\boldsymbol{r}) = 0 \tag{7.130}$$

for some incident wave $\exp(i\boldsymbol{kr})$.

7.4.1 The Fermi Pseudopotential

Now we need to find the potential (7.129), its reciprocal lattice (vectors $\boldsymbol{\kappa}$), and coefficients $v(\boldsymbol{\kappa})$. At first sight one can think of the potential itself as an array of spherical potential wells, every one of which represents a strong attractive interaction between the neutron and the nucleus at the lattice point.

However, this is not the case!Interaction of neutrons with a nucleus at a point $\boldsymbol{r}'$ in matter is described not by the negative nuclear potential, but by some artificial pseudopotential, $4\pi b\delta(\boldsymbol{r} - \boldsymbol{r}')$, introduced by E. Fermi, where b is the scattering amplitude. This pseudo potential has nothing to do with the real neutron–nucleus interaction. In most cases it is even repulsive, in contrast to strong attractive forces. However, it is related to the nuclear potential in one aspect. The scattering amplitude b calculated from the pseudo potential in perturbation theory is identical to the scattering amplitude b found accurately with the help of the real neutron–nucleus potential.

It is important to note that in previous sections we never used the notion of the potential. We needed only the scattering amplitude, so our approach was logically more self consistent than the standard approach with the potential (7.129).

The pseudopotential of the whole crystal is

$$v(\boldsymbol{r}) = 4\pi \sum_{n} \sum_{j=0}^{m} b_j \delta(\boldsymbol{r} - \boldsymbol{r}_{nj}) , \tag{7.131}$$

where coordinates $\boldsymbol{r}_{nj}$ of atoms in the crystal are represented by the sum of two vectors $\boldsymbol{r}_{nj} = \boldsymbol{r}_n + \boldsymbol{\xi}_j$, the first one being the radius vector of the origin of the nth elementary cell and the second one being the position of the jth atom in the nth elementary cell. In the case of a cubic crystal with elementary cell parameter a we have $\boldsymbol{r}_n = \boldsymbol{n}a$, where $\boldsymbol{n}$ is a vector with integer coordinates $n_{x,y,z}$.

According to the rule (7.12) the potential (7.131) is transformed to the sum

$$\begin{aligned} v(\boldsymbol{r}) &= 4\pi \sum_{\boldsymbol{m}} \int d^3 n \exp(2\pi i \boldsymbol{m}\boldsymbol{n}) \sum_{j=0}^{m} b_j \delta(\boldsymbol{r} - \boldsymbol{r}_{nj}) \\ &= 4\pi \sum_{\kappa} \int \frac{d^3 r'}{a^3} \exp(i\boldsymbol{\kappa}\boldsymbol{r}') \sum_{j=0}^{m} b_j \delta(\boldsymbol{r} - \boldsymbol{r}' - \boldsymbol{\xi}_j), \end{aligned} \tag{7.132}$$

where $\boldsymbol{m}$ is a vector with integer coordinates $m_{x,y,z}$ and in the last sum we introduced the vectors $\boldsymbol{\kappa} = 2\pi\boldsymbol{m}/a$ of the reciprocal lattice.

After integration over $\boldsymbol{r}'$ we obtain Eq. (7.129), where $v(\boldsymbol{\kappa}) = 4\pi N_c F(\boldsymbol{\kappa})$, where $N_c = 1/a^3$ is the number of elementary cells in a unit volume and

$$F(\boldsymbol{\kappa}) = \sum_{j=0}^{m} b_j \exp(i\boldsymbol{\kappa}\boldsymbol{\xi}_j) \tag{7.133}$$

is the structure factor.

7.4.2
The Standard Dynamical Diffraction Theory

Let us take our crystal as an infinite crystalline plate of thickness D perpendicular to the z-axis. Interaction with neutrons is described by the three-dimensional periodic potential (7.129). The wave function $\psi(\boldsymbol{r})$ inside a periodic potential is a product of a periodic function and the Bloch phase factor:

$$\psi(\boldsymbol{r}) = \exp(i\boldsymbol{q}\boldsymbol{r})\phi(\boldsymbol{r}), \tag{7.134}$$

where $\boldsymbol{q}$ is the Bloch wave vector and $\phi(\boldsymbol{r})$ is the periodic part of the wave function, which like the potential (7.129) is representable by a Fourier series:

$$\phi(\boldsymbol{r}) = \sum_{\kappa} \Phi(\boldsymbol{\kappa}) \exp(i\boldsymbol{\kappa}\boldsymbol{r}). \tag{7.135}$$

Comparison of Eqs. (7.134) and (7.135) shows that the components $\boldsymbol{q}_\parallel$ parallel to the crystal interface can be put equal to $\boldsymbol{k}_\parallel$. Continuity of the wave function at the interface and diffraction on crystalline planes makes the components of the wave vector parallel to the interface the same as $\boldsymbol{k}_\parallel$ or contain an additional vector $\boldsymbol{\kappa}_\parallel$ of the reciprocal lattice of the crystalline planes. However, the components $\boldsymbol{\kappa}_\parallel$ can be included in the sum (7.135), so we are left with the single component $\boldsymbol{q}_\parallel = \boldsymbol{k}_\parallel$, but $q_\perp$ are some unknown parameters.

Substitution of the wave function (7.135) into the Schrödinger equation gives

$$\exp(i\boldsymbol{q}\boldsymbol{r})\sum_{\boldsymbol{\kappa}}\left([(\boldsymbol{q}+\boldsymbol{\kappa})^2-k^2]\Phi(\boldsymbol{\kappa})+\sum_{\boldsymbol{\kappa}'}v(\boldsymbol{\kappa}')\Phi(\boldsymbol{\kappa}-\boldsymbol{\kappa}')\right)\exp(i\boldsymbol{\kappa}\boldsymbol{r})=0\,, \tag{7.136}$$

which is equivalent to an infinite set of equations for coefficients $\Phi(\boldsymbol{\kappa})$:

$$[(\boldsymbol{q}+\boldsymbol{\kappa})^2-k^2]\Phi(\boldsymbol{\kappa})+\sum_{\boldsymbol{\kappa}'}v(\boldsymbol{\kappa}')\Phi(\boldsymbol{\kappa}-\boldsymbol{\kappa}')=0\,. \tag{7.137}$$

The system is homogeneous and has a nontrivial solution only if its determinant is equal to zero. This condition is an algebraic equation of infinite order for $q_\perp$, which has an infinite number of solutions $q_{n\perp}$. For every one of them we can resolve all the coefficients $\Phi(\boldsymbol{\kappa})$ via, say, $\Phi(0)$, and further find $\Phi(0)$ from boundary conditions.

Of course all this program is not realizable and we have to use some approximations. The potential is usually small, so its components $v(\boldsymbol{\kappa})$ are also small. If, for instance, only one wave in the crystal has large amplitude $\Phi(\boldsymbol{\kappa})$, then the system (7.137) is reduced to the single equation

$$(\boldsymbol{q}+\boldsymbol{\kappa})^2-k^2+u_0=0\,, \tag{7.138}$$

where $u_0=v(0)$. The only wave (if it is the single one) created in the crystal by the incident plane wave $\exp(i\boldsymbol{k}\boldsymbol{r})$ is $\exp(i\boldsymbol{q}\boldsymbol{r})$, with $\boldsymbol{\kappa}=0$ and $\boldsymbol{q}_\parallel=\boldsymbol{k}_\parallel$. Therefore, from Eq. (7.138) we immediately find $q_\perp=\sqrt{k_\perp^2-u_0}$. If reflection at the interface at $z=0$ can be neglected, $\Phi(\boldsymbol{\kappa}=0)$ coincides with the amplitude of the incident wave, that is, $\Phi(\boldsymbol{\kappa}=0)=1$.

If two waves are large inside the crystal, say, waves $\Phi(0)=\Phi(\boldsymbol{\kappa}=0)$ and $\Phi(\boldsymbol{\kappa})$ with some $\boldsymbol{\kappa}\neq 0$, then the infinite system of equations reduces to the system of two equations:

$$\begin{aligned}&\left[q_\perp^2-k_\perp^2+u_0\right]\Phi(0)+\Phi(\boldsymbol{\kappa})v(-\boldsymbol{\kappa})=0\,,\\ &\left[(q_\perp+\kappa_\perp)^2-k_{\kappa\perp}^2+u_0\right]\Phi(\boldsymbol{\kappa})+\Phi(0)v(\boldsymbol{\kappa})=0\,,\end{aligned} \tag{7.139}$$

where $k_{\kappa\perp}^2=k^2-(\boldsymbol{k}_\parallel+\boldsymbol{\kappa}_\parallel)^2$. Equating its determinant to zero we get

$$\left[q_\perp^2-k_\perp^2+u_0\right]\left[(q_\perp+\kappa_\perp)^2-k_{\kappa\perp}^2+u_0\right]-v(\boldsymbol{\kappa})v(-\boldsymbol{\kappa})=0\,. \tag{7.140}$$

After change of variables $x=q_\perp+\kappa_\perp/2$, it is reduced to a biquadratic equation. In the standard dynamical theory (7.140) is simplified even further to a quadratic one.

7.4.2.1 Specular Bragg Diffraction

In the case of specular Bragg diffraction we have $\boldsymbol{\kappa}_\parallel=0$. Denote $\kappa_\perp=-2k_B$ and $q_\perp=k_B+x$, where x near Bragg diffraction is small. Then Eq. (7.140) becomes

$(u_0 + k_B^2 - k_\perp)^2 - 4x^2k_B^2 - v(\kappa)v(-\kappa) = 0$, and its solutions are $x_{1,2} = \pm\Delta$, where $\Delta = \sqrt{(k_B^2 - k_\perp^2 + u_0)^2 - v(\kappa)v(-\kappa)}/2k_B$. Therefore, $q_{1,2\perp} = k_B \pm \Delta$.

For both $q_{1,2\perp}$ we can express $\Phi^{(1,2)}(\kappa)$ via $\Phi^{(1,2)}(0)$:

$$\Phi^{(1,2)}(\kappa) = -\frac{q_{1,2\perp}^2 - k_\perp^2 + u_0}{v(-\kappa)}\Phi^{(1,2)}(0)\,. \tag{7.141}$$

For determinations of $\Phi^{(1,2)}(0)$ two boundary conditions are used: (1) continuity of the incident wave at the entrance surface at $z = 0$, $\Phi^{(1)}(0) + \Phi^{(2)}(0) = 1$; and (2) absence of diffracted waves at the exit surface $z = D$, $\Phi^{(1)}(\kappa)\exp(iq_1D) + \Phi^{(2)}(\kappa)\exp(iq_2D) = 0$. From these two conditions we can find $\Phi^{(1,2)}(\kappa)$ and the amplitude of the reflected wave:

$$\Psi(\kappa) = -\frac{iu\sin(\Delta D)}{i\left(k_B^2 - k_\perp^2 + u_0\right)\sin(\Delta D) + \sqrt{\left(k_\perp^2 - k_B^2\right)\left(k_\perp^2 - k_B^2 - 2u_0\right)}\cos(\Delta D)}\,. \tag{7.142}$$

The probability of the Bragg diffraction is equal to $|\Psi(\kappa)|^2$. In the case of monatomic crystals, where $v(\pm\kappa) = u_0$, inside the Bragg peak, where Δ is imaginary, we obtain the reflectivity, which completely coincides with Eq. (7.109).

7.4.2.2 **Symmetrical Laue Diffraction**

In the case of symmetrical Laue diffraction we have $\kappa_\perp = 0$, so Eq. (7.140) reduces to $\left[q_\perp^2 - k_\perp^2 + u_0\right]\left[q_\perp^2 - k_{\kappa\perp}^2 + u_0\right] - v(-\kappa)v(\kappa) = 0$, and its two solutions, which correspond to waves going away from $z = 0$, are $q_{1,2\perp} = \sqrt{k_\perp^2 + k_{\kappa\perp}^2 - 2u_0 \pm \sqrt{\left(k_\perp^2 - k_{\kappa\perp}^2\right)^2 + 4v(-\kappa)v(\kappa)}}/\sqrt{2}$. When the Bragg condition is satisfied precisely ($k_\perp = k_{\kappa\perp}$) we get $q_{1,2\perp} = \sqrt{k_\perp^2 - 2u_{1,2}}$, where $u_{1,2} = u_0 \pm \sqrt{v(-\kappa)v(\kappa)}$. In the case of a monatomic crystal they become $q_{1\perp} = k_\perp$ and $q_{2\perp} = \sqrt{k_\perp^2 - 2u_0}$.

To find $\Phi^{(1,2)}(0)$ and $\Phi^{(1,2)}(\kappa)$ from Eq. (7.141) two boundary conditions are used: (1) continuity of the incident wave at the entrance surface at $z = 0$, $\Phi^{(1)}(0) + \Phi^{(2)}(0) = 1$; and (2) continuity of the diffracted wave at the entrance surface at $z = 0$, $\Phi^{(1)}(\kappa) + \Phi^{(2)}(\kappa) = 0$. After solution of these equations we obtain at the exit surface $z = D$ the amplitudes of the transmitted (direct), $\Psi(0)(D) = \Phi^{(1)}(0)\exp(iq_1D) + \Phi^{(2)}(0)\exp(iq_1D)$, and diffracted, $\Psi(\kappa)(D) = \Phi^{(1)}(\kappa)\exp(iq_1D) + \Phi^{(2)}(\kappa)\exp(iq_1D)$, waves. Substitution of $q_{1,2}$ gives

$$\Psi(\kappa)(D) = \exp(iq^+D)\frac{i\sin(q^-D)}{\sqrt{1+\eta^2}}\,,$$

$$\Psi(0)(D) = \exp(iq^+D)\left(\cos(q^-D) + \frac{i\eta\sin(q^-D)}{\sqrt{1+\eta^2}}\right)\,, \tag{7.143}$$

where $\eta = (k_\perp^2 - k_{\kappa\perp}^2)/2u_0$ and $q^\pm = (q_{1\perp} \pm q_{2\perp})/2$.

Denote $D_0 = 1/q^-$, then the intensities of the direct and diffracted waves can be represented as

$$I(0) = 1 - \frac{1}{1+\eta^2}\sin^2(D/D_0), \qquad I(\kappa) = \frac{1}{1+\eta^2}\sin^2(D/D_0). \tag{7.144}$$

We see that they oscillate with increase of D, and the oscillation period and amplitude $1/(1+\eta^2)$ decrease with increase of $|\eta|$. Oscillations appear because the wave function inside the crystal consists of two pairs of waves as was discussed for Eq. (7.127).

7.5 Comparison of the Two Methods

Though the matrix method is more logically consistent and physically transparent, both methods describe the same processes and lead to nearly the same results. So the choice of the method for the calculation of neutron diffraction in single crystals is a matter of the researcher's taste. The Fourier expansion used by standard theory gives a simple way to find the positions and intensities of the diffraction peaks for crystals with an arbitrary complicated elementary cell. The matrix method can also be used for investigation of crystals with a complicated elementary cell [352]; however, it is better to calculate scattering on a single period by perturbation theory. With it the matrix method permits one to consider the diffraction process in more detail and to control the error at every step of the calculation.

The Fourier method is comparatively simple in the case of the two waves approximation and becomes more and more cumbersome when more waves come to action. The matrix method gives the same algorithm for calculations, independently of number of participating waves.

The standard dynamical diffraction theory presented in many textbooks is not self-consistent, because matching of the wave function at interfaces is not complete. In the case of specular Bragg diffraction this defect can be easily corrected (see, e. g., [353]), but this case is reduced to diffraction in a one-dimensional periodic system. More difficult is the case of diffraction when in the two waves approximation diffracted and incident waves have different components $\boldsymbol{k}_{\parallel}$ parallel to the interface. In that case full matching at two interfaces leads to a system of eight linear equations with eight unknown coefficients. It is clear that solution of such a system can be analyzed only numerically, which was demonstrated in [354, 355].

The advantage of the matrix method is its self-consistency. It uses only scattering amplitudes and does not require matching of the wave function at interfaces. It gives a new vision for the diffraction and because of that it helps to perceive new phenomena, which can also be perceived in the Fourier approach, but with some difficulty. As an example we can mention diffraction at forbidden reflections, which are predicted to have zero width by the Fourier theory. A more precise calculation with matrix theory shows that their width is 4–6 orders of magnitude narrower than that of nonforbidden ones, but not zero [199]. Another example is simultaneous Bragg and Laue diffraction. We also think that diffraction at small grazing incident angles will show an analogous effect, because the total reflection is the Bragg reflection of zero order. Therefore, when for neutrons or X-rays incident on a crystal at a grazing angle below the critical one a Laue condition $|\boldsymbol{k}_{\parallel}| = |\boldsymbol{k}_{\parallel} + \boldsymbol{\kappa}_{\parallel}|$ for some vector $\boldsymbol{\kappa}_{\parallel}$ of reciprocal lattice is satisfied, we can expect that half of the intensity will be transmitted through the crystal in forward and diffracted directions. This is also discussed in [356]. The matrix method opens a new area for research of very subtle effects in diffraction, which will help to improve our understanding of the neutron–crystal interaction and of the structure of the neutron wave function. For instance, since the neutron wave function is a wave packet, we can expect diffraction for every incident energy, because we can always find a wave in the packet for which the Bragg condition is satisfied exactly. The diffracted intensities in this case characterize the structure of the wave packet. Till now no diffraction pattern has been analyzed from this point of view.

We have considered only the principal questions in this book, and have not touched on real systems because the application of diffractometry has been considered well in many other books. For those who want more details on the standard theory and its applications to research in neutron, X-ray, and synchrotron radiation physics, we can recommend the books [357–362] (Textbook [I]).

7.6 Kinematical Diffraction Theory and Practical Crystallography

Dynamical diffraction theory explains the diffraction processes in single crystals. The goal of crystallography is to find structures. There are very few ideal single crystals in nature; for this reason most crystallography experiments are performed with powder, which contains small crystallites with dimension of order $0.1 \div 100\ \mu$m. Here we show some formulas used to interpret powder diffraction experiments.

For small crystals usually a perturbation theory is applied, and the result is called kinematical diffraction. The scattering amplitude of all the atoms inside the small crystal is

$$A(q) = \sum_{n,j} b_{n,j} \exp(i\boldsymbol{q}\boldsymbol{r}_{n,j}), \tag{7.145}$$

where $\boldsymbol{q} = \boldsymbol{k} - \boldsymbol{k}'$ is momentum transfer, $\boldsymbol{k}'$ is the wave vector after scattering, and $\boldsymbol{r}_{n,j}$ denotes the position of an atom j with scattering amplitude b_j within an elementary cell marked by three-dimensional number vector $\boldsymbol{n} = (n_x, n_y, n_z)$ ($n_{x,y,z}$ are integers). Note that for generality we assumed that the scattering amplitudes can be different from cell to cell. This suggestion is necessary if we want to take into account defects, deformations, and incoherent scattering. However, for now we suppose that $b_{n,j} = b_j$, that is, all the cells in the crystal are identical.

Application to Eq. (7.145) of the sum rule (7.12) gives

$$A(q) = \sum_{\kappa} \int_{V} N_c d^3 r \exp(i[\boldsymbol{q} + \boldsymbol{\kappa}]\boldsymbol{r}) F(\boldsymbol{q}) = (2\pi)^3 F(\boldsymbol{q}) \sum_{\kappa} \delta(\boldsymbol{q} + \boldsymbol{\kappa}) , \quad (7.146)$$

where N_c is number of elementary cells in a unit volume, V is the crystal volume,

$$F(\boldsymbol{q}) = \sum_{j} b_j \exp(i\boldsymbol{q}\boldsymbol{r}_j) \quad (7.147)$$

is the structure factor, and we supposed that the volume is sufficiently large to replace the integral over it by the δ function.

The differential cross section is $|A(q)|^2$, which is equal to

$$\sigma'(q) = N_c^2 |F(\boldsymbol{q})|^2 V \sum_{\kappa} (2\pi)^3 \delta(\boldsymbol{q} + \boldsymbol{\kappa}) , \quad (7.148)$$

where $\delta^2(\boldsymbol{q} + \boldsymbol{\kappa})$ was replaced by $[V/(2\pi)^3]\delta(\boldsymbol{q} + \boldsymbol{\kappa})$. After multiplication of Eq. (7.129) by the number of final states $d^3k'V/(2\pi)^3$ and integration over d^3k' around direction $\boldsymbol{q} \approx \boldsymbol{\kappa}$, we obtain the total cross section with the momentum transfered $\boldsymbol{\kappa}$:

$$\sigma(\boldsymbol{k} \to \boldsymbol{k}' = \boldsymbol{k} - \boldsymbol{\kappa}) = (V N_c)^2 |F(\boldsymbol{\kappa})|^2 = N^2 |F(\boldsymbol{\kappa})|^2 , \quad (7.149)$$

where $N = N_c V$ is the total number of elementary cells in the crystal.

This result is too simple and is proportional to square of the volume, whereas usually the intensity diffracted by a single small crystal is supposed to be linear in volume. For that reason we try to find the reflection from a small crystal within a simplified dynamical diffraction theory, noting that for small crystals there are no reasons to distinguish between Bragg and Laue diffraction. In both cases we have the reflection from some sets of crystalline planes. So we shall proceed in the same way as we did for Bragg scattering from small crystalline blocks in mosaic crystals on page 314.

Let us look at a single period of a small, but sufficiently large crystal, which can be considered as a set of parallel crystalline periods. In the case of a complicated elementary cell it is not a single crystalline plane but several of them. To find the diffraction on a single period we are to apply multiple wave scattering theory mentioned in Section 7.1.1. The positions of the atoms in the period are defined by vectors $\boldsymbol{r}_{n,j}$. In the first approximation we can suppose that every scatterer inside the period is illuminated by the same incident wave $\exp(i\boldsymbol{k}\boldsymbol{r})$. Then the reflected wave field at $z < 0$, like in Eq. (7.11), is equal to

$$\begin{aligned} \psi(\boldsymbol{r}) &= e^{i\boldsymbol{k}\boldsymbol{r}} - \frac{i}{2\pi} \int \frac{d^2 q_\parallel}{q_\perp} e^{i q_\parallel r - i q_\perp z} \sum_{n,j} b_j \exp(i[\boldsymbol{k} - \boldsymbol{q}]\boldsymbol{r}_{n,j}) \\ &= e^{i\boldsymbol{k}\boldsymbol{r}} - \frac{i}{2\pi} \int \frac{d^2 q_\parallel}{q_\perp} e^{i q_\parallel r - i q_\perp z} \sum_{n} F(\boldsymbol{k} - \boldsymbol{q}) \exp(i[\boldsymbol{k} - \boldsymbol{q}]\boldsymbol{r}_{n\parallel}) , \end{aligned} \quad (7.150)$$

where $F(\boldsymbol{k} - \boldsymbol{q})$ is the structure factor (7.147).

After transformation of the sum according to the rule (7.12) we obtain (like in Eq. (7.16))

$$\psi(\boldsymbol{r}) = \exp(i\boldsymbol{k}\boldsymbol{r}) - \sum_{\kappa} \frac{2\pi i N_2 F(\boldsymbol{\kappa})}{k_\perp(\boldsymbol{\kappa})} \exp(i\boldsymbol{k}_{\kappa\|}\boldsymbol{r}_\| - ik_\perp(\boldsymbol{\kappa})z) . \quad (7.151)$$

Therefore, the reflection matrix of a single period is

$$\hat{\rho} = -i\hat{\boldsymbol{\beta}}|\boldsymbol{n}\rangle\langle\boldsymbol{n}| , \quad (7.152)$$

where $\hat{\boldsymbol{\beta}}$ is a diagonal matrix with matrix elements $\beta_n = 2\pi N_2 F(\boldsymbol{\kappa}_n)/k_\perp(\boldsymbol{\kappa}_n)$.

The intensity reflected by a crystalline block of thickness D and area S for incident flux density I_0 is characterized, according to Eq. (7.116), by the expression

$$I(\boldsymbol{\kappa}) = I_0 S\, R(\boldsymbol{\kappa}) \approx I_0 S\, Q D \mathcal{L} = I_0 V |N_0 F(\boldsymbol{\kappa})|^2 \lambda^3 \mathcal{L} , \quad (7.153)$$

which is linear in V and contains the well-known Lorentz factor $\mathcal{L} = 1/\cos\theta_B \sin(2\theta_B)$. Sometimes it is written as $\mathcal{L} = 1/\sin\theta_B \sin(2\theta_B)$, but it depend on whether θ is the angle between the wave vector and the normal to the reflecting plane, as in our case, or it is the grazing angle.

In practical crystallography this expression is more complicated. In powder diffraction, commonly thin cylindrical samples are used, and several correction factors are included in Eq. (7.153):

$$I(\boldsymbol{\kappa}) = I_0 V Q \mathcal{L} J A E , \quad (7.154)$$

where J is the number of equivalent reflections related to the lattice symmetry, A is correction for absorption, and E is correction for the secondary extinction. Moreover, sometimes it is necessary to take into account thermal vibrations of atoms and possible deformation and defects of the crystalline lattice. After that, for extraction of atomic positions a theoretical model with guessed positions is fitted by minimization of χ^2 to the results of measurement at many points $\boldsymbol{\kappa}$. In crystallography this way of fitting is customarily called the Rietveld method. We cannot dwell on this and refer the reader to [349–351].

7.7
Unitarity and the Detailed Balance Principle

Above we considered dynamical diffraction for the simplest case of a monatomic crystal and ignored some questions, which can be important, when we generalized our methods to more complex crystals, where perturbation theory for a single elementary cell must be used. The perturbation theory always brings some errors, which, if not corrected, will accumulate and give rise to false effects such as violation of unitarity. Here we want to consider basic principles which should not be violated. They are the law of particle conservation in the absence of losses, which is unitarity, and the detailed balance condition, which means that scattering should not lead to a decrease of entropy of closed systems. Because of volume limitations, we shall consider them very briefly.

7.7.1 Unitarity

Unitarity follows from the Schrödinger equation and it means that in the case of a stationary wave function the neutron flux through any closed surface is zero, that is, the number of neutrons entering the volume inside a considered surface is equal to the number of outgoing ones. In the case of an infinite surface like a plane, the unitarity principle says that the flux density of incident neutrons must be equal to the flux density of outgoing ones. For a single fixed-in-space scatterer with scattering amplitude $b(\Omega)$, where Ω is the scattering angle, unitarity means that $\mathrm{Im}\,b(0) = k\sigma_t/4\pi$, where σ_t is the total cross section, which contains scattering and absorption. This statement is known as the "optical theorem." In the case of an isotropic scatterer and the absence of losses, the optical theorem means that the scattering amplitude can be represented as

$$b = \frac{b_0}{1 + ikb_0}, \tag{7.155}$$

where b_0 is a real number. If instead of one scatterer we have a crystalline plane with isotropic scatterers, the total wave function becomes $\psi = \psi_0 + \psi_s \equiv \Theta(z < 0)[\psi_0 + \psi_r] + \Theta(z > 0)\psi_t$. The incident flux, J_0, is determined by the incident wave ψ_0. The outgoing flux is determined by the reflected one $J_r = J_s$, produced by the reflected part ψ_r of the wave function and the transmitted part J_t produced by the transmitted part $\psi_t = \psi_s + \psi_0$ of the wave function. The last one contains J_0 and J_s and the interference part J_i. Unitarity means that $J_0 = J_0 + 2J_s + J_i$, or $2J_s + J_i = 0$. The scattered flux density, J_s according to the wave function Eq. (7.16), is

$$J_s = (2\pi N_2)^2 |bC|^2 \sum_n^{k_{n\perp}^2 > 0} \frac{1}{k_{n\perp}},$$

where the sum over n does not contain exponentially decaying waves. The interference part of the flux through the plane S_2 in the direction $\boldsymbol{k}_0$ is $J_i = -2\pi i N_2(bC - b^* C^*)$. The unitarity condition $J_i + 2J_s = 0$ means limitation on bC:

$$\mathrm{Im}(bC) = -\gamma |bC|^2, \quad \gamma = 2\pi N_2 \sum_n^{k_{n\perp}^2 > 0} \frac{1}{k_{n\perp}}. \tag{7.156}$$

According to Eq. (7.7), the number C is representable as

$$C = 1/(1 + ib\alpha), \tag{7.157}$$

where α, in general, is a complex number, $\alpha' - i\alpha''$. Taking into account Eq. (7.155) for b, which is valid in the absence of absorption, we get

$$bC = \frac{b_0}{1 + b_0\alpha'' + ib_0(k + \alpha')} = \frac{b_1}{1 + ib_1(k + \alpha')}, \tag{7.158}$$

where $b_1 = b_0/(1 + b_0\alpha'')$ is the renormalized real part of the scattering amplitude. It follows from Eq. (7.158) that $\mathrm{Im}(bC) = -(k + \alpha')|bC|^2$. Comparing this with

Eq. (7.156), we immediately obtain

$$\alpha' = -k + \gamma = -k + 2\pi N_2 \sum_n^{k_{n\perp}^2>0} \frac{1}{k_{n\perp}} . \qquad (7.159)$$

Therefore the factor C Eq. (7.157) of the amplitude b changes the bookkeeping of unitarity. The first term $-k$ in Eq. (7.159) cancels the unitarity which was valid for a single scatterer, and the second one introduces new rules for unitarity in new conditions, when we do not have a single scatterer but the whole crystalline plane of scatterers.

It is easy to guess that the complex number α is obtained from the real α' Eq. (7.159) by extension of the sum in Eq. (7.159) to all the possible $\boldsymbol{\kappa}_n$:

$$\alpha = -k + 2\pi N_2 \sum_n^{\infty} \frac{1}{k_{n\perp}} . \qquad (7.160)$$

Thus, those $\boldsymbol{\kappa}_n$ which are related to propagated waves contribute to the real part α', and those which are related to exponentially decaying waves contribute to the imaginary part α''. The series in Eq. (7.160) diverges; however, this divergence is fictive. The infinite series is a formal representation of a finite function, like the sum of the geometrical progression x^n at $x > 1$ is a formal representation of the finite function $1/(1-x)$.

In all the calculations that involve the scattering amplitude we can ignore renormalization and accept b as a real number. However, at the end of the calculations it is necessary to change b in accord with unitarity requirements, that is, for instance, to put it as $b = b_0/(1 + ikb_0)$ in the case of a single center, or

$$b = \frac{b_1}{1 + 2\pi i N_2 b_1 \sum_n (1/k_{n\perp})} \qquad (7.161)$$

in the case of diffraction on a crystalline plane.

At small $k \ll 2\pi/a$ all the diffracted waves decay exponentially, and we are left with only single specularly reflected and directly transmitted waves. The reflected wave has amplitude $-ip/(k_\perp + ip)$, and transmitted one has amplitude $k_\perp/(k_\perp + ip)$, which is in complete agreement with the case when the plane is described by the δ-function potential $u = 2p\delta(r_\perp)$, in which $p = 2\pi N_2 b_1$. Let us note again an interesting phenomenon which follows from Eq. (7.161). If the wave vector $\boldsymbol{k}_0$ is such that for some vector $\boldsymbol{\kappa}$ of the reciprocal lattice of the atomic plane the component $k_\perp(\boldsymbol{\kappa})$ of a diffracted wave is zero, the denominator becomes infinite and the amplitudes of all the diffracted waves become zero except for the single one propagating along the crystalline plane. This theoretical prediction has not yet been verified in an experiment.

7.7.2
The Detailed Balance Principle

The detailed balance is as follows: if we have a scatterer with elastic and inelastic scattering and we put it in a neutron gas at thermal equilibrium, that is, a gas with isotropic and Maxwellian distribution of neutron velocities, and the states of the scatterer are also in equilibrium with the same temperature as the neutron gas, then the neutron scattering on such a scatterer should not change the equilibrium distribution neither of the scatterer nor of the neutron gas. Since the equilibrium state has maximal entropy, the principle of detailed balance means that the scattering must not diminish the entropy.

If $w(\boldsymbol{k} \to \boldsymbol{k}')$ is the probability of scattering on a single scatterer, the detailed balance principle requires

$$k\rho_0 \exp(-k^2/2T)d^3k\,w(\boldsymbol{k} \to \boldsymbol{k}')d^3k' = \\ k'\rho_0 \exp(-k'^2/2T)d^3k'w(\boldsymbol{k}' \to \boldsymbol{k})d^3k\,, \tag{7.162}$$

where ρ_0 =const is the neutron density in phase space, T is temperature, and the factor k characterizes the flux density of neutrons incident on the scatterer. The condition (7.162) means that the number of neutrons with energy k^2 scattered per unit time from an element d^3k of the phase space into an element d^3k' at energy k'^2 is equal to the number of neutrons with energy k'^2 scattered per unit time from the element d^3k' of phase space into the element d^3k at energy k^2. From Eq. (7.162) a limitation follows: the probability of the transition $w(\boldsymbol{k} \to \boldsymbol{k}')$ must be of the form

$$w(\boldsymbol{k} \to \boldsymbol{k}') = k' \exp(-(k'^2 - k^2)/4T)w^s(\boldsymbol{k} \to \boldsymbol{k}')\,, \tag{7.163}$$

where $w^s(x \to y)$ is a function symmetrical with respect to permutation of arguments.

If instead of a single scatterer we have the whole plane of scatterers, which are in thermodynamical equilibrium with each other and with the neutron gas on both sides of the plane, then the probability of neutron scattering from the state $\boldsymbol{k}$ into state $\boldsymbol{k}'$ described by the function $w(\boldsymbol{k} \to \boldsymbol{k}')$, according to the detailed balance principle for the plane, should satisfy the requirement

$$k_\perp\rho_0 \exp(-k^2/2T)d^3k\,w(\boldsymbol{k} \to \boldsymbol{k}')d^3k' = \\ k'_\perp\rho_0 \exp(-k'^2/2T)d^3k'w(\boldsymbol{k}' \to \boldsymbol{k})d^3k\,, \tag{7.164}$$

where the factor $k_\perp$ is related to the neutron flux density incident on the plane. From Eq. (7.164) there follows some limitation on the form of the scattering function:

$$w(\boldsymbol{k} \to \boldsymbol{k}') = k'_\perp \exp(-(k'^2 - k^2)/4T)w^s(\boldsymbol{k} \to \boldsymbol{k}')\,, \tag{7.165}$$

where $w^s(x \to y)$ is a function symmetrical with respect to permutation of its arguments.

In the case of pure elastic scattering we have $w(\boldsymbol{k} \to \boldsymbol{k}') = w(\boldsymbol{\Omega} \to \boldsymbol{\Omega}')\delta(k^2 - k'^2)$; therefore the detailed balance condition (7.164) is reduced to the form $\cos\theta\, d\Omega\, w(\boldsymbol{\Omega} \to \boldsymbol{\Omega}')d\Omega' = \cos\theta' d\Omega' w(\boldsymbol{\Omega}' \to \boldsymbol{\Omega})d\Omega$, and it follows that

$$w(\boldsymbol{\Omega} \to \boldsymbol{\Omega}') = \cos\theta' w^s(\boldsymbol{\Omega} \to \boldsymbol{\Omega}')\,, \tag{7.166}$$

where $w^s(\boldsymbol{\Omega} \to \boldsymbol{\Omega}')$ is a function symmetrical with respect to permutation of $\boldsymbol{\Omega} \leftrightarrow \boldsymbol{\Omega}'$.

In the case of the diffraction on a crystalline plane, the diffraction probability $w_d(\boldsymbol{k} \to \boldsymbol{k})$, when the incident neutron has the normal component of the wave vector $k_\perp$, and the diffracted one has the normal component of the wave vector $k'_\perp$, is equal to ratio of the diffracted flux density $J(\boldsymbol{k}')$ to the incident flux density $J(\boldsymbol{k}) = k_\perp$, that is, $w_d(\boldsymbol{k} \to \boldsymbol{k}') = J(\boldsymbol{k}')/J(\boldsymbol{k}) = |p|^2/k_\perp k'_\perp$. This probability of the discrete transition is symmetrical with respect to permutation $\boldsymbol{k} \leftrightarrow \boldsymbol{k}'$, which, seems to contradict (7.166). However, Eq. (7.166) is related to a continuous distribution of scattered neutrons. Therefore, to check the detailed balance condition we need to represent the discrete probability in continuous form:

$$w(\boldsymbol{k}) = \sum_{k'} \frac{|p|^2}{k_\perp k'_\perp} \equiv \sum_{\kappa} \frac{|p|^2}{k_\perp k_{\perp\kappa}}$$
$$= \frac{|p|^2}{k_\perp} \int \sum_{\kappa} \delta(\boldsymbol{k}_\parallel - \boldsymbol{k}'_\parallel + \boldsymbol{\kappa}) \delta(k^2/2 - k'^2/2) d^3k' \,, \tag{7.167}$$

where the summation over all k' in the first equality is replaced by the summation over all the vectors of the reciprocal lattice in the second equality. The integral in the third equality is identical to $1/k_{\perp\kappa}$.

From the continuous representation (7.167) it follows that

$$w(\boldsymbol{k} \to \boldsymbol{k}') = \frac{|p|^2}{k_\perp} \sum_{\kappa} \delta(\boldsymbol{k}_\parallel - \boldsymbol{k}'_\parallel + \boldsymbol{\kappa}) \delta(k^2/2 - k'^2/2) \tag{7.168}$$

completely satisfies the detailed balance requirement $w(\boldsymbol{k} \to \boldsymbol{k}') = k'_\perp w^s(\boldsymbol{k} \to \boldsymbol{k}')$, where the symmetrical function w^s is $w^s(\boldsymbol{k} \to \boldsymbol{k}') = (|p|^2/k_\perp k'_\perp) \sum_\kappa \delta(\boldsymbol{k}_\parallel - \boldsymbol{k}'_\parallel + \boldsymbol{\kappa}) \delta(k^2/2 - k'^2/2)$. This function is symmetrical because the summation runs over all $\boldsymbol{\kappa} = (2\pi/a)(n_x, n_y, 0)$, with positive and negative integer n_x and n_y.

7.8
The Optical Potential u_0

From the very start of this book we have practically everywhere used the notion of the optical potential $u_0 = 4\pi N_0 b$, which plays a very important role in the interaction of neutrons with matter. Now we want to show how to derive it from first principles. In fact we have no potential at all. All we have is scattering on every atom characterized by the amplitude b. Nevertheless the investigation of scattering on an ordered system of atoms demonstrated that at small $k_\perp^2$ in the range $0 \le k_\perp^2 \le u_0$ the neutron is totally reflected from the substance. It is this fact which permits us to ascribe to the matter the repelling potential u_0, notwithstanding that the optical potential of the strong neutron–nucleus interaction is negative and, therefore, attractive.

The expression for u_0 obtained in Eq. (7.53) is not absolutely precise. It is important to find corrections to it. They are especially important in relation to the anomalous losses observed in storage of ultracold neutrons (textbook [G]). To estimate these corrections we again consider neutron diffraction on a semi-infinite crystal, limiting ourselves to small energies, when all the diffracted waves decay exponentially [363].

When $k \ll \pi/a$ the wave function Eq. (7.16) after scattering on a crystalline plane is reduced to

$$\psi(\boldsymbol{r}) = \exp(i\boldsymbol{k}_\| \boldsymbol{r}_\|)\left(\Theta(z<0)\left[e^{ik_{0\perp}z} - i\frac{p_c}{k_{0\perp}}e^{-ik_{0\perp}z}\right] + \left[1 - i\frac{p_c}{k_{0\perp}}\right]\Theta(z>0)e^{ik_{0\perp}z}\right) - \sum_{\kappa\neq 0}\frac{p_c}{|k_\perp(\kappa)|}\exp(i\boldsymbol{k}_{\kappa\|}\boldsymbol{r}_\| - |k_\perp(\kappa)||z|)\,, \tag{7.169}$$

where $k_{0\perp} = k_\perp(\kappa = 0)$ and $p_c = 2\pi N_2 bC$. The constant C can be always represented in the form $C = 1/(1+ib\alpha)$ (see Eq. (7.157)), where α at low energies according to Eq. (7.160) is $\alpha = -k + 2\pi N_2/k_{0\perp} - i\alpha''$, and $\alpha'' = 2\pi N_2\sum_{\kappa\neq 0}^{\infty}|k_\perp(\kappa)|^{-1}$. Therefore, in the absence of losses $p_c = p_1/(1+ip_1/k_{0\perp})$, where $p_1 = 2\pi b_1 N_2$ and $b_1 = b_0/(1+b_0\alpha'')$ is the renormalized real scattering amplitude. Substitution of p_c into Eq. (7.169) shows that the first part of the wave function is identical to the one obtained for scattering from the potential $u(z) = 2p_1\delta(z)$.

If we replace all the crystalline planes by such a potential, we obtain the Kronig–Penney potential (4.88), considered in Section (4.5.2). Reflection from such a potential described by formulas (4.96) and (4.99) becomes total when $k^2 < u_1 = 2p_1/a$, where a is the distance between crystalline planes. So we obtained a result which is identical to reflection from the optical potential, though no potential was introduced. Thus, we can deduce that at low energies, when the structure of the substance is not important, interaction of a neutron with it can be represented by the optical potential u_0.

7.8.1 Kronig–Penney Corrections to the Optical Potential

Formula (4.99) is an approximation for Eq. (4.96). We can improve the expansion in Eq. (4.96) and find corrections (call them Kronig–Penney corrections) of order u_1a^2.

Reflection from the Kronig–Penney potential can be approximately represented in the form

$$R = \frac{k - \sqrt{\frac{1+u_1a^2/12}{1+u_1a^2/4}}\sqrt{k^2 - \frac{u_1}{1+u_1a^2/12}}}{k + \sqrt{\frac{1+u_1a^2/12}{1+u_1a^2/4}}\sqrt{k^2 - \frac{u_1}{1+u_1a^2/12}}}\,. \tag{7.170}$$

It is seen that the reflection is total at $k^2 < u_1/(1+u_1a^2/12)$. The correction $u_1a^2/12 \approx b_1/a$ arises because of discreteness of the crystal in the direction of the normal to its interface, whereas the discreteness of the crystalline planes themselves is not taken into account. For most substances the correction is on the order of less than 10^{-4}, and it can be neglected.

7.8.2 Diffraction Correction to the Optical Potential

There are also corrections of another type. We call them diffraction corrections. They appear because of exponentially decaying waves in expression (7.169). Every exponentially decaying wave with $k_\perp(\boldsymbol{\kappa}) = i|k_\perp(\boldsymbol{\kappa})|$, created on one of the planes, after diffraction on the consecutive crystalline plane can transform back to the wave with original $k_\perp(\boldsymbol{\kappa} - \boldsymbol{\kappa}) = k_{0\perp}$, which further propagates along $\boldsymbol{k}_0$, or along the direction of specular reflection. If we take an exponentially decaying wave with the minimal vector $\boldsymbol{\kappa}_{1\|}$ of the reciprocal lattice, then the least decaying wave near the next plane will have magnitude $\propto \exp(-2\pi) \approx 2 \cdot 10^{-3}$. However, we have not yet taken into account the preexponential factor. This factor is $k_{0\perp}/|\boldsymbol{\kappa}_{1\|}| \approx a/\lambda \approx 200$ times less than the reflection amplitude ρ_s of a single plane because the wavelength $\lambda = 2\pi/k$ of ultracold neutrons ($k^2 < u_0$) is on the order of 600 Å. Thus, after back-diffraction with $\boldsymbol{\kappa} = -\boldsymbol{\kappa}_{1\|}$ on the neighboring crystalline plane the exponentially decaying wave gives a correction to the amplitude of the propagating wave on the order of $\approx 10^{-5}$.

Let us note that the anomalous loss coefficient of ultracold neutrons in storage bottles is of the same order. Therefore, it is important to find how this correction influences the coefficient of total reflection. However, from the very beginning we can be sure that this correction can change the height of the potential u_0 and the phase of the reflected wave, but cannot create losses or leakage of neutrons inside the matter. Indeed, at the total reflection the minimal decay constant is $q = \sqrt{u_0 - k_{0\perp}^2}$. A diffraction correction on the order of 10^{-5} can change this decay constant a little, but cannot add an imaginary part to it. If a decay constant q because of the correction were changed to $q - i\alpha$, then a flux $J \propto \alpha$ would appear inside the crystal. This flux would decay along the depth z according to the law $\exp(-2qz)$. If, because of that, reflection is not total such a decay of flux would mean violation of unitarity: the flux of reflected neutrons would be less than that of incident ones, and the deficit of the flux would disappear inside matter in the absence of absorption by nuclei. If reflection remains total, then again there is a violation of unitarity: the total flux of neutrons reflected and leaked inside the matter at the interface is larger than that of incident ones. From these considerations we can conclude that the diffraction correction can only change the phase of the total reflection by an amount on the order of 10^{-5}, and this correction can always be neglected.

If nuclei of the crystal absorb neutrons, then the reflection becomes nontotal, but the losses should be proportional to the absorption cross section, and diffraction correction can increase the losses. However, the relative increase is again on the order of 10^{-5}, and it can be neglected.

One note should be added to these considerations. The normalization of the scattering amplitude discussed on page 299 contains the correction of the real part of the scattering amplitude on the order of 1%, which is proportional to $1/\sigma_l$. The less is σ_l, the larger is the correction. It changes the height of the optical potential but changes little its imaginary part. Nevertheless this point remains subtle, and deserves deeper consideration.

7.8.3 Digression on the Optical Potential

The optical potential is some constant. What is our interest in such a constant? We know that it determines the refractive index. Can we obtain with it something else? Now we want to illustrate with some examples how important this constant is in physics.

7.8.3.1 Optics

We were talking above mainly about potential barriers $u > 0$, which repel particles; however, the scattering amplitude b can be negative (typical examples are Ti and H). In that case $u_0 = 4\pi N_0 b < 0$, and neutron–matter interaction becomes attractive. The potential u_0 defines the refractive index for neutrons $n = \sqrt{1 - u_0/k^2}$ and we see that for attractive potentials $n > 1$.

Let us notice that in the case of light the majority of substances have a refractive index larger than unity. This means that almost all matter attracts light photons. It is because of this that we have no total external reflection in optics (we do not consider here X-rays), but only total internal reflection, because for the light inside matter the vacuum looks like a potential barrier.

Indeed, interaction of matter with an electromagnetic field is described by the refractive index $n = \sqrt{\epsilon}$, where ϵ is the dielectric permeability. The wave number inside matter is $k' = nk$, which can also be represented as $k' = \sqrt{\epsilon k^2} = \sqrt{k^2 + (\epsilon - 1)k^2} = \sqrt{k^2 - u}$, where similarly to neutrons the optical potential $u = (1 - \epsilon)k^2$ is introduced.[29] For ϵ we usually write $\epsilon = 1 + 4\pi N_0 \alpha$, where α is the polarizability of atoms in the matter. Therefore, the potential u is represented in full analogy with neutrons as $u = 4\pi N_0 b$, where $b = -\alpha k^2$.

Polarizability is absolutely naturally enters the light–matter interaction. Indeed, if an atom has its own electric dipole moment $\boldsymbol{d}$, then it has interaction potential with light equal to $-\boldsymbol{d}\boldsymbol{E}$. If the atom does not have $\boldsymbol{d}$, it can acquire an induced dipole moment $\boldsymbol{d} = \alpha \boldsymbol{E}$. Therefore, its interaction with the field is $-\alpha E^2 < 0$. Since $E^2/4\pi$ is the density of electromagnetic energy, then for photons with frequency ω it can be represented as $\hbar\omega N_\gamma$, where N_γ is the density of photons. Therefore, interaction of the atom with the field is described by the potential $U_\gamma = -\alpha 4\pi\hbar\omega N_\gamma$. If $N_\gamma(\boldsymbol{r})$ is nonuniform in space, then the force $\boldsymbol{F} = -\nabla U_\gamma(\boldsymbol{r})$ appears, which pulls the atom into the space with highest density of N_γ.

29) It is worth mentioning that the speed of light in contrast to that of the neutron decreases inside an attractive potential.

In contrast to neutrons, the photon's potential depends on the frequency of the incident radiation $\omega = kc$. At high frequencies (X-ray optics) the polarizability α is inversely proportional to the square of the frequency and changes sign. As a result, the amplitude $-\alpha k^2$ ceases to depend on ω^2 and becomes similar to the neutron one, where the neutron–nucleus scattering amplitude b is replaced by the Thomson scattering length e^2/mc^2. The optical potential u for X-rays contains the density of electrons $n_e = N_0 Z$ (Z is the the charge of the atomic nucleus) instead the atomic density N_0. Thus, $u = 4\pi N_0 Z e^2/mc^2 > 0$, that is, matter for X-rays is represented by a potential barrier, like for neutrons, and X-rays can be totally reflected from it.

Self-focusing The interaction of optical photons with matter is negative. This means that the substance attracts light, and therefore the photons attract matter. This attraction results in electrostriction, which leads to the phenomenon of self-focusing.

Imagine a powerful beam of light being transmitted by a substance. The density of matter inside the beam increases because of electrostriction, and therefore the density of matter outside it, near its edges, decreases. The rarefaction at edges results in total reflection of the light from the edges to inside the beam, which leads to self-focusing.

Nonlinearity Nonlinearity in light–matter interaction is described by the dependence of ϵ on intensity I. This dependence arises because of electrostriction. The compression of matter toward the highest density of photons makes the atomic density $N = N_0 + \alpha I$ increase with light intensity, which leads to a change of ϵ with intensity: $\epsilon = \epsilon_0 + \beta I$, where β is a characteristic quantity for every given substance.

Since $I \propto E^2$, then the linear dependence of ε on I creates cubic nonlinearity. Since $E = E_0 \cos(\omega t)$, then because of nonlinearity, the interaction of light of frequency ω with matter creates waves with frequencies 3ω. Interference of these waves creates even higher harmonics with frequencies $n\omega$, where $n > 3$.

Ball lightning The optical potential can have a relation to a phenomenon such as ball lightning [364]. Imagine a point explosion in the atmosphere. Its origin, for example, can be fast evaporation of a water drop after being struck by linear lightning. As a result of the explosion, a spherical shock wave is created. The air behind the shock wave front is compressed, and the refractive index behind the front is higher than that before it. Therefore a photon, if it appears behind the front, can be totally reflected from the front at a small grazing angle.

Imagine that the gas behind the front is strongly excited, and one of the atoms emits a photon in such a mode that it is totally reflected from the front. This photon can create a laser discharge of photons in the whispering gallery mode. If the density of photons is sufficiently high, the electrostriction forces can stop expansion of the shock wave front, and we are left with a stable ball, which contains a thin spherical shell full of photons. How long this shell can survive and how it is possible in principle to ignite such a ball in a laboratory is considered in [364]; however, serious investigations in this direction have not yet been conducted. It seems such research could be important because it should also be able to shed light on the origin of hurricanes and tornadoes.

7.8.3.2 Neutrostriction in Neutron Stars

> It all can be written on the back of an envelop. I don't believe it is not known to astrophysicists, so I do not recommend publication.
>
> A referee.

> Everything here is wrong and should never be published.
>
> A referee.

> I concur with the comments of both referee. The physics discussed in this manuscript is not new and is presented in a misleading way (e. g., the comparison of gravitational and "optical" energies using the scattering length only throughout the volume of the neutron star).
>
> An editor.

We know that at $b < 0$ the optical potential becomes attractive, that is, the substance represents for neutrons a potential well. There are bound states in such a well. This means that substance holds the neutron if it is on one of the bound levels. It is understandable that the neutron also holds the substance. Indeed, let us imagine that the neutron is in the ground state of, say, a drop of liquid hydrogen (absorption is switched off), and the hydrogen starts to dilate. When it dilates its atomic density N_0 becomes lower; therefore, the depth of the well becomes shallower and the energy of the ground state should increase. This means that for dilation we need to do work on the neutron, that is, the neutron compresses the substance or resists its dilation. This compression is like electrostriction, and we can call it neutrostriction, or neutronostriction.

Of course, one neutron creates a negligible force; however, in nature there is a situation where this force is very large. This is the case of a neutron star. It is known that neutron–neutron scattering is characterized by a negative scattering amplitude $b \approx -1.8 \cdot 10^{-12}$ cm. Since the scattering takes place only in singlet state, when the total spin of two scattering neutrons is zero, then the coherent scattering amplitude is 4 times less: $b_c \approx 0.45 \cdot 10^{-12}$ cm.

We imagine the simplest model of the neutron star consisting of only neutrons. Such a star is a potential well for every neutron. The density n in the neutron star is on the order more than 10^{37} cm^{-3}; therefore, the depth of the potential well $(\hbar^2/2m)4\pi n b$ is larger than 10 MeV.

The potential energy of all the neutrons in the star is $(\hbar^2/2m)4\pi n^2 b V$, where V is the star volume. It can be larger than the gravitational energy. Indeed, if the density inside the star is constant, then the gravitational energy of the star is $U_g = -(3/5)GM^2/R$, where G is the gravitational constant, R is radius of the star, and $M = mnV$ is the total mass of the star. The optical energy is $U_0 = (\hbar^2/2m)4\pi bM^2/V = (\hbar^2/2m)3bM^2/R^3$. Therefore, $U_g < U_o$ when $R < R_0 = \sqrt{5|b|\hbar^2/2m^3G} \approx 20\,\text{km}$.

The critical radius R_0 is 20 km and does not depend on density only when $b = -4.5$ fm does not depend on energy E. However it does depend on energy, and the absolute value $|b|$ decreases with increase of E. Taking this dependence into account, we can calculate the effect of the optical potential on the density, mass, and radius of neutron stars. This is done in [366]. We shall not continue with this theme because it goes too far beyond the scope of this book. We want only to point out that the optical potential notion introduces a lot of interesting and not yet studied problems related to phenomena such as pulsations of the star and explosions in the presence of resonances in the scattering amplitude.

7.8.4 The End of the Digression

At the end it is worth recalling again that though the scattering amplitude is the result of strong but short-range interactions, the optical potential itself is a long-range one. It is seen in everyday experiments in neutron physics in phenomena such as diffraction and specular reflection, where the interaction involves many atoms far apart at distances many order of magnitude larger than the radius of a nucleus and the wavelength of a neutron, and one of the fundamental questions is: can we find a length which characterizes how long this long-range interaction is?

Scattering and optical potential are properties not only of neutrons. Scattering of atoms and molecules can also create an optical potential, which may play an important role in the evolution of ordinary stars. It can be important in exotic systems such as Bose condensates, and it may be crucial for the explanation of phenomena such as superconductivity and superfluidity.

7.9 Conclusion

The algebraic method used in Chapter 4 for one-dimensional systems was generalized here to three-dimensional periodic media. It gave us an opportunity to formulate the dynamical diffraction theory differently from the standard approach and to deduce rigorously the optical potential. However, here we limited ourselves to totally ordered systems of atoms fixed in space. The next step is to generalize the method to three-dimensional disordered systems. This will be done in the next chapter.

8 Disordered Media. Incoherent and Small-Angle Scattering

In previous chapters we were dealing with ordered media. Now we turn to disordered ones. First we shall consider waves in random media, and then the propagation of corpuscles in them. Application of the algebraic method to corpuscles will give us a new approach to diffusion and to albedo reflection from ultradispersive powders.

The investigation of random media always includes some elements of phenomenology. They are related to the so-called correlation functions. These functions should satisfy some conditions, but to a great extent they are arbitrary. The introduction of such functions is equivalent to the introduction of some phenomenological parameters, the physical meaning of which is clear, but their values can be found only from experiment.

8.1 Wave Processes in Three-Dimensional Disordered Media

Let us consider reflection of a plane wave from a wall with a disordered distribution of atoms. It is clear in advance that we can expect specular reflection and scattering in nonspecular directions. The specular reflection is determined by the average parameters of the medium, and nonspecular scattering appears because of fluctuations inherent to all the random medium. To see all that we have to use multiple wave scattering (MWS) theory.

8.1.1 Equations of MWS

The equations of MWS (see [14, 16]) are presented in Eqs. (7.3) and (7.4). If atoms are distributed randomly and (or) amplitudes b_j are random complex numbers, then the wave function, which we shall denote now $\Psi(\boldsymbol{r})$, and coefficients $\psi(\boldsymbol{r}_n)$ are also random [367]. So we can define their average $\Phi(\boldsymbol{k}, \boldsymbol{r}) = \langle\langle\Psi(\boldsymbol{r})\rangle\rangle$, $\phi(\boldsymbol{r}_n) = \langle\langle\psi(\boldsymbol{r}_n)\rangle\rangle$, and correlations $\Xi(\boldsymbol{r}, \boldsymbol{r}') = \langle\langle\Psi^*(\boldsymbol{r})\Psi(\boldsymbol{r}')\rangle\rangle - \langle\langle\Psi^*(\boldsymbol{r})\rangle\rangle\langle\langle\Psi(\boldsymbol{r}')\rangle\rangle$, where $\langle\langle\cdots\rangle\rangle$ means averaging over the elements (atoms, isotopes, i. e., amplitudes b_j), and positions of the atoms.

Neutron Optics. Masahiko Utsuro, Vladimir K. Ignatovich

ISBN: 978-3-527-40885-6

Denote $b_j = b + \delta b_j$, $\psi(\boldsymbol{r}_j) = \phi(\boldsymbol{r}_j) + \delta\psi(\boldsymbol{r}_j)$, where $b = \langle b_j \rangle$ and δb and $\delta\psi$ are fluctuations. We can take as granted that the δb_j do not correlate with each other and with $\delta\psi(\boldsymbol{r}_j)$. This comes from the fact that $\psi(\boldsymbol{r}_j)$ is a sum of all the waves coming from all the scatterers except the *j*th one. Therefore, to a good approximation, we can suggest that $\delta\psi(\boldsymbol{r}_j)$ does not depend on δb_j.

If the medium is homogeneous, its average density is a constant N_0. The average (or coherent) part of the wave function $\Phi(\boldsymbol{k}, \boldsymbol{r})$ can be represented as

$$\langle\langle \Psi(\boldsymbol{r})\rangle\rangle \equiv \Phi(\boldsymbol{k}, \boldsymbol{r}) = \exp(i\boldsymbol{k}\boldsymbol{r}) - N_0 b \int \phi(\boldsymbol{r}') \frac{\exp(ik|\boldsymbol{r}-\boldsymbol{r}'|)}{|\boldsymbol{r}-\boldsymbol{r}'|} d^3 r' , \tag{8.1}$$

where the function $\phi(\boldsymbol{r})$ satisfies the equation

$$\langle\langle \psi(\boldsymbol{r})\rangle\rangle \equiv \phi(\boldsymbol{r}) = \exp(i\boldsymbol{k}\boldsymbol{r}) - N_0 b \int g(\boldsymbol{r}-\boldsymbol{r}') d^3 r' \phi(\boldsymbol{r}') \frac{\exp(ik|\boldsymbol{r}-\boldsymbol{r}'|)}{|\boldsymbol{r}-\boldsymbol{r}'|} . \tag{8.2}$$

Here we introduced the correlation function $g(\boldsymbol{r}-\boldsymbol{r}')$, which denotes the probability of finding the second scatterer at point $\boldsymbol{r}'$, when the first one is at point $\boldsymbol{r}$. To solve Eq. (8.2) we must define $g(\boldsymbol{r})$. We know that this function is positive-definite because it is a probability. This probability should go to zero when $\boldsymbol{r} \to 0$, because the point $r = 0$ is already occupied, and to unity at $\boldsymbol{r} \to \infty$, because particles far apart are independent. Moreover, with increase of distance r, the function $g(r)$ may oscillate near $g = 1$, because there is some average distance between neighboring atoms, and the oscillation amplitude should decrease, because the distance between atoms is not certain. The function $g(r)$ can be represented as $g(r) = 1 - \gamma(r)$, and since

$$\int g(r) N_0 d^3 r = N - 1 , \tag{8.3}$$

where N is the total number of atoms in the system, we have

$$\int \gamma(r) N_0 d^3 r = 1 . \tag{8.4}$$

The function $\gamma(r)$ is equal to unity at $r = 0$, goes to zero at $r \to \infty$, and oscillates with monotonically decreasing amplitude around zero. We do not need the precise form of this function. We need only its property Eq. (8.4), and we need the positive value of the average of the square radius

$$\langle r^2 \rangle = N_0 \int \gamma(r) r^2 d^3 r > 0 . \tag{8.5}$$

For instance, we can accept the function

$$\gamma(r) = \frac{\sin(s'r)}{s'r} \exp(-sr) , \tag{8.6}$$

or

$$\gamma(r) = \cos(s'r) \exp(-sr) \tag{8.7}$$

with some acceptable parameters s and s'.

8.1.2
Solution for Coherent Values

It is natural to suppose that the averaged or "coherent" wave function inside the medium is of the form $\Phi(\boldsymbol{k}, \boldsymbol{r}) = A\exp(i\boldsymbol{k}'\boldsymbol{r})$, where A is an as yet unknown constant factor, and the wave vector $\boldsymbol{k}'$ is $(\boldsymbol{k}_{\parallel}, k'_{\perp})$. The components $\boldsymbol{k}_{\parallel}$ parallel to the interface are equal to those of the incident wave, and the normal component $k'_{\perp}$ is not equal to $k_{\perp}$.

It is natural to suppose that the factors $\phi(\boldsymbol{r})$ are proportional to $\Phi(\boldsymbol{k}, \boldsymbol{r})$:

$$\phi(\boldsymbol{r}) = C\Phi(\boldsymbol{k}, \boldsymbol{r}) = CA\exp(i\boldsymbol{k}'\boldsymbol{r}) . \tag{8.8}$$

To find A and $k'_{\perp}$, substitute ϕ and Φ into Eq. (8.1). When the point $\boldsymbol{r}$ is inside the medium we get

$$A\exp(i\boldsymbol{k}'\boldsymbol{r}) = \exp(i\boldsymbol{k}\boldsymbol{r}) - N_0 bCA\exp(i\boldsymbol{k}'\boldsymbol{r})\int \exp(i\boldsymbol{k}'(\boldsymbol{r}' - \boldsymbol{r}))\frac{\exp(ik|\boldsymbol{r} - \boldsymbol{r}'|)}{|\boldsymbol{r} - \boldsymbol{r}'|}d^3r' . \tag{8.9}$$

Use the two-dimensional Fourier expansion of the spherical function (7.10):

$$\frac{\exp(ikr)}{r} = \frac{i}{2\pi}\int d^2p_{\parallel}\frac{\exp(i\boldsymbol{p}_{\parallel}\boldsymbol{r}_{\parallel} + ip_{\perp}|z|)}{p_{\perp}} , \tag{8.10}$$

where $p_{\perp} = \sqrt{k^2 - p_{\parallel}^2}$. Substitute it into Eq. (8.9). After integration over coordinates $\boldsymbol{r}'_{\parallel}$, parallel to the interface and cancellation of the factor $\exp(i\boldsymbol{k}_{\parallel}\boldsymbol{r}_{\parallel})$ in both sides of Eq. (8.9), we get

$$A\exp(ik'_{\perp}z) = \exp(ik_{\perp}z) - i4\pi N_0 bAC\int_0^{\infty}\frac{\exp(ik_{\perp}|z - z'|)}{2k_{\perp}}\exp(ik'_{\perp}z')dz' , \tag{8.11}$$

where $k_{\perp} = \sqrt{k^2 - k_{\parallel}^2}$. After integration over z' on the right-hand side we obtain

$$A\exp(ik'_{\perp}z) = \exp(ik_{\perp}z) - 4\pi N_0 bAC \times\left(\frac{1}{2k_{\perp}(k_{\perp} - k'_{\perp})}\exp(ik_{\perp}z) + \frac{1}{k'^2_{\perp} - k^2_{\perp}}\exp(ik'_{\perp}z)\right) . \tag{8.12}$$

To get this result we supposed that $k'_{\perp}$ contains a positive imaginary part. Equating the preexponential factors of different exponents, we find

$$1 = 4\pi N_0 bAC\frac{1}{2k_{\perp}(k_{\perp} - k'_{\perp})} , \quad 1 = -4\pi N_0 bC\frac{1}{k'^2_{\perp} - k^2_{\perp}} . \tag{8.13}$$

It follows from the second equation that $k'^2 = k^2 - 4\pi N_0 b C$, and the first equation gives

$$A = 2k_\perp \frac{k_\perp - k'_\perp}{4\pi N_0 b C} = \frac{2k_\perp}{k_\perp + k'_\perp} . \tag{8.14}$$

For the space outside the medium $z < 0$, substitution of Eq. (8.8) into Eq. (8.1) gives $\Phi(\boldsymbol{k}, \boldsymbol{r}) = \exp(i\boldsymbol{k}_\parallel \boldsymbol{r}_\parallel)\Phi(k_\perp, z)$, where

$$\Phi(k_\perp, z) = \exp(ik_\perp z) + 4\pi N_0 b A C \frac{1}{2k_\perp(k_\perp + k'_\perp)} \exp(-ik_\perp z) . \tag{8.15}$$

After substitution of $A = 2k_\perp/(k_\perp + k'_\perp)$, Eq. (8.15) is reduced to the form

$$\Phi(k_\perp, z) = \exp(ik_\perp z) + \overrightarrow{r_0} \exp(-ik_\perp z) , \tag{8.16}$$

where $\overrightarrow{r_0} = (k_\perp - k'_\perp)/(k_\perp + k'_\perp)$ is the reflection amplitude.

To determine C, we have to substitute Eq. (8.8) into Eq. (8.2) inside the medium. As a result we obtain

$$\begin{aligned} A C e^{i k' r} &= e^{i k r} - A C N_0 b \exp(i\boldsymbol{k}'\boldsymbol{r}) \\ &\times \int\limits_{z'>0} w(|\boldsymbol{r}' - \boldsymbol{r}|) \frac{\exp(ik|\boldsymbol{r}' - \boldsymbol{r}|)}{|\boldsymbol{r}' - \boldsymbol{r}|} \exp(i\boldsymbol{k}'(\boldsymbol{r}' - \boldsymbol{r})) d^3 r' . \end{aligned} \tag{8.17}$$

Substitution of $w(r) = 1 - \gamma(r)$ gives

$$\begin{aligned} A C e^{i k' r} &= e^{i k r} - A C N_0 b \exp(i\boldsymbol{k}'\boldsymbol{r}) \\ &\times \int\limits_{z'>0} [1 - \gamma(|\boldsymbol{r}' - \boldsymbol{r}|)] \frac{\exp(ik|\boldsymbol{r}' - \boldsymbol{r}|)}{|\boldsymbol{r}' - \boldsymbol{r}|} \exp(i\boldsymbol{k}'(\boldsymbol{r}' - \boldsymbol{r})) d^3 r' . \end{aligned} \tag{8.18}$$

With account of Eq. (8.9), we transform Eq. (8.18) to

$$\begin{aligned} A(C - 1) e^{i k' r} &= A C N_0 b \exp(i\boldsymbol{k}'\boldsymbol{r}) \\ &\times \int\limits_{z'>0} \gamma(|\boldsymbol{r}' - \boldsymbol{r}|) \frac{\exp(ik|\boldsymbol{r}' - \boldsymbol{r}|)}{|\boldsymbol{r}' - \boldsymbol{r}|} \exp(i\boldsymbol{k}'(\boldsymbol{r}' - \boldsymbol{r})) d^3 r' . \end{aligned} \tag{8.19}$$

Therefore, for z far away from the interface we get

$$C = \frac{1}{1 - c} , \quad c = b\Gamma(k) , \quad \Gamma(k) = \int \gamma(r) \frac{\exp(ikr)}{r} \exp(i\boldsymbol{k}'\boldsymbol{r}) N_0 d^3 r . \tag{8.20}$$

Since Γ contains $k'_\perp = \sqrt{k_\perp^2 - u_0 C}$ $(u_0 = 4\pi N_0 b)$, which also depends on C, the expression (8.20) is a transcendent equation for determination of C. However, when $k^2 \gg |u_0|$ and $k_\perp^2 > |u_0|$, we can neglect u_c under the integral, and $\Gamma(k)$

determines c completely. In this case

$$\Gamma(k) = \int \gamma(r) \frac{\exp(ikr)}{r} \exp(i\boldsymbol{kr}) N_0 d^3 r \,. \tag{8.21}$$

So, with the help of MWS theory we found an average (we called it coherent) neutron wave field inside and outside the medium. It is easy to see that this field is identical to

$$\Phi(\boldsymbol{k}, \boldsymbol{r}) = e^{ik_\| r_\|} \Big\{ \Theta(z<0) \Big[\exp(ik_\perp z) + \overrightarrow{r_0}(k_\perp) \exp(-ik_\perp z) \Big] \\ + \Theta(z>0) \overrightarrow{t_0}(k_\perp) \exp(ik'_\perp z) \Big\} \,, \tag{8.22}$$

where $\overrightarrow{r_0}(k_\perp)$, and $\overrightarrow{t_0}(k_\perp)$ are the reflection and refraction amplitudes found for the medium represented by the optical potential $u_c = 4\pi N_0 b C$.

8.1.3 Renormalization of the Scattering Amplitude

The optical potential u_c contains an imaginary part even if there is no absorption in the system, and if all the scatterers have the same scattering amplitude b. This imaginary part appears because of randomness in the positions of the scatterers and enters via renormalization of b with the factor C.

It follows from Eq. (7.155) that the scattering amplitude of an atom outside the medium in the absence of absorption and incoherent scattering is representable as $b = b_0/(1 + ikb_0)$ with real b_0. Inside the medium the amplitude is renormalized by the factor C, which changes the bookkeeping of unitarity (see page 330)): $b_c = b_0/(1 + ikb_0 - b_0\Gamma)$. Since $\Gamma = \Gamma' + i\Gamma''$ is a complex number, the amplitude b_c can be represented as $b_1/\{1 + ikb_1[1 - f(k)]\}$, where $b_1 = b_0/[1 - b_0\Gamma']$ is a normalized real quantity and $f(k) = \Gamma''(k)/k$.

The function f cannot be larger than unity because the imaginary part of the scattering amplitude cannot be positive. The positive imaginary part of the amplitude would mean the creation of particles and violation of unitarity.

If $f = 1$, the imaginary part of the scattering amplitude is zero. This means that scattering in the system stops. On the other side, if $f = 0$, then the imaginary part of the scattering amplitude becomes similar to that of a single atom outside the medium. The function f can be negative. In such a case the scattering of the atom inside the medium becomes even stronger than that of the independent atom. It is possible if there is resonant enhancement of waves scattered on neighboring atoms.

The imaginary part of b_c should vanish at low energies, when the neutron wavelength becomes much larger than the interatomic distance and discreetness of the medium becomes unimportant. This means that $f(k) \to 1$ when $k \to 0$.

Indeed, at small k, we can expand $(\exp(ikr)/r)\exp(i\boldsymbol{kr})$ in Eq. (8.21) in a power series over k. In the cubic approximation we get

$$i\Gamma'' \approx ik \int \gamma(r) \left(1 - \frac{k^2}{3} r^2\right) N_0 d^3 r = ik \left(1 - \frac{k^2}{3} \langle r^2 \rangle \right) \,, \tag{8.23}$$

where $\langle r^2\rangle = \int \gamma(r) N_0 r^2 d^3 r$, and normalization (8.4) was taken into account. It follows that

$$f(k) \approx 1 - \frac{k^2}{3}\langle r^2\rangle\,, \tag{8.24}$$

and $f \to 1$ when $k \to 0$. Moreover, $f(k) < 1$, if $\langle r^2\rangle > 0$. The last inequality imposes a limitation on the form of function $\gamma(r)$, or its parameters, if it is chosen according to Eqs. (8.6) or (8.7).

In the presence of incoherent scattering, the imaginary part in the denominator of b_c should include the incoherent scattering cross section, that is, it should be $ik\{b_1[1 - f(k)] + |\tilde{b}_{\text{inc}}|^2/b_1\}$.

The quite reasonable scheme described above is destroyed when we look at the case $k^2 < u_c$. For these k^2 we have to expand $\Gamma(k)$ in Eq. (8.20) in powers of k and $\boldsymbol{k}'$ without neglecting u_c inside the integral, and we immediately arrive at a result that at some k^2 the function $f(k)$ in the imaginary part of the denominator becomes larger than unity, and unitarity is violated. This shows that something is wrong in our considerations. We can guess that it happens because we used a spherical wave under the integral of $\Gamma(k)$, which is valid in free space but not in the medium. In the medium the spherical wave has to be modified. With modified functions (see below) we find that the function $f(k)$ is never larger than unity.

8.2
Scattering

Since violation of unitarity is due to an incorrect form of the additional imaginary part of the amplitude b_c, and this imaginary part is related to scattering density fluctuations, which are ubiquitous in a random distribution of atoms and variation of their scattering amplitudes, we have to investigate the scattering in more detail.

The scattered waves at $z < 0$ go out from the medium. They can be represented as

$$\delta\Psi(\boldsymbol{r}) = \int d^2 p_\parallel A(\boldsymbol{k}, \boldsymbol{p}) \exp(i\boldsymbol{p}\boldsymbol{r})\,, \quad \boldsymbol{p} = (\boldsymbol{p}_\parallel, -p_\perp)\,, \quad p_\perp = \sqrt{k^2 - p_\parallel^2}\,, \tag{8.25}$$

with random amplitudes $A(\boldsymbol{k}, \boldsymbol{p})$. If we average over a randomness ensemble and find that

$$\langle A(\boldsymbol{k}, \boldsymbol{p})\rangle = \overline{A}(k_\perp)\delta(\boldsymbol{p}_\parallel - \boldsymbol{k}_\parallel)\,, \tag{8.26}$$

then

$$\langle \delta\Psi(\boldsymbol{r})\rangle = \overline{A}(k_\perp) \exp(i\boldsymbol{k}\boldsymbol{r})\,, \quad \boldsymbol{k} = (\boldsymbol{k}_\parallel, -k_\perp)\,, \tag{8.27}$$

and we have a correction to the specular reflection amplitude:

$$\vec{r}(k_\perp) = \vec{r_0}(k_\perp) + \overline{A}\,. \tag{8.28}$$

Besides this, we can find the total flux through any plane parallel to the interface of nonspecularly reflected neutrons. To do that we define

$$\overline{\delta}\Psi(\boldsymbol{r}) = \delta\Psi(\boldsymbol{r}) - \langle\delta\Psi(\boldsymbol{r})\rangle\,, \tag{8.29}$$

and with it we obtain

$$\begin{aligned} J_\perp &= \int_S \frac{d^2 r_\parallel}{2i}\left\langle\left[\overline{\delta}\Psi^*(\boldsymbol{r})\frac{d}{dz}\overline{\delta}\Psi(\boldsymbol{r}) - \overline{\delta}\Psi(\boldsymbol{r})\frac{d}{dz}\overline{\delta}\Psi^*(\boldsymbol{r})\right]\right\rangle \\ &= (2\pi)^2\int_S p_\perp\langle|A(\boldsymbol{k},\boldsymbol{p})|^2\rangle d^2 p_\parallel - S|\overline{A}|^2 k_\perp\,, \end{aligned} \tag{8.30}$$

where $*$ means the complex conjugate and S is some large area of the plane, over which we integrate. The ratio of this flux to the incident one $k_\perp$ gives the cross section:

$$\sigma(\boldsymbol{k}) = \frac{J_\perp(\boldsymbol{k})}{k_\perp} = (2\pi)^2\int\frac{p_\perp}{k_\perp}\langle|A(\boldsymbol{k},\boldsymbol{p})|^2\rangle d^2 p_\parallel - S|\overline{A}|^2\,. \tag{8.31}$$

Expression (8.31) can be transformed to angular variables. Since for scattered waves $\boldsymbol{p}^2 = k^2$, and propagating waves have $p_\perp = \sqrt{k^2 - p_\parallel^2}$, a real number, then

$$d^2 p_\parallel/p_\perp = 2d^3 p\,\delta(p^2 - k^2) = k\,d\Omega = 2\pi\,d p_\perp\,, \tag{8.32}$$

and the last equality is correct when we can integrate over all the other components of the scattered vector $\boldsymbol{p}$. As a result, Eq. (8.31) is transformed to

$$\begin{aligned} \sigma(\boldsymbol{k}) = \sigma(\boldsymbol{\Omega}_0) &= (2\pi)^3\int_0^k\frac{p_\perp^2}{k_\perp}\langle|A(\boldsymbol{k},\boldsymbol{p})|^2\rangle d p_\perp \\ &= (2\pi)^2\int_{2\pi}\frac{\cos^2\vartheta}{\cos\vartheta_0}\langle|A(\boldsymbol{k},\boldsymbol{p})|^2\rangle k^2 d\Omega\,, \end{aligned} \tag{8.33}$$

where $\cos\vartheta = p_\perp/k$ and $\cos\vartheta_0 = k_\perp/k$. Now we can define the differential cross section:

$$\frac{d\sigma(\boldsymbol{\Omega}_0\to\boldsymbol{\Omega})}{d\Omega} = (2\pi)^2 k^2\frac{\cos^2\vartheta}{\cos\vartheta_0}\langle|A(\boldsymbol{k},\boldsymbol{p})|^2\rangle - S|\overline{A}|^2\delta(\boldsymbol{\Omega}-\boldsymbol{\Omega}_0)\,, \tag{8.34}$$

where $\boldsymbol{\Omega}_0$ denotes the solid angle of the incident neutron and $\boldsymbol{\Omega}$ is that for the scattered ones. If we find that

$$\langle|A(\boldsymbol{k},\boldsymbol{p})|^2\rangle = S\overline{|A(\boldsymbol{k},\boldsymbol{p})|^2}\,, \tag{8.35}$$

we define the ratio σ/S, which gives the probability of nonspecular reflection or the indicatrix of nonspecular reflectivity:

$$w(\boldsymbol{\Omega}_0\to\boldsymbol{\Omega}) = \frac{d\sigma(\boldsymbol{\Omega}_0\to\boldsymbol{\Omega})}{S\,d\Omega} = (2\pi)^2 k^2\frac{\cos^2\vartheta}{\cos\vartheta_0}\overline{|A(\boldsymbol{k},\boldsymbol{p})|^2} - |\overline{A}|^2\delta(\boldsymbol{\Omega}-\boldsymbol{\Omega}_0), \tag{8.36}$$

In reflectometer experiments we do not need angles, we need normal components of the wave vectors in the incidence plane; therefore, we represent Eqs. (8.34) and (8.36) in the form

$$\frac{d\sigma(k_\perp \to p_\perp)}{dp_\perp} = (2\pi)^3 \frac{p_\perp^2}{k_\perp} \left(\langle |A(\boldsymbol{k},\boldsymbol{p})|^2\rangle - 2\pi S |\overline{A}|^2 \delta(k_\perp - p_\perp)\right), \tag{8.37}$$

$$w(k_\perp \to p_\perp) = \frac{d\sigma(k_\perp \to p_\perp)}{S\,dp_\perp} = (2\pi)^3 \frac{p_\perp^2}{k_\perp} \overline{|A(\boldsymbol{k},\boldsymbol{p})|^2} - 2\pi |\overline{A}|^2 \delta(k_\perp - p_\perp). \tag{8.38}$$

Now, let us look at the random scattered wave function

$$\delta\Psi(\boldsymbol{r}) = -\sum_j \delta(\psi(\boldsymbol{r}_j)b_j) \frac{\exp(ik|\boldsymbol{r}-\boldsymbol{r}_j|)}{|\boldsymbol{r}-\boldsymbol{r}_j|}, \tag{8.39}$$

where $\delta(\psi(\boldsymbol{r}_j)b_j) = \psi(\boldsymbol{r}_j)b_j - \phi(\boldsymbol{r}_j)b_c$. If we calculate the nonspecular reflectivity with this expression, we find that the result will not satisfy the detailed balance condition: $d\sigma(\Omega_0 \to \Omega)/d\Omega = \cos\vartheta\, \sigma^s(\Omega_0 \to \Omega)$, where the last factor should be a function symmetrical with respect to permutation of its arguments (7.166). This shows that to describe scattering we need to modify the spherical wave.

8.2.1
Propagation of Scattered Waves inside the Medium

If the medium were infinite, that is, without interfaces, then we could replace $\exp(ikr)/r$ with the similar function $\exp(ik'r)/r$ having wave number k' inside the medium instead of k. The presence of an interface distorts it, and our task is to find the distorted spherical function.

The spherical wave is the Green function $G_0(k,\boldsymbol{r}) = \exp(ik|\boldsymbol{r}|)/|\boldsymbol{r}|$, which satisfies the equation $\left(\Delta + k^2\right) G_0(k,\boldsymbol{r}-\boldsymbol{r}') = -4\pi\delta(\boldsymbol{r}-\boldsymbol{r}')$. It has the two-dimensional Fourier representation (8.10)

$$G_0(k,\boldsymbol{r}-\boldsymbol{r}') = -\int \frac{d^2 p_\parallel}{\pi} \exp\left(i\boldsymbol{p}_\parallel(\boldsymbol{r}-\boldsymbol{r}')\right) G_0(p_\perp, z, z'), \tag{8.40}$$

where $p_\perp = \sqrt{k^2 - p_\parallel^2}$ and where

$$G_0(p_\perp, z, z') = (1/2ip_\perp)\exp\left(ip_\perp|z-z'|\right) \tag{8.41}$$

satisfies the one-dimensional equation

$$\left(d^2/dz^2 + p_\perp^2\right) G_0(p_\perp, z, z') = \delta(z-z'). \tag{8.42}$$

In the presence of the potential u_c, the spherical function is to be replaced by the Green function $G(u_c, k, \boldsymbol{r}, \boldsymbol{r}')$, which satisfies the equation

$$\left(\Delta + k^2 - \Theta(z>0)u_c\right) G(u_c, k, \boldsymbol{r}, \boldsymbol{r}') = -4\pi\delta(\boldsymbol{r}-\boldsymbol{r}'). \tag{8.43}$$

Since the space along the interface is uniform, the Green function also has the two-dimensional Fourier representation shown in Eq. (8.40):

$$G\left(u_c, k, \boldsymbol{r}, \boldsymbol{r}'\right) = -\int \frac{d^2 p_{\|}}{\pi} \exp\left(i \boldsymbol{p}_{\|}\left(\boldsymbol{r}_{\|} - \boldsymbol{r}'_{\|}\right)\right) G\left(u_c, p_{\perp}, z, z'\right) . \quad (8.44)$$

Substitution of this into Eq. (8.43) shows that $G(u_c, p_\perp, z, z')$ satisfies the equation

$$\left(d^2/dz^2 + p_\perp^2 - \Theta(z > 0) u_c\right) G(u_c, p_\perp, z, z') = \delta(z - z') . \quad (8.45)$$

According to common rules, it can be constructed with the help of two linearly independent solutions $\psi_{1,2}(u_c, p_\perp, z)$ of the homogeneous equation

$$\left(d^2/dz^2 + p_\perp^2 - \Theta(z > 0) u_c\right) \psi_{1,2}(u_c, p_\perp, z) = 0 . \quad (8.46)$$

For the function $\psi_1(u_c, p_\perp, z)$ we can take $\Phi(p_\perp, z)$ (8.22), and for $\psi_2(u_c, p_\perp, z)$ it is appropriate to take the function

$$\begin{aligned} \psi_2(u_c, p_\perp, z) = {} & \Theta(z < 0) \exp(-i p_\perp z) \tau'(p_\perp) \\ & + \Theta(z > 0) \left(\exp(-i p'_\perp z) - \rho(p_\perp) \exp(i p'_\perp z)\right) , \end{aligned} \quad (8.47)$$

where

$$\rho(p_\perp) = \frac{p_\perp - p'_\perp}{p_\perp + p'_\perp} , \quad \tau'(p_\perp) = 1 - \rho(p_\perp) = \frac{2 p'_\perp}{p_\perp + p'_\perp} , \quad (8.48)$$

and $p'_\perp = \sqrt{p_\perp^2 - u_c}$. The function (8.47) contains the incident wave propagating inside the matter.

Of course, we can take an arbitrary superposition of functions (8.22) and (8.47) instead of ψ_2; however, only the function (8.47) or one proportional to it gives the causal Green function. Any other combination will produce at scattering not only waves going out from a scatterer but also waves going toward it. The Wronskian of the functions $\psi_{1,2}(u_c, p_\perp, z)$ is $w_{12}(p_\perp) = 2 i p_\perp \tau'(p_\perp)$.

With functions $\psi_1(u_c, p_\perp, z)$ from Eq. (8.22) and $\psi_2(u_c, p_\perp, z)$ from Eq. (8.47) we have

$$G(u_c, p_\perp, z, z') = \frac{1}{2 i p_\perp \tau'(p_\perp)} \times \\ \left[\Theta(z > z') \psi_1(u_c, p_\perp, z) \psi_2(u_c, p_\perp, z') + \Theta(z' > z) \psi_1(u_c, p_\perp, z') \psi_2(u_c, p_\perp, z)\right]. \quad (8.49)$$

In the limit $u_c \to 0$ this function transforms to Eq. (8.41).

8.2.2 Scattering with the Modified Green Function

8.2.2.1 Scattering on a Single Point Scatterer inside the Medium

With the Green function (8.44) we can easily find in perturbation theory the wave function for scattering on a single nucleus with scattering amplitude b_1 located at a point $\boldsymbol{r}_1$ inside the medium:

$$\Psi(\boldsymbol{k},\boldsymbol{r}) = \Phi(\boldsymbol{k},\boldsymbol{r}) - \Phi(\boldsymbol{k},\boldsymbol{r}_1) b_1 G(u_c, k, \boldsymbol{r}, \boldsymbol{r}_1) . \tag{8.50}$$

Such a perturbation theory is called the "distorted wave Born approximation".

Substitution of Φ from (8.22) and G from Eq. (8.44) into Eq. (8.50) gives scattered waves at $z < 0$:

$$\Psi_s(\boldsymbol{k},\boldsymbol{r}) = \int d^2 p_{\|} \exp(i\boldsymbol{p}\boldsymbol{r}) \frac{i\overrightarrow{t_0}(p_\perp)}{2\pi p_\perp} b_1 \overrightarrow{t_0}(k_\perp) \exp(i\boldsymbol{\kappa}\boldsymbol{r}_1) , \tag{8.51}$$

where $\boldsymbol{\kappa} = \boldsymbol{k}' - \boldsymbol{p}'$, $\boldsymbol{k}' = (\boldsymbol{k}_{\|}, k'_\perp)$, $\boldsymbol{p}' = (\boldsymbol{p}_{\|}, -p'_\perp)$, $k'_\perp = \sqrt{k_\perp^2 - u_c}$ and $p'_\perp = \sqrt{k^2 - p_{\|}^2 - u_c}$. Comparison with Eq. (8.25) shows that

$$A(\boldsymbol{k},\boldsymbol{p}) = \frac{i\overrightarrow{t_0}(p_\perp)\overrightarrow{t_0}(k_\perp)}{2\pi p_\perp} b_1 \exp(i\boldsymbol{\kappa}\boldsymbol{r}_1) \tag{8.52}$$

according to Eq. (8.34)

$$d\sigma(\boldsymbol{\Omega}_0 \to \boldsymbol{\Omega})/d\Omega = k|b_1|^2 \frac{1}{k_\perp} \left|\overrightarrow{t_0}(p_\perp)\overrightarrow{t_0}(k_\perp)\right|^2 \exp\left(-2(p''_\perp + k''_\perp)z_1\right) , \tag{8.53}$$

where $k''_\perp = u''_c/2k_\perp$, $p''_\perp = u''_c/2p'_\perp$, and we took into account that the potential u_c is a complex number $u'_c - iu''_c$ with positive u''_c. We see that Eq. (8.53) completely satisfies the detailed balance condition (7.166).

8.2.2.2 Scattering on Fluctuations of $\delta(\psi(\boldsymbol{r}_j)b_j)$

Now we can return to the problem formulated on page 346. The fluctuation part of the wave function outside the medium with account of the new Green function is

$$\delta\Psi(\boldsymbol{r}) = \int d^2 p_{\|} \exp(i\boldsymbol{p}\boldsymbol{r}) \frac{i\overrightarrow{t_0}(p_\perp)}{2\pi p_\perp} \sum_j \exp(-i\boldsymbol{p}'\boldsymbol{r}_j)\delta(\psi(\boldsymbol{r}_j)b_j) , \tag{8.54}$$

and comparison with Eq. (8.25) shows that

$$A(\boldsymbol{k},\boldsymbol{p}) = \frac{i\overrightarrow{t_0}(p_\perp)}{2\pi p_\perp} \sum_j \exp(-i\boldsymbol{p}'\boldsymbol{r}_j)\delta(\psi(\boldsymbol{r}_j)b_j) . \tag{8.55}$$

We suppose now that $\langle\delta(\psi(\boldsymbol{r}_j)b_j)\rangle = 0$; therefore, there are no first-order corrections to $\overrightarrow{r_0}$.

To find the cross section (8.34) we need to find $\langle|A(\boldsymbol{k},\boldsymbol{p})|^2\rangle$ over fluctuations $\delta(\psi_j b_j) = \psi_j b_j - \phi(\boldsymbol{r}_j)b = \phi(\boldsymbol{r}_j)\delta b_j + b\delta\psi_j$. The first term on the right side shows the contribution of the incoherent part of the scattering amplitude, and the second term shows the contribution of fluctuations of the factors ψ_j. It is natural to suppose the following correlation functions for these two random values: $\langle \delta b_j^* \delta b_{j'} \rangle = |b_{\rm inc}|^2 \delta_{j,j'}$, $\langle \delta^* \psi_j \delta \psi_{j'} \rangle = K_0 |\phi(\boldsymbol{r}_j)|^2 \delta_{j,j'}$. The last relation follows from the fact that fluctuations of ψ_j should decrease with decrease of coherent field ϕ, that is, we can accept that $\delta \psi_j \propto \phi(\boldsymbol{r}_j)$, so the fluctuations are proportional to $|\phi(\boldsymbol{r}_j)|^2$, with some proportionality factor K_0. Therefore, $\langle \delta^*(b_j \psi_j) \delta(b_{j'} \psi_{j'}) \rangle = |\phi(\boldsymbol{r}_j)|^2 \left(|b_{\rm inc}|^2 + K_0 |b|^2\right) \delta_{j,j'}$. With this result we obtain

$$
\begin{aligned}
\langle |A(\boldsymbol{k}, \boldsymbol{p})|^2 \rangle &= |\tilde{b}|^2 \left| \frac{\overrightarrow{t_0}(p_\perp)}{2\pi p_\perp} \right|^2 \sum_j e^{-2p''_\perp z_j} |\phi(\boldsymbol{r}_j)|^2 \\
&= |\tilde{b}|^2 \left| \frac{\overrightarrow{t_0}(p_\perp) \overrightarrow{t_0}(k_\perp)}{2\pi p_\perp} \right|^2 \sum_j e^{-2(p''_\perp + k''_\perp) z_j} , \qquad (8.56)
\end{aligned}
$$

where $|\tilde{b}|^2 = |b|^2_{\rm inc} + b_c^2 K_0$, and we substituted $\phi(\boldsymbol{r}_j) = \overrightarrow{t_0}(k_\perp) \exp(i \boldsymbol{k}' \boldsymbol{r}_j)$.

After substitution of Eq. (8.56) into Eq. (8.34) and transformation of the summation to integration over $N_0 d^3 r_j$ we obtain

$$
\langle |A(\boldsymbol{k}, \boldsymbol{p})|^2 \rangle = S |\tilde{b}|^2 N_0 \frac{1}{k_\perp} \frac{\left| \overrightarrow{t_0}(p_\perp) \overrightarrow{t_0}(k_\perp) \right|^2}{2(p''_\perp + k''_\perp)} , \qquad (8.57)
$$

and according to (8.36) we obtain the reflection indicatrix

$$
w(k_\perp \to p_\perp) = \frac{|\tilde{b}|^2 N_0}{k_\perp} \frac{\left| \overrightarrow{t_0}(p_\perp) \overrightarrow{t_0}(k_\perp) \right|^2}{2(p''_\perp + k''_\perp)} , \qquad (8.58)
$$

which completely satisfies the detailed balance condition. This indicatrix in the form

$$
w(k, p) = \frac{c}{k} \frac{\left| \overrightarrow{t_0}(p) \overrightarrow{t_0}(k) \right|^2}{p'' + k''} , \qquad (8.59)
$$

with some arbitrary constant c and omitted index $\perp$, and its integral,

$$
wi(k) = c \int_0^3 \frac{dp}{k} \frac{\left| \overrightarrow{t_0}(p) \overrightarrow{t_0}(k) \right|^2}{p'' + k''} , \qquad (8.60)
$$

are shown in Figure 8.1. The number 3 in the upper limit of the integral (8.60) is chosen to cover most of the scattered intensity, which can be registered by a wide stationary detector in the experiment [188]. It can be increased, but qualitatively the result will not change.

We tried to use function (8.60) to improve the fitting of the boron glass reflectivity shown in Figure 4.5a. In the fitting we used this function instead of the background parameter n_b in Eq. (4.25). However, after such a fitting we could improve χ^2 only from 29 to 23, which is not at all satisfactory.

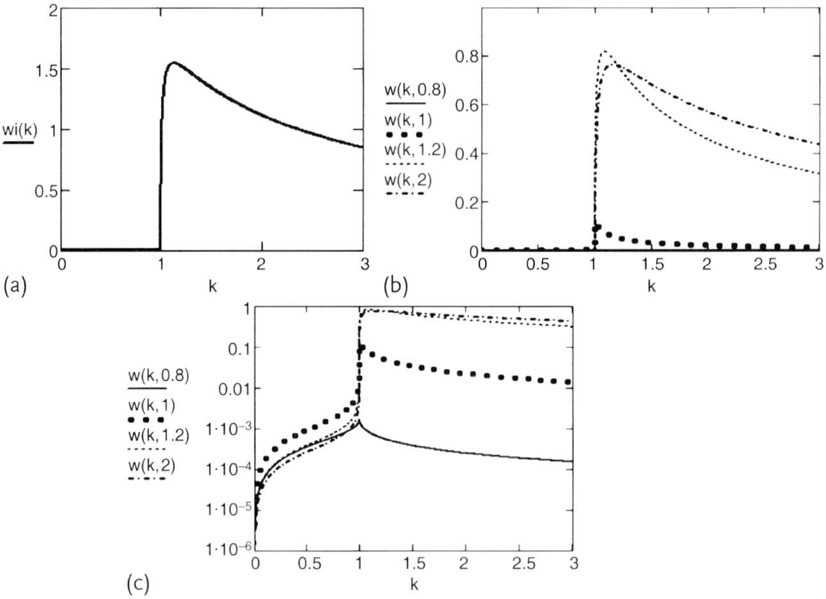

Figure 8.1 (a) Dependence of function (8.60) on k for some c and $u = 1 - i0.0002$; (b) function (8.59) on a linear scale for different p; (c) the same as (b) but on a logarithmic scale.

8.2.2.3 **Yoneda Effect [368]**

The maximum near the critical energy, seen in Figure 8.1, is known as the *Yoneda effect* [368]. Yoneda found that the incident beam of X-rays splits after reflection into two components as shown in Figure 8.2. Besides the specularly reflected beam, there appears also an anomalous one at an angle close to the critical one. This anomalous part is due to scattering on roughnesses. When the roughness of the surface increases, the specular part decreases and the anomalous one increases. This proves that the anomalous part appears because of scattering, and scattered radiation concentrates near the critical angle. Because of symmetry of the expression (8.57) the scattering increases also when the incident angle is close to the critical one. Above we considered scattering not on roughnesses, but found a similar result.

8.2.2.4 **Ultracold Neutrons**

For ultracold neutrons we have $k^2 < u'_c$, and $|k_\perp + k'_\perp|^2 = |p_\perp + p'_\perp|^2 \approx u'_c$. Therefore, Eq. (8.53) is transformed to

$$\sigma'(\boldsymbol{\Omega}_0 \to \boldsymbol{\Omega}) = 16\frac{k^4}{u'^2_c}|b_1|^2 \cos^2\vartheta \cos\vartheta_0 \exp\left(-2(|p'_\perp| + |k'_\perp|)z_1\right) , \quad (8.61)$$

which again satisfies the detailed balance condition.

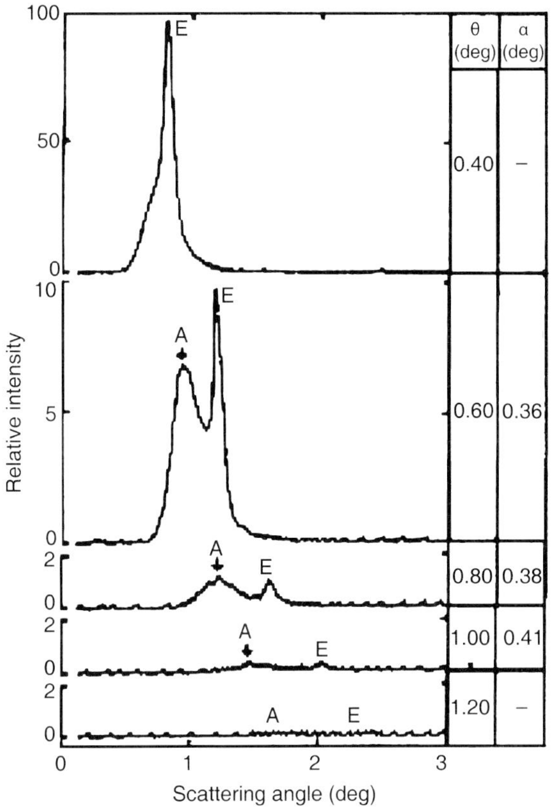

Figure 8.2 Reflection of 35-kV X-rays from a Cr surface. The reflected beam contains two components: one is specular (E) with grazing angle θ and the other one (A) is reflected at grazing angle α. When θ increases, α, which is near critical angle of Cr, remains almost constant.

For many independent atoms distributed with density N_1 we can obtain like in Eq. (8.57) the probability of reflection:

$$dw(\boldsymbol{\Omega}_0 \to \boldsymbol{\Omega})/d\Omega = 8\frac{k^4}{u_c'^2}|b_1|^2 N_1 \frac{\cos^2\vartheta\cos\vartheta_0}{|p'_\perp| + |k'_\perp|} \,. \tag{8.62}$$

From this expression it is seen that the probability of nonspecular reflection is proportional to $\cos^2\vartheta$, where ϑ is the angle between the the external normal to the interface and the direction of the flight of the reflected neutrons, and that this probability decreases with increase of the incident angle ϑ_0. This is important information for consideration of diffusion of ultracold neutrons along neutron guides [369].

Diffusion of ultracold neutrons in neutron guides with reflection indicatrix

$$[1 - w(\boldsymbol{\Omega}_0)]\delta(\boldsymbol{\Omega} - \boldsymbol{\Omega}_0) + 8\frac{k^4}{u_c'^2}|b_1|^2 N_1 \frac{\cos^2\vartheta\cos\vartheta_0}{|p'_\perp| + |k'_\perp|} \,,$$

where

$$w(\boldsymbol{\Omega}_0) = \int_{2\pi} 8\frac{k^4}{u_c'^2}|b_1|^2 N_1 \frac{\cos^2\vartheta \cos\vartheta_0}{|p'_\perp| + |k'_\perp|} d\Omega ,$$

has some peculiarity. Because of such an indicatrix the neutron random walk is not a Brownian one, but is described by richer Levy statistics [370].

8.2.2.5 Scattering on a Fixed Center outside the Medium

The wave function for scattering from a nucleus with scattering amplitude b_1 located at a point $\boldsymbol{r}_1$ outside the medium is given by the same expression (8.50), but the wave $\Phi(\boldsymbol{k}, \boldsymbol{r}_1)$, illuminating the nucleus in this case, according to Eq. (8.22) is equal to $\exp(i\boldsymbol{k}_{||}\boldsymbol{r}_{1||})\Phi(k_\perp, z_1)$, where

$$\Phi(k_\perp, z_1) = \exp(ik_\perp z_1) + \vec{r_0}(k_\perp)\exp(-ik_\perp z_1) , \quad \vec{r_0}(k_\perp) = \frac{k_\perp - k'_\perp}{k_\perp + k'_\perp} , \tag{8.63}$$

and $z_1 < 0$. The respective component $\psi_1(p_\perp, z_1)$ of the Green function is also equal to Eq. (8.63), with replacement of $k_\perp$ by $p_\perp$. As a result, according Eq. (8.37) we have

$$\frac{d\sigma(k_\perp \to p_\perp)}{dk_\perp} = \frac{|b_1|^2}{k_\perp} \left| \left[e^{ip_\perp z_1} + \vec{r_0}(p_\perp)e^{-ip_\perp z_1} \right] \times \left[e^{ik_\perp z_1} + \vec{r_0}(k_\perp)e^{-ik_\perp z_1} \right] \right|^2 , \tag{8.64}$$

where $k_\perp = k\cos\theta_0$ and $p_\perp = k\cos\theta$.

If we have many atoms with surface density N_1 over a reflecting plane at height $|z_1|$, every one of which is characterized by the scattering cross section (8.64), we find the reflection indicatrix, which like Eq. (8.57) can be represented as

$$w(k, p) = \frac{c}{k} \left| \left[e^{ipz_1} + \vec{r_0}(p)e^{-ipz_1} \right] \left[e^{ikz_1} + \vec{r_0}(k)e^{-ikz_1} \right] \right|^2 , \tag{8.65}$$

where c is some constant and index $\perp$ is omitted.

The function $w(k, p)$ on k for different p and the function

$$wi(k) = \int_0^3 w(k, p)dp = c\int_0^3 \frac{dp}{k} \left| \left[e^{ipz_1} + \vec{r_0}(p)e^{-ipz_1} \right] \left[e^{ikz_1} + \vec{r_0}(k)e^{-ikz_1} \right] \right|^2 \tag{8.66}$$

are shown in Figure 8.3. We see that in accord with the Yoneda effect, the peak of the scattered intensity has a maximum near the critical edge, but in contrast to Figure 8.1 the peak's position is below the edge. With increase of $|z_1|$ the peak widens and becomes more symmetrical.

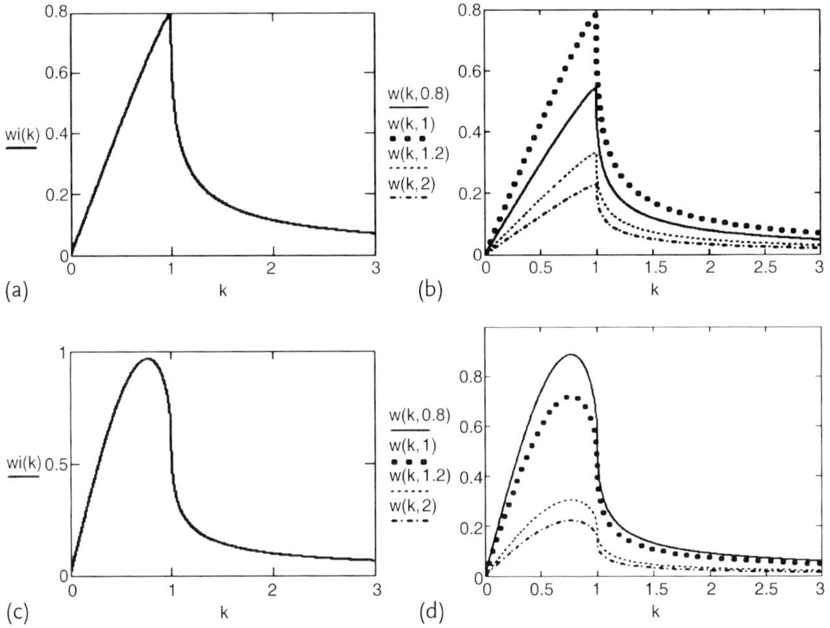

Figure 8.3 (a) Dependence of function (8.65) on k for some c and $u = 1 - i0.0002$; (b) function (8.66) on a linear scale for different p. In both cases $c = -z_1 = 0.05$. (c) and (d) are the same as (a) and (b) but for $z_1 = -0.5$.

8.2.2.6 Scattering because of Disorder

Now we can consider scattering because of disorder. This means that we assume that all $\psi(\boldsymbol{r}_j)b_j$ are equal to their average (coherent) values $\phi(\boldsymbol{r}_j)b_c$; nevertheless there is still scattering because of correlations between the positions of different scatterers. To find this scattering we again use Eq. (8.25) with

$$A(\boldsymbol{k},\boldsymbol{p}) = b_c \frac{i\overrightarrow{t_0}(p_\perp)}{2\pi p_\perp} \sum_j \exp(-i\boldsymbol{p}'\boldsymbol{r}_j)\phi(\boldsymbol{r}_j)\,. \tag{8.67}$$

Its average gives $\overrightarrow{r_0}(k_\perp)\delta(\boldsymbol{p}_\parallel - \boldsymbol{k}_\parallel)$, which can be expected according to Eq. (8.16) and should be considered as a correction to specular reflectivity. However, we must also calculate

$$\langle|A(\boldsymbol{k},\boldsymbol{p})|^2\rangle = |b_c|^2 \left|\frac{\overrightarrow{t_0}(p_\perp)\overrightarrow{t_0}(k_\perp)}{2\pi p_\perp}\right|^2 \left\langle \sum_{j,j'} e^{-i\kappa'^* r_j} e^{i\kappa' r_{j'}} \right\rangle, \tag{8.68}$$

where $\boldsymbol{\kappa}' = (\boldsymbol{k}_\parallel - \boldsymbol{p}_\parallel, k'_\perp + p'_\perp)$. The double sum has a diagonal part, which after transformation to the integral over $N_0 d^3 r_j$ gives

$$\langle|A(\boldsymbol{k},\boldsymbol{p})|_d^2\rangle_d = \frac{N_0|b_c|^2}{2(k''_\perp + p''_\perp)} \left|\frac{\overrightarrow{t_0}(p_\perp)\overrightarrow{t_0}(k_\perp)}{2\pi p_\perp}\right|^2 S\,, \tag{8.69}$$

and indicatrix

$$w_d(k_\perp \to p_\perp) = N_0|b_c|^2 \frac{\left|\overrightarrow{t_0}(p_\perp)\overrightarrow{t_0}(k_\perp)\right|^2}{2k_\perp(k''_\perp + p''_\perp)} . \tag{8.70}$$

The nondiagonal part of the sum can be transformed to a double integral:

$$\left\langle \sum_{j,j'} F^*(\boldsymbol{r}_j)F(\boldsymbol{r}_{j'}) \right\rangle = N_0 \int d^3 r_j \int g(\boldsymbol{r}_j - \boldsymbol{r}_{j'})F^*(\boldsymbol{r}_j)F(\boldsymbol{r}_{j'}) , \tag{8.71}$$

where

$$g(\boldsymbol{r}) = \delta(\boldsymbol{r}) - N_0\gamma(\boldsymbol{r}) \tag{8.72}$$

is a correlation function chosen in such a way as to exclude the diagonal part of the sum. For simplicity we shall take for $\gamma(r)$ the Gaussian

$$\gamma(\boldsymbol{r}) = (2\pi)^{-3/2}\exp(-r^2/2a^2) , \quad a = N_0^{-1/3} . \tag{8.73}$$

After substitution of $\phi(\boldsymbol{r})$ into Eq. (8.68) and integration we obtain like in Eq. (8.58) the expression

$$\begin{aligned} w_{nd}(\boldsymbol{\Omega}_0 \to \boldsymbol{\Omega}) &= |b_c|^2 N_0 \frac{1}{k_\perp} \left|\overrightarrow{t_0}(p_\perp)\overrightarrow{t_0}(k_\perp)\right|^2 \frac{1 - \exp(-a^2\boldsymbol{\kappa}^2/2)}{p''_\perp + k''_\perp} \\ &\approx \frac{c}{k_\perp} \left|\overrightarrow{t_0}(p_\perp)\overrightarrow{t_0}(k_\perp)\right|^2 \frac{1 - \exp(-a^2[p'_\perp - k'_\perp]^2/2)}{p''_\perp + k''_\perp} , \end{aligned} \tag{8.74}$$

where in the last equality we put $c = |b_c|^2 N_0$, and used the approximation $\kappa^2 \approx [p'_\perp - k'_\perp]^2$, which is valid for small $p'_\perp$ and $k'_\perp$ and a small angle between $\boldsymbol{k}$ and $\boldsymbol{p}$.

The function (8.74) represented in the form

$$w_{nd}(k, p) = \frac{c}{k} \frac{\left|\overrightarrow{t_0}(p)\overrightarrow{t_0}(k)\right|^2}{p'' + k''}[1 - \exp(-a^2[p' - k']^2/2)] \tag{8.75}$$

and its integral,

$$wi(k) = c \int_0^b \frac{dp}{k} \frac{\left|\overrightarrow{t_0}(p)\overrightarrow{t_0}(k)\right|^2}{p'' + k''}[1 - \exp(-a^2[p' - k']^2/2)] , \tag{8.76}$$

for $c = 10^{-4}$, $a^2 = 0.2$, and $b = 5$ are shown in Figure 8.4. The curves are sensitive to parameter a and the upper integration limit b. With increase of b the left maximum in Figure 8.4a increases compared with the right one, and the right peak shifts to higher k.

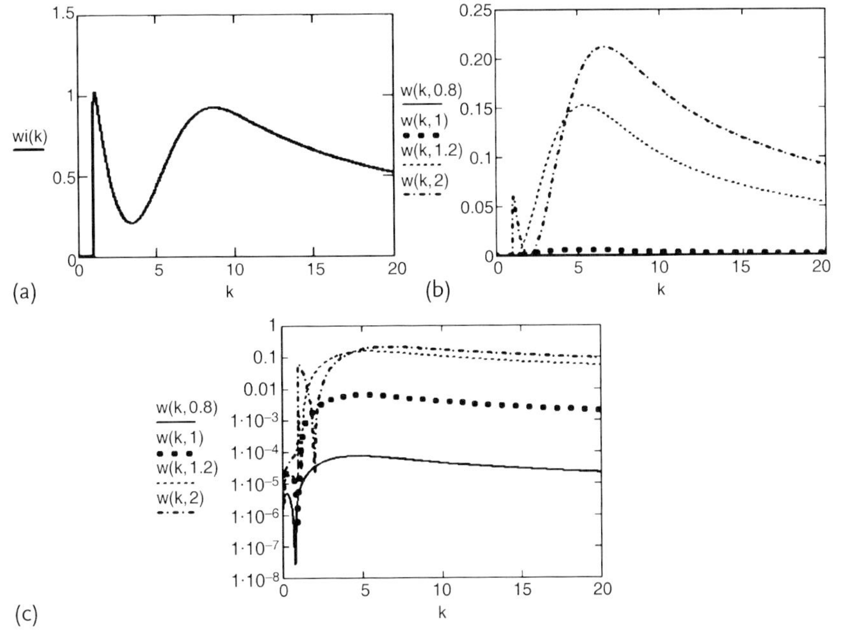

Figure 8.4 (a) Dependence of function (8.76) on k for $c = 0.0001$, $u = 1 - i0.0002$, $a^2 = 0.2$, $b = 5$; (b) function (8.75) on a linear scale for different p; (c) the same as for (b) but on a logarithmic scale.

8.2.3
Experience with an Experiment

The function (8.76) has two maxima, so we can hope to get better fitting for the glass reflectivity shown in Figure 4.5a. In the fitting we assumed the potential of Boron glass to be $u = 0.408 - 0.003i$ as was obtained before. We also included smoothing of the interface with a Debye–Waller factor, that is, instead of $\vec{r_0}(k)$ we used $r_m(k) \equiv \vec{r_0}(k)\exp(-2h^2kk')$, where $2h^2$ is a fitting parameter.

Scattering on randomness and fluctuations was represented by the function

$$w_1(k) = c\int_0^b \frac{dp}{k}\frac{\left|\vec{t_0}(p)\vec{t_0}(k)\right|^2}{p'' + k''}[2 - \exp(-a^2[p' - k']^2/2)]\,. \tag{8.77}$$

The number 2 in brackets shows that we take the sum of two functions (8.74) and (8.70). In the function (8.77) there are two fitting parameters: factor c and $a^2/2$.

The specular reflectivity was defined as $R_s(k) = |r_m(k)|^2 - w_1(k)$, because scattering decreases specular reflectivity. For the total reflectivity we used the expression

$$R(k) = \int_{k-\delta}^{k+\delta} R_s(k)\frac{dp}{2\delta} + n_b + w_1(k)\,, \tag{8.78}$$

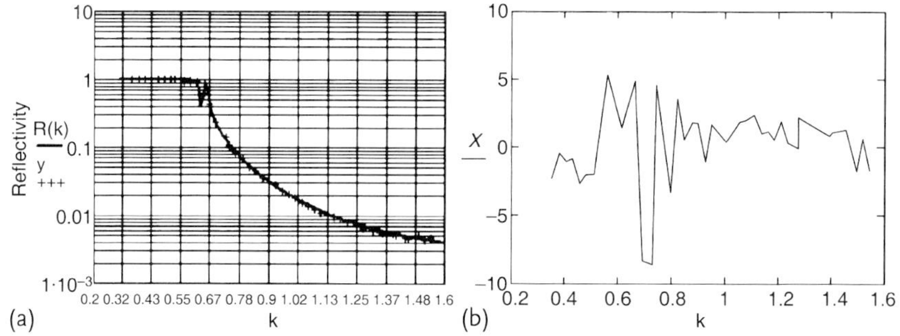

Figure 8.5 (a) Result of fitting on a logarithmic scale the reflectivity from boron glass [188] with function (8.77). The glass potential $u = 0.408 - 0.003i$ was fixed. The fitting parameters were found to be $c = 6.323 \cdot 10^{-4}$, $a^2/2 = 0.102$, $\delta = 0.022$, $n_b = 2.401 \cdot 10^{-3}$, and $h^2 = 0.048$. All the values are quite reasonable. (b) χ distribution for such fitting. We see that some fluctuations still remain near the edge, but their amplitude decreased much compared with that in Figure 4.5d. The dip in χ means that the measured quantity is larger than the theoretical one. This can be explained. In the theory we did not take into account the scattering into the glass. These scattered neutrons after going through the sample can be registered by the detector. Because of them the experimental value is higher than the theoretical one.

which contains two more fitting parameters: δ and n_b. Thus, in total we had five fitting parameters. The result of the fitting is shown in Figure 8.5. For the upper limit of the integral $b = 9$ in Eq. (8.77), χ^2 was 8.5. Though this is too high, it decreased considerably from the value of 29. The most important result is that the fitting shows a dip in reflectivity near the edge. The fitting parameters were found to be $c = 8.3 \cdot 10^{-4}$, which corresponds to randomness of positions of all the atoms in the glass; $a^2 = 0.204$, which means that the correlation length is near 40 Å; resolution $\delta = 2.2\%$, which is better than in previous fittings; background $n_b = 2.4 \cdot 10^{-3}$, which is nearly the same as before; and $2h^2 = 0.048$, that is, $h \approx 15$ Å, which means that in the preparation of the float glass some Sn atoms diffused into the glass to a depth of 15 Å. The results of the fitting shows that the dip near the edge is quite well described by nonspecular reflectivity, and is not a result of some instrumental peculiarity.

8.2.4
Small-Angle Scattering on a Single Inhomogeneity inside the Medium

Up to now we considered only point isotropic scatterers. Now we shall look at extended scatterers with nonisotropic scattering. If these scatterers are sufficiently large, they produce only small-angle scattering. In contrast to the usual small-angle scattering [378–380], we shall take into account the interface, and will treat the small-angle scattering with the help of the distorted wave Born approximation.

As an example let us consider the scattering on an inhomogeneity of a volume V. The wave scattered from it is

$$\Psi_s(\boldsymbol{k},\boldsymbol{r}) = -\int\limits_V G(u_c, k, \boldsymbol{r}, \boldsymbol{r}')\phi(\boldsymbol{k},\boldsymbol{r}')b(\boldsymbol{r}')n(\boldsymbol{r}')d^3r' , \tag{8.79}$$

where $b(\boldsymbol{r}')$ and $n(\boldsymbol{r}')$ are the coherent scattering amplitude and atomic density at the point $\boldsymbol{r}'$ inside the volume V.

Substitution of Eqs. (8.44) and (8.49) with account of Eq. (8.47) gives

$$\Psi_s(\boldsymbol{k},\boldsymbol{r}) = \int d^2p_\parallel \exp(i\boldsymbol{p}\boldsymbol{r})\frac{i\overrightarrow{t_0}(p_\perp)\overrightarrow{t_0}(k_\perp)}{2\pi p_\perp}\exp(i\boldsymbol{\kappa}\boldsymbol{r}_1)f(\boldsymbol{\kappa}) , \tag{8.80}$$

where $\boldsymbol{r}_1$ is the center of the inhomogeneity, $\boldsymbol{\kappa} = \boldsymbol{k}' - \boldsymbol{p}' + i(k''_\perp + p''_\perp)$, $\boldsymbol{k}' = (\boldsymbol{k}_\parallel, k'_\perp)$, $\boldsymbol{p}' = (\boldsymbol{p}_\parallel, -p'_\perp)$, $k'_\perp = \sqrt{k^2 - k_\parallel^2 - u'_c}$, $p'_\perp = \sqrt{k^2 - p_\parallel^2 - u'_c}$, $k''_\perp = u''_c/2k'_\perp$, $p''_\perp = u''_c/2p'_\perp$, and

$$f(\boldsymbol{\kappa}) = \int_V \exp(i\boldsymbol{\kappa}\boldsymbol{r}')b(\boldsymbol{r}')n(\boldsymbol{r}')d^3r' \tag{8.81}$$

is the structure factor of the inhomogeneity.

The scattering cross section defined as $J_\perp/k_\perp$ is

$$\begin{aligned}\sigma &= \int \frac{d^2p_\parallel}{k_\perp p_\perp}\left|\overrightarrow{t_0}(p_\perp)\overrightarrow{t_0}(k_\perp)\right|^2 e^{-2(p''_\perp+k''_\perp)z_1}|f(\boldsymbol{\kappa})|^2 \\ &= \frac{k}{k_\perp}\int d\Omega \left|\overrightarrow{t_0}(p_\perp)\overrightarrow{t_0}(k_\perp)\right|^2 e^{-2(p''_\perp+k''_\perp)z_1}|f(\boldsymbol{\kappa})|^2 ,\end{aligned} \tag{8.82}$$

where in the last equality the transformation (8.32) was used. The differential cross section is

$$d\sigma(\boldsymbol{\Omega}_0 \to \boldsymbol{\Omega})/d\Omega = \frac{k}{k_\perp}\left|\overrightarrow{t_0}(p_\perp)\overrightarrow{t_0}(k_\perp)\right|^2 e^{-2(p''_\perp+k''_\perp)z_1}|f(\boldsymbol{\kappa})|^2 . \tag{8.83}$$

This satisfies the detailed balance condition and also shows the Yoneda effect.

In the following we restrict ourselves to a spherically symmetrical inhomogeneity. If b and n in Eq. (8.81) are constant, then

$$\begin{aligned}f(\boldsymbol{\kappa}) &= bn\int_V \exp(i\boldsymbol{\kappa}\boldsymbol{r}')d^3r' = \frac{4\pi nb}{\kappa}\int\limits_0^{r_s}\sin(\kappa r)r\,dr \\ &= \frac{4\pi nb}{\kappa^3}[\sin(\kappa r_s) - \kappa r_s\cos(\kappa r_s)] .\end{aligned} \tag{8.84}$$

In the case of measurements at small angles at a reflectometer, we can replace κ with $|p'_\perp - k'_\perp|$.

If there are many similar inhomogeneities distributed with uniform density N_1, and interference of their scattered waves can be neglected, we can integrate

Eq. (8.83) over $N_1 d^3 r_1$, and as a result we obtain like in Eq. (8.59) the indicatrix of nonspecular reflection:

$$w(k \to p) = c \frac{\left| \overrightarrow{t_0}(p) \overrightarrow{t_0}(k) \right|^2}{k(p'' + k'')} |f(|p' - k'|)|^2 , \tag{8.85}$$

where c is a constant and index $\perp$ is omitted. Figure 8.6 shows the structure factor $f(\kappa) = f(k - p)_{k=0}$ and two appropriately normalized functions $f1(p) = [w(k \to p)/k]_{k=0}$ and $f2(p) = [w(k \to p)/k]_{k=1}$ on a logarithmic scale[30]. To show them on a linear scale it was necessary to multiply $f2(p)$ by an additional factor of 0.01. On a linear scale it is seen how large the difference between functions (8.84) and (8.85) is. The additional factor in Eq. (8.85) concentrates the peak near the critical edge, which explains the Yoneda effect.

30) In the definition of $f1$ we have to divide w by k so not to have identical zero at $k = 0$.

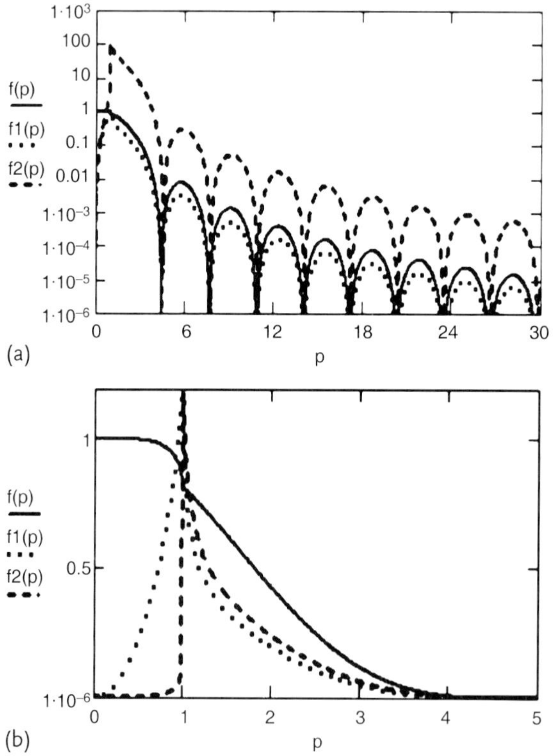

Figure 8.6 (a) Form factor $f(p)$ of a spherical inhomogeneity of radius $r = 1$ (100 Å in Ni) on a logarithmic scale and similar functions defined in the text, which take into account the interface. (b) The same functions on a linear scale. To present them all in the same range it was necessary to multiply $f2$ by 0.01.

8.2.4.1 Asymptotic of the Structure Factor at Small κ

At small κ, such that $R\kappa \ll 1$, where R is the radius of the inhomogeneity, we can approximate

$$f(\kappa) = \int_V \exp(i\kappa \mathbf{r}) b(r) n(r) d^3 r$$
$$= 4\pi \int_R \frac{\sin(\kappa r)}{\kappa r} b(r) n(r) r^2 dr \approx N b (1 - \kappa^2 \langle R^2 \rangle / 6) , \qquad (8.86)$$

where

$$N = \int_V n(\mathbf{r}) d^3 r = 4\pi \int_R n(r) r^2 dr \qquad (8.87)$$

is the total number of atoms in the inhomogeneity, $b = \int_V b(r) n(r) d^3 r / N = (4\pi/N) \int_R b(r) n(r) r^2 dr$ is the average amplitude, and $\langle R^2 \rangle = \int_V r^2 b(r) n(r) d^3 r / N b$ is its gyration radius.

The scattering intensity, $I(\kappa)$, is proportional to square of the amplitude (8.86) and at small κ it is approximated by $I(\kappa) \propto |f(\kappa)|^2 \approx |bN|^2 (1 - \kappa^2 \langle R^2 \rangle / 3)$.

8.2.4.2 Asymptotic of the Structure Factor at Large κ

At large κ, such that $R\kappa \gg 1$, integration of (8.86) by parts gives

$$f(\kappa) = 4\pi \int_R \frac{\sin(\kappa r)}{\kappa} b(r) n(r) r dr$$
$$= -4\pi r \frac{\cos(\kappa r)}{\kappa^2} b(r) n(r) \bigg|_{r=R} + O(1/\kappa^3) \approx -4\pi R \frac{\cos(\kappa R)}{\kappa^2} b(R) n(R) . \qquad (8.88)$$

So

$$|f(\kappa)|^2 \approx (4\pi R)^2 |b(R) n(R)|^2 \frac{\cos^2(\kappa R)}{\kappa^4} = 4\pi |b(R) n(R)|^2 S \frac{\cos^2(\kappa R)}{\kappa^4} , \qquad (8.89)$$

where $S = 4\pi R^2$ is the surface area of the inhomogeneity, $n(R)$ is the atomic density at $r = R$, and $b(R)$ is the coherent scattering amplitude at its surface. If $n(r)$ and $b(r)$ inside the particle are constant, then $b(R) n(R) = bn$.

If the radius of the inhomogeneity is a random number with uniform distribution inside the range $\overline{R} - \Delta R < R < \overline{R} + \Delta R$, $\overline{R} \gg \Delta R \gg 1/\kappa$, then Eq. (8.89) have to be averaged over R. As a result of such averaging, the factor $\cos^2(\kappa R)$ can be replaced by 1/2, and the asymptotic at large κ becomes $|f(\kappa)|^2 \approx 2\pi |bn|^2 \overline{S} / \kappa^4$, where $\overline{S} = 4\pi \overline{R}^2$ is the average surface area of the inhomogeneity.

8.2.4.3 Porod Invariant for the Structure Factor

As usual in small angle scattering science, we can introduce the magnitude $Q = \int |f(\kappa)|^2 d^3\kappa / 4\pi$, which is called the "Porod invariant" and defines the scattering capability of the inhomogeneity $Q = 2\pi^2 |b|^2 \int_R n^2(r) d^3 r$. At constant $n(r)$ and b

we obtain $Q = 2\pi^2 n^2 |b|^2 V$, where V is the volume of the inhomogeneity. Its value can be extracted from the scattered intensity Eq. (8.83). Direct integration of Eq. (8.83) over $d^3\kappa$ accounts for the interface and gives the dependence of the Porod function on incident angle.

8.2.5
Small-Angle Scattering on a Set of Inhomogeneities

In the presence of many inhomogeneities, formula (8.80) changes to

$$\Psi_s(\boldsymbol{k},\boldsymbol{r}) = \int d^2 p_\parallel \exp(i\boldsymbol{p}\boldsymbol{r}) \frac{i\,\overrightarrow{t_0}(p_\perp)\overrightarrow{t_0}(k_\perp)}{2\pi p_\perp} \sum_{j=1}^{\infty} \exp(i\boldsymbol{\kappa}\boldsymbol{r}_j) f_j(\boldsymbol{\kappa}) \,, \tag{8.90}$$

where $\boldsymbol{r}_j$ is the center of the jth inhomogeneity and

$$f_j(\boldsymbol{\kappa}) = \int_{V_j} \exp(i\boldsymbol{\kappa}\boldsymbol{r}) b_j(\boldsymbol{r}) n_j(\boldsymbol{r}) d^3 r \tag{8.91}$$

is its structure factor. Similarly to Eqs. (8.25) and (8.31), we obtain the cross section

$$\sigma(\boldsymbol{k}) = \frac{J_\perp(\boldsymbol{k})}{k_\perp} = \frac{k}{k_\perp} \int |\overrightarrow{t_0}(p_\perp)\overrightarrow{t_0}(k_\perp)|^2 |B(\boldsymbol{k},\boldsymbol{p})|^2 d\Omega \,, \tag{8.92}$$

where

$$|B(\boldsymbol{k},\boldsymbol{p})|^2 = \left\langle \left| \sum_{j=1}^{\infty} \exp(i\boldsymbol{\kappa}\boldsymbol{r}_j) f_j(\boldsymbol{\kappa}) \right|^2 \right\rangle , \tag{8.93}$$

and $\langle F \rangle$ means average over positions and forms of inhomogeneities. We assume that all inhomogeneities are distributed uniformly but randomly with density $\nu(\boldsymbol{r}) = \nu_0 = \text{const}$.

Again we can extract the diagonal part of double sum in Eq. (8.93) and the nondiagonal one. The diagonal part gives

$$w_d(\boldsymbol{\Omega}_0 \to \boldsymbol{\Omega}) = \frac{\nu_0 \overline{|f(\boldsymbol{\kappa})|^2}}{k_\perp} \frac{\left|\overrightarrow{t_0}(p_\perp)\overrightarrow{t_0}(k_\perp)\right|^2}{p''_\perp + k''_\perp} \,, \tag{8.94}$$

and the nondiagonal one gives

$$w_{nd}(\boldsymbol{\Omega}_0 \to \boldsymbol{\Omega}) = \frac{\nu_0 |\overline{f(\boldsymbol{\kappa})}|^2}{k_\perp} \left|\overrightarrow{t_0}(p_\perp)\overrightarrow{t_0}(k_\perp)\right|^2 \frac{1 - \exp(-a^2[p'_\perp - k'_\perp]^2/2)}{p''_\perp + k''_\perp} \,, \tag{8.95}$$

where $\overline{f(\boldsymbol{\kappa})}$ is some kind of coherent structure factor, and $\overline{|f(\boldsymbol{\kappa})|^2}$ is an analog of the total cross section, which includes coherent and incoherent parts. The expression (8.94) was derived in the same way as Eq. (8.75) with a correlation function $g(\boldsymbol{r})$ similar to Eqs. (8.72) and (8.73).

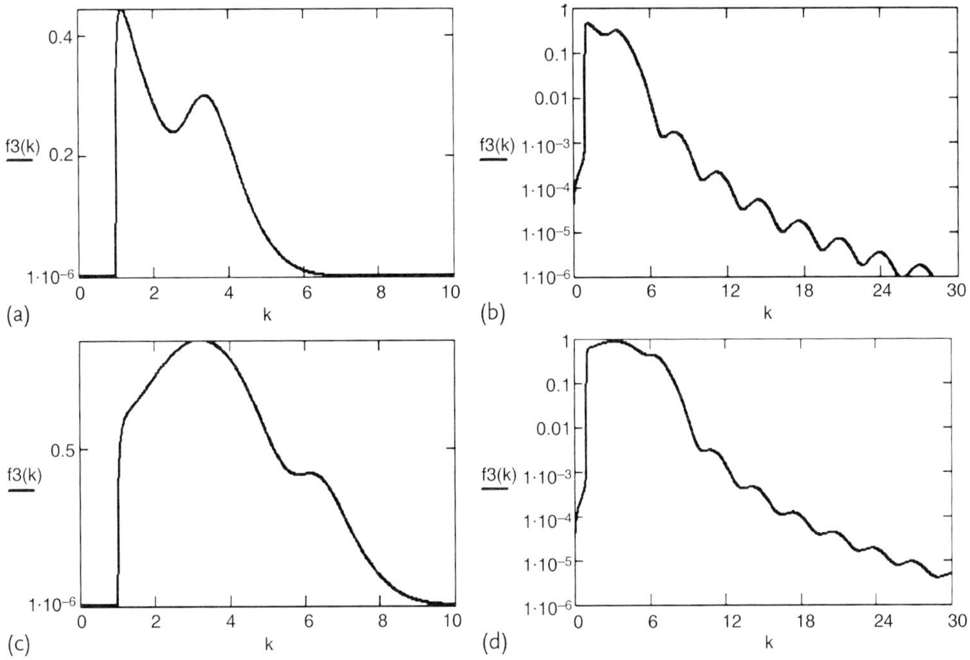

Figure 8.7 Function (8.96) (a) on a linear and (b) on a logarithmic scale for $b = 3$ in the upper limit. (c) and (d) are similar to (a) and (b) but with $b = 6$.

Figure 8.7 presents the curves for

$$f3(k) = c\int_0^b dp \frac{|f(|k' - p'|)|^2}{k}\left|\vec{t_0}(p)\vec{t_0}(k)\right|^2 \frac{1 - \exp(-a^2[p' - k']^2/2)}{p'' + k''} \tag{8.96}$$

with some normalization constant c and two upper limits of the integral: $b = 3$ and $b = 6$. In the expression (8.96) we approximated $\kappa = p'_\perp + k'_\perp$ and omitted the index $\perp$. We see that function (8.96) varies with change of the upper limit in the integral. This is very important in experiments with a large stationary detector. The upper limit depends on the size of the detector. If it is a position-sensitive one, we can find the integral count by summing the counts for different numbers of cells, and from the variation of the integral counts we can extract information on the parameters of the inhomogeneities.

It is interesting to note that if scattering becomes strong, then its contribution to the imaginary part of the potential is large. If it is the main contribution to the imaginary part, the values $k''_\perp$ and $p''_\perp$ become proportional to $|f(\kappa = 0)|^2$, and the probability of nonspecular scattering becomes of the order of unity.

On the other side, if $|f(\kappa = 0)|^2$ is large, the imaginary part of the optical potential also becomes large. However, when the imaginary part of the potential is large, the reflection amplitude $(k_\perp - k'_\perp)/(k_\perp + k'_\perp)$ at some $k_\perp$ becomes close to unity. Therefore, we have a paradox: at large scattering the probability of nonspecular reflection increases; however, the probability of specular reflection also increases, which means a decrease of nonspecular reflectivity. This paradox is worth further investigation.

8.2.6
Scattering on Roughnesses at the Interface

Roughnesses at the interface can be considered as inhomogeneities on an ideal plane interface at $z_j = 0$, and then we can proceed in the same way as in previous sections and immediately write down results similar to Eqs. (8.94) and (8.95):

$$w_d(\boldsymbol{\Omega}_0 \to \boldsymbol{\Omega}) = \frac{\nu_2 \overline{|f(\kappa)|^2}}{k_\perp} \left| \overrightarrow{t_0}(p_\perp) \overrightarrow{t_0}(k_\perp) \right|^2 , \tag{8.97}$$

$$w_{nd}(\boldsymbol{\Omega}_0 \to \boldsymbol{\Omega}) = \frac{\nu_2 \overline{|f(\kappa)|^2}}{k_\perp} \left| \overrightarrow{t_0}(p_\perp) \overrightarrow{t_0}(k_\perp) \right|^2 \left[1 - \exp(-a^2[\boldsymbol{p}_\parallel - \boldsymbol{k}_\parallel]^2/2)\right] , \tag{8.98}$$

where ν_2 is the surface density of inhomogeneities. We see that these expressions do not contain $p''_\perp + k''_\perp$ in the denominator. This is because the integral, which replaces sums, does not contain integration over z and all z_j are equal to zero. In the Gaussian we did not replace κ by $|p'_\perp + k'_\perp|$ because here $\boldsymbol{\kappa}$ contains only components $\boldsymbol{\kappa}_\parallel$.

8.2.6.1 Common Approach to Roughness Scattering [372]

The usual approach is to consider roughness as a Gaussian process. This means that all the rough surface is treated as a single inhomogeneity and the scattered wave function is written as

$$\Psi_s(\boldsymbol{k}, \boldsymbol{r}) = \int d^2 p_\parallel \exp(i\boldsymbol{p}\boldsymbol{r}) A(\boldsymbol{k}, \boldsymbol{p}) , \tag{8.99}$$

where

$$A(\boldsymbol{k}, \boldsymbol{p}) = -N_0 b \frac{i \overrightarrow{t_0}(p_\perp) \overrightarrow{t_0}(k_\perp)}{2\pi p_\perp} f(\kappa') , \tag{8.100}$$

where

$$f(\kappa) = \int d^2 r'_\parallel \int_0^{\zeta(r'_\parallel)} dz' \exp(i\kappa \boldsymbol{r}') = \int \frac{d^2 r'}{i\kappa_\perp} \exp\left(i\kappa_\parallel \boldsymbol{r}'_\parallel\right) \left[\exp\left(i\kappa_\perp \zeta\left(\boldsymbol{r}'_\parallel\right)\right) - 1\right] \Theta\left(\zeta\left(\boldsymbol{r}'_\parallel\right) > 0\right) , \tag{8.101}$$

if the roughness is a cavity (note that its scattering density is $-N_0 b$, and $\boldsymbol{\kappa}' = (\boldsymbol{\kappa}_{\|}, k'_{\perp} + p'_{\perp})$, where $q'_{\perp} = \sqrt{q_{\perp}^2 - u_0}$), and

$$A(\boldsymbol{k}, \boldsymbol{p}) = N_0 b \int d^2 r'_{\|} \exp\left(i\boldsymbol{\kappa}_{\|} \boldsymbol{r}'_{\|}\right) \int\limits_{\zeta(\boldsymbol{r}'_{\|})}^{0} dz' \Theta\left(\zeta\left(\boldsymbol{r}'_{\|}\right) < 0\right) \times \left[\exp(ik_{\perp}z') + \overrightarrow{r_0}(k_{\perp})\exp(-ik_{\perp}z')\right] \times \left[\exp(ip_{\perp}z') + \overrightarrow{r_0}(p_{\perp})\exp(-ip_{\perp}z')\right], \tag{8.102}$$

if the roughness is a bump above[31] the average interface. The parameter ζ is a random variable with probability density distribution

$$P(\zeta) = \frac{1}{\sigma\sqrt{2\pi}} \exp(-\zeta^2/2\sigma^2), \tag{8.103}$$

where σ characterizes the average height of the roughnesses. Averaging both expressions over ζ, we obtain, according to Eq. (8.27), corrections to the specular reflectivity amplitude. It will be a combination of error function $\Phi(q\sigma)$, where $q = k_{\perp} \pm p_{\perp}$ or $q = k'_{\perp} + p'_{\perp}$.

For averaging of $|A(\boldsymbol{k}, \boldsymbol{p})|^2$, which depends on random variables ζ at two different points $\boldsymbol{r}'_{\|}$, we need the density distribution of the Gaussian process:

$$P(\zeta_1, \zeta_2) = \frac{1}{2\pi\sigma^2\sqrt{1 - K^2(\boldsymbol{r}_{\|})}} \exp\left(-\frac{\zeta_1^2 + \zeta_2^2 - 2\zeta_1\zeta_2 K(\boldsymbol{r}_{\|})}{2\sigma^2(1 - K^2(\boldsymbol{r}_{\|}))}\right), \tag{8.104}$$

where K is a correlation function, which depends on the distance $r_{\|}$ between two points, where $\zeta_{1,2}$ are defined.

Small roughnesses Though calculations with these formulas can be done up to the end without principal difficulties, we shall not proceed in this complicated way and shall simplify our task by assuming that the height, σ, of the roughnesses is sufficiently small [111], that is, $\sigma k_{\perp} \ll 1$. For small grazing angles this means that $\sigma \ll 100$ Å, which is quite practical. In that case we can take $A(\boldsymbol{k}, \boldsymbol{p})$ in the form

$$A(\boldsymbol{k}, \boldsymbol{p}) = -N_0 b \frac{i\,\overrightarrow{t_0}(p_{\perp})\overrightarrow{t_0}(k_{\perp})}{2\pi k_{\perp}} \int d^2 r'_{\|} \zeta(\boldsymbol{r}'_{\|}) \exp(i\boldsymbol{\kappa}\boldsymbol{r}') \tag{8.105}$$

for all positive and negative ζ. With this function we do not have corrections to $\overrightarrow{r_0}$ because $\langle\zeta\rangle = 0$. The scattered waves are determined by

$$\int d^2 r'_{1\|} \exp\left(i\boldsymbol{\kappa}_{\|}\boldsymbol{r}'_{1\|}\right) \int d^2 r'_{2\|} \exp\left(i\boldsymbol{\kappa}_{\|}\boldsymbol{r}'_{2\|}\right) \zeta\left(\boldsymbol{r}'_{1\|}\right) \zeta\left(\boldsymbol{r}'_{2\|}\right). \tag{8.106}$$

31) In our geometry, where the medium is at $z > 0$, the bump is below the interface.

Averaging of $\zeta\left(\boldsymbol{r}'_{1\|}\right)\zeta\left(\boldsymbol{r}'_{2\|}\right)$ over Eq. (8.104) gives

$$\left\langle\zeta\left(\boldsymbol{r}'_{1\|}\right)\zeta\left(\boldsymbol{r}'_{2\|}\right)\right\rangle = \sigma^2 K\left(\boldsymbol{r}'_{1\|} - \boldsymbol{r}'_{2\|}\right) . \tag{8.107}$$

Therefore,

$$\langle|A(\boldsymbol{k},\boldsymbol{p})|^2\rangle = S\frac{|N_0 b\sigma|^2}{(2\pi p_\perp)^2}\left|\vec{t_0}(p_\perp)\vec{t_0}(k_\perp)\right|^2 \int d^2 r'_{\|} \exp\left(i\boldsymbol{\kappa}_{\|}\boldsymbol{r}'_{\|}\right) K\left(r'_{\|}\right) , \tag{8.108}$$

and the indicatrix of nonspecular scattering is

$$w(\boldsymbol{\Omega}_0 \to \boldsymbol{\Omega}) = 2\pi\frac{|N_0 b\sigma|^2}{k_\perp}\left|\vec{t_0}(p_\perp)\vec{t_0}(k_\perp)\right|^2 \int d^2 r'_{\|} \exp\left(i\boldsymbol{\kappa}_{\|}\boldsymbol{r}'_{\|}\right) K\left(r'_{\|}\right) . \tag{8.109}$$

To finish the calculations we need to define the correlation function. It is natural to suppose that

$$K(\boldsymbol{r}_{\|}) = \exp\left(-r_{\|}^2/2l^2\right) , \tag{8.110}$$

where l is the correlation length, or the average dimension of the roughnesses along the interface. With this function we have

$$K_F(\boldsymbol{\kappa}_{\|}) = \int d^2 r'_{\|} \exp\left(i\boldsymbol{\kappa}_{\|}\boldsymbol{r}'_{\|}\right) \exp\left(-r_{\|}^2/2l^2\right) = 2\pi l^2 \exp\left(-l^2\boldsymbol{\kappa}_{\|}^2/2\right) , \tag{8.111}$$

and substitution into Eq. (8.109) gives

$$w(\boldsymbol{\Omega}_0 \to \boldsymbol{\Omega}) = |N_0 b l\sigma|^2\frac{(2\pi)^2}{k_\perp}\left|\vec{t_0}(p_\perp)\vec{t_0}(k_\perp)\right|^2 \exp\left(-l^2\boldsymbol{\kappa}_{\|}^2/2\right) . \tag{8.112}$$

Some authors [374, 375] use a different correlation function:

$$K(r_{\|}) = \exp\left(-[r_{\|}/l]^{2\mu}/2\right) , \tag{8.113}$$

where parameter μ is related to the so-called self-affine fractal structure of the rough surface. We do not think it helps provide a better understanding of roughnesses if we call them a fractal with dimension not 2, like any surface, but with dimension $3-\mu$. Though the mathematics related to fractals is very nice, we think that the parameter μ is only an additional fitting parameter. Its value does not clarify any physics. Therefore, to have the smallest number of such parameters it is sufficient to use a correlation function with fixed $\mu = 1$. We can easily show, for instance, that there is no difference whether $\mu = 1$ or $\mu = 0.5$. For the last case we have

$$K_F(\boldsymbol{\kappa}_{\|}) = \int d^2 r'_{\|} \exp\left(i\boldsymbol{\kappa}_{\|}r'_{\|}\right) \exp(-|r_{\|}|/l) = \frac{2\pi l^2}{\left(1 + l^2\boldsymbol{\kappa}_{\|}^2\right)^{3/2}} . \tag{8.114}$$

This function is different from a Gaussian; however, both correlation functions give almost identical distributions over $p_{\perp}$. Even unity instead of $K_F(\kappa_{\|})$ gives almost the same distribution in the plane of incidence. This is because the function $K_F(\kappa_{\|})$ controls broadening of the scattered distribution mainly in the direction perpendicular to the incidence plane.

However, let us look at what experiments show. Experiments on scattering from surface roughnesses were performed mainly with X-rays. For example, the results of measurements [376] of X-ray scattering from the surface of Au and Ag films evaporated under different conditions on a quartz substrate are shown in Figure 8.8. In the experiment the detector was fixed at a such a position that at the incident grazing angle θ it detected specularly reflected X-rays. The sample plate was rotated around an axis perpendicular to the plane of incidence by a small angle near θ. The dependence of detector count on incident grazing angle θ_{in} reveals a specular peak together with small-angle scattering around it, and two Yoneda peaks: one for $\theta_{\text{in}} = \theta_c$, and the other one at the reflected angle $2\theta - \theta_{\text{in}} = \theta_c$. We see that in the case of Au film evaporated on the substrate at low temperature ($T = 80$ K), the surface becomes so rough that the specular peak is absent. Only small-angle scattering is seen around the specular position, and two Yoneda peaks, which characterize the intensity of the scattering, are high and broad compared with the Yoneda peaks which are seen in the rocking curve of the less rough substrate surfaces. Parameters σ, μ, and ξ are obtained by fitting the experimental data. For instance, fitting of the Au(80) data in Figure 8.8 gave $\sigma = 8.5 \pm 0.8$ Å,

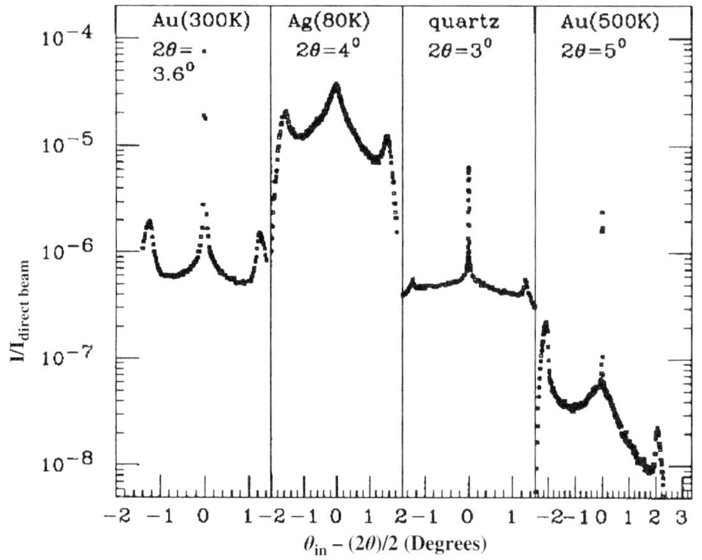

Figure 8.8 Rocking curve of different samples measured with X-rays in [376]. The detector was fixed at an angle 2θ with respect to the incident beam, and rotation of the sample changed the incident grazing angle θ_{in}.

$\mu = 0.46 \pm 0.1$, and $\xi = 1450 \pm 145$ Å. However, we cannot judge how good the fitting was because no χ^2 was reported.

With theoretical formula (8.112) we can reproduce Figure 8.8 for Au(500 K) if we describe the experimental reflectivity by the expression

$$g(x) = w(x_0 + x, x_0 - x) + |\vec{r_0}(x_0 + x)|^2 \Theta(-\delta + s < x < \delta + s)\,, \quad (8.115)$$

where

$$w(x, y) = \frac{c}{k_\perp} \left| \vec{t_0}(p_\perp)\vec{t_0}(k_\perp) \right|^2 K_F(|\kappa_\parallel| l)\,, \quad (8.116)$$

$c = 4 \cdot 10^{-8}$ is a normalization constant, $x_0 = 4$, $s = 1.5$, $\delta = 0.01$, and $|\kappa_\parallel|$ is replaced by

$$|\kappa_\parallel| = |k_x - p_x| = |\sqrt{k^2 - k_\perp^2} - \sqrt{k^2 - p_\perp^2}| \approx |k_\perp^2 - p_\perp^2|/2k\,.$$

The results of the calculations are shown in Figure 8.9. We demonstrated here results for $K_F = 1$ and two functions, $K_F(\kappa_\parallel) = \exp(-\xi(k_\perp^2 - p_\perp^2)^2)$ and $K_F(\kappa_\parallel) = 1/[1 + \xi(k_\perp^2 - p_\perp^2)^2]^{-3/2}$, with the same $\xi = 8 \cdot 10^{-4}$.

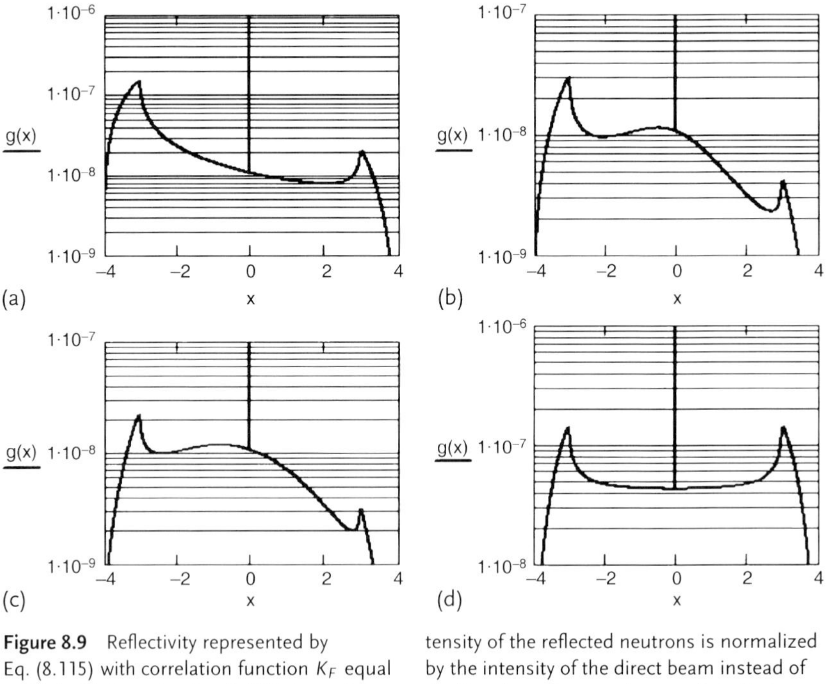

Figure 8.9 Reflectivity represented by Eq. (8.115) with correlation function K_F equal to (a) 1; (b) Gaussian Eq. (8.111) ($\mu = 1$ in Eq. (8.113)); (c) Eq. (8.114); (d) symmetrical result obtained with $K_F = 1$. This result is obtained if the experimentally measured intensity of the reflected neutrons is normalized by the intensity of the direct beam instead of the flux along the sample normal. In that case one has to exclude $k_\perp$ in the denominator of (8.116).

The experimental data for quartz in Figure 8.8 look symmetrical. This means that the measured intensity was normalized to the full intensity of the incident beam instead of its flux along the normal to the sample. This means that in Eq. (8.116) we have to omit $k_\perp$ in the denominator. The result is shown in Figure 8.9d.

8.2.6.2 Roughness of Interfaces in Multilayers

There are so many options for how to treat roughness of interfaces in multilayers that to fit experimental data on specular reflectivity with many parameters related to roughness and inhomogeneities is almost useless. Fitting becomes reasonable only after measurement of the nonspecular distribution when it shows some particular features related to special kind of imperfectness. So below we present some formulas for the effect of roughnesses at interfaces of a multilayer system, but will not try to apply them to improve χ^2 in the fitting of reflectivity of the multilayers considered in Chapter 4.

Suppose that we have a periodic system composed of N bilayers over a substrate. Every bilayer consists of a barrier layer with higher potential and a well layer with lower potential. Let us suppose that all the interfaces have similar roughnesses which do not correlate with each other. The scattered wave function created by roughnesses on the upper interface of the nth bilayer can be represented similarly to Eqs. (8.99) and (8.105):

$$\Psi_s(\boldsymbol{k}, \boldsymbol{r}) = \int d^2 p_\parallel \exp(i\boldsymbol{p}\boldsymbol{r}) A(\boldsymbol{k}, \boldsymbol{p}) , \tag{8.117}$$

where

$$A(\boldsymbol{k}, \boldsymbol{p}) = -N_0 b \overleftarrow{T}_{N-n} \frac{i \overrightarrow{T}_{n,s}(p_\perp) \overrightarrow{T}_{n,s}(k_\perp)}{2\pi p_\perp} \int d^2 r'_\parallel \zeta(\boldsymbol{r}'_\parallel) \exp(i\boldsymbol{\kappa}\boldsymbol{r}') , \tag{8.118}$$

where $\overrightarrow{T}_{n,s}(p_\perp)$ denotes the transmission amplitude of n bilayers and the substrate and $\overleftarrow{T}_{N-n}$ denotes the transmission of the remaining $N-n$ bilayers back to the vacuum.

To take into account the roughnesses on the well layer in the middle of the nth bilayer, we use the expression

$$A(\boldsymbol{k}, \boldsymbol{p}) = -N_0 b \overleftarrow{T}_{N-n,b} \frac{i \overrightarrow{T}_{w,n-1,s}(p_\perp) \overrightarrow{T}_{w,n-1,s}(k_\perp)}{2\pi k_\perp} \int d^2 r'_\parallel \zeta(\boldsymbol{r}'_\parallel) \exp\left(i\boldsymbol{\kappa}\boldsymbol{r}'\right) , \tag{8.119}$$

where

$$\overrightarrow{T}_{w,n-1,s} = \frac{t_w \overrightarrow{T}_{n-1,s}}{1 - r_w \overrightarrow{R}_{n-1,s}} , \qquad \overleftarrow{T}_{N-n,b} = \frac{\overleftarrow{T}_{N-n} t_b}{1 - r_b \overleftarrow{R}_{N-n}} . \tag{8.120}$$

Now we need to sum all that over n from 0 to N, square the modulus, and independently average over distribution (8.104). As a result we get

$$w(\boldsymbol{\Omega}_0 \to \boldsymbol{\Omega}) = 2\pi \frac{|N_0 b\sigma|^2}{k_\perp} K_F(\kappa_\parallel) \left\{ \sum_{n=0}^{N} \left| \overleftarrow{T}_{N-n} \overrightarrow{T}_{n,s}(p_\perp) \overrightarrow{T}_{n,s}(k_\perp) \right|^2 \right.$$
$$\left. + \sum_{n=1}^{N} \left| \overleftarrow{T}_{N-n,b} \overrightarrow{T}_{w,n-1,s}(p_\perp) \overrightarrow{T}_{w,n-1,s}(k_\perp) \right|^2 \right\}, \qquad (8.121)$$

which for a given multilayer structure can be compared with the measured distribution of nonspecularly reflected intensity.

8.3
Scattering on Magnetic Inhomogeneities

In Chapter 5 we studied neutron scattering mainly in homogeneous magnetic fields with some interfaces, now we shall consider scattering on magnetic inhomogeneities, when the field $\boldsymbol{B}(\boldsymbol{r})$ varies in space. According to the Maxwell theory the field should satisfy the equation

$$\nabla \cdot \boldsymbol{B}(\boldsymbol{r}) = 0\,, \qquad (8.122)$$

and its Fourier image $\boldsymbol{B}_F(\boldsymbol{\kappa}) = \int \boldsymbol{B}(\boldsymbol{r}) \exp(i\boldsymbol{\kappa}\cdot\boldsymbol{r}) d^3r$ must be tangential: $\boldsymbol{\kappa}\cdot\boldsymbol{B}_F(\boldsymbol{\kappa}) = 0$. This means that the field $\boldsymbol{B}_F(\boldsymbol{\kappa})$ is representable as $\boldsymbol{B}_F(\boldsymbol{\kappa}) = \boldsymbol{h}(\boldsymbol{\kappa}) - \boldsymbol{\kappa}(\boldsymbol{\kappa}\cdot\boldsymbol{h}(\boldsymbol{\kappa}))/\kappa^2$ via some other vector, $\boldsymbol{h}(\boldsymbol{\kappa})$, which is not orthogonal to $\boldsymbol{\kappa}$. Our task is to present the theory of the distorted wave Born approximation for scattering on magnetic inhomogeneities inside media.

Consider a semi-infinite wall at $z > 0$ with nuclear optical potential u_0, uniform magnetic field $\boldsymbol{B}_i$, and magnetic inhomogeneities $\boldsymbol{B}(\boldsymbol{r})$. Outside the wall there is a uniform magnetic field $\boldsymbol{B}_o$ which is noncollinear to $\boldsymbol{B}_i$. The lowercase letters $\boldsymbol{b}_0$, $\boldsymbol{b}_i$, and $\boldsymbol{b}$ will denote unit vectors along the respective fields.

The Schrödinger equation for a spinor wave function $|\hat{\Psi}(\boldsymbol{r})\rangle$ is

$$\left(\Delta + k^2 - \Theta(z<0)\boldsymbol{\sigma}\cdot\boldsymbol{B}_o - \Theta(z>0)[u_0 + \boldsymbol{\sigma}\cdot\boldsymbol{B}_i + \boldsymbol{\sigma}\cdot\boldsymbol{B}(\boldsymbol{r})]\right)|\hat{\Psi}(\boldsymbol{r})\rangle$$
$$= 0\,. \qquad (8.123)$$

We have included the factor 2, which was always present before magnetic fields in Chapter 5, in the definition of the magnetic fields.

The solution of Eq. (8.123) can be represented as $|\hat{\Psi}\rangle = |\hat{\psi}_0\rangle - |\hat{\Psi}_s\rangle$, where $|\hat{\psi}_0\rangle$ is a nonperturbed spinor wave function satisfying the equation

$$\left[\Delta + k^2 - \Theta(z<0)\boldsymbol{\sigma}\cdot\boldsymbol{B}_o - \Theta(z>0)(u_0 + \boldsymbol{\sigma}\cdot\boldsymbol{B}_i)\right]|\hat{\psi}_0(\boldsymbol{r})\rangle = 0\,, \qquad (8.124)$$

and $|\hat{\Psi}_s\rangle$ is the spinor state describing scattering on the inhomogeneity. The scattered wave is calculated with the help of perturbation theory:

$$|\hat{\Psi}_s(\boldsymbol{r})\rangle = \int \hat{G}(\boldsymbol{r}, \boldsymbol{r}')\boldsymbol{\sigma} \cdot \boldsymbol{B}(\boldsymbol{r}')|\hat{\psi}_0(\boldsymbol{r}')\rangle d^3 r' , \tag{8.125}$$

where $\hat{G}(\boldsymbol{r}, \boldsymbol{r}')$ is the spinor Green function. It is 2×2 matrix which satisfies the equation

$$\left[\Delta + k^2 - \Theta(z < 0)\boldsymbol{\sigma} \cdot \boldsymbol{B}_o - \Theta(z > 0)(u_0 + \boldsymbol{\sigma} \cdot \boldsymbol{B}_i)\right] \hat{G}(\boldsymbol{r}, \boldsymbol{r}') = \hat{\boldsymbol{I}}\delta(\boldsymbol{r} - \boldsymbol{r}') , \tag{8.126}$$

with the unit matrix $\hat{\boldsymbol{I}}$ at the right.

Let us note that the Green function $\hat{G}(\boldsymbol{r}, \boldsymbol{r}')$ is symmetrical with respect to permutation of $\boldsymbol{r}$ and $\boldsymbol{r}'$, and the equation for it can be represented as

$$\hat{G}(\boldsymbol{r}', \boldsymbol{r})\left[\Delta + k^2 - \Theta(z < 0)\boldsymbol{\sigma} \cdot \boldsymbol{B}_o - \Theta(z > 0)(u_0 + \boldsymbol{\sigma} \cdot \boldsymbol{B}_i)\right] = \hat{\boldsymbol{I}}\delta(\boldsymbol{r} - \boldsymbol{r}') , \tag{8.127}$$

where all the operators act to the left.

8.3.1
The Spinor Green Function

Our task now is to construct such a matrix Green function which is causal and satisfies Eq. (8.126). Because of homogeneity of the space along the interface (x, y-plane) the Green function can be represented like in Eq. (8.44) as

$$\hat{G}(\boldsymbol{r}, \boldsymbol{r}') = \int \frac{d^2 p_\parallel}{(2\pi)^2} \exp\left(i\boldsymbol{p}_\parallel \cdot (\boldsymbol{r} - \boldsymbol{r}')_\parallel\right) \hat{G}(p_\perp, z, z') , \tag{8.128}$$

where $\hat{G}(p_\perp, z, z')$ satisfies

$$\left[\frac{d^2}{dz^2} + p_\perp^2 - \Theta(z < 0)\boldsymbol{\sigma} \cdot \boldsymbol{B}_o - \Theta(z > 0)(u_0 + \boldsymbol{\sigma} \cdot \boldsymbol{B}_i)\right] \times \hat{G}(p_\perp, z, z') = \hat{\boldsymbol{I}}\delta(z - z') , \tag{8.129}$$

and $p_\perp = \sqrt{k^2 - p_\parallel^2}$.

In the scalar case this function is constructed with two linearly independent solutions, $f_{1,2}(z)$, of the homogeneous equation like in Eq. (8.49):

$$G(z, z') = \frac{1}{w}\left(\Theta(z > z') f_1(z) f_2(z') + \Theta(z < z') f_2(z) f_1(z')\right) ,$$

where w is the Wronskian $w = f_1'(z) f_2(z) - f_1(z) f_2'(z)$. In the case of the spinor Green function we should act similarly. Since $\hat{\boldsymbol{G}}$ is a 2×2 matrix it should be constructed from matrix spinor wave functions. The first of them is found in Eq. (5.17):

$$\hat{f}_1(q, z) = \Theta(z < 0)[\exp(i\hat{q}_o z) + \exp(-i\hat{q}_o z)\hat{\rho}] + \Theta(z > 0)\exp(i\hat{q}_i z)\hat{\tau} , \tag{8.130}$$

where

$$\hat{\boldsymbol{\rho}} = (\hat{\boldsymbol{q}}_o + \hat{\boldsymbol{q}}_i)^{-1}(\hat{\boldsymbol{q}}_o - \hat{\boldsymbol{q}}_i), \quad \hat{\boldsymbol{\tau}} = \hat{\boldsymbol{I}} + \hat{\boldsymbol{\rho}},$$
$$\hat{\boldsymbol{q}}_i = \sqrt{q^2 - u_0 - \boldsymbol{\sigma}\cdot\boldsymbol{B}_i}, \quad \hat{\boldsymbol{q}}_o = \sqrt{q^2 - \boldsymbol{\sigma}\cdot\boldsymbol{B}_o}. \tag{8.131}$$

The spinor wave function $|\psi_0(k_\perp, z)\rangle$ is obtained from Eq. (8.130) by multiplication on the right by an arbitrary spinor $|\xi\rangle$:

$$|\psi_0(k_\perp, z)\rangle = \hat{\boldsymbol{f}}_1(k_\perp, z)|\xi\rangle.$$

The second solution can be the matrix

$$\hat{\boldsymbol{f}}_2(q,z) = \Theta(z<0)\exp(-i\hat{\boldsymbol{q}}_o z)\hat{\boldsymbol{\tau}}' + \Theta(z>0)\left(\exp(-i\hat{\boldsymbol{q}}_i z) + \exp(i\hat{\boldsymbol{q}}_i z)\hat{\boldsymbol{\rho}}'\right), \tag{8.132}$$

where

$$\hat{\boldsymbol{\rho}}' = -\hat{\boldsymbol{\rho}}, \quad \hat{\boldsymbol{\tau}}' = \hat{\boldsymbol{I}} + \hat{\boldsymbol{\rho}}' = \hat{\boldsymbol{I}} - \hat{\boldsymbol{\rho}}. \tag{8.133}$$

However we cannot simply multiply these functions because simple multiplication will not give us the Green function which satisfies Eq. (8.127). The appropriate Green function can be guessed in the form

$$\hat{\boldsymbol{G}}(q,z,z') = \Theta(z>z')\hat{\boldsymbol{f}}_1(q,z)\hat{\boldsymbol{w}}^{-1}\hat{\boldsymbol{f}}_2^J(q,z')$$
$$+\,\Theta(z<z')\hat{\boldsymbol{f}}_2(q,z)\hat{\boldsymbol{w}}^{-1}\hat{\boldsymbol{f}}_1^J(q,z'), \tag{8.134}$$

the functions with upper index J satisfy the right-side Schrödinger equation

$$\hat{\boldsymbol{f}}_{1,2}^J(q,z)\left[\frac{\overleftarrow{d}^2}{dz^2} + q^2 - \Theta(z<0)\boldsymbol{\sigma}\cdot\boldsymbol{B}_o - \Theta(z>0)(u_0 + \boldsymbol{\sigma}\cdot\boldsymbol{B}_i)\right] = 0. \tag{8.135}$$

In this equation all the operators act from the right to the left. It is the Japanese way of writing texts, and because of that the index J is introduced.

The solutions of Eq. (8.135) are

$$\hat{\boldsymbol{f}}_1^J(q,z) = \Theta(z<0)\left(\exp(i\hat{\boldsymbol{q}}_o z) + \hat{\boldsymbol{\rho}}^J\exp(-i\hat{\boldsymbol{q}}_o z)\right)$$
$$+\,\Theta(z>0)[\hat{\boldsymbol{I}} + \hat{\boldsymbol{\rho}}^J]\exp(i\hat{\boldsymbol{q}}_i z), \tag{8.136}$$

$$\hat{\boldsymbol{f}}_2^J(q,z) = \Theta(z<0)[\hat{\boldsymbol{I}} - \hat{\boldsymbol{\rho}}^J]\exp(-i\hat{\boldsymbol{q}}_o z)$$
$$+\,\Theta(z>0)\left(\exp(-i\hat{\boldsymbol{q}}_i z) - \hat{\boldsymbol{\rho}}^J\exp(i\hat{\boldsymbol{q}}_i z)\right), \tag{8.137}$$

where $\hat{\boldsymbol{\rho}}^J = (\hat{\boldsymbol{q}}_o - \hat{\boldsymbol{q}}_i)(\hat{\boldsymbol{q}}_o + \hat{\boldsymbol{q}}_i)^{-1}$.

The matrix $\hat{w}^{-1}$ in Eq. (8.134) is the Wronskian matrix like in the scalar case, and we can guess that it is

$$\hat{w}^{-1} = \frac{1}{2i\hat{q}_o}[\hat{I} - \hat{\rho}^J]^{-1} = \frac{1}{4i}\left(\hat{q}^{-1} + \hat{q}'^{-1}\right) . \tag{8.138}$$

Let us check that the matrix (8.134) constructed in such a way does really satisfy Eq. (8.129). Substitute Eq. (8.134) into Eq. (8.129). The left-hand side becomes $\delta(z-z')\hat{Q}(z)$, where for $z < 0$

$$\begin{aligned}\hat{Q}(z) &= \left(\frac{d}{dz}\hat{f}_1(q,z)\right)\hat{w}^{-1}\hat{f}_2^J(q,z) - \left(\frac{d}{dz}\hat{f}_2(q,z)\right)\hat{w}^{-1}\hat{f}_1^J(q,z)\\ &= (\exp(i\hat{q}_o z)\hat{q}_o - \exp(-i\hat{q}_o z)\hat{q}_o\hat{\rho})\frac{1}{2\hat{q}_o}[\hat{I} - \hat{\rho}^J]^{-1}[\hat{I} - \hat{\rho}^J]\exp(-i\hat{q}_o z)\\ &+ \exp(-i\hat{q}_o z)\hat{q}_o[\hat{I} - \hat{\rho}]\frac{1}{2\hat{q}_o}[\hat{I} - \hat{\rho}^J]^{-1}\left(\exp(i\hat{q}_o z) + \hat{\rho}^J\exp(-i\hat{q}_o z)\right) .\end{aligned} \tag{8.139}$$

The second line Eq. (8.139) is reduced to

$$\begin{aligned}&\frac{1}{2}\left(\hat{I} - \exp(-i\hat{q}_o z)\hat{q}_o\hat{\rho}\hat{q}_o^{-1}\exp(-i\hat{q}_o z)\right) =\\ &\frac{1}{2}\left(\hat{I} - \exp(-i\hat{q}_o z)\hat{\rho}^J\exp(-i\hat{q}_o z)\right) ,\end{aligned} \tag{8.140}$$

where in the right-hand side we used the relation

$$\begin{aligned}\hat{q}_o\hat{\rho}\hat{q}_o^{-1} &= \hat{q}_o\left[\left(\hat{I} + \hat{q}_i\hat{q}_o^{-1}\right)\hat{q}_o\right]^{-1}\left(\hat{I} - \hat{q}_i\hat{q}_o^{-1}\right)\\ &= \left(\hat{I} - \hat{q}_i\hat{q}_o^{-1}\right)\left(\hat{I} + \hat{q}_i\hat{q}_o^{-1}\right)^{-1}\\ &= (\hat{q}_o - \hat{q}_i)(\hat{q}_o + \hat{q}_i)^{-1} = \hat{\rho}^J .\end{aligned} \tag{8.141}$$

of (8.141), the third line Eq. (8.139) becomes $\left(\hat{I} + \exp(-i\hat{q}_o z)\hat{\rho}^J\exp(-i\hat{q}_o z)\right)/2$; therefore the sum of Eqs. (8.139) and (8.139) is $\hat{I}$, and the function (8.134) satisfies the Eq. (8.129).

We considered only the case $z < 0$. It is a good exercise to check that the same result will be obtained when $z > 0$.

8.3.2
The Scattered Wave Function

Suppose that all the magnetic inhomogeneities are inside the medium $z > 0$. The matrix part of the state (8.125) in the vacuum $z < 0$ after scattering can be represented as

$$\hat{\Psi}_{\rm sc}(\boldsymbol{r}) = -i\int\frac{d^2p_{\parallel}}{(2\pi)^2}\exp(i\boldsymbol{p}_{\parallel}\boldsymbol{r}_{\parallel})\exp(i\hat{\boldsymbol{p}}_{\perp}z)(\hat{\boldsymbol{p}}_{\perp}+\hat{\boldsymbol{p}}'_{\perp})^{-1}\int\exp(i\boldsymbol{\kappa}_{\parallel}\boldsymbol{r}'_{\parallel})d^3r'$$
$$\times\exp\left(i\hat{\boldsymbol{p}}'_{\perp}z'\right)\boldsymbol{\sigma}\cdot\boldsymbol{B}(\boldsymbol{r}')\exp\left(i\hat{\boldsymbol{k}}'_{\perp}z'\right)\left(\hat{\boldsymbol{k}}_{\perp}+\hat{\boldsymbol{k}}'_{\perp}\right)^{-1}2\hat{\boldsymbol{k}}_{\perp}$$
$$= -i\int\frac{d^2p_{\parallel}}{(2\pi)^2}\exp(i\boldsymbol{p}_{\parallel}\boldsymbol{r}_{\parallel})\hat{\boldsymbol{M}}(p_{\perp},k_{\perp})\,, \tag{8.142}$$

where $\boldsymbol{\kappa}_{\parallel} = \boldsymbol{k}_{\parallel} - \boldsymbol{p}_{\parallel}$, $\hat{\boldsymbol{p}}_{\perp} = \sqrt{p_{\perp}^2 - \boldsymbol{\sigma}\cdot\boldsymbol{B}_o}$, $\hat{\boldsymbol{k}}_{\perp} = \sqrt{k_{\perp}^2 - \boldsymbol{\sigma}\cdot\boldsymbol{B}_o}$, $\hat{\boldsymbol{p}}'_{\perp} = \sqrt{p_{\perp}^2 - u_0 - \boldsymbol{\sigma}\cdot\boldsymbol{B}_i}$, $\hat{\boldsymbol{k}}'_{\perp} = \sqrt{k_{\perp}^2 - u_0 - \boldsymbol{\sigma}\cdot\boldsymbol{B}_i}$, $\hat{\boldsymbol{\kappa}} = \boldsymbol{\kappa}_{\parallel} + \boldsymbol{n}\left(\hat{\boldsymbol{p}}'_{\perp} + \hat{\boldsymbol{k}}'_{\perp}\right)$, $\boldsymbol{n}$ is a unit vector directed along the z-axis, and we introduced matrix $\hat{\boldsymbol{M}}(p_{\perp},k_{\perp})$, which is the product of three other matrices

$$\hat{\boldsymbol{M}}(p_{\perp},k_{\perp}) = \hat{\boldsymbol{M}}_1(p_{\perp})\hat{\boldsymbol{M}}_2(\hat{\boldsymbol{\kappa}})\hat{\boldsymbol{M}}_3(k_{\perp})\,,$$
$$\hat{\boldsymbol{M}}_1(p_{\perp}) = \exp\left(i\hat{\boldsymbol{p}}_{\perp}z\right)\left(\hat{\boldsymbol{p}}_{\perp}+\hat{\boldsymbol{p}}'_{\perp}\right)^{-1}\,, \tag{8.143}$$

$$\hat{\boldsymbol{M}}_2(\hat{\boldsymbol{\kappa}}) = \boldsymbol{\sigma}\boldsymbol{B}_F(\hat{\boldsymbol{\kappa}}) = \int\exp\left(i\boldsymbol{\kappa}_{\parallel}\boldsymbol{r}'_{\parallel}\right)d^3r'\exp\left(i\hat{\boldsymbol{p}}'_{\perp}z'\right)\boldsymbol{\sigma}\cdot\boldsymbol{B}(\boldsymbol{r}')\exp\left(i\hat{\boldsymbol{k}}'_{\perp}z'\right)\,, \tag{8.144}$$

$$\hat{\boldsymbol{M}}_3(k_{\perp}) = \left(\hat{\boldsymbol{k}}_{\perp}+\hat{\boldsymbol{k}}'_{\perp}\right)^{-1}2\hat{\boldsymbol{k}}_{\perp}\,. \tag{8.145}$$

8.3.2.1 Scattering on a Helicoid in Matter

To obtain a final result with these formulas, let us consider the scattering on a helicoid of radius a and length l inside a magnetic mirror. Its field is $\boldsymbol{B} = b_h\Theta(\rho < a)(\cos(2qz), \sin(2qz), 0)$, where $\rho = \sqrt{x^2+y^2}$. In that case

$$\hat{\boldsymbol{M}}_2(\hat{\boldsymbol{\kappa}}) = 2\pi(a/\kappa_{\parallel})b_hJ_1(\kappa_{\parallel}a)\int dz'\exp\left(i\hat{\boldsymbol{p}}'_{\perp}z'\right)\exp\left(-iq\sigma_zz'\right)\sigma_x$$
$$\times\exp\left(iq\sigma_zz'\right)\exp\left(i\hat{\boldsymbol{k}}'_{\perp}z'\right)$$
$$= 2\pi(a/\kappa_{\parallel})b_hJ_1(\kappa_{\parallel}a)\sigma_x\frac{I-\exp(iQ(\sigma_z)l)}{iQ(\sigma_z)}\,, \tag{8.146}$$

where $J_1(x)$ is the Bessel function and

$$Q(\sigma_z) = \sqrt{k_{0\perp}^2 - u - B_i\sigma_z} + \sqrt{p_{\perp}^2 - u + B_i\sigma_z} + 2q\sigma_z\,. \tag{8.147}$$

The matrix (8.146) has only nondiagonal matrix elements:

$$M_{2\mp\pm} = \langle z_{\mp}|\hat{\boldsymbol{M}}_2|z_{\pm}\rangle = 2\pi\frac{a}{\kappa_{\parallel}}b_hJ_1(\kappa_{\parallel}a)\frac{I-\exp(iQ(\pm1)l)}{Q(\pm1)}$$
$$= 2\pi i\frac{a}{\kappa_{\parallel}}b_hJ_1(\kappa_{\parallel}a)e^{iQ(\pm1)l/2}\frac{\sin(Q(\pm1)l/2)}{Q(\pm1)/2}\,, \tag{8.148}$$

where $|z_\pm\rangle$ are eigenspinors of matrix σ_z: $\sigma_z|z_\pm\rangle = \pm|z_\pm\rangle$. It is seen that the most important is the matrix element M_{2+-}, because its denominator is

$$Q(-1) = \sqrt{k_{0\perp}^2 - u + B_i} + \sqrt{p_\perp^2 - u - B_i} - 2q\,. \qquad (8.149)$$

To calculate the scattering amplitudes with and without spin flip with respect to the external field $\boldsymbol{B}_o$, we need to find the matrix elements of the matrix $\hat{\boldsymbol{M}}$ between states

$$|\boldsymbol{b}_{o\pm}\rangle = \frac{\hat{\boldsymbol{I}} \pm \boldsymbol{\sigma b}_o}{\sqrt{2(1 \pm b_{oz})}}|z_+\rangle\,, \qquad (8.150)$$

where $\boldsymbol{b}_o = \boldsymbol{B}_o/B_o$ is the unit vector along $\boldsymbol{B}_o$. With account of Eq. (8.148), the matrix elements of the matrix $\hat{\boldsymbol{M}}$ between any initial, $|\boldsymbol{b}_{oi}\rangle$, and final, $|\boldsymbol{b}_{of}\rangle$, states can be represented as a sum of two terms:

$$\langle \boldsymbol{b}_{of}|\hat{\boldsymbol{M}}|\boldsymbol{b}_{oi}\rangle = \langle \boldsymbol{b}_{of}|\hat{\boldsymbol{M}}_1|z_+\rangle M_{2+-}\langle z_-|M_3|\boldsymbol{b}_{oi}\rangle + \langle \boldsymbol{b}_{of}|\hat{\boldsymbol{M}}_1|z_-\rangle M_{2-+}\langle z_+|M_3|\boldsymbol{b}_{oi}\rangle\,. \qquad (8.151)$$

It is possible to calculate them rigorously for any scattering parameters; however the most interesting case is when $\boldsymbol{b}_o$ is parallel to the z-axis and when $Q(-1)$ in Eq. (8.149) is small. In that case we can retain in Eq. (8.151) only the term M_{2+-}, and we see that the scattering is accompanied by spin flip from state $|z_-\rangle$ to state $|z_+\rangle$. Therefore, the scattering amplitude is proportional to the term $\boldsymbol{\sigma q}$, which violates time parity conservation.

8.3.2.2 General Calculation of Scattering Amplitude on an Inhomogeneity

We can easily calculate all the matrix elements of matrix (8.142) with the help of the matrix elements of matrices (8.143)–(8.145). To do that we denote spin states $|\boldsymbol{b}_{o\pm}\rangle$ with respect to $\boldsymbol{B}_o$ as $|n\rangle$, with $n = 0, 1$, respectively. The spin states $|\boldsymbol{b}_{i\pm}\rangle$ with respect to $\boldsymbol{B}_i$ we denote as $|n'\rangle$, with $n' = 0, 1$, respectively. Then we insert the unit matrix $\hat{\boldsymbol{I}} = \sum_{n'}|n'\rangle\langle n'|$ between matrices $\hat{\boldsymbol{M}}_1$, $\hat{\boldsymbol{M}}_2$ and $\hat{\boldsymbol{M}}_2$, $\hat{\boldsymbol{M}}_3$. After that matrix elements of the matrix product between the initial n_i and final n_f states outside the medium are

$$\hat{\boldsymbol{M}}_{fi} \equiv \langle n_f|\hat{\boldsymbol{M}}(p_\perp, k_\perp)|n_i\rangle = \sum_{n',m'}\langle n_f|\hat{\boldsymbol{M}}_1(p_\perp)|n'\rangle\langle n'|\hat{\boldsymbol{M}}_2(\hat{\kappa})|m'\rangle\langle m'|\hat{\boldsymbol{M}}_3(k_\perp)|n_i\rangle\,. \qquad (8.152)$$

Matrix elements of $\hat{M}_1$ To calculate the matrix elements $\langle n_f|\hat{\boldsymbol{M}}_1(p_\perp)|n'\rangle$ and $\langle m'|\hat{\boldsymbol{M}}_3(k_\perp)|n_i\rangle$ we write (see Eq. (5.41))

$$\left(\hat{\boldsymbol{p}}_\perp + \hat{\boldsymbol{p}}'_\perp\right)^{-1} \equiv \left(\hat{\boldsymbol{p}}_\perp(\boldsymbol{\sigma}\cdot\boldsymbol{B}_o) + \hat{\boldsymbol{p}}'_\perp(\boldsymbol{\sigma}\cdot\boldsymbol{B}_i)\right)^{-1} = \frac{1}{N(p_\perp)}\left(\hat{\boldsymbol{p}}_\perp(-\boldsymbol{\sigma}\cdot\boldsymbol{B}_o) + \hat{\boldsymbol{p}}'_\perp(-\boldsymbol{\sigma}\cdot\boldsymbol{B}_i)\right)\,,$$

where

$$N(p_\perp) = \left(\hat{p}_\perp(-\boldsymbol{\sigma}\cdot\boldsymbol{B}_o) + \hat{p}'_\perp(-\boldsymbol{\sigma}\cdot\boldsymbol{B}_i)\right)\left(\hat{p}_\perp(\boldsymbol{\sigma}\cdot\boldsymbol{B}_o) + \hat{p}'_\perp(\boldsymbol{\sigma}\cdot\boldsymbol{B}_i)\right)$$
$$= \left(p_\perp(B_o) + p'_\perp(B_i)\right)\left(p_\perp(-B_o) + p'_\perp(-B_i)\right)$$
$$+ p_\perp^- p_\perp'^-[1 - \boldsymbol{b}_o\cdot\boldsymbol{b}_i]/2\,, \tag{8.153}$$

$p_\perp^- = \sqrt{p_\perp^2 - B_o} - \sqrt{p_\perp^2 + B_o}$, and $p_\perp'^- = \sqrt{p_\perp^2 - u - B_i} - \sqrt{p_\perp^2 - u + B_i}$. After these transformations we obtain

$$\langle n_f|\hat{\boldsymbol{M}}_1(p_\perp)|n'\rangle = \frac{p_\perp(-(-1)^{n_f}B_o) + p'_\perp(-(-1)^{n'}B_i)}{N(p_\perp)} \times \exp\left(i\hat{p}_\perp((-1)^{n_f}B_o)z\right)\langle n_f||n'\rangle\,, \tag{8.154}$$

where $p_\perp(x) = \sqrt{p_\perp^2 - x}$, $p'_\perp(x) = \sqrt{p_\perp^2 - u - x}$, and we need to find the product $\langle n_f||n'\rangle$. To calculate it we take the quantization axis along $\boldsymbol{B}_o$ and introduce unit vectors $\boldsymbol{b}_o = \boldsymbol{B}_o/B_o$ and $\boldsymbol{b}_i = \boldsymbol{B}_i/B_i$. The state $|n'\rangle$ can be represented as

$$|\boldsymbol{n}'\rangle = \frac{\hat{\boldsymbol{I}} + (-1)^{n'}\boldsymbol{\sigma}\boldsymbol{b}_i}{\sqrt{2[1 + (-1)^{n'}(\boldsymbol{b}_i\cdot\boldsymbol{b}_0)]}}|0\rangle\,, \tag{8.155}$$

where state $|0\rangle$ is an eigenstate of $(\boldsymbol{\sigma}\cdot\boldsymbol{b}_o)$: $(\boldsymbol{\sigma}\cdot\boldsymbol{b}_o)|0\rangle = |0\rangle$. With this notation we have

$$\langle n_f||n'\rangle = \frac{\delta_{n_f,0}[1 + (-1)^{n'}(\boldsymbol{b}_i\cdot\boldsymbol{b}_o)] + \delta_{n_f,1}(-1)^{n'}[\boldsymbol{b}_o\times[\boldsymbol{b}_i\times\boldsymbol{b}_o]]}{\sqrt{2\left[1 + (-1)^{n'}(\boldsymbol{b}_i\cdot\boldsymbol{b}_0)\right]}}\,. \tag{8.156}$$

Matrix elements of $\hat{M}_3$ A similar result is obtained for matrix elements of $\hat{\boldsymbol{M}}_3$:

$$\langle m'|\hat{\boldsymbol{M}}_3(k_\perp)|n_i\rangle = \frac{k_\perp(-(-1)^{n_i}B_o) + k'_\perp(-(-1)^{m'}B_i)}{N(k_\perp)} \times 2k_\perp((-1)^{n_i}B_o)\langle m'||n_i\rangle\,, \tag{8.157}$$

where

$$\langle m'||n_i\rangle = \frac{\delta_{n_i,0}\left[1 + (-1)^{m'}(\boldsymbol{b}_i\cdot\boldsymbol{b}_o)\right] + \delta_{n_i,1}(-1)^{m'}[\boldsymbol{b}_o\times[\boldsymbol{b}_i\times\boldsymbol{b}_o]]}{\sqrt{2\left[1 + (-1)^{m'}(\boldsymbol{b}_i\cdot\boldsymbol{b}_0)\right]}}\,. \tag{8.158}$$

Matrix elements of $\hat{M}_2$ The middle matrix elements are

$$\langle n'|\hat{\boldsymbol{M}}_2|m'\rangle = \langle n'|\boldsymbol{\sigma}\cdot\boldsymbol{B}_F(\boldsymbol{\kappa}_{m',n'})|m'\rangle\,, \tag{8.159}$$

where

$$\boldsymbol{B}_F(\boldsymbol{\kappa}_{m',n'}) = \int d^3r' \boldsymbol{B}(\boldsymbol{r}') \exp(i\boldsymbol{\kappa}_{m',n'}\boldsymbol{r}') , \tag{8.160}$$

$$\boldsymbol{\kappa}_{m',n'} = \boldsymbol{\kappa}_\parallel + \boldsymbol{n}\left(p'_\perp((-1)^{n'}) + k'_\perp((-1)^{m'})\right) , \quad \boldsymbol{\kappa}_\parallel = \boldsymbol{k}_\parallel - \boldsymbol{p}_\parallel . \tag{8.161}$$

After substitution of Eq. (8.155) into Eq. (8.159) we get

$$\begin{aligned}\langle n'|\boldsymbol{\sigma}\cdot\boldsymbol{B}_F(\hat{\boldsymbol{\kappa}}_{m',n'})|m'\rangle &= \delta_{m',n'}(-1)^{n'}(\boldsymbol{b}_i\cdot\boldsymbol{B}_F(\boldsymbol{\kappa}_{m',n'})) + (1-\delta_{m',n'})\\ &\times \frac{(\boldsymbol{b}_o\cdot[\boldsymbol{b}_i\times[\boldsymbol{B}_F(\boldsymbol{\kappa}_{m',n'})\times\boldsymbol{b}_i]]) + i(-1)^{n'}(\boldsymbol{b}_o\cdot[\boldsymbol{b}_i\times\boldsymbol{B}_F(\boldsymbol{\kappa}_{m',n'})])}{\sqrt{1-(\boldsymbol{b}_i\cdot\boldsymbol{b}_o)^2}} .\end{aligned} \tag{8.162}$$

8.3.2.3 Calculations of Cross Sections

With wave function (8.142) it is possible to calculate the total cross sections $\sigma(\boldsymbol{k}_i)$ like in Eq. (8.31), and after transformation (8.32) to get partial cross sections

$$\begin{aligned}\frac{d\sigma_{f,i}(\boldsymbol{k}_i\to\boldsymbol{k}_f)}{d\Omega_f} &= k_f\frac{k_{f\perp}^2 k_{i\perp}}{\pi^2}\Bigg|\sum_{n',m'=0}^{1}\langle n_f|\left(\hat{\boldsymbol{k}}_{f\perp}+\hat{\boldsymbol{k}}'_{f\perp}\right)^{-1}|n'\rangle\\ &\langle n'|\boldsymbol{\sigma}\cdot\boldsymbol{B}_F(\boldsymbol{\kappa}_{m'm'})|m'\rangle\langle m'|\left(\hat{\boldsymbol{k}}_{i\perp}+\hat{\boldsymbol{k}}'_{i\perp}\right)^{-1}|n_i\rangle\Bigg|^2 ,\end{aligned} \tag{8.163}$$

where

$$k_f = \sqrt{k^2-(-1)^{n_f}B_o} , \quad k_{f\perp} = \sqrt{k^2-\boldsymbol{k}_f^2-(-1)^{n_f}B_o} ,$$

$$k_{i\perp} = \sqrt{k^2-\boldsymbol{k}_i^2-(-1)^{n_i}B_o} ,$$

$$\hat{\boldsymbol{k}}_{f\perp} = \sqrt{k^2-\boldsymbol{k}_f^2-\boldsymbol{\sigma}\cdot\boldsymbol{B}_o} , \quad \hat{\boldsymbol{k}}_{i\perp} = \sqrt{k^2-\boldsymbol{k}_i^2-\boldsymbol{\sigma}\cdot\boldsymbol{B}_o} ,$$

$$\hat{\boldsymbol{k}}'_{f\perp} = \sqrt{k^2-\boldsymbol{k}_f^2-u-\boldsymbol{\sigma}\cdot\boldsymbol{B}_i} , \quad \hat{\boldsymbol{k}}'_{i\perp} = \sqrt{k^2-\boldsymbol{k}_i^2-u-\boldsymbol{\sigma}\cdot\boldsymbol{B}_i} .$$

If $\boldsymbol{B}(\boldsymbol{r}')$ extends all over the area of the interface, then the cross section is proportional to the interface area S, and instead of the cross section we must use the indicatrix of reflection: $w(\boldsymbol{\Omega}_i\to\boldsymbol{\Omega}_f) = d\sigma_{f,i}(\boldsymbol{k}_i\to\boldsymbol{k}_f)/S\,d\Omega_f$.

The investigation of scattering on a system of magnetic inhomogeneities is performed like on page 360. We can construct correlation functions for scattered waves with given polarization and take into account correlation of the position and direction of the magnetic fields in two different inhomogeneities. In general, the formulas are very cumbersome. So as not to overload the book, we shall not present them here. In real life we can use simpler or less simple models, which facilitate the investigations, but discussion of such models is worth a separate scientific monograph devoted to magnetism.

8.4
Neutron Albedo

Above we considered scattering of waves in disordered media. Now we shall investigate scattering of particles as corpuscles. Our main task will be the calculation of the albedo [385], which is met in reactor and radiation physics. We shall show that the methods developed for waves can also be applied to corpuscles.

Albedo is the total probability that a particle incident on a wall will be reflected in an arbitrary direction. Therefore, the problem is to find the total reflection coefficient R_L for a wall of thickness L if we have in our disposal the scattering cross section $\sigma'(\boldsymbol{\Omega} \to \boldsymbol{\Omega}')$ for a single scatterer. To solve this problem we first find scattering in a thin layer, then with separation of the thin layer from a semi-infinite wall, as shown in Figure 8.10, we find the albedo of an infinitely thick wall R_∞ and an analog of the Bloch wave vector q, which characterizes the extinction length $L_0 = 1/q$ of the neutron density in the wall. The parameter L_0 is an analog of the diffusion length in the diffusion theory.

With the help of R_∞ and q, we can find albedo R_L and transmission T_L for a layer of thickness L. For that we again do as shown in Figure 8.10, but this time we split off the infinitely thick wall not the infinitesimally thin layer ζ, but the layer of thickness L.

Besides characteristics such as R_∞ and q, it is also desirable to find the angular distribution of reflected neutrons. However, in general we cannot find an analytical solution for an arbitrary angular dependence of the cross section. The problem can be solved only in special cases. In particular, it can be solved for isotropic scattering and we shall show that the result is identical to the one obtained with the help of the invariant embedding method (see [386]) developed by Ambartsumyan and Chandrasekhar in their investigation of radiation of stars.

To apply the algebraic method to corpuscular physics, let us first define some mathematical notions which we shall work with.

8.4.1
Main Notions

We shall use the following notions:

Elementary vector states

$$|\boldsymbol{\Omega}\rangle\,. \tag{8.164}$$

They describe neutrons propagating in the direction $\boldsymbol{\Omega}$, and they are normalized according to the equation

$$\langle\boldsymbol{\Omega}'||\boldsymbol{\Omega}\rangle = \delta(\boldsymbol{\Omega} - \boldsymbol{\Omega}')\,. \tag{8.165}$$

The direction $\boldsymbol{\Omega}$ is characterized by angles with the polar axis along the internal normal $\boldsymbol{n}$ to the wall interface.

A general state vector

$$|\psi\rangle = \int d\Omega\, \psi(\boldsymbol{\Omega})|\boldsymbol{\Omega}\rangle\,. \tag{8.166}$$

It describes a neutron which flies along direction $|\boldsymbol{\Omega}\rangle$ with probability $\psi(\boldsymbol{\Omega})$. Note that the letter ψ here has no relation to quantum mechanics.

The isotropic state of incident neutrons is described by the state vector

$$|\psi\rangle = \int\limits_{n\boldsymbol{\Omega}>0} d\Omega\, \psi(\boldsymbol{\Omega})|\boldsymbol{\Omega}\rangle = \int\limits_{n\boldsymbol{\Omega}>0} d\Omega \frac{\cos\theta}{\pi}|\boldsymbol{\Omega}\rangle\,. \tag{8.167}$$

Normalization vector. To define the norm of any state we introduce the so-called normalization vector:

$$|e\rangle = \int\limits_{4\pi} d\Omega\,|\boldsymbol{\Omega}\rangle\,. \tag{8.168}$$

The norm ν of any vector state $|\boldsymbol{\phi}\rangle$ is defined with the help of $|e\rangle$ as

$$\nu_\phi = \langle e||\boldsymbol{\phi}\rangle\,. \tag{8.169}$$

In particular, the norm of states (8.164) and (8.167), calculated with account of (8.165), is unity.

Operators. Scattering, absorption, and reflection change the states, and the changes can be described by the operators

$$\hat{F} = \int\int |\boldsymbol{\Omega}\rangle F(\boldsymbol{\Omega} \leftarrow \boldsymbol{\Omega}')d\Omega\, d\Omega'\langle\boldsymbol{\Omega}'|\,, \tag{8.170}$$

where $F(\boldsymbol{\Omega} \leftarrow \boldsymbol{\Omega}')$ is a numerical function which describes the transition of the neutron flight from the direction $\boldsymbol{\Omega}'$ to direction $\boldsymbol{\Omega}$. Under the action of the operator $\hat{\boldsymbol{F}}$, the state $|\boldsymbol{\phi}\rangle = \int \phi(\boldsymbol{\Omega})d\Omega|\boldsymbol{\Omega}\rangle$ with the norm $\nu_\phi = \langle e||\boldsymbol{\phi}\rangle = \int \phi(\boldsymbol{\Omega})d\Omega$ is transformed to the state $|\chi\rangle = \hat{\boldsymbol{F}}|\boldsymbol{\phi}\rangle = \int\left[\int F(\boldsymbol{\Omega} \leftarrow \boldsymbol{\Omega}')d\Omega'\phi(\boldsymbol{\Omega}')\right]d\Omega|\boldsymbol{\Omega}\rangle$ with the norm $\nu_\chi = \langle e||\chi\rangle = \int d\Omega \int F(\boldsymbol{\Omega} \leftarrow \boldsymbol{\Omega}')d\Omega'\phi(\boldsymbol{\Omega}')$.

8.4.2
Albedo of an Infinitely Thick Wall

Our approach here is very close to that of "invariant embedding" method used, for instance, in [387] for calculation of backscattering of electrons. Let us separate virtually a layer of little thickness ζ from an infinitely thick wall as shown in Figure 8.10, and introduce the probability of reflection from the thin layer, $\hat{\boldsymbol{\rho}}_b = \zeta\hat{\boldsymbol{\Sigma}}_b$, where

$$\begin{aligned}\hat{\boldsymbol{\Sigma}}_b &= \int\limits_{n\boldsymbol{\Omega}<0}\int\limits_{n\boldsymbol{\Omega}'>0} |\boldsymbol{\Omega}\rangle d\Omega\, \Sigma_s(\boldsymbol{\Omega} \leftarrow \boldsymbol{\Omega}')\frac{d\Omega'}{\cos\theta'}\langle\boldsymbol{\Omega}'| \\ &\equiv \int\limits_{n\boldsymbol{\Omega}>0}\int\limits_{n\boldsymbol{\Omega}'<0} |\boldsymbol{\Omega}\rangle d\Omega\, \Sigma_s(\boldsymbol{\Omega} \leftarrow \boldsymbol{\Omega}')\frac{d\Omega'}{|\cos\theta'|}\langle\boldsymbol{\Omega}'|\,.\end{aligned} \tag{8.171}$$

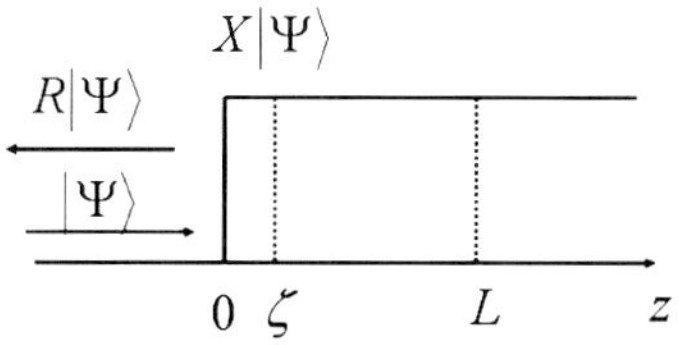

Figure 8.10 Chipping off a thin layer of matter from an infinitely thick wall for calculation of the reflection, R, and extinction, X, of the neutron flux in the wall. To find the albedo of a wall of thickness L, we chip off a layer of thickness L from the infinite wall.

The index b means that the reflection goes back to the same semispace from where the incident neutron propagated. The first equality refers to reflection from the left and the second equality refers to reflection from the right. The factor $1/\cos\theta'$ shows that at small angles the probability of scattering increases, because the number of atoms increases on the neutron flight path. The factor $\Sigma_s(\boldsymbol{\Omega} \leftarrow \boldsymbol{\Omega}')$ denotes the macroscopic differential cross section $\Sigma_s(\boldsymbol{\Omega} \leftarrow \boldsymbol{\Omega}') = N d\sigma_s(\boldsymbol{\Omega}' \rightarrow \boldsymbol{\Omega})/d\Omega$, where N is the number density of scatterers.

Let us introduce also an operator of transmission of the thin layer $\hat{\boldsymbol{\tau}} = \hat{\boldsymbol{I}} - \hat{s} + \hat{\boldsymbol{\rho}}_f$, where $\hat{\boldsymbol{I}} = \int |\boldsymbol{\Omega}\rangle d\Omega \langle\boldsymbol{\Omega}|$ is a unit operator, $\hat{s} = \zeta\Sigma_t\hat{S}$, $\hat{S} = \int |\boldsymbol{\Omega}\rangle [d\Omega/|\cos\theta|] \langle\boldsymbol{\Omega}|$, $\Sigma_t = N\sigma_t = \Sigma_a + \Sigma_s$ is the macroscopic total cross section, that is, absorption Σ_a and scattering Σ_s. The operator $\hat{\boldsymbol{\rho}}_f$ is $\hat{\boldsymbol{\rho}}_f = \zeta\hat{\boldsymbol{\Sigma}}_f$, and

$$\hat{\boldsymbol{\Sigma}}_f = \int\limits_{n\Omega>0}\int\limits_{n\Omega'>0} |\boldsymbol{\Omega}\rangle d\Omega\, \Sigma_s(\boldsymbol{\Omega} \leftarrow \boldsymbol{\Omega}') \frac{d\Omega'}{\cos\theta'} \langle\boldsymbol{\Omega}'|$$
$$\equiv \int\limits_{n\Omega<0}\int\limits_{n\Omega'<0} |\boldsymbol{\Omega}\rangle d\Omega\, \Sigma_s(\boldsymbol{\Omega} \leftarrow \boldsymbol{\Omega}') \frac{d\Omega'}{|\cos\theta'|} \langle\boldsymbol{\Omega}'| \,. \tag{8.172}$$

The index f means that after scattering the neutron is on the opposite side of the layer with respect to the incident neutron. The first equality describes scattering for the neutron incident on the layer from the left, and the second one describes scattering for the neutron incident on the layer from the right.

Let us introduce operator $\hat{\boldsymbol{X}}$, transforming the state $|\psi\rangle$ of neutrons incident on the interface into the state $\hat{\boldsymbol{X}}|\psi\rangle$ of the neutrons incident on the infinite wall after the thin layer. Let us also introduce reflection operator $\hat{\boldsymbol{R}}$ for the infinite wall. Operators $\hat{\boldsymbol{X}}$ and $\hat{\boldsymbol{R}}$ are related to each other via the system of equations

$$\hat{\boldsymbol{X}} = \hat{\boldsymbol{\tau}} + \hat{\boldsymbol{\rho}}_b\hat{\boldsymbol{R}}\hat{\boldsymbol{X}} \,, \quad \hat{\boldsymbol{R}} = \hat{\boldsymbol{\rho}}_b + \hat{\boldsymbol{\tau}}\hat{\boldsymbol{R}}\hat{\boldsymbol{X}} \,. \tag{8.173}$$

Resolve $\hat{\boldsymbol{X}}$ from the first equation

$$\hat{\boldsymbol{X}} = (\hat{\boldsymbol{I}} - \hat{\boldsymbol{\rho}}_b\hat{\boldsymbol{R}})^{-1}\hat{\boldsymbol{\tau}} \,, \tag{8.174}$$

and substitute it into the second one. As a result we obtain the equation for $\hat{\boldsymbol{R}}$: $\hat{\boldsymbol{R}} = \hat{\boldsymbol{\rho}}_b + \hat{\boldsymbol{\tau}}\hat{\boldsymbol{R}}(\hat{\boldsymbol{I}} - \hat{\boldsymbol{\rho}}_b\hat{\boldsymbol{R}})^{-1}\hat{\boldsymbol{\tau}}$. Since the thickness ζ can be chosen to be arbitrarily small, all the matrices $\hat{\boldsymbol{\rho}}_{b,f}$ and $\hat{\mathrm{s}}$ are small and the equation can be linearized:

$$\begin{aligned}\hat{\boldsymbol{R}} &= \hat{\boldsymbol{\rho}}_b + \hat{\boldsymbol{\tau}}\hat{\boldsymbol{R}}(\hat{\boldsymbol{I}} - \hat{\boldsymbol{\rho}}_b\hat{\boldsymbol{R}})^{-1}\hat{\boldsymbol{\tau}} = \hat{\boldsymbol{\rho}}_b + (\hat{\boldsymbol{I}} - \hat{\mathrm{s}} + \hat{\boldsymbol{\rho}}_f)\hat{R}(\hat{\boldsymbol{I}} + \hat{\boldsymbol{\rho}}_b\hat{R})(\hat{\boldsymbol{I}} - \hat{\mathrm{s}} + \hat{\boldsymbol{\rho}}_f) \\ &= \hat{\boldsymbol{R}} + \hat{\boldsymbol{R}}\hat{\boldsymbol{\rho}}_b\hat{\boldsymbol{R}} - \hat{\mathrm{s}}\hat{\boldsymbol{R}} - \hat{\boldsymbol{R}}\hat{\mathrm{s}} + \hat{\boldsymbol{\rho}}_f\hat{\boldsymbol{R}} + \hat{\boldsymbol{R}}\hat{\boldsymbol{\rho}}_f + \hat{\boldsymbol{\rho}}_b\,. \end{aligned} \tag{8.175}$$

It follows that

$$\hat{\boldsymbol{R}}\hat{\boldsymbol{\Sigma}}_b\hat{\boldsymbol{R}} + (\hat{\boldsymbol{\Sigma}}_f - \Sigma_t\hat{\boldsymbol{S}})\hat{\boldsymbol{R}} + \hat{\boldsymbol{R}}(\hat{\boldsymbol{\Sigma}}_f - \Sigma_t\hat{\boldsymbol{S}}) + \hat{\boldsymbol{\Sigma}}_b = 0\,. \tag{8.176}$$

Suppose that independently of the incident angle the reflected neutrons have an isotropic distribution. This means that $\hat{\boldsymbol{R}}$ is representable in the form

$$\hat{\boldsymbol{R}} = R_\infty \int\limits_{n\boldsymbol{\Omega}<0}\int\limits_{n\boldsymbol{\Omega}'>0} |\boldsymbol{\Omega}\rangle d\Omega \frac{|\cos\theta|}{\pi} d\Omega' \langle\boldsymbol{\Omega}'| = R_\infty|\psi\rangle\langle\boldsymbol{e}|\,. \tag{8.177}$$

Multiply Eq. (8.176) by $|\psi\rangle$ Eq. (8.167) on the right and by $\langle\boldsymbol{e}|$ Eq. (8.168) on the left. As a result we get

$$\begin{aligned}&\langle\boldsymbol{e}|[\hat{\boldsymbol{R}}\hat{\boldsymbol{\Sigma}}_b\hat{\boldsymbol{R}} + (\hat{\boldsymbol{\Sigma}}_f - \Sigma_t\hat{\boldsymbol{S}})\hat{\boldsymbol{R}} + \hat{\boldsymbol{R}}(\hat{\boldsymbol{\Sigma}}_f - \Sigma_t\hat{\boldsymbol{S}}) + \hat{\boldsymbol{\Sigma}}_b]|\psi\rangle \\ &\quad = R_\infty^2\Sigma_b - 2(\Sigma_t - \Sigma_f)R_\infty + \Sigma_b = 0\,,\end{aligned} \tag{8.178}$$

where

$$\begin{aligned}\Sigma_b &= \int\limits_{n\boldsymbol{\Omega}<0} d\Omega \int\limits_{n\boldsymbol{\Omega}'>0} \Sigma_s(\boldsymbol{\Omega} \leftarrow \boldsymbol{\Omega}')\frac{d\Omega'}{2\pi}\,, \\ \Sigma_f &= \int\limits_{n\boldsymbol{\Omega}>0} d\Omega \int\limits_{n\boldsymbol{\Omega}'>0} \Sigma_s(\boldsymbol{\Omega} \leftarrow \boldsymbol{\Omega}')\frac{d\Omega'}{2\pi}\end{aligned} \tag{8.179}$$

are the cross sections for scattering back and forward, respectively, averaged over incidence angles. Let us note that

$$\Sigma_b + \Sigma_f = \int\limits_{4\pi} d\Omega \int\limits_{n\boldsymbol{\Omega}'>0} \Sigma_s(\boldsymbol{\Omega} \leftarrow \boldsymbol{\Omega}')\frac{d\Omega'}{2\pi} = \Sigma_s \int\limits_{2\pi} \frac{d\Omega'}{2\pi} = \Sigma_s\,. \tag{8.180}$$

The solution of Eq. (8.178) is representable as

$$R_\infty = \frac{\sqrt{\Sigma_t - \Sigma_f + \Sigma_b} - \sqrt{\Sigma_t - \Sigma_f - \Sigma_b}}{\sqrt{\Sigma_t - \Sigma_f + \Sigma_b} + \sqrt{\Sigma_t - \Sigma_f - \Sigma_b}} = R_\infty = \frac{\sqrt{\Sigma_a + \Sigma_{\mathrm{tr}}} - \sqrt{\Sigma_a}}{\sqrt{\Sigma_a + \Sigma_{\mathrm{tr}}} + \sqrt{\Sigma_a}}\,, \tag{8.181}$$

where in the last equality we introduced the so-called transport cross section $\Sigma_{\mathrm{tr}} = \Sigma_s - \Sigma_f + \Sigma_b = 2\Sigma_b$. At $\Sigma_a = 0$ the albedo of the semi-infinite wall is 1, and at $\Sigma_{\mathrm{tr}}/\Sigma_a \to 0$ the albedo goes to zero.

8.4.3
Extinction of the Neutron Density in Matter

To find the albedo of a wall of finite thickness it is necessary to find how fast the neutron flux decays in the depth of the wall. This decay is described by the operator $\hat{\boldsymbol{X}}$. According to Eq. (8.174) it can be represented as

$$\hat{\boldsymbol{X}} = (\hat{\boldsymbol{I}} - \hat{\boldsymbol{\rho}}_b \hat{\boldsymbol{R}})^{-1} \hat{\boldsymbol{\tau}} \approx \hat{\boldsymbol{I}} + \hat{\boldsymbol{\rho}}_b \hat{\boldsymbol{R}} - \hat{\boldsymbol{s}} + \hat{\boldsymbol{\rho}}_f \,. \tag{8.182}$$

The operator $\hat{\boldsymbol{X}}$ transforms the state $|\psi\rangle$ at the interface to the state $\hat{\boldsymbol{X}}|\psi\rangle$ at the depth ζ inside the matter. The norm of this state is $\langle e|\hat{\boldsymbol{X}}|\psi\rangle = 1 - 2\zeta[\Sigma_a + \Sigma_b(1 - R_\infty)] \approx \exp(-q\zeta)$. It means that at depth z from the interface the neutron density decays as $X(z) = \exp(-qz) = \exp(-z/L_0)$, where

$$q = 2\Sigma_a + \Sigma_{\text{tr}}(1 - R_\infty) = 2\sqrt{\Sigma_a(\Sigma_a + \Sigma_{\text{tr}})}\,, \quad L_0 = \frac{1}{2\sqrt{\Sigma_a(\Sigma_a + \Sigma_{\text{tr}})}}\,. \tag{8.183}$$

Therefore, the operator $\hat{\boldsymbol{X}}$ can be represented as

$$\hat{\boldsymbol{X}}(z) = \exp(-z/L_0)|\psi\rangle\langle e| \,. \tag{8.184}$$

8.4.4
Albedo of a Wall of Thickness *L*

To find the reflection $\hat{\boldsymbol{R}}_L$ from a wall of thickness L we again consider a semi-infinite wall and write down equations similar to Eq. (8.173), but now we chip off not a thin layer, but a layer of thickness L:

$$\hat{\boldsymbol{X}}_L = \hat{\boldsymbol{T}}_L + \hat{\boldsymbol{R}}_L \hat{\boldsymbol{R}} \hat{\boldsymbol{X}}_L \,, \quad \hat{\boldsymbol{R}} = \hat{\boldsymbol{R}}_L + \hat{\boldsymbol{T}}_L \hat{\boldsymbol{R}} \hat{\boldsymbol{X}}_L \,. \tag{8.185}$$

Here $\hat{\boldsymbol{X}}_L = \hat{\boldsymbol{X}}(L)$, and $\hat{\boldsymbol{T}}_L$ is the transmission operator of the wall of thickness L.

Exclude $\hat{\boldsymbol{T}}_L$ from the second equation. As a result we get $\hat{\boldsymbol{R}} = \hat{\boldsymbol{R}}_L + [\hat{\boldsymbol{X}}_L - \hat{\boldsymbol{R}}_L \hat{\boldsymbol{R}} \hat{\boldsymbol{X}}_L] \hat{\boldsymbol{R}} \hat{\boldsymbol{X}}_L$. It follows that

$$\begin{aligned} &\hat{\boldsymbol{R}} - \hat{\boldsymbol{X}}_L \hat{\boldsymbol{R}} \hat{\boldsymbol{X}}_L = \hat{\boldsymbol{R}}_L[\hat{\boldsymbol{I}} - \hat{\boldsymbol{R}} \hat{\boldsymbol{X}}_L \hat{\boldsymbol{R}} \hat{\boldsymbol{X}}_L] \\ &\quad \to \hat{\boldsymbol{R}}_L = [\hat{\boldsymbol{R}} - \hat{\boldsymbol{X}}_L \hat{\boldsymbol{R}} \hat{\boldsymbol{X}}_L][\hat{\boldsymbol{I}} - \hat{\boldsymbol{R}} \hat{\boldsymbol{X}}_L \hat{\boldsymbol{R}} \hat{\boldsymbol{X}}_L]^{-1} \,. \end{aligned} \tag{8.186}$$

Let us note that

$$\begin{aligned} &\hat{\boldsymbol{I}} - \hat{\boldsymbol{R}} \hat{\boldsymbol{X}}_L \hat{\boldsymbol{R}} \hat{\boldsymbol{X}}_L = \int |\boldsymbol{\Omega}\rangle d\Omega \langle \boldsymbol{\Omega}| - R_\infty^2 X_L^2 |\psi\rangle\langle e| = \hat{\boldsymbol{I}} - \alpha|\psi\rangle\langle e| \,, \\ &\qquad \alpha = R_\infty^2 X_L^2 \,, \end{aligned} \tag{8.187}$$

where $X_L = \exp(-qL)$. The reciprocal matrix $(\hat{\boldsymbol{I}} - \alpha|\psi\rangle\langle e|)^{-1}$ can be represented as

$$(\hat{\boldsymbol{I}} - \alpha|\psi\rangle\langle e|)^{-1} = \hat{\boldsymbol{I}} + c|\psi\rangle\langle e| \,, \tag{8.188}$$

where c is determined from the evident equation $\hat{I} = (\hat{I} - \alpha|\psi\rangle\langle e|)(\hat{I} - \alpha|\psi\rangle\langle e|)^{-1} = (\hat{I} - \alpha|\psi\rangle\langle e|)(\hat{I} + c|\psi\rangle\langle e|) = \hat{I} - (\alpha - c + c\alpha)|\psi\rangle\langle e|$. So $\alpha - c + c\alpha = 0$, or

$$c = \frac{\alpha}{1-\alpha} = \frac{R_\infty^2 X_L^2}{1 - R_\infty^2 X_L^2} . \tag{8.189}$$

Substitution of Eq. (8.188) with account of Eq. (8.189) into Eq. (8.186) gives $\hat{R}_L = R_L|\psi\rangle\langle e|$, where

$$R_L \equiv \langle e|\hat{R}_L|\psi\rangle = R_\infty \frac{1 - \exp(-2qL)}{1 - R_\infty^2 \exp(-2qL)} . \tag{8.190}$$

In the absence of absorption we have $R_\infty = 1$, $q = 0$, and the right-hand side of Eq. (8.190) becomes an uncertain ratio 0/0. This uncertainty is resolved with the L'Hopital rule by differentiation of the numerator and the denominator of the fraction (8.190) over $\sqrt{\Sigma_a}$ at $\Sigma_a = 0$. Since, according to Eqs. (8.183) and (8.181), $q' \equiv dq/d\sqrt{\Sigma_a} = 2\sqrt{\Sigma_{\text{tr}}}$, and $R'_\infty \equiv dR_\infty/d\sqrt{\Sigma_a} = -2/\sqrt{\Sigma_{\text{tr}}}$, then

$$\lim_{\sqrt{\Sigma_a}\to 0}\left(\frac{1 - \exp(-2qL)}{1 - R_\infty^2 \exp(-2qL)}\right) = \frac{q'L}{-R'_\infty + q'L} = \frac{\Sigma_{\text{tr}} L}{1 + \Sigma_{\text{tr}} L} . \tag{8.191}$$

Besides the albedo, it is sometimes also useful to know the transmission $\hat{T}_L$ or shielding of the radiation by the wall. Substituting of $\hat{R}_L$ from the second equation of (8.185) into the first one, we get $\hat{X}_L = \hat{T}_L + (\hat{R} - \hat{T}_L\hat{R}\hat{X}_L)\hat{R}\hat{X}_L$, and we find $\hat{T}_L = (\hat{I} - \hat{R}^2)\hat{X}_L(\hat{I} - \hat{R}\hat{X}_L\hat{R}\hat{X}_L)^{-1}$. Substitution of Eqs. (8.177) and (8.184), with account of Eqs (8.187)– (8.189), gives $\hat{T}_L = T_L|\psi\rangle\langle e|$, where

$$T_L \equiv \langle e|\hat{T}_L|\psi\rangle = \exp(-qL)\frac{1 - R_\infty^2}{1 - R_\infty^2 \exp(-2qL)} . \tag{8.192}$$

8.4.5
Comparison with Diffusion Formulas for the Albedo

– You tell that diffusion theory is not correct? So now those, who all their life used this theory, should to shoot themselves?
– Oh, no! There is less radical way to improve knowledge.

From discussion with an opponent.

Expressions (8.181) and (8.190) differ from the respective albedo formulas obtained within diffusion theory. In the case of infinite wall diffusion, the albedo (see, e. g., [388], Eq. (6.3.6) and [389], Eq. (3.3)) is of the form

$$R_\infty = \frac{1 - 2D/L_D}{1 + 2D/L_D} , \tag{8.193}$$

where $D = 1/3(\Sigma_a + \Sigma_{\text{tr}})$ is the diffusion coefficient and $L_D = 1/\sqrt{3\Sigma_a(\Sigma_a + \Sigma_{\text{tr}})}$ is the diffusion length. Equation (8.193) is easily reduced to the form

$$R_\infty = \frac{\sqrt{\Sigma_a + \Sigma_{\text{tr}}} - \sqrt{(4/3)\Sigma_a}}{\sqrt{\Sigma_a + \Sigma_{\text{tr}}} + \sqrt{(4/3)\Sigma_a}} , \tag{8.194}$$

and it follows that at $\Sigma_{\text{tr}} < \Sigma_a/3$ the albedo, which is a probability and must be positive, becomes negative.

The albedo of a wall of thickness L (see, e. g., [388], Eq. (6.3.5)) is represented as

$$R_L = \frac{1 - 2(D/L_D)\coth(L/L_D)}{1 + 2(D/L_D)\coth(L/L_D)}\,, \tag{8.195}$$

and at $L \to 0$ we get an absurd result: the albedo goes to -1.

Absurdities show that the diffusion approach is valid only at $\Sigma_{\text{tr}} \gg \Sigma_a$ and $L \gg L_D$. Since our expressions (8.181) and (8.190) never give absurd results, this means that our algebraic approach is valid for a broader range of parameters than the diffusion one.

8.4.6 Albedo of Ultrathin Powdered Matter

Let us consider the albedo of ultradispersive powder, or powder consisting of small grains with radius in the range of several nanometers.

Dispersion creates additional scattering on a grain as a whole. It increases Σ_{tr} and therefore increases the albedo. It is interesting to estimate the contribution of additional scattering to the albedo. If the contribution is large, then in reactor construction elements, active core, reflector, and moderator, we can economize in terms of fuel elements and the reactor size [390, 391].

So our task is to estimate the total transport cross section, Σ_{tr}, which is the sum of two terms $\Sigma_{\text{tr}} = \Sigma_{\text{tr}}^c + \Sigma_N$. The first one, Σ_{tr}^c, relates to coherent scattering on a grain as a whole, and the second one, Σ_N, is related to scattering on separate nuclei of the grain. The second term is itself a sum of two parts: an isotropic incoherent one $\Sigma_N^{\text{ic}} = 4\pi N_0 |b_{\text{ic}}|^2$, where b_{ic} is the incoherent part of the scattering amplitude of nuclei, and a coherent cross section Σ_N^c.

Coherent scattering on nuclei is not isotropic in general. It describes diffraction on the crystalline lattice of the grain and depends on the neutron wavelength. In the case of thermal neutrons ($\lambda = 1.8$ Å), coherent scattering on nuclei can be considered isotropic and represented as $\Sigma_N^c = 4\pi N_0 |b_c|^2$, where b_c is the coherent scattering amplitude, because of random orientation of lattices in different grains.

Let us estimate Σ_{tr}^c for scattering on a grain of radius a as a whole. The scattering amplitude of a grain found in perturbation theory is

$$\begin{aligned} B &= N_0 b_c \int d^3 r \exp(i\boldsymbol{qr}) = \frac{2\pi N_0 b_c}{iq} \int_{-a}^{a} r dr e^{iqr} \\ &= -\frac{4\pi N_0 b_c a}{q^2} \left(\cos(qa) - \frac{\sin(qa)}{qa}\right)\,. \end{aligned} \tag{8.196}$$

Therefore, the differential cross section is

$$d\sigma_s^c(\Omega) = |B|^2 d\Omega = a^2 \frac{u^2}{q^4} \left(\cos(qa) - \frac{\sin(qa)}{qa}\right)^2 d\Omega\,, \tag{8.197}$$

where $u = 4\pi N_0 b_c$ is the reduced optical potential of the medium: $U = \hbar^2 u/2m \approx 10^{-7}$ eV. With it we can calculate the transport cross section, $\sigma_{\rm tr}$, $\sigma_{\rm tr}^c = \int_{4\pi} d\Omega \left[d\sigma_s^c(\Omega)/d\Omega\right](1 - \cos\theta) = \int\limits_{4\pi} |B|^2(1 - \cos\theta)d\Omega$, and the macroscopic transport cross section,

$$\Sigma_{\rm tr}^c = \gamma N_1 \sigma_{\rm tr}^c = \gamma N_0 \frac{\sigma_{\rm tr}^c}{4\pi N_0 a^3/3} = \gamma \Sigma_N^c F\,, \quad F = 6\pi \frac{N_0}{k^3} g(2ka)\,, \tag{8.198}$$

where N_1 is the number of grains in a unit volume if all the continuous material can be transformed into grains, $\gamma \leq 0.74$ is packing density, and the function

$$\begin{aligned} g(2ka) &= \int_0^{2ka} (x\cos x - \sin x)^2 \frac{dx}{2kax^3} \\ &\approx \begin{cases} \frac{\ln(2ka)+0.27}{4ka} & \text{for} \quad 2ka \gg 1 \\ 2(ka)^3/9 & \text{for} \quad 2ka \ll 1\,, \end{cases} \end{aligned} \tag{8.199}$$

has a maximum $\max(g) \approx 0.25$ at $2ka \approx 3.5$, that is, at $a \approx 0.28\lambda$.

The factor γ enters all the parameters of formula (8.181) and is excluded, but it is not excluded in the penetration depth. Therefore,

$$\begin{aligned} R_\infty &= \frac{\sqrt{\Sigma_a + \Sigma_N^{\rm ic} + \Sigma_N^c(1+F)} - \sqrt{\Sigma_a}}{\sqrt{\Sigma_a + \Sigma_N^{\rm ic} + \Sigma_N^c(1+F)} + \sqrt{\Sigma_a}}\,, \\ L_0 &= \frac{1}{2\gamma\sqrt{\Sigma_a + \Sigma_N^{\rm ic} + \Sigma_N^c(1+F)}} \end{aligned} \tag{8.200}$$

and the albedo $R_L \propto R_\infty[1 - \exp(-2L/L_0)]$ of the wall of thickness L in the case of an ultradispersive medium can be lower than that of a homogeneous medium.

8.4.7
Angular Distribution of Reflected Neutrons for Isotropic Scattering

Above we used the simple model (8.177) for the isotropic reflection operator. The real distribution cannot be isotropic. In general, the reflection operator for a semi-infinite wall can be represented as

$$\hat{R} = \int\limits_{n\boldsymbol{\Omega}<0}\int\limits_{n\boldsymbol{\Omega}'>0} |\boldsymbol{\Omega}\rangle d\Omega|\cos\theta|\mathcal{R}(\boldsymbol{\Omega} \leftarrow \boldsymbol{\Omega}')\frac{d\Omega'}{2\pi}\langle\boldsymbol{\Omega}'|\,, \tag{8.201}$$

and the albedo averaged over the isotropic distribution of the incident neutrons is

$$R_\infty = \int\limits_{n\boldsymbol{\Omega}<0}\int\limits_{n\boldsymbol{\Omega}'>0} d\Omega|\cos\theta|\mathcal{R}(\boldsymbol{\Omega} \leftarrow \boldsymbol{\Omega}')\frac{d\Omega'}{2\pi}\,. \tag{8.202}$$

From the detailed balance principle it follows that in the absence of absorption $\mathcal{R}(\boldsymbol{\Omega} \leftarrow \boldsymbol{\Omega}')$ is a symmetrical function with respect to permutation of its arguments. We shall suppose that it is a symmetrical function $\mathcal{R}(\mu,\mu')$ of $\mu = \cos\theta$ even with absorption. To find $\mathcal{R}(\mu,\mu')$ in general for an arbitrary angular dependence of scattering on a single scatterer is a pretty complicated problem. We can find $\mathcal{R}(\mu,\mu')$ only for isotropic scattering on a single atom. Here we shall report only the result without details of the derivation It was shown in [385] that $\mathcal{R}(\mu,\mu')$ can be represented as

$$\mathcal{R}(\mu,\mu') = \frac{\alpha}{2}\frac{H(\mu)H(\mu')}{\mu+\mu'}\,, \tag{8.203}$$

where $\alpha = \Sigma_s/\Sigma_t$, $H(\mu) = 1 + \mu Q(\mu)$, where $Q(\mu) = \int_0^1 \mathcal{R}(\mu,\mu')d\mu'$. From these definitions it follows that the function $H(\mu)$ satisfies the integral equation

$$\frac{1}{H(\mu)} = 1 - \frac{\alpha\mu}{2}\int_0^1 \frac{H(\mu')}{\mu+\mu'}d\mu'\,. \tag{8.204}$$

This equation completely coincides with Eq. (5) in Chapter 9.3 in [386], where it was obtained from the kinetic equation. The solution of Eq. (8.204) is also presented there, so we do not need to repeat it here.

It is important, however to mention that for calculation of the albedo R_∞ we do not need the exact solution of Eq. (8.204). Integration in (8.202) over angles with account of Eq. (8.203) gives

$$\alpha(1+R_\infty)^2 = 4R_\infty \to R_\infty = \frac{\sqrt{\Sigma_t}-\sqrt{\Sigma_a}}{\sqrt{\Sigma_t}+\sqrt{\Sigma_a}}\,. \tag{8.205}$$

The R_∞ coincides with Eq. (8.181) completely, because in the case of isotropic scattering $\Sigma_{\text{tr}} = 2\Sigma_b \equiv \Sigma_s$.

8.4.8
Angular Distribution for Anisotropic Scattering

Our approach can be applied to the general case of arbitrary scattering. Suppose we have a powdered wall and want to find angular distribution of reflected neutrons. There was an attempt to solve such a problem [392] for $\gamma =$ quanta; however, it was not fully successful. Here we only outline the way to solve the problem.

The starting point is the integral equation (8.175):

$$\hat{\boldsymbol{R}}\hat{\rho}_b\hat{\boldsymbol{R}} - \hat{s}\hat{\boldsymbol{R}} - \hat{\boldsymbol{R}}\hat{s} + \hat{\rho}_f\hat{\boldsymbol{R}} + \hat{\boldsymbol{R}}\hat{\rho}_f + \hat{\rho}_b = 0\,. \tag{8.206}$$

Multiply it by $|\boldsymbol{\Omega}\rangle$ on the right and by $\langle\boldsymbol{\Omega}|$ on the left. As a result we get

$$\langle\boldsymbol{\Omega}|[\hat{\boldsymbol{R}}\hat{\boldsymbol{\Sigma}}_b\hat{\boldsymbol{R}} + (\hat{\boldsymbol{\Sigma}}_f - \Sigma_t\hat{S})\hat{\boldsymbol{R}} + \hat{\boldsymbol{R}}(\hat{\boldsymbol{\Sigma}}_f - \Sigma_t\hat{S}) + \hat{\boldsymbol{\Sigma}}_b]|\boldsymbol{\Omega}\rangle = 0\,, \tag{8.207}$$

which after substitution of all the operators and division by $\cos\theta$ can be written as

$$\begin{aligned}
&\int\limits_{n\Omega''>0}\int\limits_{n\Omega'''<0} R(\Omega,\Omega'')\frac{d\Omega''}{2\pi}\Sigma_{\mathrm{sb}}(\boldsymbol{\Omega}''\leftarrow\boldsymbol{\Omega}''')\frac{d\Omega'''}{2\pi}R(\Omega''',\Omega')\\
&\quad+\frac{\Sigma_{\mathrm{sb}}(\boldsymbol{\Omega}\leftarrow\boldsymbol{\Omega}')}{\cos\theta\cos\theta'}-2\Sigma_t R(\Omega,\Omega')\\
&\quad+\int\limits_{n\Omega''<0}\frac{\Sigma_{sf}(\boldsymbol{\Omega}\leftarrow\boldsymbol{\Omega}'')}{\cos\theta}\frac{d\Omega''}{2\pi}R(\Omega'',\Omega')\\
&\quad+\int\limits_{n\Omega''>0}R(\boldsymbol{\Omega}\leftarrow\boldsymbol{\Omega}'')\frac{d\Omega''}{2\pi}\frac{\Sigma_{sf}(\Omega'',\Omega')}{\cos\theta'}\\
&=0\,.
\end{aligned}\tag{8.208}$$

Let us denote

$$\begin{aligned}
\Sigma_{\mathrm{sb}}(\boldsymbol{\Omega}\leftarrow\boldsymbol{\Omega}')&=\cos\theta\cos\theta'\rho(\boldsymbol{\Omega},\boldsymbol{\Omega}')\,,\\
\Sigma_{sf}(\boldsymbol{\Omega}\leftarrow\boldsymbol{\Omega}')&=\cos\theta\cos\theta' F(\boldsymbol{\Omega},\boldsymbol{\Omega}')\,.
\end{aligned}\tag{8.209}$$

Then Eq. (8.208) becomes

$$\begin{aligned}
&\int\limits_{n\Omega''>0}\int\limits_{n\Omega'''<0} R(\boldsymbol{\Omega},\boldsymbol{\Omega}'')\cos\theta'' d\Omega''\rho(\boldsymbol{\Omega}'',\boldsymbol{\Omega}''')\cos\theta''' d\Omega''' R(\boldsymbol{\Omega}''',\boldsymbol{\Omega}')\\
&\quad+\rho(\boldsymbol{\Omega},\boldsymbol{\Omega}')-2\Sigma_t R(\boldsymbol{\Omega},\boldsymbol{\Omega}')\\
&\quad+\int\limits_{n\Omega''<0}F(\boldsymbol{\Omega},\boldsymbol{\Omega}'')\cos\theta'' d\Omega'' R(\boldsymbol{\Omega}'',\boldsymbol{\Omega}')\\
&\quad+\int\limits_{n\Omega''>0}R(\boldsymbol{\Omega},\boldsymbol{\Omega}'')\cos\theta'' d\Omega'' F(\boldsymbol{\Omega}'',\boldsymbol{\Omega}')\\
&=0\,.
\end{aligned}\tag{8.210}$$

We can denote $\cos\theta = x$, and all the symmetrical functions of the type $Q(\Omega,\Omega')$ met in Eq. (8.210) as $Q(x,x',\phi-\phi')$. Then the equation becomes

$$\begin{aligned}
&\int_0^1 dy\int_0^1 dz\int_{-\infty}^{\infty}\frac{d\psi}{2\pi}\int_{-\infty}^{\infty}d\frac{d\chi}{2\pi}R(x,y,\phi-\psi)y\rho(y,z,\psi-\chi)zR(z,x',\chi-\phi')\\
&\quad+\rho(x,x',\phi-\phi')-2\Sigma_t R(x,x',\phi-\phi')+\int_0^1 y\,dy\int_{-\infty}^{\infty}\frac{d\psi}{2\pi}\\
&\quad\times[F(x,y,\phi-\psi)R(y,x',\psi-\phi')+R(x,y,\phi-\psi)F(y,x',\psi-\phi')]\\
&=0\,.
\end{aligned}\tag{8.211}$$

We supposed that all the functions of the azimuthal angle strongly decrease with increase of ϕ and because of that extended integration over this angle to infinite limits.

Multiplication of this equation by $\exp(-is\phi)$ and integration over $\phi/2\pi$ gives

$$\int_0^1 dy \int_0^1 dz\,\bar{R}(x,y,s)\,y\bar{\rho}(y,z,s)z\,\bar{R}(z,x',s) + \bar{\rho}(x,x',s) - 2\Sigma_t\bar{R}(x,x',s)$$
$$+ \int_0^1 y\,dy[\bar{F}(x,y,s)\bar{R}(y,x',s) + \bar{R}(x,y,s)\bar{F}(y,x',\psi-\phi')] = 0, \tag{8.212}$$

where

$$\bar{g}(s) = \int_{-\infty}^{\infty} \frac{d\psi}{2\pi}\exp(-is\phi)g(\phi)\,.$$

Now it is necessary to find a complete system of polynomials $P_n(x)$ orthogonal in the interval (0,1) with weight x:

$$\int_0^1 x\,dx\,P_n(x)P_m(x) = \delta_{mn}\,,$$

Represent all the functions as an expansion over them

$$g(x,y,s) = \sum_n g_n(s)P_n(x)P_n(y)\,,$$

substitute these expansions into Eq. (8.212), and find algebraic equations for the coefficients g_n,

$$R_n^2(s)\rho_n(s) + \rho_n(s) - 2\Sigma_t R_n(s) + 2F_n(s)R_n(s) = 0\,, \tag{8.213}$$

from which it follows that

$$R_n(s) = \frac{\sqrt{\Sigma_t - F_n(s) + \rho_n(s)} - \sqrt{\Sigma_t - F_n(s) - \rho_n(s)}}{\sqrt{\Sigma_t - F_n(s) + \rho_n(s)} + \sqrt{\Sigma_t - F_n(s) - \rho_n(s)}}\,, \tag{8.214}$$

and the formal solution of the problem is

$$\cos(\theta')R(\Omega' \leftarrow \Omega) = \cos(\theta')\int_{-\infty}^{\infty} ds\,e^{is(\phi'-\phi)}\sum_{n=0}^{\infty} P_n(\cos\theta')P_n(\cos\theta)$$
$$\times \frac{\sqrt{\Sigma_t - F_n(s) + \rho_n(s)} - \sqrt{\Sigma_t - F_n(s) - \rho_n(s)}}{\sqrt{\Sigma_t - F_n(s) + \rho_n(s)} + \sqrt{\Sigma_t - F_n(s) - \rho_n(s)}}\,. \tag{8.215}$$

We stop here because the rest involves the art of construction of polynomials, numerical calculation of integrals, and fitting of parameters to experimental data.

We considered only reflection from a semi-infinite medium. Reflection from a medium of finite thickness can be also found, but we do not consider it here so as not to increase the size of the book too much.

8.5 Conclusion

In this chapter we considered scattering in disordered media. This is a wide field of neutron optics. We discussed specular and nonspecular reflection from disordered matter. We showed that the coherent wave, which appears in random media, is identical to the one which is created when neutron–matter interaction is described by the optical potential. This optical potential is complex even in the absence of losses in the matter. The imaginary part of it is the result of scattering or neutron leakage from the coherent field.

The detailed balance principle and unitarity require one to include this potential in the description of the random part of the wave function, which characterizes scattering. Therefore, the perturbation theory which is used for investigation of scattering on inhomogeneities should be modified to the distorted wave approximation. With this approach we considered several problems related to small-angle scattering on magnetic and nonmagnetic inhomogeneities inside matter.

We also showed how algebraic matrix methods used for description of one-dimensional scattering on magnetic and crystalline objects can be generalized to the case of continuous distributions over angles. Here matrices are replaced by integral operators, and we demonstrated the power of this method in calculating the corpuscular problem of albedo reflection of neutrons from matter.

It is necessary to point out that the algebraic approach has a clear advantage compared with diffusion approach for solution of such problems. For instance, let us look at transport of ultracold neutrons along neutron guides. To find the transmission of a neutron guide of some length one usually solves the diffusion equation (textbook [G]) after calculation of the diffusion coefficient D. This coefficient depends on the characteristics of the neutron reflection from the guide walls. If the reflection indicatrix contains a specular component which increases with decrease of the grazing angle of the incident neutron, then the diffusion coefficient can become infinite. This case corresponds to random walk when some steps can be very long [370]. At infinite D, the diffusion approach becomes meaningless. On contrast to it, the algebraic approach is applicable at any reflection indicatrix. We do not consider here the propagation of ultracold neutrons in neutron guides, but we can promise to those who will try to solve this problem that they will encounter a lot of interesting features related to unusual statistics.

We considered the general problem of neutron albedo and angular distribution for nonisotropic scattering only schematically. All the calculations can be performed for any model with arbitrary precision and sufficient number of parameters to describe the results of any experiment. Realization of this program and comparison with existing methods (see [392]) is a subject for future work.

9
Neutron Interference and Phase Observation

The fundamental principle of experiments to observe the effects of neutron interference[32] is, simply speaking, the measurement of some interference pattern produced by superposing waves coherently after some operation on either one or more of the wave functions *split* into different optical paths in any one of coordinates for the wave function of a neutron. Therefore, many kinds of neutron interference experiments, that is, various kinds of neutron interferometry experiments, can be considered and performed in view of the principle and the experimental configuration.

For example, we can classify them by the selection of the splitting coordinate as experiments involving spatial splitting or temporal splitting into two or more partial waves, and spin splitting into two spinor components. On the other hand, we can classify them by the configuration of the interferometer, that is, by the type of essential optical element, as experiments involving a single crystal interferometer, a grating mirror interferometer, a multilayer mirror interferometer, and so on. In this chapter, various kinds of these neutron interferometers will be described in the course of explanations of the experiments on neutron interference and phase observation.

Among the many interesting and important reports already published about experiments on neutron interference and phase observations, in this chapter we will be mainly concerned with the experiments relating to the optical phenomena introduced in the previous chapters or that are highly relevant to the foundation of quantum mechanics.

32) Since we have no such source at present to *produce primarily coherent neutrons*, unlike the laser as a photon source, *neutron* in this phrase means interference of waves for *a neutron*.

Neutron Optics. Masahiko Utsuro,Vladimir K. Ignatovich

ISBN: 978-3-527-40885-6

9.1
Neutron Interferometry with a Perfect Crystal Interferometer

9.1.1
Triple Laue Diffraction Type Silicon Crystal Interferometer

The first reported experimental attempt at neutron interferometry was performed by Maier-Leibnitz and Springer by using a biprism interferometer [20] as mentioned briefly in Section 1.2. However, in that configuration it was difficult to insert a sample like a phase shifter because only a slight separation between the split beams was possible owing to the refractive index for neutrons being very near unity.

The great epoch that initiated the dramatic developments in neutron interferometry experiments was the successful observation of interference patterns by Rauch, Treimer, and Bonse [393] after their challenging and continuing efforts with a triple Laue diffraction type *silicon perfect single crystal interferometer*, with an optical configuration akin to the *Mach–Zehnder interferometer* in conventional optics and also as an analogy to the X-ray interferometer realized with a perfect single crystal. The configuration of their experiment is shown in Figure 9.1 with their experimental results using the neutron beam from the TRIGA reactor at the Atomic Institute of the Technical University of Austria [393].

As already explained in Section 3.4.2, the distances between the interferometer parts (splitter S, mirror M, and analyzer A) in Figure 9.1a have to be set very accurately to avoid defocusing effects and loss of coherence. After etching, the thickness of the crystal plates in their interferometer was 4.3954 ± 0.0008 mm, and the distances were 27.2936 ± 0.0009 mm. The crystal was tested as an X-ray interferometer, and by a suitable mounting and the application of small weights, the inherent

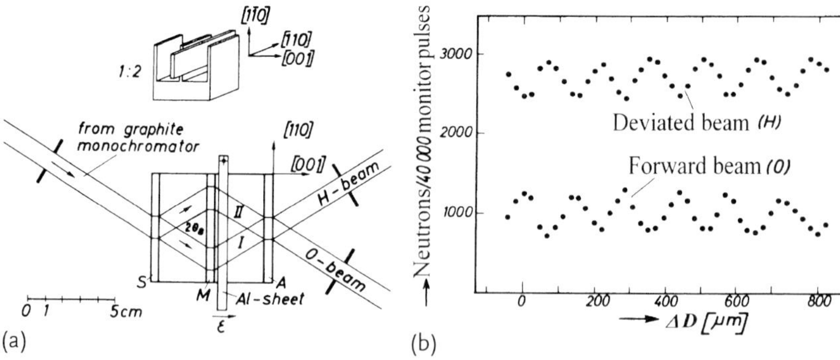

Figure 9.1 The first successful neutron interferometry experiment with a silicon perfect single crystal interferometer: (a) configuration of the interferometer; (b) measured intensity modulation of the deviated (H) and forward (O) beams as a function of the different optical paths for beams I and II due to an inserted Al sheet (The statistical error is smaller than the size of the data points.) (Rauch *et al.* [393]).

Moiré pattern could be balanced to give a homogeneous interference pattern over the whole region of interest.

According to the dynamical diffraction theory explained in Section 6.1 ([34], p. 202), the intensities of the *forward diffracted beam (O-beam)* and the *deviated diffracted beam (H-beam)* after the crystal in the configuration shown in Figure 9.1a are given by Eqs. (9.1) and (9.2), respectively:

$$P_o/P_i = (1+y^2)^{-1}\left[y^2 + \cos^2\left(A\sqrt{1+y^2}\right)\right] , \tag{9.1}$$

$$P_h/P_i = (1+y^2)^{-1}\sin^2\left(A\sqrt{1+y^2}\right) , \tag{9.2}$$

where $A = b_c|F|\lambda t/a^3 \cos\theta_B$, $y = a^3(\theta_B - \theta)\sin 2\theta_B/\lambda^2 b_c|F|$, θ_B is the Bragg angle, b_c is the coherent scattering amplitude, $|F|$ is the crystal structure factor, a is the lattice constant, and t is the thickness of the crystal.

Equations (9.1) and (9.2) satisfy $P_o/P_i + P_h/P_i = 1$. Insertion of a sheet with thickness D and refractive index n into the path induces a *phase shift* $\Delta\chi = (1-n)2\pi D/\lambda$. When the sheet is rotated by an angle ϵ, the optical path difference between the split beams becomes $\Delta D = D[\cos^{-1}(\theta_B+\epsilon) - \cos^{-1}(\theta_B-\epsilon)]$. An ordinary Al sheet with $D = 5$ mm was used in the experiment. The modulation of the superposed beam will be decreased by the effects of the wavelength width in the beam and every possible inaccuracy in configuring the interferometer.

Their first measurements were carried out with a beam after a graphite crystal monochromator with wavelength $\lambda = 2$ Å, wavelength width $\Delta\lambda/\lambda = 0.6\%$, and a cross section of $10 \times 30\,\text{mm}^2$ at the 250 kW TRIGA reactor. The modulated interference pattern was clearly demonstrated as observed in Figure 9.1b.

In the present kind of interference experiment, an index representing the degree of interference, *visibility* V, originally introduced for interference in light optics, is used:

$$V = \frac{I_{\text{max}} - I_{\text{min}}}{I_{\text{max}} + I_{\text{min}}} , \tag{9.3}$$

where I_{max} and I_{min} denote the maximum and minimum, respectively, of the interference intensity. We will also use the above terminology and the definition in this book. For example, if we roughly estimate the visibility from the pattern given in Figure 9.1b, we obtain the value for V as about 0.1–0.3. Of course, the magnitudes of the maximum and minimum depend not only on the degree of interference of neutron waves, but also on other conditions in the measurement.

An similar index is also often used to represent the degree of light and dark in an interference pattern, named *contrast*, but different definitions for *contrast* appear in the literature depending on the research fields and situations. In some cases, *contrast* is used with the same definition as the right side of Eq. (9.3), but in many cases it is used to represent simply the attenuation of the difference between light and dark parts in a pattern (i. e., attenuation of contrast), that is, $(I_{\text{max}} - I_{\text{min}})/(I_{\text{max0}} - I_{\text{min0}})$, where I_{max0} and I_{min0} are the maximum and the minimum intensity, respectively, in a reference condition. Furthermore, in a certain field, I_{max}/I_{min} is quite generally called the contrast ratio. In the present book, the word *contrast* is used with the meaning in the original reference to describe or discuss the experiment or the result.

9.1.2
Experimental Verification of the Equivalence Principle on the Gravity Effect

With the success of the silicon perfect crystal interferometer mentioned above, Overhauser and Colella immediately proposed a neutron interferometry experiment to combine quantum mechanics with gravity [394] as one of the most interesting and fundamental tests in present-day physics. The proposed principle of the interferometry experiment is shown in Figure 9.2.

Since the gravity fall of free neutrons following parabolic curve with classical gravity acceleration had already been precisely confirmed by measurements [395–397], the *Newtonian potential in the Hamiltonian* for neutrons with mass m in the interferometer shown in Figure 9.2 was assumed to be expressed by mgr on the basis of the principle of equivalence, where g is the acceleration due to gravity and r is the vertical height.

First, suppose at the outset that the paths are exactly given by the parallelogram in Figure 9.2. Then, the most naive expectation might suggest that the phase difference for the segment BC will be same as that for DE. However, in the actual situation, the phase difference for the path segment BD will exceed that for CE by $\Delta\chi = s(k_{BD} - k_{CE})$, where the wave number difference $\Delta k = k_{BD} - k_{CE}$ arises with the magnitude given by the next equation at vertical height r for the neutron with wavelength λ based on the gravity potential $m_g gr$ and the kinetic energy $\hbar^2(k - \Delta k)^2/2m_i$:

$$\Delta k \cong \frac{m_g m_i g r \lambda}{2\pi\hbar^2}, \tag{9.4}$$

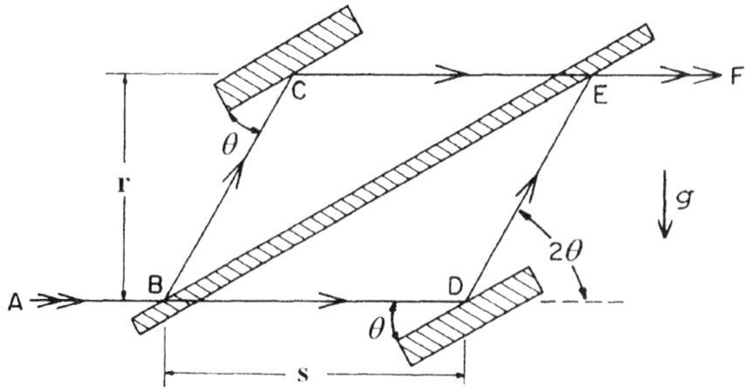

Figure 9.2 Principle of neutron interferometry to test a gravitationally induced quantum phase shift; the illustration is drawn on the plane parallel to gravity, where the neutron beams are Bragg-reflected at B, C, D, and E by protruding slabs (shaded) of a single crystal of Si. θ is the Bragg angle (Overhauser and Colella [394]).

where m_g and m_i are the gravity mass and inertia mass of the neutron, respectively, and *the principle of equivalence* predicts the equality $m_g = m_i$. The fact that Eq. (9.4) includes g and $\hbar$ indicates the combination of gravity and quantum mechanics.

As the next step, if one considers in a little more detail the actual path segments expected in the configuration shown in Figure 9.2, the neutron after the finite flight time for segment BD will strike the reflecting plane at a point a little lower than point D on the straight line in Figure 9.2, resulting in an incident angle a little larger than θ. Such a larger reflection angle will shift the reflection point E to the left, that is, to a point nearer D. However, at that point slightly shifted leftward, the horizontal partner ray passing through the same point will have an equally shorter path length and so the same phase. Therefore, the resultant phase difference due to these small deviations of the reflection point will produce no contribution to first order in g in the interference pattern by the superposition of split waves. In the same way, for segment BC, inducing a small deviation of point C to the right, that is, upward, will make the segment a little longer, and the beam will then be reflected by an angle a little smaller than θ to a point shifted a little to the right of E. This will also correspond to an equally little longer path to the same point for the rising waves from D, and so again there will be no contribution to the superposed interference pattern.

According to these considerations, the observation of the interference pattern as the first step in the configuration shown in Figure 9.2 has to be carried out. Then, as the next step, the configuration is rotated by an angle of 180° around the axis AD, and the interference measurement is repeated in the situation of the relative roles with the alternate paths interchanged. Consequently, the total shift in the interference fringe can be expressed by the number of periods as

$$\Delta N = 2\Delta\chi/2\pi = 2m_g m_i g r s\lambda/h^2 , \tag{9.5}$$

and we notice ΔN is proportional to the area, $A = rs$, enclosed by the two paths. For an interferometer with area $A = 6\,\text{cm}^2$ and neutrons with wavelength $\lambda = 1.42$ Å, the expected fringe shift becomes $\Delta N = 10.7$.

The possible effect due to elastic distortion of the interferometer during the 180° rotation can be measured by making use of X-rays, which have no mass. If the experimental result disagrees with the theoretical expectation by putting $m_g = m_i$ in Eq. (9.4), then the validity of the principle of equivalence in the quantum limit will be questioned.

The experiment was carried out by Colella, Overhauser, and Werner [398] with neutrons of wavelength 1.445 Å (silicon Bragg angle $\theta = 22.1°$) at the Ford Nuclear Reactor, University of Michigan. The configuration of their interferometer is different from that shown in Figure 9.2, which makes use of the Bragg-case reflections; instead it makes use of Laue-case reflections for both the splitter and analyzer as in the interferometer of Rauch *et al.* shown in Figure 9.1a. The interferometer consists of three protruding slabs with a thickness of 0.2 cm and a separation of 3.5 cm. The count rates I_H and I_O in the direction of the H-beam and the O-beam, respectively, were measured for angle φ by rotating the interferometer around the axis of the extension line of the incident beam. Special consideration was given to mounting the

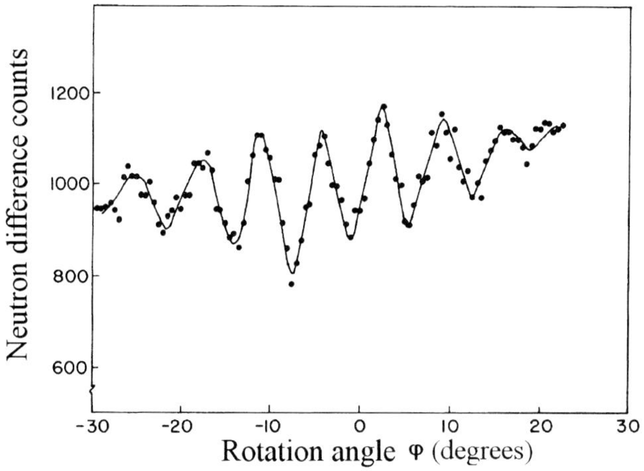

Figure 9.3 The difference counts $I_H - I_O$ for a measuring time of about 80 min as a function of interferometer rotation angle φ (Colella *et al.* [398]).

crystal to find the contribution of bending of the interferometer, which becomes constant across the transverse dimensions of the interfering beams, and such an arrangement limited rotations to $-30° < \varphi < 30°$. The sum of the count rates, $I_H + I_O$, was confirmed to be independent of rotation angle φ as theoretically expected, and therefore the interference effect was most conveniently displayed by plotting the difference $I_H - I_O$ for rotation angle φ as shown in Figure 9.3.

The periodic structure in the figure is explained as the gravity effect by putting $g \to g \sin\varphi$ in Eq. (9.5), and is connected to the phase difference $\beta = q \sin\varphi$ between the paths of the split beam in the interferometer, where $q = 2\pi\lambda g m^2 A/h^2 = 4\pi\lambda g m^2 d(d + a\cos\theta)\tan\theta/h^2$, $m = m_g = m_i$, θ is the Bragg angle, a and d are the slab thickness and separation, respectively, in the interferometer. They derived from the Fourier analyses of the experimental result the value $q_{\text{exp}} = 54.3$, whereas the theoretical value was $q_{\text{th}} = 59.6$. The discrepancy between the two values was attributed to the bending effect q_{bend} of the interferometer base during rotation, and such an explanation was supported by rotation experiments using X-rays.

Although the observed discrepancy between the theoretical expectation and the experimental result reported by Colella, Overhauser, and Werner on the gravity effect, the so-called (*COW experiment*), was mostly caused by the bending effect of the interferometer base, the possibility of some slight discrepancy remaining even after exact correction for the bending effects was assumed, and many experimental verifications and theoretical analyses were performed thereafter to investigate more accurately the reproducibility of the COW experiment.

In the gravity experiment performed by Littrell *et al.* [399, 400], special consideration was given to the effects of bending of the silicon crystal, which had a very important influence on the experimental result of the COW experiment. They thought it would be difficult to estimate exactly the bending effects from the experimental result with X-rays, since the distribution of waves propagating in the Borrmann fan

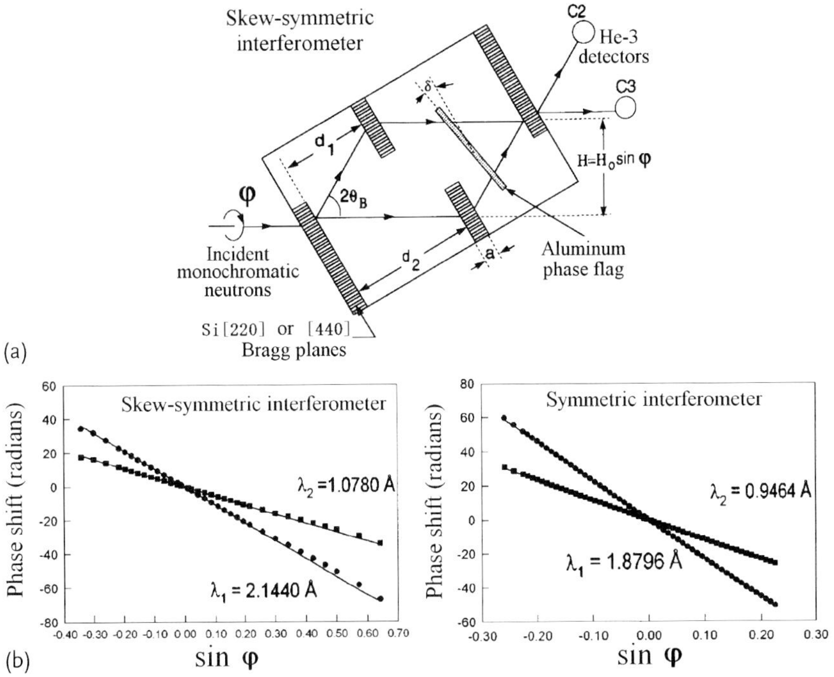

Figure 9.4 Gravity experiments with two-wavelength and two-geometry interferometer configurations. (a) The geometry of a skew-symmetrical interferometer with the blade separations $d_1 = 16.172 \pm 0.003$ mm and $d_2 = 49.449 \pm 0.003$ mm and thickness $a = 2.621 \pm 0.003$ mm. (b) Experimental results of two-wavelength measurements with the skew-symmetrical and symmetrical interferometers, The experimental points were plotted after the Sagnac effect (see the footnote on page 423) and the bending effect phase shifts had been subtracted, and the curves represent the gravity phase shifts expected from theory (Littrell *et al.* [399, 400]).

region for neutrons with a high transmission ratio should be very different from that for X-rays with high absorption in the crystal. Therefore, they employed another experimental approach to correct the bending effect without X-ray measurement, that is, by making use of two almost harmonic wavelengths, the fundamental one, λ_1, and the first higher harmonic, $\lambda_1/2$, for the silicon crystal interferometer. The beams following approximately the same paths through the interferometer and illuminated the same regions of each blade in the interferometer. Furthermore, to separate and identify possible systematic discrepancies other than the previously known bending correction, they used two interferometers of different geometry, that is, in addition to the symmetrical interferometer as used in the experiment of Colella *et al.*, analogous measurements with a skew-symmetrical one as illustrated in Figure 9.4a were also carried out. These experimental results were analyzed in detail.

The experimental results obtained at the University of Missouri research reactor are compared with the theoretical curves in Figure 9.4b. It was found the the experimental values for the wavelength-independent parameter proportional to $m_g m_i g$

for the gravity phase shift were lower than the theoretically expected value by 1.5% for the skew-symmetrical interferometer data, and by 0.8% for the symmetrical interferometer data, with relative uncertainties of 0.12 and 0.11%, respectively.

As known from Eq. (9.4), the sensitivity of the COW experiment becomes higher by making use of neutrons with longer wavelength, or of an interferometer with a larger enclosed area rs. In this respect, more recently an improved COW experiment was performed by Zouw *et al.* by employing grating mirrors for very cold neutrons; they attained an interaction time longer by two orders. Their experiment will be described in Section 9.3, the result of which gave a significantly improved accuracy of 0.4%. Their results were agreement with theory and the classical precision measurement of local gravity, which was assumed to conform to theory while the earlier anomalous results on the 1σ level were excluded [401, 402]. In these experimental situations, a future study on the principle of equivalence will be required with higher experimental accuracy.

9.1.3
Coherence Length and Dispersive/Nondispersive Arrangements

As the next task for interferometry experiments, we proceed to the investigation of the coherence length of neutrons introduced in Section 1.3. Consider how the coherence length affects the interference pattern produced by the superposition of split waves in a silicon crystal interferometer. According to textbooks on quantum mechanics (e. g., [C], Section 1, Eq. (12.20), or [D], Section 6.2.2, Eq. (6.59)), the Gaussian wave packet for a free neutron can be extended with elapsed time t in the flight to the spreading σ_x in real space as

$$\sigma_x^2(t) = \frac{1}{4\sigma_k^2} + \frac{\hbar^2 t^2}{m^2}\sigma_k^2 , \tag{9.6}$$

where σ_k is the wave number width for the amplitude of the packet.

In a silicon crystal interferometer, propagation of either of the split partial waves can be delayed by inserting a phase shifter in the path. Employing such a method, Kaiser *et al.* carried out the first experiment to measure the coherence length of neutrons in a neutron interferometer [403]. They used neutrons with wavelength $\lambda = 1.268$ Å and measured wavelength width $\Delta\lambda/\lambda \cong 0.0125$. The coherence length derived from the measured amplitudes in the interference pattern was 94 Å full width at half maximum for a Gaussian distribution. This did not correspond to the expanding width, about 5×10^6 Å, calculated from Eq. (9.6) with the flight time of neutrons in the interferometer. They concluded that the present result could be understood by supposing the minimum width $\sigma_x(0)$ of the packet is directly measured in the experiment and that there is a nonspreading coherence length for the packet arising from an asymmetrical internal structure, that is, shorter oscillation periods (i. e., faster motion) in the front part and longer periods (i. e., slower motion) in the rear part than in the packet center. Then the packet distribution spreads with time but the coherence length is invariant over the packet spread in space and in time.

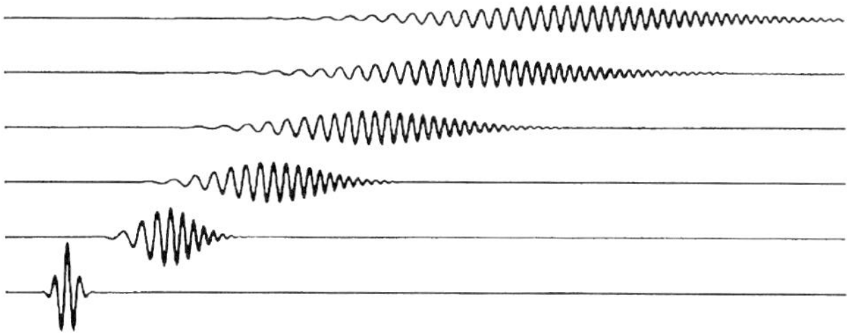

Figure 9.5 Evolution of the real part of the wave function for a propagating wave packet, indicating the effect of dispersion. The abscissa from left to right represents the direction of propagation and the time evolution is from the bottom to the top. (Klein *et al.* [404]).

At the same time, Klein *et al.* [404] also explained the phenomenon using the illustration shown in Figure 9.5 and an analysis based on the working of the phase shifter in the superposition of the split waves, both with the internal structure expanding as shown in Figure 9.5 but either of them shifted. Then the coherence length obtained from the superposition in the experiment should correspond not to the time-spreading width of the packet amplitudes, but to the magnitude $\sigma_x(0)$.

By the way, when we consider a little more carefully the configuration of the phase shifter inserted to extract an experimental value for the coherence length in the interferometer, a certain kind of attenuation effect is expected to arise owing to the working of the phase shifter in a *dispersive arrangement*, and therefore for the correct estimation of the intrinsic coherence length for a wave packet, a more well considered arrangement (belonging to the *nondispersive arrangement*) is required. In this respect, Rauch *et al.* carried out interferometry experiments [405] at the Grenoble high-flux reactor with two kinds of configuration for inserting the phase-shifting sample (Bi sample of 99.999% purity, i. e., 5N9 block) as shown in Figure 9.6.

In contrast to the conventional configuration of the crystal interferometer, or more typically in contrast to the sample configuration shown in Figure 9.6a, in the *nondispersive arrangement* shown in Figure 9.6b both the front and the rear surfaces of the sample are aligned parallel to the reflecting Bragg planes of the splitter and reflectors, that is, perpendicular to the surfaces of the plates in the interferometer. In another sense, as supposed from the refracted path through the sample, the former configuration corresponds to inducing a longitudinal (x-direction) shift of the wave packet, while the latter a transverse (y-direction) shift [406]. The attenuation ratio of *contrast* in the interference pattern can be derived by plotting the experimental results for these different arrangements as $(I_{max} - I_{min})/(I_{max0} - I_{min0})$, where I_{max} and I_{min} are the maximum and minimum, respectively, of the interference intensity, and I_{max0} and I_{min0} are those without the sample. They performed the experiments with an auxiliary Al phase shifter in addition to the Bi sample, and obtained the results shown in Figure 9.6c and d by varying the angle of the inserted Al shifter for incident neutrons with wavelength $\lambda = 1.9225$ Å and wavelength

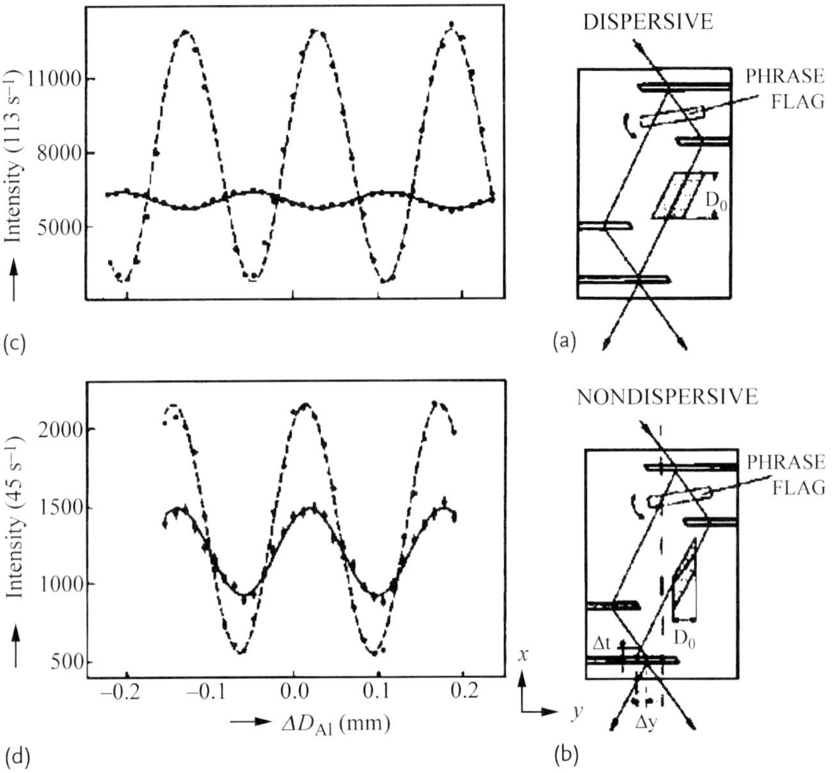

Figure 9.6 Two kinds of configuration for phase shifter insertion and the experimental results of interference contrast. (a) Dispersive arrangement (sample thickness $D_0 = 30.316$ mm) and (b) nondispersive arrangement (sample thickness $D_0 = 16.899$ mm). (c) Measured contrast of the interference pattern for the dispersive arrangement (256th period) and (d) nondispersive arrangement (248th period), with the reference pattern around the zero interference order (dashed lines) (Rauch *et al.* [406]).

width $\Delta\lambda = 0.0048$ Å. Comparison of the experimental results indicates obviously higher contrast for the case of the nondispersive arrangement.

The cause of the difference observed can be explained as follows. The wave number width of incident neutrons in the interferometer is usually broader than the dynamical reflection width mentioned in Section 6.1 of the interferometer crystal. When these neutrons are split, and further reflected in the interferometer, the wave vector component perpendicular to the crystalline plane (y-component in Figure 9.6a, b) will be reversed for the incident neutrons with wave vector within the narrow reflection width of the Bragg reflection defined by the dynamical diffraction theory, whereas the other components (x- and z-components) are unchanged during the propagation and maintain the broad wave vector distribution of the incident neutrons. According to Eq. (1.16), or referring to Eq. (1.23), the phase difference produced by the phase-shifting material is

$$\chi = -k(1-n)D_{\text{eff}} \cong -\lambda b' N D_{\text{eff}}, \tag{9.7}$$

where b' is (the real part of) the coherent scattering amplitude, and, as mentioned above, λ is strictly defined by the Bragg condition as $\lambda = 2d_{hkl} \sin \theta_B$. Therefore, such a configuration as to satisfy the condition $D_{\text{eff}} = D_0 / \sin \theta_B$ for the broad wave vector distribution in the x- and z-components of incident neutrons will be required for the nondispersive arrangement to produce a higher contrast. The arrangement of the sample in Figure 9.6b corresponds to such a configuration.

The attenuation ratio of the contrast derived from Figure 9.6d for the 248th period is 0.351. If we consider this attenuation ratio is caused by the coherence length of neutrons, it becomes about 500 Å. However, they estimated that the attenuation essentially derived from the thickness variation of the sample and the beam attenuation, with a negligible effect of the defocusing phenomenon indicated by Δt in Figure 9.6b; thus, the effect of coherency of the wave packet was only slight even if the coherency was included.

In this way, the present experimental result and the estimation of the coherence length, about 3 000 Å or more, from the Pendellösung interference experiment in a silicon perfect crystal performed by Shull [32], introduced in Section 1.3, can be considered to be compatible. As another estimation, Scheckenhofer and Steyerl reported the value for the coherence length intrinsic to a neutron, if it exists, to be larger than 10^6 Å from the diffraction pattern on an about 0.8 μm pitch grating mirror observed using the gravity diffractometer [63] shown in Section 2.1.

In the case of the sample arrangement shown in Figure 9.6a with the sample surfaces at large angles to the x-axis, the observed contrast was attenuated to a much smaller value of 0.065, which was essentially caused by the effect of the wave vector distribution width conserved on the incident neutrons as mentioned above.

Finally, if we apply these classifications of the sample arrangement to the experiment of Kaiser *et al.* introduced previously [403], it will belong to the case of the conventional dispersive arrangement for the insertion of a phase shifter.

Further discussions and experiments on the coherence length of neutrons will be reported in Section 9.6.

9.1.4
Stochastic and Deterministic Processes in the Absorbing Element

As the next interesting optical phenomenon, let us consider two typical cases for the insertion of an absorbing element for neutrons in the either of the split paths as shown in Figure 9.7a and b. In Section 1.2, the refractive index formula given by Goldberger and Seitz [15] was introduced as Eq. (1.14) for the case where absorption is neglected, but they originally derived the *refractive index formula with absorption* as follows [15]:

$$k'^2 - k^2 = (n^2 - 1)k^2 = \frac{4\pi N}{k} \left\{ \pm \left[\frac{k^2 \sigma_s}{4\pi} - \left(\frac{k^2 \sigma_a}{4\pi} \right)^2 \right]^{1/2} + i \frac{k^2 \sigma_a}{4\pi} \right\}, \quad (9.8)$$

where n is the refractive index, N is the atomic density, σ_s and σ_a are the scattering and absorption cross sections, respectively, k is the wave number without the medi-

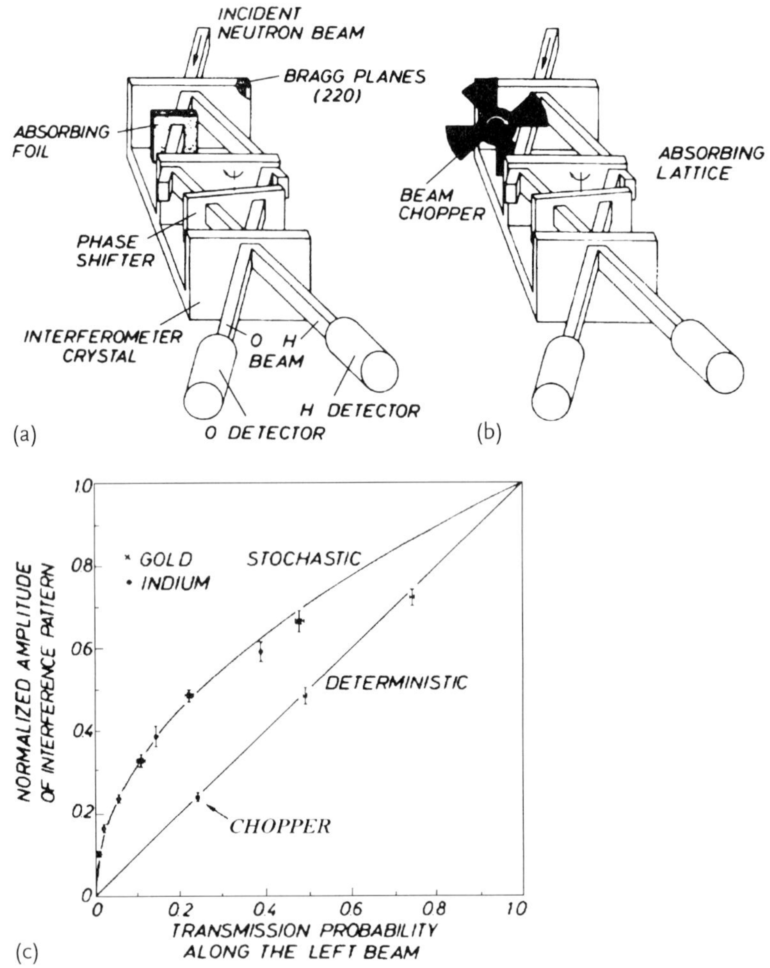

Figure 9.7 Setup for the experiments with different kinds of absorbers in the left beam path and the experimental results: (a) with gold or indium foils as absorbers; (b) with a beam chopper; (c) normalized amplitude of the interference pattern as a function of transmission probability through the left beam, which indicates the points for the foil absorbers follow closely the curve for stochastic absorption, whereas those for the chopper absorber follow the straight line for deterministic absorption. (Summhammer *et al.* [407]).

um, and k' is the wave number in the medium with absorption, which, in general, is a complex number.

Rewriting Eq. (9.8) with the coherent scattering amplitude b_{coh}, and taking the negative sign for $n < 1$ in the medium with positive coherent scattering amplitude, as mentioned in Section 2.1, one can express the forward diffracted beam (O-beam) in the superposition of split paths I and II in the interferometer as

$$\phi_{\text{O}} = \phi_{\text{OI}} + \phi_{\text{OII}} e^{i\chi} , \tag{9.9}$$

where the relative phase χ is given by the relation [407]

$$\chi = -\lambda N \Delta D \sqrt{b_c^2 - (\sigma_r/2\lambda)^2} + i\sigma_r N \Delta D/2\,, \tag{9.10}$$

where ΔD is the difference in the thickness of the phase shifter as seen along paths I and II, and σ_r is the reaction cross section including absorption and incoherent scattering processes, $\sigma_r = \sigma_a + \sigma_{\text{inc}}$.

In the case of absorbing foils, Figure 9.7a, the real part of χ can be neglected for an absorbing foil that is thin in comparison with the phase shifter, and absorption is neglected for the Al phase shifter. Then the intensity at the O-detector becomes

$$I_{Oa}(\chi_s) = |\phi_{\text{OI}} e^{-\sigma_r NL/2} + \phi_{\text{OII}} e^{i\chi_s}|^2 = |\phi_{\text{OII}}|^2 (1 + a + 2\sqrt{a}\cos\chi_s)\,, \tag{9.11}$$

where L is the absorbing foil thickness, $\chi_s = -N\lambda b_c \Delta D$ is the phase difference between paths I and II induced by the phase shifter, and $a = |e^{-\sigma_r NL/2}|^2$ is the transmission probability of the absorbing foil to the O- or H-detector along beam path I. Further, the equality $\phi_{\text{OI}} = \phi_{\text{OII}}$ was assumed for an ideal interferometer without both the absorber and the phase shifter.

In the next arrangement with a slow chopper, Figure 9.7b, since path I is either fully opened or completely blocked, with the time-averaged effective transmission probability a, the expectation for the intensity becomes

$$I_{Ob}(\chi_s) = a|\phi_{\text{OI}} + \phi_{\text{OII}} e^{i\chi_s}|^2 + (1-a)|\phi_{\text{OII}}|^2 = |\phi_{\text{OII}}|^2 (1 + a + 2a\cos\chi_s)\,. \tag{9.12}$$

The fundamental difference between these two arrangements can be considered as whether a neutron is absorbed or not at a certain instance and at a certain spot. This cannot be predicted for the foil absorber arrangement, whereas it can be predicted for the chopper with regard to the rotating blade in the spot or not. Therefore, the former situation was named stochastic absorption, and the latter was named deterministic absorption [407, 408].

The experimental results of Summhammer *et al.* obtained at the Grenoble high-flux reactor are shown in Figure 9.7c, with the transmission probability dependences of the interference amplitude closely following the corresponding theoretical lines. A remarkable point is the much larger amplitude of interference for the foil absorber compared with the chopper absorber in the region of very small transmission probability. Making use of the present characteristics enabled the coherent scattering amplitude of strongly absorbing ^{3}He gas to be successfully measured with the silicon crystal interferometer [409].

Furthermore, an *intermediate absorber element* is also possible and represents a gradual transition from deterministic to stochastic absorption. As practical devices to realize such an element, one can suppose a lattice absorber or a very fast chopper. Actually, Summhammer *et al.* [407] carried out such an experiment by using a lattice absorber, and obtained the result for interference amplitude in the transition region between the two extreme cases given in Figure 9.7c.

In the limit of a further lower transmission probability, a theory [410] predicts an additional reduction of the interference pattern, or visibility, compared with the purely stochastic case, owing to the effect of statistical fluctuations in the absorbing element. Rauch *et al.* performed an interferometry experiment in the region of such transmission probabilities by inserting Gd solutions of different concentrations in one of the paths [411]. The values obtained at the very low transmission probabilities lay slightly below the purely stochastic curve, which they thought might be caused by experimental errors or by any fundamental phenomena as mentioned above, such as fluctuations of the absorption probability [410].

9.1.5
Study of Quantum Entangled States with a Skew-Symmetrical Perfect Crystal Interferometer

To end this section, we will look at an interesting study of the so-called *postselection procedure* in which an additional spectral selection is applied to the beam behind the crystal interferometer. The results from such an experiment lead us to consider the fundamental concept of *quantum entanglement*[33] on two coherent beams split and separated from each other by a significant distance through the operation of a phase shifter in an interferometer.

The experiment was carried out by Jacobson *et al.* [412] in the configuration with a skew-symmetrical interferometer and a postselection analyzer crystal shown in Figure 9.8 by using neutrons with wavelength $\lambda_0 \cong 2.3$ Å and wavelength width $\delta\lambda \cong 0.036$ Å at the University of Missouri research reactor. The interference patterns for several kinds of bismuth phase-shifter thickness D were plotted by changing the phase rotator angle α in a fine step. Especially, the forward diffracted beam was operated by spectral filtering with a few kinds of silicon mosaic crystals and the interference patterns after the spectral selection were investigated in detail.

The experimental results given in Figure 9.9 indicate that the visibility decreased seriously at a large phase shift in the usual overall beam interference (Figure 9.9b) but can be clearly enhanced and almost recovered in the narrow spectral beam interference (Figure 9.9a). Further, comparison of the measured spectra in Figure 9.9c reveals that the spectrum in the case of severe visibility decrease was strongly modulated after the transmission through a thick phase shifter.

On the basis of this evidence, they pointed out that interference has to be treated in phase space rather than in ordinary space, and further that the plane-wave components of the wave packet, that is, narrow-bandwidth components, interact over a much larger distance than the size of the packets. Thus, *spatially separated packets could remain entangled* in phase space and nonlocality will appear as a result of this entanglement.

33) A quantum mechanical phenomenon that two or more quantum states must be described by connection to each other even though they are separated far apart from each other in real space.

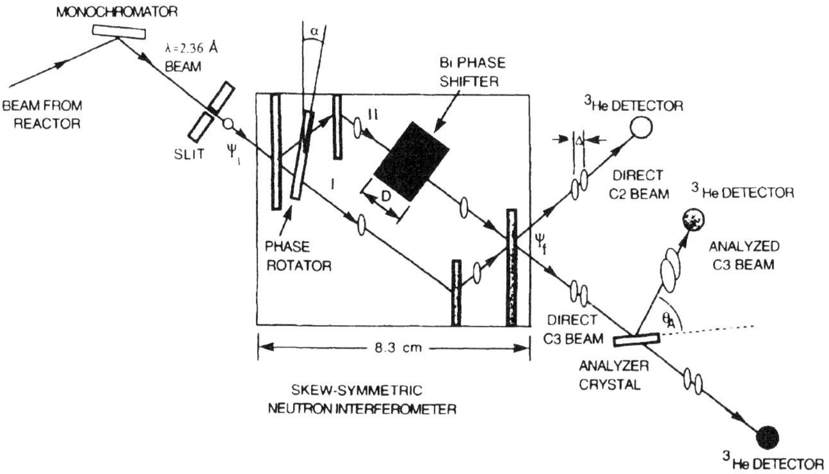

Figure 9.8 Experimental arrangement of a skew-symmetrical perfect crystal interferometer and a postselection analyzer crystal (Jacobson *et al.* [412]).

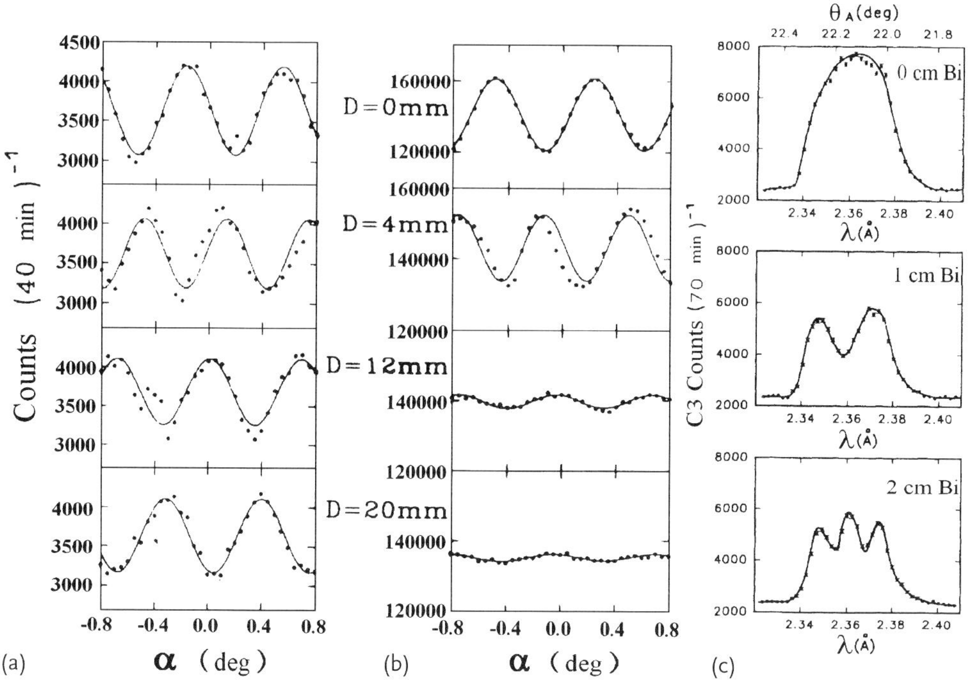

Figure 9.9 Experimental results for the interference pattern: (a) Interference pattern of the beam in a very narrow band reflected from a nearly perfect crystal analyzer ($\delta k/k_0 = 0.0003$) in the antiparallel position; (b) that of the overall beam without spectral selection ($\delta k/k_0 = 0.012$); (c) spectral modulation of the forward diffracted outgoing beam measured by angular scan of the analyzer crystal with resolution 0.002 Å for different bismuth phase-shifter thicknesses (Jacobson *et al.* [412]).

However, it must be remarked that a theoretical work on wave function collapse [413] led to the conclusion that the coherence has to be lost at a *statistical level* when many interfering particles interact with a macroscopic object. Therefore, these states coherently separated in ordinary space seem to be notoriously fragile and sensitive to dephasing effects, and are forced to transfer to a statistical mixture, owing to interactions with a macroscopic object. Further discussions on quantum entanglement will also appear in the next section.

9.2 Experiments on Spin Behavior with a Perfect Crystal Interferometer

9.2.1 4π Periodicity of Spin Precession in a Magnetic Field

The phase of the Larmor precession and the sign reversal of the wave function for a neutron by rotation through 2π radians in a magnetic field were shown by the Fresnel diffraction experiment in Section 3.4. In the present subsection, the experiment of Werner *et al.* [414] which proved for the first time with a neutron interferometer the *Larmor precession phase and the 4π periodicity* for neutrons will be described. The experimental arrangement is essentially the same as that of the COW experiment explained in the previous section, and an unpolarized neutron beam with wavelength $\lambda = 1.445$ Å was used. Instead of the Al phase shifter in Figure 9.1a, the magnetic device illustrated in Figure 9.10a to produce a variable magnetic field was inserted in split path II, and the neutron count rates I_H and I_O for the H- and O-beam, respectively, were measured for different values of the magnetic field strength. The intensity difference $I_H - I_O$ plotted for the change of the magnetic field gave an obvious interference pattern, shown in Figure 9.10b.

Inserting into Eq. (3.35) for the Larmor precession of a neutron spin in a magnetic field, or into the half of the phase difference for the transmissions of the magnetic fields $\pm B$ given by Eq. (3.54), the value of the magnetic dipole moment for a neutron; $\bar{\mu}_n = \mu/\mu_N = -1.913$, we obtain the equation for the optical path length l in the magnetic field B to produce a 4π phase difference (complete period) between the paths with and without the magnetic field in an interferometer as follows:

$$Bl = 272/\lambda \, , \tag{9.13}$$

where B is expressed in gauss, the path length l in centimeters, and the wavelength in angstroms.

On the other hand, the magnetic field measurement on the present magnet in the interferometer gave the value for the effective magnetic path integral Bl as

$$\langle Bl \rangle = 2.7 B_{\text{gap}} \, (\text{G cm}) \, , \tag{9.14}$$

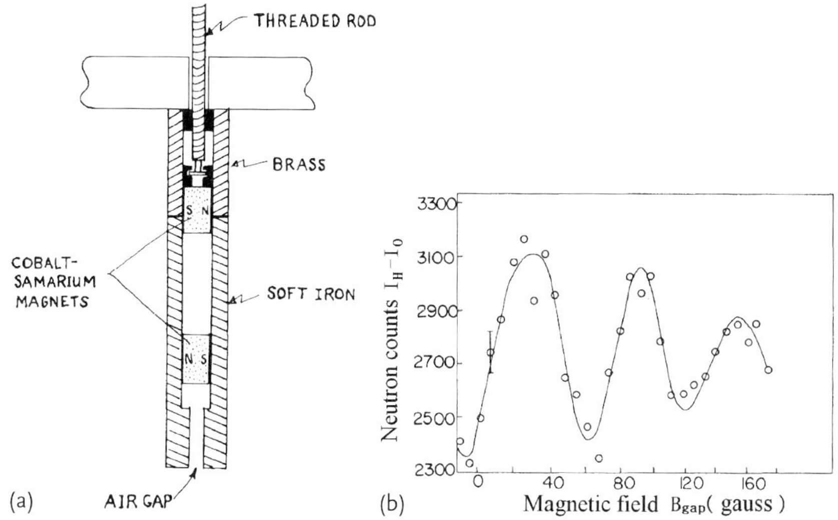

Figure 9.10 Observation of the phase shift of a spin-precessing neutron in a magnetic field with a silicon crystal interferometer: (a) configuration of the magnetic device used in the experiment; (b) neutron count difference $I_H - I_O$ between the deviated and forward diffracted beams plotted as a function of the magnetic field in the magnet air gap. The counting time was approximately 40 min per point (Werner *et al.* [414]).

where B_{gap} is the magnetic field in the magnet air gap. The oscillation period obtained by interferometry, Figure 9.10b, is 62 ± 2 G. Thus,

$$\langle Bl \rangle = 242/\lambda \,, \tag{9.15}$$

which agreed with the theoretical value within the estimated experimental errors.

Furthermore, Grigorie *et al.* and Kraan *et al.* observed the neutron interference and 4π periodicity of the spinor by making use of spin rotation in a magnetic field gradient [415, 416].

9.2.2
Complementary Principle in a Double-Resonance Experiment

The *complementary principle* introduced by Bohr in 1928 is one of the most important principles in quantum mechanics. It requires the some of the elements in complementary relation to each other to make up a complete classical description are actually mutually exclusively, and yet these complementary elements are all necessary for the description of various aspects of the phenomenon. The principle states that atomic phenomena cannot be described with the completeness demanded by classical dynamics. For example (referring to textbook [C]; Chapter 1, Section 3), in the case of a double-slit diffraction experiment, performing a measurement to obtain the interference pattern at a sufficiently high resolution to precisely determine the angular (i. e., momentum) distribution of particles and setting the position indicator behind the slits with sufficiently high resolution to definitely determine which slit a particle passed through are not compatible, since the momentum and the position of a particle are just such fundamental elements as required by the complementary principle to be complementary.

As another expression (referring to textbook [A]; Chapter I, Section 1), if the use of a classical concept excludes that of another, we call both concepts (e. g., position and momentum coordinates of a particle) *complementary* (to each other), following Bohr.

The result of the time-dependent experiment introduced below accompanied by the energy transfer within a resonance coil in an interferometer has to be discussed in terms of the mutually exclusive relation of two elements, the energy and the time on a particle, in the sense of the complementary principle, or in other words the energy–time uncertainty relation and the particle number–phase uncertainty relation [417, 418].

Badurek *et al.* carried out a series of experiments demonstrating the time-dependent superposition of spinors by making use of a radiofrequency spin-flipping device inserted into split paths in a silicon crystal neutron interferometer [417, 418]. Especially, the *double-resonance experiment* [418] is very interesting in connection with the complementary principle mentioned above and so will be introduced below.

In the experimental arrangement shown in Figure 9.11a, the spin-reversal process caused by the radiofrequency spin flip coil is associated with a change of the total energy of neutrons according to emission or absorption of photons. The energy transfer according to emission or absorption of a photon was directly measured in a separate experiment by Alefeld *et al.* by means of a high-resolution neutron backscattering spectrometer [419]. The findings indicated two obviously split peaks with energy separation $4\mu B_0$ after energy transfer of $\pm 2\mu B_0$ owing to the spin reversal of $\mp$spin components, respectively, of the unpolarized incident neutrons, when the unpolarized monochromatic incident neutrons pass through a resonance spin flip device satisfying the resonance frequency condition $\omega_{\mathrm{HF}} = \omega_L = 2\mu B_0/\hbar$ and the resonance amplitude condition $B_1 = \pi\hbar/\mu l$ for the *Zeeman energy splitting*. Thus, the energy of a neutron detected in the backscattering spectrometry experiment was either given or taken off the energy $2\mu B_0$.

Badurek *et al.* employed in their double-resonance experiment with the configuration shown in Figure 9.11a the method of *stroboscopic measurement*, in which the beat period $T = 1/(\nu_{r1} - \nu_{r2})$ for two slightly different radiofrequencies ν_{r1} and ν_{r2} was subdivided synchronously into eight consecutive time intervals of equal width τ_0 to gate the inputs of eight separate scalers. Then, the stroboscopic registration transformed the interference beating into a stationary intensity distribution as a function of the time shift τ; $\tau_n = n\tau_0$, $n = 1, 2, 3 \ldots 8$, with respect to an arbitrarily chosen reference time point.

Prior to the double-resonance experiment, a single-resonance experiment with the radiofrequency coil inserted into only one of the split paths had been performed [417]. In this case, at the resonance condition of the radiofrequency, $\omega_{\mathrm{rf}} =$

$2\mu B_0/\hbar$, the interference pattern of *stroboscopic measurement* synchronized to the single resonance frequency indicated the Larmor precession on the x-y-plane for the spin of the superposed beam, but no counting difference was observed for the z-component, that is, there was no z-component of the polarization. In this way, the energy change by the amount $2\mu B_0$ for one of the split beams for the incident neutrons polarized in the z-direction resulted in the spin lying on the x-y-plane of the superposed beam at the exit of the interferometer.

In both arrangements of these resonance experiments, the interference pattern should become time-dependent if the energy difference is induced between two split paths as a result of energy change by photon transfers; then stroboscopic measurement is required. In the case of the double-resonance experiment, a quantum beat condition from the difference of two radiofrequencies in synchronous operation arises between the spins of partial waves reversed in both of the split paths as indicated by arrows in Figure 9.11a.

The experiment was performed at the Grenoble high-flux reactor by using neutrons with wavelength $\lambda = 1.893$ Å, wavelength width $\Delta\lambda/\lambda = 1.2\%$, and polarized by a couple of magnetic prisms (illustrated in Figure 3.15a) in series and then incident in the skew-symmetrical interferometer shown in Figure 9.11a. There is a finite path length of several centimeters for two coherent subbeams propagating in parallel directions. This provides the advantage of facilitating the installation of devices in one of the subbeams without them influencing each other. Each of the radiofrequency flip coils consists of a single layer of 103 turns and connected to the output stages of a high-performance stereo amplifier. The calculated resonance amplitude was 0.55 A, somewhat lower than the experimentally determined value of 0.60 ± 0.03 A. An initial degree of polarization $P_i = 0.91 \pm 0.02$ and a maximum spin-flip probability $W_{max} = 0.99 \pm 0.02$ of both coils were found at resonance fre-

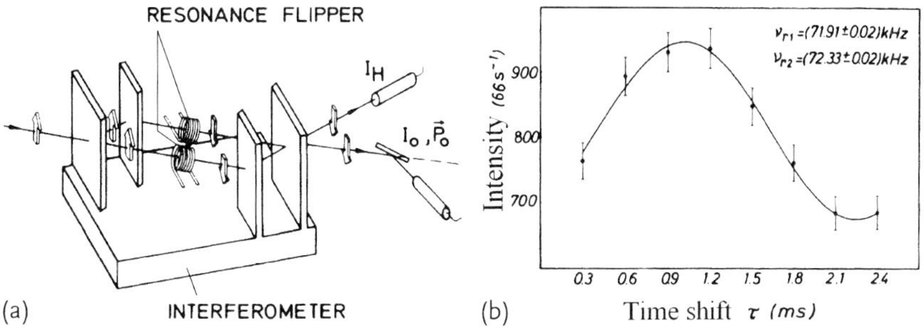

Figure 9.11 Double-resonance experiment with a silicon crystal neutron interferometer: (a) Radiofrequency flip coils separately but operated synchronously are inserted in each subbeam within the skew-symmetrical interferometer, and a Heusler crystal analyzes the polarization of the O-beam; (b) *stroboscopically measured* interference pattern at a frequency difference of 420 Hz between the two radiofrequency fields (Badurek *et al.* [418]).

quency $f_r \sim 72\,\text{kHz}$. There exists a large Bloch–Siegert shift[34] with respect to the Larmor frequency $f_L = 58.3\,\text{kHz}$ of the neutrons in the static field $B_0 = 2.0\,\text{mT}$, which was produced by a Helmholz solenoid covering the whole region between the polarizing prisms and the Heusler crystal analyzer. By means of an electronic phase-shift circuit, 24 discrete values of phase differences within the interval 0, 2π could be established between the radiofrequency coils. Any time dependence of the polarization behind the interferometer which varied synchronously with either the frequencies of the radiofrequency coils or their difference, respectively, could be discovered by means of a *stroboscopic detection method* based on an electronic phase-locked-loop circuitry. The efficiencies of both flip coils with the rather large frequency difference of 420 Hz remained as high as about 98% because of the extremely broad resonance curve.

Considering the neutron flux used, we can suppose only one neutron exists at most within the interferometer. Therefore, if we could obtain from the stroboscopic measurement any interfering beat pattern corresponding to the difference between two radiofrequencies, we could say that a neutron emits or absorbs two photons, each one for the radiofrequency in each of the two split paths[35], while at the same time the neutron shows a wave nature as the self-interference pattern in the superposition of the split beams.

The result of the stroboscopic measurement shown in Figure 9.11b indicates an obvious interference pattern with the period corresponding to the frequency difference of 420 Hz. Furthermore, they also performed a similar measurement with an exceedingly small frequency difference of 0.002 Hz between the two resonance coils and hence obtained a macroscopically observable beat effect in real time with a period of about 500 s without any special stroboscopic electronics.

A series of these experiments provided some fundamental insight into the remarks by Vigier *et al.* [420] relating to the complementary principle in quantum mechanics which had stimulated the present double-resonance experiment.

In the first experiment with a single resonance coil in one of the split paths, the exchange of energy in one of subbeams caused a time dependence of the interference pattern, but in order to convert the time-dependent interference pattern into a stationary one by means of stroboscopic neutron registration, the phase of the radiofrequency field had to be known with accuracy $\Delta\varphi < 2\pi$, and because of the particle number–phase uncertainty relation, $\Delta N \Delta\varphi \geq 2\pi$ [417], the mean number of photons would be known only to $\Delta N > 1$. Therefore, it was in principle impossible to detect single-photon transitions simultaneously with the interference pattern.

In this context of the energy transfer not representing a measuring process, Vigier *et al.* [420] proposed an experiment with two separate resonance coils inverting

34) The shift of the magnetic resonance frequency by about [(rotating magnetic field)/(axial direction static magnetic field)]^2 induced by a magnetic field with frequency twice that of the Larmor frequency.

35) Then, one would further want to assure with an ultra-high resolution time-of-flight spectroscopy of neutrons behind the interferometer whether the neutron could actually exchange the energy for flight by the net amount of the difference or summation of those of two photons.

the spin state of each subbeam within an interferometer. According to the result of the double-resonance experiment introduced above, Vigier *et al.* claimed that because of the indivisibility of each exchanged photon, the energy transfer had to be associated with the particle properties of the neutron, and concluded that if one accepts Einstein's postulate on energy–momentum conservation in all individual microprocesses, the individual neutrons inside the interferometer behaved as waves *and* particles simultaneously [421], whereas in the Bohr–Heisenberg interpretation of quantum mechanics the complementary principle permits a manifestation of particle *or* wave properties only. On the other hand, Badurek *et al.* reported [418] that the theoretical predictions for the outcome of the double-resonance experiment they performed were identical for both interpretations and that all of the observed interference phenomena can be easily explained in terms of wave mechanics, but that at the same time well-defined particle properties such as mass, spin, and the associated magnetic moment can also be attributed to the neutron.

9.2.3
Experimental Test of the Einstein–Podolsky–Rosen Paradox

In this subsection, we proceed to the next fundamental subject of quantum mechanics: to test experimentally the *Einstein–Podolsky–Rosen paradox (EPR paradox)* [422]. As one of the typical products in the very fundamental discussions between Einstein and Bohr at the early stage of quantum mechanics, Einstein, Podolsky, and Rosen [422] submitted in 1935 a criticism of quantum mechanics in the form of a paradox based on the analysis of a *gedankenexperiment.* In this criticism, they insisted that the following requirements seem to be necessary for a correct and complete physical theory:

1. *Every element of the physical reality must have a counterpart in the physical theory.* (The condition of completeness.)
2. *If, without in any way disturbing a system, we can predict with certainty (i.e., with probability equal to unity) the value of a physical quantity, then there exists an element of physical reality corresponding to this physical quantity.* (The condition of reality.)

To examine whether quantum mechanics satisfies both of these criteria, they performed a *gedankenexperiment* in which they supposed two systems interacted from time $t = 0$ to $t = T$, then after that time there was no longer any interaction between them. The conclusion they arrived at from the gedankenexperiment was that the description of reality as given by a wave function is not complete.

As further work to advance the EPR paradox by testing it through a practical physical experiment, Bell developed the theoretical analysis [423] with the variant gedankenexperiment advocated by Bohm and Aharonov [424] on the EPR paradox, as an argument that quantum mechanics could not be a complete theory but should be supplemented by additional variables. He formulated his idea mathematically and presented an inequality, so-called *Bell's inequality*, to be satisfied by the correlation in the experimental results if there exists such an additional variable, that is, any local, and not being quantum mechanical, *hidden variable*, which up to now could not be observed. These additional variables are thought to restore to the theory causality and locality, and be incompatible with the statistical predictions of quantum mechanics. Therefore, it was pointed out to be crucial for the definite test experiment in which there is no allowance for the measuring devices to reach mutual agreement by exchange of signals with the velocity of light.

A few years after Bell's work, Clauser, Horne, Shimony, and Holt (CHSH) further extended the theorem of Bell which proves that certain predictions of quantum mechanics are inconsistent with the entire family of local hidden-variable theories to a more practical formula applicable to realizable experiments [425]. To test the EPR paradox based on the locality assumption that the setting of one measuring device does not influence the reading of another remote instrument, they considered an experiment consisting of an ensemble of correlated pairs of particles which move so that one enters apparatus I_a located in direction $\boldsymbol{a}$ and the other enters apparatus I_b in direction $\boldsymbol{b}$. In each apparatus, a particle must select one of two channels labeled +1 and −1. As a result of the analysis, they presented an advanced version of Bell's inequality, known as *CHSH's inequality*, as follows:

$$-2 \leq S \leq 2\,, \tag{9.16}$$

where

$$S = E(\boldsymbol{a}, \boldsymbol{b}) - E(\boldsymbol{a}, \boldsymbol{b}') + E(\boldsymbol{a}', \boldsymbol{b}) + E(\boldsymbol{a}', \boldsymbol{b}')\,, \tag{9.17}$$

and $E(\boldsymbol{a}, \boldsymbol{b})$ is the correlation coefficient in the detection of the pair of particles defined by

$$E(\boldsymbol{a}, \boldsymbol{b}) = P_{++}(\boldsymbol{a}, \boldsymbol{b}) + P_{--}(\boldsymbol{a}, \boldsymbol{b}) - P_{+-}(\boldsymbol{a}', \boldsymbol{b}) - P_{-+}(\boldsymbol{a}, \boldsymbol{b})\,, \tag{9.18}$$

where $P_{\pm\pm}(\boldsymbol{a}, \boldsymbol{b})$ is the probability of obtaining the results $\pm$ in the measurement along $\boldsymbol{a}$ for particle 1 and at the same time the results $\pm$ in the measurement along $\boldsymbol{b}$ for particle 2.

The experimental realization of the situation proposed by Bell and CHSH to verify whether the local hidden variables exist or not was actually carried out about 10 years later by Aspect *et al.* for the case of photons [426]. They measured the linear polarization correlation of the photons emitted in a radiative atomic cascade of calcium, and derived the value for S in Eq. (9.17) as $S_{\text{obs}} = 2.697 \pm 0.015$, strongly violating CHSH's inequality, Eq. (9.17), that is, the generalized Bell inequality, and in excellent agreement with the quantum mechanical predictions. Therefore, the argument mentioned above based on the *realistic local theories* to indicate the incomplete features of quantum mechanics is considered at present became difficult for a system such as photons in atomic cascade decays. As explained in Section 9.1.5, such a quantum phenomenon with correlated states in a certain coordinate in phase space between the particles or between the split parts of a particle in spite of being located far apart from each other in real space is called *quantum entanglement*.

With regard to these situations, testing Bell's inequality or its generalized CHSH inequality by making use of neutrons as matter waves is a very interesting and important task. Recently, Hasegawa *et al.* performed a neutron-spin correlation experiment with a perfect crystal interferometer [427]. In the case of the neutron interferometry experiment, the *quantum entanglement* realizable in actual experiments is related not to the correlation between two particles as in the case of photons, but to the different freedoms of a particle, that is, the *quantum entanglement between different freedoms*, say, real space and spin in this case, of a single particle.

They employed a Bell-like inequality for single-neutron interferometry by making use of the quantity S':

$$-2 \leq S' \leq 2\,, \tag{9.19}$$

where S' is expressed with expectation value E',

$$S' = E'(\alpha_1, \chi_1) - E'(\alpha_1, \chi_2) + E'(\alpha_2, \chi_1) + E'(\alpha_2, \chi_2)\,, \tag{9.20}$$

and $E'(\alpha, \chi)$ is the expectation value in a spin-path joint measurement in neutron interferometry with the configuration shown in Figure 9.12a. They used this to find neutrons with spin state $(|\uparrow\rangle + e^{i\alpha}|\downarrow\rangle)/\sqrt{2}$, and path state $(|\mathrm{I}\rangle + e^{i\chi}|\mathrm{II}\rangle)/\sqrt{2}$, analogously to the expectation value of a joint measurement for correlated-photon-pair experiments.

In the experimental configuration shown in Figure 9.12a, a spin-turner is inserted between the first and second plates of the interferometer. The spin-turner is a soft-magnetic MuMetal sheet of 0.5-mm thickness in an oval ring form and provided with two DC coils for magnetization. It induces an additional relative phase shift between two beams, I and II, in the interferometer, working as the *spin-phase shifter*, the induced relative phase difference being adjusted by rotation [428], as for the spatial phase shifter based on the nuclear scattering amplitude inserted between the second and third plates of the interferometer. In all interference oscillations in the present experiment given in Figure 9.12, the relative spin phase was shifted by about π radians.

By making use of the formula for the superposed beam given in Section 3.4, the expectation value $E'(\alpha, \chi)$ for the joint measurement of neutron count rates with a single detector for the spinor and the path is given by the formula [427]

$$E'(\alpha, \chi) = \frac{N'(\alpha, \chi) + N'(\alpha + \pi, \chi + \pi) - N'(\alpha, \chi + \pi) - N'(\alpha + \pi, \chi)}{N'(\alpha, \chi) + N'(\alpha + \pi, \chi + \pi) + N'(\alpha, \chi + \pi) + N'(\alpha + \pi, \chi)}\,, \tag{9.21}$$

tand is calculated from the measured count rates, where $N'(\alpha_j, \chi_k)$ is the count rate with spin rotation α_j and spatial phase shift χ_k. Thus, the present expectation value was thought to correspond to the correlation coefficient $E(\boldsymbol{a}, \boldsymbol{b})$ in the conventional EPR argument.

Quantum theory predicts a sinusoidal behavior for the count rate, $N'_{\mathrm{qm}}(\alpha.\chi) = [1 + \cos(\alpha + \chi)]/2$, and the same behavior is also derived for the expectation value, $E'_{\mathrm{qm}}(\alpha.\chi) = \cos(\alpha + \chi)$. These functions will lead to the violation of Bell-like inequality for various sets of polarization analysis and phase shift. The maximum violation is expected, for instance, for the following set: $\alpha_1 = 0$, $\alpha_2 =$

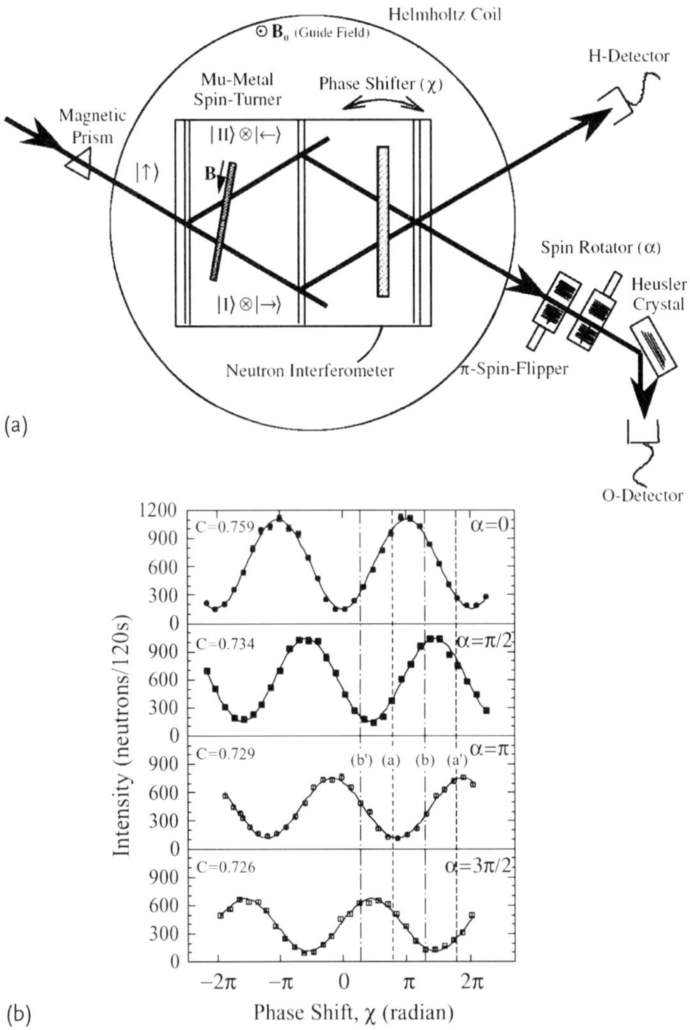

Figure 9.12 Experiment to demonstrate the violation of a Bell-like inequality in single-neutron interferometry: (a) experimental setup consisting of three stages, that is, preparation of the entangled state using a spin-turner, manipulation of the spinor rotation angle α and the spatial phase shift χ, and detection; (b) typical interference oscillations with contrasts of 76% for one and about 73% for the other three, where the dotted and broken lines indicate (a) $\chi = 0.78\pi$, (a$'$)$\chi = 1.78\pi$, (b) $\chi = 1.29\pi$, and (b$'$)$\chi = 0.29\pi$ for deriving the expectation values (Hasegawa *et al.* [427]).

$\pi/2$, $\chi_1 = -\pi/4$, $\chi_2 = \pi/4$, and if the experiment is ideal, the result should indicate $S' = 2\sqrt{2} = 2.82 > 2$. In an actual experiment, however, the visibility can be reduced owing to various causes, and when the visibility is reduced, the expectation value E'_{obs} is reduced as well. Therefore, to demonstrate the violation of the Bell-like inequality, a visibility higher than $1/\sqrt{2} = 70.7\%$ is required for the experiment.

The experiment of Hasegawa *et al.* was carried out with the silicon perfect crystal interferometer at the S18 beamline of the Grenoble high-flux reactor. The neutron beam was monochromatized to have a mean wavelength of $\lambda_0 = 1.92\,\text{Å}$ by the use of a silicon perfect crystal monochromator. The interference pattern obtained indicated a visibility higher than 70% around the measuring condition mentioned above, as shown in Figure 9.12b. From these experimental data, they derived the following results for the expectation values: $E'_{\text{obs}}(0, 0.79\pi) = 0.542 \pm 0.007$, $E'_{\text{obs}}(0, 1.29\pi) = 0.4882 \pm 0.012$, $E'_{\text{obs}}(0.5\pi, 0.79\pi) = -0.538 \pm 0.006$, and $E'_{\text{obs}}(0.5\pi, 1.29\pi) = 0.438 \pm 0.012$. Inserting these values led to the result $S'_{\text{obs}} = 2.051 \pm 0.019$.

Although this experimental result is not so distinct in violating the Bell-like inequality as the previously mentioned result from the two photons in atomic cascade decays, the value derived from the single-neutron experiment slightly exceeds the critical value of 2. Since it does not seem possible at present to undertake such a two-neutron correlation experiment as performed with photons, it will be meaningful to carry out further a single-neutron experiment with different experimental configurations or different physical situations to verify whether Bell-like inequality is violated or not.

9.3
Very Cold Neutron Interferometry with Grating Mirrors

9.3.1
Development of a Grating Mirror Interferometer and Experiment on the Gravity Effect

There are various advantages in using a perfect single crystal interferomerter, which is made as a whole of a single crystal. On the other hand, there are a couple of serious restrictions: the size of the interferometer, and as a result the spatial spitting distance between the subbeams is limited, and that it can only be applied to thermal neutrons capable of being Bragg-scattered by the crystal lattices. Therefore, other types of interferometers are required for application of neutrons with longer wavelengths, such as cold or ultracold neutrons, since they have very attractive properties for optical experiments as already described in previous chapters. Actually, various development works for such kinds of *interferometers for long-wavelength neutrons* have been performed up to now.

For example, Ioffe *et al.* [429] prepared and tested a *diffraction grating interferometer for cold neutrons* manufactured by vacuum deposition of an about 1500 Å ^{58}Ni film onto a diffraction grating with a symmetrical photolithographed rectangular profile 70-mm square with a depth of about 1000 Å and a period of 21 μm on a glass plate, and discussed the application to ultracold neutrons.

On the other hand, Pruner *et al.* [430] developed a *volume-phase grating interferometer for cold and very cold neutrons* utilizing holographical production of the gratings from photosensitized slabs of deuterated poly(methyl methacrylate).

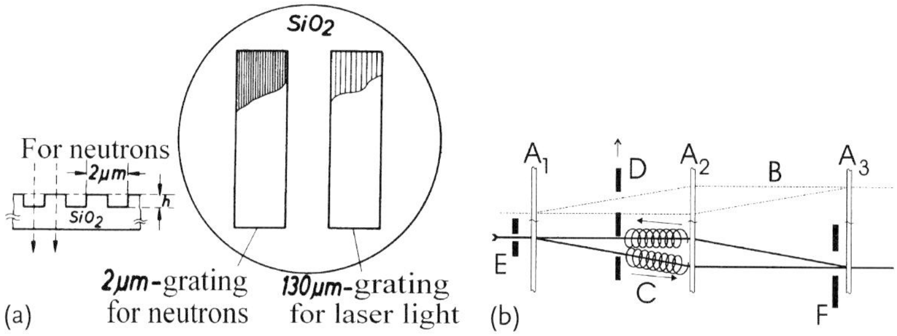

Figure 9.13 A very cold neutron interferometer. (a) Phase gratings; quartz glass plate with a phase grating for neutrons and a light-absorption grating (right, not to scale) and a cross section of the phase grating (left), where the height *h* determines the phase shift (Guber *et al.* [431]). (b) Top view of the neutron interferometer; solid lines, neutron paths; dotted lines, laser light paths; A_1, coherent beam splitter; A_2, deflector; A_3 analyzer; C, two antiparallel solenoids; D, chopper selecting the neutrons to be subjected to the field pulse; E, second slit of the collimator; F, shielding slit; C and D were both inserted for the Aharonov–Bohm experiment (Zouw *et al.* [401]).

In this subsection, an *interferometer for very cold neutrons using phase gratings in quartz-glass plates* developed by collaboration of Zeilinger's group and Gähler *et al.* will be described by introducing an experimental example applied for study of the gravity effect [431].

A sketch of the structure of their *diffraction gratings* is shown in Figure 9.13a. Two gratings are arranged side by side on a piece of 2.5-mm-thick quartz glass. The grating used for neutrons was sputter-etched into the quartz, whereas the grating for light was an absorption grating produced by evaporation of chromium onto the quartz substrate. Both gratings are parallel to each other to better than 0.1 μm over the whole length of the line. The grating constant for the absorption grating was chosen to be 130 μm to obtain the same diffraction angle (5 mrad) for He–Ne laser light as for the $\lambda = 102$ Å neutrons at the 2-μm phase gratings. The interferometer employs three diffraction gratings, A_1, A_2, and A_3 in parallel geometry as shown in Figure 9.13b, and the entire setup was located for the test at the exit of a very cold neutron guide as one branch bypassing the ultracold neutron turbine (see Figure 2.18a) on the level-D floor at the Grenoble high-flux reactor. The result of the test measurements indicated an obvious interference pattern with a maximum contrast of over 60% in the O-beam count rate as a function of the first grating position [431].

As mentioned previously at the end of Section 9.1, Zouw *et al.* carried out an experiment to measure the phase shift due to the gravity effect by tilting the interferometer as illustrated in the inset in Figure 9.14. The experiment was performed on the top floor of the Grenoble high-flux reactor by making use of the incident very cold neutrons with an average wavelength of 9.5 nm and a wavelength width of about 4 nm full width at half maximum.

The experimental results [401] are given by solid squares in Figure 9.14. The least-squares fit of the theoretical formula with the gravitational phase shift, $\beta = q_0 A\lambda g \sin\varphi$, where $q_0 = 2\pi m_i m_g/h^2$, A is the area enclosed by the interferometer

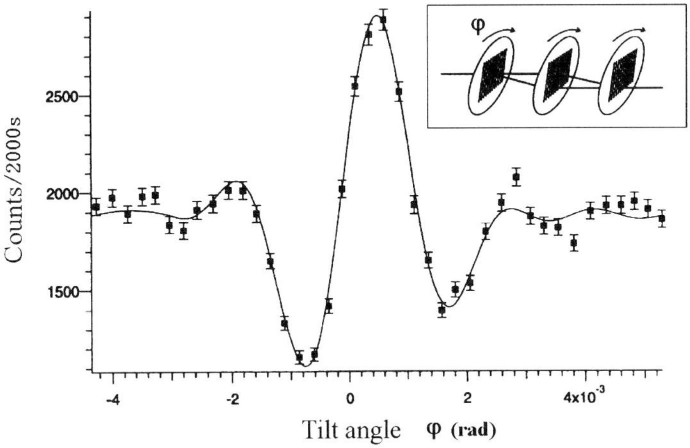

Figure 9.14 Experimental result of the gravitational phase shift (solid squares) as a function of the tilt angle φ (refer to the inset). The solid line represents a least-squares fit to the theoretical formula based on the measured neutron spectrum (Zouw *et al.* [401]).

paths, and φ is the angle between the normal on A and the direction of g, to the experimental data points based on the measured wavelength spectrum and the local gravitational acceleration $g = 9.80507(2)$ measured by a precision spring balance at the experimental site gave the value of

$$q_{0\text{exp}} = 4.012(36) \times 10^{13}\,\text{rad}\,\text{s}^2\,\text{m}^{-4}\,, \tag{9.22}$$

The value obtained is consistent within the measurement accuracy of 1% with the theoretical value independent of experimental parameters:

$$q_{0\text{theor}} = 4.014763681(98) \times 10^{13}\,\text{rad}\,\text{s}^2\,\text{m}^{-4}\,, \tag{9.23}$$

derived from the precise ratio of the Planck constant to the neutron mass and the assumption that $m_i = m_g$.

Zouw carried out an improved measurement with a significantly higher accuracy of 0.4% by making use of a tailored incoming neutron spectrum through a filter, and obtained the result

$$q_{0\text{exp}} = (4.023 \pm 0.015[\text{stat.}] \pm 0.007[\text{syst.}]) \times 10^{13}\,\text{rad}\,\text{s}^2\,\text{m}^{-4}\,, \tag{9.24}$$

in agreement with the theoretical value to the 1σ level [402], that is, much better than the result obtained by Littrell *et al.* discussed in Section 9.1.

As mentioned in Section 9.1, the experimental result for the gravity effect obtained by Littrell *et al.* indicated a disagreement of around 1% in magnitide with the theoretical value, but the result obtained by Zeilinger's group given in this subsection removed the disagreement by such a significant magnitude. In this way, the disagreements of the earlier experimental results with the theoretical expectation by putting $m_g = m_i$ in Eq. (9.4), which might have suggested the validity of the principle of equivalence should be questioned in the quantum limit, a long-standing issue in the field of neutron interferometry since the experiment with thermal neutrons by Colella *et al.*, were much improved to support the principle of equivalence to the level of 10^{-3} in the very cold neutron experiment. Therefore, a higher experimental accuracy would be required to undertake a further study on the principle of equivalence[36].

9.3.2 Experimental Verification of Aharonov–Bohm Effect

As a starting point for the present problem, let us consider the wave function of a particle acted on by forces. It is given, for example, by Pauli in his textbook [A], Section 4, as follows.

The description of the states of a system of particles subject to forces emerged as a generalization of that for free particles. Evidently, the corresponding concepts and laws should be consistent with one another and contain the laws of classical particle mechanics as limiting cases.

First, we shall give the modifications necessary in the wave equation in the presence of an external magnetic field. If A_k are the components of the *vector potential*, e the charge of the particle, and c the velocity of light, the magnetic field strength is given by[37]

$$B_{kl} = \frac{\partial A_l}{\partial x_k} - \frac{\partial A_k}{\partial x_l} . \tag{9.25}$$

The electric field strength has an additional term

$$E_k = -\frac{1}{c}\frac{\partial A_k}{\partial t} \tag{9.26}$$

if A_k depends explicitly on the time.

The fact that the classical *Hamiltonian* is obtained from the one without the magnetic field by replacing p_k by $p_k - \frac{e}{c}A_k$ makes it reasonable to write the wave equation of the particle in a magnetic field by replacing the operator $\frac{\hbar}{i}\frac{\partial}{\partial x_k}$ by $\frac{\hbar}{i}\frac{\partial}{\partial x_k} - \frac{e}{c}A_k$ in the wave equation without the magnetic field.

It is an important fact that the potentials A_k are determined only up to an additional gradient term, since by such an addition the magnetic field strengths B_{kl} are not altered. Hence, it is permitted to make the substitution

$$A'_k = A_k + \frac{\partial f}{\partial x_k} , \tag{9.27}$$

36) In this connection, in the gravity effect study by making use of an ultracold atom interferometer with laser-cooled atoms, the experimental accuracy of $\Delta g/g \cong 3 \times 10^{-9}$ was reported as being attained [432].

37) We prefer to write $\boldsymbol{B}$ as an antisymmetrical tensor ($B_{kl} = -B_{lk}$) so that B_{23}, B_{31} and B_{12} denote, respectively, the 1-, 2-, and 3-components of $\boldsymbol{B}$. The vector product $[\dot{x}, \boldsymbol{B}]$ then has the 1-component $\dot{x}_2 B_{12} - \dot{x}_3 B_{31}$ and this is, in fact, equal to $\Sigma_l B_{1l}\dot{x}_l$, since $B_{31} = -B_{13}$, $B_{11} = 0$.

where f is an arbitrary function of the position coordinates. Since f can contain the time explicitly, one has to set, at the same time,

$$U' = U - \frac{e}{c}\frac{\partial f}{\partial t} \tag{9.28}$$

to keep the expression for the force.

In the wave equation, not only the magnetic and electric field strengths but also the potentials U and A_k enter. If ϕ is a solution of the wave equation with the potentials U and A_k, we obtain a solution ϕ', with the potentials U' and A'_k given by Eqs. (9.27) and (9.28) if we make the substitution

$$\phi' = \phi \exp\left(\frac{ie}{\hbar c} f\right) . \tag{9.29}$$

The group of substitutions defined by Eqs. (9.27)–(9.29) is called the *gauge group*. Quantities which do not change under these substitutions are called *gauge-invariant* quantities. It is worth noting that not only the probability density $\phi\phi^*$ but also the current $\boldsymbol{i}$ and the stress tensor T_{kl} are gauge-invariant quantities. From this point of view the form of the wave equation and in particular the special choice of the Hamiltonian operator in the wave equation must be considered as quite natural. On the other hand, the wave equation depends essentially on the assumption that the field quantities U and A_k themselves can be considered as classical quantities (given space–time functions) of such a kind that the possible influence of the quantum of action on the definition of these field quantities can be disregarded.

In 1959, Aharonov and Bohm [433] pointed out very interesting properties of the electromagnetic potentials in the quantum domain as follows. In classical electrodynamics, the *vector* and *scalar potentials* were first introduced as a convenient mathematical aid for calculating the fields, whereas the fundamental equations of motion can always be expressed directly in terms of the fields alone. In quantum mechanics, on the other hand, the potentials cannot be eliminated from the basic equations in the canonical formalism. Nevertheless, these equations, as well as the physical quantities, are all gauge-invariant, so it has been considered that even in quantum mechanics the potentials themselves have no independent significance. However, Aharonov and Bohm showed that the above conclusions are not correct and that a further interpretation of the potentials, that is, consideration of the so-called *Aharonov–Bohm effect* (AB effect), is needed in the quantum mechanics, and further they described possible experimental configurations to test the effects they predicted.

The principle of their experiment to verify the predicted effect is described as follows. Consider a single coherent electron beam to be split into two parts, and each part is then allowed to enter a long cylindrical metal tube, as shown in Figure 9.15a. After the beams pass through the tubes, they are combined to interfere coherently at a location F. By means of time-determining shutters, the beam is chopped into wave packets W_1 and W_2 with size long compared with the wavelength but short compared with the length of the tubes M_1 and M_2. The potential in each tube is controlled to be nonzero only while the electrons are well inside the tube (region

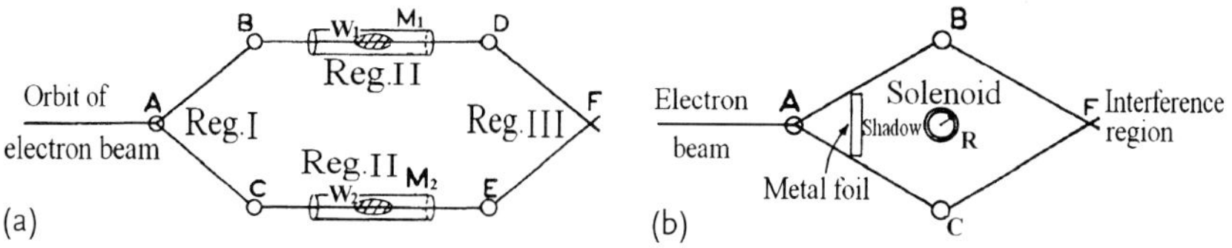

Figure 9.15 Configuration of experiments proposed by Aharonov and Bohm to demonstrate the effect of interference interpreted by potentials: (a) experiment on time-dependent scalar potential; (b) experiment on time-independent vector potential (Aharonov and Bohm [433]).

II), and zero while the electrons are in region I or III; thus, it is ensured that the electron is in a time-varying potential without ever being in a field. In such a situation, a term for electric potential $V_i(t) = e\varphi_i(t)$, $i = 1, 2$, a function of time only, is added to the Hamiltonian of the particle in each part for the region inside tubes 1 and 2, respectively. Then, the solution will differ from that without the potential just by a phase factor. Thus, the total solution of the wave function ϕ' for the electron at the interfering position F after the superposition in region III will be given by

$$\phi' = \phi_1^0 e^{-iS_1/\hbar} + \phi_2^0 e^{-iS_2/\hbar} ,$$
$$S_1 = e \int \varphi_1 dt , \quad S_2 = e \int \varphi_2 dt , \tag{9.30}$$

and it is evident that the interference of the two parts at F will depend on the phase difference $(S_1 - S_2)/\hbar$, inducing a physical effect even though no force is ever actually exerted on the electron.

Another special case of two paths in space only and independent of time, but inducing an associated phase shift is illustrated in Figure 9.15b. It corresponds to the situation including a current flowing through a very closely wound cylindrical solenoid of radius R, center at the origin and axis in the z-direction, and creating a magnetic field $\boldsymbol{H}$. The magnetic field is essentially confined within the solenoid, but the vector potential $\boldsymbol{A}$ is evidently nonzero everywhere outside the solenoid, because the total flux through every circuit containing the origin is equal to a constant:

$$\phi_m = \int \boldsymbol{H} \cdot d\boldsymbol{s} = \oint \boldsymbol{A} \cdot d\boldsymbol{x} . \tag{9.31}$$

In this way, the interference pattern between the two beams enclosing the vector potential $\boldsymbol{A}$, that is, passing on opposite sides of the solenoid, as illustrated in Figure 9.15b, will evidently depend on the phase difference:

$$(S_1 - S_2)/\hbar = (e/\hbar c) \oint \boldsymbol{A} \cdot d\boldsymbol{x} = (e/\hbar c) \int \boldsymbol{H} \cdot d\boldsymbol{s} = (e/\hbar c)\phi_m , \tag{9.32}$$

where ϕ_m is the total magnetic flux inside the circuit. This effect is called the *vector AB effect*.

Concerning the effect in the present argument, they also pointed out as the essential result of the present discussion that in quantum theory an electron, for example, can be influenced by the potentials even if all the field regions are excluded from it. Furthermore, more recently an extension of the AB effect was presented by Aharonov and Casher [434], who predicted that a neutral particle possessing a magnetic dipole moment should experience an analogous phase shift when diffracted around a line of electric charge (Aharonov–Casher effect, AC effect). A possible experimental configuration to verify the AC effect corresponds to first replacing the solenoid in Figure 9.15b with an equivalent line of electrically neutral magnetic dipoles, and next interchanging the roles of charge and magnetic dipole; thus, as a result one has two split paths of an electron enclosing the solenoid to be replaced by two paths of a particle with a magnetic dipole moment enclosing the static electric charge.

The AB effect predicted by Aharonov and Bohm was observed and measured for the first time in 1986 in texperiments with electrons by Tonomura *et al.* [435]. The experimental verification of the AC effect with a neutron interferometer had already been performed, for example, by Cimmino *et al.* [436]. Furthermore, the experimental demonstration of the *scalar AB effect* for a neutral particle with a magnetic moment in a time-dependent magnetic field was carried out by making use of an atomic interferometer in the time domain [437]. In addition to these studies, various experimental investigations on the AB and AC effects by using electrons and neutrons have been reported.

In the following, the experimental results for the scalar AB effect obtained by Zouw *et al.* with the diffraction grating type of very cold neutron interferometer shown in Figure 9.13a will be introduced [401, 402].

The whole configuration of the interferometer was arranged as shown in Figure 9.13b, in accord with Aharonov and Bohm's proposal in Figure 9.15a, but with the replacement of electrons by neutrons, and also the role of the term $e\varphi$ (φ being electric potential) in the Hamiltonian for the electric AB effect was replaced with that of the term $\boldsymbol{\mu} \boldsymbol{H} = \mu \boldsymbol{\sigma} \boldsymbol{H}$ in the Hamiltonian for neutrons.

The experiment was performed on the top floor of the Grenoble high-flux reactor by making use of incident very cold neutrons with an average wavelength of 9.5 nm and a wavelength full width at half maximum of about 4 nm. In advance of the experiment on the AB effect, the dispersivity of the phase shift from a static magnetic field was demonstrated as shown in Figure 9.16a with the count rate pattern as function of the DC current. The interference pattern was clearly washed out after a few fringes owing to the phase shift being dependent on the neutron velocity, and agreed well with the theoretical expectation assuming a Gaussian wavelength distribution.

The next measurement to demonstrate the nondispersivity of the scalar Aharonov–Bohm phase shift was carried out by applying switched magnetic fields to two beam paths in opposite directions in such a way that the neutrons never experience a classical force, that is, the field was switched on only while the chopper-pulsed neutron was completely inside its homogeneous region. In the present experiment using an unpolarized neutron beam, the spin-independent phase of the interferometer had to be adjusted to multiples of π for the spin-dependent Aharonov–Bohm fringes to have maximum visibility. This requirement was achieved by translating

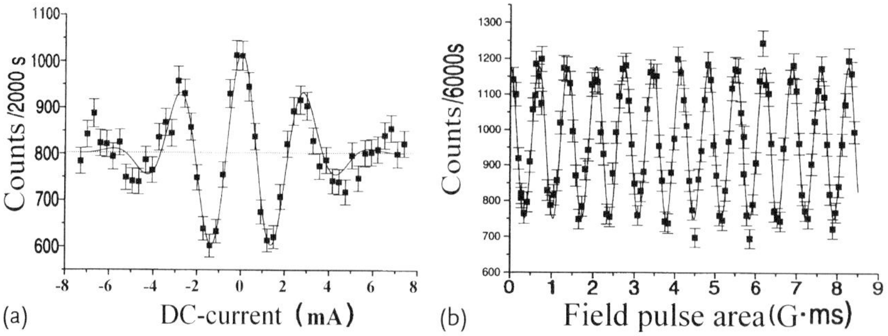

Figure 9.16 Experimental results for the scalar Aharonov–Bohm effect with a diffraction grating type of very cold neutron interferometer. (a) Phase shift pattern as a function of a DC current through the coil; the disappearance of the fringes demonstrates the dispersivity of a phase shift from a static magnetic field. The solid line represents a fit to an analytical model assuming a Gaussian wavelength distribution and no adiabatic effects. (b) Aharonov–Bohm-type phase shift as a function of the field pulse integral over time; the fringe visibility remains constant over the entire range, demonstrating the nondispersivity of the phase shift. The solid line is a fit to a sinusoid with frequency $\mu/\hbar = 9.14\,\text{rad}\,\text{G}^{-1}\,\text{ms}^{-1}$ (theoretical expectation, $9.16\,\text{rad}\,\text{G}^{-1}\,\text{ms}^{-1}$) (Zouw *et al.* [401]).

the first grating, as mentioned previously concerning the present diffraction grating type of interferometer. The experimental result is shown in Figure 9.16b with the count rate as a function of the field integral over time, as calculated from the current-pulse integral and the coil geometry. The interference contrast remained constant over the entire range and far more fringes were visible than in the static case. The frequency of the interference pattern agreed well with the expected value. The present result clearly demonstrated the verification and the nondispersivity of the scalar Aharonov–Bohm phase shift for neutrons.

9.4 Time-Dependent Interferometry

Some examples of *time-dependent interferometry* by making use of diffraction grating interferometers were introduced in the previous section. The experimental approach was illustrated in Figure 9.13b; the incident neutrons were pulsed by a chopper and the magnetic field controlled to synchronize with the pulse was exerted only while the neutrons were completely within the magnetizing solenoid. The experimental result shown in Figure 9.16b was obtained to verify the Aharonov–Bohm phase shift. In that experiment using unpolarized incident neutrons, the neutron beam was spatially split and switched magnetic fields had to be applied to two beam paths in opposite directions to verify the scalar Aharonov–Bohm phase shift (scalar AB effect).

In contrast to the experimental approach mentioned above, Badurek *et al.* [233] made use of polarized neutrons for an experiment with the same purpose but earlier than the experiment described in the previous section, that is, to demonstrate

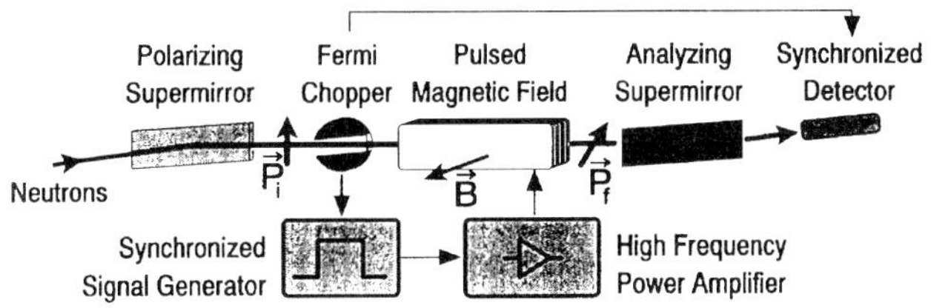

Figure 9.17 Experimental setup of time-dependent neutron interferometry. A rotating Fermi chopper imposes a time structure on the neutrons polarized through reflection from a supermirror. The operation of the subsequent magnetic field oriented in a direction orthogonal to the neutron polarization could be selected according to the experimental purpose, either constant in time or synchronously switched on and off so as to exert the field only while the neutrons are inside the field region (Badurek *et al.* [233]).

for the first time the nondispersivity of the AB effect, by using the configuration shown in Figure 9.17. In their configuration, the incident polarization vector $\boldsymbol{P}$ is perpendicular to the beam direction, and a homogeneous time-dependent magnetic field $\boldsymbol{B}(t)$ is exerted along the direction perpendicular to both of these directions, and is switched on and off in such a way that it is zero both when the neutron enters the magnetic field region and when it leaves that region again. Then, the two spin eigenstates experience just a different change in frequency $\Delta\omega = \pm\mu B/\hbar$, and the particle momentum remains a constant of motion. Evidently, such a magnetic field leads to a phase shift on the neutron wave function predicted as the AB effect with the time integral of $\Delta\omega$, and thus it is nondispersive as required for a topological effect and of different sign for the two spin eigenstates.

The experiment was performed with neutrons of a wide wavelength range from 2 to 6 Å at the experimental neutron guide laboratory of the FRM-I research reactor of the Technical University of Munich.

In advance of the time-dependent interferometry, measurement with a constant magnetic field was performed and the result is shown in Figure 9.18a. No intensity oscillations behind the analyzer can be observed, since the phase difference accompanied by a relative displacement of two eigenstates exceeds the limited length of coherent spread of the neutrons used with the wide spectrum.

On the other hand, in the case of the time-dependent measurement, the switching times of the magnetic field were chosen appropriately according to the requirement of the force-free scalar Aharonov–Bohm situation as explained in the figure caption. Then, the interference pattern after the magnetic field can be measured as the intensity of the neutrons reflected by the supermirror analyzer by triggering the detector synchronously to the chopper burst. The experimental result shown in Figure 9.18b demonstrates unambiguously that the phase shift caused by the magnetic field is not accompanied by a reduction of contrast and hence there is essentially no limit on the number of observable intensity oscillations whatever the spectral width of the incident beam may be, provided one can prepare adequate pulsed magnetic fields to meet the requirement mentioned in the figure caption.

The present experiment demonstrated for the first time the nondispersivity of the AB effect, as a spin-rotation analog of the scalar AB effect, and indicated the

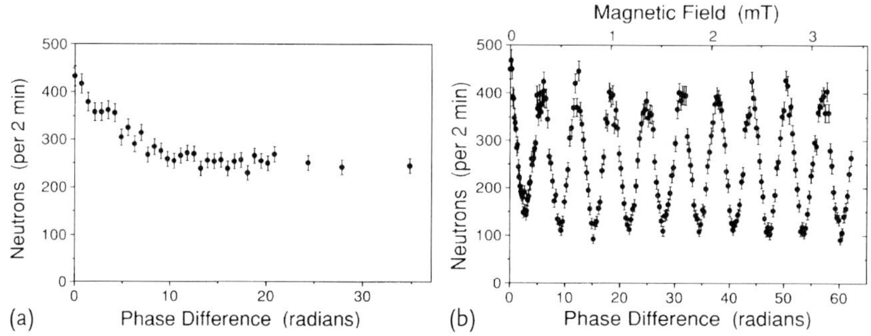

Figure 9.18 Number of neutrons counted after the second supermirror plotted against the phase difference between the spin eigenstates as introduced by the magnetic field: (a) a static magnetic field exerted in the region between the polarizer and the analyzer; (b) as in (a) but now with the magnetic field switched on and off to exert a field only while the neutrons are inside the magnetic field region (Badurek *et al.* [233]).

essential feature of the *topological nature of the Aharonov–Bohm-type phase*, that is, independent of the velocity (wavelength) of the interfering particles.

In addition to the experiments already introduced in Chapter 3 and in the present chapter, further various experimental studies with neutron interferometers have been reported relating to the neutron spin as investigations on the properties of the topological phase.

We would like to add at the end of this section in relation to time-dependent interferometry that Felber *et al.* [266] presented a little later than Badurek's experimental report the concept of using neutron time interferometry as a very attractive counterpart of conventional spatial interferometry and performed some theoretical analyses on the concept.

9.5
Neutron Interferometry with Single-Layer and Multilayer Mirrors

As mentioned at the beginning of Section 9.3, the perfect single crystal type of interferometer is essentially restricted applications for thermal neutrons, and therefore various types of *interferometers with single-layer and multilayer mirrors* have been developed and applied for neutrons with wavelength longer than that of thermal neutrons. A single-layer mirror could be used for ultracold neutrons, whereas multilayer mirrors have a variety of applicabilities covering cold neutrons and also very cold neutrons.

As an example of a proposal for an ultracold neutron interferometer, Steyerl *et al.* [438] considered a *Michelson-type interferometer* with the arrangement illustrated in Figure 9.19, and analyzed the performance as well as the expected neutron intensity in the case of it being inserted in place of a set of the sample mirrors in Figure 2.3 in the gravity diffractometer introduced in Chapter 2.

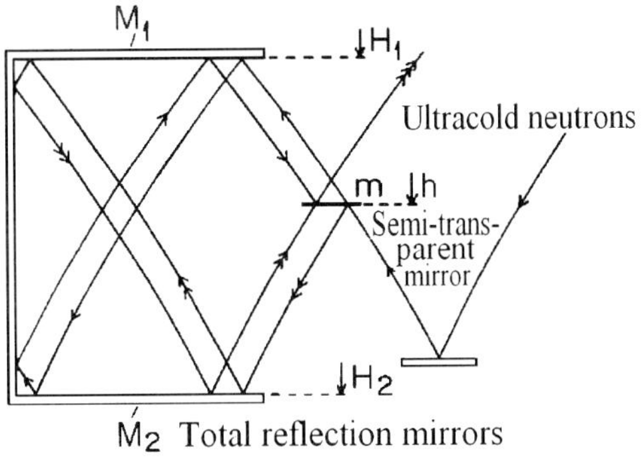

Figure 9.19 Confuguration of a version of a Michelson interferometer to be applied for ultracold neutrons: A single semitransparent mirror m acts both as a beam splitter and an analyzer, and two horizontal mirrors, M_1, M_2, and one vertical mirror are arranged such that the area enclosed by each slit path of an ultracold neutron beam is nearly zero, thus eliminating shifts due to the Coriolis force (Steyerl *et al.* [438]).

In the present geometry of the split paths, a significantly wide band width of horizontal neutron velocities may be admitted for experiments requiring a high resolution only in the vertical direction. The geometry is arranged such that the effective area enclosed by the coherent beams is virtually zero, thus rendering the interferometer insensitive to the Coriolis force, which is thought to give rise to a small fringe shift proportional to the enclosed area, as known from the phenomenon called the Sagnac effect[38)] in conventional optics.

For the realization of the ultracold neutron interferometer described above, the manufacturing precision for a mirror size of $10 \times 10\,\text{cm}^2$ with an allowable angular deviation due to misadjustment or long-range surface waviness must not exceed about 10^{-7} rad. In addition precautions must be taken to keep temperature variations below tolerable limits.

Another approach for the development of an interferometer for cold and very cold neutrons was carried out by making use of *multilayer mirrors* by Ebisawa *et al.* at the Kyoto University Research Reactor Institute [440]. In the case of single-crystal interferometers introduced in earlier sections of this chapter, all optical components of the splitter *S*, the reflector for subbeams *M*, and the analyzer *A* consist of a common single crystal as illustrated in Figure 9.1a. In contrast, in the case of a multilayer interferometer composed of two or more sheets of artificial

38) Suppose one introduces two split continuous coherent light beams traveling in opposite directions along the trajectory of a closed loop enclosing an area, and then superpose them at the joining point on the loop. If a rotation is applied to the closed loop, a phase difference is induced in the interference pattern owing to the difference between the distances traveled by each beam. This effect reported by Sagnac [439] is named the Sagnac effect, and is nowadays applied for practical uses such as in optical gyroscopes to measure the phase difference proportional to the angular velocity of the optical system containing a closed loop, as, for example, in ring laser gyroscopes in global positioning systems.

mirrors, the relative positions and angular settings of these components prepared independently must be technically arranged and controlled with the precision of the neutron wavelength to be used in the experiments. Ebisawa *et al.* arranged these optical components on a common flat baseplate, and developed an effective approach for high-precision control of the mirror positions and the setting angles by making use of the combination of precision induction sensors and piezo actuators. These components and technologies were further applied for the realization of multilayer mirror spin interferometers for cold and very cold neutrons, to be described in detail in the following and also in the next section.

As a more practical arrangement for the multilayer mirror interferometer, Funahashi *et al.* employed the geometrical configuration of a *Jamin type of interferometer* arrangement shown in Figure 9.20 consisting of a couple of multilayer mirrors with an intermediate monolayer between them. This moderated very effectively the severe technical requirement of parallelism and relative positions between four optical elements, a splitter, two reflectors, and an analyzer, compared with the configuration setting these four elements independently. They successfully observed interference patterns as shown in Figure 9.21 in multilayer interferometry with cold neutrons [441].

The experiment was performed by using beamline MINE at the cold neutron guide tube of the JRR-3M reactor, Japan Atomic Energy Research Institute, Tokai. The neutrons in the incident beam with a wavelength of 12.6 Å and a wavelength width of 3.5% full width at half maximum were coherently split into Φ_b and Φ_c by the reflection from the first pair. The relative phase difference induced for the incident angle θ_1 and the wavelength λ can be expressed as

$$\Delta\chi_1 = 2\pi \frac{2D \sin\theta_1}{\lambda}, \tag{9.33}$$

where D denotes the effective distance between the two multilayer mirrors in a single pair, and is given by the relation $D = T + Nd$ for the intermediate monolayer thickness T and the multilayer of layer number N with unit bilayer thickness d. In the experimental results shown in the inserts (i) in Figure 9.21 for the measurements with only a single pair of multilayers, the interference fringes are

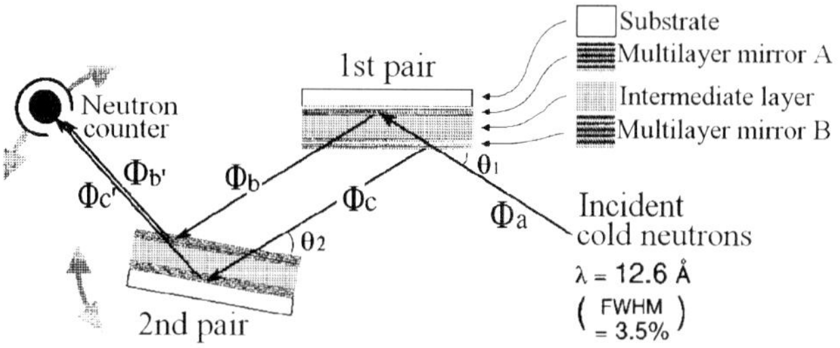

Figure 9.20 Arrangement of the multilayer mirror interferometer. The angles θ_1 and θ_2 denote the grazing angles for the first and second pair of multilayer mirrors, respectively. FWHM full width at half maximum. (Funahashi *et al.* [441]).

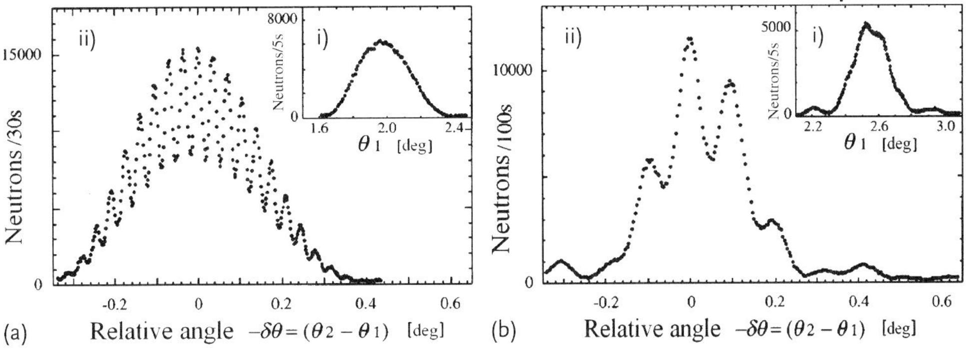

Figure 9.21 Experimental results with a multilayer interferometer. (a) Intensity registered by the neutron counter for the case with a multilayer pair consisting of a unit bilayer thickness *d* of 180 Å and an effective distance between the two multilayer mirrors *D* of 10 100 Å. (b) The same as (a) but for a pair consisting of a unit bilayer thickness *d* of 140 Å and an effective distance between the two multilayer mirrors *D* of 3 700 Å. in both (a) and (b), (i) is the reflected intensity from a single pair, and (ii) is the interference fringes observed by changing the relative angle between the first and second pairs. (Funahashi *et al.* [441]).

not obviously observed, since the phase difference $\Delta\chi_1$ becomes much larger than $2\pi(\lambda/\Delta\lambda)$ for the wavelength width $\Delta\lambda$ of the beam.

With the addition of the second pair as illustrated in Figure 9.20, the phase differences induced within each pair mostly cancel each other and the resultant phase difference remaining between $\Phi_{b'}$ and $\Phi_{c'}$ is given by

$$\Delta\chi_1 - \Delta\chi_2 = 2\pi\frac{2D}{\lambda}(\sin\theta_1 - \sin\theta_2) \cong 2\pi\frac{2D}{\lambda}\delta\theta\,, \tag{9.34}$$

where $\delta\theta = \theta_1 - \theta_2$.

In this way, scanning $\delta\theta$ over the region from $-(\lambda/2D)(\lambda/\Delta\lambda)$ to $(\lambda/2D)(\lambda/\Delta\lambda)$, one observes interference fringes with the period $\lambda/2D$ according to Eq. (9.34), as obviously indicated in the main plots (ii) in Figure 9.21. From these results, the present approach of the multilayer mirror interferometer can be applied to much wider wavelength regions of cold and very cold neutrons, compared with the single crystal type of interferometer, the application of which is restricted to thermal neutrons. Those neutrons with longer wavelength are expected to be advantageous for precision measurements to study various phenomena in fundamental physics. Especially, as introduced in the next section, a variety of experimental studies have been performed by making use of the present type of interferometers consisting of magnetic multilayers and/or polarized incident neutrons.

9.6
Spin Interferometry Experiments with Magnetic Multilayer Mirrors

9.6.1
Development of a Phase-Spin-Echo Interferometer

In this subsection, we would like to study some details of the *phase-spin-echo interferometer* with *magnetic multilayers* developed by Ebisawa *et al.* at the Kyoto University Research Reactor Institute. A short description of the principle of the device was given in Section 3.4.

With regard to their spin interferometer, Figure 3.16 illustrates the schematic configuration of the optical elements, the typical interference pattern observed, and the outline of the magnetic multilayer used as the spin splitter and analyzer, respectively. The overall setup is shown in Figure 9.22a [158]. In the present configuration, two identical multilayer spin splitters (MSS1 and MSS2) are mounted on the goniometer in the respective precession fields, and work with simultaneous occurrence of phase-echo and spin-echo.

The incident neutrons are polarized in the $(x$-$y)$-plane as illustrated by an arrow in a circle in Figure 3.16c, perpendicular to the quantization z-axis defined by the guide field, by the combination of a polarizer and $\pi/2$ flipper 1 in Figure 9.22a. According to the rule of spin superposition described in Section 3.4, the present spin state $|S_{xy}\rangle$ can be expressed as the superposition of the spin eigenstates $|\uparrow_z\rangle$ and $|\downarrow_z\rangle$ as follows:

$$|S_{xy}(\chi)\rangle = \frac{1}{2}\{|\uparrow_z\rangle + e^{i\chi}|\downarrow_z\rangle\} = \frac{e^{i\chi/2}}{\sqrt{2}}\left\{\cos\left(\frac{\chi}{2}\right)|\uparrow_x\rangle - i\sin\left(\frac{\chi}{2}\right)|\downarrow_x\rangle\right\}, \tag{9.35}$$

where χ is the phase difference between the spin eigenstates. Therefore, the expectation value of the x-component $\langle S_x\rangle$ is given by

$$\langle S_x(\chi)\rangle = \cos\chi\,. \tag{9.36}$$

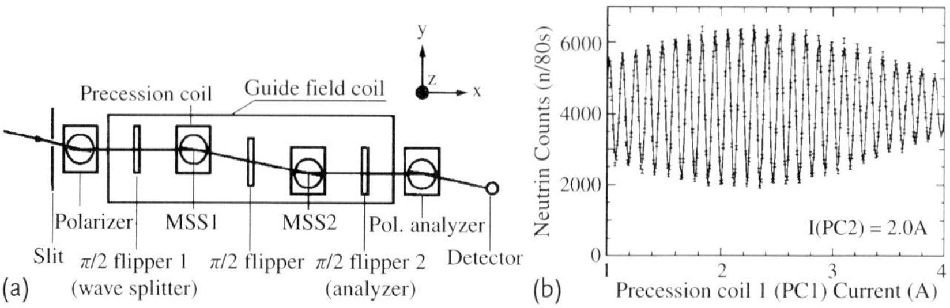

Figure 9.22 Cold neutron phase-spin-echo interferometer with magnetic multilayers: (a) arrangement of the interferometer with two identical multilayer spin splitters, MSS1 and MSS2, mounted on respective goniometers (dotted lines) in the precession fields; (b) measured spin-echo profile for a phase-spin-echo interferometer composed of PG/G5/NT multilayers for MSS1 and MSS2 at a splitter setting angle of about 1.7° (Ebisawa *et al.* [158]).

Then, the neutrons are reflected by the first multilayer pair element as a spin splitter MSS1 in the precession coil.

In a similar way, the neutrons coming out of the second multilayer spin splitter MSS2, positioned parallel to MSS1, but now in the role to superpose split spins, are analyzed by the combination of $\pi/2$ flipper 2 and the polarization analyzer, and the final expectation value of the x-component, $\langle S_x \rangle$, can be determined from the neutron counts registered by the detector. Thus, the conditions of the *spin echo* of the split beam and also *phase echo* by Eq. (9.34) as described below can be simultaneously attained, and this results in a neutron spin-phase-echo interferometer with a couple of magnetic multilayer pairs.

The present interferometer was developed and tested by making use of a cold neutron beam with a wavelength of 8.0 Å and a wavelength width of 4.5% from the CN-1 guide at the cold neutron source facility of the Kyoto University research reactor, Kumatori. It was used for experiments on neutron optics studies. Furthermore, the configuration and elements of the spin interferometer were duplicated and provided for various neutron optics experiments with incident neutrons with a longer wavelength of 12.6 Å and a wavelength width of 3.5% as one of the joint-use instruments at beamline C3-1-2 in the cold guide hall at the JRR-3M reactor, Japan Atomic Energy Agency (JAEA), Tokai.

In the multilayer pair shown in Figure 3.16a, the parallel spin component $|\uparrow_z\rangle$ among the incident neutrons with wavelength λ satisfying the Bragg condition $\lambda = 2d \sin\theta$ is reflected by the multilayer at the surface side, whereas the antiparallel component $|\downarrow_z\rangle$ passes through both the magnetic layer and the intermediate gap layer, and is then Bragg-scattered by the inner nonmagnetic multilayer. Thus, the spin state $|S_{xy}\rangle$ is split into parallel and antiparallel eigenstates, and according to the optical path difference depending on the thickness of the gap layer, an additional phase difference χ_a is induced as

$$\chi_a = \frac{4\pi D n_\perp(\theta)\sin\theta}{\lambda}, \tag{9.37}$$

where $n_\perp(\theta)$ is the effective refractive index of the gap layer for a neutron with wavelength λ at grazing angle θ given by

$$n_\perp(\theta) = \left[1 - \frac{Nb_c}{\pi}\left(\frac{\lambda}{\sin\theta}\right)^2\right]^{1/2}. \tag{9.38}$$

Therefore, the phase-spin-echo interferometer with simultaneous occurrence of *spin-echo and phase-echo* conditions is realized by arranging a couple of identical spin splitters in a parallel geometry and a π flipper between them in the flight path between a couple of $\pi/2$ flippers in the guide field as illustrated in Figure 9.22a.

Several kinds of multilayer pairs were prepared with different layer thickness and layer numbers, and as an example, the experimental result for the interference pattern for a multilayer element consisting of a magnetic multilayer (seven bilayers of permalloy-45 ($Fe_{55}Ni_{45}$; PA) with thickness 100 Å and titanium with the same thickness, and a gap layer with thickness 2000 Å) and a nonmagnetic multilayer (seven bilayers of nickel with thickness 100 Å and titanium with the same thickness), was given in Figure 3.16b. The experimental result for an extended region of the precession current is shown in Figure 9.22b for the case of the same magnetic and nonmagnetic multilayers, but the gap layer was 5000-Å germanium. Both of these interference patterns indicate high contrast obtained as a result of simultaneous occurrence of the spin-echo and phase-echo interference.

9.6.2
Observation of the Transverse Coherence Length

The concept and theoretical formula for coherence length and coherence time were introduced for the interference of light. From the viewpoint of experiment, the coherence length or coherence time can be estimated in relation to the *visibility* mentioned in Section 9.1 to attenuation down to $1/e$ of the original value, where the visibility V can be defined by the expression

$$V(\boldsymbol{\Delta}\,;\tau) = \frac{I_{\max}(\boldsymbol{\Delta}\,;\tau) - I_{\min}(\boldsymbol{\Delta}\,;\tau)}{I_{\max}(\boldsymbol{\Delta}\,;\tau) + I_{\min}(\boldsymbol{\Delta}\,;\tau)}\,, \tag{9.39}$$

with the intensities $I_{\max}(\boldsymbol{\Delta}\,;\tau)$ and $I_{\min}(\boldsymbol{\Delta}\,;\tau)$ as the maximum and minimum, respectively, in the interference pattern for a spatial shift $\boldsymbol{\Delta}$ and a temporal shift τ. Although there are some common features in such concepts and formula used in light optics that are also applicable to neutron optics, we notice important differences and essential difficulties in applying the framework in light optics to neutron optics, which handles matter waves. For example, every spectral component of neutrons propagates with a different velocity (e. g., refer to textbook [D], pp. 433–437).

Concerning matter waves described by the Schrödinger equation, and assuming a quasi-stationary state or a sufficiently slow time dependence, Rauch represented these related fundamental relations by using the *coherence function* $\Gamma(\boldsymbol{\Delta})$ as a function of the three-dimensional spatial shift $\boldsymbol{\Delta}$ [406]. Especially, for the case of Gaussian momentum distributions having widths $\delta k_i (i = x, y, z)$ in each of the three orthogonal directions x, y, z, he obtained a Gaussian coherence function for continuous wave experiments:

$$\Gamma(\boldsymbol{\Delta}) = \prod_{i=x,y,z} e^{-[(\Delta_i \delta k_i)^2/2]}\,. \tag{9.40}$$

The coherence function and the visibility of the interference pattern from the superposition of split beams I_1 and I_2 can be related as

$$V(\boldsymbol{\Delta}) = \frac{2\sqrt{I_1 I_2}}{I_1 + I_2}|\Gamma(\boldsymbol{\Delta})|\,. \tag{9.41}$$

Thus, the *coherence length* Δ_{cx} for a one-dimensional spatial shift is directly related to δk_x in the present theoretical framework as

$$\Delta_{cx} = \frac{\sqrt{2}}{\delta k_x}\,. \tag{9.42}$$

Therefore, it would be an interesting task to evaluate the visibility to be observed in a certain spatial direction in neutron interferometry experiments and to investigate the consistency with the results from the Gaussian description of the coherence function. Especially, it would be very stimulating if any discrepancy, more or less, against the results from the wave function description using the Schrödinger equation might come out from an interferometry experiment with long-wavelength neutrons.

In Section 9.1 we studied measurements of longitudinal coherence length where the possibility of a coherence length inherent to a neutron was discussed. In the present subsection, two different experimental approaches for measuring the *transverse coherence length*, that is, in the direction perpendicular to that of the wave propagation, of a neutron and the experimental results will be described.

The first experiment was carried out by Hino *et al.* [442] by making use of the multilayer spin interferometer described in the previous subsection. In their experiment, the mutual geometrical arrangement of the spin splitter (MSS1) and the spin analyzer (MSS2) in Figure 9.22a was reversed from that in Figure 9.23a to that in Figure 9.23b. Then, in contrast to the superposition of two subbeams in Figure 9.23a with the transverse separations canceling each other by the reflection by the spin splitters in parallel arrangement on both sides of π flipper, in the case of the symmetrical arrangement shown in Figure 9.23b the transverse shifts are additionally summed and the possible resultant effects due to the transverse shift dependent on the gap layer thickness and on the mirror angle should be observed in the interference pattern with the spin analyzer. The spin polarization visibility to be observed in the spin interferometry is expected to indicate mostly the effect of the transverse coherence length of a neutron since the present arrangement induces a transverse spatial separation but spin convergence of $\uparrow$ and $\downarrow$ components. The experimental results are plotted in Figure 9.23c for an incident angle to the splitter of $1.15°$, where the divergence angle of the incident neutron was varied by changing the slit width in the incident neutron line [442].

Assuming the visibility decrease in the observed results as the difference between the dotted line and the solid line for a gap layer thickness in Figure 9.23c is caused by the transverse separation in the subbeam superposition, one can relate the results to Eqs. (9.40)–(9.42) in the coordinate of the transverse direction. The estimated transverse coherence length based on Eq. (9.42) has a value larger than 10 μm.

As the next viewpoint, the intersections of these lines extrapolated to the condition of no incident divergence gave values smaller than unity, in contrast to the expectation from Eqs. (9.40) and (9.41) of unity. The present observed effect might be related to some intrinsic coherent separation of a neutron, which should diminish somewhat the observed visibility depending on the transverse separation. However, the intersections observed experimentally do not seem to be consistent with such a supposed dependence on the transverse separation, and only the intersection for the symmetrical spin splitters with a thicker gap layer of 1.0 μm decreased to significantly less than unity. Anyway, they considered that one of the most probable causes of the present visibility decrease without the incident neutron divergence is

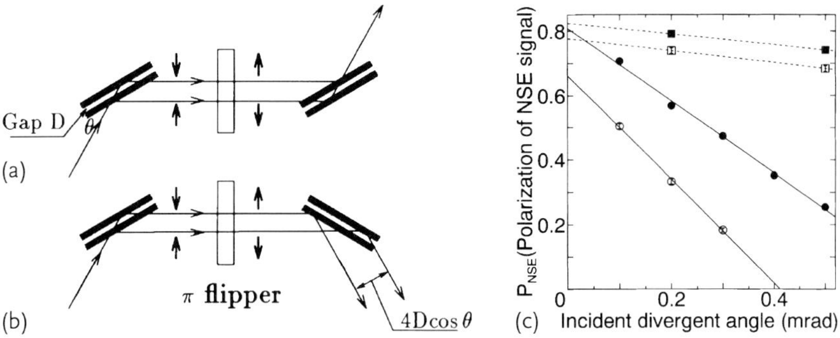

Figure 9.23 Arrangements for transverse spatial separation but spin convergence of ↑ and ↓ components. (a) and (b) Two kinds of arrangements for a couple of multilayer spin splitters, in which each splitter element consists of, from the surface side, a magnetic layer (80-nm-thick permalloy-45), a gap layer (germanium with thickness *D*), and a nonmagnetic layer (80-nm-thick nickel). (c) Observed visibility of the spin-echo signal as a function of the incident neutron divergence angle. The solid and open symbols indicate P_{NSE} reflected by a couple of spin splitters with gap layer thickness 0.5 and 1.0 μm, respectively. The squares and circles correspond to the results for the parallel, that is, (a), and symmetrical, that is, (b), arrangements of the couple of spin splitters, respectively (Hino *et al.* [442]).

the effect of surface roughness in the spin splitters, since there is a tendency for evaporation of a thicker gap layer to induce a greater roughness of the interface and surface, or a greater thickness inhomogeneity, on the upper magnetic layer. From these considerations, they concluded conservatively that the present experimental results could not convincingly indicate the existence of an intrinsic maximum coherent separation of a neutron.

The second experiment on the transverse coherence length was carried out by Kitaguchi *et al.* by making use of slightly different optical elements and interferometry [443]. They employed etalons[39] as the beam splitter and the analyzer, as shown in the plan view in Figure 9.24a, after vacuum-evaporating a magnetic multilayer of PA and germanium on the surface of one side, and a nonmagnetic multilayer of nickel and titanium on the surface of the other side. Further, to observe the visibility variation, the relative arrangement of the splitter and the analyzer was twisted as illustrated in the vertical section in Figure 9.24a, and the visibility measurements were performed for a stepwise change of the tilting angle α, instead of changing the incident beam divergence as in the previous experiments of Hino *et al.*

The multilayer in the beam splitter and analyzer consists of eight bilayers with an effective lattice constant of about 240 Å, and the spacing in the etalon is 9.75 μm. The measurements were performed by using cold neutrons with a wavelength 8.8Å and a wavelength width 2.4% at beamline MINE2 of the JRR-3M reactor. As

39) An etalon is an optical element consisting of two reflecting surfaces facing each other and separated by certain gap. Various kinds of high-precision optical etalons are available commercially.

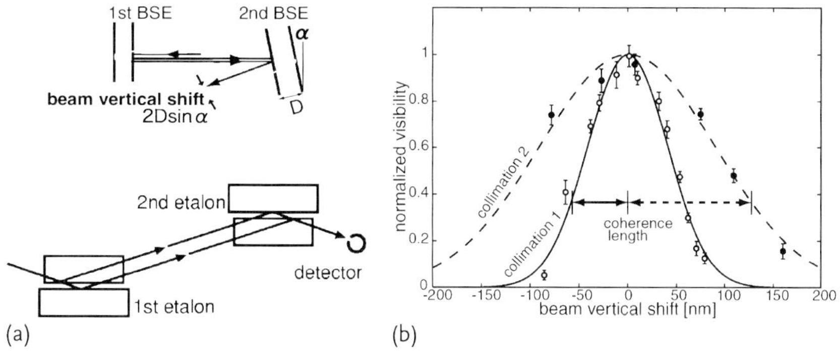

Figure 9.24 Multilayer spin interferometry with optical elements of etalons for the measurement of the transverse coherence length. (a) Setup of the beam splitter, first etalon (1st BSE) and analyzer, second etalon (2nd BSE), and the vertical section indicating the vertical shift of subbeams to be superposed after the reflection by 2nd BSE. (b) Visibility change measured as a function of the vertical shift between superposed subbeams. The open circles correspond to the case of the sharper incident beam with standard deviation of the vertical wave number distribution $\delta k_z = 0.024\,\text{nm}^{-1}$, and the solid circles correspond are that of the incident beam with standard deviation of the vertical wave number distribution $\delta k_z = 0.011\,\text{nm}^{-1}$. The solid and broken lines are the results calculated with Eq. (9.40), and the horizontal arrows show the coherence length estimated with Eq. (9.42) (Kitaguchi *et al.* [443]).

shown by the experimental results in Figure 9.24b, the coherence lengths of 58 ± 2 and $129 \pm 10\,\text{nm}$ were obtained for the incident beam collimations 1 and 2, respectively, in good agreement with the theoretical values of 59 and 130 nm from applying Eq. (9.42) to the vertical direction. In this way, the transverse coherence length derived from the visibility decrease in the present experiment gave reasonable agreement with that estimated from the standard deviation in the momentum distribution of the incident neutron. There was thus no obviously positive indication of a finite intrinsic transverse coherence length of the neutron.

9.6.3
Observation of Quasi-Bound States of the Neutron in a Magnetic Resonator

As described in Figures 2.20 and 2.21 for unpolarized neutron experiments, multilayers with a double-hump potential are capable of working as resonators to bind neutrons temporarily in the potential. Applying the present principle to a spin-dependent potential as illustrated in Figure 9.25a, we can design such a magnetic resonator such that $\uparrow$ spin neutrons with energy lower than the height of the potential barrier in a magnetic resonator temporarily compose a *quasi-bound state*, and $\downarrow$ spin neutrons feel a nearly flat potential distribution in the resonator and so are easily transmitted. It would be a very interesting task to investigate experimentally the Larmor precession of neutrons in such a magnetic resonator as well as to analyze theoretically the behavior of spinors in the condition of magnetic resonance.

As an extension of the study given in the previous section on the neutron optical element of magnetic multilayers, Hino *et al.* prepared a *Fabry–Pérot magnetic*

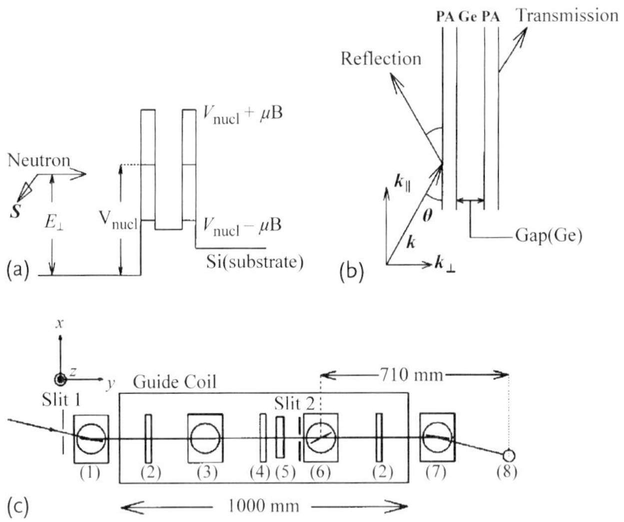

Figure 9.25 Experiments involving neutron magnetic resonance and quasi-bound states with a multilayer spin interferometer. (a) Potential energy structure and (b) outline of the neutron optical path for an incident neutron with grazing angle θ in a Fabry–Pérot magnetic resonator. PA permalloy-45. (c) Layout of a neutron spin interferometer for the magnetic resonance experiment, where (1) is a polarizer, (2) is a $\pi/2$ spin flipper coil, (3) is a precession coil 1,(4) is a π spin flipper, (5) is an accelerator coil, (6) is the second precession coil and the experimental sample (Fabry–Pérot magnetic resonator), (7) is a polarization analyzer, and (8) is a ^{3}He neutron detector (Hino *et al.* [267]).

resonator consisting of potential barriers of PA and gap layers of germanium as illustrated in Figure 9.25b, and investigated in detail the experimental characteristics of the magnetic resonator in comparison with theoretical analyses [267, 444].

The wave functions for neutrons reflected from and transmitted through a multilayer can be expressed by using the reflection and transmission coefficients r and t on the multilayer given by Eq. (2.19). These coefficients are further represented by Eq. (3.14) by making use of the transfer matrix $\underline{\boldsymbol{M}}$ as explained in Section 3.2. Furthermore, the *reflection and transmission coefficients for a magnetic multilayer,* $r_\pm$ and $t_\pm$, can be derived from the 4×4 transfer matrix $\underline{\boldsymbol{M}}$, given by Eq. (3.20). Therefore, the wave $|\varphi_{\mathrm{tr}}\rangle$ transmitted through a magnetic multilayer becomes

$$|\varphi_{\mathrm{tr}}\rangle = \begin{pmatrix} t_+ e^{iky} \\ t_- e^{iky} \end{pmatrix} , \tag{9.43}$$

where $t_\pm$ are the transmission coefficients for ↑ and ↓ spins, respectively, and can be decomposed to the form

$$t_\pm = T_\pm^{1/2} e^{i\Delta\chi_\pm} e^{-ikd} e^{\pm i\delta/2} , \tag{9.44}$$

in which $\Delta\chi$ is the *additional phase due to a resonator,* T and d are the transmission probability and the total thickness, respectively, of the multilayer, and δ is the precession angle of an incident neutron at the resonator surface.

In this way, the expectation values for the spin components of the transmitted neutron are given by the next relations:

$$\langle S_x; \mathrm{tr}\rangle = \frac{\hbar}{2}\langle\phi_{\mathrm{tr}}|\sigma_x|\phi_{\mathrm{tr}}\rangle = \hbar\cos(\Delta\chi_+ - \Delta\chi_- - \delta)\frac{\sqrt{T_+T_-}}{T_+ + T_-}, \tag{9.45}$$

$$\langle S_y; \mathrm{tr}\rangle = \frac{\hbar}{2}\langle\phi_{\mathrm{tr}}|\sigma_y|\phi_{\mathrm{tr}}\rangle = -\hbar\sin(\Delta\chi_+ - \Delta\chi_- - \delta)\frac{\sqrt{T_+T_-}}{T_+ + T_-}, \tag{9.46}$$

$$\langle S_z; \mathrm{tr}\rangle = \frac{\hbar}{2}\langle\phi_{\mathrm{tr}}|\sigma_z|\phi_{\mathrm{tr}}\rangle = \frac{\hbar}{2}\frac{T_+ - T_-}{T_+ + T_-}, \tag{9.47}$$

where $\Delta\chi_+ - \Delta\chi_-$ corresponds to the *additional precession angle* Ω induced by the transmission of the magnetic multilayer.

The experimental arrangement shown in Figure 9.25c employed the so-called spin-echo configuration [55] in which the Larmor precession number before (4) in the midpoint is subtracted from that after the π flipper and thus their difference is observed. Inserting a Fabry–Pérot magnetic resonator at position (6) in the present arrangement, the resultant precession number δN of neutrons becomes

$$\delta N = N_0 - N_1 - N_{\mathrm{acc}} - \Delta N = \beta\left(\frac{H_1 l_1}{v_1} - \frac{H_2 l_2}{v_2} - \frac{H_{\mathrm{acc}} l_{\mathrm{acc}}}{v_2}\right) - \frac{\Omega}{2\pi}, \tag{9.48}$$

where $\beta = 2|\mu|/h = 29.16\,\mathrm{kHz/mT}$, N is the Larmor precession number, ΔN is the additional precession number due to the sample, l is the optical path length for the magnetic field H, v is the neutron velocity, and the subscripts 1, 2, and acc denote the first and second precession coils and the acceleration coil, respectively.

In the case of the transmission geometry shown in Figure 9.25c, $v_1 = v_2$, and H_1, H_2, l_1, l_2, and l_{acc} are constant. Therefore, the difference between the Larmor precession numbers with and without the sample in the conditions of a constant grazing angle θ and a constant acceleration field H_{acc} indicates the precession number due to the sample at that grazing angle.

The experiment was carried out by using neutrons with a wavelength of 1.26 ± 0.044 nm, a divergence angle of 1.0 mrad, and a magnetic field strength of 2 mT at the sample position in the cold neutron spin interferometer located in beamline C3-1-2 at the JRR-3M reactor, JAEA, Tokai.

The experimental results compared with the calculated ones are shown in Figure 9.26 for the sample of a double-hump Fabry–Pérot magnetic resonator. The theoretical calculations were performed using Eqs. (9.43)–(9.46), taking the wavelength distribution of neutrons into consideration. The experimental results for the transmission probability and the precession number agree well with the corresponding theoretical calculations. The solid vertical line in Figure 9.26c at a grazing angle of 1.41° represents the critical angle of the potential barrier for ↑ spin neutrons.

The dotted vertical line in Figure 9.26d indicates the grazing angle giving the maximum transmission probability for ↑ spin neutrons. We notice in the figure that the Larmor precession number at that grazing angle becomes the same for a sample with a gap layer thickness of 40 nm and for a sample with a gap layer

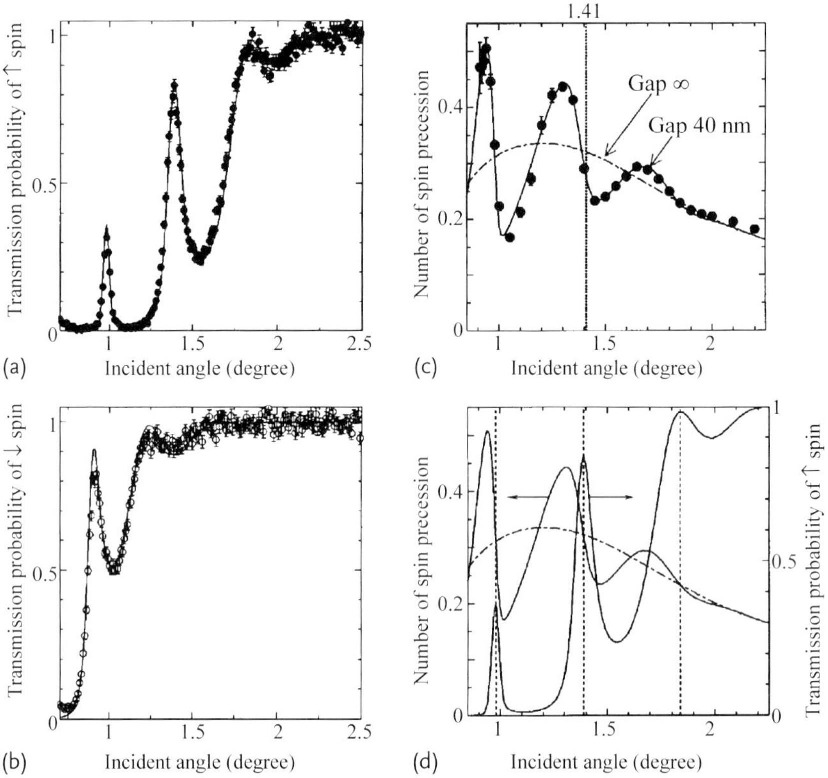

Figure 9.26 Experimental results compared with the theoretical calculations on the magnetic resonance and quasi-bound states of neutrons as function of the grazing angle to a PA(20 nm)–Ge(40 nm)–PA(20 nm) Fabry–Pérot magnetic resonator. (a) and (b) Experimental transmission probabilities (solid and open circles) and calculated results (solid lines), for (a) ↑ spin and (b) ↓ spin neutrons. (c) Comparison between the experimental result (solid circles) and calculated ones (solid and dot-dashed lines) on the precession number for a multilayer with Ge gap layer thickness X; $X = 40$ nm (solid line) and $X = \infty$ (dot-dashed line), and (d) the calculated precession number compared with the transmission probability distribution (Hino *et al.* [267]).

with infinite thickness, that is, at that condition it seems that the reflections by the potential barrier have almost no contribution to add to the Larmor precession. In contrast, at a grazing angle a little below that of the maximum transmission probability but above that of the minimum transmission probability, a remarkable increase of the additional Larmor precession is induced. The present relation is very important with regard to the estimation on the quasi-bound states based on the resonant peak in the neutron transmission probability in the unpolarized neutron resonance experiment mentioned in Section 2.5.

The experiment using a multiple-hump Fabry–Pérot magnetic resonator with a larger number of potential barriers will be introduced in the next section. It will indicate the steep growth of the additional precession number around the transmission probability peak according to the increase of the barrier number, or in a more exact expression, the precession number increases in proportion to the number of the potential valley.

Hino *et al.* considered the increase of the precession number as the result of the increase of the phase difference between ↑ spin neutrons and ↓ spin ones due to the *dwell time of the ↑ spin neutrons in the magnetic resonator,* although the Larmor precession might not be so simply supposed to be a real clock in the neutron behavior. They estimated [267, 444] the possible dwell time in their experiment to be the on the order of 10 ns ×(gap layer number). On the other hand, the time length as the product of the additional Larmor precession number at the resonant peak and the precession period for neutrons in the nonmagnetic gap layer is on the order of several microseconds per gap layer. The dwell time length of neutrons estimated from the resonance width in the reflection and transmission curves is between these two estimates.

9.6.4 Observation of Larmor Precession on Resonant Tunneling Neutrons

We introduce in this subsection two extensions by Hino *et al.* for the experiment using the setup shown Figure 9.25c to observe the Larmor precession in the resonant tunneling transmission of neutrons through a Fabry–Pérot magnetic resonator. The first extension is the resonator with further increased layer number [267], and typical results from spin interferometry of the spin precession number compared with the theoretical calculation for a resonator with the potential structure of [PA(20 nm)–Ge(40 nm)]^{10}–PA(20 nm) are given in Figure 9.27a. The dot-dashed line indicates, like that in Figure 9.26b, the theoretical calculation for a gap layer with infinite thickness.

In Figure 9.27a, the transmission probability of the resonator for ↑ spin neutrons is also plotted as a dotted line. As shown in the figure, even in the present case of a resonator with a many more potential barriers, the experimental results for the precession number and the transmission probability agreed well with the corresponding calculations with the Schrödinger equation.

Another extension was provided in a study on a resonator sample with a large absorption cross section for neutrons. As we learned in Section 2.2, the contribution of neutron absorption in optics was assumed to be expressed in the formalism with a complex scattering amplitude, as Eq. (2.16). According to such a formalism, the contributions of the absorption for ↑ and ↓ spin components of neutrons in a magnetic multilayer resonator are also expressed here by the *complex potential,* to be inserted into Eq. (1.19):

$$U = U_{\text{nucl}} \pm \mu B + i\, U_{\text{ab}} \,, \tag{9.49}$$

where U_{nucl} and U_{ab} are the optical potentials calculated from the coherent scattering amplitude b_{coh} and the absorption cross section σ_{a} with insertion of Eqs. (2.17) and (2.18), respectively. The double signs correspond to the ↑ spin and ↓ spin components, respectively, of a neutron.

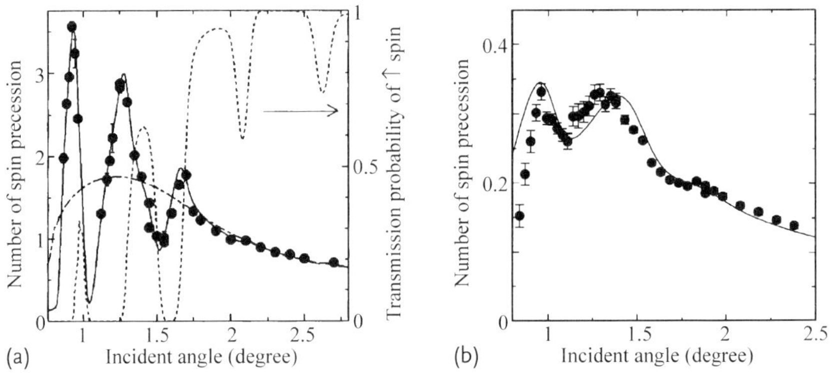

Figure 9.27 Results of spin interferometry similar to those in the previous subsection but with further extended cases of the resonator sample: (a) much increased layer number, that is, a resonator with the potential structure of $[PA(20\,nm){-}Ge(40\,nm)]^{10}{-}PA(20\,nm)$ (Hino *et al.* [267]); (b) Fabry–Pérot magnetic resonator including a strongly absorbing gap layer for neutrons, that is, a resonator with the potential structure of SSD(20 nm)–Ge/Gd(26 nm)–SSD(20 nm) PA permalloy-45, SSD Supersendust (Hino *et al.* [445]).

Hino *et al.* [445] prepared a Fabry–Pérot magnetic resonator of SSD–(Ge/Gd)–SSD (where SSD is Supersendust; $Fe_{86.8}Si_6Al_4\ Ni_{3.2}$) with a gap layer of a neutron absorber by evaporating a mixture of Ge(70%) and Gd(30%) with a thickness of 26.0 nm between the magnetic layers of SSD with a thickness of 20 nm. They performed the spin interferometry experiment with the SSD–(Ge/Gd)–SSD resonator by using the spin interferometer with the C3-1-2 beamline cold neutrons of wavelength 12.6 Å ± 0.44 Å and a divergence angle of 0.7×10^{-3} rad at the JRR-3M reactor, JAEA, Tokai. In the preparation of and the theoretical calculation on the resonator, the optical potentials of SSD were assumed to be 190 and 78.4 neV for ↑ and ↓ spin components, respectively, and the optical potentials U_{nucl} and U_{ab} of the (Ge/Gd) layer were estimated as 80 and 30 neV, respectively. With the present condition of the resonator potentials, the critical grazing angle of total reflection for the ↑ spin component of 12.6-Å-wavelength neutrons becomes 1.31°.

The experimental results compared with the calculated curve by using the one-dimensional Schrödinger equation are shown in Figure 9.27b. We notice the trend for the experimental results to follow rather well the distribution of the calculated curve, thus indicating the expression with the complex optical potential for the effects of neutron absorption is acceptable in the present case of resonant transmission. However, we also notice an obvious quantitative discrepancy between results of the present experiment and the calculation, in contrast to the more exact agreement obtained in the case of a negligible absorption shown in Figure 9.27a. The present result and the remaining discrepancy will stimulate further study on the optical effects of strong neutron absorption in such a case of resonant transmission of neutrons.

9.6.5
Experimental Test of Wave Packet Theories with Very Cold Neutron Interferometry

In this book, the neutron wave function in free space is very often expressed simply by plane waves as the fundamental solution of the Schrödinger equation. Such formulations appeared, for example, in Eqs. (1.22), (2.19), (2.20), (3.9), and (6.18). In the actual situations where incident neutrons have a wavelength distribution with a finite wavelength width, the incident waves have to be expressed by an incoherent superposition of plane waves over the wavelength distribution. We denote here such a formalism the plane wave model, or simply *plane waves* (PW).

On the other hand, in Section 9.1 we considered a *wave packet* as a coherent superposition of elementary waves in order to investigate the coherence length of a neutron. Especially in interferometry experiments, it is very important to consider the phase of waves exactly based on the practical geometry and the experimental situation, and therefore the concept of the wave packet was fundamentally necessary as described in Figure 9.5. One of the standard expressions of the neutron wave packet is a coherent assembly of waves which satisfy the Schrödinger equation, that is, the coherent superposition of plane waves, as given by Rauch [406] for the neutron wave packet in free space:

$$\Phi(\boldsymbol{r}, t) = \int a(\boldsymbol{k}) e^{i(\boldsymbol{k}\cdot\boldsymbol{r}-\omega t)} d^3\boldsymbol{k} , \tag{9.50}$$

where $a(\boldsymbol{k})$ is the amplitude factor for the wave vector component $\boldsymbol{k}$. We denote here such an expression *Schrödinger's packet* (SP). It is often pointed out in textbooks (e. g., as a formulation in textbook [D], pp. 357–362), or as described previously in Figure 9.5, that such a wave packet will disperse in space and time according to propagation owing to the wave number distribution $a(\boldsymbol{k})$.

Furthermore, fundamentally different types of wave packet can be supposed. For example, de Broglie proposed in 1927, that is, at the very beginning of the development of quantum mechanics, such a wave packet for matter waves as possessing a singular center inside the packet and propagating by maintaining a certain spatial distribution around the center. His concept on matter waves requires a theoretical framework fundamentally different from that of the Schrödinger equation, and so could not be accepted in the course of the successful development of quantum mechanics. Even late in his life, that is, a few decades after the establishment of quantum mechanics, he again insisted on his concept of wave mechanics in his book [446]. Around these years, a slightly modified interpretation of the Schrödinger equation was also proposed (e. g., [447]), based on a similar concept to de Broglie's theory on the mechanics of matter waves. With regard to these fundamental interest in the optics of matter waves, some theoretical analyses and experimental investigations relating to de Broglie's wave packet will be included in this subsection .

The de Broglie wave packet in free space can be expressed as follows:

$$\Psi_0(\boldsymbol{k}_0, \boldsymbol{r}, t; s_0) = C_0 \exp[i(\boldsymbol{k}_0\boldsymbol{r} - \omega_0 t)] \frac{\exp(-s_0|\boldsymbol{r} - \boldsymbol{v}_0 t|)}{|\boldsymbol{r} - \boldsymbol{v}_0 t|} , \tag{9.51}$$

where $\boldsymbol{k}_0$ is the wave vector of the packet, $\boldsymbol{v}_0 = \hbar \boldsymbol{k}_0/m$, and s_0 denotes a parameter determining the spread of the packet. In contrast to PW and SP satisfying the homogeneous Schrödinger equation, the present wave packet, Eq. (9.51), is the solution of the nonhomogeneous equation with a singular source term proportional to $C_0\delta(\boldsymbol{r} - \boldsymbol{v}_0 t)$ [448, 449], where $\delta(x)$ is Dirac's delta function. We denote such a wave packet *de Broglie's packet* (DB).

We cannot distinguish between these three kinds of wave formalism, PW, SP, and DB, from the results of any conventional neutron optical experiment, including the experiments introduced up to here in this book. Even with the neutron interferometry experiments introduced up to now in this book, evaluation of neutron waves as SP or DB was not possible. In such a situation, a specially arranged spin interferometry experiment with neutrons in *tunneling transmission* was performed by Utsuro and the results were compared with the calculated results from these wave packet theories, that is, SP and DB [449], in order to examine the possibility to distinguish between these two wave packet theories.

According to the theoretical analysis [449], the solution for the *tunneling-transmitted wave packet*, that is, the wave function of the neutron transmitted through a layer with finite thickness l with the grazing angle below the critical grazing angle for total reflection, can be expressed for DB as follows in the Cartesian coordinates with the z-axis perpendicular to the layer surface and the x-axis along the direction of the incident wave vector projected on the surface:

$$\begin{aligned}\Psi_{0\mathrm{T}}(r_{xy}, z, t; s_0) = &\ D_{0\mathrm{TF}} \times e^{-k_{Bz}l} \times e^{i\omega_{s0}t} e^{ik_{0x}(x - v_{0x}t)} e^{ik_{0z}(z - l - v_{0z}t)} \\ &\times \Lambda_0(r_{xy}, z, t; s_0; l)\,,\end{aligned} \tag{9.52}$$

$$\Lambda_0(r_{xy}, z, t; s_0; l) = \frac{\exp\left[-s_0\varrho(r_{xy}, z, t; l)^{1/4}\cos\alpha(r_{xy}, z, t; l)\right]}{\varrho(r_{xy}, z, t; l)^{1/4}}\,, \tag{9.53}$$

$$\begin{aligned}\varrho(r_{xy}, z, t; l) = &\left\{r_{xy}(t)^2 + (z - l - v_{0z}t)^2 - \frac{v_{0z}^2 l^2}{v_{Bz}^2}\right\}^2 \\ &+ 4(z - l - v_{0z}t)^2 \frac{v_{0z}^2 l^2}{v_{Bz}^2}\,,\end{aligned} \tag{9.54}$$

$$\alpha(r_{xy}, z, t; l) = 1/2 \arctan \xi(r_{xy}, z, t; l)\,, \tag{9.55}$$

$$\xi(r_{xy}, z, t; l) = \left\{\frac{2(z - l - v_{0z}t)\frac{v_{0z}}{v_{Bz}} l}{r_{xy}(t)^2 + (z - l - v_{0z}t)^2 - \frac{v_{0z}^2}{v_{Bz}^2} l^2}\right\}\,, \tag{9.56}$$

where $D_{0\mathrm{TF}}$ is the complex amplitude coefficient, and the notation $r_{xy}(t)^2 = (x - v_{0x}t)^2 + y^2$, $k_{Bz} = (2mU/\hbar^2 - k_{0z}^2)^{1/2}$, $v_{Bz} = \hbar k_{Bz}/m$, and $\omega_{s0} = \hbar(k_0^2 + s_0^2)/2m$ was used.

The present solution for the tunneling-transmitted packet, in contrast to the wave packet of Eq. (9.51) with the singularity only at a single point $\boldsymbol{r} = \boldsymbol{v}t$, has the singularity on the surface of an elliptic cylinder expressed by

$$\left.\begin{array}{ll} r_{xy}(t) & = \frac{v_{0z} l}{v_{Bz}}, \\ z & = l + v_{0z} t \end{array}\right\} , \tag{9.57}$$

which indicates that the wave packet propagates along the broad surface of a cylinder. As the result of such a characteristic propagation, the tunneling-transmitted DB is expected to show a phase distribution fundamentally different from that of SP and also from that of the refractively transmitted DB.

Furthermore, the optical propagation path of the tunneling-transmitted DB diverges, as given by the radius in Eq. (9.57), in proportion to the layer thickness l, in contrast to the cases of SP and of refractively transmitted DB for which we can follow a single optical path for the packet center after the tunneling. The present character of the tunneling-transmitted DB will have a much broader phase distribution with different values depending on the transverse location in the propagating space.

The experiment to test the DB model represented by the above-mentioned formalism was carried out by using a *spin interferometer for very cold neutrons* recently arranged at the high-flux reactor, Grenoble [450], with insertion of a couple of Fabry–Pérot magnetic resonators with a number of magnetic layers. In such an arrangement, two spin components will behave through the first resonator at a certain condition of the grazing angle as the parallel ($\uparrow$) spin component in tunneling transmission and as the antiparallel ($\downarrow$) spin component in refractive transmission, and then after the π flipper their transmission conditions through the second resonator will be exchanged, and they are again superposed and analyzed after the second resonator to observe the spin interference pattern. If the above-mentioned analytical model for DB is correct, the resultant spin interference pattern should show the effects of the diverged phase (i.e., spin precession) distribution for the tunneling-transmitted component through a number of the magnetic layers, as the *visibility* decrease in the spin interference pattern, and most of other spin-precession effects through the resonators will compensated each other in the present arrangement of the spin-phase-echo geometry.

According to the theoretical relations of Eqs. (9.52)–(9.56), it can be expected [451] that the measurement for the present effects will become more advantageous by making use of neutrons with a longer wavelength giving a larger critical grazing angle. From the viewpoint of a practically available neutron intensity, the use of very cold neutrons supplied at the very cold neutron beam port on the top floor of the high-flux reactor, Grenoble, was considered to be the best condition for the experiment.

The experiment was performed by using the very cold neutron spin interferometer with the configuration shown in Figure 9.28. A couple of the magnetic multilayer resonators with 11 magnetic layers (i.e., 10 gap layers) with a layer structure of [PA(20 nm)–Ge(40 nm)]^{10}–PA(20 nm) were prepared specially at the same time for the experiment. The main point of the experiment was to determine whether the visibility of the measured spin interference pattern decreases or not in the cas-

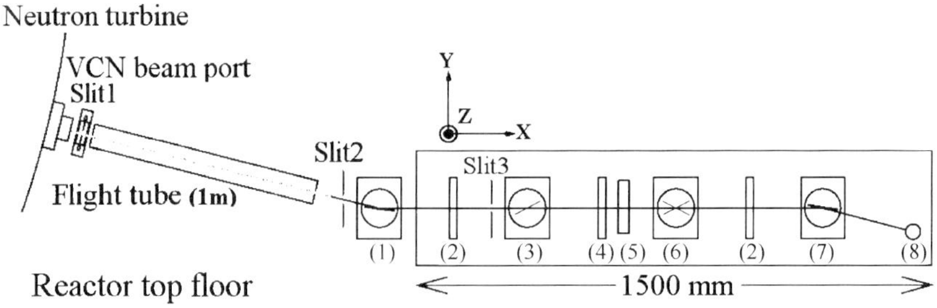

Figure 9.28 Experimental setup of the very cold neutron (VCN) spin interferometer arranged at the Grenoble high-flux reactor: (1) multilayer monochromator and polarizer; (2) $\pi/2$ spin flipper coils; (3) first Fabry–Pérot magnetic resonator (solid line in the circle) on the first goniometer; (4) π spin flipper coil; (5) accelerator coil; (6) second Fabry–Pérot magnetic resonator (solid line in the circle for the parallel twin arrangement, and dotted line for the symmetrical twin arrangement, to the first resonator) on the second goniometer; (7) polarization analyzer; (8) ^{3}He neutron detector (Hino *et al.* [450]).

es of tunneling transmission for the parallel spin component and of the refractive transmission for the antiparallel component, compared with that in the case of refractive transmission for both spin components. Further, the experimental *net visibility*[40)] derived from the experimental results can be compared with the theoretically calculated values based on the three kinds of wave formalism, PW, SP, and DB, as explained above.

To enhance the effects due to tunneling transmission in the experiment, the potential structure of the magnetic resonator with 10 bilayers, $[\mathrm{PA}(20\,\mathrm{nm})\text{–}\mathrm{Ge}(40\,\mathrm{nm})]^{10}\text{–}\mathrm{PA}(20\,\mathrm{nm})$, similar to the sample in the experiment reported in Figure 9.27a, was employed, and a couple of the same resonator samples specially prepared in the same evaporation procedure were named as samples A and B for identification. Very cold neutrons with an average wavelength of 5.7 nm and a wavelength width of 10% were used, and these wavelength values were also used in the theoretical calculations for comparison with experiment.

In advance of the interferometry experiment, a numerical simulation calculation on the slit geometry was performed to determine the possible interference effects of the three slits (all with the same width of 3 mm) in a serial arrangement as shown in Figure 9.28. The effect was found to be minor for the exact measurement of the net visibility [452].

Further, in advance of the twin resonator measurements, experiments with a single resonator setup were carried out by inserting the resonator in either the first or the second goniometer, and the experimental results were used for the determination of the subcritical and transcritical resonant angles for the parallel spin component, and also for determining the magnitude (1.40 T) of the magnetic potential of the PA layers. These are important conditions for realizing tunneling transmission

40) The value for the visibility calculated from the count rates for the measured interference amplitudes by using Eq. (9.39) after the correction for the background count rates with the resonator samples, and further divided by the corresponding value for the visibility without the samples. The appropriateness of the present procedure was supported by the theoretical consideration [451].

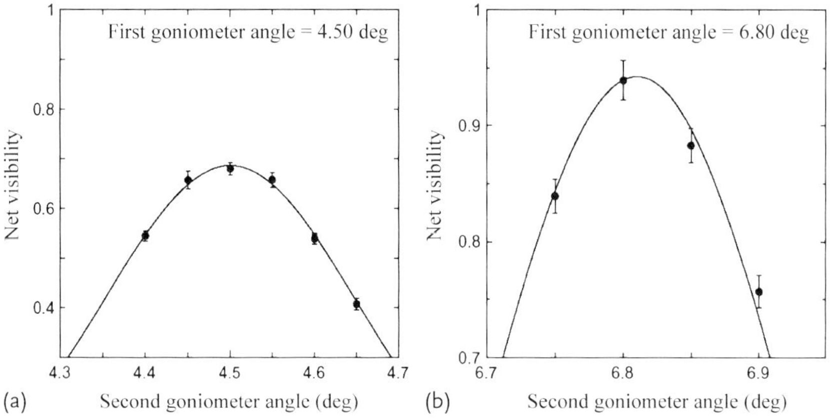

Figure 9.29 Gauss fit procedure to derive the exact value for the net visibility corresponding to the condition of precise matching of twin resonator angular settings from the measured visibility distributions for parallel twin resonators; (a) at the subcritical resonant grazing angle; (b) at the transcritical resonant grazing angle (Utsuro [451]).

or refractive transmission and also for accurate theoretical simulations. The experimental results for the single resonator setup were also used for the evaluation of the wave packet models by comparing them with the theoretical calculations.

In the case of the twin resonator experiments, a Gauss fit procedure on the net visibility distribution for parallel twin resonators was employed as shown in Figure 9.29 to derive the exact value for the net visibility corresponding to the condition of precise matching of twin resonator angular settings.

The final results of the neutron spin interference visibility in the grazing angle regions below and around the critical angle for total reflection of parallel spin neutrons are plotted on the two-dimensional graphs, taking the visibility in the subcritical condition as the ordinate and that in the transcritical condition as the abscissa, in Figure 9.30a and b, for the cases of the single and twin resonator setups, respectively [451]. In addition, the experimental results are compared with the calculated ones based on the three kinds of wave packet models.

The experimental results are expressed by marks with error bars as the root-squared sum of the statistical errors and the standard deviations in the sinusoidal and Gauss fitting procedures. The letters A and B identify the sample for the single resonator setup, and the marks with error bars for SA or SB and those for $Twin1$ and $Twin2$ represent the experimental reproducibility, since at the end of the first experiment the resonator sample arrangement was completely disassembled. After about a half year or more, fine adjustments of the interferometer as well as the sample setting were made and then the second measurements were performed. The experimental result SB' corresponds to the single resonator setup in the symmetrical sample direction as indicated by the dotted line on the second goniometer in Figure 9.28.

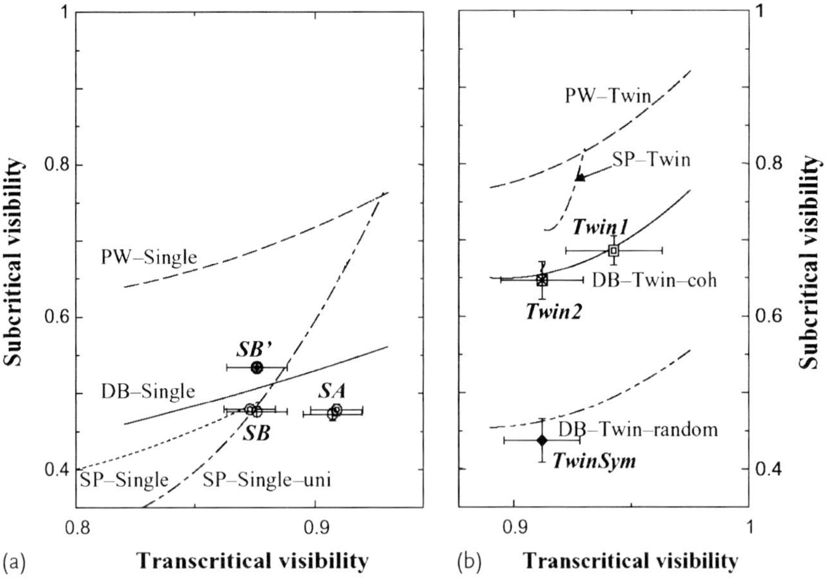

Figure 9.30 Resultant net visibility derived from spin interferometry experiments (marks with error bars) compared with the results of theoretical calculations from three kinds of wave packet models (curves): (a) single resonator measurements; (b) twin resonator measurements. PW, plane waves; SP, Schrödinger wave packet model; DB, de Broglie wave packet model; SP-Single-uni, Schrödinger wave packet model for a single resonator without consideration of layer thickness distributions; DB-Twin-coh, de Broglie wave packet model for parallel twin resonators with a strong coherency between the phases induced by tunneling through twin resonators; DB-Twin-random, the same model but with a complete incoherency between the phases induced by tunneling through twin resonators (Utsuro [451]).

The theoretical curves were plotted by changing the uniformity of the layer thickness, that is, the width of the layer thickness distribution, for the single resonator setup (0–7% from the right to the left along the curves in Figure 9.30a) for a fixed distribution width of 7% of the magnetic potential for the magnetic layers and a fixed wavelength resolution of 10%, or by changing the uniformity of the magnetic potential, that is, the distribution width of the magnetic potential for magnetic layers, for the twin resonator setup (0–10% from the right to the left along the curves in Figure 9.30b) for a fixed wavelength resolution of 10%.

In the case of the SP model for the single resonator setup, two curves as typical cases were plotted by changing the distribution width of the layer thickness (0–5% from the right to the left in Figure 9.30a) for a fixed packet wavelength width of 2% as the dotted line with the notation SP-Single, and by changing the packet wavelength width (0.1–3% from the right to the left in Figure 9.30a) for the completely uniform thickness of layers as the dot-dashed line with the notation SP-Single-uni. Furthermore, in the case of the SP model for the twin resonator setup, the calculated curve for the most typical case was plotted by changing the packet wavelength width (0.1–3% from the right to the left in Figure 9.30b) for a fixed magnetic uniformity, that is, the distribution width of 5% for the magnetic potential.

We can derive several conclusions from these comparisons between the experiments and the theoretical calculations. In the first place, the comparisons for the case of the single resonator setup indicate there is a significant discrepancy in the PW model, which suggests the important role of the coherent natures in the spectral, spatial, or spin structures of the wave packet in the visibility measured in the neutron spin interferometry experiment.

Next, in the case of the twin resonator setup, theoretical calculations for both the PW and the SP models could not explain well the experimental results of the interference visibility in the subcritical condition for the parallel spin component. On the other hand, the theoretical calculation with the DB model assuming strong coherency between the tunneling-transmitted packet phases agrees well with both results of the twice-repeated experiments for the parallel twin resonator arrangement. Further, the calculated result of the DB model assuming complete incoherency between the tunneling-transmitted packet phases seems to correspond to the symmetrical twin resonator arrangement indicated by a dotted line for the second sample in Figure 9.28. We can understand these relative correspondences to be reasonable in view of the possible matching between the additional phases due to the tunneling transmission through twin resonators in the case of the parallel arrangement, and the complete nonmatching between those through the two resonators with the geometry of the second sample reversed compared with the first one with regard to the optical paths on the elliptic cylinder mentioned previously in the case of the symmetrical twin arrangement.

Thus, the present results raise a fundamental question on the description of the interference phenomena by the Schrödinger equation for neutrons in tunneling transmission, and suggest the possibility of a certain distributed phase being induced additionally with characteristics such as those introduced above by the DB model in tunneling transmission.

10
Contradictions of the Scattering Theory, and Quantum Mechanics

In this chapter we discuss contradictions of the present-day quantum scattering theory [453] and some hot questions of standard quantum mechanics. It is addressed to those who are not satisfied with their level of understanding.

First we show that at present there are three scattering theories, and every one contains contradictions. We try to resolve them, but arrive at new ones. We shall be able to overcome all contradictions except one. The resolution of the remaining contradiction requires nonlinearity of interaction. We have no nonlinear scattering theory, and are able to introduce nonlinearity only implicitly [453].

First we formulate problems and give their solutions without detailed discussion. We decided to do so for readers to be able to grasp the whole picture of contradictions without entering into the details. However, the details are important and they are given in Section 10.5.

Since we found contradictions in the scattering theory, we looked critically at other issues of quantum mechanics such as uncertainty relations, the Einstein, Podolsky, and Rosen (EPR) paradox, and the interpretation of quantum mechanics.

10.1
How many Scattering Theories Do We Have?

> – I could easily show that the author is not right by pointing out to any standard textbook. The author does not want textbooks, but it is necessary to have something to start with.
> – You can start with the Holy Bible, but why not to start with my arguments?
>
> From discussions with a referee.

We discuss three-dimensional scattering of slow scalar particles, which is the main tool for research in condensed matter physics with neutrons. No relativistic concepts will be required. First we want to show that we have three scattering theories.

Neutron Optics. Masahiko Utsuro, Vladimir K. Ignatovich

ISBN: 978-3-527-40885-6

10.1.1 Theory of Spherical Waves

The simplest theory is borrowed from the scattering theory of classical waves such as sound and light. The scattering on a point scatterer fixed in space is described by the wave function

$$\psi = \exp(i\boldsymbol{k}\cdot\boldsymbol{r}) - b\frac{\exp(ikr)}{r}\,, \tag{10.1}$$

which contains the plane $\exp(i\boldsymbol{k}\cdot\boldsymbol{r})$, and the spherical $\exp(ikr)/r$ waves corresponding to the incident and scattered particles, respectively. The parameter b has dimension of length and is called the scattering amplitude[41]. The amplitude b gives the differential cross section $d\sigma = |b|^2 d\Omega$ for elastic scattering into an element of solid angle $d\Omega$ and the total, integrated over angles, cross section $\sigma = 4\pi|b|^2$. We see that the scatterer can be interpreted as an obstacle of radius $2b$ for the incident wave.

If the scatterer is not a pointlike one, the amplitude b becomes a function of the scattering angle Ω (see, e. g. textbook [B]), then $d\sigma(\Omega) = |b(\Omega)|^2\, d\Omega$, where $\sigma = \int |b(\Omega)|^2\, d\Omega$. The angular dependence appears after expansion of the total wave function of scattered particles over spherical harmonics (see formula (123.11) in textbook [B] with account of our notation):

$$b(\Omega) = \frac{i}{2k}\sum_{l=0}^{\infty}(2l+1)\,P_l(\vartheta)(S_l - 1)\,,$$

where ϑ is the polar angle with respect to the direction of the incident beam, P_l are the Legendre polynomials, and S_l is the scattering matrix of the lth harmonic.

If the scatterer can be excited by an incident particle, the amplitude b can also depend on energy, and the wave number k in spherical wave (10.1) can differ from $|\boldsymbol{k}|$ of the incident plane wave [454].

In classical physics such a description characterizes perturbation of the wave because of an obstacle. In quantum mechanics it is the change of the wave function caused by the scatterer. However, such a description of scattering is contradictory. In quantum mechanics we measure not a wave function after scattering, but particles with their energies and momenta. The spherical wave in Eq. (10.1) does not describe such particles.

41) As a distinction from all textbooks, we choose a sign minus before spherical waves, which is more convenient because in this case positive b corresponds to a positive, or repulsive, optical potential, while negative b corresponds to a negative, or attractive, optical potential.

10.1.2 The Standard Scattering Theory

The theory of spherical waves is a particular case of the standard scattering theory (textbook [B]) [455, 456]. In fact, the standard theory is not a theory at all, but a set of recipes for calculation of scattering cross sections.

The main recipe is the "Fermi golden rule". According to it, one must take a matrix element $\langle f|V|i\rangle$ of the interaction V between the initial, $|i\rangle$, and final, $|f\rangle$, states of the neutron and the scattering system, construct an expression for the so-called probability, w_t, of scattering per unit time, and divide it by the incident flux density J of the single particle. This ratio is then averaged over initial states and summed over final states of the scattering system.

A scattering amplitude, which is squared to get the cross section, is not calculated[42]. The recipes listed give directly the cross sections.

10.1.3 The Fundamental Scattering Theory

The fundamental theory is used mainly for justification of the standard one. It attempts to formulate the scattering process rigorously (see, e. g., [457, 458]). Three types of states of the scattering system are distinguished in it: the current state $\Psi(t)$ of the particle interacting with the scatterer, and asymptotical "really free" states in infinite past, $|\mathrm{in}\rangle$, and infinite future, $|\mathrm{out}\rangle$, where interaction is absent (switched off). The current state at $t = 0$ is related to asymptotical states by the Möller operators $U(+)$, $U(-)$: $\Psi(t = 0) = U(+)|\mathrm{in}\rangle$, and $\Psi(t = 0) = U(-)|\mathrm{out}\rangle$. These operators help to relate two asymptotical states

$$|\mathrm{out}\rangle = U^{-1}(-)\,U(+)|\mathrm{in}\rangle \,,$$

and to define the S matrix: $S = U^{-1}(-)\,U(+)$. Matrix elements of the S matrix give the probability amplitudes of scattering, which with the help of some tricks are transformed into scattering cross sections. Below we show that the fundamental theory does not provide a proof of the validity of the standard one.

10.2 Analysis of the Three Theories

10.2.1 Scattering of Spherical Waves

Spherical waves $\exp(ikr)/r$ do not satisfy the free Schrödinger equation and therefore they do not describe free particles. They are the solution of the inhomogeneous

42) There is a description with the T matrix, but calculation with it requires squaring the generalized δ functions.

equation

$$(\Delta + k^2)\frac{\exp(ikr)}{r} = -4\pi\delta(\boldsymbol{r})\,, \tag{10.2}$$

and characterize the current state (in the terminology of the fundamental theory) when the neutron is still interacting with the scatterer located at the point $r = 0$.

Usually it is said that the spherical wave is an asymptotical wave function far away from the scatterer where all the powers $1/r^n$, $n \geq 2$, can be neglected, and at these distances the spherical wave does satisfy the free Schrödinger equation.

Indeed at $r \neq 0$ the right-hand side of Eq. (10.2) is zero; therefore, the spherical wave satisfies the free equation. However, it is not sufficient to describes a free particle there. A bound-state wave function outside the potential well also satisfies the free equation, but it does not represent a free particle there, because it decays exponentially at infinity.

Let us accept the exponential decay as a feature of nonfree particles. The spherical wave contains two parts: one is a superposition of exponentially decaying waves (they bind the neutron to the scattering center) and the other one is a superposition of free propagating waves. The correct asymptotic of the spherical wave is the second part.

To find it we use a two-dimensional Fourier expansion of the spherical wave:

$$b\frac{\exp(ikr)}{r} = \frac{ib}{2\pi}\int \frac{d^2 p_{\parallel}}{\sqrt{k^2 - p_{\parallel}^2}} \exp\left(i\boldsymbol{p}_{\parallel}\cdot\boldsymbol{r}_{\parallel} + i\sqrt{k^2 - p_{\parallel}^2}\,|z|\right)\,, \tag{10.3}$$

where the z-axis is chosen in the direction from the scattering center at $\boldsymbol{r} = 0$ to the observation point $\boldsymbol{r}$. The two-dimensional integration runs over an infinite plane of vectors $\boldsymbol{p}_{\parallel}$ parallel to a plane which is perpendicular to the z-axis. It includes such $\boldsymbol{p}_{\parallel}$, for which $p_{\parallel} \geq k$. Waves with these $\boldsymbol{p}_{\parallel}$ decay exponentially along the z-axis. To get the correct asymptotics of the spherical wave we must neglect them, and restrict the integration range to $p_{\parallel} \leq k$. This part of the integral (10.3) is

$$\begin{aligned} b\frac{\exp(ikr)}{r} &\approx \frac{ib}{2\pi}\int\limits_{p_{\parallel}\leq k} \frac{d^2 p_{\parallel}}{\sqrt{k^2 - p_{\parallel}^2}} \exp\left(i\boldsymbol{p}_{\parallel}\cdot\boldsymbol{r}_{\parallel} + i\sqrt{k^2 - p_{\parallel}^2}\,|z|\right) \\ &= \frac{ib}{2\pi}\int d^3p\,\delta(p^2/2 - k^2/2)\exp(i\boldsymbol{p}\cdot\boldsymbol{r})\Theta(p_{\perp}z > 0) \\ &= \frac{ikb}{2\pi}\int\limits_{4\pi} \Theta(\boldsymbol{k}_{\Omega}\cdot\boldsymbol{r} > 0)\,d\Omega\,\exp(i\boldsymbol{k}_{\Omega}\cdot\boldsymbol{r})\,, \end{aligned} \tag{10.4}$$

where we used the relation $d^2 p_{\parallel}/\sqrt{k^2 - p_{\parallel}^2} = d^3p\,\delta(p^2/2 - k^2/2)$ valid at $p_{\parallel}^2 \leq k^2$, introduced the step Θ function, which warrants that only waves propagating away from the scattering center will arrive at the observation point, and in the last equality integrated over dp. The result Eq. (10.4) represents the correct asymptotic of the spherical wave. It contains a superposition of the scattered plane waves of free particles with wave vectors $\boldsymbol{k}_{\Omega}$ of length k pointing in the direction of the unit vector $\boldsymbol{\Omega}$

in the interval $d\Omega$ of the solid angle Ω. The amplitudes $f(\Omega) = ikb/2\pi = b/\lambda$ of the scattered waves represent probability amplitudes for scattering into the angle Ω.

So, we corrected the logical deficiency of the theory of spherical waves; however, we immediately arrived at a contradiction.

10.2.1.1 Contradiction 1

Instead of the cross section we obtained the dimensionless probability of scattering, which seems to be

$$dw_1(\Omega) = |f(\Omega)|^2 \, d\Omega = |b/\lambda|^2 \, d\Omega \, . \tag{10.5}$$

10.2.1.2 The Need for a Cross Section

We need a cross section because without it we cannot use such an evident formula as $T = \exp(-N_0\sigma_t d)$ for neutron transmission through a sample of thickness d and atomic density N_0. It contains the total cross section σ_t, which together with $N_0 d$ gives a dimensionless exponent. The factor σ cannot be replaced by the dimensionless probability!

10.2.2 Cross Section in the Standard Theory

The standard theory is a set of rules or recipes for calculation of cross sections (see, e. g., [456]). The main recipe is the Fermi golden rule, which defines the *probability of scattering per unit time*:

$$dw_t = \frac{2\pi}{\hbar} \left|\langle \lambda_f, \boldsymbol{k}_f | V | \lambda_i, \boldsymbol{k}_i \rangle\right|^2 \rho(E_{fk}) \, , \tag{10.6}$$

where $\boldsymbol{k}_{i,f}$ are the initial and final momenta of the neutron, and $\lambda_{i,f}$ are the initial and final quantum numbers of a scatterer. The factor $\rho(E_{fk})$ in (10.6) denotes the density of final neutron states:[43)]

$$\rho(E_{fk}) = \delta(E_{ik} + E_{i\lambda} - E_{fk} - E_{f\lambda}) \frac{L^3 \, d^3 k_f}{(2\pi)^3} \, , \tag{10.7}$$

where the δ function represents the energy conservation law, $E_{i,fk}$ and $E_{i,f\lambda}$ are the initial and final energies of the neutron and scatterer, respectively, and L^3 is some arbitrary volume.

The golden rule itself contains a contradiction.

10.2.2.1 Contradiction 2

The probability of scattering per unit time Eq. (10.6) does not depend on time; therefore, its integral over time, which must give the total probability, is meaningless.

43) The presentation given in [456] is slightly rationalized here.

In the standard theory the cross section is defined as the ratio of the probability Eq. (10.6) and the flux density of a single incident neutron: $d\sigma = dw_t/J_i$. To define the flux density J_i for the single neutron, its wave function $|\boldsymbol{k}_{i,f}\rangle$ is represented in the form of a normalized plane wave: $|\boldsymbol{k}\rangle = L^{-3/2}\exp(i\boldsymbol{k}\cdot\boldsymbol{r})$, where L is the linear dimension, which is the same as in Eq. (10.7). The flux density with this wave function is defined as $J_i = \hbar k_i/mL^3$. The ratio of Eq. (10.6) to this J_i gives the cross section:

$$d\sigma(\boldsymbol{k}_i,\lambda_i \to \boldsymbol{k}_f,\lambda_f) \equiv \frac{dw_t}{J_i} = \frac{2\pi m}{\hbar^2 k_i}\left|\langle\lambda_f,\boldsymbol{k}_f|V|\lambda_i,\boldsymbol{k}_i\rangle\right|^2 \times \delta(E_{ik}+E_{i\lambda}-E_{fk}-E_{f\lambda})\frac{L^6 d^3k_f}{(2\pi)^3}\,. \tag{10.8}$$

With account of $E_{fk} = \hbar^2 k_f^2/2m$ and $d^3k_f = mk_f\,dE_f\,d\Omega_f/\hbar^2$, we define the double-differential cross section as

$$\frac{d^2\sigma}{dE_f d\Omega_f}(\boldsymbol{k}_i,\lambda_i \to \boldsymbol{k}_f,\lambda_f) = \frac{m^2 k_f}{\hbar^4 k_i}\left|\langle\lambda_f,\boldsymbol{k}_f|V|\lambda_i,\boldsymbol{k}_i\rangle\right|^2 \times \delta(E_{ik}+E_{i\lambda}-E_{fk}-E_{f\lambda})\frac{L^6}{(2\pi)^2}\,. \tag{10.9}$$

To compare this with experiment it is necessary to average Eq. (10.9) over the initial states and to sum over all the final states of the scatterer:

$$\frac{d^2\sigma}{dE_f d\Omega_f}(\boldsymbol{k}_i \to \boldsymbol{k}_f,\mathcal{P}) = \sum_{\lambda_i,\lambda_f}\mathcal{P}(\lambda_i)\frac{m^2 k_f}{\hbar^4 k_i}\left|\langle\lambda_f,\boldsymbol{k}_f|V|\lambda_i,\boldsymbol{k}_i\rangle\right|^2 \times \delta(E_{ik}+E_{i\lambda}-E_{fk}-E_{f\lambda})\frac{L^6}{(2\pi)^2}\,, \tag{10.10}$$

where $\mathcal{P}(\lambda_i)$ is the probability for the scatterer to be initially in the state $|\lambda_i\rangle$.

The notion of the flux density for a single particle and the arbitrary volume L^3 are artificial features of the standard scattering theory. The introduction of L is not a harmless step. Sometimes [366] it leads to false physical effects.

The rules of the standard theory are used to calculate directly the cross section; however this cross section contains coherent and incoherent parts. The coherent part can be defined with the coherent amplitude, which is an average of the scattering amplitude not only over chemical, isotopic compositions and spin states of nuclei, but also over the initial states of the scatterer. The standard theory does not consider it, though such a coherent part of the scattering cross section can give additional information about the scattering process.

10.2.3
Fundamental Scattering Theory [457, 458]

In the fundamental theory the incident particle is described by a wave packet $|\phi\rangle$. When it is far from the scatterer (target), its dynamics is governed by the free

Hamiltonian H_0: $|\phi(t)\rangle = \exp(-iH_0t)|\phi\rangle$. The wave packet is represented by the Fourier expansion:

$$|\phi\rangle \equiv |\phi(\boldsymbol{k})\rangle = \int d^3p\, a(\boldsymbol{k}-\boldsymbol{p})|\boldsymbol{p}\rangle\,, \tag{10.11}$$

where $\boldsymbol{k}$ is the average momentum of the wave packet, and $|\boldsymbol{p}\rangle = \exp(i\boldsymbol{p}\boldsymbol{r})$ is a plane wave.

Because of scattering, the wave function (10.11) transforms to the function

$$|\phi_s\rangle \equiv |\phi_s(\boldsymbol{k})\rangle = \int \hat{S}|\boldsymbol{p}\rangle d^3p\, a(\boldsymbol{k}-\boldsymbol{p})\,, \tag{10.12}$$

where $\hat{S}$ is the scattering matrix. Substitution of the unity matrix $\hat{I} = \int |\boldsymbol{q}\rangle\, d^3q\, \langle\boldsymbol{q}|$ before $\hat{S}$ in Eq. (10.12) gives

$$|\phi_s\rangle = \int \hat{I}\hat{S}|\boldsymbol{p}\rangle\, d^3p\, a(\boldsymbol{k}-\boldsymbol{p}) = \int d^3q\, b(\boldsymbol{q},\boldsymbol{k})|\boldsymbol{q}\rangle\,, \tag{10.13}$$

where $b(\boldsymbol{q},\boldsymbol{k}) = \int \langle\boldsymbol{q}|\hat{S}|\boldsymbol{p}\rangle\, d^3p\, a(\boldsymbol{k}-\boldsymbol{p})$. The probability of scattering is defined as

$$dw = d^3q\,|b(\boldsymbol{q},\boldsymbol{k})|^2 = d^3q\left|\int d^3p\,\langle\boldsymbol{q}|\hat{S}|\boldsymbol{p}\rangle a(\boldsymbol{k}-\boldsymbol{p})\right|^2\,. \tag{10.14}$$

However, *this definition is faulty!* According to it the scattering probability is defined as the transformation of the wave packet of the incident neutron $|\phi\rangle$ into a plane wave $|\boldsymbol{q}\rangle$. Such a transformation violates unitarity completely because it transforms a normalized wave function into a nonnormalizable plane wave.

According to unitarity the wave function after scattering is to be represented as $\psi_s = \hat{S}|\phi(\boldsymbol{k})\rangle = \int f(\boldsymbol{k}\to\boldsymbol{k}')d^3k'\,|\phi(\boldsymbol{k}')\rangle$, where $f(\boldsymbol{k}\to\boldsymbol{k}')\,d^3k'$ is the probability amplitude of scattering to the state with the average wave vector $\boldsymbol{k}'$ in the interval d^3k'. It is dimensionless because the incident and scattered wave packets have the same dimensionality.

Thus, the fundamental theory introduces dimensionless probabilities, like the self-consistent theory of spherical harmonics, which we can consider as the fundamental theory of scattering of plain waves instead of wave packets.

In standard textbooks the initial wave packet is considered not as a property of particles but as a result of the preparation of the initial state. In such a case it is logically more consistent to consider scattering of plane waves into plane waves and to average the scattering probability over the distribution of initial plane waves in the incident beam.

10.2.3.1 Contradiction 3

The fundamental scattering theory contains an error and does not justify the recipes of the standard one.

10.3
General Definition of the Scattering Cross Section

In all textbooks, the scattering cross section is defined as the ratio of the number of events per unit time to the flux density of the incident particles. This is a phenomenological definition because it is equivalent to the definition via the experiment shown schematically in Figure 10.1. If we have a small sample s in the neutron flux from a source S, then the ratio of the detector count rate $\dot{N}_D$ to the flux density J of the incident neutrons is the effective area Σ of the whole sample which is able to intercept neutrons from the incident flux and direct them to the detector. The ratio of Σ to the total number of particles in the sample $N_s = N_0 V$, where N_0 is the number density of atoms and V is the sample volume, gives the effective cross section of a single atom: $\sigma = \dot{N}_D / J N_0 V$. It is clear that this definition contains no information about the interaction of a single neutron with a single atom. It is only a parameter used for description of the experimental results.

10.3.1
Phenomenological Definition of the Cross Section in the Self-Consistent Theory of Spherical Waves

With the asymptotical wave function (10.4) we can obtain the phenomenological cross section for a single atom. According to the common definition, the total flux of scattered particles through an infinite plane $S(\boldsymbol{n})$ with normal $\boldsymbol{n}$, which points to the observation point $\boldsymbol{r}$ from the scatterer, is

$$J_S = \int\limits_{S(\boldsymbol{n})} \frac{d^2 r_{\parallel}}{2i} \left[\psi_s^*(\boldsymbol{r}, t) \nabla_{\perp} \psi_s(\boldsymbol{r}, t) - \psi_s(\boldsymbol{r}, t) \nabla_{\perp} \psi_s^*(\boldsymbol{r}, t)\right] , \tag{10.15}$$

where $\nabla_{\perp}$ is a derivative along the normal to this plane. Substitution of the wave function (10.4) in the form of a two-dimensional Fourier expansion gives

$$J_S = |b|^2 \int\limits_{k_{\parallel}^2 < k^2} \frac{d^2 k_{\parallel}}{k_{\perp}} = k|b|^2 \int\limits_{2\pi(\boldsymbol{n})} d\Omega \, . \tag{10.16}$$

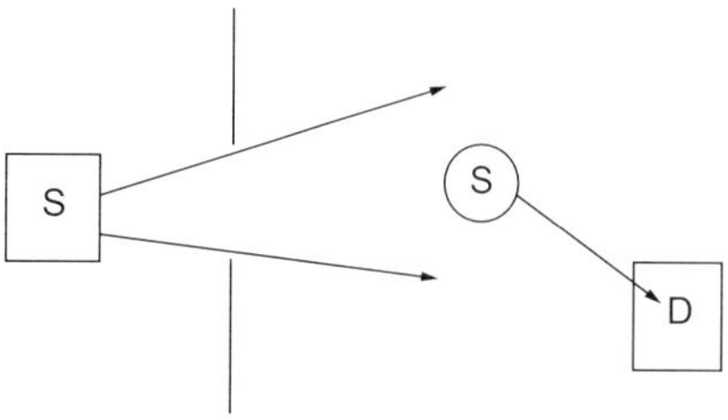

Figure 10.1 Phenomenological definition of a cross section as a ratio of the detector count rate to the flux density of the incident neutrons. S source, s sample, D detector.

If the flux density for the incident neutrons is accepted as k, then the phenomenological differential cross section is

$$d\sigma/d\Omega = |b|^2 . \tag{10.17}$$

We see that this cross section does not have the meaning of the scatterer's cross area, but it means an area pierced by scattered neutrons in a plane perpendicular to a given direction.

10.3.2 Relation between Cross Section and Probability

To see how to relate the scattering probability to experimental data, let us consider the experiment shown schematically in Figure 10.2. Our sample is now sufficiently wide to embrace all the incident neutrons, its thickness d is sufficiently small for us to neglect multiple scattering, and we suppose that all the atoms scatter incoherently. If the count rate of the incident neutrons is $\dot{N}_i$ (it is measured without the sample) and the detector count rate of the scattered neutrons in the presence of the sample is $\dot{N}_D$, then the ratio $W = \dot{N}_D/\dot{N}_i$, which is the sole outcome of such an experiment, gives the probability of scattering of a single neutron in the whole sample.

To find probability w_1 of a single neutron scattering on a single atom, we must divide the total probability W by the number of atoms N_a the neutron meets on its way through the sample. If we suppose that $N_a = \sigma N_0 d$, then $w_1\sigma = W/N_0 d$, which is possible only when $w_1 = 1$, because $W/N_0 d$ is just the experimentally defined σ. However, according to Eq. (10.5) we can expect that $w_1 \propto (b/\lambda)^2 \ll 1$; therefore, we cannot accept $N_a = \sigma N_0 d$ and must introduce the area of the neutron wave function $A \gg \sigma$ such, that $N_a = N_0 A d$, and $w_1 A = \sigma$. The area A includes not only the wave front of the neutron but also the size of the nucleus;

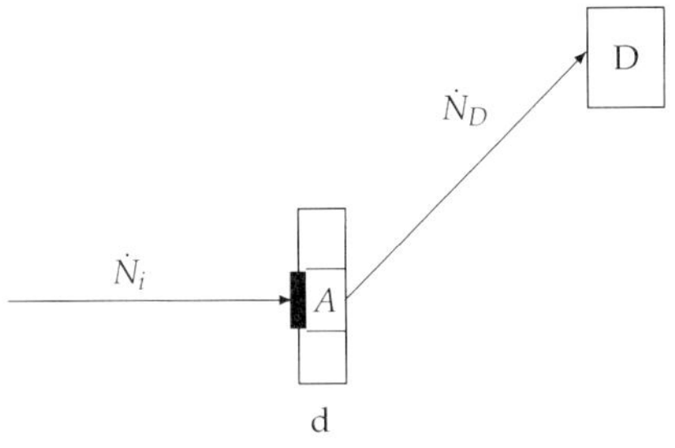

Figure 10.2 The definition of the scattering cross section of a single atom. A thin sample of thickness d and large area is pierced by $\dot{N}_i$ neutrons per unit time. The count rate of the detector is $\dot{N}_D$. The total probability for scattering of a single neutron in the sample is $W = \dot{N}_D/\dot{N}_i$. If the cross area of the single neutron's wave function front is A, then the probability of scattering of a single neutron on a single atom is $w_1 = W/N_0 A d$, where N_0 is the number density of atoms in the sample and $N_0 A d$ is the number of atoms swept by the wave function front during flight of the neutron through the sample. It follows that the cross section is $\sigma = A w_1 = \dot{N}_D/\dot{N}_i N_0 d$.

however, in the following we suppose that the size of the nucleus can be neglected. Comparison of theory with experiment means comparison of $\sigma_{\text{theor}} = Aw_1$ and $\sigma_{\text{exp}} = \dot{N}_D / N_0 d\dot{N}_i$. We suppose we are able to calculate probability w_1, but we can say nothing about A.

10.3.2.1 A Consequence of the Introduction of the Parameter *A*

> The only thing, I am absolutely sure in my life, is that wave packets do not exist!
>
> Roland Gähler

Introduction of area A means that the free neutron's wave function is supposed to be a wave packet. We can consider area A as a fitting parameter for experimental data; however, it is more fruitful to find it from some different considerations and experiments. Usually the wave packets are not considered because in the fundamental theory, which operates with wave packets in the initial state, it is proven that the wave packet structure and dimensions do not matter, that is, the parameter A is excluded from the cross section. In some respect this appears trivial because we can accept $\sigma_{\text{exp}} = 4\pi|b|^2$, and not worry about w_1 and A; however, with w_1 and A we can hope to gain insight into the scattering processes beyond the phenomenology.

Let us look at how the fundamental theory does relate cross section to the scattering probability.

10.3.2.2 Relation between the Cross Section and the Probability in the Fundamental Theory

The transition from probability to cross section is a little bit different in different books. In [457], the authors suppose "for the sake of simplicity" that at $t = 0$ the scatterer coincides with the center of the wave packet of the initial particle as is shown in Figure 10.3, and they find the cross section as the ratio of the probability and neutrons density in the packet. Taylor [458] introduces the impact parameter of the scatterer with respect to the center of the wave packet and integrates over the impact parameter. However, both of these methods are equivalent to multiplication of the scattering probability by the packet cross area A, and acceptance of

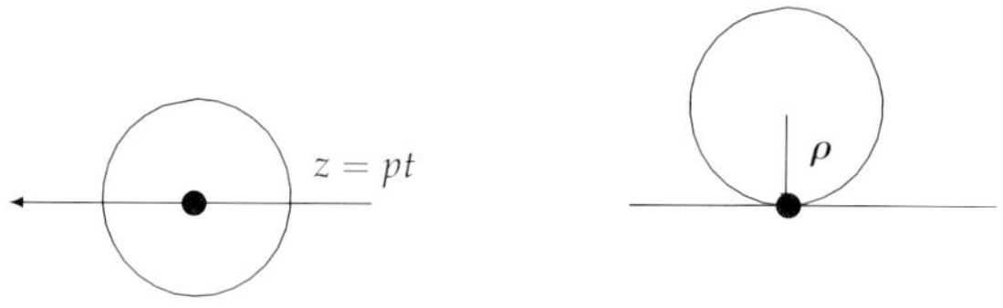

Figure 10.3 Position of the scatterer at $t = 0$ with respect to the center of the wave packet of the incident particle. In (a) they coincide. This is used in the derivation of Eq. (10.45). In (b) the target is at the impact parameter ρ from the packet center. This geometry is used in the derivation of Eq. (10.47).

an implicit assumption that there are no scattering if the scatterer is outside the wave packet. This assumption is in accord with common sense, and above we also followed it. However, the scattering amplitude does not depend on any trajectory, so the implicit assumption is not verified, and in the next paragraph it is shown to be even wrong.

10.3.2.3 Scattering of Wave Packets

To avoid the error of the fundamental theory and to derive correctly the cross section from the probability and the wave packet form, we need to study the process of scattering of wave packets when the final state is the same wave packet as the initial one. This is done in Section 10.5. The result of this study shows that the scattering of wave packets does not depend on their structure and proceeds like the scattering of plane waves. Moreover, in the case of scattering on a point scatterer, the probability of scattering into a given angle does not depend on the impact parameter.

This result looks puzzling, but it is a consequence of linearity of the theory. A wave packet is a superposition of plane waves, which are enhanced inside and cancel each other outside the wave packet. However, they exist in the whole space and scatter independently of each other. So we arrive at a new contradiction.

10.3.2.4 Contradiction 4

We introduced wave packets to transform the probability into a cross section. However, this does not help because scattering on a target does not depend on how large the impact parameter of the target is with respect to the wave packet center.

To resolve this contradiction we need nonlinearity of interaction to ensure that the scattering of a separate plane wave at a point takes place only when the full sum of the plane waves at this point is nonzero. We do not have nonlinearity, therefore we do not have a scattering theory. Nevertheless we have to use the parameter A if we want to do something! So we assume that introduction of the parameter A is an implicit introduction of nonlinearity into the interaction.

10.4 Wave Packets

The parameter A is unavoidable for the deduction of the cross section; therefore, it cannot be a result of an arbitrary preparation of the initial state of the particle. It must be a property of the neutron itself. However, if it is a property, the wave packet should not spread. In other case we should observe an increase of the scattering cross section when a sample shifts away from the source. Nobody has ever observed such an increase, though, in fact, no one has tried to look for it.

A packet has finite area A only if it is normalizable; however, in quantum mechanics all normalizable wave packets spread. So we arrive again at a contradiction.

10.4.1 Contradiction 5

The wave packet is to be normalizable, but then it spreads, and area A cannot be a property of a particle.

10.4.1.1 The Singular de Broglie Wave Packet

The last contradiction can be resolved if we use the singular de Broglie wave packet [446]:

$$\psi_{dB}(s, \boldsymbol{k}, \boldsymbol{r}, t) = C \exp(i\boldsymbol{k} \cdot \boldsymbol{r} - i\omega t) \frac{\exp(-s|\boldsymbol{r} - \boldsymbol{k}t|)}{|\boldsymbol{r} - \boldsymbol{k}t|} , \tag{10.18}$$

where $\boldsymbol{k} = \boldsymbol{v}$ is the neutron velocity (in units $\hbar = m = 1$), $\omega = (k^2 - s^2)/2$, $C = \sqrt{s/2\pi}$ is the normalization constant, defined via relation $\int d^3r\, |\psi(s, \boldsymbol{k}, \boldsymbol{r}, t)|^2 = 1$, and s is a parameter which gives the packet width in momentum space and reciprocal width in configuration space. The value of ω is less than that of the kinetic energy $v^2/2$ by the amount $s^2/2$, which can be considered as a bound energy of the packet. The size A of the packet is proportional to $1/s^2$.

The singular de Broglie wave packet satisfies the inhomogeneous Schrödinger equation

$$[2i\partial/\partial t + \Delta]\, \psi_{dB}(s, \boldsymbol{k}, \boldsymbol{r}, t) = -4\pi C e^{i(v^2+s^2)t/2} \delta(\boldsymbol{r} - \boldsymbol{k}t) , \tag{10.19}$$

with a source on the right-hand side, which is zero everywhere except at a single point. It looks as if we have arrived at a new contradiction: we discussed the theory with free initial and final states, which are solutions of the homogeneous equation, whereas Eq. (10.19) is an inhomogeneous one. However, it is not a contradiction. It is in our will and right to redefine the notion of a free particle.

10.4.2 Estimation of *s*

There is no theory which can give us the size A of the packet, so we must find it experimentally. At present we have only one kind of experiment from which we can deduce the size s, that is, experiments on storage of ultracold neutrons (see Chapter 1), where an anomaly was found.

Anomalous losses at every collision with the walls can be attributed to the wave packet form of the neutron wave function [173]. Indeed, the wave packet is a superposition of plane waves with different wave numbers $\boldsymbol{p}$. Some of them have the normal component, $p_\perp$, with respect to the wall, larger than $p_c = \sqrt{u_0}$, where $u_0 = 4\pi N_0 b$ is the optical potential of the vessel walls. We can suppose that the fraction of such wave vectors in the full spectrum of the wave packet determines the probability of nontunneling transmission through the wall.

In the linear theory such an explanation is not valid because reflection from a wall works like a filter. The part of the wave packet with $p^2 > p_c^2$ goes out through the wall, and the part with $p^2 < p_c^2$ remains inside the bottle, and it is these components that constitute ultracold neutrons. In the nonlinear theory we can suppose that the wave packet is a property of the neutron, so the reflected and transmitted parts are the same as the incident packet.

If anomalous losses of ultracold neutrons in bottles are related to this mechanism, then they give information on the value of s. An estimations made in [173] shows that the width s for ultracold neutrons must be on the order $3 \cdot 10^{-5}k$, or the area A of the neutron with ultracold neutron energy should be on the order of several square millimeters and it should decrease with the neutron energy[44]! This conclusion about the dimensions of the wave packet leads immediately to a new contradiction.

10.4.3 Contradiction 6

The size of the wave packet must be large but it cannot be large.

Indeed, since the scattering probability is proportional to $|b|^2k^2$, and $A \sim 1/s^2$, then the cross section is $\sigma \sim |b|^2(k^2/s^2)$. The ratio $k^2/s^2 \approx 10^9$ is large. Therefore, to get a cross section σ on the order of 1 b, or 10^{-24} cm^2, we must put the amplitude b to be less than 10^{-16} cm. However, then the optical potential $4\pi N_0 b$ becomes more than 4 orders of magnitude less than the experimentally measured value.

10.4.4 Resolution of Contradiction 6. Quantization of the Scattering Angle

We rejected spherical harmonics and arrived at contradiction 1. We resolved it by the introduction of a wave packet and its cross area A. We rejected recipes of the standard scattering theory as logically inconsistent, and the rejection is equivalent to resolution of contradiction 2. We turned toward the fundamental theory, but even there we found contradiction 3. To resolve contradiction 3, we considered scattering of a wave packet into wave packets, and immediately arrived at contradiction 4. We found that its resolution requires nonlinearity. We do not have a nonlinear scattering theory, and must accept nonlinearity implicitly. Only with nonlinearity can we accept resolution of contradiction 1 and neglect contradiction 5. Finally, we are left with contradiction 6.

To resolve contradiction 6, we must return to contradiction 1 and analyze what probabilities we found after abandoning the spherical waves. The analysis shows that the definitions of probabilities (10.5) and (10.14) from wave functions (10.4) and (10.13), respectively, are wrong. Such definitions give nonunique scattering probabilities [453]. In particular, the scattering cross section of neutrons on an atom

44) The results of the two experiments reported in [459, 460] contradict each other. This contradiction can be resolved only by further experiments.

becomes different if we calculate it directly in the laboratory frame or if we calculate it first in the center-of-mass reference system and then return to the laboratory frame.

Indeed, suppose we have a one-dimensional wave function of the form $\psi(x) = \int dp\, f(p)\exp(ipx)$. It is wrong to say that the probability of finding the particle in the state with momentum p is $dw(p) = |f(p)|^2 dp$, because this probability changes with transformation of the coordinates. For instance, change of coordinates $x = y/2$, $p = 2q$ transforms the integral to $\psi(y) = \int dq 2f(2q)\exp(iqy)$. If we accept the probability for the particle to have momentum q as $dw(q) = 4|f(2q)|^2 dq$, then after back-transformation $y = 2x$, $q = p/2$ we get $dw(p) = 2|f(p)|^2 dp$, which is twice as large as before.

Therefore, to get the probability of scattering for the wave function

$$\psi_s = \int\limits_{2\pi} g(\Omega)\, d\Omega \exp(i\boldsymbol{k}_\Omega \cdot \boldsymbol{r}) , \tag{10.20}$$

we must replace the integral (10.20) with the integral sum:

$$\psi_s = \sum_{j=1}^{N} g(\Omega_j)\, \delta\Omega_j \exp(i\boldsymbol{k}_{\Omega_j} \cdot \boldsymbol{r}) = \sum_{j=1}^{N} F_j \exp(i\boldsymbol{k}_{\Omega_j} \cdot \boldsymbol{r}) , \tag{10.21}$$

where discrete values $F_j = g(\Omega_j)\delta\Omega_j$ are introduced. With this representation the probability of scattering in the direction Ω_j is $|F_j|^2$, and the probability of scattering into $2n$ states with index j, $j_1 - n \le j \le j_1 + n$, around some average direction Ω_{j1} is

$$dw_n(\Omega_{j1}) = \sum_{j=j_1-n}^{j_1+n} |F_j|^2 \approx W_{j1}\Delta\Omega ,$$

where we used the notation

$$\Delta\Omega = \sum_{j=j_1-n}^{j_1+n} \delta\Omega_j \approx 2n\delta\Omega_{j1} , \quad W_{j1} = |g(\Omega_{j1})|^2 \delta\Omega_{j1} ,$$

and supposed that probabilities W_j change negligibly in the interval $j_1 - n \le j \le j_1 + n$.

In the approximation of the integral by the integral sum, the directions Ω_j and intervals $\delta\Omega_j$ can be chosen arbitrarily. In the spirit of quantization, we can suppose that all the amplitude elements $F_j = g(\Omega_j)\,\delta\Omega_j$ are equal, that is, there is a quantum of area $\int_{\delta\Omega} g(\Omega)\, d\Omega$. In the case of isotropic scattering, the function $g(\Omega)$ is a constant; therefore, quantization of the area is equivalent to quantization of the element of the solid angle $\delta\Omega$. The value of the quantum $\delta\Omega$ is very important because the correct choice of it helps to resolve contradiction 6.

Change of variables changes the function $g(\Omega)$ and the interval $\delta\Omega$, but the amplitude quantum, $F(\Omega) = g(\Omega)\,\delta\Omega$, and the number of quanta in the integral $\int_{4\pi} g(\Omega)\, d\Omega$ are invariant.

10.4.4.1 The Value of the Angular Quantum and Resolution of Contradiction 6

The probability amplitude of scattering according to Eq. (10.4) is $g(\Omega) = ibk/2\pi$; therefore, the amplitude quantum is $F(\Omega) = (ibk/2\pi)\delta\Omega$. The probability of scattering into some interval of solid angle $\Delta\Omega = 2n\delta\Omega$ is $\Delta w = 2n|F(\Omega)|^2 = |bk/2\pi|^2\delta\Omega\,\Delta\Omega$, and the cross section found with the help of area A is

$$\frac{d\sigma}{d\Omega} = \frac{\Delta w}{\Delta\Omega}A = \left|\frac{bk}{2\pi}\right|^2 A\,\delta\Omega\,. \tag{10.22}$$

It follows immediately that to have agreement with an experimental $d\sigma/d\Omega = |b|^2$ we must put

$$\left|\frac{k}{2\pi}\right|^2 A\,\delta\Omega = 1\,. \tag{10.23}$$

This means that the size of the angular quantum $\delta\Omega$ correlates with the size of the wave packet. This means that we can have a large area A and it will not lead to contradiction 6 because of compensation by the smallness of the angular quantum $\delta\Omega$. Moreover, we see that if the quantum $\delta\Omega$ does not depend on energy, then $A \sim 1/k^2$, that is, cross area A decreases with increase of the particle's energy. We can also estimate the value of the angular quantum $\delta\Omega$. If we estimate A from the ultracold neutron anomaly as $\pi 10^9/k^2$, then $\delta\Omega \approx 10^{-8}$.

10.4.5 Final Remarks

Thus, we have resolved all the contradictions except one, which requires nonlinearity. All these contradictions are logical. They do not invalidate the results obtained in physics up to now. However, the interpretation of some phenomena can change. Moreover, new directions for research can be opened and new fine effects can be found which in the traditional approach are invisible.

The next section contains the details, which are addressed to those who want to check the results presented in this section. After that we touch briefly on some principal points of quantum mechanics, such as uncertainty relations the EPR paradox, and diffraction of a particle by two slits, and discuss some experiments to investigate the wave packet property of the neutron wave function. But before proceeding, we want to note that the contradictions listed do not invalidate the results of other chapters of the book, because there we almost everywhere studied the problems of reflectometry and diffractometry, which deal with probabilities and not with cross sections.

10.5 Details

Here we first briefly recall main points of the fundamental theory, and how it arrives at the probability of scattering. Then we discuss how the probability is transformed to the cross section, and show how to calculate the scattering of a wave packet rigorously.

10.5.1
Main Points of the Fundamental Scattering Theory According to [457, 458]

There are several common points in all the books devoted to the fundamental scattering theory (we shall limit ourselves to [457, 458]):

A wave packet $|\phi\rangle$ for the incident particle far away from the scatterer (target) is introduced. Its dynamics is described by the free Hamiltonian $H_0 : |\psi_{\text{in}}\rangle \equiv |\phi(t)\rangle = \exp(-iH_0 t)|\phi\rangle$. The wave packet can be represented by the Fourier expansion (10.11).

The current wave function of the particle $|\Psi\rangle$ at time $t \approx 0$, when the particle interacts with the target, is introduced. Its dynamics is described by the full Hamiltonian H, which contains the interaction potential $V : |\Psi(t)\rangle = \exp(-iHt)|\Psi\rangle$.

The limiting Möller operator [458] Ω_+, defined as $\Omega_+ = \lim_{t\to-\infty} U(0, t)$, where $U(0, t) = e^{iHt}e^{-iH_0 t}$, is introduced. It relates $\exp(-iHt)|\Psi\rangle$ to the asymptotics $\exp(-iH_0 t)|\phi\rangle$ at $t \to -\infty$, so $\Psi(0) \leftarrow \Omega_+|\phi\rangle$.

The Möller operator depends on the interaction potential. From the time derivative of the operator $i(\partial/\partial t)U(0, t) = -e^{iHt}Ve^{-iH_0 t}$ we find $\Omega_+ = 1 - i\int_{-\infty}^{0} e^{iHt'}Ve^{-iH_0 t'}dt'$, and it follows that

$$
\begin{aligned}
|\Psi\rangle &= \left(1 - i\int_{-\infty}^{0} dt' e^{iHt'}Ve^{-iH_0 t'}\right)|\phi\rangle \\
&= \int d^3p\, a_i(\boldsymbol{k}-\boldsymbol{p})\left[1 - i\int_{-\infty}^{0} dt'\, e^{i(H-E_p)t'}V\right]|\boldsymbol{p}\rangle\,, \qquad (10.24)
\end{aligned}
$$

where $a_i(\boldsymbol{k}-\boldsymbol{p})$ are the Fourier components of the incident wave packet, and we used the relation $\exp(-iH_0 t)|\boldsymbol{p}\rangle = \exp(-iE_p t)|\boldsymbol{p}\rangle$. Integration over t' in Eq. (10.24) leads to

$$
\int d^3p\, a_i(\boldsymbol{k}-\boldsymbol{p})\left[1 - \frac{1}{H - E_p - i\epsilon}V\right]|\boldsymbol{p}\rangle = \int d^3p\, a_i(\boldsymbol{k}-\boldsymbol{p})|\psi_{\boldsymbol{p}}\rangle\,, \qquad (10.25)
$$

where the functions $|\psi_{\boldsymbol{p}}\rangle$ replace the plane waves $|\boldsymbol{p}\rangle$ in the packet.

The intermediate function, which in the wave packet expansion (10.25) during interaction replaces the plane wave $|\boldsymbol{p}\rangle$, is

$$
|\psi_{\boldsymbol{p}}\rangle = \left[1 - \frac{1}{H - E_p - i\epsilon}V\right]|\boldsymbol{p}\rangle\,. \qquad (10.26)
$$

It satisfies the stationary Schrödinger equation with the full Hamiltonian, $(H - E_p)|\psi_{\boldsymbol{p}}\rangle = 0$, and, according to the standard theory, contains incident plane and scattered spherical waves.

The asymptotical state $|\chi\rangle$ at an infinite future after scattering is introduced. Its dynamics is governed again by the free Hamiltonian H_0: $|\chi(t)\rangle = \exp(-iH_0t)|\chi\rangle$. This state is also a wave packet of the type $|\chi\rangle = \int d^3p\,|\boldsymbol{p}\rangle a_s(\boldsymbol{k}-\boldsymbol{p})$.

The asymptotical relation between states $|\psi\rangle$ and $|\chi\rangle$: $\exp(-iHt)|\Psi\rangle \to \exp(-iH_0t)|\chi\rangle$ or $\Psi(0) \to \Omega_-|\chi\rangle$ is established at $t \to +\infty$, from which it follows that

$$|\chi\rangle = \Omega_-^{-1}\Psi(0) = \lim_{t\to\infty} e^{iH_0t}e^{-iHt}\int d^3p\, a_i(\boldsymbol{k}-\boldsymbol{p})|\psi_{\boldsymbol{p}}\rangle$$
$$= \lim_{t\to\infty} e^{iH_0t}\int d^3p\, a_i(\boldsymbol{k}-\boldsymbol{p})e^{-iE_pt}|\psi_{\boldsymbol{p}}\rangle\,. \tag{10.27}$$

Substitution of unity $I = \int |\boldsymbol{p}'\rangle d^3p'\langle\boldsymbol{p}'|$ after $\exp(-iH_0t)$ gives

$$|\chi\rangle = \lim_{t\to\infty}\int d^3p\, a_i(\boldsymbol{k}-\boldsymbol{p})\int d^3p'\, e^{iE_{p'}t}|\boldsymbol{p}'\rangle\langle\boldsymbol{p}'||\psi_{\boldsymbol{p}}\rangle e^{-iE_pt}\,. \tag{10.28}$$

The relation between two asymptotical states is established by substitution of Eq. (10.26) into Eq. (10.28):

$$|\chi\rangle = \lim_{t\to\infty}\int d^3p'|\boldsymbol{p}'\rangle\int d^3p\, a_i(\boldsymbol{k}-\boldsymbol{p})$$
$$\times\left[\delta(\boldsymbol{p}-\boldsymbol{p}') - \langle\boldsymbol{p}'|\left(\frac{e^{i(E_{p'}-E_p)t}}{H-E_p-i\epsilon}V\right)|\boldsymbol{p}\rangle\right]\,. \tag{10.29}$$

A matrix $\mathcal{T}$ is introduced. It is defined with the help of the simple relation

$$\frac{1}{H-E_p-i\epsilon}V = \frac{1}{H_0-E_p-i\epsilon}\left(1-V\frac{1}{H-E_p-i\epsilon}\right)V$$
$$= \frac{1}{H_0-E_p-i\epsilon}\mathcal{T}\,. \tag{10.30}$$

The scattering S matrix is introduced as follows. Substitution of Eq. (10.30) into Eq. (10.29) gives

$$|\chi\rangle = \lim_{t\to\infty}\int d^3p'|\boldsymbol{p}'\rangle\int d^3p\, a_i(\boldsymbol{k}-\boldsymbol{p})$$
$$\times\left[\delta(\boldsymbol{p}-\boldsymbol{p}') - \frac{e^{i(E_{p'}-E_p)t}}{E_{p'}-E_p-i\epsilon}\langle\boldsymbol{p}'|\mathcal{T}|\boldsymbol{p}\rangle\right]\,. \tag{10.31}$$

At $t \to \infty$ we can use the relation

$$\lim_{\epsilon\to 0}\frac{\exp(ixt)}{x-i\epsilon} \equiv i\int_{-\infty}^{t}\exp(ixt')\,dt'\,, \tag{10.32}$$

which transforms (10.31) to $|\chi\rangle = \int d^3p'|\boldsymbol{p}'\rangle\int d^3p\,\langle\boldsymbol{p}'|\hat{S}|\boldsymbol{p}\rangle a_i(\boldsymbol{k}-\boldsymbol{p})$, where the $\hat{S}$ matrix is defined via its matrix elements

$$\langle\boldsymbol{p}'|\hat{S}|\boldsymbol{p}\rangle = \delta(\boldsymbol{p}-\boldsymbol{p}') - 2\pi i\delta(E_{p'}-E_p)T(\boldsymbol{p}',\boldsymbol{p})\,, \tag{10.33}$$

with $T(\boldsymbol{p}',\boldsymbol{p}) = \langle\boldsymbol{p}'|\mathcal{T}|\boldsymbol{p}\rangle$.

In fact, the fundamental theory considers scattering of plane waves because the matrix elements of the S matrix are calculated between plane waves.

Let us note that in the whole formalism described above there is no dependence of matrix elements on the position of the target with respect to the wave packet. We must remember this in discussing the tricks used in [457, 458]. But first let us consider scattering of a wave packet not to plane waves but to wave packets of the same form as the incident one.

10.5.2
Elastic Scattering of a Wave Packet on a Fixed Center

Every spherically symmetrical wave packet $\psi(s, \boldsymbol{k}, \boldsymbol{r}, t)$ can be represented by the Fourier integral

$$\psi(s, \boldsymbol{k}, \boldsymbol{r}, t) = \int d^3 p\, a(\boldsymbol{p}, \boldsymbol{k}, s) \exp[i\boldsymbol{p}\cdot\boldsymbol{r} - i\omega(\boldsymbol{p}, \boldsymbol{k})t]\,, \tag{10.34}$$

where the Fourier coefficients $a(\boldsymbol{p}, \boldsymbol{k}, s)$ and the frequency $\omega(\boldsymbol{p}, \boldsymbol{k}, s)$ are functions of terms which are invariant with respect to rotation in the coordinate reference frame. Scattering of a wave packet is scattering of its plain waves. Every plane wave $\exp(i\boldsymbol{p}\cdot\boldsymbol{r})$ after scattering on a center fixed at a point $\boldsymbol{r}_1$, which can be identified with the impact parameter $\boldsymbol{\rho}$, transforms to the new superposition of plane waves:

$$\exp(i\boldsymbol{p}\cdot\boldsymbol{r}) - e^{i\boldsymbol{p}\cdot\boldsymbol{\rho}}\frac{ipb}{2\pi}\int\limits_{2\pi} d\Omega\, \exp(i\boldsymbol{p}_\Omega\cdot(\boldsymbol{r}-\boldsymbol{\rho}))\,. \tag{10.35}$$

Substitution of Eq. (10.35) into Eq. (10.34) gives the scattered field in the form

$$\begin{aligned}\psi_s(s, \boldsymbol{k}, \boldsymbol{r}, t) = &\int d^3 p\, a(\boldsymbol{p}, \boldsymbol{k}, s)\exp(i\boldsymbol{p}\cdot\boldsymbol{\rho})\\ &\times\int d\Omega\; f(p, \Omega)\exp[i\boldsymbol{p}_\Omega\cdot(\boldsymbol{r}-\boldsymbol{\rho}) - i\omega(\boldsymbol{p}, \boldsymbol{k}, s)t]\,,\end{aligned} \tag{10.36}$$

where $f(p, \Omega) = -ipb/2\pi$. If the spectrum of $\boldsymbol{p}$ of the packet is very narrow and is concentrated near $\boldsymbol{k}$, we can replace p in $f(p, \Omega)$ by k, and represent the scattered wave field in the form

$$\begin{aligned}\psi_s(s, \boldsymbol{k}, \boldsymbol{r}, t) = &\int d\Omega\; f(k, \Omega)\int d^3 p\, a(\boldsymbol{p}, \boldsymbol{k}, s)\exp(i\boldsymbol{p}\cdot\boldsymbol{\rho})\\ &\times\exp[i\boldsymbol{p}_\Omega\cdot(\boldsymbol{r}-\boldsymbol{\rho}) - i\omega(\boldsymbol{p}, \boldsymbol{k}, s)t]\,.\end{aligned} \tag{10.37}$$

We see that every wave vector $\boldsymbol{p}$ after scattering rotates by the angle Ω.

Since $a(\boldsymbol{p}, \boldsymbol{k}, s)$, $\boldsymbol{p}\cdot\boldsymbol{\rho}$, and $\omega(\boldsymbol{p}, \boldsymbol{k}, s)$ are all invariant with respect to rotation, we can write them in the form $a(\boldsymbol{p}_\Omega, \boldsymbol{k}_\Omega, s)$, $\boldsymbol{p}_\Omega\cdot\boldsymbol{\rho}_\Omega$, and $\omega(\boldsymbol{p}_\Omega, \boldsymbol{k}_\Omega, s)$. After that we can change the integration variable $\boldsymbol{p}\to\boldsymbol{p}_\Omega$, and omit the index Ω of $\boldsymbol{p}_\Omega$. As a result, Eq. (10.37) is transformed to

$$\begin{aligned}\psi_s(s, \boldsymbol{k}, \boldsymbol{r}, t) = &\int\limits_{2\pi} d\Omega\; f(k, \Omega)\int d^3 p\, a(\boldsymbol{p}, \boldsymbol{k}_\Omega, s)\\ &\times\exp[i\boldsymbol{p}\cdot(\boldsymbol{r}-\boldsymbol{\rho}+\boldsymbol{\rho}_\Omega) - i\omega(\boldsymbol{p}, \boldsymbol{k}_\Omega, s)t]\,,\end{aligned} \tag{10.38}$$

which is equivalent to

$$\psi_s(s, \boldsymbol{k}, \boldsymbol{r}, t) = \int d\Omega\ f(k, \Omega)\psi_0(\boldsymbol{r} - \boldsymbol{\rho} + \boldsymbol{\rho}_\Omega, \boldsymbol{k}_\Omega, t, s)\,, \tag{10.39}$$

where ψ_0 denotes the wave packet of the same form as the incident one.

So, we see that the packet is scattered as a whole with probability amplitude $f(k, \Omega) = -ibk/2\pi$, which, surprisingly, does not depend on the impact parameter $\boldsymbol{\rho}$ like in the case of plane waves. The independence of the probability amplitude of the impact parameter is the result of the linearity of the scattering theory. The plain waves comprising the wave packet are present in the whole space, and they are scattered independently of each other. To make them feel like the sum of all the waves at the scattering point, we have to introduce nonlinearity into the interaction. We do not have a nonlinear theory, but we want to get the scattering cross section; therefore, we multiply the probability by the wave packet area A and this step means that we implicitly, together with authors of [457, 458], introduce nonlinearity at such a step.

10.5.3
Transition from Probabilities to Cross Sections

Now we show how the probability is transformed to a cross section in [457, 458], but start with steps related to transformations of probability (10.14) that are common to both books.

According to Eqs. (10.14), (10.31), and (10.33), the scattering probability is

$$\begin{aligned} dw(\boldsymbol{k} \to \boldsymbol{p}') = \\ &= d^3p'\,(2\pi)^2 T(\boldsymbol{p}', \boldsymbol{p}_i) T^*(\boldsymbol{p}', \boldsymbol{p}'_i) a_i(\boldsymbol{k} - \boldsymbol{p}_i) a_i^*(\boldsymbol{k} - \boldsymbol{p}'_i) \\ &\times d^3p_i\, d^3p'_i\, \delta\left(p'^2/2 - p_i'^2/2\right) \delta\left(p'^2/2 - p_i^2/2\right)\,. \end{aligned} \tag{10.40}$$

The spectrum of $\boldsymbol{p}_i$ in the wave packet is supposed to be sufficiently narrow around $\boldsymbol{k}$, so $\boldsymbol{p}_i$ and $\boldsymbol{p}'_i$ in the matrix elements $T(\boldsymbol{p}', \boldsymbol{p}'_i)$ and $T(\boldsymbol{p}', \boldsymbol{p}_i)$ can be replaced by $\boldsymbol{k}$. Therefore,

$$T(\boldsymbol{p}'; \boldsymbol{p}_i) T^*(\boldsymbol{p}', \boldsymbol{p}'_i) \approx |T(\boldsymbol{p}', \boldsymbol{k})|^2\,. \tag{10.41}$$

The product $\delta(p'^2/2 - p_i'^2/2)\,\delta(p'^2/2 - p_i^2/2)$ in Eq. (10.40) is identical to $\delta(p_i^2/2 - p_i'^2/2)\,\delta(p'^2/2 - p_i^2/2)$. Momentum p_i in $\delta(p'^2/2 - p_i^2/2)$ is approximated by k, and $d^3p'\,\delta(p'^2/2 - k^2/2)$ is replaced with $k\,d\Omega'$, where Ω' is the scattering angle.

The factor $\delta(p_i^2/2 - p_i'^2/2)$ is represented as

$$\delta\left(p_i^2/2 - p_i'^2/2\right) = \frac{1}{2\pi} \int_{-\infty}^{\infty} dt \exp\left(it\left(p_i^2 - p_i'^2\right)/2\right)\,. \tag{10.42}$$

The difference in the exponent is approximated by $p_i^2 - p_i'^2 = (\boldsymbol{p}_i + \boldsymbol{p}'_i) \cdot (\boldsymbol{p}_i - \boldsymbol{p}'_i) \approx 2\boldsymbol{k} \cdot (\boldsymbol{p}_i - \boldsymbol{p}'_i)$, and as a result the product $a_i(\boldsymbol{k} - \boldsymbol{p}_i) a_i^*(\boldsymbol{k} - \boldsymbol{p}'_i) d^3p_i\, d^3p'_i\, \delta(p_i^2/2 -$

$p_i'^2/2)$ is reduced to

$$\int d^3p_i\, a_i(\boldsymbol{k}-\boldsymbol{p}_i)\exp(i\boldsymbol{k}\cdot\boldsymbol{p}_i t)\int d^3p_i' a_i^*(\boldsymbol{k}-\boldsymbol{p}_i')\exp(-i\boldsymbol{k}\cdot\boldsymbol{p}_i' t) = |\phi(\boldsymbol{r})_{r=kt}|^2, \tag{10.43}$$

where the Fourier expansion (10.11) of the wave packet is used. If we direct the z-axis along $\boldsymbol{k}$ and change the integration variable dt to $dz = k dt$, then Eq. (10.40) is reduced to

$$dw = d\Omega' 2\pi|T(\boldsymbol{p}', \boldsymbol{k})|^2 \int_{-\infty}^{\infty} dz\, |\phi(0, 0, z)|^2 . \tag{10.44}$$

10.5.3.1 Transition to the Cross Section According to [457]

Since the wave packet is normalized to unity, $\int |\phi(0, 0, z)|^2\, dz$ has dimensionality $1/\text{cm}^2$. It can be interpreted as a density of incident particles per unit area. So the ratio of probability (10.44) to the density of incident particles is the cross section:

$$d\sigma = \frac{dw}{\int |\phi(0, 0, z)|^2\, dz} = d\Omega' 2\pi|T(\boldsymbol{p}', \boldsymbol{k})|^2 . \tag{10.45}$$

10.5.3.2 Transition to the Cross Section According to [458]

In [457] the wave packet was $\int d^3p\, a_i(\boldsymbol{p}-\boldsymbol{k})\exp(i\boldsymbol{p}\cdot\boldsymbol{r} - ip^2t/2)$, and the scatterer was at the point $\boldsymbol{r} = 0$. Therefore, the center of the wave packet at $t = 0$ coincides with the scatterer. In [458] the wave packet was chosen to be $\int d^3p\, a_i(\boldsymbol{p}-\boldsymbol{k})\exp(i\boldsymbol{p}\cdot(\boldsymbol{r}-\boldsymbol{\rho}) - ip^2t/2)$, and the scatterer was again at the point $\boldsymbol{r} = 0$; therefore, at $t = 0$ the packet is at the impact parameter ρ from the scatterer and Taylor instead of Eq. (10.44) obtained the probability

$$dw(\boldsymbol{\rho}) = d\Omega' 2\pi|T(\boldsymbol{p}', \boldsymbol{k})|^2 \int_{-\infty}^{\infty} dz\, |\phi(\boldsymbol{\rho}, z)|^2 . \tag{10.46}$$

The cross section is obtained by integration over the impact parameter:

$$d\sigma = \int d^2\rho\, dw(\boldsymbol{\rho}) = d\Omega' 2\pi|T(\boldsymbol{p}', \boldsymbol{k})|^2 \times \int_{-\infty}^{\infty} d^3r\, |\phi(\boldsymbol{r})|^2 = d\Omega' 2\pi|T(\boldsymbol{p}', \boldsymbol{k})|^2 , \tag{10.47}$$

where normalization of the incident wave packet was taken into account.

10.5.3.3 Analysis of Procedures in [457, 458]

The approximation (10.41) is equivalent to the assumption that scattering of the wave packet is identical to the scattering of a plane wave with momentum $\boldsymbol{k}$ into a plane wave with momentum $\boldsymbol{p}'$. The division of the probability by $|\phi(0, 0, z)|^2$

in Eq. (10.45) or integration over the impact parameter in Eq. (10.47) is absolutely equivalent to multiplication of the probability by the packet's cross area, that is, $d\sigma = A dw$, as was deduced from the experiment outlined in Figure 10.2. The considerations about the position of the wave packet with respect to the scatterer are absolutely not related to scattering itself. They only show that there is believed to be no scattering if the wave packet does not overlap with the scatterer. This description of scattering is equivalent to the suggestion that the point scatterer is illuminated not by every plane wave but by the full wave packet. In other words, in Eq. (10.35), describing scattering of the plane wave $\exp(i\boldsymbol{p}\boldsymbol{r})$, the scatterer at the point $\boldsymbol{\rho}$ is illuminated not by the wave $\exp(i\boldsymbol{p}\boldsymbol{\rho})$ but by $\phi(\boldsymbol{\rho})\exp(i\boldsymbol{p}\boldsymbol{\rho})$, which means that the component $\exp(i\boldsymbol{p}\boldsymbol{\rho})$ enters twice, that is, the interaction is nonlinear.

10.6 The Hot Problems of Quantum Mechanics

> A manuscript must convey new physics. The existing work on the subject must be briefly reviewed and the author must indicate in what way existing theory is insufficient to solve certain specific problems, then it must be shown how the proposed new theory resolves the difficulty. Your paper does not satisfy these requirements, hence we cannot accept it for publication.
>
> A Physical Review editor

To discuss the hot problems of quantum mechanics, it is best to start with the famous EPR paradox [422], where the weak points are especially clearly seen, though they were never discussed. The paper [422] triggered a flood of theoretical and experimental works proving the existence of so-called spooky action at a distance in quantum mechanics, which by EPR themselves was considered as the most unacceptable solution of their paradox.

10.6.1 The EPR Paradox, Its Logic, Error, and Consequences

The paper shows that two particles have a joint ("entangled") wave function, and one of the two particles has exact position and momentum simultaneously, which contradicts uncertainty relations. This contradiction shows that quantum mechanics is incomplete theory. There are two ways to resolve the contradiction: (1) the theory must be completed by the introduction of new, not yet known (hidden) parameters or (2) one has to admit that quantum mechanics is a nonlocal theory, where manipulation of one particle instantly affects the state of another one notwithstanding how far apart the particles are.

In this section we review the entanglement and the uncertainty relations. We then attack the problem of the paradox in quantum mechanics from a different prospective – instead of trying to resolve the paradox, we a take closer look at it and find that the source of the paradox is a matter of simple definition, which has no solid underlying reason. Redefining such a definition has a profound and far-reaching consequence. It eliminates the paradox itself and takes on the foundations of quantum mechanics, which are unquestionably familiar in our minds. We will see that particles, in fact, can have exactly defined positions and momentum simultaneously and that the uncertainty relations have a completely different physical meaning. For entangled states, we will show that a joint wave function is simply a list of possible product states; therefore, the nonlocality or action at a distance does not exist. Although these statements may sound absurd to both experienced and inexperienced physicists, I invite the reader to find a flaw in my reasoning (no one has found one yet). We will also review some of the key experiments that were specifically designed to prove the existence of the paradox, and show that these experiments in fact do not prove anything.

10.6.1.1 **Entangled States**

Consider two particles that interacted in the past and then flew apart far away. So they do not interact any more. However, because of the past interaction they have a joint wave function $\Psi(x_1, x_2)$ which is not a product, $\psi_1(x_1)\psi_2(x_2)$, of two independent functions.

The joint function $\Psi(x_1, x_2)$ can be expanded over eigenfunctions, $u_n(x_1)$, of, say, the first particle

$$\Psi(x_1, x_2) = \sum_n u_n(x_1) A_n(x_2) , \tag{10.48}$$

where $A_n(x_2)$ are expansion coefficients. From this expansion it follows that if one measures the first particle and finds it in a state $u_k(x)$, then the state of the second particle is discovered to be $A_k(x_2)$. However, the second particle does not interact with the first one; therefore, its state $A_k(x_2)$ existed before measurement of the first particle. If this were not so, then the measurement of the first particle would instantly affect the state of the second one notwithstanding how far apart the particles are, which is absolutely unacceptable because no action can propagate faster than the speed of light.

The EPR logic, however, can be continued. If the state of the second particle before measurement of the first one is $A_k(x_2)$, then the state of both particles before any measurement is $A_k(x_2)u_k(x_1)$, that is, all the wave function is not Eq. (10.48), but the simple product $\psi_1(x_1)\psi_2(x_2)$ of two independent functions, and the measurement only discovers what these functions $\psi_1(x_1)$ and $\psi_2(x_2)$ are. The sum (10.48) is only a list of potentially realizable product states.

Such a clear-cut logical conclusion, however, is not accepted by many physicists. The state (10.48) of two particles is considered as a real quantum mechanical object, and the measurement of the first particle is believed to change the state of the second particle immediately. The existence of such a "spooky" action at a distance is even "supported" by many experiments. Later we shall consider some of them, and propose a different experiment, which can prove that there is no action at a distance, but now we shall discuss the second point of the EPR paper.

10.6.1.2 Error of the EPR Paper and Definition of Physical Values

> Ha! We needed to wait 73 years for some one ... would open our eyes to an Einstein's error!
>
> A referee

The joint function $\Psi(x_1, x_2)$ can be expanded over another set of eigenfunctions, $v_n(x_1)$, of the first particle:

$$\Psi(x_1, x_2) = \sum_n v_n(x_1) B_n(x_2) , \tag{10.49}$$

where $B_n(x_2)$ are expansion coefficients. If the first particle after measurement is found to be in a state $v_k(x)$, then the state of the second particle is $B_k(x_2)$, and it must exist before any measurement. Thus we conclude that the second particle should have two states: $A_k(x_2)$ from Eq. (10.48), and $B_k(x_2)$ from Eq. (10.49) simultaneously. It may happen that $A_k(x_2)$ corresponds to a state with a precisely defined position, and $B_k(x_2)$ corresponds to a state with a precisely defined momentum. Therefore the second particle has simultaneously precisely defined position and momentum, which contradicts the uncertainty relations. Therefore, there is a paradox, which shows that quantum mechanics is not complete.

However, what does it mean for a particle to have a precisely defined, say, momentum? According to EPR it means that the particle should be in a state which is described by an eigenfunction of the momentum operator. Such a function can be only a plane wave $\exp(ikx_2)$.

Considering a plane wave state $\psi(x) = \exp(ikx)$ of a particle, the authors [422] wrote that the particle has a momentum, k, but does not have a position. About position one can

> only say that the relative probability that a measurement of the coordinate will give a result lying between a and b is
>
> $$P(a, b) = \int_a^b |\psi(x)|^2 dx = b - a . \tag{[189]6}$$

It is immediately seen that $P(a, b)$ in ([422]6) is not a probability, because it is not dimensionless and not normalizable. So, the relation ([422]6) *is an error*. To define a relative probability, one needs a normalizable wave function $\psi(x)$. The plane wave is not normalizable and must be rejected.

The standard textbooks offer normalized plane waves such as $L^{-1/2} \exp(ik_0x)$, where L is some arbitrary dimension of the space. However, introduction of L means either imposition of periodic boundary conditions or restriction of the space by two impenetrable walls. The first recipe does not help, because the wave function remains not normalizable, and the second recipe changes the plane wave into a combination of $\sin(kx)$ and $\cos(kx)$ functions, which are not eigenfunctions of the momentum operator.

For EPR only a plane wave is acceptable because according to their definition the particle has a momentum only [422] if its state is described by an eigenfunction of the momentum operator $-i d/dx$:

> If ψ is an eigenfunction of the corresponding operator A, that is, if
>
> $$\psi' \equiv A\psi = a\psi \,, \tag{[189]1}$$
>
> where a is the number, then the physical quantity A has with certainty the value a whenever the particle is in the state given by ψ.

This definition in general is not right. With it no physical system has a momentum because no normalizable function is an eigenfunction of the momentum operator. According to [470], the correct definition for the value a of a physical quantity described by an operator $\hat{A}$ is the expectation value

$$a = \int \psi^+ \hat{A}\psi \, dx \,. \tag{10.50}$$

With this definition all the Hermitian operators, even noncommuting ones, have simultaneously defined values for any state ψ, and uncertainty relations do not matter.

So, after correction of the error ([422]6), which unambiguously leads to the definition (10.50), we arrive at the conclusion that the paradox does not exist.

For Eq. (10.50) to be acceptable as a definition of the value of a physical quantity, $\hat{A}$ we have to discuss the meaning of uncertainty relations and of dispersion

$$\Delta^2 a = \int \psi^+ (\hat{A} - a)^2 \psi \, dx \,. \tag{10.51}$$

10.6.1.3 The Meaning of the Uncertainty Relations and of Dispersion

The uncertainty relations have only a trivial meaning [462]. The inequality

$$\langle (x - x_0)^2 \rangle \langle (k - k_0)^2 \rangle \geq 1/4 \tag{10.52}$$

shows how the range of a normalizable function $f(x)$ (the size of the interval, where $f(x)$ is essentially concentrated) is related to the range of its Fourier image:

$$f_F(k) = \int f(x) \exp(-ikx) dx/\sqrt{2\pi} \,. \tag{10.53}$$

Equation (10.52) is valid in all branches of physics where functions are used, and it is nothing special in quantum mechanics. To show this, let us recall the proof of Eq. (10.52).

Take an arbitrary function $f(x)$ of finite range, and its Fourier image, $f_F(k)$, Eq. (10.53). The functions have the norm

$$\int\limits_{-\infty}^{+\infty} |f(x)|^2 dx = \int\limits_{-\infty}^{+\infty} |f_F(k)|^2 dk \equiv N < \infty \,. \tag{10.54}$$

With the function $f(x)$ we can calculate the positively defined integral

$$I = \frac{1}{N}\int |(\alpha(x-x_0) + d/dx - ik_0)f(x)|^2 = \alpha^2 A + \alpha B + C > 0\,, \quad (10.55)$$

where α is an arbitrary real parameter,

$$x_0 = \frac{1}{N}\int x|f(x)|^2 dx\,, \quad k_0 = \frac{1}{N}\int k|f_F(k)|^2 dk\,, \quad (10.56)$$

$$A = \frac{1}{N}\int (x-x_0)^2|f(x)|^2 dx = \frac{1}{N}\int (x^2 - x_0^2)|f(x)|^2 dx \equiv \langle(\Delta x)^2\rangle, \quad (10.57)$$

$$B = \frac{1}{N}\int x\frac{d}{dx}|f(x)|^2 dx = \int dx\frac{d}{dx}(x|f(x)|^2) - \int dx|f(x)|^2$$
$$= -\frac{1}{N}\int dx|f(x)|^2 = -1\,, \quad (10.58)$$

$$C = \frac{1}{N}\int (k-k_0)^2|f_F(k)|^2 dk = \frac{1}{N}\int (k^2 - k_0^2)|f_F(k)|^2 dk \equiv \langle(\Delta k)^2\rangle\,. \quad (10.59)$$

Substitution of Eqs. (10.57)–(10.59) into Eq. (10.55) gives

$$\alpha^2\langle(\Delta x)^2\rangle - \alpha + \langle(\Delta k)^2\rangle \geq 0\,. \quad (10.60)$$

This inequality is valid for arbitrary α only when Eq. (10.52) is satisfied.

We see that uncertainty relations do not depend on the Schrödinger equation, and therefore they contain nothing, specific to quantum mechanics.

A particle in quantum mechanics is described by a wave packet. The wave packet has a dimension, and the dispersion Δx^2 is not a statistical uncertainty, but a parameter characterizing the size of the packet. It is easily checked, for instance, in the case of Gaussian packets.

10.6.1.4 Locality, Nonlocality, Measurements, and Hidden Parameters

The EPR paradox does not exist because particles can have position and momentum simultaneously. However, there is a question about entangled states in the EPR paper which, though logically resolved, also needs experimental confirmation. The problem in a simplified version is as follows. If a wave function $\Psi(x_1, x_2)$, of two far-apart particles is

$$\Psi(x_1, x_2) = u_1(x_1)v_1(x_2) + u_2(x_1)v_2(x_2)\,, \quad (10.61)$$

does it mean that in an every experimental run it is reduced to one of the products $u_i v_i (i = 1, 2)$ and the experiment only recovers which product is it, or is the wave function in every run (10.61), and measuring the first particle in the state, say, u_1, immediately adjusts the state of the second particle to v_1? The first version corresponds to local quantum mechanics, whereas the second one corresponds to a nonlocal theory where action propagates instantly to arbitrarily long distances or to the Copenhagen interpretation of quantum mechanics with instant reduction of wave packets.

To be specific we shall consider only two of the most popular systems discussed in the literature: two spin-1/2 particles in an *s*-state, and two photons with zero total electric field.

10.6.1.5 A Pair of Spin-1/2 Particles in an *s*-State

Bohm and Aharonov [424] considered the decay of a scalar molecule into two atoms with spin 1/2. Because of Fermi statistics, the spin wave function of the two atoms in the molecule is

$$|\psi(1,2)\rangle = \frac{1}{\sqrt{2}}[|\psi_+(1)\psi_-(2)\rangle - |\psi_-(1)\psi_+(2)\rangle] , \tag{10.62}$$

where $\psi_\pm(1,2)$ refers to states of atom A or B with spin projection $\pm 1/2$ on an arbitrarily selected, say, z-axis.

According to [424] after atoms have been separated far apart

> enough so that they cease to interact, any desired component of the spin of the first particle (A) is measured. Then, because the total spin is still zero, it can be concluded that the same component of the spin of the other particle (B) is opposite to that of A.

They continue

> If this were a classical system there would be no difficulties, in interpreting the above result, because all components of the spin of each particle are well defined at each instant of time.

However

> In quantum theory, a difficulty arises, in the interpretation of the above experiment, because only one component of the spin of each particle can have a definite value at a given time. Thus, if the x component is definite, then y and z components are indeterminate and we may regard them more or less as in a kind of random fluctuation.

This means that there must be an action at a distance because if there were not how could the second particle know which one of the spin components should cease to fluctuate when we measure a spin component of the first particle. However, let us look at whether the spin components really do fluctuate and at what we really do measure.

In fact every particle has an absolutely defined polarization, and we can attach to it a classical unit vector – a spin arrow. We can consider it as a hidden parameter. When we know the direction of the spin arrow, as in the case of polarized beams, we are able to manipulate the particle at our will in the framework of quantum mechanics without violation of causality.

A particle polarized along a unit vector $\boldsymbol{\Omega} = (\sin\theta\cos\varphi, \sin\theta\sin\varphi, \cos\theta)$, where θ and φ are polar angles, is described by the spinor

$$|\psi_{\Omega}\rangle = e^{i\varphi}\frac{1+\boldsymbol{\sigma}\cdot\boldsymbol{\Omega}}{\sqrt{2(1+\Omega_z)}}\begin{pmatrix}1\\0\end{pmatrix}, \tag{10.63}$$

where $\boldsymbol{\sigma} = (\sigma_x, \sigma_y, \sigma_z)$ are Pauli matrices, and φ is an arbitrary numerical phase. The values of the spin projection to any axis are well defined simultaneously, and there are no fluctuations.

Let us look at a molecule in a state (10.62). The question is: will the spinor function of two atoms after decay of the molecule remain the same as given in Eq. (10.62), or will the decay change it to a state described by a product of two independent spinors? An answer to this question can be found only in an experiment, but first we should answer another question: what should we measure, and what does the measurement mean?

According to [424] we can measure a spin component. That is wrong. We cannot measure a spin component. We can only count a particle after its transmission through a filter (analyzer). If the analyzer is aligned along a unit vector $\boldsymbol{a}$, the probability of transmission for the particle in the state (10.63) is

$$w(\boldsymbol{\Omega}) = \langle\psi_{\Omega}|\frac{1+\boldsymbol{\sigma}\cdot\boldsymbol{a}}{2}|\psi_{\Omega}\rangle = \frac{1+\boldsymbol{a}\cdot\boldsymbol{\Omega}}{2}, \tag{10.64}$$

but that does not mean that we measure projection $\boldsymbol{a}\cdot\boldsymbol{\Omega}$, because in a single event we cannot measure a probability. In an experiment with a single particle we either count it or do not count it. Only if $\boldsymbol{a} = \boldsymbol{\Omega}$ can we be sure that we will count the particle. The seekers of hidden parameters ask if it is possible in principle to find an additional parameter which predicts exactly if a particle will be counted after such an analyzer or not, that is, whether it is possible to use a classical theory instead of quantum mechanics. Later we shall show that in principle it is possible, but now we are quite satisfied with quantum mechanics itself, and only want to know whether entangled states exist or should be abandoned as being nonphysical, like we reject exponentially growing solutions of the Schrödinger equation.

According to the nonlocality, the spinor function of two particles after decay remains the same Eq. (10.62) as before it, and the probability of detecting both atoms with analyzers directed along $\boldsymbol{a}$ and $\boldsymbol{b}$ is

$$w_q = \langle\psi(1,2)|\left(\frac{1+\boldsymbol{a}\cdot\boldsymbol{\sigma}}{2}\right)_1\left(\frac{1+\boldsymbol{b}\cdot\boldsymbol{\sigma}}{2}\right)_2|\psi(1,2)\rangle = \frac{1-\boldsymbol{a}\cdot\boldsymbol{b}}{4}. \tag{10.65}$$

We see that because of interference (contribution of cross terms), the probability of detecting both particles at parallel analyzers ($\boldsymbol{a} = \boldsymbol{b}$) is zero.

According to local realism, the atoms flying apart after decay of a singlet molecule are in a product state $|\psi_{-\Omega}(2)\rangle|\psi_{\Omega}(1)\rangle$ with opposite polarizations along some unknown direction $\boldsymbol{\Omega}$. If an analyzer for particle 1 is aligned along $\boldsymbol{a}$, particle 1 after filtering will be in the state $\psi_{\boldsymbol{a}}$, but particle 2 will remain in the state $\psi_{-\Omega}(2)$, which is not $\psi_{-\boldsymbol{a}}$. This means that if an analyzer for particle 2 is aligned along a unit vector $\boldsymbol{b}$, particle 2 will be registered with probability $w_2 = (1 - \boldsymbol{b} \cdot \boldsymbol{\Omega})/2$. The probability for both particles to be registered is $w(\boldsymbol{a}, \boldsymbol{b}, \boldsymbol{\Omega}) = (1-\boldsymbol{b}\cdot\boldsymbol{\Omega})(1+\boldsymbol{a}\cdot\boldsymbol{\Omega})/4$. Therefore, if $\boldsymbol{b} = \boldsymbol{a}$, the probability for detection of both particles with parallel analyzers is $w(\boldsymbol{a}, \boldsymbol{a}, \boldsymbol{\Omega}) = [1-(\boldsymbol{a}\cdot\boldsymbol{\Omega})^2]/4 > 0$. Since the direction of the polarization vector $\boldsymbol{\Omega}$ is random, averaging of $w(\boldsymbol{a}, \boldsymbol{b}, \boldsymbol{\Omega})$ over a uniform distribution (look, here $\boldsymbol{\Omega}$ is a hidden parameter!) gives

$$\langle w(\boldsymbol{a}, \boldsymbol{b})\rangle = \frac{1}{4}\left(1 - \frac{\boldsymbol{a}\cdot\boldsymbol{b}}{3}\right) = \frac{1}{6} + \frac{1 - \boldsymbol{a}\cdot\boldsymbol{b}}{12} . \tag{10.66}$$

The difference between the two predictions apart from the normalization factor is in fact only a constant, which in measurements can easily be mistaken for a background and subtracted, after which the results of an experiment can be reported as a proof of nonlocality. We shall see how this can happen in experiments with photons.

10.6.1.6 Nonlocality with Photon Pairs

Most of the experiments proving nonlocality were done with photons. The first such experiment was presented in [472]. Correlation of the polarizations of two photons from the cascade $6^1 S_0 \to 4^1 P_1 \to 4^1 S_0$ in a calcium atom was measured. The authors wrote ($\boldsymbol{\epsilon}_{1,2}$ are polarization vectors of two photons):

> Since the initial and final atomic states have zero total angular momentum and the same parity, the correlation is expected to be of the form $(\boldsymbol{\epsilon}_1 \cdot \boldsymbol{\epsilon}_2)^2$.

This means that the wave function of the photons inside the atom before the decay was supposed to be

$$|\psi\rangle = \frac{1}{\sqrt{2}}(C^x C^x + C^y C^y)|\psi_0\rangle , \tag{10.67}$$

where $|\psi_0\rangle$ is the vacuum state and $C^{x,y}$ are creation operators for photons with polarizations along x or y directions. If we take into account that the total electric field of the two photons is zero, the wave function (10.67) of the two photons has to be represented in a form similar to Eq. (10.62):

$$|\psi\rangle = \frac{1}{\sqrt{2}}[|x, -x\rangle + |y, -y\rangle] , \tag{10.68}$$

where the first letter is related to the polarization direction of the photon going to the left and the second letter is related to the polarization direction of the photon going to the right.

If a photon is polarized along a unit vector $\boldsymbol{\Omega}$ and the analyzer is directed along a unit vector $\boldsymbol{a}$, then the probability of registration of the photon is $|\boldsymbol{a} \cdot \boldsymbol{\Omega})|^2$. According to nonlocality, the wave function after decay remains (10.68), and the probability for detection of both photons after analyzers aligned along unit vectors $\boldsymbol{a}$ and $\boldsymbol{b}$ is

$$w(\boldsymbol{a}, \boldsymbol{b}) = |\langle \boldsymbol{a}, \boldsymbol{b}||\psi\rangle|^2 = \frac{(\boldsymbol{a}\boldsymbol{b})^2}{2} = \frac{1 + \cos(2\xi)}{4} , \tag{10.69}$$

where ξ is the angle between vectors $\boldsymbol{a}$ and $\boldsymbol{b}$. We see that the probability of detecting both particles with orthogonal analyzers is zero.

According to local realism, the two photons after decay fly away in the product state

$$|\psi(1,2)\rangle = |\psi_\phi(1), \psi_{-\phi}(2)\rangle \tag{10.70}$$

with opposite polarizations along some direction $\boldsymbol{\phi}$ in the plane perpendicular to the momenta of the photons. The probability of detect both photons after analyzers aligned along unit vectors $\boldsymbol{a}$ and $\boldsymbol{b}$ is

$$w(\boldsymbol{a}, \boldsymbol{b}, \boldsymbol{\phi}) = |\langle \boldsymbol{a}, \boldsymbol{b}||\psi_\phi(1), \psi_{-\phi}(2)\rangle|^2 = (\boldsymbol{a}\boldsymbol{\phi})^2(\boldsymbol{b}\boldsymbol{\phi})^2 . \tag{10.71}$$

Since the direction of the polarization vector $\boldsymbol{\phi}$ is random, averaging $w(\boldsymbol{a}, \boldsymbol{b}, \boldsymbol{\phi})$ over a uniform distribution over angle ϕ gives

$$\begin{aligned}\langle w(\boldsymbol{a}, \boldsymbol{b})\rangle &= \frac{1}{2\pi}\int_0^{2\pi} d\phi (\boldsymbol{a}\boldsymbol{\phi})^2(\boldsymbol{b}\boldsymbol{\phi})^2 = \frac{1}{2\pi}\int_0^{2\pi} d\phi \cos^2\phi \cos^2(\phi+\xi) \\ &= \frac{1}{4}\left(\frac{1}{2} + \cos^2\xi\right) = \frac{1}{8} + \frac{1+\cos(2\xi)}{8} \equiv \frac{1}{8} + \frac{(\boldsymbol{a}\cdot\boldsymbol{b})^2}{4} .\end{aligned} \tag{10.72}$$

Again we see that apart from a normalization factor, the two angular dependences (10.72) and (10.69) differ only by a constant, which can be mistaken as a background. According to Eq. (10.72), the ratio of the count rate at $\xi = 90°$ to that at $\xi = 0°$ is 1/3 instead of 0, as follows from Eq. (10.69).

10.6.1.7 Results of the First Experiment

Let us look at the first experiment [472], where the dependence of the coincidence counts of two photons on the coincidence window were measured at parallel and orthogonal analyzers. The experimental setup is shown in Figure 10.4, and the results are demonstrated in Figure 10.5. We clearly see that the background can be accepted at a level of 40 counts; therefore, the coincidence count with parallel filters is equal to $w(0) \approx 100$, and the coincidence count with orthogonal filters is $w(\pi/2) \approx 30$. At the precision level presented there is good agreement with $w(\pi/2)/w(0) \approx 1/3$.

However, the authors, though absolutely honest, were very reluctant to accept such interpretation. Their conclusion is

> The results of a 21-h run, shown in Figure 10.5 indicate clearly the difference between the coincidence rates for parallel and perpendicular orientations. They are consistent with a correlation of the form $(\epsilon_1 \cdot \epsilon_2)^2$.

They attributed the discrepancy at the level of 30% to 6% transmission of crossed filters, as is documented in the caption to Figure 10.5. This is the EPR hypnosis!

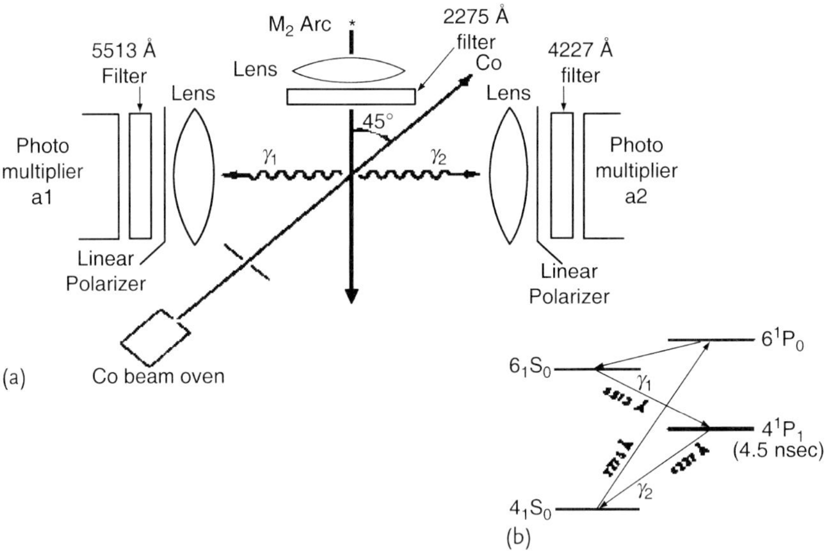

Figure 10.4 The first experiment with cascade photons [472]. (a) Schematic diagram of apparatus; (b) Level scheme for calcium.

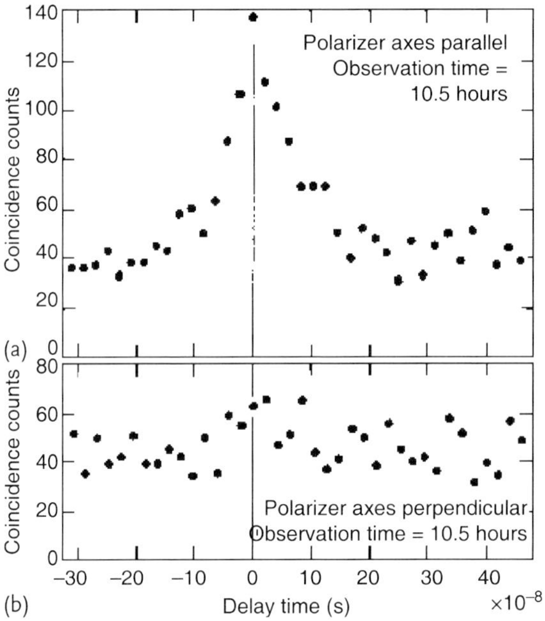

Figure 10.5 Results for coincidence counts with different delays obtained in the experiment reported in [472].

10.6.1.8 The Experiments of Aspect *et al.*

The concept of entanglement has been thought to be in the heart of quantum mechanics. The seminal experiment of Aspect *et*

> *al.* [463] has proved the "spooky" nonlocal action of quantum mechanics by observing violation of Bell inequality with entangled photon pairs.
>
> From [469]

We shall not dwell on Bell's inequalities because they are not relevant to our clearly formulated problem: do we really have nonlocality in quantum mechanics, or are entangled states to be rejected as nonphysical ones? Bell's inequalities serve as a thick mist covering all the activity aimed at proof of nonlocality in quantum mechanics, because most frequently they are discussed in the context of hidden parameters. For instance, the paper [463] is titled "Experimental test of realistic local theories via Bell's theorem". With such a title it is possible only to applaud the positive result that Bell's inequalities are violated, which proves that quantum mechanics is a good theory not needing those notorious hidden parameters. However, in fact the experiment is considered as proving nonlocality, which is exactly pointed out in [469]. We shall show, however, that it provcs nothing.

The experiment reported in [463] is very similar to the one reported in [472] but instead of $6^1 S_0$ the state $4p^{21} S_0$ was excited, and the coincidence count rates were measured at different angles of two analyzers. Let us look at the experimental results, shown in Figure 10.6.

This result can be described as Eq. (10.69) and not as Eq. (10.72); however, the difference is only in a constant 1/8 term and in normalization. We shall not discuss the normalization, because it is more or less arbitrary, but subtraction of the constant term is very important. Let us look how it was done.

> Typical coincidence rates without polarizers are 240 coincidences per second in the null delay channel and 90 accidental coincidences per second; for a 100-s counting period we thus obtain 150 true coincidences per second with a standard deviation less than 3 coincidences per second.

Suppose that in 90 coincidences they subtracted, 50 were indeed accidental ones, but the remaining 40 were not accidental. If you add these counts you will find that at angle $\varphi = \pi/2$ you have 40 counts and at $\varphi = 0$ you have 115 counts $(150/2+40)$. So their ratio is approximately 1/3. In that case there is no contradiction to the local

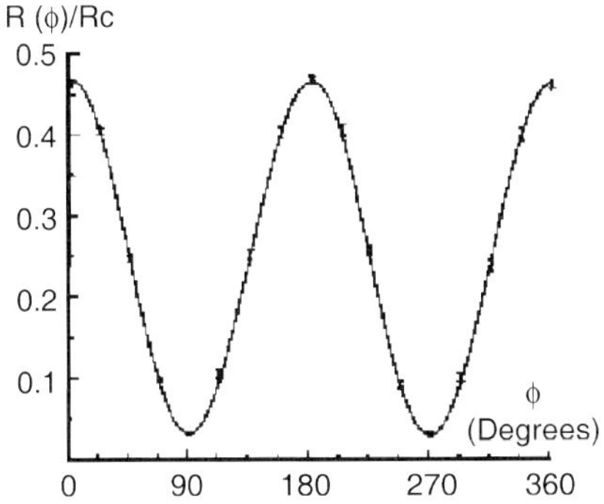

Figure 10.6 Results of the experiment reported in [463].

quantum mechanics, and there is no spooky action at a distance. Everything is in accord with our calculations and local realism.

Thus, we see that the experiments, and between them "the most seminal" one, did not reject local realism. Because of volume restrictions we do not consider here many experiments on "Bell's inequalities" with downconversion of photons. They have become very popular in the last few years, and some of them were considered in [466]. Here we can finish with a proposal for a crucial experiment. With photons it is necessary to measure the coincidence count rates of two particles with orthogonal analyzers. Entangled states give zero. Local realism predicts $1/3$ of the count rate with respect to parallel analyzers.

Finally, it is necessary to say that entangled states do exist for particles which are at the same space point or in the vicinity of it. They also exist for particles which fly apart and then meet at a point. In that respect we can say that there is a wonderful relation between quantum mechanics and geometry. This relation is not yet clear and it is worth researching.

10.6.2
Interpretation of Quantum Mechanics and Hidden Variables

There are so many interpretations of quantum mechanics, and none of them help to understand what in reality happens in an individual quantum process. One model, so-called Bohmian mechanics (see, e. g., [473]), attempts to do that. In Bohmian mechanics, the complex wave function of a particle is represented as $\psi(\boldsymbol{r}, t) = R(\boldsymbol{r}, t)\exp(i S(\boldsymbol{r}, t)$ with real R and S functions. The gradient, $\nabla S/m$, of S, where m is the particle mass, is interpreted as the particle's speed $\boldsymbol{v}$. The Schrödinger equation is transformed to the Newton one $d\boldsymbol{v}/dt = -\nabla(V + Q)$, where V is the potential of the system which enters the Schrödinger equation and Q is the so-called quantum potential: $Q = (\hbar^2/2m)\Delta R^2/R^2$. However, $\boldsymbol{v}(\boldsymbol{r}, t)$ depends on variables $\boldsymbol{r}$, so it is not the speed of a particle, but some field of speeds. So Bohmian mechanics is not an interpretation, but a method to solve the Schrödinger equation. One separates the complex equation into real and imaginary parts, and solves the system of new real equations instead of a single complex one.

We believe that there is a possibility to interpret the Schrödinger equation as an approximation to a nonlinear classical system of equations for the dynamics of a single particle if its wave function is interpreted as a field. To explain the idea we shall demonstrate the interference of a classical electron at two slits in a screen. Then we will derive the nonlinear classical system of equations which we believe can replace the Schrödinger equation.

10.6.2.1 Interference in Classical Physics

Let us consider the well-known experiment with two slits, shown in Figure 10.7. We have a source of classical electrons which gives a beam of electrons passing only through the upper slit. The screen is supposed to be a thin metallic foil. The electrons have their Coulomb field, which interacts with the screen through the bound-

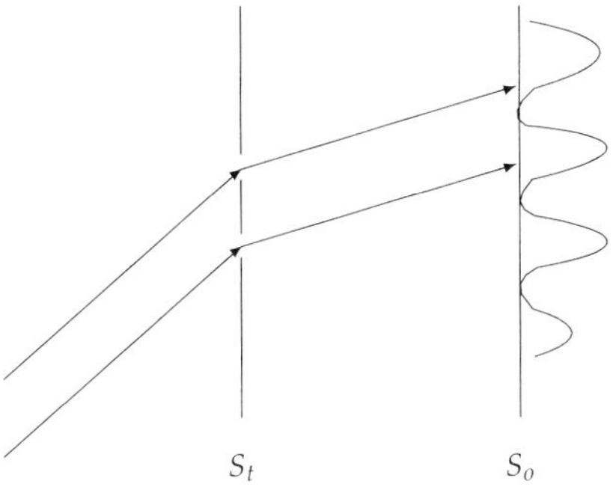

Figure 10.7 The quantum mechanical experiment with two slits. According to quantum mechanics, the particle goes through both slits in the screen S_t, and superposition of two parts of the wave function after S_t gives the interference pattern on the observation screen S_o.

ary condition on its surface. The field reflected from the screen interacts with the electron itself, so the electron on its trip changes its trajectory. This change depends on the distance between the electron and the edges of the slit. At the observation screen we can see the distribution of the electrons.

When the second slit is closed, we see one distribution, when the second slit is opened, we see another one though all the electrons go through the same upper slit. The distribution changes because the boundary condition on the screen S_t changes on opening of the second slit, and the change affects the electron's trajectory even when it goes through the same upper slit. The perturbation created by the second slit is the interference. In the case of the pure Coulomb field we cannot expect to observe a diffraction pattern on the screen S_o, because we do not have a parameter such as wavelength. However, such a parameter can appear if our calculations are performed in discrete steps, or if we take into account the retardation because of special relativity effects. It is interesting to see whether it is possible to introduce the Plank constant by requiring that the electron trajectory changes only after such a step Δl along its trajectory, where $p\Delta l = h$, where p is the electron momentum at the beginning of the interval Δl.

10.6.2.2 **Nonlinear System of Classical Equations Instead of the Schrödinger Equation**

The discussion in the section shows that to describe the electron trajectory $\boldsymbol{r}(t)$ we need to solve the Newton equation:

$$\frac{d^2\boldsymbol{r}}{dt^2} = \boldsymbol{F}(\Phi(\boldsymbol{r}(t)))\,, \tag{10.73}$$

where the force on the right-hand side depends on the electron's field $\Phi(\boldsymbol{r}(t))$ at the point of the electron position. To find this force we need to find the field, that is, to solve the field equation

$$\hat{L}\Phi(\boldsymbol{r}, t) = j(\boldsymbol{r}(t))\,, \tag{10.74}$$

where the source of the field depends on the particle trajectory, which is not yet known without solution of the first equation. To solve (10.74) and to find the field at the point $\boldsymbol{r} = \boldsymbol{r}(t)$,it is necessary to take into account the initial and boundary conditions. We see that to predict the pattern on the observation screen we need to solve the highly nonlinear system of equations.

This example shows that the wave function in quantum mechanics, especially de Broglie's singular wave packet, can be interpreted as a some kind of field, and the linear Schrödinger equation can be replaced by the system of two nonlinear classical equations. Quantum mechanics is an ingenious theory, which circumvented solution of such nonlinear system, and replaced it with the single linear Schrödinger equation. However, the price for it is the appearance of probability and the loss of causality. It is very interesting to check whether it is possible with Eqs. (10.73) and (10.74) to explain the reflection and transmission of a particle at a semitransparent mirror. We think that the reflection and transmission is a bifurcation of the particle trajectory, similar to that studied in stochastic dynamics, and the reason for bifurcation can be found in the dependence of the motion on the initial conditions.

10.6.3 Discussion of some Experiments with Wave Packets

If the neutron wave function is a wave packet, we have to find the properties of the packet. The theory tells us nothing about it. So the properties should be found from experiments. From experiments, we must find the size A of the wave packet and its dependence on neutron momentum k.

The size of the packet was estimated from experiments with ultracold neutrons, where it was also found that $A \propto 1/k^2$. This estimation was used to predict some transmission over a potential barrier with neutrons of higher energies [465] when the grazing angle of the neutrons incident on a mirror is below the critical angle. It was expected that because of the dependence on k^2, the transmission probability would increase with increase of neutron energy. However, experiments [465] could not find transmission at the level 10^{-4}, so it was concluded that the size of the packet, if it exists, does not depend on the neutron energy.

A consequence of this result [453] was that if we measure the transmission of neutrons through ^{4}He gas, its dependence on temperature T will be proportional to $T^{3/2}$. The standard theory without wave packets predicts a dependence proportional to $T^{1/2}$, which is supported by the experiment reported in [467], where neutron transmission through noble gases was measured. However, in that experiment only the dependence on the neutron energy near the thermal point was measured, and the temperature dependence was deduced. It was desirable to perform an experiment at low neutron energies and to measure directly the temperature dependence without energy variation. Such an experiment performed at the Institut Laue–Langevin reactor [460] showed that the temperature dependence is in accord with the predictions of the standard scattering theory. Wave packet theory can give such a dependence only if $A \propto 1/k^2$. But then we are in contradiction with the results of the experiments reported in [465]. It is possible, however, that the size of the packet, which determines the transmission, depends on the relative speed between the neutron and the interface. In that case we cannot expect an increase of transmission with increase of the neutron energy, because subcritical transmission remains the same as for ultracold neutrons, that is, it can be expected at the level of 10^{-5}. We also think that it is necessary to repeat the experiments reported in [465] with more careful examination of the angular distribution of transmitted neutrons at a subcritical incidence.

Besides these experiments, it is possible to search for wave packet properties in experiments of the Goos–Hänchen type [182]. We have two types of Goos–Hänchen experiments. In the first type, the shift of the wave packet along surface mirror is measured at total reflection. The closer the grazing angle of the incident neutrons is to the critical one, the larger is the shift. In the case of multilayered systems we can achieve a sufficiently large shift at one of the internal interfaces and look for interference of wave functions of the packets reflected at the first and at deeper interfaces (see [182]).

Another type of Goos–Hänchen effect is deviation of the reflected beam from the specular direction when the grazing angle between incident neutron and the mirror surface is above but close to the critical one. Again this deviation is larger when the grazing angle is closer to the critical one.

It is also possible to study wave packet properties with diffraction in ideal crystals. Analysis of the intensities of different reflections at a distance from the Bragg conditions can give information about the amplitudes of plane waves that constitute the wave packet.

All the experiments described except the one with ^{4}He are very difficult because they look for very small effects. However, they are feasible and can be performed if some of the funds given to the search for parity nonconservation and precise measurement of the neutron lifetime are directed to them.

10.7
Modern Art in Neutron Optics

> This medicine is highly effective because it acts at a quantum level. It was processed in torsion fields, and a test after processing had shown that its Bell index reached the value $BI = 2.82$.
>
> From a commercial

We would like to finish this chapter with a discussion of an experiment with a neutron interferometer [427], the scheme of which is shown in Figure 10.8. In this experiment the interferometer is considered as a device producing entangled state of two commuting degrees of freedom of a single particle:

$$|\psi\rangle = A_1|\xi_y\rangle \exp(ikx_1 + \varphi_1) \oplus A_2|\xi_{-y}\rangle \exp(ikx_2 + i\varphi_2) . \qquad (10.75)$$

One degree of freedom is spin and the other one is the path the neutron propagates along after the first beam splitter. The first term in Eq. (10.75) describes the neutron

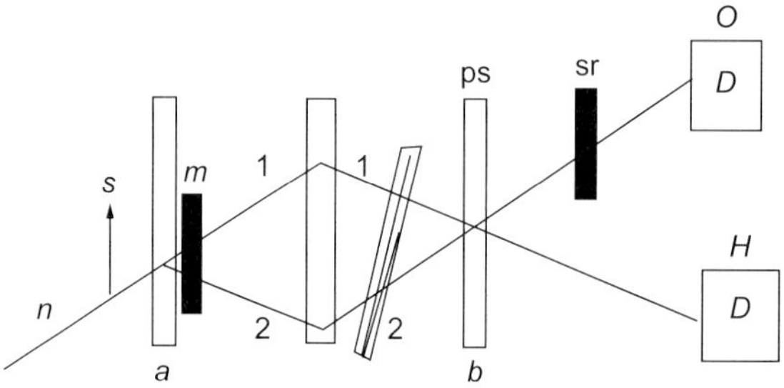

Figure 10.8 Generation of an entangled state in an interferometer.

propagating along path 1 after the beam splitter in spin state $|\xi_y\rangle$, and the second term describes the same neutron propagating after the beam splitter along path 2 with spin state $|\xi_{-y}\rangle$. To achieve different spin states along the two paths after the beam splitter $\boldsymbol{a}$ there is a magnet **m** with a strong inhomogeneous field, which according to the Stern–Gerlach effect directs the neutrons in state $|\xi_y\rangle$ along path 1 and the neutrons in state $|\xi_{-y}\rangle$ along path 2. The incident neutrons before the first splitter $\boldsymbol{a}$ are polarized along the z direction.

At beam splitter b the two terms of Eq. (10.75) recombine and give two new beams, one of which (beam H) propagates toward detector H and the other one (beam O) propagates toward detector O. The spin state of beam O can be represented as

$$|\psi_O\rangle = A\left[|\xi_y\rangle + r|\xi_{-y}\rangle \exp(i\vartheta)\right] , \tag{10.76}$$

where A and r are some constants (r is real), and phase ϑ can be varied by the phase shifter **ps** (see Figure 10.8). The state (10.76) is a spinor state $|\psi_O(\boldsymbol{b})\rangle$ polarized along some direction $\boldsymbol{b}$:

$$|\psi_O(\boldsymbol{b})\rangle = C\frac{I + \boldsymbol{b\sigma}}{\sqrt{2(1+b_z)}}|\xi_z\rangle , \tag{10.77}$$

where C is a normalization constant, determined from

$$\begin{aligned} |C|^2 &= \langle\psi_O(\boldsymbol{b})||\psi_O(\boldsymbol{b})\rangle \\ &= |A|^2\left[\langle\xi_y| + r\langle\xi_{-y}|\exp(-i\vartheta)\right]\left[|\xi_y\rangle + r|\xi_{-y}\rangle\exp(i\vartheta)\right] \\ &= |A|^2(1+r^2) . \end{aligned} \tag{10.78}$$

The vector $\boldsymbol{b}$ is

$$\begin{aligned} \boldsymbol{b} &= \frac{1}{|C|^2}\langle\psi_O(\boldsymbol{b})|\boldsymbol{\sigma}|\psi_O(\boldsymbol{b})\rangle \\ &= \frac{1}{1+r^2}\left[\langle\xi_y| + r\langle\xi_{-y}|\exp(-i\vartheta)\right]\boldsymbol{\sigma}\left[|\xi_y\rangle + r|\xi_{-y}\rangle\exp(i\vartheta)\right] \\ &= \frac{(2r\sin\vartheta, 1-r^2, 2r\cos\vartheta)}{1+r^2} , \end{aligned} \tag{10.79}$$

and $|\boldsymbol{b}| = 1$.

In the experiment reported in [427] there was also a spin rotator **sr** (see Figure 10.8), which rotated the spin around the x-axis to any desirable angle ϕ, and after the spin rotator there was also an analyzer, which transmitted neutrons polarized along the z-axis. Therefore, the count rate of detector O is

$$\begin{aligned} I_O &= \frac{|C|^2}{4(1+b_z)} |\langle \xi_z| \exp(i\phi\sigma_x)(I + \boldsymbol{\sigma b})|\xi_z\rangle|^2 \\ &= |C|^2 \frac{|(1+b_z)\cos\phi + i(b_x + ib_y)\sin\phi|^2}{4(1+b_z)} \\ &= \frac{|A|^2}{4}\left[1 + r^2 + \sqrt{1 + r^4 + 2r^2\cos(2\vartheta)}\cos(2\varphi + \alpha)\right], \\ \alpha &= \operatorname{arccot}\left(\frac{2r\cos\vartheta}{1-r^2}\right). \end{aligned} \tag{10.80}$$

We see that the intensity in detector O oscillates proportionally to $\cos(2\varphi + \alpha)$, its phase depends on phase shift ϑ, and the contrast of oscillations[45)]

$$V = \frac{\sqrt{1 + r^4 + 2r^2\cos(2\vartheta)}}{1 + r^2} \tag{10.81}$$

also depends on ϑ.

This relation is obtained for the ideal case of a monochromatic beam, ideal incident polarization, and the absence of a background. If we include only the background β, the expression for the contrast will change to

$$V = \frac{\sqrt{1 + r^4 + 2r^2\cos(2\vartheta)}}{1 + r^2 + 2\beta}. \tag{10.82}$$

It is interesting to check the dependence of I_O on different parameters and to apply the device to the investigation of physical phenomena. However, there seems to be a deficit of ideas. Therefore, the goal of the experiment reported in [427] was different. It was checked whether the entangled state (10.75) can be represented as a product state of the wave function of spin and the wave function of paths:

$$|\psi(\boldsymbol{r})\rangle = \left[\exp(i\boldsymbol{k}_1\boldsymbol{r}) + e^{i\theta}\exp(i\boldsymbol{k}_2\boldsymbol{r})\right]\left[|\xi_{+y}\rangle + |e^{i\phi}\xi_{-y}\rangle\right], \tag{10.83}$$

where θ is the phase produced by the phase shifter and ϕ is the phase produced by spin rotator. Such a product supposedly follows from the local realism.

It is so strange to check all these "local realistic classical" theories that flourished starting from John Bell! All these theories give predictions contradicting quantum mechanics; therefore, they were rejected from the very start as no-go theorems. However, they are not. John Bell invented his inequalities in an attempt to get a

45) I have a chance to correct the equations in *Concepts of Physics* [466] similar to Eqs. (10.80) and (10.81).

local realistic description of entangled states of separated particles. His failure is considered as a proof of nonlocality. So every experimental demonstration of the violation of Bell's inequalities is claimed as the experimental evidence of nonlocality, though this is wrong.

In fact we need an absolutely different approach. We need first to prove that entangled states for separated particles are nonphysical ones and should be excluded from the list of solutions. They exist only for particles which are at the same place or meet after a temporal separation. How it occurs is an exciting problem of quantum mechanics, which is worth investigating. Instead of this, a lot of researchers prefer to prove again and again that Bell's inequalities are violated. It is strange to read that instead of some physical research with an excellent experimental device such as the neutron interferometer the goal of the experiment is to shed light on noncontextuality [427]:

> The concept of quantum noncontextuality represents a straightforward extension of the classical view: the result of a particular measurement is determined independently of previous (or simultaneous) measurements on any set of mutually commuting observables. Local theories represent a particular circumstance of noncontextuality, in that the result is assumed not to depend on measurements made simultaneously on spatially separated (mutually noninteracting) systems. In order to test noncontextuality, joint measurements of commuting observables that are not necessarily separated in space are required.

The concepts of noncontextuality and contextuality are similar to the notions of classical and quantal (a good discussion of these concepts can be found in [474]). To check persistently that the neutron interferometer operates according to quantal laws [475–477] seems very strange, because it operates only owing to a quantal phenomenon such as interference.

However, if somebody does not understand Modern Art, he should not dare even to suspect that the "King is bare".

10.8 Conclusion

Neutron optics is more logically consistent than neutron physics on the whole, because it deals mainly with probabilities which are a result of the interaction of the neutron with many atoms simultaneously. However, we also have to investigate the interactions of the neutron with individual atoms. Scientific practice worked out the rules for how to do this. They are useful and must be used. However, it is also necessary to see the limitations of these rules and to try to overcome these limitations.

We showed that the EPR paradox does not exist. Its logical solution leads to the conclusion that entangled states for separated systems are only a list of simple product states. The inconsistency of quantum mechanics pointed out by EPR is the result of the error. Correction of this error leads inevitably to a change of the definition of the momentum and the position of a particle. The momentum and position are not eigenvalues of the respective operators. They must be defined as expectation values of the operators. After the change of definition, the uncertainty relations cease to forbid particles from having position and momentum simultaneously. The dispersions of momentum and position resulting from the new definition are not statistical uncertainties, but are characteristics of the corresponding wave packets.

Rejection of entangled states for separate particles leads to rejection of nonlocality. The absence of nonlocality can be proved by experiments like that reported in [463] but with correct measurement of the background. Experiments demonstrating violation of Bell's inequalities are absolutely useless. They reject the classical description of quantal phenomena, which was already rejected many times.

It is also necessary to point out that a commonly defined notion such as an observable has a semantic error, because no observable has ever been observed. For instance, the value of the spin component σ_x of a particle has never been measured. In all experiments we count particles after transmission through a filter. If we were able to guess correctly the direction of the particle's spin arrow, we could arrange our filter parallel to this arrow and then we could be sure that the particle would pass the filter with 100% probability. We cannot guess. If we polarize particles with the help of some filter, we change their state. Again, we do not measure the spin components of the polarized particles. If we put the second filter in the particle's path, we only check how well the two filters are adjusted.

All I said above does not mean that I do not see problems in quantum mechanics. They really exist. It is very important to understand what the meaning of the neutron spin precession wave, described in Chapter 5, is. How can the neutron exist in a superposition of states with different energies? Is there really an entangled state with an absorbed or emitted photon? What are the properties of a particle's wave packet? How can we observe and measure them? Is it possible to replace the Schrödinger equation by a nonlinear system of classical equations with account for some kind of stochastic dynamics? We formulated these problems with the hope that the scientific community will pay attention to them.

11
Application of Neutron Optics to Material Science

11.1
Nuclear Scattering Density Profile Analyses of Layered Materials

In Section 2.3, we studied the developments of various kinds of neutron mirrors and their performances. On the other hand, in Section 2.1 the possibility to investigate the structure of a multilayer sample from the analysis of the measured neutron reflectivity was pointed out. For example, the complicated layer structure produced with copolymers and the process occurring in the multilayer formation of copolymers have been experimentally studied by neutron reflectometry, augmenting the experimental results from small-angle scattering of neutrons and X-rays. The important procedure in these neutron reflectometry experiments is the deduction of the correct *distribution for the scattering amplitude density for neutrons*, $\rho = N b_{\text{coh}}$, from the measured reflectivity distributions.

Concerning the requirement mentioned above, Majkrzak and Berk [137, 138, 478] investigated at the National Institute of Standards and Technology (NIST) reactor, United States, how we can deduce the exact and unique scattering amplitude profile from the experimental results of neutron reflectivity, according to the 2×2 matrix formalism with Croce and Pardo's optical approach [92]. The importance of such an investigation is indicated in the example shown in Figure 11.1.

We can notice that at least two kinds of completely different scattering amplitude density profiles as illustrated in Figure 11.1a become compatible with the measured data for a titanium oxide layer shown in Figure 11.1b. This is caused by the fact that we measure only the absolute value of the reflectivity for the data shown in Figure 11.1b, with respect to the reflection coefficient, which is actually a complex quantity with real and imaginary parts, or with the amplitude and the phase.

Therefore, it will become possible to obtain the scattering amplitude density profile without such an ambiguity by taking into consideration the information on the phase in addition to the absolute magnitude of the reflection coefficient. De Haan *et al.* considered such a measurement on a sample by including a known magnetic layer in addition to the unknown layer, and reported the analytical approach as well as the numerical result [479]. Majkrzak *et al.* undertook a similar procedure to take into consideration the phase which is lacking in the example described above, by increasing the measured information [480]. For the present procedure, they pro-

Neutron Optics. Masahiko Utsuro, Vladimir K. Ignatovich

ISBN: 978-3-527-40885-6

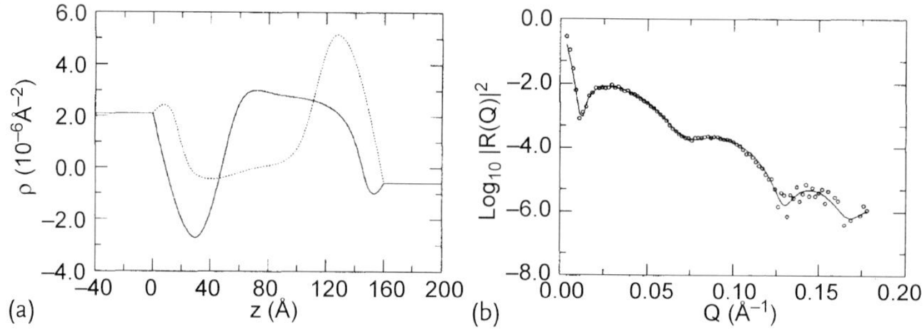

Figure 11.1 Correspondence of the scattering amplitude density profiles with the measured neutron reflectivity: (a) two kinds of scattering amplitude density profile (solid and dotted lines); (b) an example of the actually measured data on a titanium oxide layer (circles) and essentially the same reflectivity profiles (solid and dotted lines) corresponding to the two cases illustrated in (a) (Berk *et al.* [478]).

posed and practiced two kinds of approaches. The first one is named the *reference layer method*, in which, in addition to the usual measurement on an original sample with an unknown layer, similar measurements are repeated on samples with supplemented known layers in front of, or on backside of, the original layer sample. The measured data on two or three kinds of composite samples are used to deduce a correct and unique scattering amplitude density profile for the original sample [481, 482].

Figure 11.2 shows their results of the reference layer method applied to a sample with a deuterated polystyrene layer on a silicon substrate. The sample was prepared

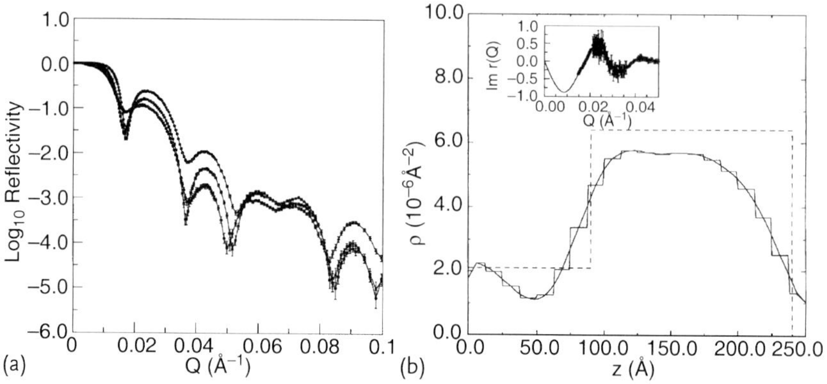

Figure 11.2 Structure analyses on measured neutron reflectivity with the reference layer method applied to a deuterated polystyrene layer on a silicon substrate. (a) Measured reflectivity data of composite films with an unknown layer provided by Ni, Cu, and Mo references (top to bottom at $Q = 0.03\,\text{Å}^{-1}$). (b) Inverted scattering amplitude density profiles from (a) extracted by applying the reference layer method for a gap and unknown parts of the potential. The insert is the resultant Im $r(Q)$ for the unknown part. (The curves are spline-smoothed, and the dashed line is the ideal expected profile.) (Majkrzak *et al.* [480]).

by sputtering on a substrate first a reference layer with a thickness of about 100 Å, next a gap layer of silicon, and finally a sample layer. The experimental result shown in Figure 11.2a was obtained with the neutron beam incident on the substrate side of silicon. The analyzed result applying the reference layer method for the gap and the sample layers as unknown ones gave the unique scattering amplitude density profile shown in Figure 11.2b and indicated gap layer and sample layer thicknesses of about 90 Å and about 150 Å, respectively.

However, to employ the present reference layer method, it is required to prepare two or three samples by a rather laborious procedure. Another approach, the polarized neutron method, is comparatively simple for the preparation of the sample, and was introduced in Chapter 3.

11.2 Polarized Neutron Reflectometries

11.2.1 Penetration Depth in Superconductors

One of the experimental fields where important information was derived from the application of polarized neutron reflectometry is the study of the *magnetic field penetration depth* in *superconductors*. As mentioned in textbooks on superconductors [483], superconductivity was discovered 1911 by Kamerlingh Onnes in Leiden. The *London penetration depth* Λ_L was introduced as one of the fundamental length scales in the equation of H. and F. London which describes the electromechanical properties of superconductors. The theory of superconductivity was epitomized in a successful direction by Ginzburg and Landau (GL) in 1950, and the GL theory introduces a characteristic length called the *GL coherence length* ξ, which represents the distance over which the wave function of superconducting electrons can vary without undue energy increase. In typical superconductors, $\Lambda_L \cong 500$ Å and $\xi \cong 3000$ Å.

In 1957, Abrikosov investigated the situation where the GL coherence length ξ becomes comparable to or smaller than the penetration depth Λ. Defining the *GL parameter* κ as $\kappa = \Lambda/\xi$, the situation $\kappa \ll 1$ holds for typical classic pure superconductors, and in this case there is a positive surface energy associated with a domain wall between normal and superconducting material. Abrikosov considered the case where $\kappa > 1/\sqrt{2}$ and found that this should lead to a negative surface energy, so the process of subdivision into domains would proceed until it is limited by the microscopic length ξ. This behavior is thought to be radically different from the classic intermediate-state behavior considered earlier. He showed that the exact breakpoint between these two regimes is at $\kappa = 1/\sqrt{2}$, and he named the material he considered *type II superconductors* to distinguish them from the earlier *type I superconductors*.

Furthermore, in 1986, another new type of superconductor, the so-called high-temperature superconductor, was discovered by Bednorz and Müller.

The experimentally measured length Λ of magnetic screening, that is, the *magnetic screening length*, is an important parameter for estimating Λ_L, and among various possible experimental approaches, polarized neutron reflectometry has two advantages [484]. First, direct measurement of the temperature dependence $\Lambda(T)$ is possible, and therefore it is unnecessary to introduce a model for the temperature dependence. Second, polarized neutron reflectometry is directly sensitive to the profile of the penetrated field itself, in contrast to approaches that are only sensitive to the penetrated field being an integrated quantity on the profile.

To utilize these advantages, it is crucial to reduce a unique inherent profile independent of any model from measured reflectivity data including the effect of relatively weak magnetic scattering compared with that of nuclear scattering. As typical examples applying polarized neutron reflectometry to the study of superconductors, two kinds of experiments on niobium will be introduced below.

In the first example of the experiments, Felcher *et al.* [485] arranged a niobium film with a thickness of 5 μm sputtered on the substrate of a polished silicon single crystal in the sample position indicated in Figure 3.2b, and maintained the sample at a temperature lower than the superconducting transition temperature T_c in the magnetic field H parallel to the surface and less than the critical magnetic flux entry field in the polarized neutron reflectometry experiment. The magnetic induction at depth z in the material is $B(z) = H\exp(-z/\Lambda)$. The next relation for the refractive index can be derived from Eq. (3.6):

$$n_{\pm}^2 = 1 - \frac{\lambda^2}{\pi}\left\{N b_{\mathrm{coh}} \mp cH\left[1 - \exp\left(-\frac{z}{\Lambda}\right)\right]\right\} . \tag{11.1}$$

In such a situation with the refractive index dependent on the depth, the reflectivity should be decided by the result of reflections inside the film, in addition to the contributions from the surfaces and interfaces. Although no simple and exact formula as a general solution for this case was obtained, Felcher *et al.* gave the next approximate formula for the spin-dependent reflectivity of a medium with the refractive index given by Eq. (11.1):

$$R_{\pm} = |r|^2\left\{1 \mp \frac{2\theta_i}{p_0}\frac{cH}{N b_{\mathrm{coh}}}\frac{1}{1 + (4\pi p_0 \Lambda/\lambda)^2}\right\} , \tag{11.2}$$

where $r = (\theta_i - p_0)/(\theta_i + p_0)$ is the reflection coefficient in the case without the magnetic potential, and $p_0 \cong (\theta_i^2 - \lambda^2 N b_{\mathrm{coh}}/\pi)^{1/2}$ is the effective angle of the neutron beam in the medium.

From Eq. (11.2), it can be understood that the region in which the effect of the magnetic potential becomes predominant is at small p_0 in the experimental condition of $\lambda \ll \Lambda$, that is, the wavelength region near the critical wavelength, $\lambda \sim \lambda_c = \theta_i(\pi/N b_{\mathrm{coh}})^{1/2}$.

In advance of the spin-dependent experiments, the measurement was carried out at a temperature of $T = 10$ K, that is, higher than $T_c = 9.2$ K, and from the result the effects of the surface roughness were checked. Possible oscillations in the measured intensity due to the interference terms in the finite sample thickness are averaged over the beam divergence angle $\Delta\theta_i = 0.02°$. An example of

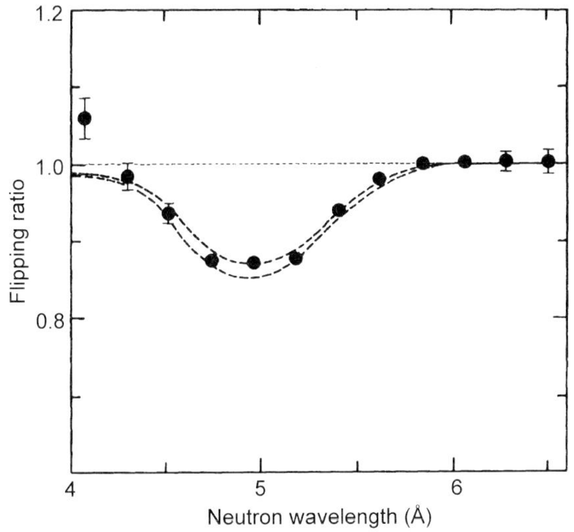

Figure 11.3 Experimental results (circles) of the polarized neutron reflectivity ratio at grazing angle $\theta_i = 0.34 \pm 0.02°$ to a superconducting niobium film at a temperature of 4.6 K and with a magnetic field of 500 Oe compared with the calculated results (broken lines: upper; magnetic penetration depth $\Lambda = 480$ Å, lower; $\Lambda = 380$ Å). (Felcher *et al.* [485]).

the results thus obtained on the reflectivity ratio $F_R = R_+/R_-$ from the measured spin-dependent reflectivity is shown in Figure 11.3. As expected, the minimum of the reflectivity ratio is obvious around the critical wavelength $\lambda_c^- = 5.1$ Å. From comparison with the theoretically calculated curves, the value for the magnetic field penetration depth $\Lambda = 430 \pm 30$ Å was obtained at the temperature and the magnetic field of the present measurements. The theoretical estimation from the present experimental result on the zero-temperature value gives $\Lambda(0) = 410 \pm 40$ Å, which was concluded as reasonable with magnitude similar to that of the London penetration depth $\Lambda_L(0)$ obtained by other experimental methods.

As another study, Zhang *et al.* [484] carried out polarized neutron reflectometry experiments similarly on superconductive niobium film by making use of the NIST research reactor, and compared the results with the numerical analyses on the polarization ratio of reflected neutrons. The exact solution for the magnetic field penetration profile with the boundary condition $H = H_a$ at both surfaces $z = 0$ and d of a superconductive film based on the local electromagnetic equations of London (H. and F. London) can be written as $B(z) = H_a \cosh[(2z - d)/2\Lambda]/\cosh(d/2\Lambda)$, and the numerical analyses on the reflected neutron polarization ratio were performed with the *density profile of the neutron scattering amplitude*

$\rho(z)$ given by the equation

$$\rho_{\pm}(z) = N b_{\text{coh}} \mp c H_a \frac{\cosh\left(\frac{2z-d}{2\Lambda}\right)}{\cosh\left(\frac{d}{2\Lambda}\right)}, \quad 0 \leq z \leq d \tag{11.3}$$

where Λ is the *magnetic screening length.*

Two kinds of samples were prepared by sputtering a niobium film with a thickness of 600 nm (sample 1) and 300 nm (sample 2), respectively, on a silicon substrate. Three kinds of angular resolutions were used by exchanging the neutron beam slit. The neutron beam was monochromatized with a wavelength of $\lambda = 2.35$ Å and a wavelength resolution of $\Delta\lambda/\lambda \sim 0.1$ by using pyrolytic graphite. Then the critical angle for the wavelength is $\theta_c = 0.15°$. The measured data on the reflected intensities I_+ and I_- for the upward and downward spin neutrons, respectively, were converted to the *polarization function,*

$$PF = \frac{I_+ - I_-}{I_+ + I_-}, \tag{11.4}$$

where PF is the polarization function, and were compared with the numerical calculations based on Eq. (11.3).

In the course of these calculations, in addition to the magnetic screening length Λ, the dead layer thicknesses x_1 (air side) and x_2 (Si side) were also adjusted to obtain the best fit. The angular dependence of the reflected intensity for sample 1 varied smoothly, as the polarization function derived from the reflected intensities for the sample 1 at 4.5 K are shown in Figure 11.4 for each resolution, in comparison with the numerical calculations. Averaging with equal weight on these three kinds of data resulted in $\Lambda = 1100 \pm 10$ Å. In contrast the reflected intensities from sample 2 gave results including oscillating components due to the effects of interferences, but also led to a similar value for the magnetic screening length by applying a comparison with the numerical calculations.

If the electron mean free path l_e becomes shorter owing to possible effects of impurities in the sample or for any other reason, then the magnetic screening length becomes longer than the London penetration length Λ_L. Therefore, if we correct the value for the magnetic screening length Λ derived from the experiments on both samples for the present effect as $\Lambda = \Lambda_L(1 + \xi_0/l_e)^{1/2}$, where ξ_0 is the *GL coherence length* introduced previously, then for the London penetration depth this gives $\Lambda_L(4.5\text{K}) = 494$ and 369 Å for samples 1 and 2, respectively, the averaged value becoming 430 ± 80 Å, in reasonable agreement with the result of Felcher *et al.* at 4.6 K.

As indicated by these two studies, the two advantages mentioned above of polarized neutron reflectometry, that is, (1) direct measurement of the absolute magnitude of the magnetic screening length Λ, compared with the indirect measurements through, for example, the temperature dependence in other experimental methods, and (2) measurement sensitive to the penetrated magnetic field profile, compared with the integral measurements of the penetrated magnetic field in other experimental methods, will be most utilized for the study of superconductors.

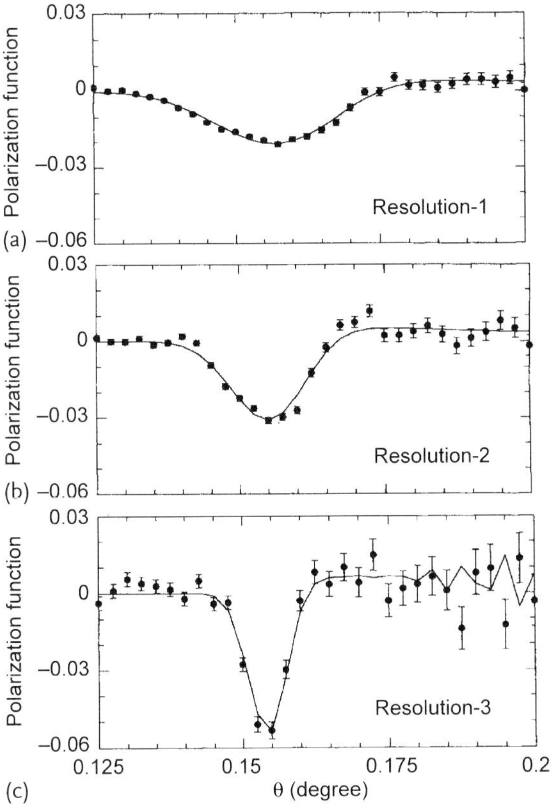

Figure 11.4 Experimental results (circles) for the polarization function $(I_+ - I_-)/(I_+ + I_-)$ of neutrons with wavelength $\lambda = 2.35$ Å for superconducting niobium film sample 1 at a temperature of 4.5 K and with a magnetic field of 423 G, compared with the calculation (curve: magnetic screening length $\Lambda = 1092$ Å (a), 1107 Å (b), 1091 Å (c)) (Zhang *et al.* [484]).

Actually, various kinds of measurements have been performed to investigate the characteristics of a variety of samples. As an example, the measurement of the magnetic profile of multilayers with sandwiched superconductor layers was carried out with the polarized neutron reflectometer at the KENS Facility, the High Energy Accelerator Research Organization, Tsukuba [486]. Further, the effectiveness of the present reflectometry for study of the ordered lattice of magnetic vortices parallel to the interface was also demonstrated with the reflectometer at the Missouri University research reactor [487].

Similarly to the description in Section 2.1, in addition to the specular reflection measurements for the studies on the scattering amplitude density profile perpendicular to the surface or interface, measurements on nonspecular reflections at the angles different from the incident angle to the surface provide information on the magnetic structure parallel to the surface.

As an example, nanometer-scale structures can be prepared on magnetic materials owing to recent progresses in lithography and evaporations. Temst *et al.* observed the nonspecular reflection peaks from the two-dimensional regular structure of dots with width 1 μm and period 10 μm on a polycrystalline cobalt sample by making use of the polarized neutron reflectometer at the Hahn–Meitner Institute, Berlin [488]. For large-angle measurements including nonspecular reflection angles, a position-sensitive detector was used, and for the neutron polarization and polarization analysis, Si–FeCo supermirrors were used.

On the other hand, Chen *et al.* carried out a measurement on a cobalt film with a lithography pattern by making use of a specially developed polarized neutron nonspecular reflectometer with the combination of a polarized ^{3}He analyzer and a position-sensitive detector at the NIST reactor [489].

11.2.2
Reference Layer Methods for Multilayers

An example of the practical application of the present polarized neutron reference layer method is illustrated in Figure 11.5 for the wave vector transfer $Q = 2k_\perp$, to evaluate the applicability of the present method to reflection phase analyses.

The experiment was performed with the polarized neutron reflectometer NG-1 at the NIST reactor for the sample structure indicated in Figure 11.5a, where the Fe layer was saturatedly magnetized with a magnetic field of about 200 G perpendicular to the wave vector surface. The neutron wavelength and the wavelength width ratio used were 4.75 Å and 1%, respectively. The neutron polarization ratio was kept between 98 and 99%, and the depolarization ratio of the reflected neutrons was negligible. Further, a nonspecular reflection component was not observed.

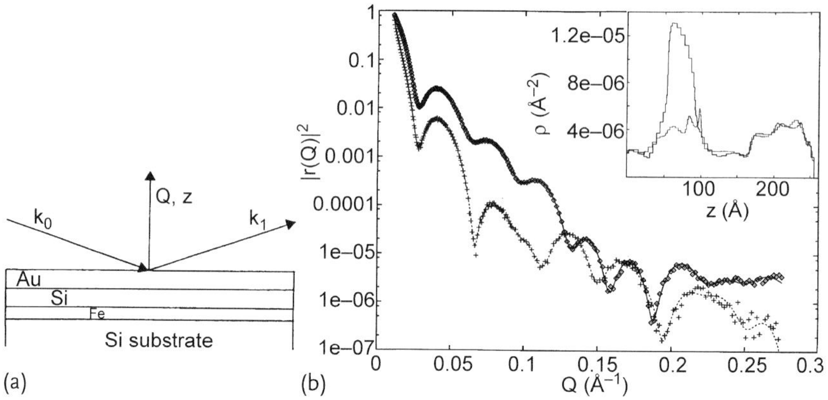

Figure 11.5 Polarized neutron reflection experiment: (a) structure of the multilayer; (b) neutron reflectivity data for + (diamonds) and − (crosses) spin states; the insert shows the distribution of the scattering amplitude density profile ρ derived from each set of data (Majkrzak *et al.* [490]).

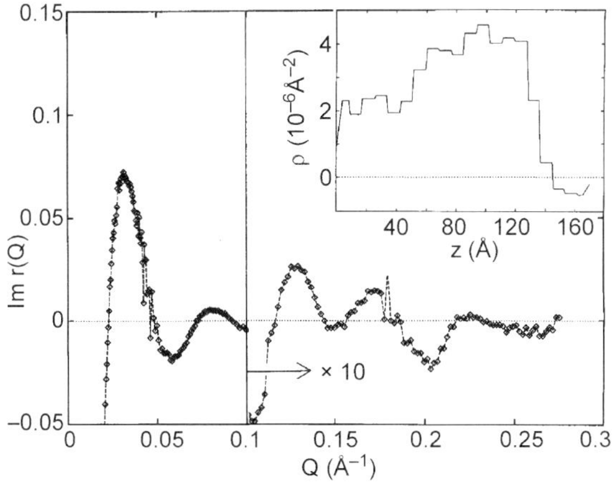

Figure 11.6 Im *r* versus *Q* derived from the data shown in Figure 11.5b; the insert shows the scattering amplitude density profile for "unknown" Si and Au layers as obtained by inversion of Im *r* (Majkrzak *et al.* [490]).

The insert in Figure 11.5b shows the scattering amplitude density profile obtained by the application of the present method of data analyses to the experimental results. For a further examination on the phase determination derived by the present method, the imaginary part of *r* thus derived and the scattering amplitude density profile calculated from Im *r* with the numerical integral method are plotted in Figure 11.6 and in the insert, respectively. The scattering amplitude density profiles given by the inserts in Figures 11.5 and 11.6 are in good agreement, thereby demonstrating the consistency of the present polarized neutron reference layer method with the use of known reference layers in addition to the unknown layer to be studied.

11.2.3 Nonspecular Reflection from Magnetic Multilayers

Practical polarized neutron experiments on a sample with a significant noncollinearity similar to that supposed in the previous calculation were carried out by Nickel *et al.* [491]. The polarized neutron reflectivities with and without spin flip were measured over a wide reflection angle including the nonspecular reflection region on a sample of a FeCr multilayer; the experimental results are shown in Figure 11.7.

The diffractometer EVA at the high-flux reactor of the Institut Laue–Langevin was used in an experimental setup fundamentally similar to that shown in Figure 3.2. For the polarization analyzer covering the wide reflection angle, a polarized ^{3}He gas filter was used. A multilayer sample with 200 periods of 19-Å Fe and 42-Å Cr was grown in the molecular beam epitaxy system at 300 °C on a Nb buffer layer on a crystalline Al_2O_3 substrate. There was a terrace structure with a period of 50 Å

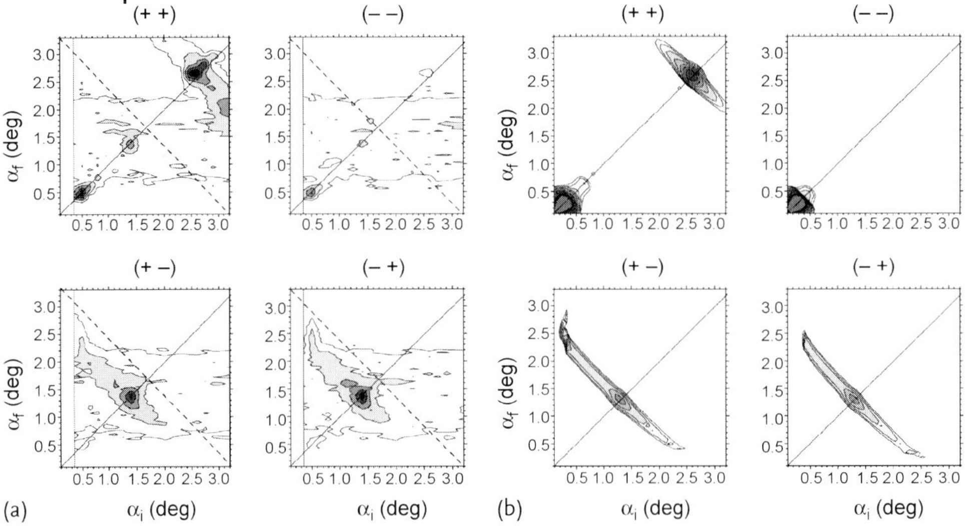

Figure 11.7 Reflected intensity contours on a logarithmic scale with neutron spin analyses for a FeCr multilayer sample, where α_i is the incident angle and α_f is the reflected angle: (a) experimental results for a 200-bilayer sample; (b) modeled calculation results for 16 bilayers with the supermatrix method (Nickel *et al.* [491]).

and an inclination angle of 2.3° on the upper interface of the buffer layer owing to the growth characteristics of the Nb layer, and this further propagated to Fe and Cr layers with a similar structure. The results of measurements and of numerical simulations were plotted in two-dimensional maps with the incident angle as the abscissa and the reflection angle as the ordinate for each reflection component with and without spin flips. In the figure, reasonable correspondence can be seen between the experiments and the calculations with the supermatrix method, not only in the non-spin-flip specular reflections, but also in the spin-flip reflections, over a wide angular region including the nonspecular reflections.

11.2.4
Magnetic Materials with Artificial Properties

Polarized neutron reflectometry helps to investigate artificial magnetic systems – their magnetization profile, domain structure, and magnetic fluctuations – and their correlations and behavior with variation of external conditions – fields and temperature.

Recently a lot of attention has been paid to layered magnetic systems and also to lateral nanostructures such as stripes and islands on a smooth surface. Their properties depend on coupling between elements of the structure such as the proximity effect, that is, interaction of magnetic parts through a thin nonmagnetic isolator, exchange bias, that is, interaction between neighboring ferromagnetic

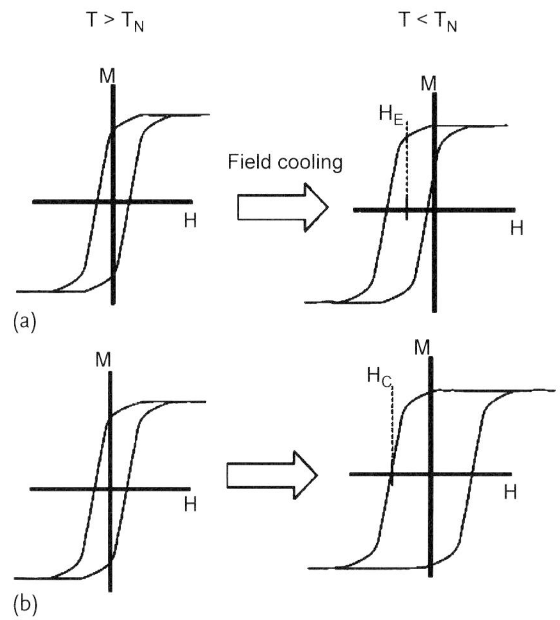

Figure 11.8 Some effects induced by the ferromagnetic–antiferromagnetic exchange coupling, that is, (a) loop shift and (b) coercivity enhancement after field cooling below Neel, T_N, or Curie, T_C, temperatures [492].

and antiferromagnetic parts, and spring exchange, that is, interaction between neighboring soft and hard ferromagnetic layers. The artificial systems have very unusual properties, which can be illustrated by their hysteresis curves. In the case of exchange bias [492, 493], the most spectacular is the shift, H_E, of the hysteresis curve after field cooling below the Neel temperature, as shown in Figure 11.8a, or an increase of the coercivity, H_C, as shown in Figure 11.8b. Which of the two effects happens depends on the strength of the antiferromagnetic anisotropy.

In the case of spring exchange, the main interest is related to the possibility to increase the value of $B\,H_{max}$ and to improve the performance of magnets. To increase $B\,H_{max}$ it is necessary to have high coercivity, H_C, and high saturation magnetization M_{sat}. It can be achieved [494] by combining soft and hard ferromagnetic films. The properties of such a system depend on the thickness of the soft film, as shown in Figure 11.9.

If the system is magnetized in a saturation field in one direction, and then the external field decreases and turns to the other direction, the soft magnetic becomes separated into layers with different magnetization, which turns from the direction almost parallel to the external field to the direction for a hard magnetic. This turn can go along a spiral in a plane parallel to the interface, and then we obtain a helicoidal structure, or this turn can go in a plane perpendicular to the interface, and then we have a fanlike system. The rotation can go the other way and mea-

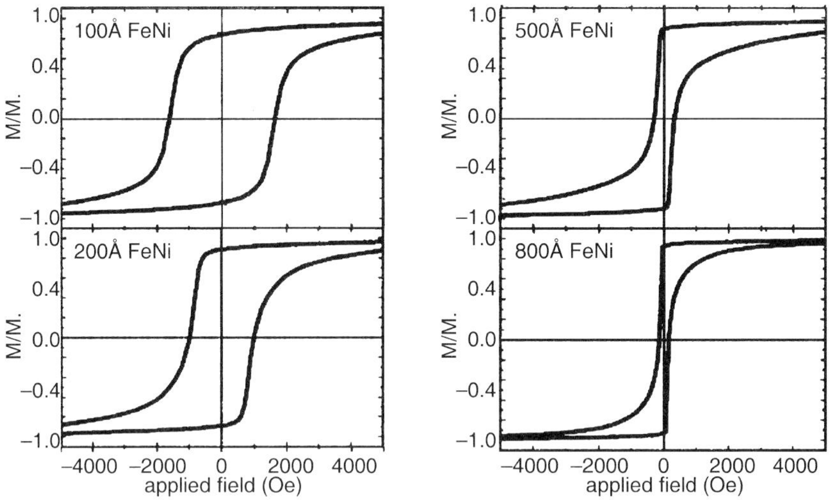

Figure 11.9 Hysteresis loops for FePt/NiFe spring-exchange films with 200-Å FePt (hard magnetic) layers and increasing FeNi (soft magnetic) thickness values as given in each panel [494].

surements of reflectivities of polarized neutrons with and without spin flip help to reveal the distribution of magnetization in such a system.

The reflectivity curves from one such measurement [244] performed using neutrons of wavelength 0.475 nm with the NG-1 reflectometer at the NIST Center for Neutron Research are shown in Figure 11.10. The sample contained hard magnetic $Fe_{55}Pt_{45}$ of thickness of 20.0 nm evaporated on a glass substrate, above it the soft magnetic $Ni_{80}Fe_{20}$ (permalloy) of thickness 50.0 nm, and all the system was capped with Pt.

The neutron beam was polarized along the x-axis. The sample was first magnetized by a field of 890 mT along the $-x$-axis. After that a smaller field H was applied along $+x$. The data in Figure 11.10 were obtained at $H = 16$ mT. The magnetic moment in the soft ferromagnet was expected to rotate uniformly around $z||Q$.

The reflectivity curves were also measured at other values of H: 5, 10, 16, 20, 26, 50, 100, 260, 350, and 630 mT. With a change of the external field, the directions of magnetization in the soft magnetic also change as shown in Figure 11.11 for three points along the depth: on the surface, in the middle, and near the surface of the hard magnetic.

A helical type of magnetization can be found in nature [495] (holmium single crystal at a temperature below 133 K), and can be prepared artificially [496]. One such artificial construction is shown in Figure 11.12. It consists of three one-domain-thickness ferromagnetic (Co) disks separated by 3-nm nonmagnetic spacers. Such a system can be stable even at room temperature. Neutron scattering on a helical system was considered in Chapter 6, and we shall not repeat the discussion here.

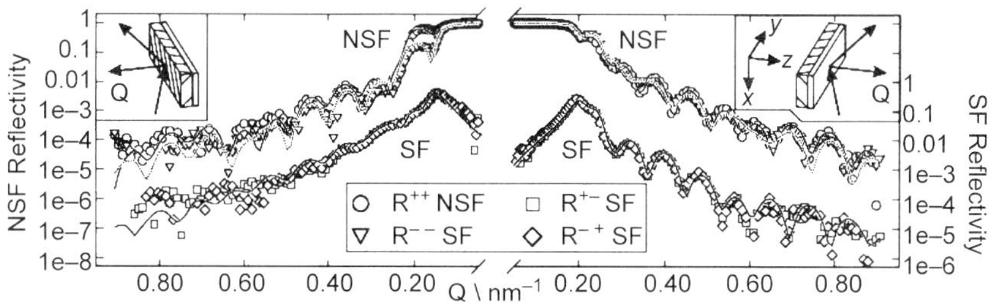

Figure 11.10 Reflectivity at 16 mT. Data from the front (back) surface are shown on the right (left) with $Q = 2k_{\perp}$ increasing toward the right (left). The non-spin-flip (NSF) data are plotted against the left axis. Differences in the two front NSF reflectivities are masked by the size and density of the symbols in the plot. The spin-flip (SF) data are plotted against the right axis, which is shifted by 2 orders of magnitude. For the back surface, when $Q > 0.62\,\text{nm}^{-1}$, the uncertainty in R is equal to or greater than R. Fits are shown as a dashed line for R^{--} and solid lines for R^{++} and R^{SF}. The insets show the scattering geometry for the two experimental configurations, with the incident and exit wave vectors unlabeled, and the momentum transfer Q labeled [244].

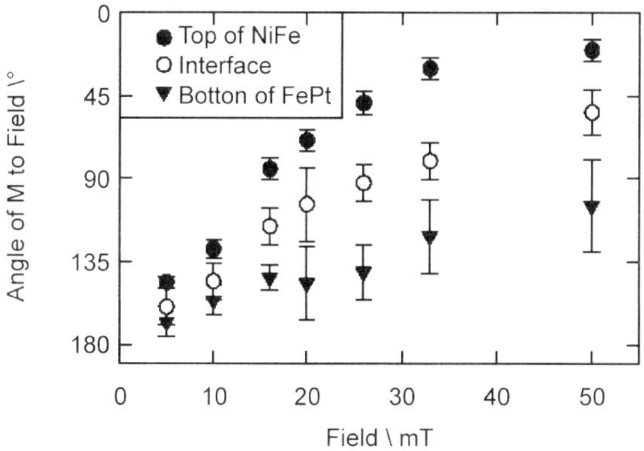

Figure 11.11 Dependence of angle of the magnetization at the top of the NiFe layer (black circles), at the interface (open circles), and at the bottom of the FePt layer (triangles) on the applied field [244].

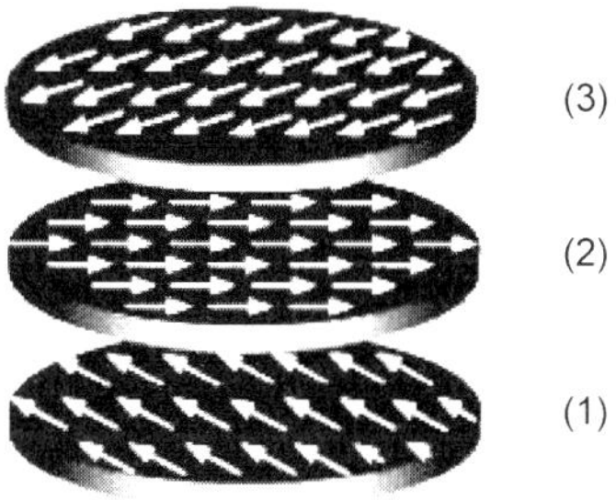

Figure 11.12 A noncollinear magnetic helix in the three disks system [496].

Before closing this section we also want to add that application of a periodic radiofrequency field to magnetics gives an additional opportunity to investigate of the distribution of the constant (DC) field B because of resonant enhancement of the spin flip processes at the points, where the frequency ω of the RF field is equal to B [270].

11.3 Neutron Scattering Experiments in Various Wavelength Regions

11.3.1 Graphite Inhomogeneity Structure Observed with Very Cold Neutrons

Various kinds of scattering experiments in material science with long-wavelength neutrons were carried out by making use of the very slow neutron spectrometer at the research reactor of the Technical University of Munich described in Section 6.2. As the first example indicating the typical inhomogeneity structure, the experimental study on graphite performed by Steyerl *et al.* [497, 498] will be described below. The experimental arrangement and the data analyses are essentially the same as those given in Section 6.2. In their experiment, two kinds of typically different carbon materials were studied. One is more generally used, *electrographite* produced in an electric furnace, and the other is a kind of crystalline carbon, named *pyrographite crystal*. The electrographite sample had density $\rho = 1.715\,\text{g/cm}^2$, and a thickness of 0.10 mm or more, whereas the pyrographite crystal sample had density $\rho = 2.24\,\text{g/cm}^2$ (i. e., 98.6% of the perfect crystalline density), and a thickness of 3.8 mm. In the case of the measurement on the electrographite, to diminish erroneous counting of the small angle scattered component as the effect of inhomogeneity scattering, both the angle of the incident beam to the sample and the suspended angle to the detector were more severely restricted than in the case in Figure 6.7b, that is, the former was 0.026 sr and the latter 0.095 sr.

As obviously recognized in Figure 11.13, the result for the electrographite includes a very distinguished scattering component steeply increasing at lower velocity, that is, at longer wavelength, in the measured velocity region. The observed velocity dependence indicates gradients of v'^{-2} and v'^{-4} below and above, respectively, the neutron velocity corresponding to the *characteristic wave number* k_0. It was also determined in the experiment that the present tendency does not vary essentially on changing the sample temperature. From these observations, the present distinguished scattering component can be attributed to the typical effect of inhomogeneity scattering. The characteristic wave number k_0 corresponding to the corner of the velocity dependence gives an approximate measure for the size of the inhomogeneity regions or particles in the case of the system of inhomogeneity particles with low packing density (the fraction of the total volume occupied by the regions or particles), through the relation $k_0 \cong 1/(r\theta_1)$, where r is the size of the inhomogeneity regions or particles, and θ_1 is the angle extending from the de-

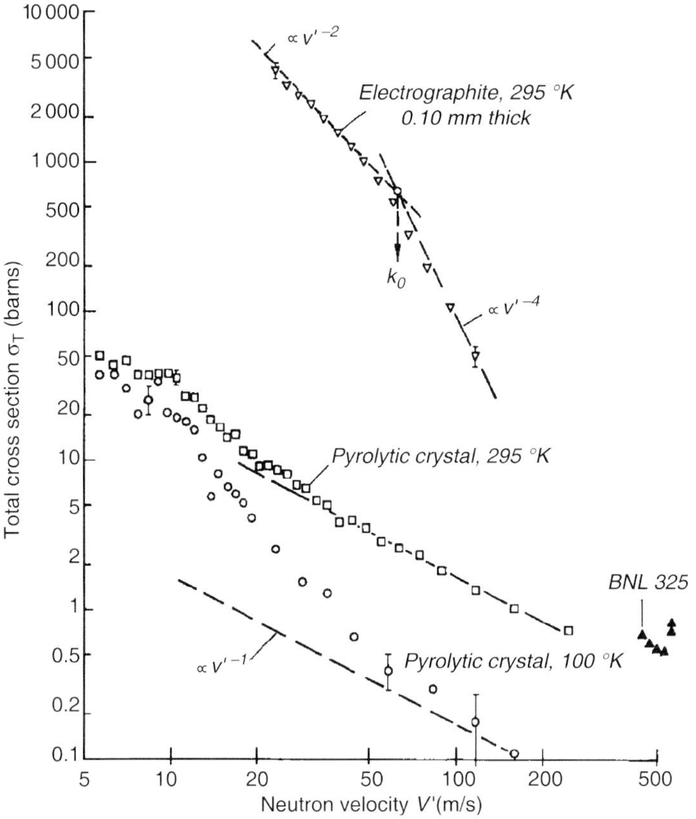

Figure 11.13 Experimental results of total cross sections for very slow neutrons on electrographite and a pyrographite crystal. The steep increase of σ_T for the electrographite in the lower-velocity region is caused by the predominant density fluctuations in the material. (Steyerl [498]).

tector window radius to the sample. However, in the case of high packing density, the present simple estimation leads to underestimated size for the inhomogeneity regions or particles.

In contrast to the result for electrographite, the experimental results for the pyrographite crystal vary nearly proportionally to $1/v'$ in the velocity region of $v' > 50\,\text{m/s}$, as expected from the absorption and inelastic cross sections, and indicate an effect of inhomogeneity scattering in the region $v' < 50\,\text{m/s}$.

As already explained in Section 6.3, this kind of the *diffractive scattering of waves in an inhomogeneous sample* has been analyzed since the early years of the field of X-ray scattering, and for the simplest case to be analyzed theoretically, *spherical particles* with radius r distributed in space sufficiently far apart from each other, leads to the angular distribution of the scattering per particle, $d\sigma/d\Omega$, in the Born approximation given by Rayleigh's formula [300]; Eq. (6.10). Thereafter, many kinds of materials such as those of colloidally dispersed particles and of various kinds of nonspherical particles were also studied [301–304], and such analyses can be applied to the inhomogeneity scattering of neutrons.

The macroscopic cross section for a sample is given by Eq. (6.11). Furthermore, to compare the theoretical solution with the experimental result shown in Figure 11.13, integrating Eq. (6.11) over the solid angle of neutrons, $d\Omega = 2\pi \sin\theta \, d\theta = 2\pi\kappa \, d\kappa / k^2$, with direction out of the detector can be performed for the case of Eq. (6.11) as

$$\Sigma(k) = \nu \int_{k\theta_1}^{2k} \frac{d\sigma(k)}{d\Omega} d\Omega$$
$$= 6\pi^2 \eta \rho_d^2 \left[j_0^2(kr\theta_1) + j_1^2(kr\theta_1) - j_0^2(2kr) - j_1^2(2kr) \right] / k^2 , \quad (11.5)$$

where the parameter η (packing volume ratio) regarding the packing density of the inhomogeneity zones is defined as the fraction of the total volume occupied by the regions or particles and is given by $\eta = 4\pi r^3 \nu / 3$. In the present equations, the packing ratio was assumed to be so low that the zones are sufficiently far apart from each other. In contrast, in the case of a high packing ratio, interference effects between zones must be taken into consideration.

The velocity dependence, that is, the wave number dependence of the neutron cross section given by Eq. (11.5), varies successively from the saturated region in $k < 3/(2r\sqrt{2})$, to the intermediate region proportional to k^{-2} in $3/(2r\sqrt{2}) < k < 1/(r\theta_1)$, and finally to the much steeper region proportional to k^{-4} in $k > 1/(r\theta_1)$. By comparing the experimental results with the theoretical curve, one can derive information on the effective radius r of the zone and the product of the packing ratio and the coherent scattering power squared, $\eta \rho_d^2$. Therefore, we apply the analytical formulas derived above to the experimental results on the electrographite shown in Figure 11.13, supposing the electrographite structure to be the spherical microcrystals of graphite packed in a medium of amorphous carbon, and further take the effects of interferences between the microcrystalline zones into consideration on the basis of the *correlation function from the Percus–Yevick equation* for hard spheres [308, 309]. Then, we arrive at the final results [497] for the particle radius $r = 110$ Å and the packing ratio $\eta = 0.26$. The derived value for the effective radius is considered to be reasonable as the microcrystal size in electrographite.

11.3.2
Solid Water Inhomogeneity Structures Observed with Cold and Very Cold Neutrons

As the next example of studies on solid-state structures with regard to the inhomogeneity and possible deformation from simple crystalline structures, very cold and cold neutron scattering experiments by Utsuro *et al.* on heavy water ice at low temperatures will be described below.

First, the very cold neutron experiments on solid heavy water at very low temperatures were carried out [499] with same experimental approach as that described in the the previous subsection by using the spectrometer shown in Figure 6.7a in Munich.

Solid water (H_2O ice) is one of the most widely used materials in our daily life and also in nature, but in spite of its wide use and importance, the microscopic structure of ice is not so easy to determine. Many investigations have been performed on the solid-state properties and the structure of solid water, and a number of neutron scattering studies have been reported. For the structure studies with neutron scattering, it is well known to be advantageous to substitute hydrogen (H) in the water for deuterium (D), that is, to use heavy water (D_2O) as the measuring sample, since the former nucleus has a very large incoherent scattering amplitude for neutrons, whereas the latter is essentially a coherent scattering nucleus. For the inhomogeneity studies also, such a substitution, that is, the employment of solid heavy water, is considered to be advantageous.

The experiments and analyses of the inhomogeneity studies on the solid heavy water (heavy water ice) were carried out using the same procedure for electrographite described in the previous subsection by making use of the very cold neutron spectrometer at the Munich reactor and also of the theoretical formulas mentioned previously. For comparison with regard to the inhomogeneity in the solid state, two kinds of samples were prepared: one was a *a single crystal of heavy water* and the other was a sample prepared by *simply frozen solidification* from liquid heavy water.

The single-crystal sample was prepared by cutting it out very carefully, preventing possible contamination of hydrogen impurities, to a thickness of 10 mm from a very large block of a single crystal of D_2O [500] stored in the refrigeration room at the Physics Laboratory of the Technical University of Munich [500]. Immediately after it had been cut out, the solid heavy water specimen was fixed at the sample position in the cryostat in the precooled state of the measuring assembly shown in Figure 6.7a. Thereafter it was cooled down gradually to the measuring temperature with the refrigerant liquid nitrogen.

On the other hand, the sample of simply frozen solid heavy water was prepared by simple cooling down in the cryostat from the liquid heavy water enclosed in a vacuum-tight aluminum capsule prepared by welding to a sample thickness of about 20 mm. It was then cooled down to the measuring temperature with the refrigerant liquid nitrogen, and then with liquid helium. Furthermore, in the case of the latter sample, the outer thickness of the capsule containing heavy water was again measured with vernier calipers to check for a possible thickness change due to the heavy water freezing by taking the capsule out of the cryostat after a series of measurements. The result was taken into consideration for the estimation of the actual thickness during the neutron measurements.

The experimental results shown in Figure 11.14 were fitted with the next relation in the low-energy region of neutrons:

$$\Sigma_T(k) = \Sigma_{\text{inc}}^{el} + A/k + \Sigma(k)\,, \tag{11.6}$$

where $\Sigma_T(k)$ is the total cross section and $k = mv'/\hbar$ is the wave number for the velocity v' of the neutrons inside the sample. The first, second, and last terms on

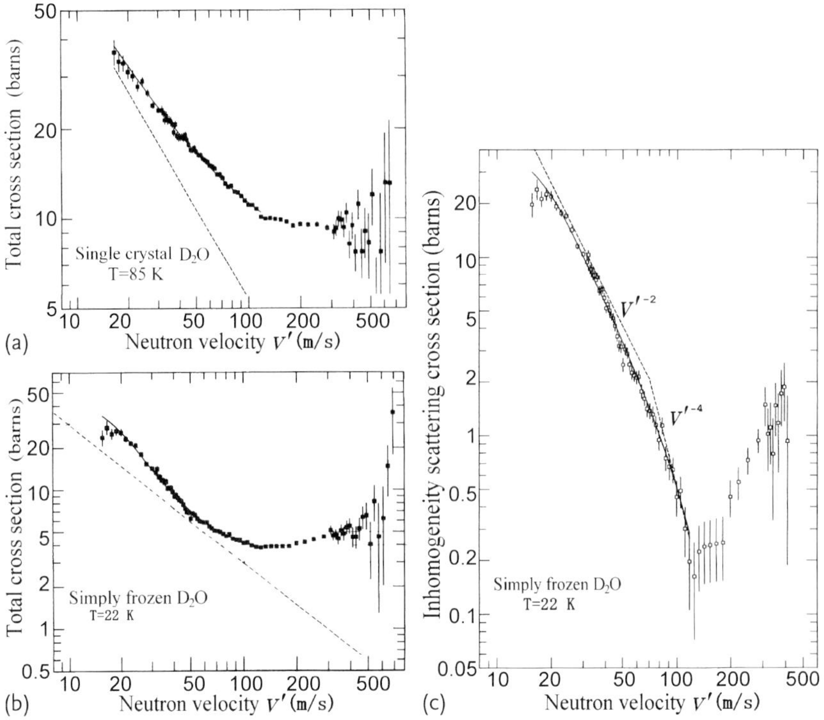

Figure 11.14 Experimental results of neutron total cross section experiments on a single crystal and a simply frozen sample of heavy water at low temperatures, provided by the theoretical analyses on inhomogeneity scattering: (a) measured total cross section for a sample of a single crystal of heavy water; (b) measured total cross section for a sample of simply frozen solid heavy water; (c) inhomogeneity scattering component extracted from experimental result (b). The solid lines in each graph indicate the corresponding results calculated with a procedure similar to that in previous subsection based on the theory introduced in Section 6.2 (Utsuro *et al.* [499]).

the right side of Eq. (11.6) correspond to the incoherent elastic scattering, the summation of absorption and inelastic scattering, and the coherent elastic scattering cross sections, respectively.

As obviously observed from the differences between the experimental results and the $1/v'$ formula given by the dotted lines in Figure 11.14a and b, the result for the single crystal essentially consists of the first and second terms on the right side of Eq. (11.6), whereas that for the simply frozen solid includes a predominant contribution of the $k^{-n}, n > 1$ type of component in the velocity region $v' < 50\,\mathrm{m/s}$, which indicates a significant fraction of packing by inhomogeneity zones with a size of several tens of angstroms. The inhomogeneity scattering component extracted from Figure 11.14b is illustrated in Figure 11.14c. The application of the same procedure for the electrographite, that is, the correction of Eq. (11.5) for the effects of interzone interference with the Percus–Yevick model for hard spheres [308, 309], gives the solid line in Figure 11.14c, where the parameters for the zone radius $r \cong 47 \pm 2$ Å, the density ratio of the medium layer to the zone $\rho_m/\rho_z = 0.97$, and the packing ratio of zones $\eta \cong 0.14$ were used. By the way, the isotope purity of the sample used in these measurements could be estimated from the magnitude of the first term on the right side of Eq. (11.6), and the results indicated a hydrogen impurity of less than 1% in the case of the simply frozen sample and about 4% for the sample of the single crystal. The latter hydrogen impurity is supposed to be caused by the handling process during the cutting out of the sample and its installation in the cryostat.

These measurements on the inhomogeneity correspond to the structure studies with the neutrons with a wavelength of several tens of angstroms, and therefore should reveal microscopic structures on the scale of several tens of angstroms. As indicated by the analyses, these samples studied with the inhomogeneity have partially crystalline character, or include microcrystalline zones. Therefore, it is considered to be an interesting task for understanding the nature of the present inhomogeneity to investigate such samples by using neutrons with a little shorter wavelength, that is, to observe the microscopic structures with atomic or molecular scales. From such a viewpoint, Morishima *et al.* carried out the total cross section measurements using time-of-flight spectroscopy with pulsed neutrons with a wavelength of about 1–20 Å extracted from a cold source, and compared the experimental results with the theoretical calculation to deduce a microscopic structure deformed from the perfect crystalline structure [501, 504]. Their results are shown in Figure 11.15 for the solid heavy water temperatures of 77 and 15 K. In their experiments, the pulsed cold source with a solid mesithylene cold moderator installed in the Kyoto University Research Reactor Institute linac facility, which was introduced in Section 2.1, was used, and the sample heavy water was simply frozen in the process of cooling down in a helium cryostat.

In the present case of the sample in which the neutron absorption can be neglected and the inelastic scattering cross section is sufficiently small owing to the very low sample temperature and very small incident neutron energy, the total cross section can be expressed by the next equation for the neutron energy E and the wavelength λ;

$$\sigma_T(E) = \sigma_{\text{inc}} + 8\pi b_{\text{coh}}^2 \left(\frac{\lambda}{4\pi}\right)^2 \int_0^{4\pi/\lambda} QS(Q)dQ, \tag{11.7}$$

where $S(Q)$ is the *structure factor* of the sample for the wave vector transfer Q. Putting $\xi = 4\pi/\lambda$ into the present equation and differentiating the equation, one can derive the following relation to extract the structure factor from the measured total cross section:

$$S(Q) = \frac{1}{8\pi b_{\text{coh}}^2} \frac{1}{Q} \left[\frac{d}{d\xi} \{(\sigma_T(\xi) - \sigma_{\text{inc}})\xi^2\}\right]_{\xi=Q}. \tag{11.8}$$

By application of this relation to the experimental results for the solid heavy water at 77 and 15 K given in Figure 11.15a and b, respectively, the structure factors for these temperatures were derived as shown by the solid and dotted lines, respectively, in Figure 11.15c. Commonly for both temperatures, several characteristic

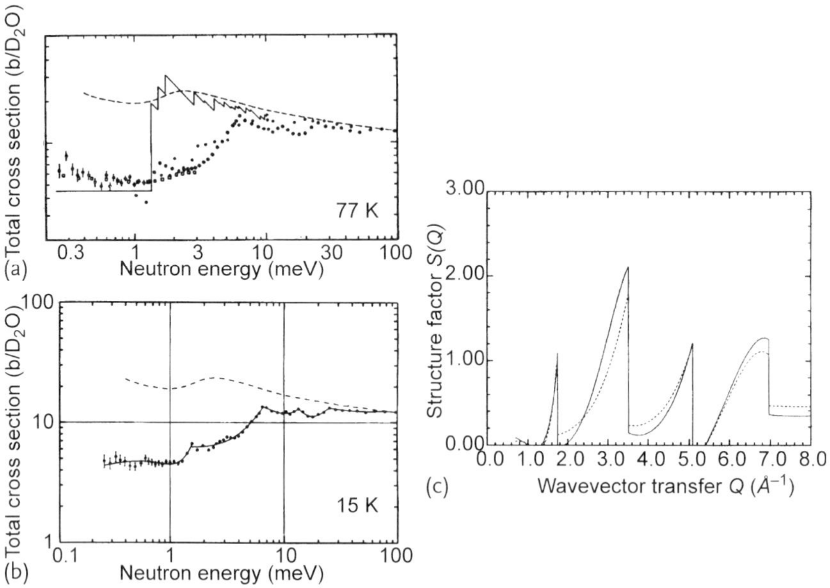

Figure 11.15 Experimental results and analyses on the cold neutron total cross section of simply frozen solid heavy water at low temperatures. (a) Experimental results at 77 K (open circles with the statistical error bars represent the results obtained by Utsuro and Morishima [501], solid circles represent the results obtained by Whittemore and McReynolds [502], and open squares represent the results obtained by Gissler *et al.* [503]) compared with the calculated result (solid line) for a hexagonal crystal of solid heavy water. (b) Experimental results at 15 K (solid circles represent the results obtained by Utsuro and Morishima [501]) compared with the calculated total cross section (solid curve) from the structure factor of the dotted line in (c). Broken lines in (a) and (b) represent liquid heavy water at room temperature. (c) Structure factors for simply frozen solid heavy water deduced from the experimental results at 77 K (solid line) and at 15 K (dotted line) (Morishima *et al.* [501, 504]).

peaks corresponding to the crystalline structure are recognized, and at the same time the apparent dispersion of the crystalline peaks to respective low-wave-number sides is also observed. This typical feature is considered to be essentially related to the inhomogeneity in the solid sample of simply frozen water indicated in Figure 11.14b.

11.3.3
Mechanically Alloyed Graphite Structure Observed with Neutron Diffraction

As the last example of the neutron studies on the inhomogeneity in solids, the neutron scattering experiment on a solid with a crystalline structure containing artificially dispersed atoms will be described.

Fukunaga *et al.* performed an artificial treatment of so-called *mechanical alloying* on typical crystalline solids and compounds, and carried out structure studies with neutron diffraction experiments on samples prepared in such a way [505]. The

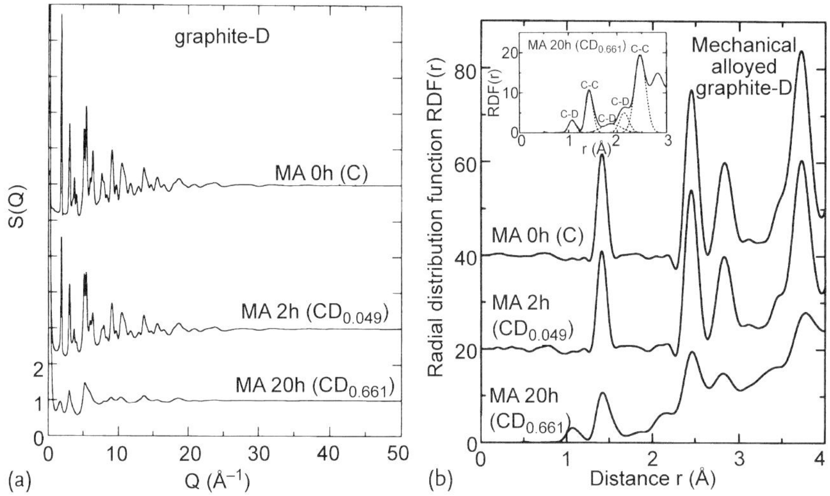

Figure 11.16 Structure factor $S(Q)$ and radial distribution function RDF(r) measured by neutron diffraction for graphite samples (CD_x) after mechanical alloying in a deuterium gas atmosphere: (a) variation of the structure factor with the milling time of mechanical alloying (MA); (b) radial distribution functions derived from the data in (a); (inset; one of the details with the results of least-squares Gaussian fitting in broken curves) (Fukunaga *et al.* [505]).

sample preparation procedure in the case of graphite is as follows. Pure hexagonal graphite powder (purity 99.999%, average particle radius 25 μm) was obtained by milling in an atmosphere of deuterium gas in a planetary ball mill. According to the increase of the milling time, the graphite sample density gradually decreased to a value (about 1.85 g/cm^3 for a milling time of 36 h) lower than that for the hexagonal graphite crystal (2.2 g/cm^3). Then, the structure study with neutron scattering was carried out with the high-intensity total scattering spectrometer HIT-II at KENS, the High Energy Accelerator Research Institute KEK, Tuskuba. The measured variation of the structure factor $S(Q)$ according to the progress of the milling is shown in Figure 11.16a. The radial distribution function RDF(r) can be derived from the Fourier transformation of the structure factor:

$$\mathrm{RDF}(r) = \frac{2r}{\pi} \int_0^{Q_{\max}} Q[S(Q) - 1] \sin(Qr)\, dQ\,. \tag{11.9}$$

The results are also shown in Figure 11.16b.

The detailed inspection of the distribution function reveals that, as shown in the inset in Figure 11.16b at a milling time of 20 h, deuterium atoms are bonded partially at the characteristic sites, dangling to carbons or between the layers in the graphite crystal lattice. Further, the slight expansion of the lattice between the graphite layers which increased according to the milling time was also observed by X-ray diffraction.

As illustrated by these experimental studies, the wave nature of neutrons in wavelength regions of various magnitudes, from thermal neutrons to very cold neutrons, plays a very useful and important role in studies of characteristic structure analyses of crystalline and noncrystalline solids by making use of nuclear scattering of slow neutrons, and provides unique information on the physics and engineering in the field of material science.

11.4 Small Angle Scattering Experiments on Metallic Alloys

11.4.1 Observation of Spinodal Decomposition

In Section 6.3 we studied the small angle neutron scattering (SANS) experimental method applied to an example of a binary alloy, Al–Zn, in which a typical case of Guinier–Preston zone formation was demonstrated. A further detailed study on Guinier–Preston zone formation in a Al–Zn alloy with the SANS experiment was carried out by Komura *et al.* [506] in view of the aging effect in the alloy sample observed with the temporal variation in the angular distribution of SANS intensity. The experiment was carried out with the neutron guide tube facility at the Kyoto University reactor with the SANS configuration shown in Figure 11.17a consisting of a helical slot neutron velocity selector operating for a selected wavelength in the region $\lambda = 3.4$–8 Å and wavelength resolution $\Delta\lambda/\lambda = 23\%$ or 14%, for incident neutrons, and a one-dimensional position-sensitive detector with an active length of 30 cm and a position resolution of 1 cm, or the length 50 cm and the resolution 0.5 cm [507]. A rectangular parallelepiped sample of Al–6.8 at% Zn alloy was homogenized at 300 °C, quenched into brine at −20 °C and kept at 77 K. After the sample had been aged for periods of 0, 1, 2, 5, 10, 30, 50, 100, 200, 500, and 800 (or 1000) min at 40 °C, SANS measurements were performed. The dependences of the differential cross section on the wave vector transfer similar to those in Figure 6.9b were measured for a series of samples after these aging periods, and the results were analyzed in relation to the aging time for the wave vector transfer, as plotted in Figure 11.17b.

As mentioned previously in Chapter 6, investigations of the aging effects on the material properties and the related mechanism are important not only from the viewpoint of the practical use of materials, but also for scientific interest in related nonlinear and nonequilibrium processes in a cooperative macrosystem. As early as 1875–1878, Gibbs [509, 510] defined two kinds of processes in equilibrating phenomena from the analyses on inhomogeneous systems. One of them was already known as the nucleation of the phase transition, but no attention was paid to the other process until Guinier [304] in 1961 reported the results of analyses on the precipitation process in an aluminum alloy from a small angle X-ray scattering experiment. The importance of the second process was first recognized when Cahn's historic paper, "On spinodal decomposition" [511], was published, in which the ex-

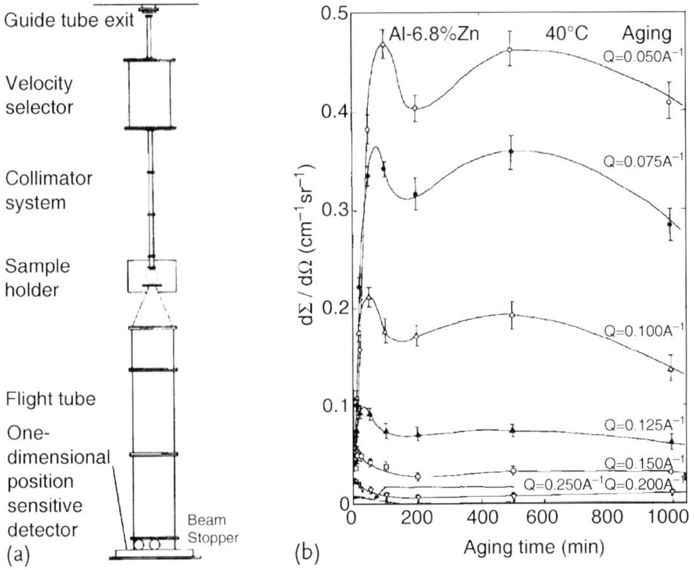

Figure 11.17 Aging time dependence of the small angle neutron scattering (SANS) differential cross section on a binary alloy sample for studying the mechanism of Guinier–Preston zone formation: (a) experimental arrangement of SANS spectrometry [507]; (b) aging time dependence of the SANS differential cross section on an Al–6.8 at% Zn sample for respective values of the wave vector transfer Q (Komura *et al.* [506]).

perimental data of Guinier were related to the definition of the second process by Gibbs. The process defined as *spinodal decomposition*[46] indicates the initiation of phase separation due to the occurrence of the unstable situation when a solid solution as a homogeneous mixture of two different components at a high temperature in the initial condition is quenched rapidly to a low temperature.

With regard to the dynamics of spinodal decomposition, Saito [512] presented his formula which explains the phenomenon approaching the equilibrium state with two-step inhomogeneity growth consisting of the earlier-stage growth of incoherent fluctuations and the later-stage growth of coherent conditions, by making use of a model on the probability distribution of the concentration represented by a summation of Gaussian distributions.

Komura *et al.* compared their aging-period-dependent results from a SANS experiment indicating obviously two peaks as shown in Figure 11.17b with Saito's formula, and explained the results with the two-step inhomogeneity structure as the anomalous growth of incoherent fluctuation and the retarded growth of a coherent ordered structure [506]. They carried out further extended measurements

46) The present terminology comes from the word spinode as the figuration on the locus curve of the boundary of the unstable region in the two-component phase diagram defined by $(\partial^2 G/\partial c^2)_{T,P} = 0$ for the free energy G and the concentration c [511].

and theoretical analyses on the time evolution of structures in Al–Zn and Al–Zn–Mg alloys [508].

11.4.2
Precipitation and Segregation in an Alloy

Concerning the separation formulas, Eqs. (6.13)–(6.15), Osamura *et al.* [513] carried out a SANS experiment and the analyses on a quarternary alloy, Fe–Cu–Ni–Mn, and derived two very different values for the Guinier radii corresponding to nuclear and magnetic scattering, respectively, which indicated the characteristic feature of precipitation and segregation in the alloy.

The sample with the chemical composition given in Figure 11.18 was prepared after forging and hot-rolling in small pieces for SANS measurements. The specimen was sealed in an evacuated ampoule and solution-treated in the α primary phase region. The ampoule was quenched and broken in ice water. The specimen was then sealed in another ampoule and then isothermally aged at 773 or 823 K. The SANS measurements were carried out at room temperature in the wave vector transfer region $Q = 0.06-3\,\text{nm}^{-1}$ with the KENS-SAN spectrometer at the KEK, Tsukuba, with a two-dimensional position-sensitive detector. From the measured two-dimensional scattered intensity distribution $d\Sigma/d\Omega(Q_x, Q_y)$, the values for $(d\Sigma/d\Omega)_{\text{nuc}}$ and $(d\Sigma/d\Omega)_{\text{mag}}$ were extracted by making use of the separation formulas, Eqs. (6.13)–(6.15). The results are plotted in Figure 11.18a for the nuclear and magnetic components as open squares and circles, respectively.

The Guinier radii for nuclear and magnetic scattering, respectively, estimated from the Q-dependence of SANS intensity distributions are shown as the correlation plot in Figure 11.18b, indicating the obvious tendency that the former goes to

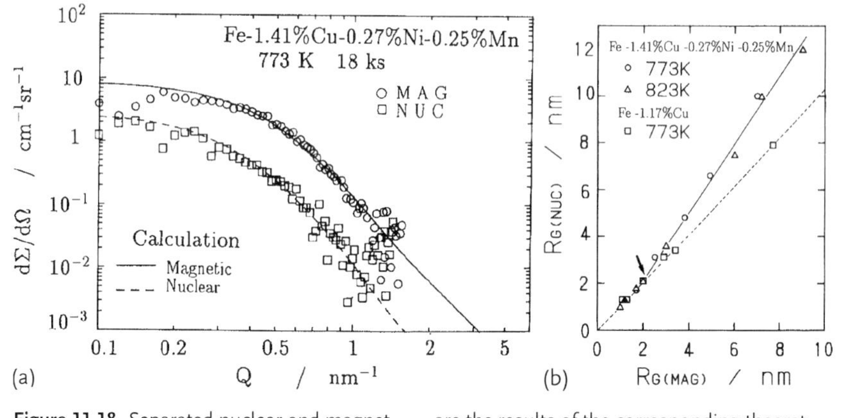

Figure 11.18 Separated nuclear and magnetic scattering components from experimental results of SANS on an aged sample of Fe–Cu–Ni–Mn quarternary alloy: (a) separated scattering intensities for the specimen aged at 773 K for 18 ks, solid and broken curves are the results of the corresponding theoretical calculations; (b) correlation between the Guinier radii from nuclear and magnetic scattering for the specimens aged at 773 or 823 K, compared with the result for a binary alloy (Osamura *et al.* [513]).

a somewhat larger value than the latter. On the basis of these results, they arrived at a model for the formation of the segregation layers as, in the supersaturated solid solution with body centered cubic (bcc) Fe, first initiation and growth of the precipitation of Cu-rich bcc clusters, then with Ni and Mn components swept out from the clusters during their growth, and finally a phase transition from the bcc cluster to the face centered cubic (fcc) Cu phase occurs surrounded by a fcc segregated layer with enriched Ni and Mn components.

In the case of such an inhomogeneity structure, the magnetic scattering amplitude density takes similar values for the matrix of almost pure Fe and for the Ni·Mn segregated layer, whereas that for the Cu-rich cluster takes a much lower value because there is no magnetic scattering by Cu. Thus, the magnetic inhomogeneity scattering will essentially see only the Cu-rich clusters. On the other hand, the nuclear scattering amplitude density for the Ni·Mn segregated layer takes a certain intermediate value, depending on the Ni·Mn concentration, between the larger value for the Fe matrix and the smaller one for the Cu-rich clusters. Therefore, the nuclear inhomogeneity scattering will also see the Ni·Mn segregated layer with a certain weight, in addition to the Cu-rich clusters. As a result, the effective zone radius for nuclear scattering will become larger than that for magnetic scattering.

Osamura *et al.* [513] performed such a model calculation by optimizing the Ni·Mn concentration, and obtained the solid and broken curves shown in Figure 11.18a, giving satisfactory agreement with the experimental results. In this way, the advantage of neutron scattering with two kinds of completely different, that is, nuclear and magnetic, scattering amplitudes played a very important role in the present case in investigating the inhomogeneity structure analyses in a complicated alloy.

11.5 Defects in Material

Here we consider the application of neutrons to investigate materials – their microstructure and macrostructure such as strain, texture, voids, precipitates, and cracks near the surface. All these are important for industry, safety, and research in geological science.

11.5.1 Matter under Pressure. Stress and Strain

Pressure applied to a substance produces strain, which in crystalline matter leads to a change of the crystalline parameters. This change can be registered by neutron diffraction methods.

Figure 11.19 shows the scheme of an experiment [514] for investigation of residual stress in samples of two grades of martensitic stainless steel called here ESR and Supreme. These samples being heating up to 1020 °C in a vacuum were cooled at three different rates: furnace cooling in a vacuum, or quenching by inert gas at pressures of 2 and 6 bar. The experiment was performed with the Australian Strain Scanner (TASS) diffractometer of the high-flux isotope reactor of the Australian Nuclear Science and Technology Organization.

The residual stress σ_i ($i = x, y, z$ in the reference frame of the sample as shown in Figure 11.19) was calculated according to the generalized Hook law:

$$\sigma_i = \frac{E}{1+\nu}\left[\varepsilon_i + \frac{\nu}{1-2\nu}(\varepsilon_x + \varepsilon_y + \varepsilon_z)\right] , \qquad (11.10)$$

where $E = 208$ GPa is the elastic modulus, $\nu = 0.28$ is the Poisson ratio, and ε_i are strains along coordinate axes. To find strains along three different directions, one uses the lattice constant of crystalline grains of α-Fe with the help of (112) Bragg diffraction.

The results of the experiment are shown in Figure 11.20. It was observed that in the ESR grade the internal stress increases with increasing cooling rate, but in the Supreme grade a considerable drop of residual stress under fast quenching at 6-bar pressure was observed; this was interpreted in terms of the high level of retained austenite.

Another example of the application of neutron diffractometry is presented in Figure 11.21. The measurements were performed at Canadian National Research Council reactor at Chalk River. The figure shows deformation along the circumference positions of a twist tube near the outer and inside surfaces and along the medial line. The strain directly after manufacturing is large and, as expected, of opposite sign near the two surfaces. After the annealing process the strain is considerably reduced. This investigation helps to optimize the technology of tube production.

11.5.2
Texture of Matter

Texture is a property of a polycrystalline material. The global texture characterizes the distribution of sizes, forms, and orientations of crystalline grains. The simple texture characterizes the distribution of only orientations, assuming that in other respects the matter is homogeneous. The texture is said to be absent if the distribution of orientations is isotropic. Thus, the texture characterizes the preferred orientation of crystallites. Texture influences the properties of materials and depends on the processing of materials by industry. Therefore, controlling the texture helps to improve the technology to get materials with the desired properties.

Texture also exists in geological rocks. In them, texture is related to the history of the formation processes. Therefore, knowledge of the texture of rocks helps to reveal the history of geological processes.

The texture, as was said above, is the three-dimensional orientation distribution function, $F(\mathbf{g})$, which is defined as volume fraction of crystallites with orientation $\mathbf{g}$:

$$F(\mathbf{g}) = \frac{1}{V}\frac{dV(\mathbf{g})}{d\mathbf{g}} . \qquad (11.11)$$

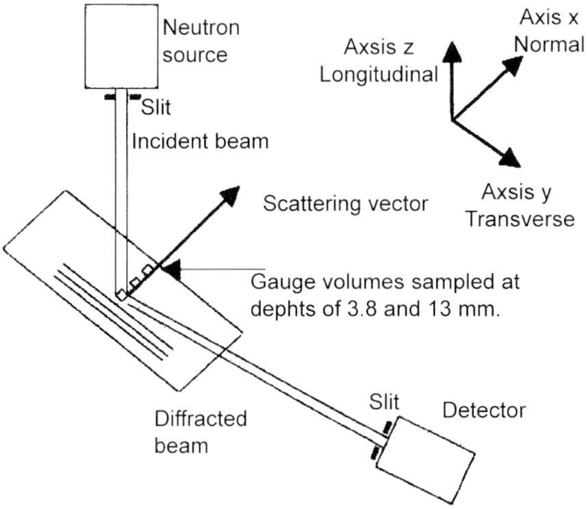

Figure 11.19 Experimental arrangement for the detection of material strain. The incident neutrons with a wavelength of 1.4 Å were selected by a Ge monochromator. The slits at the incident and diffracted beams defined the sampling volume $3 \times 3 \times 8\,\text{mm}^3$ at depths of 3, 8, and 13 mm under the surface. Diffraction angle $2\theta \sim 73^\circ$ was determined with resolution 0.075°. Neutrons were registered by a 32-wire position-sensitive detector.

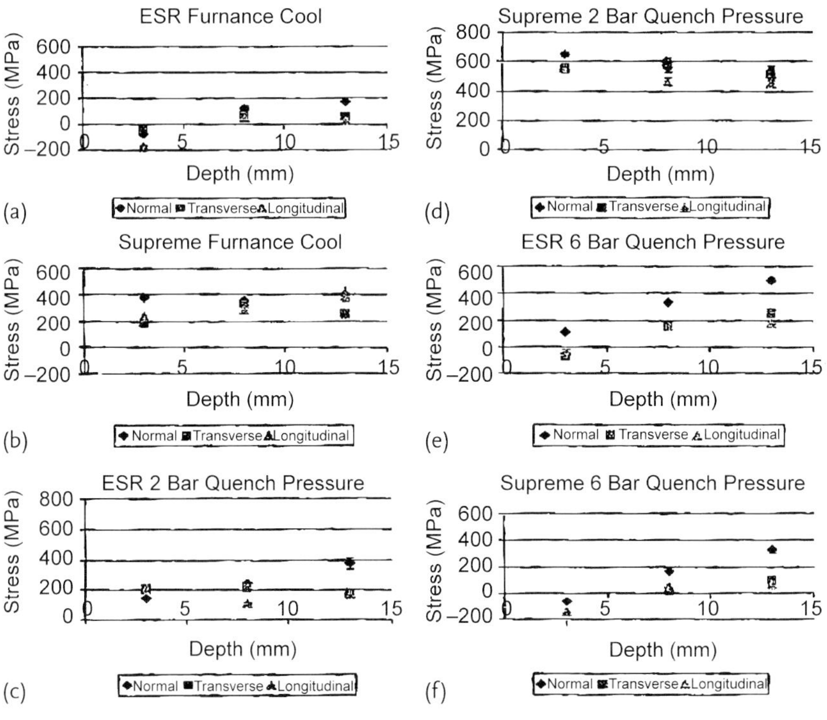

Figure 11.20 Plots in the normal, transverse, and longitudinal directions within the blocks of ESR and Supreme steels for quench conditions of furnace cool and 2 and 6 bar of a neutral gas.

This function, however, cannot be measured directly. What can be measured are the pole figures. Two o such pole figures (see [516]) are shown in Figure 11.22.

To understand what the pole figure is, imagine a crystalline grain and some planes in it. The normal to the planes in a spherical reference system has polar angle θ and azimuthal angle ϕ. These angles define a point inside the circle of the pole figure. Angle θ defines a point on the radius and angle ϕ coincides with the azimuthal angle of the circle.

Such pole figures were, in particular, measured using the high-resolution neutron diffractometer at reactor IBR-2. One measurement covers an area of 7.2° × 7.2° on a pole figure. One sample requires 714 spectra and 26 h of measurement time.

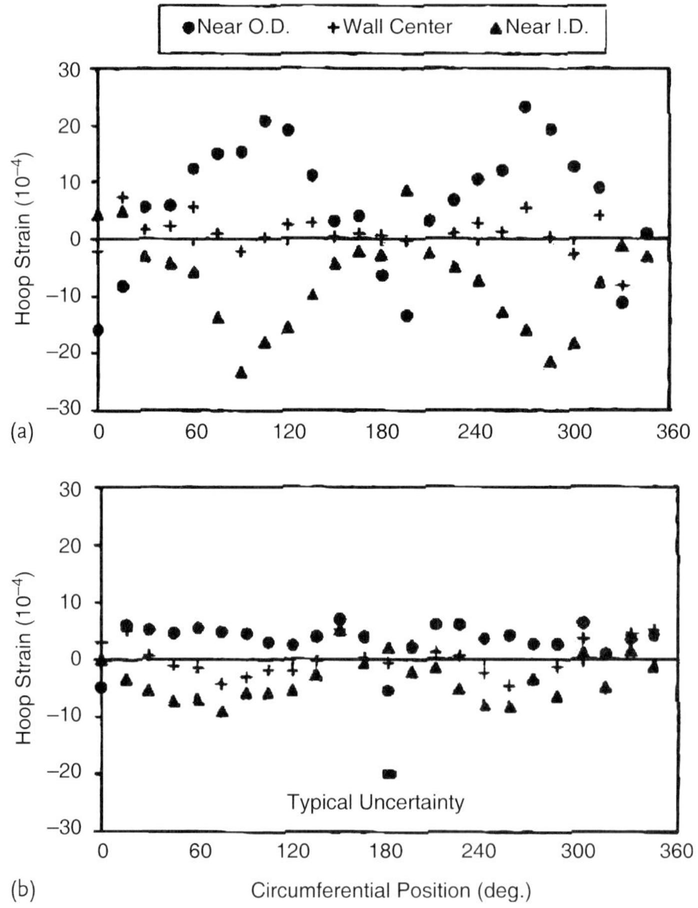

Figure 11.21 Hoop component of strain versus circumference position in the twist tube for three through-wall positions in the (a) as-manufactured and (b) annealed conditions (Source Rogge *et al.* [515]).

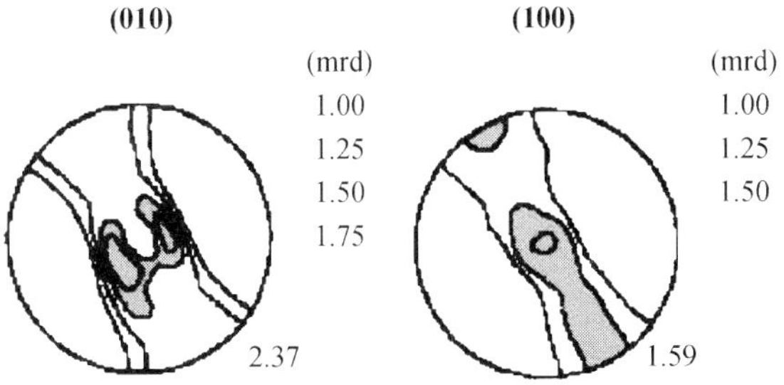

Figure 11.22 Pole figures for two reflections of the mineral plagioclase. The numbers on the right of each figure the show values of the density levels. Unity corresponds to the density of an isotropic distribution (1 mrd is multiple of random distribution). The lowest numbers denote the maximum density.

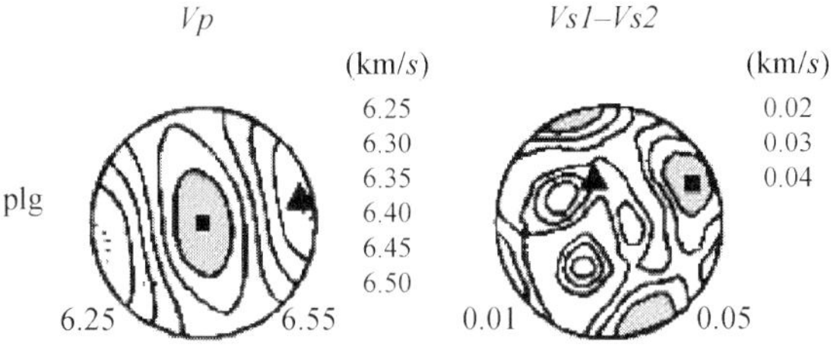

Figure 11.23 The model spatial distributions of velocities Vp and Vs1-Vs2 of the mineral plagioclase. The velocity diagrams are aligned with the system of coordinates of the sample.

After measurements of several pole figures have been performed, it is possible to find the orientation distribution function, and with it one can calculate the elastic properties of the sample. Calculations show that sound speeds are different in different directions. In Figure 11.23 the speeds of the longitudinal mode, Vp and speed difference of two shear modes Vs1-Vs2 are shown by level lines in the polar coordinates similar to those of the pole figures in Figure 11.22. The numbers on the right show the speed on the level lines. The two lowest numbers show the maximal and minimal speeds.

The calculated speeds were checked in an experiment. The results of the experiment showed that there is a disagreement with theory on the level of several percent. The difference between the theoretical and the experimental anisotropy of the sound speeds in some cases reaches a factor of three. The reasons for such a difference are discussed in [516], but that is another story.

11.5.3 Small-Angle Neutron Scattering

Small-angle scattering will be in considered detail in Chapter 12 in applications to biology. Here we only present an example of an application of SANS to investigate defects in materials.

An investigation of chromium electroplating with and without ultrasonic agitation applied to a deposition cell was reported in [517]. It is understandable that deposition is always accompanied by the formation of defects because of hydrogen evolution and abnormal growth of columnar Cr grains. Ultrasound agitation was found to improve the stability of the electroplated surface against corrosion, and SANS helps to find out what happens, which is important for the technology.

The experiment was performed at the SANS spectrometer at the HANARO reactor of the Korea Atomic Energy Research Institute. The results of measurements are shown in Figure 11.24. It is seen from the data that the number of defects decreases with ultrasonic agitation. The distribution of defect sizes is a maximum at the same point as without ultrasound, but shifts toward larger sizes. It was concluded in [517] that because of ultrasonic agitation the chromium film contains fewer and more isolated defects. This result is a qualitative one, but it is the first step on the way to technology perfection.

11.5.4 Neutron Radiography

Neutron radiography like X-ray radiography is a means to get a visible image of the invisible structure of objects [518, 519]. It is achieved by registration of transmission, $T(\boldsymbol{r}_{\|}) = |t(\boldsymbol{r}_{\|})|^2 \exp[-\mu(\boldsymbol{r}_{\|}) + i\varphi(\boldsymbol{r}_{\|})]$, of the neutrons through the object, where $t(\boldsymbol{r}_{\|})$ is the transmission amplitude and $\boldsymbol{r}_{\|}$ is a coordinate point along a photo plate registering the transmitted neutrons, which we will consider to be perpendicular to the z-axis.

The transmission amplitude can be represented as

$$t(\boldsymbol{r}_{\|}) = \exp[-\mu(\boldsymbol{r}_{\|}) + i\varphi(\boldsymbol{r}_{\|})] , \tag{11.12}$$

where $\mu(\boldsymbol{r}_{\|})$ is the extinction coefficient proportional to the total (extinction) cross section σ_t, which includes scattering and absorption, and $\varphi(\boldsymbol{r}_{\|})$ is the phase. Thus, if the incident neutrons are represented by a plane wave $\exp(ikz)$, then the transmitted wave function with account of Eq. (11.12) becomes

$$\psi(\boldsymbol{r}) = \exp(ikz)\exp[-\mu(\boldsymbol{r}_{\|}) + i\varphi(\boldsymbol{r}_{\|})] . \tag{11.13}$$

If the detector plate, which consists of a scintillator and a photo plate, is placed immediately after the object, then the intensity of the black points on the photo plate after sufficient exposition time will be proportional to $\exp(-2\mu(\boldsymbol{r}_{\|}))$. The larger is the variation of $\mu(\boldsymbol{r}_{\|})$ along $\boldsymbol{r}$, the better is the contrast of the image obtained.

This way of obtaining an image is called absorption or extinction radiography. It is based on the difference of the extinction cross sections of different substances. A lot of examples of the application of this principle can be found on the following Web pages:

`http://www-llb.cea.fr/neutrono/nr1.html`
`http://www.andor.com/learn/applications/?docID=194`

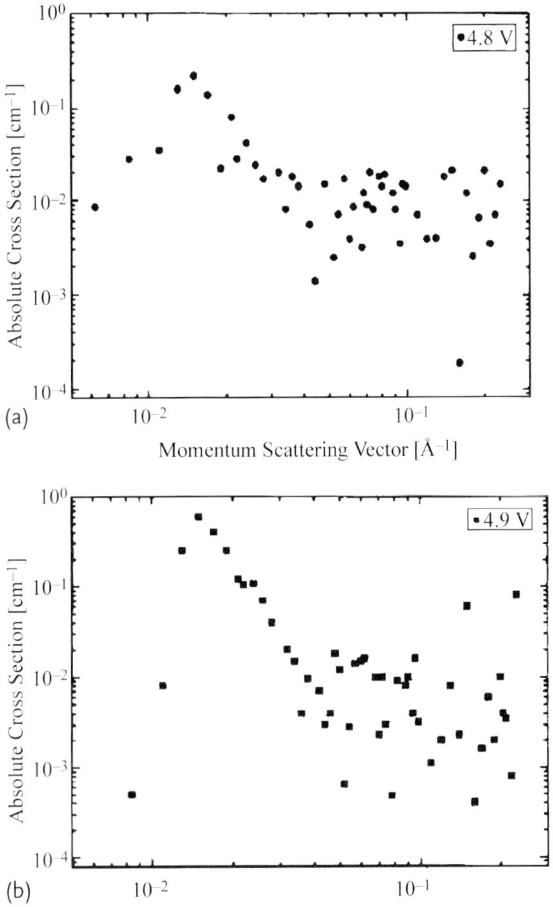

Figure 11.24 Small-angle scattering from a chromium layer (a) with and (b) without ultrasonic agitation.

The exposure time in radiography depends on the intensity of the neutron beam and its spectrum. The last point is important because the absorption cross section increases with increase of wavelength. With a high intensity it is possible to register the dynamics of processes. Measurement of the speed, v, of a Gd ball falling in melted silicate was reported in [520] (see Figure 11.25). This experiment helps to measure the viscosity, η, and the density, ρ_{melt}, of the silicate. It is based on Stokes law:

$$\eta = 2gr^2(\rho_{\mathrm{Gd}} - \rho_{\mathrm{melt}})/9v \,. \tag{11.14}$$

This law is usually corrected by a factor which depends on the radius of the vessel. To check this correction, the cylindrical vessel in the experiment had two diameters as shown in Figure 11.25. The measured speed and its change with the

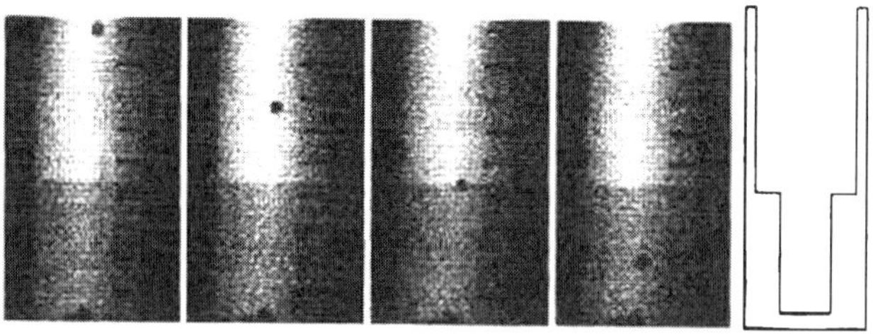

Figure 11.25 Neutron radiographic recording of a falling Gd sphere. The diameter of the sphere was 3.97 mm and the falling time was about 7 min. The inner crucible diameter in the upper part (lighter area) is 39 mm and in the lower part is only 20 mm. The lower part is darker because of an increase of neutron absorption and scattering caused by the thicker crucible walls. Right: cross section of the corresponding one-step crucible used in this experiment [520].

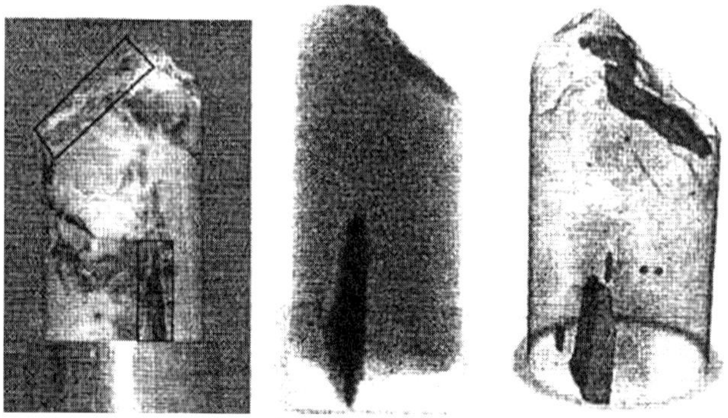

(a) (b) (c)

Figure 11.26 Neutron tomography allows one to visualize individual mineral grains or pores in rocks. (a) photo of a pegmatite sample (35 mm in diameter); (b) neutron radiograph; (c) tomographic image [520].

change of the vessel diameter show that the correction to the Poisson law is less than predicted.

In this experiment the motion was slow. However, it is also possible to investigate the dynamics of fast processes. For instance, neutron radiography of the processes in a working engine was reported in [521]. There the motion was fast but cyclic, and with a stroboscopic technique it was possible at the Institut Laue–Langevin to make a full film of a working engine with a total exposure time of 17 min.

Another example of neutron extinction radiography is tomography [520], which is radiography of rotating samples. The set of radiograms obtained for a sample at different angles allows one to reproduce the three-dimensional image of the elemental structure of the sample. As an example, in Figure 11.26 the structure of a pegmatite rock is shown. This rock contains hydrogen-bearing mica grains, which are well visible in the radiogram and especially well in tomogram.

11.5.5
Phase-Contrast Neutron Radiography

In the case where the extinction part of the transmission amplitude Eq. (11.12) is small or varies too little across $\boldsymbol{r}_{\|}$ (this means that the extinction coefficients of the sample constituents are nearly equal), the phase φ can be used for visualization of the internal structure. The phase φ is due to the real part of the refractive index, which can vary greatly for different substances.

Consider an experiment whose scheme is shown in Figure 11.27. The flux of well-collimated and monochromatized neutrons goes along the z-axis through the sample (screen 1) to a position-sensitive detector (screen 2). The transmission amplitude of the sample is $t(\boldsymbol{r}) = \exp[i\varphi(\boldsymbol{r}_{\|})]$, where vectors with index $\|$ are parallel to the sample plane. We suppose that the z-axis is parallel to the sample plane normal, so the wave function of the neutrons before the sample is the plane wave $\exp(ikz)$.

We can also consider an experiment whose scheme is shown in Figure 11.28. In this case we have a point source of monochromatic neutrons. Their wave function before screen 1 is a spherical wave:

$$\psi(\boldsymbol{r}) = \frac{\exp(ikr)}{r}, \tag{11.15}$$

where radius $r = \sqrt{\boldsymbol{r}_{\|}^2 + (z+\zeta)^2}$ has its origin at the point $z = -\zeta$. With a point source besides the variation of the neutron intensity on screen 2, which corresponds to the variation of transmission phase, can also expect magnification of the image on screen 2.

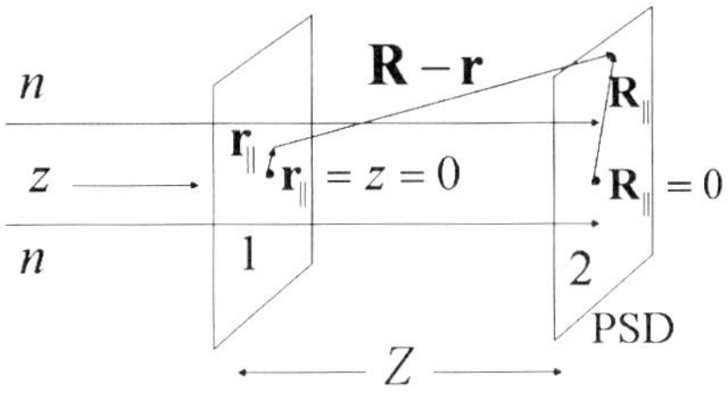

Figure 11.27 The experiment on the detection of phase contrast. PSD position-sensitive detector.

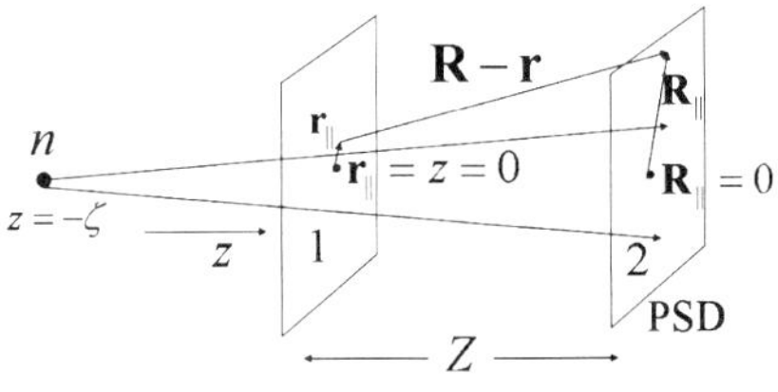

Figure 11.28 The experiment on the detection of phase contrast with a point neutron source.

Below we first present the theory of the transformation of the phase to intensity (phase contrast) for both types of neutron sources: the ideal plane wave and the point source. Though the theory already exists (see, e. g., [522]), we propose here our own original approach. After that some examples of the application of phase contrast to neutron radiography will be given.

11.5.5.1 The Theory of Phase Contrast

If the wave function on the exit plane of screen 1 is equal to $\psi(\boldsymbol{r})$, then at screen 2 after the sample the wave function, according to the Kirchhoff theorem, is representable as

$$\Psi(\boldsymbol{R}) = \frac{1}{4\pi}\int d^3r\left[\left(\psi(\boldsymbol{r})\overrightarrow{\frac{d}{dz}}\frac{\exp(ik|\boldsymbol{R}-\boldsymbol{r}|)}{|\boldsymbol{R}-\boldsymbol{r}|}\right)\right.$$
$$\left.-\left(\psi(\boldsymbol{r})\overleftarrow{\frac{d}{dz}}\frac{\exp(ik|\boldsymbol{R}-\boldsymbol{r}|)}{|\boldsymbol{R}-\boldsymbol{r}|}\right)\right]\delta(z)\,, \tag{11.16}$$

where the arrows show to which function the derivative should be applied, and because of the last δ function the integration in fact goes only over the exit surface of the sample. The spherical wave $\exp(ik|\boldsymbol{R}-\boldsymbol{r}|)/|\boldsymbol{R}-\boldsymbol{r}|$ can be represented by the two-dimensional Fourier integral:

$$\frac{\exp(ik|\boldsymbol{R}-\boldsymbol{r}|)}{|\boldsymbol{R}-\boldsymbol{r}|} = \frac{i}{2\pi}\int\frac{d^2p_{\|}}{p_z}\exp[i\boldsymbol{p}_{\|}\cdot(\boldsymbol{R}_{\|}-\boldsymbol{r}_{\|})+ip_z|Z-z|]\,, \tag{11.17}$$

where $p_z = \sqrt{k^2-p_{\|}^2}$. Equations (11.16) and (11.17) can be used for arbitrary $\psi(\boldsymbol{r})$ and now we will specify it.

11.5.5.2 The Phase Contrast for a Plane Incident Wave

With the plane incident wave $\exp(ikz)$ on the left side of the sample, the wave function on the exit plane ($z = 0$) of screen 1 is $\psi(\boldsymbol{r}) = \exp(ikz)\exp(i\varphi(\boldsymbol{r}_{\|}))$. Substitution of it and Eq. (11.17) into Eq. (11.16) gives

$$\Psi(\boldsymbol{R}) = \int\frac{d^2r_{\|}\,d^2p_{\|}}{4\pi(2\pi)}\frac{k+p_z}{p_z}\exp[i\varphi(\boldsymbol{r}_{\|})]\exp[i\boldsymbol{p}_{\|}\cdot(\boldsymbol{R}-\boldsymbol{r})_{\|}+ip_zZ]\,. \tag{11.18}$$

It is easy to check this expression: if the transmission of the sample is unity, that is, $\varphi = 0$, then after integration over $d^2r_{\|}$ we obtain under the integral $(2\pi)^2\delta(\boldsymbol{p}_{\|})$, and after integration over $d^2p_{\|}$ we obtain

$$\Psi(\boldsymbol{R}) = \exp(ikZ)\,, \tag{11.19}$$

which can be expected.

Periodic phase change Let us suppose that $\varphi(\boldsymbol{r}_{\|}) = a\sin(\boldsymbol{q}_{\|}\boldsymbol{r}_{\|})$ and $a \ll 1$. Then in a linear approximation $\exp(i\varphi(\boldsymbol{r}_{\|})) \approx 1 + ia\sin(\boldsymbol{q}_{\|}\boldsymbol{r}_{\|})$. After substitution of this approximation into Eq. (11.18) and integration first over $d^2r_{\|}$ and then over $d^2p_{\|}$, we obtain $\Psi(\boldsymbol{R})$ in the form

$$\Psi(\boldsymbol{R}) = \exp(ikZ) + ia\frac{k+q_z}{2q_z}\sin(\boldsymbol{q}_{\|}\boldsymbol{R}_{\|})\exp(iq_zZ)\,, \tag{11.20}$$

where $q_z = \sqrt{k^2 - q_{\|}^2}$. With such a wave function, the intensity on screen 2 is

$$|\Psi(\boldsymbol{R})|^2 = 1 + a\frac{k+q_z}{q_z}\sin(\boldsymbol{q}_{\|}\boldsymbol{R}_{\|})\sin[(k-q_z)Z] + a^2\left(\frac{k+q_z}{2q_z}\right)^2\sin^2(\boldsymbol{q}_{\|}\boldsymbol{R}_{\|})\,. \tag{11.21}$$

The most important term is the second one. It shows that the phase variation of the transmission amplitude at screen 1 is transformed into the amplitude variation at screen 2.

If $q \ll k$, then $\sin((k-q_z)Z) - \sin(q_{\|}^2 Z/2k)$, and at the distances $Z_n = (2n+1)\pi k/q_{\|}^2$, where n is an integer, we can see a variation of the intensity with amplitude $\sim 2a$ even with a moderate resolution. When $a \to 0$ or $q \to 0$, the image decays. It is interesting to check in the experiment that shifting of screen 2 by the distance $\Delta Z = 2\pi k/q_{\|}^2$ is accompanied by a change of the phase of the image by π.

The above approach can in fact be applied to an arbitrary transmission amplitude, because every function $t(\boldsymbol{r}_{\|})$ with a compact support can be expanded in a Fourier integral:

$$t(\boldsymbol{r}_{\|}) = \int d^2q_{\|}\,\tau(\boldsymbol{q}_{\|})\exp(i\boldsymbol{q}_{\|}\boldsymbol{r}_{\|})\,.$$

According to Eq. (11.20), we obtain for every Fourier component

$$\Psi_q(\boldsymbol{R}) = \tau(\boldsymbol{q}_{\|})\frac{k+q_z}{2q_z}\exp(i\boldsymbol{q}_{\|}\boldsymbol{R}_{\|})\exp(iq_zZ)\,, \tag{11.22}$$

where $q_z = \sqrt{k^2 - q_{\|}^2}$. Therefore, the total wave function is

$$\Psi(\boldsymbol{R}) = \int d^2q\,\Psi_q(\boldsymbol{R}) = \int \frac{k+q_z}{2q_z}d^2q\,\tau(\boldsymbol{q}_{\|})\exp(i\boldsymbol{q}_{\|}\boldsymbol{R}_{\|})\exp(iq_zZ)\,. \tag{11.23}$$

The next step depends on the form of amplitudes $\tau(\boldsymbol{q}_{\|})$, and we do not proceed further.

Parabolic phase change Let us suppose now that $\varphi(\boldsymbol{r}_{\|}) = \boldsymbol{r}_{\|}^2/2a^2$. Substitution into Eq. (11.18) and integration over $d^2r_{\|}$ gives

$$\Psi(\boldsymbol{R}) = \int \frac{a^2d^2p_{\|}}{-2\pi i}\frac{k+p_z}{2p_z}\exp\left(-ia^2\boldsymbol{p}_{\|}^2/2 + i\boldsymbol{p}_{\|}\boldsymbol{R}_{\|} + ip_zZ\right)\,, \tag{11.24}$$

where $p_z = \sqrt{k^2 - \boldsymbol{p}_\|^2}$. For sufficiently large a we can approximate $p_z = k - \boldsymbol{p}_\|^2/2k$. Then the integral becomes

$$\Psi(\boldsymbol{R}) = \int \frac{a^2 d^2 p_\|}{-2\pi i}\left(1 + \frac{p_\|^2}{4k^2}\right)\exp\left[-i\frac{(a^2k + Z)\boldsymbol{p}_\|^2}{2k} + i\boldsymbol{p}_\|\boldsymbol{R}_\| + ikZ\right]. \tag{11.25}$$

The stationary point in the exponent is $\boldsymbol{p}_\| = k\boldsymbol{R}_\|/(ka^2 + Z)$, and integration over $d^2 p_\|$ gives

$$\Psi(\boldsymbol{R}) = \frac{ka^2}{ka^2 + Z}\left[1 + \frac{\boldsymbol{R}_\|^2}{4(ka^2 + Z)^2}\right]\exp\left[i\frac{k\boldsymbol{R}_\|^2}{2(ka^2 + Z)} + ikZ\right]. \tag{11.26}$$

Therefore, the intensity on screen 2 is

$$|\Psi(\boldsymbol{R})|^2 = \frac{ka^2}{ka^2 + Z}\left[1 + \frac{\boldsymbol{R}_\|^2}{4(ka^2 + Z)^2}\right]^2 \approx \frac{ka^2}{Z}\left(1 + \frac{\boldsymbol{R}_\|^2}{2Z^2}\right), \tag{11.27}$$

where in the last expression we supposed that $Z \gg a^2k$. When $a \to \infty$, then $|\Psi(\boldsymbol{R})|^2 \to 1$.

11.5.5.3 The Phase Contrast for a Point Monochromatic Source

For the incident spherical wave (11.15) emitted from the point $z = -\zeta$ we also use a two-dimensional Fourier representation:

$$\psi_0(\boldsymbol{r}) = \frac{\exp(ikr)}{r} = \frac{i}{2\pi}\int \frac{d^2 p'_\|}{p'_z}\exp\left[i\boldsymbol{p}'_\|\boldsymbol{r}_\| + ip'_z(z + \zeta)\right], \tag{11.28}$$

where $p'_z = \sqrt{k^2 - \boldsymbol{p}'^2_\|}$. After substitution of $\psi(\boldsymbol{r}) = \psi_0(\boldsymbol{r})\exp[i\varphi(\boldsymbol{r}_\|)]$ into Eq. (11.16) we obtain

$$\Psi(\boldsymbol{R}) = -i\int \frac{d^2 r_\| d^2 p_\| d^2 p'_\|}{4\pi(2\pi)^2}\frac{p'_z + p_z}{p_z p'_z}\exp[i\varphi(\boldsymbol{r}_\|)] \times \exp\left[i\boldsymbol{p}_\| \cdot (\boldsymbol{R} - \boldsymbol{r})_\| + i\boldsymbol{p}'_\| \cdot \boldsymbol{r}_\| + ip_zZ + ip'_z\zeta\right]. \tag{11.29}$$

In the case $\exp(i\varphi(\boldsymbol{r}_\|)) = \exp(i\boldsymbol{q}_\|\boldsymbol{r}_\|)$ we obtain

$$\Psi(\boldsymbol{R}) = -i\int \frac{d^2 p_\|}{4\pi}\frac{p''_z + p_z}{p_z p''_z}\exp\left(i\boldsymbol{p}_\| \cdot \boldsymbol{R}_\| + ip_zZ + ip''_z\zeta\right), \tag{11.30}$$

where $p''_z = \sqrt{k^2 - (\boldsymbol{p}_\| - \boldsymbol{q}_\|)^2}$. We suppose that Z and ζ are large and use expansions $p_z = k - p^2/2k$ and $p'_z = k - (\boldsymbol{p} - \boldsymbol{q})^2/2k$. The stationary point in the exponent is determined from the equation $k\boldsymbol{R}_\| - \boldsymbol{p}_\| Z - (\boldsymbol{p}_\| - \boldsymbol{q}_\|)\zeta = 0$, from which it follows that $\boldsymbol{p}_{\parallel} = (k\boldsymbol{R}_{\parallel} + \boldsymbol{q}_{\parallel}\zeta)/(Z+\zeta)$. After integration over this stationary point we obtain

$$\Psi(\boldsymbol{R}) = \frac{1}{Z+\zeta}\exp\left\{i\left[\frac{k\boldsymbol{R}_{\parallel}^2}{2(Z+\zeta)} + k(Z+\zeta) - \frac{q_{\parallel}^2\zeta Z}{2k(Z+\zeta)} + \frac{\boldsymbol{R}_{\parallel}\cdot\boldsymbol{q}_{\parallel}\zeta}{Z+\zeta}\right]\right\}$$
$$= \frac{1}{R_\zeta}\exp\left[i\left(kR_\zeta - q_{\parallel}^2 ZM/2k + \boldsymbol{R}_{\parallel}\cdot\boldsymbol{Q}_{\parallel}\right)\right], \tag{11.31}$$

where we denoted $R_\zeta = \sqrt{(Z+\zeta)^2 + \boldsymbol{R}_{\parallel}^2}$, replaced $Z+\zeta$ in the preexponential factor by R_ζ, and introduced $\boldsymbol{Q}_{\parallel} = \boldsymbol{q}_{\parallel}M$ with magnification coefficient $M = \zeta/(\zeta+Z)$. Now it is evident that if $\varphi(\boldsymbol{r}_{\parallel}) = a\sin(\boldsymbol{q}_{\parallel}\cdot\boldsymbol{r}_{\parallel})$ with $a \ll 1$ we obtain at screen 2 the wave function

$$\Psi(\boldsymbol{R}) = \frac{\exp(ikR_\zeta)}{R_\zeta}\left[1 + ia\sin(\boldsymbol{R}_{\parallel}\cdot\boldsymbol{Q}_{\parallel})\exp(iq_{\parallel}^2 ZM/2k)\right], \tag{11.32}$$

and the intensity will be

$$|\Psi(\boldsymbol{R})|^2 \approx \frac{1}{R_\zeta^2}\left[1 + 2a\sin(\boldsymbol{R}_{\parallel}\cdot\boldsymbol{Q}_{\parallel})\sin(q_{\parallel}^2 ZM/2k)\right]. \tag{11.33}$$

We see the periodic change of the intensity on screen 2, but with the period enlarged $(Z+\zeta)/\zeta$ times. The same can be done with parabolic phase, but we leave it as an exercise for the interested reader.

11.5.5.4 Application of Phase-Contrast Neutron Radiography

Application of the phase contrast for crack detection and characterization in fatigued Al alloy specimens has been reported [523]. This alloy is largely used in various structural parts of commercial aircraft. Two specimens (A and B) were previously subjected to a mechanical fatigue test, as shown in Figure 11.29, to induce cracks at the notch. In specimen A the crack reached 1 mm in length, and in specimen B it reached 3 mm in length. A neutron radiography experiment with these specimens was performed at the CONRAD facility, situated at the cold source of the BER-II reactor, at the Hahn–Meitner Institute, Berlin.

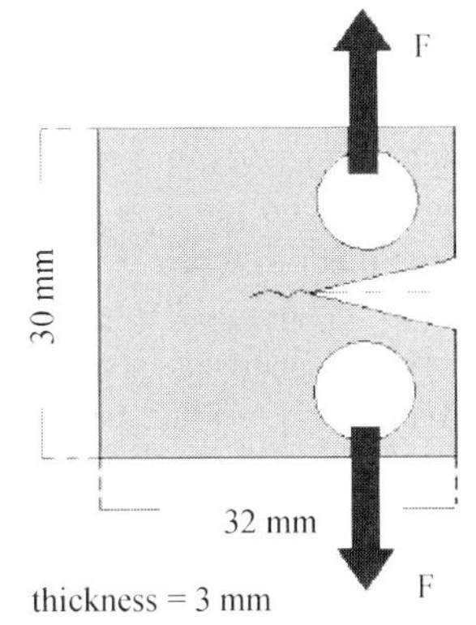

Figure 11.29 Cyclic mechanical fatigue test of specimens.

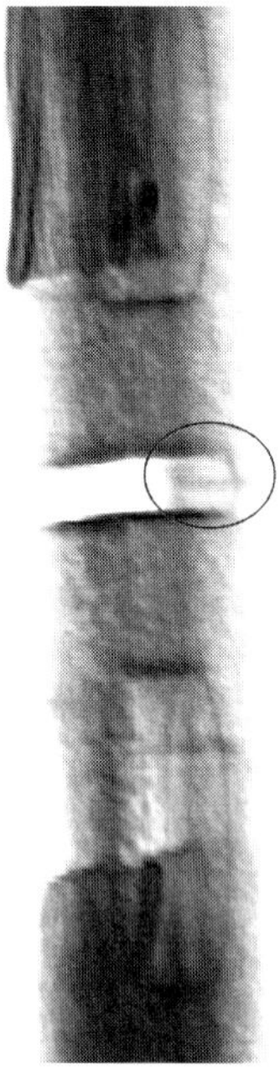

Figure 11.30 Three-dimensional image of specimen B (3-mm crack), showing the crack extension through the whole thickness of the sample.

At this facility with the phase-contrast setup, three-dimensional tomography was also performed. For the tomography the pinhole of the source was 5 mm in diameter, and 300 radiographs were obtained covering 180° with a step of 0.6°. The exposure time for every projection was 120 s, so the complete tomogram was obtained in 10 h. The reconstructed three-dimensional structure is shown in Figure 11.30.

Phase-contrast radiography was performed with a 1-mm pinhole source, $\zeta = 4\,\mathrm{m}$, and $Z = 60\,\mathrm{cm}$. For comparison, an extinction radiography measurement was also performed, with the sample just in front of the detector. The results are shown in Figure 11.31. The estimated resolution in both cases was 113 μm.

11.6
An Example of Novel Applications of Spin Echo Spectrometry – Study on Dairy Products

Neutron spin echo spectrometry has been applied to study many materials, including polymers, glasses, fluids, magnetic systems, microemulsions, and superconductors. The original neutron spin echo method and the extended version, the neutron resonance spin echo method, introduced in Section 4.6 are usually used for studies on dynamical processes in condensed matter. On the other hand, another extension of the spin-echo method, the *spin echo SANS* (SESANS) mentioned in Section 6.3, provides information on the spatial correlations of scattering centers in materials, with the range of length scale accessible lying between 50 nm and 10 μm. There are a number of reports, reviews, and also several instructive books

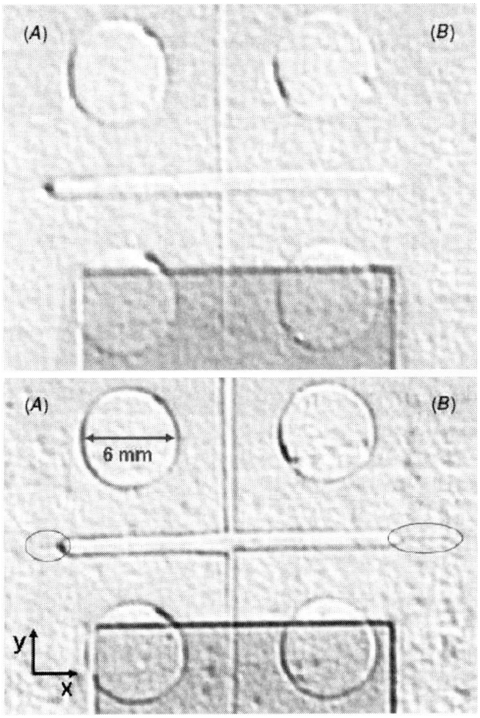

Figure 11.31 Neutron radiographs of specimens A and B. Top: conventional radiography (exposure time 10 s); bottom: phase-contrast radiography (exposure time 3600 s). The cracks are clearly visible.

on the progress and practical results of the applications of neutron spin echo spectrometries. In the present section, an example of novel applications of the neutron spin echo method for structure study with SESANS spectrometry will be briefly introduced.

One of the striking differences between SESANS and conventional SANS is that the results are obtained in real space, and we can determine the structure of materials without collimation of the neutron beam. As explained in Section 6.3, the present method is based on the Larmor precession of polarized neutrons in magnetic fields with inclined faces. The work presented below is an exploration of the possibilities of SESANS to give information on the structure of various dairy foods performed by Tromp and Bouwman [524].

As introduced by Eq. (6.16), the measured polarization is a Fourier transform of the *scattering amplitude density correlation function*. The sensitivity can be tuned by varying the applied magnetic field, the wavelength, the length of the setup, and the *tilt angle of the interfaces*.

An alternative method to explain the technique is to use the splitting of the wave function of the polarized neutrons in the applied magnetic field into two eigenstates, one parallel to the field and one antiparallel as shown in Figure 11.32. A po-

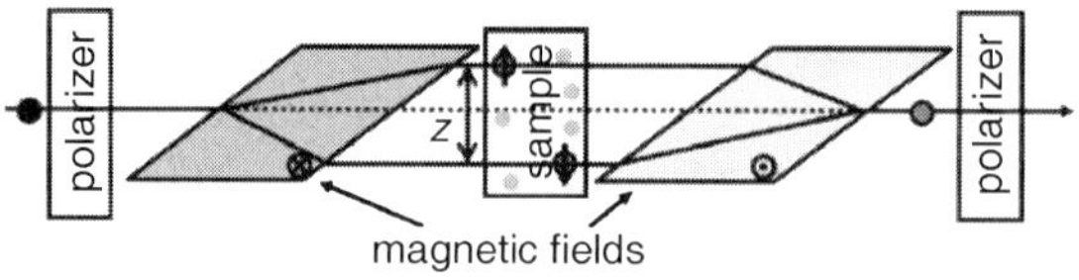

Figure 11.32 The core of the spin-echo SANS (SESANS) setup. The first tilted magnetic field splits the eigenfunction of the neutron into two eigenstates over the spin-echo length z. The second magnetic field focuses the eigenstates into its original wave function (Tromp and Bouwman [524]).

larized neutron beam passes through a tilted magnetic field that is perpendicular to the initial polarization. The wave function of each neutron can be decomposed into its two eigenstates, the up and down states with the spin, respectively, parallel and antiparallel to the magnetic field. The up state will gain kinetic energy in the direction perpendicular to the magnetic field interface and will be refracted away from its initial trajectory. The down state will be refracted in the other direction. This gives a splitting of the eigenstates over a distance identical to the *spin-echo length*. A second magnetic field with an opposite field will focus the eigenstates again to the original wave function with its original polarization. However, if one state goes through material with a density different from that of the other state, then its phase will change and the polarization will decrease. Thus, if the spin-echo length is shorter than the *inhomogeneities* in the sample, the polarization will be one and if it is longer than the inhomogeneities it will decrease. Once the spin-echo length is longer than the maximum size of the inhomogeneities, the polarization will stay at its minimum value, determined by the probability of interaction of the neutrons with the material.

The measured signal, $P(z)/P_0$, is closely related to the *SESANS correlation function* $G(z)$, which in turn is a projection of the *density autocorrelation function* $\gamma(R)$ of the particles [525]:

$$\frac{P(z)}{P_0} = \exp^{[G(z)-G(0)]} , \tag{11.34}$$

where $G(0)$ is the average number of scattering events a neutron experiences upon traversing the sample. The SESANS correlation function

$$G(z) = \int_z^{\infty} dR \frac{\gamma(R)R}{\sqrt{R^2 - z^2}} \tag{11.35}$$

is a projection of the scattering amplitude density distribution function onto a plane and thus contains all the information about the structure in the sample.

SESANS experiments [524] were carried out at Delft University of Technology in the Netherlands. Dairy samples provided by a food research company were used, in which "D_2O milk" was made by suspending 10% (w/w) of fat-free milk powder in D_2O. It turned out that yoghurt cannot be made from milk with D_2O serum, because the culture of lactic acid bacteria does not grow in D_2O. Therefore, the acidification process, which is the essence making yoghurt from milk, was simulated

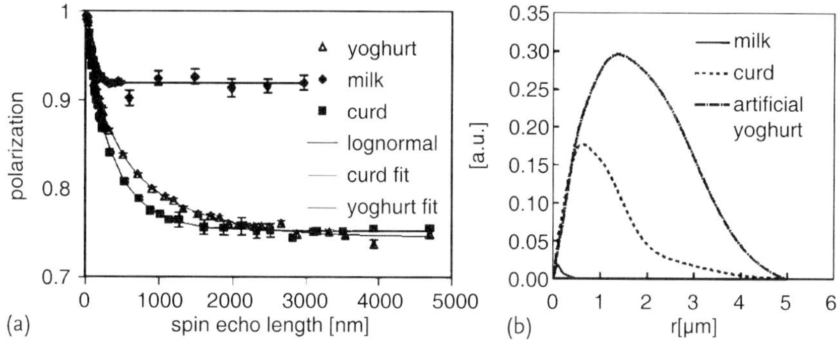

Figure 11.33 Results of SESANS measurements (a) and the derived distributions (b) on dairy products. (a) The line drawn through the data points milk describes the fit with a size distribution, whereas the lines drawn through the data points for yoghurt and curd are the results of fits with the sum of several Gaussian curves for their $G(z)$ function, giving a smooth curve. (b) Radial distribution functions derived from the fits given in (a), taking a model for the SESANS known analytically as the counterpart in the radial distribution function (Tromp and Bouwman [524]).

by adding 2.5% glucono-δ-lactone. Glucono-δ-lactone slowly, over about 4 h, hydrolyzes in water and D_2O, lowering the pH. Curdling was set off by adding 2.3 mg of a 10% chymosin solution to 100 ml of milk. Chymosin does not contribute to the SESANS signal.

The experimental results are shown in Figure 11.33a. The initial slopes for all three samples are identical, which one would expect since the compositions are the same. However, whereas the polarization of the D_2O milk decreases until a spin echo length of $z = 300$ nm is reached, the signal of the yoghurt continues to decrease up to a spin echo length of 3 μm, from which one can conclude that the micelles have aggregated to form larger structures than the original size of the casein micelles. The curd has a structure that also extends to 3 μm, but the decay at short length scales is faster, which reduces to the structure characterized by the distribution functions for curd and yoghurt as described below. This quantitative information on the bulk of yoghurt and cheese is unique and cannot be obtained by any other method.

They transformed the SESANS measurements to the scattering amplitude density correlation function by making use of the fact that for a Gaussian-shaped SESANS correlation function $G(Z)$ its counterpart as the density correlation function $\gamma(r)$ is again a Gaussian with the same width, but with an amplitude divided by this width. They fitted the measurements as shown in Figure 11.33a by taking for $G(Z)$ the sum of five Gaussians with logarithmically separated widths. From the amplitudes of the Gaussians the density correlation function was calculated, as shown in Figure 11.33b. The difference between the density correlation functions of milk, on the one hand, and curd and yoghurt, on the other hand, is the result of the reduction of the solubility of casein micelles. The curding and acidification processes weaken the steric stabilization of the micelles, resulting in an attractive interaction dominating. Both processes appear to form similar structures of casein micelles, considering the small difference between these correlation functions. This is in contrast with the different macroscopic and sensory properties of yoghurt and curd.

This novel application of SESANS measurements revealed a wide use capability and characteristic features of neutron spin echo spectrometry in spatial correlation function studies on various kinds of materials.

12
Application of Neutron Optics in Biophysical/Biological Research

First, we mention some theoretical viewpoints on neutron optics in biology. In fact, all the theoretical considerations are equally well applicable also to X-rays. Therefore, though we will speak about neutrons, some references will be given to the results obtained with X-rays. There are many excellent reviews on this topic (see, e. g., [526, 527] and textbooks [378]) and we will follow their lines.

The main data obtained in the majority of small angle scattering experiments are exemplified by the scattering curve presented in Figure 12.1, which describes the dependence of the scattering cross section (or intensity I) on the momentum transferred, which we will call the scattering vector $q = |\boldsymbol{k}_0 - \boldsymbol{k}| = 4\pi \sin\theta/\lambda$. Here $\boldsymbol{k}_0$ and $\boldsymbol{k}$ are wave vectors of the incident and scattered particles (we will deal only with elastic scattering, therefore $k_0 = k = 2\pi/\lambda$), and the scattering angle is 2θ.

The main goal of neutron biology is the measurement of the structure of biological objects by scattering methods. Since we limit ourselves only to elastic scattering, we have to consider the application to biology of methods such as reflectometry, crystallography, reflectometry, and small-angle scattering.

Crystallography is a high-resolution spectrometry, where the structure is determined with a precision of 1–2 Å, that is, on the atomic level. We will consider it in a later section of this chapter. The main problem is how to grow biological crystals of sufficient size above 1 μm; this is a problem of biology and chemistry, but not of physics. There is also a phase problem: how to determine unambiguously the atomic positions from the measured modulus of the structure factor $|F(q)|$, but its solution is well known in crystallography. We refer readers who wish to learn more about this to relevant textbooks [349–351].

Here we will consider spectrometry with low resolution of the order 10 Å, which permits us to determine the positions of small constituents in a large biological molecule. With some exceptions, we will not use biological terminology, and will call these constituents particles and the large biological molecule a construction. Usually the structure of constructions is determined by (X-ray or neutron) scattering from biological material dissolved in water. One feature of neutron scattering is the strong dependence of the scattering amplitude on the spin state of atomic nuclei and the isotope composition of the matter. It gives us an opportunity to improve the structure determination of a construction by varying the scattering of its particles by a contrast-variation method. This is method we will consider first.

Neutron Optics. Masahiko Utsuro, Vladimir K. Ignatovich

ISBN: 978-3-527-40885-6

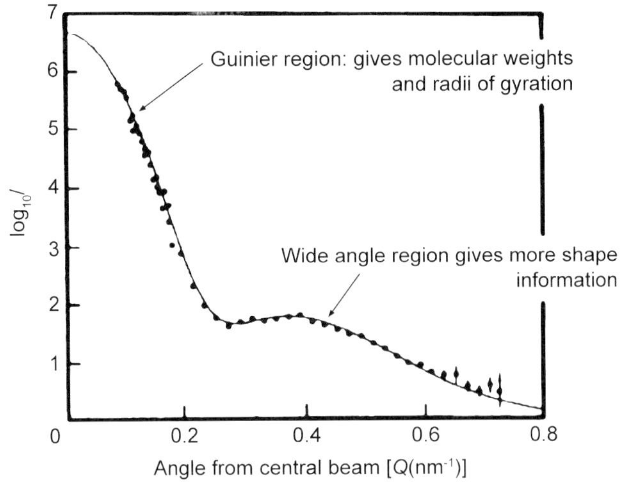

Figure 12.1 General features of a solution scattering curve. (Here the scattering vector is denoted Q. See [526].)

Further, in several experimental chapters up to now in this book, we have studied a variety of neutron optical principles and experimental approaches to investigate physical phenomena in nature and in matter. Therefore, in the later sections in this chapter, some examples of the application of such methods of the neutron optics to biophysical and biological research will be introduced. Biology as well as biophysics are rapidly growing important fields which deserve great effort to find new applications of advanced technologies such as neutron optics. The sample structures and dynamical properties to be studied are very complicated and in most cases require a well-considered combination of multidisciplinary approaches. The future scopes of such developments will be very encouraging since several high-intensity modern neutron sources and instrument facilities, such as SNS in the United States and J-PARC in Japan, are just starting operation, and the related user communities are also growing (as their base they have 50 years of neutron science history). In the later sections in this chapter, some typical examples such as the elements available for such wide and complicated applications will be described rather briefly.

We refer readers wanting more details on neutron scattering in biology or on macromolecular materials to a few textbooks [528–531].

12.1 Contrast Variation

Contrast variation is a method of change scattering intensity of different parts of an investigated object. It can be achieved by isotope substitution or by polarization of different atoms and the use of polarized neutrons. We consider both methods.

12.1.1 Isotope Substitution

Neutron scattering on a single nucleus firmly attached to point $\boldsymbol{r} = 0$ in a vacuum (imagine it to be very heavy) is described by the wave function

$$\psi(\boldsymbol{r}) = \exp(i\boldsymbol{k}_0 \cdot \boldsymbol{r}) - \frac{b}{r}\exp(ikr) , \tag{12.1}$$

where $\boldsymbol{k}_0$ is the incident neutron wave vector, $k = |\boldsymbol{k}_0|$, and b is the neutron–nucleus coherent scattering amplitude.

If the nucleus attached to point $\boldsymbol{r} = 0$ is surrounded by some medium (imagine water), then the wave function becomes

$$\psi(\boldsymbol{r}) = \exp(i\boldsymbol{k}'_0 \cdot \boldsymbol{r}) - \frac{b - b_m}{r}\exp(ik'r) , \tag{12.2}$$

where b_m is the coherent scattering amplitude of the atoms of the medium, $k' = \sqrt{k_0^2 - 4\pi N_0 b_m}$ is the neutron wave number in the medium, and N_0 is the atomic density of the medium. Usually $4\pi N_0 b_m \ll k_0^2$; therefore, we can and will replace k' by k. From Eq. (12.2) it follows that at $b_m = b$ the medium hides the atom, that is, the atom does not scatter neutrons any more. The scattering takes place only if there is a contrast (difference) between the scattering amplitudes of the atom and the medium.

If we have two atoms at points $\boldsymbol{r}_{1,2}$ with different scattering amplitudes $b_{1,2}$, the neutron wave function without rescattering between atoms becomes

$$\begin{aligned}\psi(\boldsymbol{r}) = \exp(i\boldsymbol{k}_0 \cdot \boldsymbol{r}) &- \exp(i\boldsymbol{k}_0 \cdot \boldsymbol{r}_1)\frac{b_1 - b_m}{|\boldsymbol{r} - \boldsymbol{r}_1|}\exp(ik|\boldsymbol{r} - \boldsymbol{r}_1|) \\ &- \exp(i\boldsymbol{k}_0 \cdot \boldsymbol{r}_2)\frac{b_2 - b_m}{|\boldsymbol{r} - \boldsymbol{r}_2|}\exp(ik|\boldsymbol{r} - \boldsymbol{r}_2|) .\end{aligned} \tag{12.3}$$

If we can vary b_m, then at $b_m = b_1$ we hide the first scatterer and at $b_m = b_2$ we hide the second scatterer. This result does not change if we take into account rescattering between atoms. However, because in that case the wave function is

$$\begin{aligned}\psi(\boldsymbol{r}) = \exp(i\boldsymbol{k}_0 \cdot \boldsymbol{r}) &- \frac{e^{i\boldsymbol{k}_0\cdot\boldsymbol{r}_1} + \eta e^{i\boldsymbol{k}_0\cdot\boldsymbol{r}_2}(b_2 - b_m)}{1 - \eta^2(b_1 - b_m)(b_2 - b_m)}\frac{b_1 - b_m}{|\boldsymbol{r} - \boldsymbol{r}_1|}\exp(ik|\boldsymbol{r} - \boldsymbol{r}_1|) \\ &- \frac{e^{i\boldsymbol{k}_0\cdot\boldsymbol{r}_2} + \eta e^{i\boldsymbol{k}_0\cdot\boldsymbol{r}_1}(b_1 - b_m)}{1 - \eta^2(b_1 - b_m)(b_2 - b_m)}\frac{b_1 - b_m}{|\boldsymbol{r} - \boldsymbol{r}_2|}\exp(ik|\boldsymbol{r} - \boldsymbol{r}_2|) ,\end{aligned} \tag{12.4}$$

where $\eta = \exp(ikd)/d$ and $d = |\boldsymbol{r}_1 - \boldsymbol{r}_2|$ is the distance between atoms, measurement of the dependence of the scattering cross section on b_m can give information also on the distance d.

According to perturbation theory, a single particle of n atoms in a vacuum scatters neutrons with scattering amplitude

$$f(\boldsymbol{q}) = \sum_{j=1}^{n} b_j \exp(i\boldsymbol{q}\cdot\boldsymbol{r}_j)\,, \tag{12.5}$$

where b_j is the coherent scattering amplitude of an atom j positioned at point $\boldsymbol{r}_j$ about some chosen center $\boldsymbol{r} = 0$ of the particle, and $\boldsymbol{q}$ is the scattering vector equal to the difference $\boldsymbol{k}_0 - \boldsymbol{k}$ of the incident $\boldsymbol{k}_0$ and final $\boldsymbol{k}$ neutron wave vectors. We consider here only elastic scattering; therefore, $k = k_0$ and $q = 2k\sin(\theta/2)$, where θ is the scattering angle, that is, the angle between vectors $\boldsymbol{k}$ and $\boldsymbol{k}_0$.

In the case of sufficiently large particles and small q (so as not to take into account interatomic structure), we can represent Eq. (12.5) in the form of the integral

$$f(\boldsymbol{q}) = \int_V n(\boldsymbol{r})b(\boldsymbol{r})\exp(i\boldsymbol{q}\cdot\boldsymbol{r})d^3r = \int_V \rho(\boldsymbol{r})\exp(i\boldsymbol{q}\cdot\boldsymbol{r})d^3r\,, \tag{12.6}$$

where $n(\boldsymbol{r})$ and $b(\boldsymbol{r})$ are the atomic density and the atomic scattering amplitude at the point $\boldsymbol{r}$ inside the volume V of the particle, and in the last equality we introduced the scattering density $\rho(\boldsymbol{r}) = n(\boldsymbol{r})b(\boldsymbol{r})$.

When the particle is in a solvent, its volume V replaces the same volume of the solvent with uniform scattering density ρ_s. Therefore, in the solvent the scattering amplitude Eq. (12.6) changes to

$$f_{\text{ins}}(\boldsymbol{q}) = \int_V [\rho(\boldsymbol{r}) - \rho_s]\exp(i\boldsymbol{q}\cdot\boldsymbol{r})d^3r\,. \tag{12.7}$$

The scattering density of the particle can be represented as a sum $\rho(\boldsymbol{r}) = \bar{\rho} + \rho_f(\boldsymbol{r})$ of the averaged over volume part $\bar{\rho} = \int_V \rho(\boldsymbol{r})d^3r/V$ and fluctuation $\rho_f(\boldsymbol{r})$, for which $\int_V \rho_f(\boldsymbol{r})d^3r = 0$. Substitution into Eq. (12.7) gives

$$f(\boldsymbol{q}) = f_c(\boldsymbol{q}) + f_f(\boldsymbol{q})\,, \tag{12.8}$$

where

$$f_c(\boldsymbol{q}) = V(\bar{\rho} - \rho_s)A(\boldsymbol{q})\,, \quad A(\boldsymbol{q}) = \int_V \exp(i\boldsymbol{q}\cdot\boldsymbol{r})\frac{d^3r}{V}\,,$$

$$f_f(\boldsymbol{q}) = \int_V \rho_f(\boldsymbol{r})\exp(i\boldsymbol{q}\cdot\boldsymbol{r})d^3r\,. \tag{12.9}$$

Therefore,

$$|f_{\text{ins}}(\boldsymbol{q})|^2 = |V(\bar{\rho} - \rho_s)|^2|A(\boldsymbol{q})|^2 + 2V(\bar{\rho} - \rho_s)B(\boldsymbol{q}) + |f_f(\boldsymbol{q})|^2\,, \tag{12.10}$$

where

$$B(\boldsymbol{q}) = \text{Re}\left(f_f^*(\boldsymbol{q})A(\boldsymbol{q})\right)\,, \tag{12.11}$$

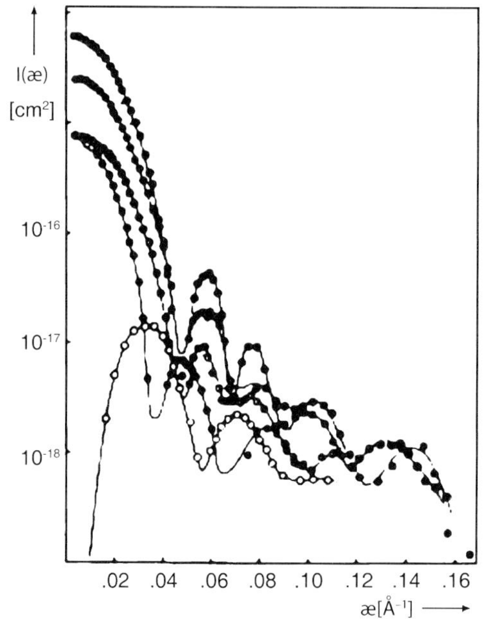

Figure 12.2 Contrast variation of the cross section for low-density lipoprotein with change of D_2O concentration. (See [532]).

and Re $(f(x))$ denotes the real part of the complex function $f(x)$. It is seen that measurements with different ρ_s permit one to find separately $\bar{\rho}$ and $|f_f(q)|^2$.

Usually biological matter is dissolved in a mixture of light and heavy water: $cD_2O + (1-c)H_2O$, where c is the heavy water concentration. The coherent scattering density of such a mixture (in $10^{12}\,\text{cm}^{-2}$) is

$$\rho_s = N_s[(1-c)b_l + cb_h] = -0.56 + 6.96c\,, \tag{12.12}$$

where $b_{l,h}$ are coherent scattering amplitudes of the light and heavy water, respectively. Thus, variation of the contrast is achieved with the help of variation of the D_2O concentration of the solvent. The typical variation with D_2O concentration of the small angle scattering cross section for low-density lipoprotein is shown in Figure 12.2.

There is though some difficulty if deuterium atoms replace hydrogen atoms in the molecule. Then the scattering amplitude of the biological object changes not only because of the contrast variation but also because of partial deuteration of the object.

12.1.2
Spin Contrast Variation

Neutron scattering on a nucleus with spin I is described by two scattering amplitudes: b_+ for the total spin state $I + 1/2$ and b_- for the total spin state $I - 1/2$. In general, we do not know in which state the neutrons are at a given scattering event, so we can only write their scattering amplitude in the form

$$b = b_0 + (\boldsymbol{I} \cdot \boldsymbol{s}) b_1 \,, \tag{12.13}$$

where $b_{0,1}$ are linear combinations of $b_{\pm}$, which are found in the following way. If in the scattering the total spin of the system is $J = I + 1/2$, then because on one side $\boldsymbol{J}^2 = J(J+1) = (I + 1/2)(I + 3/2)$ and on the other side

$$\boldsymbol{J}^2 = (\boldsymbol{I} + \boldsymbol{s})^2 = I(I+1) + 3/4 + 2\boldsymbol{I} \cdot \boldsymbol{s} \,, \tag{12.14}$$

we obtain

$$(I + 1/2)(I + 3/2) = (\boldsymbol{I} + \boldsymbol{s})^2 = I(I+1) + 3/4 + 2\boldsymbol{I} \cdot \boldsymbol{s} \,; \tag{12.15}$$

therefore,

$$\boldsymbol{I} \cdot \boldsymbol{s} = I/2 \,. \tag{12.16}$$

However, if in the scattering the total spin of the system is $J = I - 1/2$, then because on one side $\boldsymbol{J}^2 = J(J+1) = (I + 1/2)(I - 1/2) = I^2 - 1/4$ and on the other side

$$I^2 - 1/4 = I(I+1) + 3/4 + 2\boldsymbol{I} \cdot \boldsymbol{s} \,, \tag{12.17}$$

we obtain

$$\boldsymbol{I} \cdot \boldsymbol{s} = -(I+1)/2 \,. \tag{12.18}$$

From Eqs. (12.16) and (12.18) it follows that

$$b_+ = b_0 + I b_1/2 \,, \quad b_- = b_0 - (I+1) b_1/2 \,; \tag{12.19}$$

therefore,

$$b_1 = \frac{2(b_+ - b_-)}{2I + 1} \,, \quad b_0 = \frac{b_+(I+1) + b_- I}{2I + 1} \,. \tag{12.20}$$

The dot product $\boldsymbol{I} \cdot \boldsymbol{s}$ can be also represented as

$$\boldsymbol{I} \cdot \boldsymbol{s} = \boldsymbol{I}_z s_z + \frac{1}{2}(\boldsymbol{I}_x + i\boldsymbol{I}_y)(s_x - i s_y) + \frac{1}{2}(\boldsymbol{I}_x - i\boldsymbol{I}_y)(s_x + i s_y) \,, \tag{12.21}$$

where the z-axis is directed along the neutron polarization. The scattering amplitude without spin flip is

$$b_{nsf} = b_0 + P_I p_s b_1 \,, \tag{12.22}$$

where P_I is the nucleus polarization and p_s is the neutron polarization. After reversal of the neutron spin polarization we obtain

$$b_{nsf} = b_0 - P_I p_s b_1 \,. \tag{12.23}$$

The difference between the two amplitudes (the spin contrast) is $\Delta b = 2 P_I p_s b_1$.

The effect of spin contrast on the small angle scattering cross section from a solution of the homopolymer polystyrene in orthoterphenyl with 50% deuteration is demonstrated in Figure 12.3.

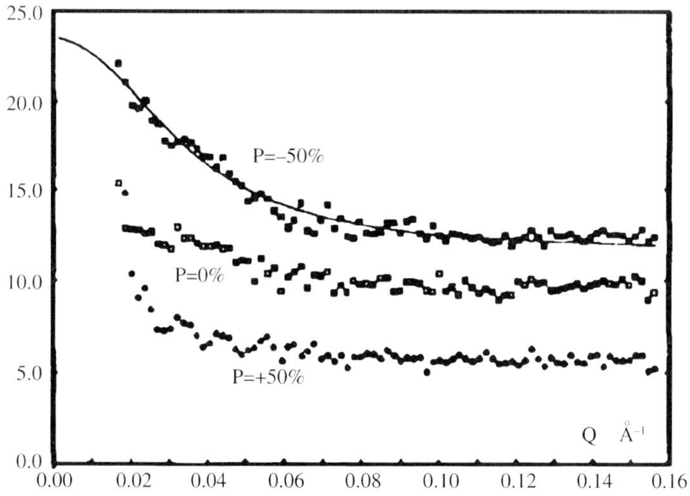

Figure 12.3 Spin contrast variation for small-angle scattering from a polymer in a 50% deuterated solvent. The nuclear polarization of ^{1}H was −50, 0, and +50% along the neutron polarization [533].

12.2
Scattering from a Dilute Ensemble of Monoparticles

First we consider a general problem of neutron scattering from a dilute solution of biomolecules of a single kind. We can write the scattering amplitude of a molecule in the form

$$f(\boldsymbol{q}) = \int_V \rho(\boldsymbol{r}) \exp(i\boldsymbol{q}\cdot\boldsymbol{r}) d^3r\,. \tag{12.24}$$

We can also find the coherent amplitude $f(q) = \overline{f(\boldsymbol{q})}$, which is an average over all directions of the scattering vector $\boldsymbol{q}$. This averaging is achieved not because of variation of the direction of $\boldsymbol{q}$ but because of the isotropic distribution of particle orientations. The change of a particle's orientation changes the direction of $\boldsymbol{q}$ in the particle's reference frame. After averaging, we get

$$f(q) = \int_V \rho(\boldsymbol{r}) \frac{\sin(qr)}{qr} d^3r\,. \tag{12.25}$$

However, we measure the cross section, and must average over all the orientations of the molecules, not the amplitude $f(\boldsymbol{q})$ but $|f(\boldsymbol{q})|^2$. This averaging gives

$$\frac{d\sigma(q)}{d\Omega} \equiv I(q) = \overline{|f(\boldsymbol{q})|^2} = \int_V \rho(\boldsymbol{r}_1) d^3r_1 \int_V \rho(\boldsymbol{r}_2) d^3r_2 \frac{\sin(q|\boldsymbol{r}_1 - \boldsymbol{r}_2|)}{q|\boldsymbol{r}_1 - \boldsymbol{r}_2|}\,, \tag{12.26}$$

which is more complicated than $|f(q)|^2$.

After change of variables $\boldsymbol{r}_2 = \boldsymbol{r} + \boldsymbol{r}_1$ Eq. (12.26) is transformed to

$$I(q) = \int_0^\infty \gamma(r) r^2 dr \frac{\sin(qr)}{qr}, \tag{12.27}$$

where

$$\gamma(r) = \int_V \rho(\boldsymbol{r}_1) d^3 r_1 \int_{4\pi} \rho(\boldsymbol{r} + \boldsymbol{r}_1) d\Omega_r, \tag{12.28}$$

where Ω_r is the solid angle of the vector $\boldsymbol{r}$. The function $\gamma(\boldsymbol{r})$ can be found from $I(q)$ by the reciprocal Fourier transformation. To show it let us represent Eq. (12.27)

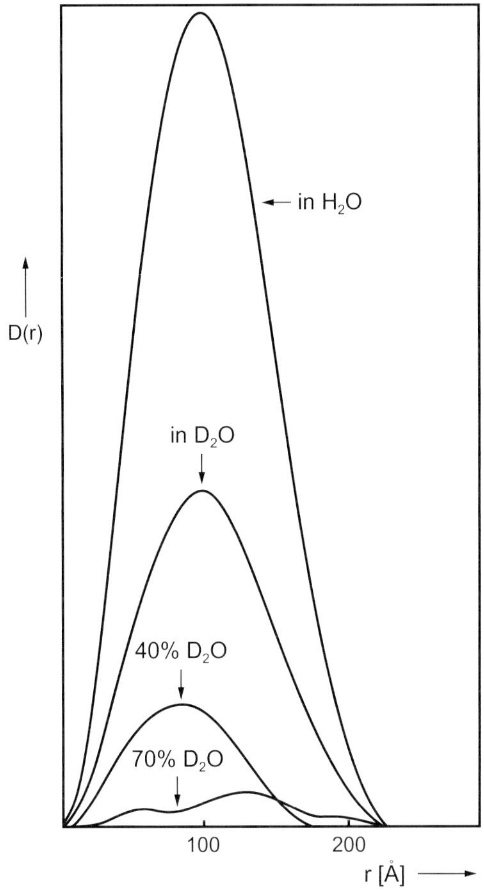

Figure 12.4 The distance distribution $D(r)$ of the 50S subunit in some H_2O–D_2O mixtures. (See [534]).

in the form

$$I(q) = \int_0^\infty \gamma(r')dr'\delta(r' - r'')r''^2 dr'' \frac{\sin(qr'')}{qr''}$$
$$= \int_0^\infty \gamma(r')dr'\delta(r' - r'')\frac{d^3 r''}{4\pi} \exp(\boldsymbol{q} \cdot \boldsymbol{r}'') . \quad (12.29)$$

The reciprocal Fourier transformation gives

$$\int \frac{d^3 q}{(2\pi)^3} I(q) \exp(\boldsymbol{q} \cdot \boldsymbol{r}) = \frac{\gamma(r)}{4\pi} , \quad (12.30)$$

and after integration over $d\Omega_q$ on the left side we obtain

$$\frac{2}{\pi} \int q I(q) dq \sin(qr) = r\gamma(r) . \quad (12.31)$$

The function $D(r) = r \int q I(q) dq \sin(qr) = \pi r^2 \gamma(r)/2$ is called the distribution of distances in the molecule.

As an example we show in Figure 12.4 the function $D(r)$ found from scattering intensities of the 50S subunit of *E. coli* ribosomes at different contrasts.

12.2.1
Properties of $\gamma(r)$

Now we can find some properties of the function $\gamma(\boldsymbol{r})$.

1. According to Eq. (12.28) it consists of the product of two densities shifted by vector $\boldsymbol{r}$. Since the volume of the molecule is finite, the product becomes zero when r is larger than the largest size D of the volume V; therefore, we can put $\gamma(r \geq D) = 0$.
2. At $\boldsymbol{r} = 0$

$$\gamma(0) = 4\pi\overline{\rho^2} V , \quad \text{where} \quad \overline{\rho^2} = \int_V \rho^2(\boldsymbol{r}) \frac{d^3 r}{V} . \quad (12.32)$$

3. Its derivative at $r = 0$ is

$$\gamma'(0) = \left.\frac{d\gamma(r)}{dr}\right|_{r=0} = \int_{4\pi} d\Omega_r \int_V \rho(\boldsymbol{r}_1) d^3 r_1 \frac{d\rho(\boldsymbol{r} + \boldsymbol{r}_1)}{dr}$$
$$= \int_{4\pi} d\Omega_r \int_V \rho(\boldsymbol{r}_1) d^3 r_1 \frac{\boldsymbol{r} \cdot \boldsymbol{\nabla}_1}{r} \rho(\boldsymbol{r}_1) \quad (12.33)$$

because the derivative over r at $\boldsymbol{r} = 0$ is the gradient over $\boldsymbol{r}_1$ along the unit vector $\boldsymbol{r}/r$. The combination $\rho(\boldsymbol{r}_1)(\boldsymbol{r} \cdot \boldsymbol{\nabla}_1)\rho(\boldsymbol{r}_1)$ can be represented as

$\nabla_1 \cdot \boldsymbol{A}(\boldsymbol{r}_1)$, where $\boldsymbol{A}(\boldsymbol{r}_1) = \boldsymbol{r}\rho^2(\boldsymbol{r}_1)/2$, and according to the Gauss theorem the integral over volume is transformed to the surface integral:

$$\gamma'(0) = \int_{4\pi} d\Omega_r \int_V d^3 r_1 \frac{\nabla_1 \cdot \boldsymbol{r}}{2r} \rho^2(\boldsymbol{r}_1) = -\int_{4\pi} d\Omega_r \int_S ds \frac{\boldsymbol{n} \cdot \boldsymbol{r}}{2r} \rho^2(\boldsymbol{r}_s) , \tag{12.34}$$

where $\boldsymbol{n}$ is the vector of the normal, pointing inside the volume V. If $\rho(\boldsymbol{r}_s) = \rho_{ms}$ is the same along the full surface of the molecule, then

$$\gamma'(0) = -\pi \rho_{ms}^2 S , \tag{12.35}$$

where S is the area of the surface. In a solution with scattering density ρ_s of the solvent we have to replace ρ_{ms} by $\rho_{ms} - \rho_s$.

4. With $\gamma(r)$ it is possible to define an average diameter of the molecule

$$l = \frac{2}{\gamma(0)} \int_0^\infty \gamma(r) dr , \tag{12.36}$$

where $\gamma(0)$ according to Eqs. (12.28) and (12.38) is $4\pi\overline{\rho^2}V$.

12.2.2
Properties of $I(q)$

The function $\gamma(r)$ defines the properties of the function Eq. (12.27):

Its asymptotic at large q is $\sim S/q^4$. Integrating Eq. (12.27) over r by parts three times we obtain

$$\begin{aligned} I(q) &= \frac{1}{q^2}[-r\gamma(r)]\cos(qr)\Big|_0^D + \frac{1}{q^3}[r\gamma(r)]' \sin(qr)\Big|_0^D \\ &+ \frac{1}{q^4}[r\gamma(r)]'' \cos(qr)\Big|_0^D - \frac{1}{q^4}\int_0^D [r\gamma(r)]''' dr \cos(qr) \\ &\approx \frac{D\gamma'(D)}{q^3}\sin(qD) + \frac{D\gamma''(D) + 2\gamma'(D)}{q^4}\cos(qD) - \frac{2\gamma'(0)}{q^4} . \end{aligned}$$

If we neglect oscillating terms, we obtain

$$I(q) \approx -\frac{2\gamma'(0)}{q^4} = \frac{2\pi\rho_{ms}^2 S}{q^4} , \tag{12.37}$$

which is the asymptotic we looked for.

Its integral $\int I(q)q^2 dq$ gives Porod invariant $2\pi^2 \overline{\rho^2} V$. Indeed

$$Q = \int I(q)q^2 dq = \int_0^\infty \gamma(r)r^2 dr \frac{\sin(qr)}{qr} q^2 dq$$
$$= \int_0^\infty \gamma(r)r^2 dr \exp(i\boldsymbol{q}\boldsymbol{r}) \frac{d^3q}{4\pi} = 2\pi^2 \int_V \rho^2(\boldsymbol{r}_1) d^3r_1 d^3r \delta(\boldsymbol{r}) = 2\pi^2 \overline{\rho^2} V , \tag{12.38}$$

where we used definition Eq. (12.28) of $\gamma(r)$ and notation Eq. (12.32) for $\overline{\rho^2}$, which for a homogeneous density is equal to ρ_0^2.

The integral $2\int_0^\infty I(q)q dq/\gamma(0)$ gives the average size l of the molecule. Indeed

$$\frac{2}{\gamma(0)} \int_0^\infty I(q)q dq = \frac{2}{\gamma(0)} \int_0^\infty \gamma(r) r dr \int_0^\infty \sin(qr) dq = \frac{2}{\gamma(0)} \int_0^\infty \gamma(r) dr$$
$$= l . \tag{12.39}$$

Integration over q is not well defined; however, according to Eq. (12.37), $I(q)$ at large q decays proportionally to q^{-4} and because of that we omitted $\cos(qr)$ at the upper limit $q = \infty$.

12.2.3
Some Special Forms of Particles

In addition to general forms we can consider three special forms of particles: spherical ones, spikes, and lamellar ones.

12.2.3.1 Spherical Particles: $\rho(\boldsymbol{r}) = \rho(r)$

In spherical particles $I(q)$ does not need averaging over directions of $\boldsymbol{q}$, because integration over the particle density does already make the scattering amplitude independent of directions of $\boldsymbol{q}$. Therefore,

$$I(q) = |f(p)|^2 = \int_V \rho(r_1) d^3r_1 \int_V \rho(r_2) d^3r_2 \exp(q|\boldsymbol{r}_1 - \boldsymbol{r}_2|)$$
$$= \left| 4\pi \int_V \rho(r) \frac{\sin(qr)}{qr} r^2 dr \right|^2 . \tag{12.40}$$

If $\rho(r) = \rho_0$ constant, then

$$f(q) = 4\pi \int_V \rho(r) \frac{\sin(qr)}{qr} r^2 dr = \frac{4\pi\rho_0}{q^3} [\sin(qR) - qR\cos(qR)] , \tag{12.41}$$

where R is the radius of the particle.

12.2.3.2 **Spikes**

These particles like long rods have a large size, L, along the rod axis. If $\pi/L \ll q_{\min}$, the minimal q achievable in an experiment, then $I(q)$ is measured in the asymptotical region. The asymptotic can be found as follows. Equation (12.27) can be put in the form

$$I(q) = \int_0^\infty \rho(\mathbf{r}')d^2r'_\perp dx' \int_0^\infty \rho(\mathbf{r}' + \mathbf{r})d^2r_\perp dx \frac{\sin(qr)}{qr} , \tag{12.42}$$

where coordinate x is along the particle axis, coordinates $\mathbf{r}_\perp$ are perpendicular to it, and $r = \sqrt{x^2 + \mathbf{r}_\perp^2}$. After change of variables $x = Lu$, we can represent $r \approx Lu$. Then we can integrate over u:

$$I(q) = \frac{L}{q} \int_0^\infty d^2r'_\perp \int_0^1 du' \rho(\mathbf{r}') \int_0^\infty d^2r_\perp \int_0^1 du \rho(\mathbf{r}' + \mathbf{r}) \frac{\sin(qLu)}{u} . \tag{12.43}$$

Since $qL \gg 1$, we can extend the integral over u to infinity. If the molecule is homogeneous along the x-axis, then we obtain

$$\int_0^\infty du \frac{\sin(qLu)}{u} = \int_{-\infty}^\infty \frac{du}{2i(u - i\epsilon)} \exp(iqLu) = \pi . \tag{12.44}$$

Therefore,

$$I(q) = \frac{\pi L}{q} S_\perp^2 \bar{\rho}^2 , \tag{12.45}$$

where $S_\perp$ is the rod's cross section and $\bar{\rho}$ is defined in Eq. (12.32).

12.2.3.3 **Lamellar Particles**

If the particles are of platelet type with small thickness and large size L along the platelet surface such that $Lq_{\min} \gg 1$, then we can use the asymptotical form of $I(q)$, which is obtained by integration in Eq. (12.27) over the surface of the platelets. To do that we represent $I(q)$ in the form

$$I(q) = \int_0^\infty d^2r'_\| dx' \rho(\mathbf{r}') \int_0^\infty d^2r_\| dx \rho(\mathbf{r}' + \mathbf{r}) \frac{\sin(qr)}{qr} , \tag{12.46}$$

and consider the integral

$$\int_0^\infty d^2r_\| \frac{\sin(qr)}{qr} = \pi \int_0^\infty dr_\|^2 \frac{\sin(qr)}{qr} = 2\pi \int_x^\infty dr \frac{\sin(qr)}{q} = 2\pi \frac{\cos(qx)}{q^2} , \tag{12.47}$$

where the value of the integral at the upper limit is neglected. Substitution in Eq. (12.46) gives

$$I(q) = \frac{2\pi S_{\|}}{q^2} I_1(q) , \tag{12.48}$$

where

$$I_1(q) = \int \gamma_1(x) \cos(qx) ,$$

$$\gamma_1(x) = \int dx' \int_{S_{\|}} \frac{d^2 r'_{\|}}{S_{\|}} \rho(\boldsymbol{r}')\rho(\boldsymbol{r}' + \boldsymbol{r})_{r_{\|}=0} ,$$

and $S_{\|}$ is the area of the platelets.

12.3 Multipole Expansion of the Scattering Density

The goal of measurements in a scattering experiment is to obtain the shape and structure of the scatterer, that is, the scattering density $\rho(\boldsymbol{r})$. Ideally, to get $\rho(\boldsymbol{r})$ it is sufficient to do the inverse Fourier transform of the scattering amplitude $f(\boldsymbol{q})$

$$\rho(\boldsymbol{r}) = \int \frac{d^3 q}{(2\pi)^3} \exp(-i\boldsymbol{q}\cdot\boldsymbol{r}) f(\boldsymbol{q}) . \tag{12.49}$$

However in real life this is impossible. A direct approach *ab initio* to find $\rho(\boldsymbol{r})$ was proposed in [535, 536]. The idea is to use multipole expansion:

$$\bar{\rho}(\boldsymbol{r}) = \sum_{l=0}^{\infty} \sum_{m=-l}^{l} \bar{\rho}_{lm}(r)\, Y_{lm}(\boldsymbol{\Omega}_r) , \tag{12.50}$$

where $Y_{lm}(\boldsymbol{\Omega})$ are spherical functions of orbital moment l and its projection m, $\boldsymbol{\Omega}_r$ is the solid angle of the vector $\boldsymbol{r}$ in a fixed reference frame of the scatterer, and $\rho_{lm}(r)$ are radial functions

$$\bar{\rho}_{lm}(r) = i^l \int \rho(\boldsymbol{r})\, Y^*_{lm}(\boldsymbol{\Omega}_r) d\Omega_r . \tag{12.51}$$

With such an expansion the scattering amplitude Eq. (12.6) (the derivation is postponed for a while) is representable as

$$f(\boldsymbol{q}) = \sum_{l=0}^{\infty} \sum_{m=-l}^{l} A_{lm}(q)\, Y_{lm}(\boldsymbol{\Omega}_q) , \tag{12.52}$$

where $\boldsymbol{\Omega}_q$ is the solid angle of the scattering vector $\boldsymbol{q}$ in the reference frame of the scatterer, $A_{lm}(q)$ are

$$A_{lm}(q) = \int \bar{\rho}_{lm}(r) j_l(qr) r^2 dr\,, \tag{12.53}$$

and $j_l(qr)$ are spherical Bessel functions. With such an expansion the result of measurement $I(q) \equiv |f(\boldsymbol{q})|^2$ of scattering from many identical isotropically oriented scatterers is representable as

$$I(q) = 4\pi \sum_{l=0}^{\infty} \sum_{m=-l}^{l} |A_{l,m}(q)|^2\,. \tag{12.54}$$

To use this expression one models the scattering density $\rho(\boldsymbol{r})$ with as many ν parameters as needed, calculates $A_{lm}(q)$ with the help of Eqs. (12.53) and (12.51), calculates the theoretical $I_t(q)$ according to Eq. (12.54), and by fitting (minimization of χ^2)

$$\chi^2 = \frac{1}{n-\nu-1} \sum_{j=1}^{N} \frac{|I_t(q_j) - I_e(q_j)|^2}{I_e(q_j)} \tag{12.55}$$

over all N experimentally measured scattering vectors q_j finds the parameters. Of course, in calculations only a few angular momenta l in summation Eq. (12.54) can be used. Moreover, in theoretical data we must include the resolution of the experimental device or deconvolute (desmear) the experimental data from resolution functions.

Figure 12.5 serves as a good illustration of the capability of such an approach. It was obtained with a different experimental technique [537], not with neutrons, however we demonstrate it because the principles for deciphering of a structure for neutron scattering experiments are very similar.

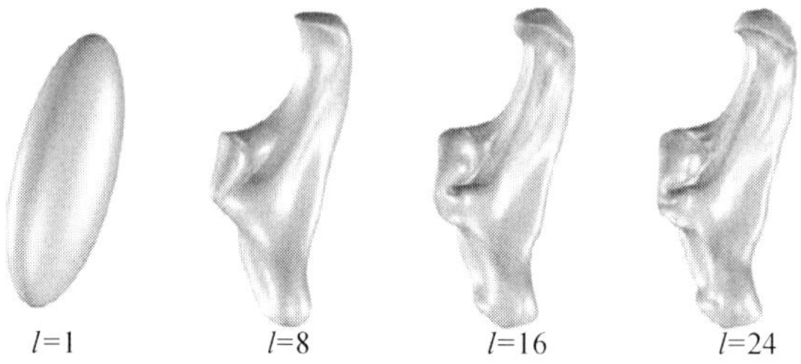

Figure 12.5 Reproduction of surface of a biological object [537] with the help of different number of spherical harmonics.

12.3.1 Derivation of Eq. (12.54)

The exponent $\exp(i\boldsymbol{q}\cdot\boldsymbol{r})$ in

$$f(\boldsymbol{q}) = \int \rho(\boldsymbol{r})\exp(i\boldsymbol{q}\cdot\boldsymbol{r})d^3r \tag{12.56}$$

can be represented as the expansion (textbook [B])

$$\exp(i\boldsymbol{q}\cdot\boldsymbol{r}) = 4\pi\sum_{l=0}^{\infty}\sum_{m=-l}^{l} i^l j_l(qr)Y^*_{lm}(\boldsymbol{\Omega}_r)Y_{lm}(\boldsymbol{\Omega}_q), \tag{12.57}$$

where $\boldsymbol{\Omega}_{q,r}$ are solid angles of vectors $\boldsymbol{q}$ and $\boldsymbol{r}$ with respect to some not yet specified reference frame. Substitution of Eq. (12.57) into Eq. (12.56) gives

$$\begin{aligned} f(\boldsymbol{q}) &= 4\pi\sum_{l=0}^{\infty}\sum_{m=-l}^{l}\int d^3r i^l\rho(\boldsymbol{r})j_l(rq)Y^*_{l,m}(\boldsymbol{\Omega}_r)Y_{l,m}(\boldsymbol{\Omega}_q) \\ &= 4\pi\sum_{l,m}A_{l,m}(q)Y_{l,m}(\boldsymbol{\Omega}_q), \end{aligned} \tag{12.58}$$

where

$$A_{l,m}(q) = i^l\int d^3r\rho(\boldsymbol{r})j_l(rq)Y^*_{l,m}(\boldsymbol{\Omega}_r). \tag{12.59}$$

With amplitude Eq. (12.58) we can find the differential cross section for scattering into a solid angle $d\Omega_k$ defined with respect to the laboratory reference frame:

$$\begin{aligned} d\sigma(q)/d\Omega_k = (4\pi)^2 &\sum_{l=0}^{\infty}\sum_{m=-l}^{l}A^*_{l,m}(q)Y^*_{l,m}(\boldsymbol{\Omega}_q) \\ &\times\sum_{l'=0}^{\infty}\sum_{m'=-l}^{l'}A_{l',m'}(q)Y_{l',m'}(\boldsymbol{\Omega}_q). \end{aligned} \tag{12.60}$$

If we could average Eq. (12.60) over directions of vector $\boldsymbol{q}$ over a 4π solid angle, then, because of orthonormality of functions $Y_{lm}(\Omega)$, we would get

$$d\sigma(q)/d\Omega_k = |f(q)|^2 = 4\pi\sum_{l=0}^{\infty}\sum_{m=-l}^{l}|A_{l,m}(q)|^2. \tag{12.61}$$

We cannot average over $d\boldsymbol{\Omega}_q$, but we have to remember that solid angles $\boldsymbol{\Omega}_{r,q}$ are to be defined with respect to some reference frame. We can define the reference frame with the help of some coordinate axes attached to a scatterer. If the orientations of different (but identical) scatterers have a uniform distribution over all the 4π solid angle, then we should average over their orientations, and this averaging for a fixed direction $\boldsymbol{q}$ is equivalent to averaging over $d\Omega_q$ which leads to Eq. (12.61).

12.3.2
Isometric Particles

Sometimes, when particles have an approximately spherical shape, $\rho(\boldsymbol{r})$ can be represented as

$$\rho(\boldsymbol{r}) = \rho_0(r)\Theta(r < R) + \sum_{l=0}^{L} \sum_{m=-l}^{l} \rho_{lm}\delta(r - R) Y_{lm}(\boldsymbol{\Omega}_r), \tag{12.62}$$

where R is an average radius of the particle, Θ is the step function, and ρ_{lm} are some constant parameters. Then the scattering intensity will be proportional to

$$I(q) = 4\pi|A_{00}|^2 + 4\pi \sum_{l=0}^{L} |a_l|^2 R^4 j_l^2(qR), \tag{12.63}$$

where

$$a_l = \sum_{m=-l}^{l} \rho_{lm}. \tag{12.64}$$

12.4
Small-Angle Scattering

Here we will consider small-angle scattering from particles and their systems in solution. First we introduce the radius of gyration for a simple particle in a solution. After that we define the radius of gyration for a system of two particles separated by a distance D, and finally we consider a system of particles and a triangulation method, which permits us find the position of different particles in a complicate construction.

12.4.1
Guinier Region. Scattering from a Dilute Monoparticle Solution

The most popular method for the description of small-angle scattering on biological structures is based on a simple consideration by Guinier [304] in 1939, that $\exp(i\boldsymbol{qr})$ at small q in the amplitude Eq. (12.56) can be expanded in powers of $\boldsymbol{q}\cdot\boldsymbol{r}$ and averaged over orientations of scatterers, which is equivalent to averaging over directions of the vector $\boldsymbol{q}$:

$$\begin{aligned}\overline{f(\boldsymbol{q})} &= \int \rho(\boldsymbol{r})\overline{\exp(i\boldsymbol{q}\cdot\boldsymbol{r})}d^3r = \int \rho(\boldsymbol{r})[1 + i\overline{\boldsymbol{q}\cdot\boldsymbol{r}} - \overline{(\boldsymbol{q}\cdot\boldsymbol{r})^2}/2 + \cdots]d^3r \\ &\approx f(0)\left[1 - q^2R_G^2/6\right],\end{aligned} \tag{12.65}$$

where

$$f(0) = \int \rho(\mathbf{r}) d^3 r = V\bar{\rho} \,, \tag{12.66}$$

and

$$R_G^2 = \frac{1}{V\bar{\rho}} \int \rho(\mathbf{r}) r^2 d^3 r \tag{12.67}$$

is square of the so-called radius of gyration R_G. The term linear in $\boldsymbol{q} \cdot \boldsymbol{r}$ is zero because of averaging. The same averaging over $\boldsymbol{q}$ gives $\overline{(\boldsymbol{q} \cdot \boldsymbol{r})^2} = q^2 r^2/3$.

The average amplitude Eq. (12.65) is some kind of coherent amplitude which helps us to construct a coherent differential scattering cross section

$$\begin{aligned} d\sigma^c(q)/d\Omega_k &\equiv I(q) = |\overline{f(\boldsymbol{q})}|^2 \\ &\approx |f(0)|^2 [1 - q^2 R_G^2/3] \approx |f(0)|^2 \exp\left(-q^2 R_G^2/3\right) \,. \end{aligned} \tag{12.68}$$

The intensity corresponding to this equation is shown in Figure 12.6, and the dependence of its logarithm on q^2 is shown in Figure 12.7.

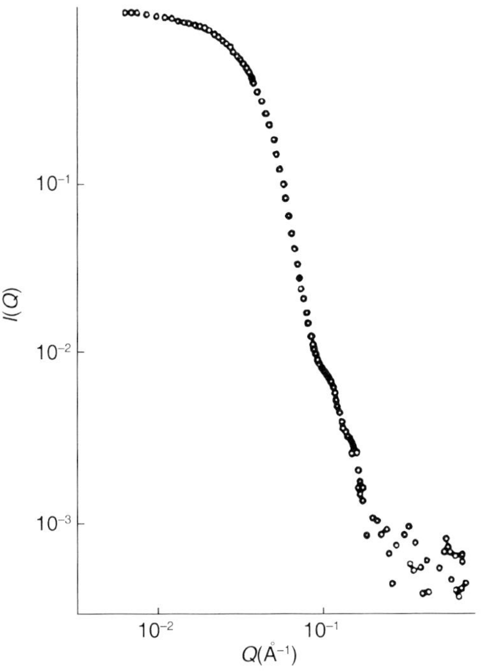

Figure 12.6 Neutron scattering data for the 50S subunit of *E. coli* ribosomes in H_2O at the Guinier region (source [527]).

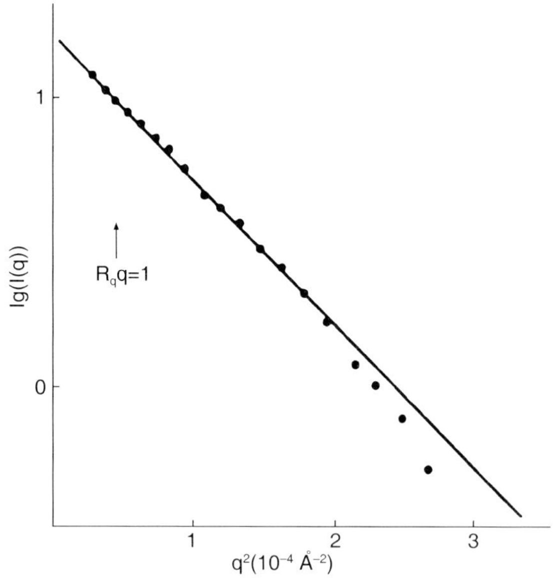

Figure 12.7 Logarithm of the intensity scattered at very small q (Guinier region) by a spherical virus (turnip yellow mosaic virus). According to Eq. (12.68) this gives a linear plot against q^2. The radius of gyration R_G obtained from the slope of the line is 123 Å (source [527]).

However we need the average of not the amplitude but of the scattering cross section. Its average is

$$\overline{d\sigma^c(q)/d\Omega_k} \equiv \bar{I}(q) = \overline{|f(\boldsymbol{q})|^2}\,. \tag{12.69}$$

Substitution of Eq. (12.56) into Eq. (12.69) gives

$$\begin{aligned}\bar{I}(\boldsymbol{q}) &= \int d^3r\rho(\boldsymbol{r})\int d^3r'\rho(\boldsymbol{r}')\overline{\exp(i\boldsymbol{q}\cdot(\boldsymbol{r}-\boldsymbol{r}'))}\\ &\approx I(0) - \frac{1}{6}\int d^3r\rho(\boldsymbol{r})\int d^3r'\rho(\boldsymbol{r}')q^2(\boldsymbol{r}-\boldsymbol{r}')^2\,.\end{aligned} \tag{12.70}$$

This expression is reduced to Eq. (12.68) only if

$$\int d^3r\rho(\boldsymbol{r})\boldsymbol{r} = 0\,. \tag{12.71}$$

This relation defines the center of gyration. Definition (12.67) without Eq. (12.71) is ambiguous. After definition of the center of gyration, Eqs. (12.70) and (12.68) become identical.

If the scattering density $\rho(\boldsymbol{r})$ is represented as the sum $\bar{\rho} + \rho_f(\boldsymbol{r})$ of the homogeneous density, $\bar{\rho} = \int_V \rho(\boldsymbol{r})d^3r/V$, and the fluctuation density, $\rho_f(\boldsymbol{r})$, such that $\int_V \rho_f(\boldsymbol{r})d^3r = 0$, then in a solvent with scattering density ρ_s the result Eq. (12.70) after substitution of $\rho_{\text{ins}}(\boldsymbol{r}) = \bar{\rho} - \rho_s + \rho_f(\boldsymbol{r})$ is

$$\int d^3r[\bar{\rho} - \rho_s + \rho_f(\boldsymbol{r})] \int d^3r'[\bar{\rho} - \rho_s + \rho_f(\boldsymbol{r}')](r^2 + r'^2) = 2I(0)R_G^2 \,. \quad (12.72)$$

$$I(\boldsymbol{q}) = I(0)\exp\left(-q^2 R_G^2/3\right) \,, \quad \text{where} \quad I(0) = V^2(\bar{\rho} - \rho_s)^2 \,. \quad (12.73)$$

The gyration center should be common to the total density $\rho_{\text{ins}}(\boldsymbol{r})$ and therefore it changes with variation of the contrast $\bar{\rho} - \rho_s$. If this change is neglected, then the radius of gyration is given by

$$R_G^2 = R_{Gc}^2 + R_{Gf}^2 \,, \quad (12.74)$$

where

$$R_{Gc}^2 = \int_V r^2 \frac{d^3r}{V} \,,$$

$$R_{Gf}^2 = \frac{1}{\bar{\rho} - \rho_s} \int_V \rho_f(\boldsymbol{r}) r^2 \frac{d^3r}{V} \,.$$

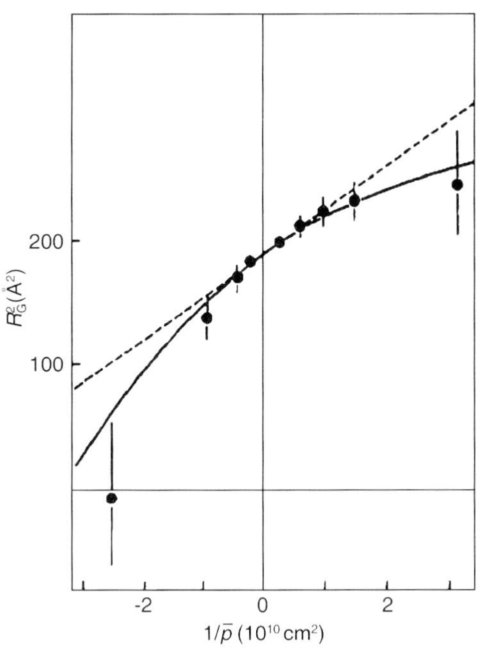

Figure 12.8 The square of the radius of gyration against $l/\bar{\rho}$, where $\bar{\rho}$ is the contrast, which in our notation is $\bar{\rho} - \rho_s$. The sample is hen egg-white lysozyme. The curve shows a quadratic variation following Eq. (12.75). This indicates that the center of gravity of the density fluctuation varies with contrast. The general slope of the curve is due to the higher scattering density of the amino acid at the surface of the protein (source [527]).

After accounting for the change of the gyration center gives [534]

$$R_G^2 = R_{Gc}^2 + \frac{1}{\bar{\rho} - \rho_s} \int_V \rho_f(\boldsymbol{r}) r^2 \frac{d^3 r}{V} + \frac{1}{(\bar{\rho} - \rho_s)^2} \int_V \rho_f(\boldsymbol{r}) \int_V \rho_f(\boldsymbol{r}')(\boldsymbol{r} \cdot \boldsymbol{r}') \frac{d^3 r d^3 r'}{V^2} . \tag{12.75}$$

This dependence is demonstrated in Figure 12.8 borrowed from [527].

12.4.2
Small-Angle Scattering from a System of Two and more Different Particles

The scattering amplitude for a system of two different particles separated by distance D is

$$f(\boldsymbol{q}) = \sum_j b_j e^{i \boldsymbol{q} \boldsymbol{r}_j} = \int_{V_1} d^3 r_1 \rho(\boldsymbol{r}_1) e^{i \boldsymbol{q} \boldsymbol{r}_1} + e^{i \boldsymbol{q} \boldsymbol{r}_{12}} \int_{V_2} d^3 r_2 \rho(\boldsymbol{r}_2) e^{i \boldsymbol{q} \boldsymbol{r}_2} = f_1(\boldsymbol{q}) + \exp(i \boldsymbol{q} \boldsymbol{r}_{12}) f_2(\boldsymbol{q}) , \tag{12.76}$$

and its cross section averaged over all orientations of the system is represented similarly to Eq. (12.26) as

$$I(q) = \overline{|f(\boldsymbol{q})|^2} = \int_{V_1} \rho_1(\boldsymbol{r}_1) d^3 r_1 \int_{V_1} \rho_1(\boldsymbol{r}_1') d^3 r_1' \frac{\sin(q|\boldsymbol{r}_1 - \boldsymbol{r}_1'|)}{q|\boldsymbol{r}_1 - \boldsymbol{r}_1'|} + \int_{V_2} \rho_2(\boldsymbol{r}) d^3 r \int_{V_2} \rho_2(\boldsymbol{r}') d^3 r' \frac{\sin(q|\boldsymbol{r} - \boldsymbol{r}'|)}{q|\boldsymbol{r} - \boldsymbol{r}'|} + 2 \int_{V_1} \rho_1(\boldsymbol{r}) d^3 r \int_{V_2} \rho_2(\boldsymbol{r}') d^3 r' \frac{\sin(q|\boldsymbol{r}' + \boldsymbol{r}_{12} - \boldsymbol{r}|)}{q|\boldsymbol{r}' + \boldsymbol{r}_{12} - \boldsymbol{r}|} . \tag{12.77}$$

It can also be represented similarly to Eq. (12.27) as

$$I(q) = \int_0^\infty [\gamma_1(r) + \gamma_2(r)] r^2 d r \frac{\sin(q r)}{q r} + 2 \gamma_{12}(r) r^2 d r \frac{\sin(q|\boldsymbol{r} + \boldsymbol{r}_{12}|)}{q|\boldsymbol{r} + \boldsymbol{r}_{12}|} , \tag{12.78}$$

where the meaning of $\gamma_{1,2}(r)$ is self-evident, and

$$\gamma_{12}(r) = \int_{V_1} \rho_1(\boldsymbol{r}_1) d^3 r \int_{V_2} \rho_2(\boldsymbol{r}_2 + \boldsymbol{r}) d \Omega_r . \tag{12.79}$$

If we calculate the distribution function of distances $D(r)$: $D(r) = r \int q d q I(q) \sin(q r)$ and calculate $\overline{r^2} = \int D(r) r^2 d r$, we will find

$$\overline{r^2} = R_{G1}^2 + R_{G2}^2 + |\boldsymbol{r}_{12}|^2 . \tag{12.80}$$

Therefore, if we know gyration radiuses of separate particles, we can find the distance between them.

We can extract the interference term and determine the distance r_{12} the other way by labeling the particles by isotope substitution (e.g., by deuteration). Suppose we can prepare first a solution containing the same concentration c of nonlabeled normal molecules and of molecules in which both components are labeled. And the second time we prepare the solution containing concentration c of molecules where only one or the other component is labeled. In the first case we measure scattered intensity $I_{00}(q) + I_{11}(q)$, and in the second case we measure the intensity $I_{01}(q) + I_{10}(q)$. Then the difference $\Delta I = I_{00} + I_{11} - (I_{01} + I_{10})$ gives the contribution to the scattering of the interference term:

$$\Delta I(q) = 2\Delta\gamma_{12}(r)r^2 dr \frac{\sin(q|\boldsymbol{r} + \boldsymbol{r}_{12}|)}{q|\boldsymbol{r} + \boldsymbol{r}_{12}|}, \tag{12.81}$$

where

$$\Delta\gamma_{12}(r) = \int_{V_1} \Delta\rho_1(\boldsymbol{r}_1) d^3 r \int_{V_2} \Delta\rho_2(\boldsymbol{r}_2 + \boldsymbol{r}) d\Omega_r, \tag{12.82}$$

and $\Delta\rho$ denotes the difference of scattering densities of labeled and nonlabeled components. The intensity $\Delta I(q)$ should contain a structure related to the distance r_{12}, and the distance can be extracted from the experimental data.

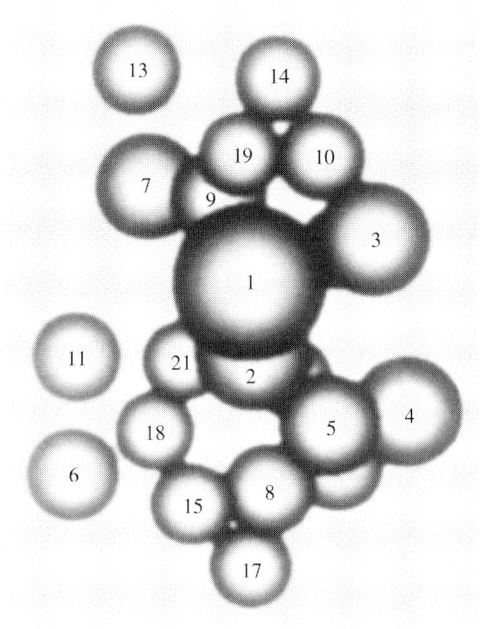

Figure 12.9 The 21-protein neutron map of the 30S ribosomal subunit from *E.coli* [538].

The two particle systems can be straightforwardly generalized to many particle ones. By labeling different particles, we can find distances between them and by geometrical method find full the construction. The task is simplified if the constituents of the construction are considered to be spherical particles with homogeneous density, or parallelepipeds. A result of such calculations is shown in Figure 12.9 .

12.5
Reflectometry on Protein and Biomaterials

As we learned in Chapters 2 and 4, the specular reflectivity for neutrons as a function of the grazing incident angle provides very useful and convenient information because the data can be interpreted in a straightforward manner with respect to the scattering amplitude density profile across the surface and interfaces. The structure of a multilayered film can, in principle, be determined by fitting the observed reflectivity curve with a histogram of scattering amplitude density, thickness, and roughness parameters. The advantages of neutron reflectometry for biophysical and biological studies include the applicability to both solid and liquid substrates, good depth resolution (about 1 nm), visibility of internal interfaces, and wide availability of contrast-matching techniques as explained in Section 12.1.

12.5.1
A New Tool to Study Protein Interactions

The reason why biophysicists are very interested in neutron reflectometry is described, for example, in a resume given by Sackmann [538] at the Technical University of Munich. During the past few years, it has been demonstrated that neutron reflectometry is a new and versatile tool to study:

1. the molecular aspects of the interaction of water-soluble proteins with lipid membranes,
2. protein–protein interaction processes at membranes;
3. the domain formation in two-dimensional lipid alloys. The method can be applied both to half-membranes at the air–water interface and to supported membranes.

The major advantages of neutron scattering techniques compared with X-ray techniques are, first, that they are free of radiation damage and, second, that by application of contrast-matching procedures parts of the stratified films may be highlighted or screened out. This is illustrated in Figure 12.10 for the case of supported membranes.

Thus, the substrate (or the air) and water may be mutually contrast matched by changing the H_2O/D_2O ratio of the subphase, enabling the observation of the adsorbed film alone. The head-group region of a supported membrane can be observed alone by matching the scattering contrast of the hydrocarbon chains to that

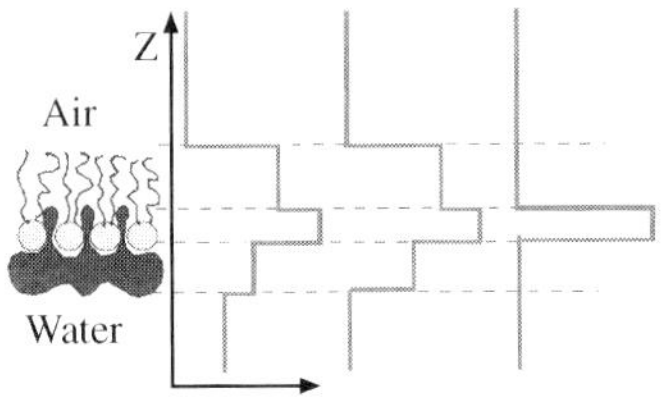

Figure 12.10 Contrast matching of supported membranes. Matching of the contrast of the solid surface and water (by changing the H_2O to D_2O ratio) allows one to observe the bilayer alone. By using mixtures of lipids with deuterated and protonated chains, one can match the contrast of the hydrophobic inner layer of the membrane to both water and the substrate to observe the head-group region of the membrane alone. The structure determination can be finally refined by analyzing intermediate states of contrast (Sackmann [538]).

of the substrate and the aqueous phase by using lipid mixtures with protonated and deuterated hydrocarbon chains.

This trick enables one to measure the partial volume fractions of each component within each plane of complex and soft self-organized interfaces.

12.5.2
Membrane Investigations

A membrane is a lipid bilayer as shown in Figure 12.11. It consists of two layers of lipid molecules with hydrophilic heads outside and hydrophobic tails inside. The total thickness of the membrane, d, is approximately 50 Å, and the total thickness of its internal part, D is approximately 35 Å. Cell membranes have a more complicated structure because they include many different proteins. Certain membranes occur naturally in stacks. We know (see Chapter 4) how to find the reflectivity if $\rho(x)$ (x is the coordinate across the membrane) is known Now we need to solve the inverse problem and find $\rho(x)$ from the reflectivity. For solution of the inverse problem we can choose a model, calculate the reflectivity, $R_c(k)$, and then find the parameters of the model by fitting $R_c(k_i)$ to experimentally measured data $R_e(k_i)$ at all points $k_i = q_i/2$.

For arbitrary $\rho(x)$ reflection amplitude, $r(k)$, where k is the normal component of the incident wave vector, can be found numerically, by a matrix method, or by

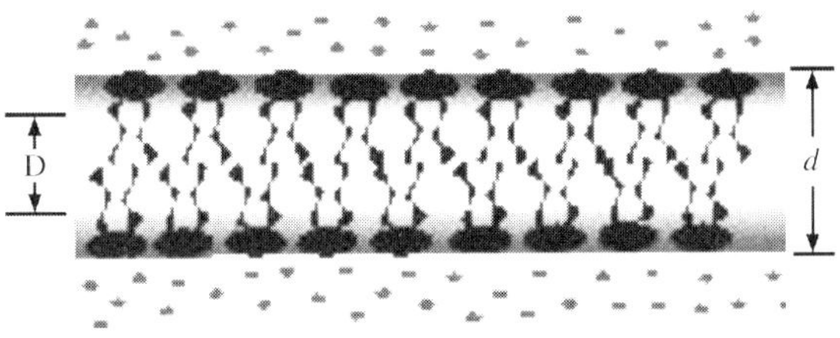

Figure 12.11 A bilayer membrane consisting of two lipid layers. Every lipid molecule contains a hydrophilic head outside the membrane and a hydrophobic tail inside it.

perturbation theory. In the last case

$$r(k) = \int_0^\infty \frac{4\pi\rho(x')}{2ik} \exp(2ikx')dx' . \tag{12.83}$$

Since $k = q/2$, we can write Eq. (12.83) as

$$r(k) = \int_0^\infty \frac{4\pi\rho(x)}{iq} \exp(iqx)dx . \tag{12.84}$$

Integrating by parts we obtain

$$r(k) = \frac{4\pi\rho(0)}{q^2} + \frac{4\pi}{q^2} \int_0^\infty \frac{d\rho(x)}{dx} \exp(iqx)dx . \tag{12.85}$$

Which expression to use is a matter of taste.

The majority of experiments have been carried out on multi bilayers of pure lipids or lipid/protein complexes. These complexes are periodic structures, and their scattering density along the x-axis can be represented by a Fourier series:

$$\rho(x) = \sum_n \pm F(n) \cos(2\pi n x/d) . \tag{12.86}$$

The reflectivity of such systems is characterized by the diffraction peaks shown, for instance (see [539]), in Figure 12.12.

From the position of the peaks it is possible to deduce that the thickness of the membrane is $d = 45.68 \pm 0.05$ Å. The resolution can be found from the number n

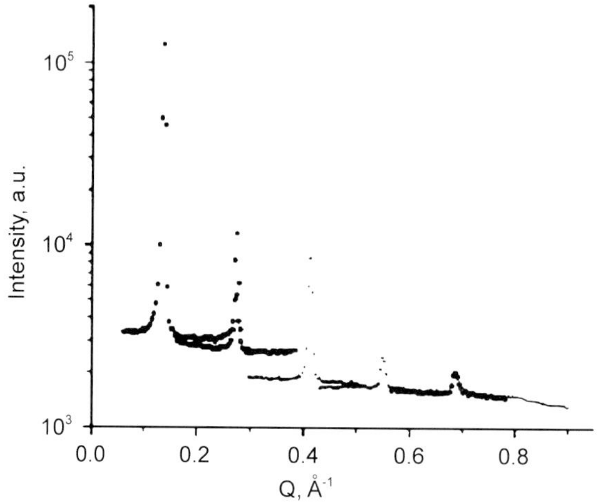

Figure 12.12 Rocking curves measured at six different detector angles (7, 11.2, 16.8, 22.4, 28, 33.6)° for a membrane.

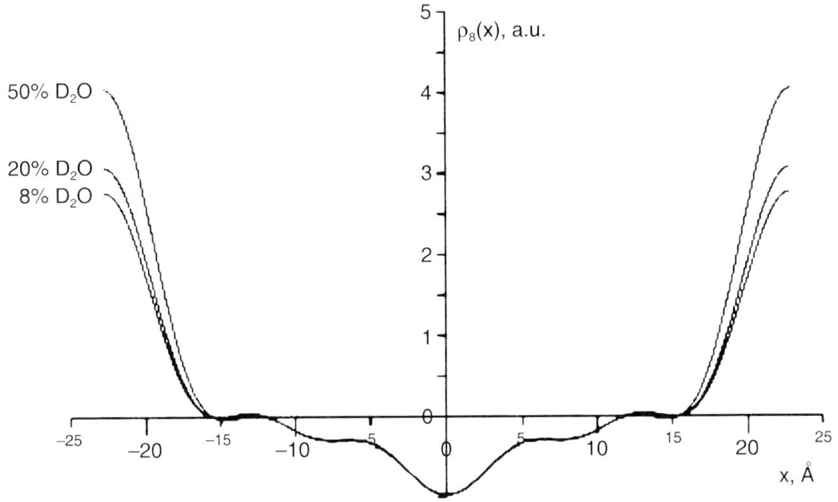

Figure 12.13 Scattering density inside a bilayer and its variation with contrast.

of observable peaks. In this case we have $n = 5$, and the resolution is given by $\Delta d = 0.6d/n = 5.5$ Å.

From five peaks and their intensities it was possible to find five Fourier coefficients in the series Eq. (12.86). The signs of the coefficients were found from contrast variation of the reflectivity. The scattering density obtained and its dependence on contrast are shown in Figure 12.13.

12.6 Diffraction on Biological Macromolecules

We already learned the short history of the first Bragg reflection experiments of neutrons in Chapter 2. Now, neutron diffraction has been developed to the level of studying the structure of the most complicated systems: protein crystals. An brief introductory description will be given here on the neutron diffraction of protein crystals from our viewpoint of neutron optics according to the review [540] by Ostanevich and Serdyuk at the Joint Institute for Nuclear Research, Dubna, and the Protein Institute of the Academy of Sciences, Pushchino.

The interest in this field to apply neutrons for structural analysis of monocrystals of biological macromolecules was based on a number of advantages that neutron studies have over X-ray studies. They include the sensitivity of neutrons substantially differing from that of X-rays, especially in the case of the presence in the structure of light atoms, and primarily of hydrogen; the unique sensitivity of neutrons to isotopic substitution of atoms, which is entirely lacking for X-rays; and the possibility, in principle, of solving the phase problem by using one isomorphous derivative of a protein containing atoms having anomalous absorption of neutrons. Finally, the radiation load on the crystal in neutron irradiation for the structure analyses is 5–6 orders of magnitude smaller than for X-rays.

At the same time, the neutron method of study possesses certain disadvantages. We would include among them the relatively high level of incoherent scattering of neutrons in hydrogen-containing materials and the presence of an appreciable inelastic scattering of neutrons. These two factors give rise to a background scattering that impedes the measurement of diffractions. Another fundamental disadvantage of neutrons is the low intensity and restricted accessibility to neutron beams. However, these two restrictions will be expected to be substantially improved by the recent developments of high-intensity accelerator neutron facilities with their user-oriented policy and at the same time by the effective collaborations in the user communities.

As the next viewpoint on the practical approaches of the experiments, we can classify them into two categories by the resolution d of the diffraction experiments; that is, *high-resolution studies* (say, arbitrarily defined as $d \leq 2$ Å) and *low-resolution studies* (say, $d \geq 10$ Å), and we can say both of them contribute to the progress in understanding the macromolecular structures and properties of biological materials. High-resolution neutron diffraction has been used to identify unambiguously the region in which individual atoms are distinguished in the Fourier maps. The procedure to replace light water with heavy water effectively reduces the incoherent background and thus improves the accuracy of the measurement of diffraction intensities. Low-resolution neutron diffraction, on the other hand, has a different role in these studies on biological macromolecules. We can consider simply that the exposure time for a study increases with increasing resolution no more slowly than V/d^3, where V is the volume of the unit cell. Hence, a relatively large number of studies can be performed with substantially poorer resolution, which shortens the duration of the experiment. In the case of biological macromolecules with complicated structures, it is also a very important approach that, not treating the object of study on the atomic level, still provides the possibility to distinguish components having different mean coherent scattering amplitude densities. In many practical cases, such coarse structural information also proves to be important and interesting.

Another important point in the diffraction studies is the *phase problem*[47], owing to the fact that one needs the complex coefficients (structure factors) for a Fourier synthesis in real space of the structure studied, while experiment allows one to measure only the square of the modulus of these coefficients. We also learned about this problem in Chapter 2 in relation to the inverse process in reflectometry. The solution of the phase problem in X-ray structure analysis is based on the *isomorphous replacement method*, in which one must study the protein of interest and several of its derivatives with heavy atoms. In the case of neutron study, this derivative can contain either paramagnetic atoms or atoms having nuclei show-

47) This is the central task in the crystal structure analysis to obtain an estimate of the structure factor phase angles which cannot be directly determined from the observed intensity data.

ing anomalous absorption for neutrons (neutron resonances). The scattering amplitude of neutrons in these cases is an experimentally variable quantity that in principle enables one to solve the phase problem.

Finally, the most important characteristic feature in biological studies will be the role of hydrogen. Actually, the roles of hydrogen in protein are very numerous. Then, along with the laborious experiments on the localization of hydrogen, low-resolution ($d = 10$–20 Å) studies are of independent interest. For such problems, neutrons prove to be an effective instrument owing to the sufficiently great differences in the scattering amplitude densities of these components. Moreover, the contrast variation method proves to be an effective instrument also for systems with periodic structure for diffraction studies. This enables one to identify unambiguously the contribution of the individual components to the intensity of the observed reflections, and thus to facilitate the solution of the phase problem in low-resolution studies.

12.7 Neutron Microscope for Biological Samples

The structure analyses with neutron diffraction explained in the previous section generally have the spatial resolution of atomic and molecular levels, or to the extent of a number of groups of macromolecules, by making use of the matter wave nature of the neutron. In contrast, there will be another direction in which developments can proceed: from macroscopic observation to magnified imaging with microscopic resolution for structure studies on biophysical and biological materials. Here we will introduce one such development carried out by Cremer *et al.* [541, 542].

As we already learned in Chapter 1, the index of refraction is very close to unity for the thermal and cold neutrons, which are available with the substantial intensity needed for high-precision beam experiments. Therefore, the focusing of these neutrons in a short distance with a single set of material lenses will be very difficult. The technique of magnetic focusing we learned about in Chapter 3 seems to be relatively more powerful, and the application of magnetic focusing to structure studies in biology will also be one of the interesting future tasks. However, Cremer and his collaborators actually employed a very attractive technique of a compound refractive lens (CRL) composed of a large number (practically 100) of biconcave lenses for imaging a biological sample with a magnification of 10–20 by making use of cold neutrons with a wavelength 8.5 Å. The configuration of the experiment they employed is shown in Figure 12.14.

A series of N biconcave lenses functioning as a CRL form a simple microscope when the CRL acts a single lens of focal length f and the object distance o and image distance i are set according to the lens equation, $1/f = 1/o + 1/i$. As in ordinary light optics, the magnification is given by $M = i/o$. A single biconcave lens can image with cold neutrons since the refractive index n of the lens material is less than unity with a focal length typically greater than 100 m. By stacking lenses,

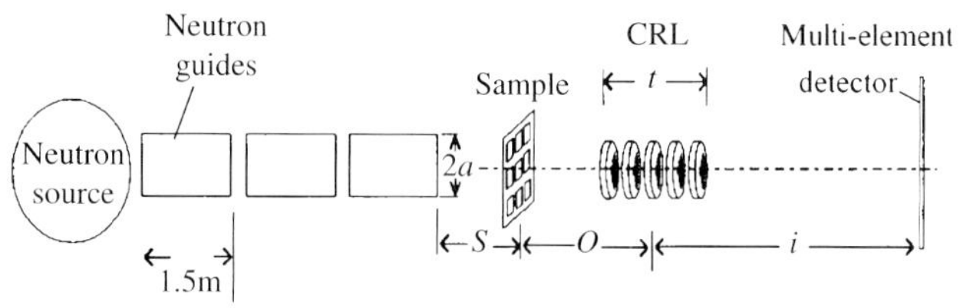

Figure 12.14 Experimental configuration for neutron imaging carried out at the NG-7 beamline of the National Institute of Standards and Technology reactor. Neutron guides can be removed to change the effective distance to the source. CRL compound refractive lens. (Cremer *et al.* [542]).

one reduces the focal length in inverse proportional to N, permitting a microscope of modest length to be fabricated. They employed both MgF_2 and Al for fabricating biconcave lenses. These materials were selected on the basis of their index of refraction, absorption, and inelastic scattering.

The decrement δ of the refractive index n from unity, that is, $\delta = 1 - \mathrm{Re}\{n\}$, determines the focal length, $f = R/2N\delta$, where R is the radius of curvature of each of the N lenses. The linear attenuation coefficient $\mu = \rho(\sigma_a + \sigma_s)$ of the lens material determines the limiting absorption aperture radius defined as the position where the transmission attenuates to e^{-2}, $r_a = \sqrt{2R/N\mu} = \sqrt{4f\delta/\mu}$, where σ_a and σ_s are absorption and scattering cross sections, respectively, and ρ is the CRL atom density.

The imaging experiments were performed on the NG-7 small angle neutron scattering instrument with the configuration shown in Figure 12.14. The exit of the rectangular neutron guide tube acted as a $50 \times 50\,\mathrm{mm}^2$ uniform source of 8.5-Å neutrons monochromatized to a 10% wavelength width. The two-dimensional detector image with a sample was normalized relative to the corresponding image with no sample. First, three plastic meshes of fine and coarse polyethylene and polypropylene grids were used as the object in an imaging experiment. The normalized line intensity profiles taken along a horizontal line are plotted in Figure 12.15. In each figure part the measured object line intensity profile (thick curve) is compared with the calculated one (thin curve) along the indicated line. The calculated intensity profile is obtained by convolution of the mesh attenuation profile with the detector system resolution, a Gaussian profile with full width at half maximum $\sigma_{tot} = 864\,\mu\mathrm{m}$. The minimum transmissions of the grid filaments were 0.44, 0.36, and 0.72 for Figure 12.15a–c, respectively. Thus, the image intensity profile yielded the approximate gap and filament dimensions of the respective meshes.

Then, with the same setup, they imaged a preserved scorpion and a leaf at 10.0 Å with a 300-s exposure time. Figure 12.16a shows a photo of the scorpion and a leaf with a superimposed 10-mm scale. Figure 12.16b shows a stitched neutron image of the scorpion with the leaf removed and the scorpion reoriented. The limited field of view, 24 mm, of the CRL required stitching two partial neutron images of the scorpion, which were displayed with a superimposed 4.4-mm scale. The figure

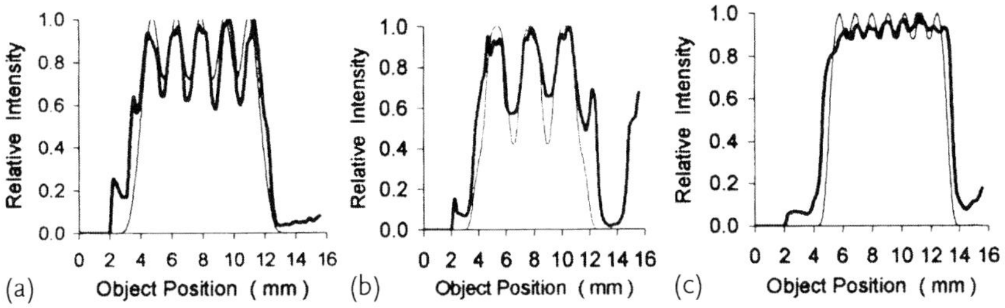

Figure 12.15 The 100-lens MgF_2 CRL neutron normalized line profile of measured intensity (thick curve) at 8.5 Å and 300-s exposure images and the theoretical intensity (thin curve) for three kinds of meshes with the grid dimensions and filament width: (a) polypropylene square grid 0.76 mm × 0.94 mm with 0.63 mm filament; (b) polyethylene diamond grid 1.52 mm × 1.52 mm with 0.51 mm filament; (c) same grid as (a) but with 0.36 mm filament (Cremer *et al.* [542]).

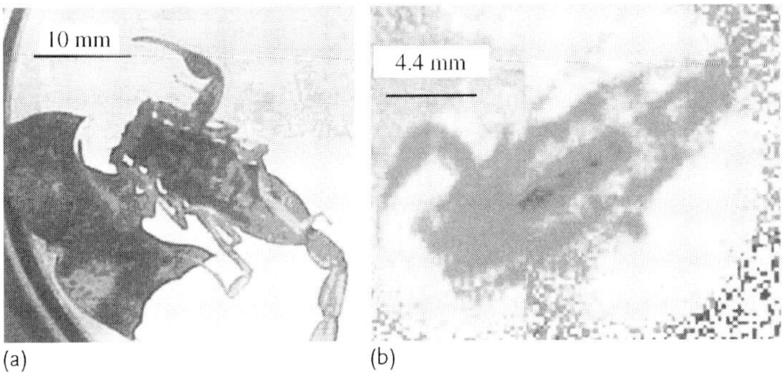

Figure 12.16 (a) Photograph showing a scorpion and a leaf. (b) Neutron microscope image of a scorpion composed from two stitched partial scorpion images at 10.0 Å and 300-s exposure with ×22.5 magnification (Cremer *et al.* [542]).

shows a variation in the density, indicating some areas of the scorpion contain more hydrogen than others.

Possible future developments to realize a higher magnification and a larger field of view with the neutron microscope in the present approach of the CRL and also in other approaches might lead to various applications of neutron microscopy with the characteristic properties in transmission and reflection of neutrons, fundamentally different from electromagnetic beams as light and X-rays and also from electrons, and will serve as a very important tool for biophysical and biological studies.

Basic Formulas in Neutron Optics

(to be given on the backside pages of the Book cover)

Neutron Optics. Masahiko Utsuro,Vladimir K. Ignatovich

ISBN: 978-3-527-40885-6

References

For the most of these references, online abstracts of the articles are available through the respective link to the journal indicated. The figures used from these journals were reproduced after copyright permission had been received from the respective journals or societies.

1 de Broglie, L. (1925) Recherches sur la theorie des quanta. *Ann. de Phys.*, 10e ser. **3**, 22–128. de Broglie, L. (2004) On the Theory of Quanta (translated by A.F. Kracklauer), Available from: URL: http://www.ensmp.fr/aflb/LDB-oeuvres/De_Broglie_Kracklauer.pdf

2 de Broglie, L. (1929) The wave nature of the electron. Nobel Lecture, The Nobel Foundation: Available from: URL:http://nobelprize.org/nobel_prizes/physics/laureates/1929/broglie-lecture.pdf.

3 Davisson, C., Germer, L.H. (1927) Diffraction of electrons by a crystal of nickel, *Phys. Rev.*, **30**(6), 705–740.

4 Thomson, G.P., Reid, A. (1927) Diffraction of cathode rays by a thin film, *Nature*, **119**, 890.

5 Kikuchi, S. (1928) Diffraction of cathode rays by mica. *Jap. J. Phys.*, **5**, 83–96.

6 Heisenberg, W. (1930) *Die physikalischen Principlen der Quantentheorie*, Verlag von S. Hirzel, Leipzig; Translated into English by Eckart, C., Hoyt F.C. (1949) *The Physical Principles of the Quantum Theory*, Dover Books on Physics, Dover Publications, New York.

7 Chadwick, J. (1932) Possible existence of a neutron, *Nature*, **129**, 312.

8 Chadwick, J. (1932) The existence of a neutron, *Proc. Roy. Soc. London A*, **136**, 692–708.

9 Mitchell, D.P., Powers, P.N. (1936) Bragg reflections of slow neutrons. *Phys. Rev.*, **50**(5), 486–487.

10 Zinn, W.H. (1947) Diffraction of neutrons by a single crystal. *Phys. Rev.*, **71**(11), 752–757.

11 Birr, M., Heidemann, A., Alefeld, B. (1971) A neutron crystal spectrometer with extremely high energy resolution. *Nucl. Instrum. Methods*, **95**, 435–439.

12 Fermi, E. (1936) Sur moto dei neutroni nelle sostanze idrogenate. *La Ricerca Scientifica*, **7**(2), 13–52.

13 Cohen, M.L. (1984) The Fermi atomic pseudopotential. *Am. J. Phys.*, **52**(8), 695–703.

14 Foldy, L.L. (1945) The multiple scattering of waves. *Phys. Rev.*, **67**(3 and 4), 107–119.

15 Goldberger, M.L., Seitz, F. (1947) Theory of the refraction and the diffraction of neutrons by crystals. *Phys. Rev.*, **71**(5), 294–310.

16 Lax, M. (1951) Multiple scattering of waves. *Rev. Mod. Phys.*, **23**(4), 287–302.

17 Fermi, E., Marshall, L. (1947) Interference phenomena of slow neutrons. *Phys. Rev.*, **71**(10), 666–677.

18 Hughes, D.J., Burgy, M.T., Ringo, G.R. (1950) Coherent neutron-proton scattering by liquid mirror reflection. *Phys. Rev.*, **77**(2), 291–292.

19 McReynolds, A.W. (1951) Neutron refraction in O_2, N_2, He, A gases. *Phys. Rev.*, **84**(5), 969–972.

20 Maier-Leibnitz, H., Springer, T. (1962) Ein Interferometer für langsame Neutronen. *Z. Phys.*, **167**, 386–402.

Neutron Optics. Masahiko Utsuro, Vladimir K. Ignatovich

ISBN: 978-3-527-40885-6

21 Landkammer, F.J. (1966) Beugungsversuche mit langsamen Neutronen. *Z. Phys.*, **189**, 113–137.

22 Friedrich, H., Heintz, W. (1978) Neutron diffraction experiments with macroscopic objects. *Z. Phys. B*, **31**, 423–427.

23 Shull, C.G. (1969) Single-slit diffraction of neutrons. *Phys. Rev.*, **179**(3), 752–754.

24 Zeilinger, A., Gähler, R., Shull, C.G., Treimer, W., Mampe, W. (1988) Single- and double-slit diffraction of neutrons. *Rev. Mod. Phys.*, **60**(4), 1067–1073.

25 Klein, A.G., Kearney, P.D., Opat, G.I., Cimmino, A. (1981) Neutron interference by division of wavefront. *Phys. Rev. Lett.*, **46**(15), 959–962.

26 Steyerl, A., Schütz, G. (1978) Zone mirror for image formation with neutrons. *Appl. Phys.*, **17**, 45–47.

27 Schütz, G., Steyerl, A. (1978) Image formation with ultracold-neutron waves. *Phys. Rev. Lett.*, **44**(21), 1400–1403.

28 Alianelli, L. (2002) Characterization and modelling of imperfect crystals for thermal neutron diffraction. Thesis, Univ. Joseph Fourier de Grenoble.

29 Moore, A.W., Poovici, M., Stoica, A.D. (2000) Neutron reflectivity and lattice spacing spread of pyrolytic graphite. *Physica B*, **276–278**, 858–859.

30 Kaiser, H., Clothier, R., Werner, S.A., Rauch, H., Wölwitsch, H. (1992) Coherence and spectral filtering in neutron interferometry. *Phys. Rev. A*, **45**(1), 31–42.

31 Keller, T., Rekveldt, M.Th., Habicht, K. (2002) Neutron Larmor diffraction measurement of the lattice-spacing spread of pyrolytic graphite. *Appl. Phys. A*, **74** [Suppl], S127–S119.

32 Shull, C.G. (1968) Observation of Pendellsung fringe structure in neutron diffraction. *Phys. Rev. Lett.*, **21**(23), 1585–1589.

33 Squires, G.L. (1996) *Introduction to the Theory of Thermal Neutron Scattering*, Dover Pub. Inc., Mineola, New York.

34 Kato, N. (1978) Diffraction and scattering. in *Solid State Physics Series*, Vol. 6. Asakura Books, Tokyo (in Japanese).

35 Dunning, J.R., Pegram, G.B., Fink, G.A., Mitchell, D.P., Segrè, E. (1935) Velocity of slow neutrons by mechanical velocity selector. *Phys. Rev.*, **48**(8), 704.

36 Alvarez, L.W. (1938) Production of collimated beams of monochromatic neutrons in the temperature range 300°–10°. *Phys. Rev.*, **54**, 609–617.

37 Baker, C.P., Bacher, R.F. (1941) Experiments with a slow neutron velocity spectrometer. *Phys. Rev.*, **59**, 332–348.

38 Felber, J., Gähler, R., Rausch, C., Golub, R. (1996) Matter waves at a vibrating surface: Transition from quantum-mechanical to classical behavior. *Phys. Rev. A*, **53**(1), 319–328.

39 Halpern, J., Estermann, I., Simpson, O.C., Stern, O. (1937) The scattering of slow neutrons by liquid ortho- and parahydrogen. *Phys. Rev.*, **52**(2), 142.

40 Schwinger, J., Teller, E. (1937) The scattering of neutrons by ortho- and parahydrogen. *Phys. Rev.*, **52**, 286–295.

41 Gttinger, P. (1931) Das Verhalten von Atomen im magnetischen Drehfeld. *Z. Phys.*, **73**, 169–184.

42 Rabi, I.I. (1937) Space quantization in a gyrating magnetic field. *Phys. Rev.*, **51**, 652–654

43 Alvarez, L.W., Bloch, F. (1940) A quantitative determination of the neutron moment in absolute nuclear magnetons. *Phys. Rev.*, **57**(2), 111–122.

44 Estermann, I., Stern, O. (1933) Über die magnetische Ablenkung von Wasserstoffmolekülen und das magnetische Moment des Protons, II. *Z. Phys.*, **85**(1–2), 17–24.

45 Stern, O. (1934) Bemerkung zur Arbeit von Herrn Schüler: Über die Darstellung der Kernmomente der Atome durch Vektoren. *Z. Phys.*, **89**(9–10):665.

46 Rogers, E.H., Staub, H.H. (1949) The signs of the magnetic moments of neutron and proton. *Phys. Rev.*, **76**(7), 980–981.

47 Ramsey, N.F. (1950) A molecular beam resonance method with separated oscillating fields. *Phys. Rev.*, **78**(6), 695–699.

48 Corngold, N.R. (1954) A measurement of the magnetic moment of the neutron. Ph.D. Thesis, Harvard University.

49 Cohen, V.W., Corngold, N.R., Ramsey, N.F. (1956) Magnetic moment of the neutron. *Phys. Rev.*, **104**(2), 283–291

50 Greene, G.L., Ramsey, N.F., Mampe, W., Pendlebury, J.M., Smith, K., Dress, W.D. *et al.* (1979) Measurement of the neutron magnetic moment. *Phys. Rev. D*, 20(9), 2139–2153.

51 Egorov, A.I., Lobashev, V.M., Nazarenko, V.A., Porsev, G.D., Serebrov, A.P. (1974) Production, storage, and polarization of ultracold neutrons. *Sov. J. Nucl. Phys.*, **19**(2), 147–152.

52 Ezhov, V.F., Ivanov, S.N., Lobashev, V.M., Nazarenko, V.A., Porsev, G.D., Serdyuk, O.V. *et al.* (1977) Adiabatic method of separated oscillating fields. *Sov. Phys. JETP-Lett.*, **24**(1), 34–37.

53 Herdin, R., Steyerl, A., Taylor, A.R., Pendlebury, J.M., Golub, R. (1978) Experiment on the efficient polarization of ultracold neutrons. Nucl. Instrum. Methods., **148**, 353–357.

54 Mezei, F. (1988) Zeeman energy, interference and neutron spin echo: a minimal theory. *Phys.*, **151**, 74–81.

55 Mezei, F. (1972) Neutron spin echo: A new concept in polarized thermal neutron techniques. *Z. Phys.*, **255**, 146–160.

56 Maier-Leibnitz, H., Springer, T. (1963) The use of neutron optical devices on beam-hole experiments. *J. Nucl. Energ. Parts A&B, Reactor Science and Technol.*, **17**, 217–225.

57 Maier-Leibnitz, H. (1966) Grundlagen für die Beurteilung von Intensitäts- und Genauigkeitsfragen bei Neutronenstreumessungen, *Nukleonik*, **8**(2), 61–67.

58 Utsuro, M., Okumura, K. (1982) Production and guide tube transmission of very cold neutrons from pulsed cold source. *J. Nucl. Sci. Technol.*, **19**(11), 863–872.

59 Kawabata, Y., Utsuro, M., Ebisawa, T. (1992) *Proc. SPIE B*, **173**, 448–453.

60 Soyama, K., Suzuki, M., Kodaira, T., Ebisawa, T., Kawabata, Y., Tasaki, S. (1995) Transmission characteristics of a supermirror bender. *Physica B*, **213&214**, 951–953.

61 Chen, H., Mildner, D.F.R., Xiao, Q.F. (1994) Neutron focusing lens using polycapillary fibers. *Appl. Phys. Lett.*, **64**(16), 2068–2070.

62 Soyama, K., Minakawa, N., Atsumi, T., Kodaira, T. (1995) Application of multicapillary fiber to neutron microguide tube. *J. Nucl. Sci. Technol.*, **32**(1), 78–80.

63 Scheckenhofer, H., Steyerl, A. (1977) Diffraction and mirror reflection of ultracold neutrons. *Phys. Rev. Lett.*, **39**(21), 1310–1312.

64 Steyerl, A. (1975) A "neutron turbine" as an efficient source of ultracold neutrons. *Nucl. Instrum. Methods*, **125**, 461–469.

65 Lanford, W.A., Golub, R. (1977) Hydrogen surface contamination and the storage of ultracold neutrons. *Phys. Rev. Lett.*, **39**(24), 1509–1512.

66 Kawabata, Y., Utsuro, M., Hayashi, S., Yoshiki, H. (1988) Hydrogen concentration on mirror surfaces of neutron bottles. *Nucl. Instrum. Methods Phys. Res. B*, **30**, 557–566.

67 Penfold, J., Thomas, R.K. (1990) The application of the specular reflection of neutrons to the study of surfaces and interfaces. *J. Phys.: Condensed Matter*, **2**, 1369–1412.

68 Ebisawa, T., Tasaki, S., Otake, Y., Funahashi, H., Soyama, K., Torikai, N. *et al.* (1995) The neutron reflectometer (C3-1-2) at the JRR-3M reactor at JAERI. *Physica B*, **213&214**, 901–903.

69 Cubitt, R. (2000) Instrument rebuild – The D17 reflectometer. *ILL Ann. Rep.*, 91.

70 Shapiro, F.L. (1972) Int. Conf. on Nuclear Structure Study with Neutrons, Budapest, Poc. Nuclear Structure Study with Neutrons (eds J. Erö and J. Szücs), Plenum, New York, London (1975).

71 Novopoltsev, M.I., Panin, Y.N., Pokotilovskii, Y.N., Rogov, E.V., Stepanchikov, V.A., Shelkova, I.G. (1989) Measurement of the absorption coefficient for UCN at a copper surface. *Z. Phys. B Cond. Matter*, **70**, 133–139.

72 Zel'dovich, Ya.B. (1959) Storage of cold neutrons. *Sov. Phys. JETP*, **36**, 1952–1953.

73 Lushchikov, V.I., Pokotilovsky, Yu.N., Strelkov, A.V., Shapiro, F.L. (1969) Observation of ultracold neutrons. *Sov. Phys. JETP-Lett.*, **9**(1), 23–26.

74 Groshev, L.V., Dvoretsky, V.N., Demidov, A.M., Panin, Yu.N., Lushchikov, V.I., Pokotilovsky, Yu.N. *et al.* (1971) Experiments with ultracold neutrons. *Phys. Lett.*, **38B**(4), 293–295.

75 Utsuro, M., Okumura, K. (1998) Ultracold neutron storage experiments with an ultra-fine polished bottle of nonmagnetic stainless steel. *KURRI Prog. Rep.*, **1999**, 16.

76 Baba, Y., Sato, K. (1990) Super mirror-flat finishing of metal surfaces by electrochemical buffing. *Surf. Sci.*, **11**(6), 1–7 (in Japanese).

77 Utsuro, M., Ebisawa, T., Okumura, K., Shirahama, S., Kawabata, Y. (1988) A supermirror turbine for an ultracold neutron source. *Nucl. Instrum. Methods Phys. Res. A*, **270**, 456–461; Utsuro, M., Kawabata, Y., Okumura K. (1994) Ultracold neutron spectra produced by an upgraded supermirror turbine. *Ann. Rep. Res. Reactor Inst. Kyoto Univ.*, **27**, 1–11.

78 Alfimenkov, V.P., Strelkov, A.V., Shvetsov, V.N., Nesvizhevskii, V.V., Serebrov, A.P., Tal'daev, R.R. *et al.* (1992) Anomalous interaction of ultracold neutrons with the surface of a beryllium trap. *Sov. Phys. JETP-Lett.*, **55**, 92–95.

79 Pokotilovski, Yu.N. (2005) UCN anomaly and the possibility for further decreasing neutron losses in traps. *Nucl. Instrum. Methods Phys. Res. A*, **554**, 356–362.

80 Kitagaki, T., Konno, O., Higuchi, M., Sato, M., Kawashima, T., Sato, E. *et al.* (1996) A new type of solid-state detector for ultra-cold neutrons. *J. Phys. Soc. Jpn.* 65, (Suppl.A), 163–168; Proc. of Internat. Symp. Advances in Neutron Optics and Related Research Facilities, (Neutron Optics in Kumatori '96; NOK'96), 163–168.

81 Kitagaki, T., Sakai, K., Hino, M., Utsuro, M., Higuchi, M., Geltenbort, P. *et al.* (2004) An abnormal ultra-cold-neutron absorption in solid UCN-detectors. *Nucl. Instrum. Methods Phys. Res. A*, **529**, 425–428.

82 Gurevich, I.I., Nemirovsky, P.E. (1999) "Metalic" reflection of neutrons. *Sov. Phys. JETP*, **14**(4), 838–839.

83 Rauch, H., Zawisky, M., Stellmach Ch, Geltenbort, P. (1999) Giant absorption cross section of ultracold neutrons in gadolimium. *Phys. Rev. Lett.*, **83**(24), 4955–4958.

84 Varlamov, V.E., Nesvizhevskii, V.V., Serebrov, A.P., Tal'daev, R.R., Kharitonov, A.G., Geltenbort, P. *et al.* (1997) Observation of the penetration of subbarrier ultracold neutrons through beryllium foils and coatings. *Sov. Phys. JETP-Lett.*, **66**(5), 336–343.

85 Geltenbort, P., Nesvizhevsky, V.V., Kartashov, D.G., Lychagin, E.V., Muzychka, A.Yu., Nekhaev, G.V. *et al.* (1999) A new escape channel for ultracold neutrons in traps. *Sov. Phys. JETP-Lett.*, **70**(3), 170–175.

86 Nesvizhevsky, V.V., Lychagin, E.V., Muzychka, A.Yu., Nekhaev, G.V., Strelkov, A.V. (2000) About interpretation of experiments on small increase in energy of UCN in traps. *Phys. Lett. B*, **479**, 353–357.

87 Pokotilovski, Yu.N. (2006) Small energy transfer at the ultra-cold neutron reflection from solid surface. *Phys. Lett. A*, **353**, 236–240.

88 Brenner, Th., Butterworth, J., Geltenbort, P., Hino, M., Malik, S.S., Nesvizhevsky, V.V. *et al.* (2000) Looking for surface states of ultra-cold neutrons. *Nucl. Instrum. Methods Phys. Res. A*, **440**(3), 722–728.

89 Schoenborn, B.P., Caspar, D.L.D., Kammerer, O.F. (1974) A novel neutron monochromator. *J. Appl. Cryst.*, **7**, 508–510.

90 Mezei, F. (1976) Novel polarized neutron devices: supermirror and spin component amplifier. *Commun. Phys.*, **1**, 81–85.

91 Mezei, F., Dagleish, P.A. (1977) Corrigendum and first experimental evidence on neutron supermirrors. *Commun. Phys.*, **2**, 41–43.

92 Croce, P., Pardo, B. (1970) Sur l'application des couches interférentielles a l'optique des rayons X et des neutrons. *Nuovo. Rev. d'Optique Applique*, **1**(4), 229–232.

93 Yamada, S., Ebisawa, T., Achiwa, N., Akiyoshi, T., Okamoto, S. (1978) Neutron-optical properties of multilayer system. *Ann. Rep. Res. Reactor Inst. Kyoto Univ.*, **11**, 8–27.

94 Ebisawa, T., Achiwa, N., Yamada, S., Akiyoshi, T., Okamoto, S. (1979) Neutron reflectivities of Ni-Mn and Ni-Ti multilayers for monochromators and supermirrors. *J. Nucl. Sci. Technol.*, **16**(9), 647–659.

95 Saxena, A.M. (1986) High-reflectivity multilayer monochromators for neutrons. *J. Appl. Cryst.*, **19**, 123–130.

96 Soyama, K., Ishiyama, W., Murakami, K. (1999) Enhancement of reflectivity of multilayer neutron mirrors by ion polishing: optimization of the ion beam parameters. *J. Phys. Chem. Solids*, **60**, 1587–1590.

97 Soyama, K., Tsunoda, H., Murakami, K. (2004) Reflectivity enhancement of large *m*-Qc supermirror by ion polishing. *Nucl. Instrum. Methods Phys. Res. A*, **529**, 73–77; Soyama K. (2005) Development of high Qc supermirror by ion beam sputtering technique in combination with ion beam polishing. Proc. Int. Symp. Res. Reactor and Neutron Sci., Daejeon, Korea, pp. 556–560.

98 Hino, M., Sunohara, H., Yoshimura, Y., Maruyama, R., Tasaki, S., Yoshino, H. *et al.* (2004) Recent development of multilayer neutron mirror at KURRI. *Nucl. Instrum. Methods in Phys. Res. A*, **529**, 54–58.

99 Hino, M., Hayashida, H., Kitaguchi, M., Kawabata, Y., Takeda, M., Maruyama, R. *et al.* (2006) Development of large-*m* polarizing neutron supermirror fabricated by using ion beam sputtering instrument at KURRI. *Physica B*, **385–386**, 1187–1189.

100 Maruyama, R., Yamazaki, D., Ebisawa, T., Hino, M., Soyama, K. (2006) Development of neutron supermirror with large-scale ion-beam sputtering instrument. *Physica B*, **385&386**, 1256–1258.

101 Utsuro, M., Shirahama, S., Okumura, K., Ishikawa, Y., Ebisawa, T. (1981) Construction of KUR ultra cold neutron source with a supermirror neutron turbine. Proc. 4th Meeting ICANS 1980, KEK Tsukuba, pp. 743–757.

102 Utsuro, M. (1996) Production of ultracold neutron beam. *Radioisotopes*, **45**(11), 727–732 (in Japanese).

103 Steyerl, A., Nagel, H., Schreiber F-X, Steinhauser K-A, Gähler, R., Ageron, P. *et al.* (1986) A new source of cold and ultracold neutrons. *Phys. Lett. A*, **116**(7), 347–352.

104 Dombeck, T.W., Lynn, J.W., Werner, S.A., Brun, T., Carpenter, J., Krohn, V. *et al.* (1978) Production of ultra-cold neutrons using Doppler-shifted Bragg scattering and an intense pulsed neutron spallation source. *Nucl. Instrum. Methods*, **165**, 139–155.

105 Brun, T.O., Carpenter, J.M., Krohn, V.E., Ringo, G.R., Cronin, J.W., Dombeck, T.W. *et al.* (1980) Measurement of ultracold neutrons produced by using Doppler-shifted Bragg reflection at a pulsed-neutron source. *Phys. Lett.*, **75A**(3), 223–224.

106 Utsuro, M., Shima, T., Okumura, K., Soyama, K. (2001–2002) Development of ultra-cold neutron source by means of super-mirror Doppler shifter. *KENS Report-XIV*, **2003-6**, 298–299.

107 Shima, T., Utsuro, M., Okumura, K., Soyama, K. (2002, 2003) Development of ultracold neutron source by means of super-mirror Doppler shifter. Osaka Univ. RCNP Ann. Rep., Section 3.2.

108 Klein, A.G., Opat, G.I., Cimmino, A., Zeilinger, A., Treimer, W., Gähler, R. (1981) Neutron propagation in moving matter: The Fizeau experiment with massive particles. *Phys. Rev. Lett.*, **46**(24), 1551–1554.

109 Hamilton, W.A., Klein, A.G., Opat, G.I., Timmins, P.A. (1987) Neutron diffraction by surface acoustic waves. *Phys. Rev. Lett.*, **58**(26), 2770–2773.

110 Kagan, Yu. (1970) Concerning a bound neutron in matter. *Pis'ma Zh. Eksp. Teor. Fiz.*, **11**(4), 235–240; (1970) *Sov. Phys. JETP-Lett.*, **11**, 147–151.

111 Steinhauser, K.-A., Steyerl, A., Scheckenhofer, H. (1980) Malik SS. Observation of quasibound states of the neutron in matter. *Phys. Rev. Lett.*, **44**(20), 1306–1309.

112 Steyerl, A., Ebisawa, T., Steinhauser, K.-A., Utsuro, M. (1981) Experimental study of macroscopic coupled resonators for neutron waves. *Z. Phys. B-Cond. Matt.*, **41**, 283–286.

113 Zhang Huai, Gallaghe, P.D., Sajita, S.K., Lindstrom, R.M., Paul, R.I., Russell, T.P. *et al.* (1994) Grazing incidence prompt gamma emissions and resonance-enhanced neutron standing waves in a thin film. *Phys. Rev. Lett.*, **72**(19), 3044–3047.

114 Aksenov, V.L., Nikitenko, Yu.V., Radu, F., Gledenov, Yu.M., Sedyshev, P.V. (2000) Observation of resonance enhanced neutron standing waves through (n, α) reaction. *Physica B*, **276–278**, 946–947.

115 Nesvizhesky, V.V., Börner, H., Gagarski, A.M., Petrov, G.A., Petukhov, A.K., Abele, H. *et al.* (2000) Search for quantum states of the neutron in a gravitational field: gravitational levels. *Nucl. Instrum. Methods in Phys. Res. A*, **440**, 754–759.

116 Nesvizhesky, V.V., Börner, H., Petukhov, A.K., Abele, H., Bäßler, S., Rueß, F.J. *et al.* (2002) Quantum states of neutrons in the earth gravitational field. *Nature*, **415**, 297–299.

117 Schwarzschild, B. (2002) Ultracold neutrons exhibit quantum states in the earth's gravitational field. *Phys. Today*, **55**[3], 20–22, (translated to Japanese by Otake Y. Parity 2003 Jun; 18).

118 Bloch, F. (1936) On the magnetic scattering of neutrons. *Phys. Rev.*, **50**, 259–260; (1937) On the magnetic scattering of neutrons. II. *Phys. Rev.*, **51**, 944.

119 Schwinger, J.S. (1937) On the magnetic scattering of neutrons. *Phys. Rev.*, **51**, 544–552.

120 Halpern, O. (1949) Double refraction and polarization of neutron beam. *Phys. Rev.*, **75**, 343.

121 Hughes, D.J., Burgy, M.T. (1949) Reflection and polarization of neutrons by magnetized mirrors. *Phys. Rev.*, **76**, 1413–1414.

122 Ignatovich, V.K. (1978) Depolarization of ultracold neutrons in refraction and reflection by magnetic-film surfaces. *Pis'ma Zh. Eksp. Teor. Fiz.*, **28**(5), 311–314; (1978) *Sov. Phys. JETP-Lett.*, **28**, 286–288.

123 Felcher, G.P., Hilleke, R.O., Crawford, R.K., Haumann, J., Kleb, R., Ostrowski, G. (1987) Polarized neutron reflectometer: A new instrument to measure magnetic depth profiles. *Rev. Sci. Instrum.*, **58**(4), 609–619.

124 Felcher, G.P., te Velthuis, S.G.E., Rühm, A., Donner, W. (2001) Polarized neutron reflectometry: recent developments. *Physica B*, **297**, 87–93.

125 Felcher, G.P. (2000) Neutron reflectometry as a tool to study magnetism. *J. Appl. Phys.*, **87**(9), 5431–5436.

126 Vladimirskii, V.V. (1961) Magnetic mirrors, channels and bottles for cold neutrons. *Sov. Phys. JETP*, **12**(4), 740–746.

127 Abov, Yu.G., Borovlv, S.P., Vasil'ev, V.V., Vladimirskii, V.V., Mospan, E.E. (1983) Measurement of the time of storage of ultracold neutrons in a magnetic trap. *Sov. J. Nucl. Phys.*, **38**(1), 70–73.

128 Kügler, K.-J., Paul, W., Trinks, U. (1978) A magnetic storage ring for neutrons. *Phys. Lett. B*, **72**(3), 422–424.

129 Inoue, N., Nihei, H., Akiyama, N., Kinoshita, K. (1996) Confinement of UCN in a multiple cusp magnetic field. Proc. Int. Symp. Advance in Neutron Optics and Related Res. Facilities (NOK'96), held at Kumatori, *J. Phys. Soc. Japan*, **65**(Suppl. A), (Textbooks [J]) 155–158.

130 Inoue, N., Nihei, H., Akiyama, A., Utsuro, M., Kawabata, Y., Okumura, K. *et al.* (1997, 1998) Transport and confinement of ultracold neutrons with burnished stainless steel guide tube and permanent magnets. KURRI Prog. Rep. 1997, 1998, p. 5.

131 Ezhov, V.F., Andreev, A.Z., Glushkov, A.A., Glushkov, A.G., Groshev, M.N., Knyazkov, V.A. *et al.* (2005) First ever storage of ultracold neutrons in a magnetic trap made of permanent magnets. *J. Res. Natl. Inst. Stand. Technol.*, **110**(4), 345–350.

132 Geltenbort, P., Bondarenko, L., Ezhov, V.F., Huffman, P., Masuda, Y., Paul, S. *et al.* (2005) Neutron lifetime measurements using cold (CN) or ultracold neutrons (UCN). Present. Int. Conf. TPFNP, Univ. South Carolina; 14–15 October 2005.

133 Paul, W., Anton, F., Paul, L., Paul, S., Mampe, S. (1989) Measurement of the neutron lifetime in a magnetic storage ring. *Z. Phys. C-Particles and Fields*, **45**, 25–30.

134 Kügler, K.J., Moritz, K., Paul, W., Trinks, U. (1985) Nestor–A magnetic storage ring for slow neutrons. *Nucl. Instrum. Methods A*, **228**(2–3), 240–258.

135 Blundell, S.J., Bland, J.A.C. (1992) Polarized neutron reflection as a probe of magnetic films and multilayers. *Phys. Rev. B*, **46**(6), 3391–3400.

136 Radu, F., Ignatovich, V.K. (1999) Generalized matrix method for the transmission of neutrons through multilayer magnetic systems with noncollinear magnetization. *Physica B*, **267–268**, 175–180.

137 Majkrzak, C.F., Berk, N.F. (1995) Exact determination of the phase in neutron reflectometry. *Phys. Rev. B*, **52**, 10827–10830.

138 Berk, N.F., Majkrzak, C.F. (1996) Inverting specular neutron reflectivity from symmetric, compactly-supported potentials. *J. Phys. Soc. Japan*, **65**(Suppl. A), 107–112.

139 Rühm, A., Toperverg, B.P., Dosch, H. (1999) Supermatrix approach to polarized neutron reflectivity from arbitrary spin strctures. *Phys. Rev. B*, **60**(23), 16073–16077.

140 Utsuro, M. (1984) New applications of ultra cold neutron physics in magnetic and gravity fields, *J. Phys. (Paris)*, **45**(C3), 269–277.

141 Weinfurter, H., Badurek, G., Rauch, H., Schwahn, D. (1988) Inelastic action of a gradient radio-frequency neutron spin flipper. *Z. Phys. B-Cond. Matt.*, **72**, 195–201.

142 Sherwood, J.E., Stephenson, T.E., Bernstein, S. (1954) Stern–Gerlach experiment on polarized neutrons. *Phys. Rev.*, **96**, 1546–1548.

143 Bloch, F. (1946) Nuclear induction. *Phys. Rev.*, **70**(7 and 8), 460–474.

144 Summhammer, J., Niel, L., Rauch, H. (1985) Focusing of pulsed neutrons by traveling magnetic potentials. *Z. Phys. B-Cond. Matt.*, **62**, 269–278.

145 Shimizu, H.M., Suda, Y., Oku, T., Nakagawa, H., Kato, H., Kamiyama, T. *et al.* (1999) Measurement of cold neutron-beam focusing effect of a permanent sextupole magnet. *Nucl. Instrum. Methods Phys. Res. A*, **430**, 423–434.

146 Shimizu, H.M. (2006) Present. Int. Conf. Present Status and Future of VCN Applications, Villigen, 14 February 2006.

147 Oku, T., Shimizu, H.M. (2000) A neutron prism. *Physica B*, **276–278**, 112–113.

148 Zimmer, O., Felber, J., Shärpf, O. (2001) Stern–Gerlach effect without magnetic-field gradient. *Europhys. Lett.*, **53**(2), 183–189.

149 Rose, M.E. (1957) *Elementary Theory of Angular Momentum*. John Wiley & Sons, Inc., New York.

150 Aharonov, Y., Susskind, L. (1967) Observability of the sign change of spinors under 2π rotations. *Phys. Rev.*, **158**, 1237–1238.

151 Bernstein, H.J. (1967) Spin precession during interferometry of fermions and the phase factor associated with rotations through 2π radians. *Phys. Rev. Lett.*, **318**, 1102–1103.

152 Klein, A.G., Opat, G.I. (1975) Observation of 2π rotations: A proposed experiment. *Phys. Rev. D*, **11**(3), 523–528.

153 Klein, A.G., Opat, G.I. (1976) Observation of 2π rotations by Fresnel diffraction of neutrons. *Phys. Rev. Lett.*, **37**(5), 238–240.

154 Wigner, E.P. (1963) The problem of measurement. *Am. J. Phys.*, **31**, 6–15.

155 Summhammer, J., Badurek, G., Rauch, H., Kischko, U. (1982) Explicit experimental verification of quantum spin-state superposition. *Phys. Lett. A*, **90**(3), 110–112.

156 Summhammer, J., Badurek, G., Rauch, H., Kischko, U., Zeilinger, A. (1983) Direct observation of fermion spin superposition by neutron interferometry. *Phys. Rev. A*, **27**(5), 2523–2532.

157 Badurek, G., Rauch, H., Summhammer, J., Kischko, U., Zeilinger, A. (1983) Direct verification of the quantum spin-state superposition law. *J. Phys. A: Math. Gen.*, **16**, 1133–1139.

158 Ebisawa, T., Tasaki, S., Kawai, T., Hino, M., Achiwa, N., Otake, Y. *et al.* (1998) Quantum precession of cold neutron spin using multilayer spin splitters and a phase-spin-echo interferometer. *Phys. Rev. A*, **57**(6), 4720–4729.

159 Gähler, R., Golub, R. (1987) A high resolution neutron spectrometer for quasielastic scattering on the basis of spin-echo and magnetic resonance. *Z. Phys. B-Cond. Matt.*, **65**, 269–273.

160 Gähler, R., Golub, R. (1987) A neutron resonance spin echo spectrometer for quasi-elastic and inelastic scattering. *Phys. Lett. A*, **123**, 43–48.

161 Klimko, S., Stadler, C., Böni, P., Currat, R., Demmel, F., Fåk, B. (2003) Implementation of a zero-field spin-echo option at the three-axis spectrometer

IN3 (ILL, Grenoble) and first application for measurements of phonon line widths in superfluid ^{4}He. *Physica B*, **335**, 188–192.

162 Ebisawa, T., Maruyama, R., Tasaki, S., Hino, M., Kawabata, Y., Yamazaki, D. (2004) Neutron resonance spin echo methods for pulsed source. *Nucl. Instrum. Methods Phys. Res. A*, **529**, 28–33.

163 Maruyama, R., Tasaki, S., Hino, M., Takeda, M., Ebisawa, T., Kawabata, Y. (2004) Performance test of neutron resonance spin echo at a pulsed source. *Nucl. Instrum. Methods in Phys. Res. A*, **530**, 505–512.

164 Yamazaki, D., Soyama, K., Ebisawa, T., Takeda, M., Torikai, N., Tasaki, S. (2005) Resonance spin-echo option on neutron reflectometers for the study of dynamics of surfaces and interfaces. *Physica A*, **356**, 229–233.

165 Yoshimi, A., Asahi, K., Sakai, K., Tsuda, M., Yogo, K., Ogawa, H. *et al.* (2002) Nuclear spin maser with an artificial feedback mechanism. *Phys. Lett. A*, **304**, 13–20.

166 Yoshimi, A., Asahi, K., Yogo, K., Sakai, K., Ogawa, H., Suzuki, T. *et al.* (2001) Novel spin maser mechanism studied for high-precision measurement of neutron electric dipole moment. *SPIN 2000, AIP Conf. Proc.*, **570**, 353–357.

167 Yoshimi, A., Asahi, K., Sakai, K., Yogo, K., Ogawa, H., Nagakura M *et al.* (2000, 2001) Application of nuclear spin maser method to neutrons–a new approach to search for neutron electric dipole moment. Proceedings of the Research Meeting on Fundamental Physics Using Neutrons held at KURRI, 28–29 August 2000; KURRI-KR-67, 18–25 December 2001 (in Japanese).

168 Berry, M.V. (1984) Quantal phase factors accompanying adiabatic changes. *Proc. Roy. Soc. London A*, **392**, 45–57.

169 Bitter, T., Dubbers, D. (1987) Manifestation of Berry's topological phase in neutron spin rotation. *Phys. Rev. Lett.*, **59**(3), 251–254.

170 Fermi, E. (1950) *Conferenze di Fisica atomica*, Accademia Nazionale dei Lincei, Roma.

171 Fermi, E., Zinn, W. (1946) *Phys. Rev.*, **70**, 103.

172 Rutherford, E. (1920) Nuclear constitution of atom. Bakerian lecture. *Proc. Roy. Soc. A*, **97**, 395.

173 Ignatovich, V.K., Utsuro, M. (1997) Tentative solution of UCN problem. *Phys. Lett. A*, **225**, 195–202.

174 Serebrov, A. (1997) Proc. of ISINN-5 14–17 May 1997, JINR E3-97-213, Dubna, 1997, p. 67.

175 Nesvizhevskii, V.V., Strelkov, A.V., Geltenbort, P., Yaidzhiev, P.S. (1998) Observation of a new loss mechanism for UCN in bottles. JINR P3 98-79, Dubna, 1998. (in Russian), see also contribution to proceedings of ILL Workshop on Low Energy Neutron Physics, 21–25 October 1998, Grenoble.

176 Pendlebury, J.M., Geltenbort, P., Nesvizhevsky, V.V., Schreckenbach, K., Serebrov, A.P., Strelkov, A.V. *et al.* (2000) Identification of a new escape channel for UCN from traps. *Nucl. Instrum. Methods. Phys. Res. A*, **440**(3), 695–703.

177 Ignatovich, V.K., Utsuro, M. (2000) Review of inelastic losses of UCN and quantum mechanics of the de Broglie wave packet. JINR E4-98-327, Dubna, 1998; *Nucl. Instrum. Methods Phys. Res. A*, **440**(3), 709–16.

178 Atchison, F., Brys, T., Daum, M., Fierlinger, P., Geltenbort, P., Henneck, R., Heule, S. *et al.* (2007) Loss and spin-flip probabilities for ultracold neutrons interacting with diamondlike carbon and beryllium surfaces. *Phys. Rev. C*, **76**, 044001-1-18.

179 Pokotilovskii, Yu.N. (2005) UCN anomaly and the possibility for further decreasing neutron losses in traps. *Nucl. Instrum. Methods A*, **554**, 356–62.

180 Bokun, R.C., Kistovich, Yu.V. (2005) Surface wave functions of non relativistic particles created by absorption. *Cryst. Rep.*, **50**(5), 862–8.

181 Ignatovich, V.K. (2009) On Neutron Surface Waves. *Cryst. Rep.* Kristallografiya, **54**(1), 116–121.

182 Ignatovich, V.K. (2004) Neutron reflection from condensed matter, the Goos–Hänchen effect and coherence. *Phys. Lett. A*, **322**, 36–46.

183 Goos, F., Hänchen, H. (1947) Ein neuer und fundamentaler Versuch zur Totalreflexion. *Ann. Phys.*, **1**(6), 333–46.

184 Goos, F., Hänchen, H. (1949) Neumessung des Strahlversetzungseffekten bei Totalreflexion. *Ann. Phys.*, **5**(5), 251–2.

185 Kukharchik, P.D., Serdyuk, V.M., Titovitskiy, I.A. (1999) Total reflection of a Gaussian light beam. *ZhTF*, **69**(4), 74 (see Sov. Tech. Phys.).

186 Matthews, A., Kivshar, Yu. (2008) Tunable Goos–Hänchen shift for self-collimated beams in two-dimensional photonic crystals. *Phys. Lett. A*, **372**(17), 3098–3101.

187 Yang, Y., Chen, Xi, Chun-Fang, Li (2007) Large and negative lateral displacement in an active dielectric slab configuration. *Phys. Lett. A*, **361**, 178–81.

188 Bodnarchuck, V., Czer, L., Ignatovich, V.K., Veres, T., Yaradaykin, S. (2009) Investigation of periodic multilayers. Communication of JINR 14-2009-127, Dubna, Russia.

189 Godfrey, G.H. (1957) Recurrence formulae for the reflectance and transmittance of multilayer films with applications. *Austral. J. Phys.*, **10**, 1–15.

190 Korneev, D.A.,Ignatovich, V.K., Yaradaykin, S.P., Bodnarchuk, V.I. (2005) Specular reflection of neutrons from potentials with smooth boundaries. *Physica B: Phys. Cond. Mat.*, **364**(1–4), 99–110.

191 Ne'vot, L., Croce, P. (1980) Caracte'risation des surfaces par re'flexion rasante de rayons X. Application a' l'e'tude du polissage de quelques verres silicates. *Rev. Phys. Appl.*, **15**, 761–80.

192 Korneev, D.A., Bodnarchuk, V.I., Yaradaikin, S.P., Peresedov, V.F., Ignatovich, V.K., Menelle, A., Gähler, R. (2000) Refelectometry study of the coherent properties of neutron. *Physica B*, **276–278**, 973–974.

193 Ignatovich, V.K. (1999) Principle of invariance, or splitting, in neutron optics and basic properties of the neutron. *Phys. At. Nucl.*, **62**(5), 738–753.

194 Darwin, C.G. (1914) The theory of X-ray reflexion. *Phil. Mag.*, **27**, 315–333, 675–690.

195 Ignatovich, V.K. (1977) Scattering of waves and particles from one dimensional periodic potentials. JINR P4-10778, Dubna.

196 Ignatovich, V.K. (1986) Etude on one dimensional periodic potential *Sov. Phys. Usp.*, **29**(9), 880–887.

197 Ignatovich, V.K. (1989) The remarkable capabilities of recursive relations. *Am. J. Phys.*, **57**(10), 873–78.

198 Ignatovich, V.K. (1991) An algebraic approach to the propagation of waves and particles in layered media. *Physica B*, **175**(1–3), 33–8.

199 Ignatovich, V.K., Protopopescu, D., Utsuro, M. (1996) Darwin table width for forbidden reflections. *Phys. Rev. Lett.*, **77**(20), 4202.

200 Ignatovich, V.K., Ignatovitch, F.V. (2000) Multilayered Systems with Forbidden Reflections. Proc. Int. Conf. "Thin films deposition of oxide multilayers. Industrial-scale processing". Vilnius, Lithuania, 28–29 September 2000, Vilnius University Press, pp. 103–107.

201 Carron, I., Ignatovich, V.K. (2003) Algorithm for preparation of multilayer systems with high critical angle of total reflection. *Phys. Rev.* , **67**, 043610.

202 Turchin, V.F. (1967) Diffraction of slow neutrons on multilayer systems. *At. En.*, **22**(3), 119 (in Russian).

203 Turchin, V.F. (1997) Scattering of Slow Neutrons on Layered Media. *Phys. At. Nucl.*, **60**(12), 1946.

204 Hayter, J.B., Mook, H.A. (1989) Discrete thin-film multilayer design for X-ray and neutron supermirrors. *J. Appl. Cryst.*, **22**, 35–41.

205 Maruyama, R., Yamazaki, D., Ebisawa, T., Hino, M., Soyama, K. (2007) Development of neutron supermirror with large critical angle. *Thin Solid Films*, **515**, 5704–6.

206 Ignatovich, V.K., Ignatovitch, F.V., Andersen, D.R. (2000) Algebraic description of multilayer systems with resonances. *Part. Nucl. Lett.*, **3**, 48–61.

207 Aksenov, V.L., Ignatovich, V.K., Nikitenko, Yu.V. (2006) Neutron standing waves in multilayered systems. *Cryst. Rep.*, **51**(5), 734–753.

208 Ignatovich, V.K., Radu, F. (2001) Theory of neutron channeling in resonant layer of multilayer systems. *Phys. Rev. B*, **64**, 205408-1–6.

209 Feng, Y.P., Majkrzak, C.F., Sinha, S.K., Wiesler, D.G., Zhang, H., Deckman, H.W. (1994) Direct observation of neutron-guided wave in a thin-film waveguide. *Phys. Rev. B*, **49**(21), 10814–17.

210 Pogossian, S.P., Menelle, A. (1996) Experimental observation of guided polarized neutrons in magnetic thin-film waveguides. *Phys. Rev.*, **53**(21), 14359.

211 Bloch, J.-F., Ignatovich, V.K. (2001) A new approach to bound states in potential wells. *Am. J. Phys.*, **69**, 1177.

212 Golub, R., Yoshiki, H. (1989) Ultracold anti-neutrons. 1. The approach to the semiclassical limit. *Nucl. Phys. A*, **501**, 869–876.

213 Ignatovich, V.K. (1987) New method for solution of onedimensional Schrödinger equation. JINR P4-87-878, Dubna.

214 Parratt, L.G. (1954) Surface studies of solids by total reflection of X-rays. *Phys. Rev.*, **95**(2), 359–369.

215 Ignatovich, V.K., Terekhov, G.I. (1976) Storage of UCN in a plane gravimagnetic trap. JINR, P4-10102, Dubna.

216 Ignatovich, V.K., Nikitenko, Yu.V. (1987) Physical properties of gravity spectrometers for UCN. JINR P3-87-832, Dubna.

217 Ignatovich, V.K. (1987) Diffusion of UCN along neutron guides in presence of gravity field. JINR P4-87-402, Dubna.

218 Golikov, V.V., Ignatovich, V.K., Nikitenko, Yu.V. (1988) Diffusion of UCN in branched neutron guides. JINR P3-88-48, Dubna.

219 Rauch, H., Werner, S. (2000) *Neutron Interferometry: Lessons in Experimental Quantum Mechanics*, Oxford University Press, Oxford.

220 Klein, A.G., Opat, G.I., Cimmino, A., Zeilinger, A., Treimer, W., Gähler, R. (1981) Neutron propagation in moving matter: The Fiseau experiment with massive particles. *Phys. Rev. Lett.*, **46**(24), 1551–4.

221 Nosov, V.G., Frank, A.I. (1998) Interaction of slow neutrons with moving matter. *Phys. At. Nucl.*, **61**(4), 613–23.

222 Frank, A.I., Geltenbort, P., Kulin, G.V., Kustov, D.V., Nosov, V.G., Strepetov, A.N. (2006) Effect of accelerating media in neutron optics. *JETP. Lett.*, **84**(7), 363–7.

223 Frank, A.I., Geltenbort, P., Jentshel, M., Kulin, G.V., Kustov, D.V., Nosov, V.G. *et al.* (2007) New gravitational experiment with UCN. *JETP. Lett.*, **86**(4), 225–9.

224 Balashov, S.N., Bondarenko, I.V., Frank, A.I., Geltenbort, P., Hoghoj, P., Kulin, G.V., Masalovich, S.V., Nosov, V.G., Strepetov, A.N. (2004) Diffraction of ultracold neutrons on a moving grating and neutron focusing in time. *Physica B: Cond. Matt.*, **350**(1–3), 246–9.

225 Radu, F., Leiner, V., Wolff, M., Ignatovich, V., Zabel, H. (2005) Quantum state of neutrons in magnetic thin films. *Phys. Rev. B*, **71**, 214423-1–6.

226 Ignatovich, V.K. (1978) Depolarization of ultracold neutrons in refraction and reflection by magnetic-film surfaces. *JETP Lett.*, **28**(5), 286–7.

227 Korneev, D.A., Bodnarchuk, V.I., Ignatovich, V.K. (1996) Off-specular neutron reflection from magnetic media with nondiagonal reflectivity matrices. *JETP Lett.*, **63**, 944–51. Proc. of the Int. symp. on advance in neutron optics and related research facilities. (Neutron Optics in Kumatori '96) (1996) *J. Phys. Soc. Japan.*, **65**(Suppl.A), 7–12.

228 Felcher, G.P., Adenwalla, S., De Haan, V.O., Van Well, A.A. (1995) Zeeman splitting of surface-scattered neutrons. *Nature*, **377**(5 October), 409–10; Felcher, G.P., Adenwalla, S., De Haan, V.O., Van Well, A.A. (1996) Observation of the Zeeman splitting for neutrons reflected by magnetic layers. *Physica B*, **221**(1–4):494–9.; Felcher, G.P. (1999) Polarized neutron reflectometry – a historical perspective. *Physica B*, **267–268**(June), 154–61.

229 Aksenov, V.L., Dokukin, E.B., Kozhevnikov, S.V., Nikitenko YuV, Petrenko, A.V., Schreiber, J. (1997) Refraction of polarized neutrons in a magnetically non-collinear layer. *Physica B*, **234–236**(2 June), 513–5.

230 Fredrikze, H., Rekveldt, T., vanWell, A., Nikitenko, Y., Syromyatnikov, V. (1998) Non-specular spin-flipped neutron reflectivity from a cobalt film on glass. *Physica B*, **248**(1), 157–62.

231 Aksenov, V.L., Nikitenko, Yu.V., Kozhevnikov, S.V. (2001) Spin-flip spatial neutron beam splitting in magnetic media. *Physica B*, **297**(1–4), 94–100.

232 Aksenov, V.L., Lauter-Pasyuk, V.V., Lauter, H., Nikitenko, Yu.V., Petrenko, A.V. (2003) Polarized neutrons at pulsed sources in Dubna. *Physica B*, **335**(1–4), 147–52.

233 Badurek, G., Weinfurther, H., Gähler, R., Kollmar, A., Wehinger, S., Zeilinger, A. (1993) Nondispersive phase of the Aharonov–Bohm effect. *Phys. Rev. Lett.*, **71**, 307–311.

234 Calvo, M. (1978) Quantum theory of neutrons in helical magnetic fields. *Phys. Rev. B*, **18**, 5073–7.

235 Aksenov, V.L., Ignatovich, V.K., Nikitenko, Yu.V. (2006) Neutron reflection from a helicoidal system. *JETP Lett.*, **84**(9), 473–8.

236 Asti, G., Ghidini, M., Pellicelli, R., Pernechele, C., Solzi, M., Albertini, F., Casoli, F., Fabbrici, S., and Pareti, L. (2006) Magnetic phase diagram and demagnetization processes in perpendicular exchange-spring multilayers. *Phys.Rev. B*, **73**, 094406.

237 Victora R.H., and Shen X. (2005) Composite Media for Perpendicular Magnetic Recording. *IEEE Trans. Magn.*, **41**, 537-42.

238 Ignatovich, V.K., Radu, F. Reflection of neutrons from fan-like magnetic systems. Submitted to Phys.Rev.B.

239 Ignatovich, V.K., Nikitenko, Yu.V. A Time-odd correlation in a neutron reflectometry experiment. *JETP* To be published.

240 Ignatovich, V.K., Nikitenko, Yu.V., Fraerman, A.A. Transmission of unpolarized neutrons through noncomplanar magnetic multilayer systems. *JETP* To be published.

241 Buzdin, A.I. (2005) Proximity effects in superconductor – ferromagnet heterostructures. *Rev. Mod. Phys.*, **77**, 935–76.

242 Leighton, C., Fitzsimmons, M.R., Yashar, P., Hoffmann, A., Nogue's, J., Dura, J., Majkrzak, C.F., Schuller, I.K. (2001) Two-stage magnetization reversal in exchange biased bilayers. *Phys. Rev. Lett.*, **86**(19), 4394–7; Fitzsimmons, M.R., Yashar, P., Leighton, C., Schuller, I.K., Nogue's, J., Majkrzak, C.F., Dura, J.A. (2000) Asymmetric magnetization reversal in exchange-biased hysteresis loops. *Phys. Rev. Lett.*, **84**(17), 3986–9.

243 Kazansky, A., Uzdin, V.M. (1995) Nodeling of the magnetic properties of the Cr-Fe interface *Phys. Rev. B*, **52** (13), 9477–85.

244 O'Donovan, K.V., Borchers, J.A., Majkrzak, C.F., Hellwig, O., Fullerton, E.E. (2002) Pinpointing chiral structures with front-back polarized neutron reflectometry. *Phys. Rev. Lett.*, **88**(6), 067201-1–4.

245 Radu, F., Ignatovich, V.K. (1999) Generalized matrix method for the transmission of neutron through multilayer magnetic system with noncollinear magnetization. *Physica B.*, 267–268, 175–80.

246 Radu, F. (2004) Polarized Neutron Reflectometry Software "POLAR" http://www.ep4.rub.de/radu/welcome/polar.html.

247 Aksenov, V.L., Nikitenko, Yu.V. (2001) Neutron interference at grazing incidence reflection. Neutron standing waves in multilayered structures: applications, status, perspectives. *Physica B*, **297**(1–4), 101–12.

248 Krupchitskiy, P.A. (1985) Fundamental research with polarised neutrons. – M., Energoatomizdat.

249 Jones, T.J.L., Williams, W.G. (1977) Spin flippers for thermal neutrons. RL-77-079/A.

250 Dabbs, J.W.T., Roberts, L.D., Bernstein, S. (1955) Report ORNL-CF-55-5-126.

251 Drabkin, G.M., Zabidarov, E.I., Kasman, Ya.A., Okorokov, A.I. (1969) *Sov. Phys. JETP.*, **29**, 261.

252 Korneev, D.A. (1979) A new spin-flipper with a prolonged working area for non-monochromatic neutron beams. JINR, P13-12362. Dubna.

253 Korneev, D.A. (1980) A new spin-flipper with a prolonged working area for non-monochromatic neutron beams. *Nucl. Instrum. Methods*, **169**, 65–8.

254 Korneev, D.A., Kudrtashov, V.A. (1980) Method of determination of flipping probability for neutron transmission through spin-flipper. JINR, P3-80-65, Dubna.

255 Korneev, D.A., Kudriashov, V.A. (1980) Experimental determination of the characteristics of a spin-flipper with a prolonged working area. JINR, P3-80-350, Dubna.

256 Korneev, D.A., Kudriashov, V.A. (1981) Experimental determination of the characteristics of a spin-flipper with a prolonged working area. *Nucl. Instrum. Methods*, **179**, 509–13.

257 Schmidt, U., Abele, H., Boucher, A., Klein, M., Stellmach, C., Geltenbort, P. (2000) Neutron polarization induced by radio frequency radiation. *Phys. Rev. Lett.*, **84**(15), 3270–3.

258 Abele, H., Boucher, A., Geltenbort, P., Klein, M., Schmidt, U., Stellmach, C. (2000) Radio frequency-induced polarisation of ultra-cold neutrons, or how to pump a two level system. *Nucl. Instrum. Methods. Phys. Res. A*, **440**, 760–3.

259 Grigoriev, S.V., Okorokov, A.I., Runov, V.V. (1997) Peculiarities of the construction and application of a broadband adiabatic flipper of cold neutrons. *Nucl. Instrum. Methods Phys. Res. A*, **384**, 451–6.

260 Ignatovich, V.K. (2008) The Berry phase: A simple derivation and relation to the electric dipole moment experiments with ultracold neutrons. *Am. J. Phys.*, **76**(3):258–64.

261 Baker, C.A., Doyle, D.D., Geltenbort, P., Green, K., van der Grinten, M.G.D., Harris, P.G., Iaydjiev, P., Ivanov, S.N., May, D.J.R., Pendlebury, J.M., Richardson, J.D., Shiers, D., Smith, K.F. (2006) Improved experimental limit on the electric dipole moment of the neutron. *Phys. Rev. Lett.*, **97**, 131801-1–4.

262 Krüger, E. (1980) Acceleration of polarized neutrons by rotating magnetic field. *Nukleonika*, **25**, 889–93.

263 Drabkin, G.M., Zhitnikov, P.A. (1960) *ZhETF*, **38**, 1013.

264 Golub, R., Gähler, R., Keller, T. (1994) A plane wave approach to particle beam magnetic resonance. *Am. J. Phys.*, **62**, 779–88.

265 Ignatovich, V.K., Ignatovich, F.V. (2003) The Krüger problem and neutron spin games. *Am. J. Phys.*, **71**, 1013–24.

266 Felber, J., Gähler, R., Golub, R., Hank, P., Ignatovich, V., Keller, T., *et al.* (1999) Neutron time interferometry. *Foundation of Phys.*, **29**, 381–396

267 Hino, M., Achiwa, N., Tasaki, S., Ebisawa, T., Kawai, T., Yamazaki, D. (2000) Measurement of spin-precession angles of resonant tunneling neutrons. *Phys. Rev. A*, **61**, 013607-1–8.

268 Gähler, R., Ignatovich, V. (2007) Neutron holography without reference beam. *Phys. Lett.*, **362**(5–6), 393–400.

269 Aksenov, V.L., Dokukin, E.B., Kozhevnikov, S.V., Nikitenko, Yu.V. (2004) Spin-precessor intended for microstructure investigations at ultrasmall-angle neutron spectrometer. *Physica B*, **345**(1–4), 254–7.

270 Ignatovich, V.K., Nikitenko, Yu.V., Radu, F. (2009) Experimental opportunity to investigate layered magnetic structures with the help of oscillating magnetic field. *Nucl.Instr&Meth. in Phys. Res. A* 604(3) 653-661.

271 Lauter, H.J., Toperverg, B.P., Lauter-Pasyuk, V., Petrenkod, A., Aksenov, V. (2004) Larmor precession reflectometry. *Physica B*, **350**(1–3)(Suppl. 1), E759–62.

272 Ebisawa, T., Yamazaki, D., Tasaki, S., Kawai, T., Hino, M., Akiyoshi, T., Achiwa, N., Otake, Y. (1998) Quantum beat experiments using a cold neutron spin interferometer. *J. Phys. Soc. Jap.*, **67**, 1569–73.

273 Achiwa, N., Shirozu, G., Ebisawa, T., Hino, M., Tasaki, S., Kawai, T., Yamazaki, D. (2001) Time beat neutron spin interferometry before and after analyzer. *J. Phys. Soc. Japan*, **70**(Suppl. A), 436–8.

274 Golub, R., Gähler, R., Habicht, K., Klimko, S. (2006) Bunching of continuous neutron beams. *Phys. Lett. A*, **349**(1–4), 59–66.

275 Ignatovich, V.K., Ostanevich, Yu.M., Podgoretskiy, M.I. (1979) A method for obtaining holograms without reference beam. Author certificate N. 745 271 with priority on 8 January 1979. Registered in USSR 7 March 1980.

276 OIPOTZ: Bulletin (1980) Discoveries, Inventions, industrial designs and commodity signs, **24**, 351.

277 Gähler, R., Golub, R. (1996) Neutron spin optics; a thought experiment with applications. *Phys. Lett. A*, **213**(5–6), 239–44.

278 Cser, L., Török, G., Krexner, G., Sharkov, I., Faragó, B. (2002) Holographic imaging of atoms using thermal neutrons. *Phys. Rev. Lett.*, **89**, 175504-1–4.

279 Sur, B., Anghel, V.N.P., Rogge, R.B., Katsaras, J. (2005) Diffraction pattern from thermal neutron incoherent elastic scattering and the holographic reconstruction of the coherent scattering length distribution. *Phys. Rev. B*, **71**, 014105-1–12.

280 v. Laue, M., Friedrich, W., Knipping, P. (1912) Interferenzen-Erscheinungen bei Röntgenstrahlen. Sitzungsberichte der mathematisch-physikalischen Klasse der Bayer. Akademie der Wissenschaften zu München, Jahrgang 1912, Verlag der Königlich Bayerischen Akademie der Wissenschaften in Kommission des Franz'schen Verlags (J. Roth), München, pp. 303–322; v. Laue, M. (1915) Nobel Lecture, Concerning the detection of X-ray interference. pp. 303–322 and 5 Tables.

281 Darwin, C.G. (1922) The reflexion of X-rays from imperfect crystals. Part, I., II. *Phil. Mag.*, **43**, 800–829.

282 Ewald, P.P. (1916) Zur Begründung der Kristalloptik, Teil, I., II. *Ann. Phys. Lpz.*, **49**, 1–38, 117–143.

283 Ewald, P.P. (1917) Zur Begründung der Kristalloptik, Teil III. *Ann. Phys. Lpz.*, **54**, 519–556, 557–597.

284 Borrmann, G. (1941) Über Extinktionsdiagramme von Quarz. *Phys. Z.*, **42**(Nr. 9/10), 157–160.

285 Borrmann, G. (1950) Die Absorption von Röntgenstrahlen im Fall der Interferenz. *Z. Phys.*, **127**, 297–323.

286 Batterman, B.W. (1964) Dynamical diffraction of X-rays by perfect crystals. *Rev. Mod. Phys.*, **36**(3), 681–717.

287 Knowles, J.W. (1956) Anomalous absorption of slow neutrons and X-rays in nearly perfect single crystals. *Acta. Cryst.*, **9**, 61–69.

288 Sippel, D., Kleinstück, K., Schulze, G.E.R. (1964) Neutron diffraction of ideal crystals using a double crystal spectrometer. *Phys. Lett.*, **8**, 241–242.

289 Kikuta, S., Ishikawa, I., Kohra, K., Hoshino, S. (1975) Studies on dynamical diffraction phenomena of neutrons using properties of wave fan. *J. Phys. Soc. Japan*, **39**, 471–478.

290 Klein, A.G., Prager, P., Wagenfeld, H., Ellis, P.J., Sabine, T.M. (1967) Diffraction of neutrons and X-rays by a vibrating quartz crystal. *Appl. Phys. Lett.*, **10**, 293–295.

291 Mikula, P., Michalec, R., Čech, J., Chalupa, B., Sedláková, L., Petržílka, V. (1974) Secondary reflexions of neutrons diffracted by a single-crystal bar vibrating at high frequency. *Acta. Cryst. A*, **30**, 560–564.

292 Mikula, P., Michalec, R., Chalupa, B., Sedláková, L., Petržílka, V. (1975) A vibrating perfect crystal assumed to be a real one. *Acta. Cryst. A*, **31**, 688–693.

293 Kohra, K., Matsushita, T. (1972) Some characteristics of dynamical diffraction at a Bragg angle of about $\pi/2$. *Z. Naturforsch.*, **27a**, 484–487.

294 Treimer, W., Strobl, M., Hilger, A. (2001) Development of a tunable channel cut crystal. *Phys. Lett. A*, **289**, 151–154.

295 Treimer, W., Strobl, M., Hilger, A. (2002) Observation of edge refraction in ultra small angle neutron scattering. *Phys. Lett. A*, **305**, 87–92.

296 Treimer, W., Strobl, M., Hilger, A. (2003) On lateral coherence. Experimental Report BENSC Instrum. V12b.

297 Steyerl, A. (1972) A time-of-flight spectrometer for ultracold neutrons. *Nucl. Instrum. Methods*, **101**, 295–314.

298 Steyerl, A., Vonach, H. (1972) Total cross-section of various homogeneous substances for ultracold neutrons. *Z. Phys.*, **250**, 166–178.

299 Hughes, D.J., Schwartz, R.B. (Eds) (1958) Neutron Cross Sections. BNL-325 (second edn.), Brookhaven National Laboratory, Upton, New York.

300 Rayleigh Lord (1914) On the duffraction of light by spheres of small relative index. *Proc. Roy. Soc. London A*, **90**, 219–225.

301 Guinier, A. (1937) Comptes Rendus hebdomadaires des Séance de l'Academie des Sciences, **204**, 1115–1117.

302 Guinier, A. (1939) La diffraction des rayons X aux trs petits angles: Application a l'tude de phnomnes ultramicroscopiques. *Ann. Phys. Paris*, 11e série, **12**, 161–237.

303 Porod, G. (1947) Die Abhängigkeit der Röntgen-Kleinwinkelstreuung von Form und Größe der kolloiden Teilchen in verdünnten Systemen, IV. *Acta Phys. Austr.*, **2**, 255–292.

304 Guinier, A., Fournet, G. (1955) *Small-Angle Scattering of X-Rays*, John Wiley & Sons, Inc., New York.

305 Hughes, D.J., Burgy, M.T., Heller, R.B., Wallace, J.W. (1949) Magnetic refraction of neutrons at domain boundaries. *Phys. Rev.*, **75**, 565–569.

306 Ginie, A. (1980) *La structure de la matire-du ciel bleu à la matire plastique*, Hachette, Paris.

307 Engelmann, G., Steyerl, A., Heidemann, A., Kostorz, G., Mughrabi, H. (1979) Comparison between very-slow-neutron transmission and small-angle neutron-scattering experiments. *Z. Phys. B*, **35**, 345–349.

308 Egelstaff, E.A. (1967) *An Introduction to the Liquid State*, Academic Press, London, New York.

309 Toda, M., Matsuda, H., Hiwatari, Y., Wadachi, M. (1976) *Structure and Properties of Liquids*, Iwanami Books, Tokyo (in Japanese).

310 Kunitomi, N. (1976) Analyses of magnetic structures. in *Neutron Diffraction* (ed. S. Hoshino). Lectures on Experimental Physics, Vol.22, Chapter 10. p. 224, Kyouritsu Pub. Co., Tokyo (in Japanese).

311 Wiedenmann, A. (2006) Small angle neutron scattering investigations of magnetic nanostructures. in *Neutron Scattering from Magnetic Materials* (ed. T. Chatterji), 1st edn., Chapter 10. p. 480, Elsevier, Amsterdam, London, Oxford, San Diego.

312 Suzuki, J. (2000) Three-dimensional small-angle neutron scattering (3D-SANS) anallysis. *J. Appl. Cryst.*, **33**, 785–787.

313 Fratzl, P., Langmayr, F., Paris, O. (1993) Evaluation of 3D small-angle scattering from non-spherical particles in single crystals. *J. Appl. Cryst.*, **26**, 820–826.

314 Bouwman, W.G., Kruglov, T.V., Plomp, J., Rekveldt, M.T. (2005) Spin-echo methods for SANS and neutron reflectometry. *Physica B*, **357**, 66–72.

315 Moshinsky, M. (1952) Diffraction in time. *Phys. Rev.*, **88**, 625–631.

316 Jeffrey, A., Dai H.-H. (2008) *Handbook of Mathematical Formulas and Integrals*, 4th edn, Chapter 22, Section 22.17, Academic Press, London, New York, San Diego.

317 Jeffrey, H., Swirles, B. (1972) *(Lady Jeffreys) Methods of Mathematical Physics*, 3rd edn., Chapter 23, University Press, Cambridge.

318 Moriguchi, K., Udagawa, K., Hitotsumatsu, S. (2002) *Mathematical Formulas III*, 18th Print, Iwanami Books, Tokyo, p. 69 (in Japanese).

319 Jeffrey, A., Dai, H.-H. (2008) *Handbook of Mathematical Formulas and Integrals*, 4th edn, Chapter 13, Section 13.2, Academic Press, London, New York, San Diego.

320 Jeffrey, H., Swirles, B. (1972) *(Lady Jeffreys) Methods of Mathematical Physics*, 3rd edn, University Press, Cambridge, p. 403, p. 569.

321 Moriguchi, K., Udagawa, K., Hitotsumatsu, S. (2003) *Mathematical Formulas I*, 23rd Print., Iwanami Books, Tokyo, p. 155 (in Japanese).

322 Moriguchi, K., Udagawa, K., Hitotsumatsu, S. (2002) *Mathematical Formulas III*, 18th Print, Iwanami Books, Tokyo, p. 21 (in Japanese).

323 Jeffrey, A., Dai, H.-H. (2008) *Handbook of Mathematical Formulas and Integrals*, 4th edn., Chapter 14, Section 14.1, Subsection 14.1.1, Academic Press, London, New York, San Diego.

324 Jeffrey, H., Swirles, B. (1972) *(Lady Jeffreys) Methods of Mathematical Physics*, 3rd edn., University Press, Cambridge, p. 473.

325 Moriguchi, K., Udagawa, K., Hitotsumatsu, S. (2002) *Mathematical Formulas III*, 18th Print, Iwanami Books, Tokyo, p. 22, p. 261 (in Japanese).

326 Moriguchi, K., Udagawa, K., Hitotsumatsu, S. (2002) *Mathematical Formulas III*, 18th Print, Iwanami Books, Tokyo, p. 23 (in Japanese).

327 Jeffrey, H., Swirles, B. (1972) *(Lady Jeffreys) Methods of Mathematical Physics*, 3rd edn., University Press, Cambridge, pp. 506–507.

328 Gähler, R., Golub, R. (1984) Time dependent neutron optics-quantum mechanical effects on beam chopping and a new type of high resolution neutron spectrometer (FOTOF). *Z. Phys. B-Cond. Matt.*, **56**, 5–12.

329 Felber, J.,Gähler, R., Golub, R. (1988) Test of the time dependent Schrödinger equation with very slow neutrons. *Physica B*, **151**, 135–139.

330 Hils, Th., Felber, J., Gähler, R., Glaser, W., Golub, R., Habicht, K. *et al.* (1998) Matter-wave optics in the time domain: Results of a cold-neutron experiment. *Phy. Rev. A*, **58**, 4784–4790.

331 Felber, J., Gähler, R., Golub, R., Hil, Th. (2000) Some aspects of a time-domain neutron optics experiment. *Nucl. Instrum. Methods in Phys. Res. A*, **440**, 585–590.

332 Gabor, D. (1948) A new microscopic principle. *Nature*, **161**, 777–778.

333 Gabor, D. (1949) Microscopy by reconstructed wave-fronts. *Proc. Roy. Soc. (London) A*, **197**, 454–487.

334 Gabor, D. (1951) Microscopy by reconstructed wave fronts: II. *Proc. Phys. Soc. (London)*, **64**(pt. 6)378B, 449–469.

335 Gabor, D. () Holography, 1948–1971. http://nobelprize.org/nobel_prizes/physics/laureates/1971/gabor-lecture.pdf.

336 Szke, A. (1986) in *Short Wavelength Coherent Radiation: Generation and Applications* (eds D.T Attwood, J. Baker), AIP Conf. Proc. No. 147, AIP, New York.

337 Gog, T., Len, T.M., Materlik, G., Bahr, D., Fadley, C.S., Sanchez-Hanke, C. (1996) Multiple-energy X-ray holography: Atomic images of hematite (Fe_2O_3). *Phys. Rev. Lett.*, **76**(17), 3132–3135.

338 Bartell, L.S., Ritz, C.L. (1974) Atomic images by electron-wave holography. *Science*, **185**, 1163–1164.

339 Tonomura, A., Matsuda, T., Endo, J., Todokoro, H., Komoda, T. (1979) Development of a field emission electron microscope. *J. Electron Microsc.*, **28**, 1–11.

340 Harp, G.R., Saldin, D.K., Tonner, B.P. (1990) Atomic-resolution electron holography in solids with localized sources. *Phys. Rev. Lett.*, **65**, 1012–1015.

341 Tegze, M., Faigel, G. (1996) X-ray holography with atomic resolution. *Nature*, **380**, 49–51.

342 Sur, B., Rogge, R.B., Hammond, R.P., Anghel, V.N.P., Katsaras, J. (2001) Atomic structure holography using thermal neutrons. *Nature*, **414**, 525–527.

343 Cser, L., Faragó, B, Krexner, G., Sharkov, I., Török, G. (2004) Atomic resolution neutron holography (principles and realization). *Physica B*, **350**, 113–119.

344 Cser, L., Krexner, G., Török, G. (2001) Atomic-resolution neutron holography. *Europhys. Lett.*, **54**(6), 747–752.

345 Ignatovich, V.K. (1992) Multiwave algebraic Darwin method in dynamic theory of diffraction. *Sov. Phys. Crystallogr.*, **37**(5), 588–600.

346 Glasser, M.L. (1973) The evaluation of lattice sums. I. Analytic procedures. *J. Math. Phys.*, **14**(3), 409–413.

347 Hardy, G.H. (1919) *Mess Math.*, **49**, 85.

348 Abramovitz, M., Stegun, I.A. (1964) *Handbook of Mathematical Functions*, NBS circular AMS v.55, (Natl. Bur. Stds.), Washington, Chapter 23.

349 Massa, W., Gould, R.O. (1999) *Crystal Structure Determination*, Berlin, Heidelberg, New York, Springer-Verlag.

350 Dinnebier, R.E., Billinge, S.J.L. (Eds) (2008) *Powder Diffraction: Theory and Practice*, RSC Publishing, Cambridge, p. 604.

351 Pecharsky, V., Zavalij, P. (2009) *Fundamentals of Powder Diffraction and Structural Characterization of Materials*, 2nd edn. (Recent Results in Cancer Research). New York, Springer Science+Business Media, LLC.

352 Ignatovich, V.K. (1990) Algebraic approach to the dynamical diffraction theory for polyatomic ideal monocrystals. *Sov. Phys. JETP*, **70**(5), 913–7.

353 Mittleman, D.M., Bertone, J.F., Jiang, P., Hwang, K.S., Colvin, V.L. (1999) Optical properties of planar colloidal crystals: Dynamical diffraction and the scalar wave approximation. *J. Chem. Phys.*, **111**, 345–54.

354 Ulyanenkov, A.P., Stepanov, S.A., Pietsch, U., Köhler, R. (1995) A dynamical diffraction approach to grazing-incidence X-ray diffraction by multilayers with lateral lattice misfits, *J. Phys. D: Appl. Phys.*, **28**, 2522–8.

355 Stepanov, S., Kohler, R. (1994) A dynamical theory of extremely asymmetric X-ray diffraction taking account of normal lattice strain. *J. Phys. D: Appl. Phys.*, **27**, 1922–8.

356 Stepanov, S.A. (1997) X-ray diffraction of multilayers and superlattices at grazing incidence and/or exit. 3th Autumn School on X-ray Scattering from Surfaces and Thin Layers, Smolenice, Slovakia. 1–4 October 1997, http://sergey.gmca.aps.anl.gov/Publications.html.

357 Bacon, G.E. (1962) *Neutron Diffraction* (Monographs on the Physics and Chemistry of Materials), Clarendon Press, Oxford.

358 Turchin, V.F. (1965) Slow neutrons. Israel Program for Scientific Translations.

359 Pinsker, Z.G. (1978) *Dynamical Scattering of X-Rays in Crystals (Reactivity and Structure)*, Springer.

360 Belyakov, V.A. (1992) *Diffraction Optics of Complex-Structured Periodic Media (Partially Ordered Systems)*, 1st edn, Springer.

361 Hughes, D.J. (1954) *Neutron Optics*, Interscience Publishers.

362 Aleksandrov, Yu.A., Sharapov, E.I., Cser, L. (1981) Diffraction methods in neutron physics. M.: Energoizdat (in Russian).

363 Ignatovich, V.K., Utsuro, M. (1997) Optical potential and dispersion law for long-wavelength neutrons. *Phys. Rev. B*, **55**(22), 14774–83.

364 Ignatovich, V.K. (1992) The ball lightning. *Laser Phys.*, **2**(6), 991–6.

365 Silbar, R.R., Reddy, S. (2004) Neutron stars for undergraduates. *Am. J. Phys.*, **72**, 892–905.

366 Ignatovich, V.K. (2007) Neutronstriction in neutron stars. *Concepts of Phys. Old and new*, **4**(4), 575–607.

367 Klyatskin, V.I. (1985) Ondes et quations stochastiques dans les milieux alatoiremeint non homognes. Monographie. Les Éditions de physique de Besanson.

368 Yoneda, Y. (1963) Anomalous surface reflection of X-rays. *Phys. Rev.*, **131**(5), 2010–3.

369 Berceanu, I., Ignatovich, V.K. (1973) Monte Carlo calculation of the transmission of straight tubes without account of UCN absorption in the walls. *Vacuum*, **23**(12), 441–5.

370 Bouchaud, J.-Ph., Georges, A. (1990) Anomalous diffusion in disordered media: statistical mechanisms, models, and physical applications. *Phys. Rep.*, **195**(4–5), 127–293.

371 Ciccariello, S., Goodisman, J., Brumberger, H. (1988) On the Porod law. *J. Appl. Cryst.*, **21**(Part 2), 117–28.

372 Pynn, R. (1992) Neutron scattering by rough surfaces at grazing incidence. *Phys. Rev. B*, **45**(2), 602–12.

373 Steyerl, A. (1972) Effect of surface roughness on the total reflection and transmission of slow neutrons. *Z. Phys.*, **254**, 169–88.

374 Sinha, S.K., Sirota, E.B., Garoff, G., Stanley, H.B. (1988) X-ray and neutron scattering from rough surfaces. *Phys. Rev. B*, **38**(4), 2297–311.

375 Robbins, M.O., Andelman, D., Joanny, J.-F. (1991) Thin liquid films on rough or heterogeneous solids. *Phys. Rev. A* **43**(8), 4344–54.

376 Chiarello, R., Panella, V., Krim, J., Thompson, C. (1991) X-ray reflectivity and adsorption isotherm study of fractal scaling in vapor-deposited films. *Phys. Rev. Lett.*, **67**(24), 3408–11.

377 Palasantzas, G., Krim, J. (1993) Effect of the form of the height-height correlation function on diffuse X-ray scattering from a self-affine surface. *Phys. Rev. B*, **48**(5), 2873–7.

378 Feigin, L.A., Svergun, D.I. (1987) *Structure Analysis by Small-Angle X-Ray and Neutron Scattering*, 1st edn. Springer.

379 Brumberger, H. (ed.) (1995) *Modern Aspects of Small Angle Scattering*, Kluwer Academic Press.

380 Sinha, S.K. (1998) Small angle and surface scattering from porous and fractal materials. in *Proc. of the Sixth Summer School on Neutron Scattering. Complementarity between neutron and synchrotron X-ray scattering* (ed. A. Furrer), World Scientific, Singapore, New Jersey, London, Hong Kong, pp. 251–81.

381 Ignatovich, V.K. (2000) Measuring surface roughness in transmission geometry. in *Proc. SPIE. Scattering and surface roughness III* (eds Gu

Zu-Han, A.A. Maradudin), V. 4100, pp. 190–198.
382 Radu, F., Vorobiev, A., Major, J., Humblot, H., Westerholt, K., Zabel, H. (2003) Spin-resolved off-specular neutron scattering from agnetic domain walls using the polarized ^{3}He gas spin filter. *Physica B*, **335**, 63–7.
383 Radu, F., Etzkorn, M., Leiner, V., Schmitte, T., Schreyer, A., Westerholt, K., Zabel, H. (2002) Polarised neutron reflectometry study of Co/CoO exchange-biased multilayers. *Appl. Phys. A*, **74**(Suppl), S1570–2.
384 Zabel, H. (2006) Neutron reflectivity of spintronic materials. *Materials Today*, **9**(1–2), 42–9.
385 Ignatovich, V.K., Shabalin, E.P. (2007) Algebraic method for calculating of neutron albedo. *Phys. At. Nucl.*, **70**(2), 265–72.
386 Case, K.M., Zweifel, P.F. (1967) *Linear Transport Theory*, Adison-Wisley Publishing Company.
387 Dashen, R.F. (1964) Theory of electrons backscttering. *Phys. Rev. A*, **34**(4), 1025–31.
388 Beckurts, K.H., Wirtz, K. (1964) *Neutron Physics*, Springer Verlag, Berlin.
389 Germogenova, T.A. *et al.* (1973) Neutron Albedo. M.: Atomizdat.
390 Artem'ev, V.A. (2003) Estimation of critical parameters of thermal neutrons reactors with active core made of nanostructured material. *VANT ser. Phys. At. React.*, **(1–2)**:7–12 (in Russian).
391 Artem'ev, V.A. (2003) Estimation of critical parameters of thermal neutrons reactors with active core made of nanostructured material. *At. Energ.*, **94**(3), 231–3 (in Russian).
392 Remizovich, V.S. (1984) Theoretical description of elastic reflection of particles (photons) incident at grazing angles without the ude of the diffusion approximation. *Sov. Phys. JETP*, **60**(2), 290–299.
393 Rauch, H., Treimer, W., Bonse, U. (1974) Test of a single crystal neutron interferometer. *Phys. Lett. A*, **47**, 369–371.
394 Overhauser, A.W., Colella, R. (1974) Experimental test of gravitationally induced quantum interference. *Phys. Rev. Lett.*, **33**, 1237–1239
395 McReynolds, A.W. (1951) Gravitational acceleration of neutrons. *Phys. Rev.*, **83**, 172–173.
396 Dabbs, J.T., Harvey, J.A., Paya, D., Horstmann, H. (1965) Gravitational acceleration of free neutrons. *Phys. Rev. B*, **139**, B756–B760.
397 Koester, L. (1967) Absolutmessung der kohärenten Streulängen von Wasserstoff, Kohlenstoff und Chlor sowie Bestimmung der Schwerebeschleunigung für freie Neutronen mit dem Schwerkraft-Refraktometer am FRM. *Z. Phys.*, **198**, 187–200.
398 Colella, R., Overhauser, A.W., Werner, S.A. (1975) Observation of gravitationally induced quantum interference. *Phys. Rev. Lett.*, **34**, 1472–1474.
399 Littrell, K.C., Allman, B., Hoag, K., Werner, S. (1996) Two wavelength difference measurement of gravitationally-induced quantum interference phases. Textbooks [J]: or *J. Phys. Soc. Japan*, **65**(Suppl. A), 86–89.
400 Littrell, K.C., Allman, B.E., Werner, S.A. (1997) Two-wavelength-difference measurement of gravitationally induced quantum interference phases. *Phys. Rev. A*, **56**, 1767–1780.
401 van der Zouw, G., Weber, M., Felber, J., Gähler, R., Geltenborrt, P., Zeilinger, A. (2000) Aharonov–Bohm and gravity experiments with the very-cold-neutron interferometer. *Nucl. Instrum. Methods Phys. Res. A*, **440**, 568–574.
402 van der Zouw, G. (2000) Gravitational and Aharanov-Bohm phases in neutron interferometry. Ph. Doctor Diss., Univ. Wien.
403 Kaiser, H., Werner, S.A., George, E.A. (1983) Direct measurement of the longitudinal coherence length of a thermal neutron beam. *Phys. Rev. Lett.*, **50**, 560–563
404 Klein, A.G., Opat, G.I., Hamilton, W.A. (1983) Longitudinal coherence in neutron interferometry. *Phys. Rev. Lett.*, **50**, 563–565.
405 Rauch, H., Seidl, E., Tuppinger, D., Petrascheck, D., Scherm, R. (1987) Nondispersive sample arrangement in neutron interferometry. *Z. Phys. B-Cond. Matt.*, **69**, 313–317.

406 Rauch, H., Wölwitsch, H., Kaiser, H., Clothier, R., Werner, S.A. (1996) Measurement and characterization of the three-dimensional coherence function in neutron interferometry. *Phys. Rev. A*, **53**, 902–908.

407 Summhammer, J., Rauch, H., Tuppinger, D. (1987) Stochastic and deterministic absorption in neutron-interference experiments. *Phys. Rev. A*, **36**, 4447–4455.

408 Rauch, H., Summhammer, J. (1992) Neutron-interferometer absorption experiments in the quantum limit. *Phys. Rev.*, **46**, 7284–7287

409 Kaiser, H., Rauch, H., Baduek, G., Bauspiess, W., Bonse, U. (1979) Measurement of coherent neutron scattering lengths of gases. *Z. Phys. A*, **291**, 231–238.

410 Namiki, M., Pascazio, S. (1990) On a possible reduction of the interference term due to statistical fluctuations. *Phys. Lett. A*, **147**, 430–434.

411 Rauch, H., Summhammer, J., Zawisky, M., Jericha, E. (1990) Low-contrast and low-counting-rate measurements in neutron interferometry. *Phys. Rev. A*, **42**, 3726–3732.

412 Jacobson, D.J., Werner, S.A., Rauch, H. (1994) Spectral modulation and squeezing at high-order neutron interferences. *Phys. Rev. A*, **49**, 3196–3200.

413 Namiki, M., Pascazio, S. (1991) Wave-function collapse by measurement and its simulation. *Phys. Rev. A*, **44**, 39–53.

414 Werner, S.A., Colella, R., Overhauser, A.W., Eagen, C.F. (1975) Observation of the phase shift of a neutron due to precession in a magnetic field. *Phys. Rev. Lett.*, **35**, 1053–1055.

415 Grigoriev, S.V., Kreuger, R., Kraan, W.H., Mulder, F.M., Rekvedt, M.Th. (2001) Neutron wave-interference experiments with adiabatic passage of neutron spin through resonant coils. *Phys. Rev. A*, **64**, 013614-1–11.

416 Kraan, W.H., Grigoriev, S.V., Rekvedt, M.Th. (2004) Observation of 4π-periodicity of the spinor using neutron resonance interferometry. *Europhys. Lett.*, **66**[2], 164–170.

417 Badurek, G., Rauch, H., Summhammer, J. (1983) The time-dependent superposition of spinors. *Phys. Rev. Lett.*, **51**, 1015–1018.

418 Badurek, G., Rauch, H., Tuppinger, D. (1986) Neutron interferometric double-resonance experiment. *Phys. Rev. A*, **34**, 2600–2608.

419 Alefeld, B., Badurek, G., Rauch, H. (1981) Observation of the neutron magnetic resonance energy shift. *Z. Phys. B*, **41**, 231–235.

420 Dewedney, C., Gueret, P., Kyprianidis, A., Vigier, J.P. (1984) Testing wave-particle dualism with time-dependent neutron interferometry. *Phys. Lett. A*, **102**, 291–294.

421 Vigier, J.P. (1988) New theoretical implications of neutron interferometric double resonance experiments. *Physica B*, **151**, 386–392.

422 Einstein, A., Podolsky, B., Rosen, N. (1935) Can quantum-mechanical description of physical reality be considered complete?. *Phys. Rev.*, **47**, 777–780

423 Bell, J.S. (1964) On the Einstein Podolsky Rosen paradox. *Physics*, **1**, 195–200; (1966) *Rev. Mod. Phys.*, **38**, 447–452.

424 Bohm, D., Aharonov, Y. (1957) Discussion of experimental proof for the paradox of Einstein, Rosen, and Podolsky. *Phys. Rev.*, **108**, 1070–1076

425 Clauser, J.F., Horne, M.A., Shimony, A., Holt, R.A. (1969) Proposed experiment to test local hidden-variable theories. *Phys. Rev. Lett.*, **23**, 880–884.

426 Aspect, A., Grangier, P., Roger, G. (1982) Experimental tests of realistic local theories via Bell's theorem. *Phys. Rev. Lett.*, **49**, 91–94.

427 Hasegawa, Y., Loidl, R., Badurek, G., Baron, M., Rauch, H. (2004) Violation of Bell-type inequality in single-neutron interferometry: quantum contextuality. *Nucl. Instrum. Methods in Phys. Res. A*, **529**, 182–186; Hasegawa, Y., Loidl, R., Badurek, G., Baron, M., Rauch H. (2004) Violation of a Bell-like inequality in single-neutron interferometer experiments: quantum contextuality. *J. Mod. Opt.*, **51**, 967–972; Hasegawa, Y., Loidl, R., Badurek, G., Baron, M., Rauch H. (2003) Violation of a Bell-like inequality in single-neutron interferometry. *Nature*, **425**, 45–48

428 Rauch, H., Wilfing, A., Bauspiess, W., Bonse, U. (1978) Precise determination of the 4π-periodicity factor of a spinor wave function. *Z. Phys. B*, **29**, 281–284.

429 Ioffe, A.I., Zabiyakin, V.S., Drabkin, G.M. (1985) Test of a diffraction grating neutron interferometer. *Phys. Lett. A*, **111**, 373–375.

430 Pruner, C., Fally, M., Rupp, R.A., May, R.P., Vollbrandt, J. (2006) Interferometer for cold neutrons. *Nucl. Instrum. Methods Phys. Res. A*, **560**, 598–605.

431 Gruber, M., Eder, K., Zeilinger, A., Gähler, R., Mampe, W. (1989) A phase-grating interferometer for very cold neutrons. *Phys. Lett. A*, **140**, 363–367.

432 Peters, A., Chung, K.Y., Chu, S. (1999) Measurement of gravitational acceleration by dropping atoms. *Nature*, **400**, 849–852.

433 Aharonov, Y., Bohm, D. (1959) Significance of electromagnetic potentials in the quantum theory. *Phys. Rev.*, **115**, 485–491

434 Aharonov, Y., Casher, A. (1984) Topological quantum effects for neutral particles. *Phys. Rev. Lett.*, **53**, 319–321.

435 Tonomura, A., Osakabe, N., Matsuda, T., Kawasaki, T., Endoh, J., Yano, S., *et al.* (1986) Evidence for Aharonov–Bohm effect with magnetic field completely shielded from electron wave. *Phys. Rev. Lett.*, **56**, 792–795.

436 Cimmino, A., Opat, G.I., Klein, A.G., Kaiser, H., Werner, S.A., Arif, M. *et al.* (1989) *Phys. Rev. Lett.*, **63**, 380–383.

437 Shinohara, K., Aoki, T., Morinaga, A. (2002) Scalar Aharonov–Bohm effect for ultracold atoms. *Phys. Rev. A*, **66**, 042106-1–4.

438 Steyerl, A., Malik, S.S., Steinhauser, K.-A., Berger, L. (1979) A Michelson interferometer for ultracold neutrons. *Z. Phys. B*, **36**, 109–112.

439 Sagnac, M.G. (1913) L'ther lumineux dmontr par l'effet du vent relatif d'ther dans un interfromtre en rotation uniforme. *Comptes Rendus de l'Academie des Sciences (Paris)*, **157**, 708–710; (1913) Sur la preuve de la ralit de l'ther lumineux par l'exprience de l'interfrographe tournant. *Comptes Rendus de l'Academie des Sciences (Paris)*, **157**, 1410–1413.

440 Ebisawa, T., Tasaki, S., Kawai, T., Akiyoshi, T., Utsuro, M., Otake Y *et al.* (1994) Multilayer mirror interferometer for very cold neutrons. *Nucl. Instrum. Methods Phys. Res. A*, **344**, 597–606.

441 Funahashi, H., Ebisawa, T., Haseyama, T., Hino, M., Masaike, A., Otake Y *et al.* (1996) Interferometer for cold neutrons using multilayer mirrors. *Phys. Rev. A*, **54**, 649–651.

442 Hino, M., Ebisawa, T., Tasaki, S., Otake, Y., Tahata, H., Hashimoto M *et al.* (1999) Measurement of transverse coherent separation of spin precessing neutron using spin splitters. *J. Phys. Chem. Solids*, **60**, 1603–1605.

443 Kitaguchi, M., Funahashi, H., Nakura, T., Taketani, K., Hino, M., Otake, Y. *et al.* (2003) Non-dispersive measurement of the transverse coherence length of a cold neutron beam. *J. Phys. Soc. Japan*, **72**, 3079–3081

444 Hino, M., Achiwa, N., Tasaki, S., Ebisawa, T., Kawai, T., Akiyoshi, T. (1998) Observation of quasibound states of neutron in Fabry–Pérot magnetic thin-film resonator using Larmor precession. *Physica B*, **241–243**, 1083–1085.

445 Hino, M., Tasaki, S., Ebisawa, T., Kawai, T., Utsuro, M., Achiwa, N. *et al.* (2000) Experimental study of neutron-optical potential with absorption using Fabry–Pérot magnetic resonator. *Physica B*, **276/278**, 981–982.

446 de Broglie, L. (1960) *Non-Linear Wave Mechanics: A Causal Interpretation*, Elsevier, Amsterdam.

447 Bohm, D. (1952) A suggested interpretation of the quantum theory in terms of "hidden" variables, I. and II. *Phys. Rev.*, **85**, 166–179 and 180–193.

448 Ignatovich, V.K., Utsuro, M. (1997) A tentative solution for the UCN anomaly problem. *Phys. Lett. A*, **225**, 195–202.

449 Utsuro, M. (2002) Analytical solution for the de Broglie wave packet description of the neutron in subcritical transmission through a mirror. *Phys. Lett. A*, **292**, 222–232.

450 Hino, M., Tasaki, S., Kawabata, Y., Ebisawa, T., Geltenbort, P., Brenner T *et al.* (2003) Development of a very cold

neutron spin interferometer at the ILL. *Physica B*, **335**, 230–233.
451 Utsuro, M. (2005) Neutron spin interference visibility in tunneling transmission through magnetic resonators. *Physica B*, **358**, 232–246.
452 Utsuro, M., Hino, M., Geltenbort, P., Butterworth, J. (2005) Observation on the visibility decrease in a VCN spin resonator interferometry. *J. Res. Nat. Inst. Standard Technol.*, **110**[3], 245–249.
453 Ignatovich, V.K. (2004) Contradictions in scattering theory. *Concepts Phys.*, **1**, 51–104.
454 Messiah, A. (1958) *Quantum Mechanics*, Wiley & Sons, Ltd, New York.
455 Marshall, W., Lovesey, S.W. (1971) *Theory of Thermal Neutron Scattering*, Clarendon Press, Oxford.
456 Lovesey, S.W. (1984) *Theory of Neutron Scattering from Condensed Matter*, Clarendon Press, Oxford.
457 Goldberger, M.R., Watson, K.W. (1964) *Collision Theory*, John Wiley & Sons, New York, London, Sydney, Toronto.
458 Taylor, J.R. (1972) *Scattering Theory. The Quantum Theory of Nonrelativistic Collisions*, John Wiley & Sons, New York, London, Sydney, Toronto.
459 Utsuro, M., Ignatovich, V.K. (1998) Experimental test of the de Broglie wavepacket description of the neutron. Phys. Lett. A, **246**, 7–15.
460 Ignatovich, V.K. (2007) Temperature dependence of neutron scattering in He-4 gas. in: Neutron Spectroscopy, Nuclear Structure, Related topics, Proceedeings of XIV International Seminar on Interaction of Neutrons with Nuclei. Held in Dubna 24–27 May 2006, JINR, Dubna, p. 41–63.
461 Fetter, A.L., Walecka, J.D. (1971) *Quantum Theory of Many-Particle Systems*, McGrow-Hill, Boston.
462 Ignatovich, V.K. (2006) On uncertainty relations and interference in quantum and classical mechanics. *Concepts of Phys.*, **III**(1), 11–22.
463 Aspect, A., Grangier, P., Roger, G. (1981) Experimental tests of realistic local theories via Bell's theorem. *Phys. Rev. Lett.*, **47**, 460–3.
464 Sanz, A.S., Borondo, F., Miret-Artes, S. (2002) Particle diffraction studied using quantum trajectories. *J. Phys.: Condens. Matt.*, **14**, 6109–45.
465 Utsuro, M., Ignatovich, V.K., Geltenbort, P., Brenner Th, Butterworth, J., Hino, M., Okumura, K., Sugimoto, M. (1999) An experimental search of subcritical transmission of very cold neutrons (VCN) described by the de Broglie wavepacket. Proc. of ISINN-7. Dubna, 1999, 110–25.
466 Ignatovich, V.K. (2009) On EPR paradox, Bell's inequalities and experiments which prove nothing. arxiv:quant-phys/0703192v4; (2008) *Concepts Phys.*, **5**(2), 227–78.
467 Genin, R., Beil, H., Signarbieux, C., Carlos, P., Joly, R., Ribrag, M. (1963) Détermination des sections efficaces d'absorption et de diffusion des gaz rares pour les neutrons thermique. *Le Journal de Physique et le Radium*, **24**, 21–6.
468 Bell, J.S. (1964) *Physics*, **1**, 195; Bell, J.S. (2004) *Speakable and Unspeakable in Quantum Mechanics*, Cambridge University Press, p. 14.
469 Hayashi, M. *et al.* (2006) Hypothesis testing for an entangled state produced by spontaneous parametric down-conversion, *Phys Rev. A*, **74**, 062321-1–8.
470 von Neumann, J. (1955) *Mathematical Foundations of Quantum Mechanics*, Chapter IV, Sections 1 and 2, Princeton University Press, Princeton, New Jersey.
471 Albertson, J. (1961) Von Neumann's hidden-parameter proof. *Am. J. Phys.*, **29**, 478–83.
472 Kocher, C.A., Commins, E.D. (1967) Polarization correlation of photons emitted in an atomic cascade. *Phys. Rev. Lett.*, **18**, 575–7.
473 Sanz, A.S., Borondo, F., Miret-Artes, S. (2002) Particle diffraction studied using quantum trajectories. *J. Phys.: Condens. Matt.*, **14**(24), 6109–45.
474 Nikolić, H. (2007) Quantum Mechanics: Myths and Facts. *Found. Phys.*, **37**, 1563–611.
475 Rauch, H., Baron, M., Filipp, S., Hasegawa, Y., Lammel, H., Loidi, R. (2006) Hidden observables in neutron

quantum interferometry. *Physica B*, **385–386**(Part II), 1359–64.

476 Hasegawa, Y., Loidi, R., Badurek, G., Baron, M., Rauch, H. (2006) Quantum contextuality in neutron interferometer experiment. *Physica B*, **385–386**(Part II), 1377–80.

477 Hasegawa, Y., Loidi, R., Badurek, G., Baron, M., Rauch, H. (2006) Quantum contextuality in a single-neutron optical experiment. *Phys. Rev. Lett.*, **97**, 230401-1–4.

478 Berk, N.F., Majkrzak, C.F. (1995) Using parametric *B* splines to fit specular reflectivities. *Phys. Rev. B*, **51**, 11296–11309.

479 de Haan, V.-O., van Well, A.A., Adenwalla, S., Felcher, G.P. (1995) Retrieval of phase information in neutron reflectometry. *Phys. Rev. B*, **52**, 10831–10833.

480 Majkrzak, C.F., Berk, N.F., Dura, J.A., Satija, S.K., Karim, A., Pedulla, J. *et al.* (1998) Phase determination and inversion in specular neutron reflectometry. *Physica B*, **248**, 338–342.

481 Majkrzak, C.F., Berk, N.F., Silin, V., Meuse, C.W. (2000) Experimental demonstration of phase determination in neutron reflectometry by variation of the surrounding media. *Physica B*, **283**, 248–252.

482 Majkrzak, C.F., Berk, N.F. (2003) Phase sensitive reflectometry and the unambiguous determination of scattering length density profiles. *Physica B*, **336**, 27–38.

483 Tinkham, M. (1996) *Introdution to Superconductivity*, 2nd edn, McGraw-Hill, New York.

484 Zhang, H., Lynn, J.W., Majkrzak, C.F., Satija, S.K., Kang, J.H., Wu, X.D. (1995) Measurements of magnetic screening lengths in superconducting Nb thin films by polarized neutron reflectometry. *Phys. Rev. B*, **52**, 10395–10404.

485 Felcher, G.P., Kampwirth, R.T., Gray, K.E., Felici, R. (1984) Polarized-neutron reflections: A new technique used to measure the magnetic field penetration depth in superconducting niobium. *Phys. Rev. Lett.*, **52**, 1539–1542.

486 Kobayashi, S., Oike, H., Takeda, M., Itoh, F. (2002) Central peak position in magnetization hysteresis loops of ferromagnet/superconductor/ferromagnet trilayered films. *Phys. Rev. B*, **66**, 214520.

487 Han S-W, Ankner, J.F., Kaiser, H., Miceli, P.F., Paraoanu, E., Greene H. (1999) Spin-polarized neutron reflectivity: A probe of vortices in thin-film superconductors. *Phys. Rev. B*, **59**, 14692–14696.

488 Temst, K., Van Bael, M.J., Swerts, J., Buntinx, D., Van Haesendonck, C., Bruynseraede, Y. *et al.* (2004) In-plane vector magnetometry on rectangular Co dots using polarized neutron reflectivity. *J. Vac. Sci, Technol. B*, **21**, 2043–2047.

489 Chen, W.C., Gentile, T.R., O'Donovan, K.V., Borchers, J.A., Majkrzak, C.F. (2004) Polarized neutron reflectometry of a patterned magnetic film with a ^{3}He analyzer and a position-sensitive detector. *Rev. Sci. Instrum.*, **75**, 3256–363.

490 Majkrzak, C.F., Berk, N.F. (1999) Inverting neutron reflectivity from layered film structures using polarized beams. *Physica B*, **267–268**, 168–174.

491 Nickel, B., Rühm, A., Donner, W., Major, J., Dosch, H., Schreyer A *et al.* (2001) Spin-resolved off-specular neutron scattering maps from magnetic multilayers using a polarized ^{3}He gas spin filter. *Rev. Sci. Instrum.*, **72**, 163–172.

492 Nogues, J., Sort, J., Langlais, V., Skumryev, V., Surinach, S., Munoz, J.S., Baro, M.D. (2005) Exchange bias in nanostructures. *Phys. Rep.*, **422**, 65–117.

493 Berkowitz, A.E., Takano, K. (1999) Exchange anisotropy – a review. *JMMM*, **200**, 552–70.

494 Hellwig, O., Kortright, J.B., Takano, K., Fullerton, E.E. (2000) Switching behavior of Fe-Pt/Ni-Fe exchange-spring films studied by resonant soft-X-ray magneto-optical Kerr effect. *Phys. Rev. B*, **62**(17), 11694–8.

495 Koehler, W.C., Cable, J.W., Wilkinson, M.K., Wollan, E.O. (1966) Magnetic structures of holmium. I. The virgin state. *Phys. Rev.*, **151**, 414–24.

496 Fraerman, A.A., Gribkov, B.A., Gusev, B.A., Hjorvarsson, B., Klimov AYu, Mironov, V.L., Nikitushkin, D.S., Rogov, V.V., Vdovichev, S.N., Zabel, H.

(2007) Artificial helical nanomagnets. arXiv:0705.2445.

497 Steyerl, A., Truestedt, W.-D. (1974) Experiments with a neutron bottle. *Z. Phys.*, **267**, 379–388

498 Steyerl, A. (1974) Very cold neutrons – a new tool in condensed matter research. *Proc. 2nd Intern. School Neutron Physics*, Alushta. JINR, Dubna, D3-7991, 42–90

499 Utsuro, M., Steyerl, A. (1978) Total cross section of solid heavy water for very slow neutrons. *Ann. Meet. Atomic Energy Soc. Japan B*, **62**, 136, Preprint (in Japanese).

500 Haltenorth, H. (1974) Messung des Debye–Waller-Faktors und des Gitterparameters a an H_2O-Eis-Ih-Einkristallen zwischen 90° und 230° K″. Thesis, Tech. Univ. München.

501 Utsuro, M., Morishima, N. (1981) Total neutron cross section in heavy ice at 77 and 15 K for neutron energies between 0.3 and 100 meV. *J. Nucl. Sci. Technol.*, **18**[9], 739–741.

502 Whittemore, W.L., McReynolds, A.W. (1961) Proc. IAEA Symp. on Inelastic Scattering of Neutrons in Solid and Liquids. IAEA Vienna, p. 511.

503 Gissler, V.W., Reinsch, C., Springer, T., Wiedemann, W. (1961) Untersuchung des Neutronenstreuquerschnittes von schwerem Eis in der Umgebung der Bragg'schen Grenzwellenlänge. *Z. Kristallogr.*, **116**, 328–344; (1963), **118**, 149–157.

504 Morishima, N., Utsuro, M., Komori, E. (1981) Structure of heavy ice formed by rapid cooling of heavy water to a temperature of 77 or 15 K. Memoirs Fac. Eng., Kyoto Univ, XLIII: Part 4, pp. 350–359.

505 Fukunaga, T., Itoh, K., Orimo, S., Aoki, M., Fujii, H. (2001) Location of deuterium atoms absorbed in nanocrystalline graphite prepared by mechanical alloying. *J. Alloys Compound.*, **327**, 224–229.

506 Komura, S., Osamura, K., Fujii, H., Takeda, T., Murakami, Y. (1981) Small-angle neutron scattering from Al-Zn and Al-Zn-Mg alloys during Guinier-Preston zone formation. *Colloid Polym. Sci.*, **259**, 670–674.

507 Komura, S., Takeda, T., Fujii, H., Toyoshima, Y., Osamura, K., Mochiki, K., and Hasegawa, K. (1983) A 6-meter neutron small-angle spectrometer at KUR. *Jap. J. Appl. Phys.*, **22**, 351–356.

508 Komura, S., Osamura, K., Fujii, H., and Takeda, T. (1985) Time evolution of the structure of quenched Al–Zn and Al–Zn–Mg alloys. *Phys. Rev. B*, **31**, 1278–1301.

509 Gibbs, J.W. (1878) *On the Equilibrium of Heterogeneous Substances*. American J. Science and Arts (eds D. James, E.S. Dana, B. Silliman), Third Series, **XVI**, No. 6: original print. New Haven, 1878.

510 Gibbs, J.W. (1928) *The Collected Works*, Vol. 1, Longmans, Green and Co., New York, London, Toronto, pp. 105–115, pp. 252–258.

511 Cahn, J.W. (1961) On spinodal decomposition. *Acta Cryst.*, **9**, 795–801.

512 Saito, Y. (1976) Relaxation in a bistable system. *J. Phys. Soc. Japan*, **41**, 388–393; Anomalous fluctuation and mode selection in dynamics of spinodal decomposition. *ibid.*: 1129–1136.

513 Osamura, K., Okuda, H., Asano, K., Furusaka, M., Kishida, K., Kurosawa F *et al.* (1994) SANS study of phase decomposition in Fe-Cu alloy with Ni and Mn addition. *Iron and Steel Inst. of Japan International*, **34**, 346–354.

514 Doyle, E.D., Wong, Y.C., Ripley, M.I. (2006) Residual stress evaluation in martensitic stainless steel as a function of gas quenching pressure using thermal neutrons. *Physica B*, **385-386**, 897–9.

515 Rogge, R.B., Root, J.H., Donaberger, R.I. (2006) Applied neutron diffraction for industry (ANDI). *Physica B*, **385-386**, 883–9.

516 Nikitin, A.N., Ivankina, T.I., Ignatovich, V.K. (2009) The wave field patterns of the propagation of longitudinal and transverse elastic waves in grain-oriented rocks. *Izvestiya, Phys. Solid Earth.*, **45**(5), 424–36.

517 Choi, Y., Hahn, Y.S., Seong, B.S., Kim, M. (2006) Study of the effect of ultrasonic agitation the defects size in electro-deposited chromium layer by small angle neutron scattering. *Physica B*, **385-6**, 911–3.

518 Harms, A.A., Wyman, Dr. (1986) Mathematics and Physics of Neutron Radiography. in *A Reidel Texts in the Mathematical Sciences*, D. Reidel Publishing Company. Springer.

519 Hussein, E.M.A. (2003) *Handbook on Radiation Probing, Gauging, Imaging and Analysis*. Springer, ISBN 1402012942, 9781402012945.

520 Winkler, B., Kahle, A., Hennion, B. (2006) Neutron radiography of rocks and melts. *Physica B*, **385–386**, 933–4.

521 Schillinger, B., Brunner, J., Calzada, E. (2006) A study of oil lubrication in a rotating engine using stroboscopic neutron imaging. *Physica B*, **385-386**, 921–4.

522 Cowley, J.M. (1975,1987) *Diffraction Physics*, North-Holland, Amsterdam, p. 206.

523 Fiori, A., Hilger, A., Kardjilov, N., Albertini, G. (2006) Crack detection in Al alloy using phase-contrast neutron radiography and tomography. *Meas. Sci. Technol.*, **17**, 2479–84.

524 Tromp, R.H., Bouwman, W.G. (2007) A novel application of neutron scattering on dairy products. *Food Hydrocolloids*, **21**, 154–158.

525 Krouglov, T., de Schepper, I.M., Bouwman, W.G., Rekveldt, M.T. (2003) Real-space interpretation of spin-echo small angle neutron scattering. *J. Appl. Crystallogr.*, **36**, 117–124.

526 Perkins, S.J. (1988) Structural studies of proteins by high-flux X-ray and neutron solution scattering. *Biochem. J.*, **254**, 313–27.

527 Jacrot, B. (1976) The study of biological structures by neutron scattering from solution. *Rep. Prog. Phys.*, **39**, 911–53.

528 Fitter, J., Gutberlet, T., Katsaras, J. (2006) *Neutron Scattering in Biology: Techniques and Applications*, Springer, Berlin, New York.

529 Schoenborn, B.P., Knott, R.B. (Eds) (1996) *Neutrons in Biology. (Proc. Workshop on Neutrons in Biology)*, Plenum Press, New York.

530 Roe, J. (2000) *Methods of X-Rays and Neutron Scattering in Polymer Science*, Oxford Univ. Press, New York, Oxford.

531 Higgins, J.S., Benoît, H.C. (1994) *Polymers and Neutron Scattering*, Clarendon Press, Oxford.

532 Stuhrmann, H.B., Tardieu, A., Mateu, L., Sardet, C., Luzzati, V., Aggerbeck, L. *et al.* (1975) Neutron scattering study of human serum low density lipoprotein. *Proc. Nat. Acad. Sci. USA*, **72**(6), 2270–3.

533 Van Der Grinten, M.G.D., Glttli, H. (1993) Spin contrast in SANS of polymers in solution. *J. Phys.*, **3**, **C8**, 427–30.

534 Stuhrmann, H.B., Haas, J., Ibel, K., De Wolfo, B., Koch, M.H.J., Parfait, R., Crichton, R.R. (1976) New low resolution model for 50s subunit of escherichia coli ribosomes. *Proc. Nat. Acad. Sci. USA*, **73**(7), 2379–83.

535 Sturmann, H.B. (1970) Interpretation of small angle scattering of dilute solutions and gases. *Acta. Cryst. A*, **26**, 297–306.

536 Sturmann, H.B. (1970) Eine neues Verfahren zur Bestimmung der Oberflächenform und der inneren Struktur von gelösten globulären Proteinen aus Röntgenkleinwinkelmessungen. *Z. Phys. Chem. N. Folge.*, **72**, 177–98.

537 Shen, Li, Farid, H., and McPeek, M.A. (2009) Modeling three-dimensional morphological structures using spherical harmonics. *Wiley InterScience, Evolution* (International journal of organic evolution), **63**(7), 1003–1016. Published Online: 17 Oct 2008. Journal compilation © 2008 The Society for the Study of Evolution. doi:10.1111/j.1558-5646.2008.00557.x

538 Sackmann, E. (2003) Neutron reflectivity and surface scattering techniques: a new tool to study protein-lipid interaction mechanisms and protein-protein recognition processes at supported membranes. http://cell.e22.physik.tu-muenchen.de/research/pdf/NeutronReflectivity.pdf; 28 August 2003, 196K.

539 Kiselev, M.A., Ryabova, N.Y., Balagurov, A.M., Dante, S., Hauss, T., Zbytovska, J. *et al.* (2005) Hydration of a stratum corneum lipid model membrane by neutron diffraction. *Eur. Biophys. J.*, **34**, 1030–40.

540 Ostanevich, Yu.M., Serdyuk, I.N. (1982) Neutron-diffraction studies of the structure of biological macromolecules. *Sov. Phys. Usp.*, **25**(5), 323–339.

541 Cremer, J.T., Piestrup, M.A., Gary, C.K., Pantell, R.H., Glinka, C.J. (2004) Biological imaging with a neutron microscope. *Appl. Phys. Lett.*, **85**, 494–496.

542 Cremer, J.T., Piestrup, M.A., Park, H., Gary, C.K., Pantell, R.H., Glinka, C.J. *et al.* (2005) Imaging hydrogenous materials with a neutron microscope. *Appl. Phys. Lett.*, **87**, 161913-1–3.

Index

a
absorption 49
- deterministic — 399
- giant — cross section 50
- stochastic — 399
adiabatic
- — following the magnetic field 27, 91
- — spin flip condition 27
aging 271
Aharonov–Bohm effect 109, 417
Aharonov–Casher effect 419
analyzers of polarization 216
angular width of the Bragg peak 306
asymmetrical potentials 133, 136
asymptotical wave function
- in future
  - — stationary approach 448

b
ball lightning 336
Bell's inequality 409
Berry's geometrical phase 108
beryllium bottle 48
Bloch
- phase factor 139, 300
- — wall 97
- wave vector 139, 322
  - in Kronig–Penney potential 143
Bloch–Siegert shift 408
Bohr magneton 192
Borrmann
- — effect 258, 319
- — fan 258
bottle experiment
- ultracold neutron — 45
bound states
- rectangular well 166
- zones in periodic potentials 166

Bragg
- — case 258
- — reflection 3
- —'s law 5
- — scattering 3
- symmetrical — case 261
- — wave number 304

c
channel cut crystals 263
Clauser–Horne–Shimony–Holt inequality 410
coefficient
- reflection — 40
- transmission — 40
coherence
- — function 428
- — length 15–16, 428
- transverse — length 429
coherent 13
- — background (coherent reference wave) 284
- — illumination 284
- — scattering
  - — —— power 273
- wave function 340–341, 343
Colella–Overhauser–Werner experiment 394
complementary principle 405
continuous-fractions method 174
contrast 391, 397
convolution 264
Cornu spiral 278
correlation function 340
critical wave number 117
cross section
- definition 452
- relation to probability 453
crystalline sum 295
current model 72

Neutron Optics. Masahiko Utsuro, Vladimir K. Ignatovich

ISBN: 978-3-527-40885-6

d
Darwin
– — region 260
– — table 144, 260, 304
– width 260
de Broglie 1
– —'s fundamental formula 2
density autocorrelation function 524
detailed balance principle 328, 330
detector
– helium gas — 49
– position-sensitive — 271
– solid state type of — 48
diffracted beam
– deviated —(H-beam) 391
– forward —(O-beam) 391
diffraction
– Bragg 140, 303
– — angular width 305
– — Kronig–Penney potential 144
– — nonspecular 306
– — specular 303
– Bragg and Laue 319
– dynamical 322
– dynamical — theory 257
– — fan 261
– — grating
– —— for laser light 414
– very cold neutron interferometer with —— 414
– kinematical 326
– kinematical — theory 257
– Laue 318
– on a crystalline plane 294
– on single crystal 299, 307
diffractometer
– double-crystal — 263
– gravity — 35, 64
dipole
– — electric — moment 108
– — magnetic — moment 71
– — model 72
– — moment
– — electric —— 25
– — magnetic —— 22–25
Doppler shifter
– crystal balde — 62
– velocity focusing type of supermirror — 62
double-slit interference in time 280
dyadic matrix 300
dynamical
– — diffraction 322
– — diffraction theory 257
– — phase 108–109

e
Eckart potential 168
Einstein–Podolsky–Rosen paradox 409, 465
elastic recoil detection analysis 38
electron diffraction
– discovery of — 2
electrostriction 336
Ewald
– —'s formula 260

f
Fermi pseudopotential 321
Fizeau experiment 62
flight
– time-of— experiment 18
– time-of— method 267
– — velocity 17
flipper 216
fm(fermi) 44
Fraunhofer diffraction 9
Fresnel
– — biprism 8
– — concave zone mirror 12
– — diffraction 11, 97
– — integral 278
– — zone plate 11
fundamental question 194, 200

g
gauge
– — group 417
– — invariance 109, 417
– — invariant
– — — quantity 417
Gaussian distribution
– two— 14
gedankenexperiment 409
geometrical phase 109
Ginzburg–Landau
– — coherence length 487
– — parameter 487
graphite
– electro— 498
– pyro— crystal 498
– pyrolytic — 13
grating
– diffraction —
– —— cold neutron interferometer 413
– —— ultracold neutron interferometer 413

– volume-phase —
– —— very cold neutron interferometer 413
gravity
– — acceleration guide 61
– — deceleration guide 61
– — diffractometer 35, 64
Green function 346
– causal 347
– Fourier expansion 347
– one-dimensional equation 177
– nonstationary 158
– spinorial 369
group velocity 2
Guinier–Preston zone 271
Guinier radius 271

h
Hamiltonian 75, 85, 416
– Newtonian potential in — 392
hidden variable 410
high-resolution study 552
hologram 248
holography 283
– — principles of — 247
Huygen's principle 9
hydrogen
– — impurity at the mirror surface 44
– — impurity on mirror surface 37

i
imaginary part of scattering amplitude 343
incoherent 13
– — scattering 6, 344
index of refraction 6
– phase shift due to — 12
inhomogeneity 524
– — scattering 273, 499
– — characteristic wave number for —— 499
interference 476
– — experiment
– — double-slit —— 9
– — single slit —— 9
– — —— with split-zone plate 12
– — fringe 16
– Pendellösung — 15
interferometer
– cold neutron —
– — diffraction grating —— 413
– experiment with a split lens — 12
– — for long-wavelength neutrons 413
– grating mirror — for very cold neutrons 414
– — in dispersive arrangement 397
– — in nondispersive arrangement 397
– — with a silicon perfect single crystal 99
– Jamin type — 424
– Mach–Zehnder 390
– Michelson — 422
– mulilayer mirror type of very cold neutron — 422
– multilayer mirror — 423
– phase-spin-echo — with magnetic multilayers 426
– silicon perfect single crystal — 390
– single layer mirror type of ultracold neutron — 422
– time-dependent — 420
– ultracold neutron —
– — diffraction grating —— 413
– very cold neutron — 439
– — volume-phase grating —— 413
internal-detector concept 286–287, 289
internal-source concept 286–288
ion-polishing 55
isomorphous replacement method 552

k
Kagan–Afanasiev effect 319
Kikuchi
– — pattern 2
– Seishi — 18
kinematical
– diffraction 326
– — diffraction theory 257
Kronig–Penney potential 141
– scattering on one period 142
Krüger problem 229–230

l
Larmor precession 90
– — frequency 22
– — motion 22
– 4π periodicity of — 404
– 2π rotation in — 96
lattice structure in time–space plane 283
Laue
– — case 258
– symmetry — diffraction 15
linear oscillating field 220
Liouville's theorem 58, 92
long-range action of optical potential 118
low-resolution study 552

m
magnetic
– — deceleration 88
– — dipole moment 22–25, 71
– — domain 269
– — domain walls 269
– Fabry–Pérot — resonator 432
– — additional phase due to —— 432
– — dwell time in —— 435
– — field penetration depth 487
– London —— 487
– — focusing 90
– gyro— ratio 90, 95
– — moment orientation 192
– — multilayer
– reflection coefficient of —— 432
– transmission coefficient of —— 432
– — potential 74–75
– — prism 94
– pseudo— field 96
– — screening length 488, 490
magnetic field
– mutually orthogonal — 24
– precession — 102
– quadrupole — 94
– reciprocating — 23
– rotating — 22–23
– sextupole — 91
– static – 22
– total — 22
– — with tilted interface 523
matrix
– — method 175
– — —— for solving magnetic systems 213
– transfer —— 54, 81, 85
– propagation — 82–83
– reflectance — 85
– super— 85
– — transfer — 54, 82
– transfer super— 86
– transmission — 82–83
– transmittance — 85
– unimodular — 84
matter wave
– quantum mechanical analysis of — 19
mechanical alloying 504
method of stationary phase 279
Miller indices 5
mosaic crystal 13, 314
multilayer 51
– high-Q — development 55
– — monochromator 51
– — polarizer mirror 53
– — spin splitter 101
multiple wave scattering theory 293, 339

n
neutron
– albedo 376
– cold 11
– coupled resonator for — 67
– — electric dipole moment of — 25
– electric dipole moment of — 108
– epithermal — 4
– — guide tube 33
– — bent —— 33
– — characteristic energy of bent —— 34
– — characteristic velocity of bent —— 34
– — characteristic wavelength of —— 33
– — garland component in bent —— 34
– — zigzag component in bent —— 34
– — magnetic dipole moment 23–25
– pulsed — resonance spin echo method 105
– quasi-bound state — 431
– quasi-bound states of — 65
– — reaction rate 67
– — reflectometry 40
– — resonance spin echo method 103
– — spin echo 29, 522
– — spin echo method 103
– — spin maser 107
– split of — resonance peak 67
– — star 337
– — storage
– — —— with electromagnets 77, 80
– — —— with material mirrors 45
– — —— with permanent magnets 77
– thermal — 3
– — turbine 35, 59
– — Steyerl's —— 59
– — supermirror —— 47, 59
– ultracold — 12, 35, 45, 88
– —— anomaly 51
– —— bottle experiment 45
– ultracold neutrons 28
– very cold — 9, 11, 19
– transmission experiment 48
neutronostriction 337
Nishina
– Yoshio 18

nonadiabatic spin flip 22
noncollinear magnetization 87
nonlinearity 336
nuclear magnetic resonance method 23
nuclear magneton 22, 24

o
optical
– — period
– — asymmetrical 141
– potential 332
– — corrections 333
– — potential 117
– —— theorem 328

p
partial waves 13, 19
particle beam
– classical mechanical analysis of — 19
Pauli matrices 96, 191
– properties 195
Pendellösung 319
– — interference 15, 258
Percus–Yevick correlation function 273, 500
periodic systems
– algebraic approach 138
perturbation theory
– scattering on a single period 176
phase
– dynamical — 109
– — problem 552
– — shifter 391
– —-spin-echo interferometer 427
– topological — 109, 422
– —velocity 2
phases of *r* and *t* 137
plane wave
– — model 437
– of particle with spin 193
polarization
– — analysis 23
– — function 490
polarizer 215
– magnetic reflection — 28, 73
– magnetic scattering — 23, 73
Porod invariant 359
postselection procedure 402
potential
– complex — 49, 67, 435
– Fermi's pseudo — 5
– gravity — 35
– optical — 7, 62
– scalar — 417
– smoothed — 36
– soft — 36
– — splitting 133
– — step 115
– step — 19, 36
– time-dependent — 19
– vector — 416–417
– well type — 64
– — with permanent magnets 78
precession
– additional — angle 433
– — field 102
precipitation 508
primary extinction coefficient 316

q
quantization of scattering angle 459
quantum
– — chopper 282
– — entanglement 402, 410
– —— between different freedoms 411
– —— of separated wave packets 402
quasi-potential 118
quench 271

r
Rabi formula 219
reaction
– (n, α) — 69
– (n, γ) — 67
– neutron — in a resonator 67
realistic local theories 410
reciprocal lattice vector 5
rectangular potential barrier 129
reduced mass factor 7
reflection
– — coefficient 40, 52
– — —— for magnetic multilayer 83
– — —— for multilayer 82
– — —— of magnetic multilayer 432
– experiment with a vibrating mirror
— 20
– from a semi-infinite periodic potential
– — symmetrical 139
– from Kronig–Penney potential 143
– from magnetic mirror 193, 201
– from magnetic periodic system 203
– from magnetized mirror
– — reflection matrix amplitude 198
– from *N* periods
– — symmetrical 140
– from semi-infinite periodic potential
– — asymmetrical 141

– from semiperiodic potential
– — symmetrical 140
– from two mirrors
– — asymmetrical 133
– — magnetic 202
– in external field 177
– — loss coefficient 43–44
– — loss probability 43
– off-specular — 41
– — phase analysis
– polarized neutron method for —— 85
– polarized neutron reference layer method for —— 85
– —— with reference layer method 486
– specular — 40–41
reflection and transmission amplitudes
– properties 136
reflectometer
– chopper and time-of-flight method of — 42
– horizontal type of — 41
– vertical type of — 41
reflectometry
– — method 40
– polarized neutron — 74
refractive index
– phase change due to —
– —— in a rotating rod 63
– — with absorption 400
renormalization of the scattering amplitude 294, 343
reproduction of image from hologram 249
resolution restriction of electron microscope 283
resonance
– double— experiment 406
– — frequency 24
– — magnetic field strength 24
resonances
– at total reflection 159
– Breit–Wigner formula 157
– decay 158
– two-barrier systems 155
resonator
– Fabry–Pérot — 65
– Fabry–Pérot magnetic — 432
– macroscopic neutron — 65
rocking curve 258
rotating
– — frame 26
– — radiofrequency field 218
rotation
– — by an angle 2π radian 96
– operation of — 96
roton 105
roughness 362
Rutherford forward scattering 38

s

Sagnac effect 423
Saito's formula 507
scalar
– — Aharonov–Bohm effect 201
– — optics 12
scattering
– — amplitude 293
– — coherent —— for a proton 32
– — complex —— 43, 49
– — amplitude (scattering length) 5
– bound atom —— 7
– — coherent —— 6
– free atom —— 7
– — amplitude density
– — —— correlation function 523
– — distribution of — for neutrons 485
– — —— profile 490
– on a crystalline plane 293, 333
– Rutherford forward — 38
scattering theory 445
– fundamental 447, 450
– — grounds 454
– — justification 460, 464
– spherical waves 446
– standard 447
– — critics 449
– — on an arbitrary system 449
secondary extinction coefficient 316
segregation 508
self-focusing 336
semi-transparent mirrors 132
singlet scattering 337
small-angle neutron scattering 269, 356, 360
– spin-echo — 276
– three-dimensional — 275
small-angle X-ray scattering 269
smoothing of boundaries 168
solid heavy water
– simply frozen — 501
– single crystal of — 501
— space density 58
spectrometer
– backscattering — 262
– double crystal — 90
– frame-overlapping time-of-flight — 280

spectrometry
– neutron spin echo 29
spectroscopy
– backscattering — 5
– crystal — 5
spherical
– — particle 273, 499
– — wave 293, 346
– —— two-dimensional Fourier representation 247, 341
spin
– —— arrow direction 191
– Drabkin — rotator
– — with constant fields 216
– —-echo length 524
– — echo small angle neutron scattering 522
– —-echo small-angle neutron scattering 276
– — flip
– — adiabatic —— method with a magnetic gradient 26
– — nonadiabatic —— 22
– — resonance condition for —— 22
– —— time 26
– — —— with Ramsey's separated oscillating magnetic field method 24
– — maser 107
– —-phase shifter 411
– precession wave 236, 250
– rotator 216
– current foil —— with constant fields 216
– — Korneev —— with constant fields 216
– — Mezei ——with constant fields 217
– — with constant fields 216
– rotators
– — —— with radiofrequency fields 218
– superposition of — states 99
spin echo small angle neutron scattering correlation function 524
spinodal decomposition 507
spinor 191
– —— states
– — —— properties 195
splittered waves 9, 389
sputter method 55
standing waves 203
Stern–Gerlach
– — effect without a magnetic field gradient 94
– experiment of magnetic bending by — 90
– — magnetic deflection experiment 24
stroboscopic
– — detection method 408
– — measurement 406
structure factor 503
– crystal — 15
superconductor 487
– type I 487
– type II 487
supermirror 144
– — polarizer 53
superposition of states 194
surface hydrogen concentration
– — analyzed with elastic-recoiled atom detection 38
– — with resonant nuclear reaction γ-rays 37
surface waves 126

t
the matter wave 1
the principle of equivalence 393
time–space reciprocal lattice 283
topology 422
– —ical phase 109
total reflection 117
– critical angle for — 6
– critical wavelength for — 32
– formula of — 7
– — from gas-liquid interface 8
– limiting velocity for — 33, 35, 44, 49
– — contribution of absorption to —— 49
transferred
– — energy 29
– — wave number vector 54
transmission
– — coefficient 40, 52
– — —— for magnetic multilayer 83
– —— for multilayer 82
– —— of magnetic multilayer 432
– through N periods
– symmetrical 140
– through two mirrors
– — asymmetrical 133
– — symmetrical 133
triplet ray reflection 198
two-step process 284

u
ultracold neutrons 117
– anomaly 118
– — anomaly 119
– — losses 119
ultrafine electromechanical composite polishing 46
unitarity 136, 328, 451
– — condition 116
– — on interface 116

v
vacuum evaporation method 51
vector Aharonov–Bohm effect 418
velocity selection
– electrical — 18
– mechanical — 17
visibility 391, 428, 439
– net — 440

w
wave front reconstruction 285
wave nature 3
wave packet 14, 437, 454
– de Broglie
– — singular 456
– de Broglie's — 438
– scattering 455
– — on a center 462
– Schrödinger's — 437
– tunnel-transmitting — 438
wave–particle duality 238
wave vector transfer 41
wavelength 2

y
Yoneda effect 350
Young double slit 8

z
Zeeman energy splitting 406